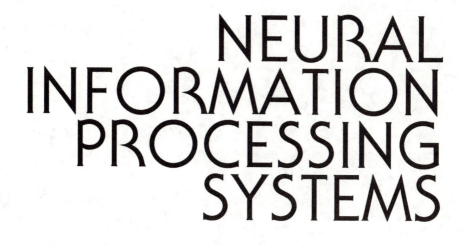

NEURAL INFORMATION PROCESSING SYSTEMS

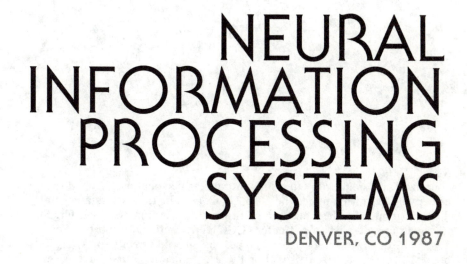

NEURAL INFORMATION PROCESSING SYSTEMS

DENVER, CO 1987

EDITOR:

DANA Z. ANDERSON
UNIVERSITY OF
COLORADO

AMERICAN INSTITUTE OF PHYSICS   NEW YORK 1988

L.C. Catalog Card No. 88-71208
ISBN 0-88318-569-5

Printed in the United States of America.

INTRODUCTION

1987 witnessed not the birth of neural networks, but a rebirth. Network processing was as much in the minds of the early computer conceivers as digital processing was. For the digital computer, everything went right: first relays, then tubes, then transistors, then integrated circuits. It has become an extremely high-class technology—one based largely on the silicon wafer. It can do anything; who needs anything else and why bother with networks? With such a promising future for the digital computer, network processing became a mere distraction. Enthusiasm for neural networks waned for many during the late sixties and early seventies. But for a few the enthusiasm never died. These were the people who were trying to understand brain function and the nervous system, and for them what the modern computer was bore little relevance to wetware. But for most of us the whole business of the neural networks is a novel idea seemingly born of the eighties. It is, in fact, a very seductive idea that has led an enormous number of people, including myself, to toy with this paradigm.

As though the field really had died (or perhaps was killed) the question is asked, "What is different now from the way it was then?" with a tacit reply, "It didn't work in the sixties, why should it work now?" I have heard it many times now. If some of the science we are seeing now has appeared before, like children we can be pleased to learn to walk even though we find a group of taller people who already know how. Reinvention is a first course for learning. However, things *are* different now than they were then: it isn't simply that our understanding of neural networks has progressed, or that electronic technology has become a superb processing medium. What has changed is our perspective: we have begun to see the limitations of the digital computer. Indeed, we have begun to run into them.

In the fifties and sixties electronic computers were popularly called electronic brains. Sometime between then and now that phrase seems to have faded. We thought we could make brains. Time has cast some doubt. There is now more of a recognition that we cannot do some of the things that are "brain-like" (I use the term loosely) with conventional electronic computers and a purely algorithmic approach to processing. At least, the conventional paradigm may not be a *practical* means for these goals. Early on, one had a choice of (at least) two paths to processing: the digital computing path and the neural network path (though the latter had no such name then). Digital electronics was the more promising path. Now that we see some tasks are not so easy the one way, it's time to explore the other one. It might also be true that neural networks are incapable of anything that proves better—there is that risk. This is, however, exactly the kind of risk that makes the field truly one of research rather than one of development. The potential is there, the payoff is high, and it demands inspection.

The explosive rise of interest in neural networks in recent years is amazing. One cannot help but feel as though we are in the midst of a revolution. If there is a scientific revolution taking place here, it is not in the mold of Thomas Kuhn's *The Structure of Scientific Revolution*. The field of neural networks has not experienced a normal evolution. There was no breakthrough, as there has recently been in superconductivity, to drive the community size from a few hundred to a few thousand in a year or two. Is there enough substance to uphold all this activity? It remains to be seen. Outside observers have high expectations for this field. It is fortunate that the community is becoming self-conscious and introspective. What can neural networks really do? This book takes that question as its primary context.

The papers comprising this volume were presented at the first IEEE Conference on Neural Information Processing Systems held in Denver, Colorado, on November 8–12, 1987. There is one exception. Professor Carver Mead of Caltech made the rather peculiar request that I include the paper by J. L. Wyatt, Jr., and D. L. Standley in lieu of his own paper. His argument was that his own contribution could be found elsewhere, but that Wyatt and Standley had solved a serious problem concerning electronic lateral inhibition networks using techniques from control theory that have not yet permeated the neural network community. I readily agreed to his request: it is on cross-fertilization that this field thrives!

My intention in assembling this book was to make it available as soon as possible to the community so that the material might prove timely. Thus I apologize to those authors who, upon rereading their own contribution would like to have made some revisions; this was a one-shot deal. At the same time, I am extremely grateful that my guillotine deadline was observed (well, at least by 95% of the authors!).

I am deeply indebted to Kris Mickus, without whose efforts the entropy of this project would have diverged. She is unbelievable!

February, 1988

Dana Z. Anderson
Boulder, Colorado

CONNECTIVITY VERSUS ENTROPY

Yaser S. Abu-Mostafa
California Institute of Technology
Pasadena, CA 91125

ABSTRACT

How does the connectivity of a neural network (number of synapses per neuron) relate to the complexity of the problems it can handle (measured by the entropy)? Switching theory would suggest no relation at all, since all Boolean functions can be implemented using a circuit with very low connectivity (e.g., using two-input NAND gates). However, for a network that learns a problem from examples using a *local* learning rule, we prove that the entropy of the problem becomes a lower bound for the connectivity of the network.

INTRODUCTION

The most distinguishing feature of neural networks is their ability to spontaneously learn the desired function from 'training' samples, i.e., their ability to program themselves. Clearly, a given neural network cannot just learn any function, there must be some restrictions on which networks can learn which functions. One obvious restriction, which is independent of the learning aspect, is that the network must be big enough to accommodate the circuit complexity of the function it will eventually simulate. Are there restrictions that arise merely from the fact that the network is expected to *learn* the function, rather than being purposely designed for the function? This paper reports a restriction of this kind.

The result imposes a lower bound on the connectivity of the network (number of synapses per neuron). This lower bound can only be a consequence of the learning aspect, since switching theory provides purposely designed circuits of low connectivity (e.g., using only two-input NAND gates) capable of implementing any Boolean function [1,2]. It also follows that the learning mechanism must be restricted for this lower bound to hold; a powerful mechanism can be designed that will find one of the low-connectivity circuits (perhaps by exhaustive search), and hence the lower bound on connectivity cannot hold in general. Indeed, we restrict the learning mechanism to be local; when a training sample is loaded into the network, each neuron has access only to those bits carried by itself and the neurons it is directly connected to. This is a strong assumption that excludes sophisticated learning mechanisms used in neural-network models, but may be more plausible from a biological point of view.

The lower bound on the connectivity of the network is given in terms of the *entropy* of the environment that provides the training samples. Entropy is a quantitative measure of the disorder or randomness in an environment or, equivalently, the amount of information needed to specify the environment. There are many different ways to define entropy, and many technical variations of this concept [3]. In the next section, we shall introduce the formal definitions and results, but we start here with an informal exposition of the ideas involved.

The environment in our model produces patterns represented by N bits $\mathbf{x} = x_1 \cdots x_N$ (pixels in the picture of a visual scene if you will). Only h different patterns can be generated by a given environment, where $h < 2^N$ (the entropy is essentially $\log_2 h$). No knowledge is assumed about which patterns the environment is likely to generate, only that there are h of them. In the learning process, a huge number of sample patterns are generated at random from the environment and input to the network, one bit per neuron. The network uses this information to set its internal parameters and gradually tune itself to this particular environment. Because of the network architecture, each neuron knows only its own bit and (at best) the bits of the neurons it is directly connected to by a synapse. Hence, the learning rules are local: a neuron does not have the benefit of the entire global pattern that is being learned.

After the learning process has taken place, each neuron is ready to perform a function *defined by what it has learned*. The collective interaction of the functions of the neurons is what defines the overall function of the network. The main result of this paper is that (roughly speaking) if the connectivity of the network is less than the entropy of the environment, the network cannot learn about the environment. The idea of the proof is to show that if the connectivity is small, the final function of each neuron is independent of the environment, and hence to conclude that the overall network has accumulated no information about the environment it is supposed to learn about.

FORMAL RESULT

A neural network is an undirected graph (the vertices are the neurons and the edges are the synapses). Label the neurons $1, \cdots, N$ and define $K_n \subseteq \{1, \cdots, N\}$ to be the set of neurons connected by a synapse to neuron n, together with neuron n itself. An environment is a subset $e \subseteq \{0, 1\}^N$ (each $\mathbf{x} \in e$ is a sample from the environment). During learning, $x_1, \cdots, x_N$ (the bits of $\mathbf{x}$) are loaded into the neurons $1, \cdots, N$, respectively. Consider an arbitrary neuron n and relabel everything to make K_n become $\{1, \cdots, K\}$. Thus the neuron sees the first K coordinates of each $\mathbf{x}$.

Since our result is asymptotic in N, we will specify K as a function of N; $K = \alpha N$ where $\alpha = \alpha(N)$ satifies $\lim_{N \to \infty} \alpha(N) = \alpha_o$ ($0 < \alpha_o < 1$). Since the result is also statistical, we will consider the ensemble of environments $\mathcal{E}$

$$\mathcal{E} = \mathcal{E}(N) = \left\{ e \subseteq \{0,1\}^N \mid |e| = h \right\}$$

where $h = 2^{\beta N}$ and $\beta = \beta(N)$ satifies $\lim_{N \to \infty} \beta(N) = \beta_o$ ($0 < \beta_o < 1$). The probability distribution on $\mathcal{E}$ is uniform; any environment $e \in \mathcal{E}$ is as likely to occur as any other.

The neuron sees only the first K coordinates of each $\mathbf{x}$ generated by the environment e. For each e, we define the function $n : \{0,1\}^K \to \{0,1,2,\cdots\}$ where

$$n(a_1 \cdots a_K) = |\{\mathbf{x} \in e \mid x_k = a_k \text{ for } k = 1, \cdots, K\}|$$

and the normalized version

$$\nu(a_1 \cdots a_K) = \frac{n(a_1 \cdots a_K)}{h}$$

The function ν describes the relative frequency of occurrence for each of the 2^K binary vectors $x_1 \cdots x_K$ as $\mathbf{x} = x_1 \cdots x_N$ runs through all h vectors in e. In other words, ν specifies the projection of e as seen by the neuron. Clearly, $\nu(\mathbf{a}) \geq 0$ for all $\mathbf{a} \in \{0,1\}^K$ and $\sum_{\mathbf{a} \in \{0,1\}^K} \nu(\mathbf{a}) = 1$.

Corresponding to two environments e_1 and e_2, we will have two functions ν_1 and ν_2. If ν_1 is not distinguishable from ν_2, the neuron cannot tell the difference between e_1 and e_2. The distinguishability between ν_1 and ν_2 can be measured by

$$d(\nu_1, \nu_2) = \frac{1}{2} \sum_{\mathbf{a} \in \{0,1\}^K} |\nu_1(\mathbf{a}) - \nu_2(\mathbf{a})|$$

The range of $d(\nu_1, \nu_2)$ is $0 \leq d(\nu_1, \nu_2) \leq 1$, where '0' corresponds to complete indistinguishability while '1' corresponds to maximum distinguishability. We are now in a position to state the main result.

Let e_1 and e_2 be independently selected environments from $\mathcal{E}$ according to the uniform probability distribution. $d(\nu_1, \nu_2)$ is now a random variable, and we are interested in the expected value $E(d(\nu_1, \nu_2))$. The case where $E(d(\nu_1, \nu_2)) = 0$ corresponds to the neuron getting no information about the environment, while the case where $E(d(\nu_1, \nu_2)) = 1$ corresponds to the neuron getting maximum information. The theorem predicts, in the limit, one of these extremes depending on how the connectivity (α_o) compares to the entropy (β_o).

Theorem.

1. If $\alpha_o > \beta_o$, then $\lim_{N \to \infty} E\left(d(\nu_1, \nu_2)\right) = 1$.

2. If $\alpha_o < \beta_o$, then $\lim_{N \to \infty} E\left(d(\nu_1, \nu_2)\right) = 0$.

The proof is given in the appendix, but the idea is easy to illustrate informally. Suppose $h = 2^{K+10}$ (corresponding to part 2 of the theorem). For most environments $e \in \mathcal{E}$, the first K bits of $\mathbf{x} \in e$ go through all 2^K possible values approximately 2^{10} times each as $\mathbf{x}$ goes through all h possible values once. Therefore, the patterns seen by the neuron are drawn from the fixed ensemble of all binary vectors of length K with essentially uniform probability distribution, i.e., ν is the same for most environments. This means that, statistically, the neuron will end up doing the same function regardless of the environment at hand.

What about the opposite case, where $h = 2^{K-10}$ (corresponding to part 1 of the theorem)? Now, with only 2^{K-10} patterns available from the environment, the first K bits of $\mathbf{x}$ can assume at most 2^{K-10} values out of the possible 2^K values a binary vector of length K can assume in principle. Furthermore, which values can be assumed depends on the particular environment at hand, i.e., ν does depend on the environment. Therefore, although the neuron still does not have the global picture, the information it has says something about the environment.

ACKNOWLEDGEMENT

This work was supported by the Air Force Office of Scientific Research under Grant AFOSR-86-0296.

APPENDIX

In this appendix we prove the main theorem. We start by discussing some basic properties about the ensemble of environments $\mathcal{E}$. Since the probability distribution on $\mathcal{E}$ is uniform and since $|\mathcal{E}| = \binom{2^N}{h}$, we have

$$\Pr(e) = \binom{2^N}{h}^{-1}$$

which is equivalent to generating e by choosing h elements $\mathbf{x} \in \{0, 1\}^N$ with uniform probability (without replacement). It follows that

$$\Pr(\mathbf{x} \in e) = \frac{h}{2^N}$$

while for $x_1 \neq x_2$,

$$\Pr(x_1 \in e \,, \, x_2 \in e) = \frac{h}{2^N} \times \frac{h-1}{2^N - 1}$$

and so on.

The functions n and ν are defined on K-bit vectors. The statistics of $n(a)$ (a random variable for fixed a) is independent of a

$$\Pr(n(a_1) = m) = \Pr(n(a_2) = m)$$

which follows from the symmetry with respect to each bit of a. The same holds for the statistics of $\nu(a)$. The expected value $E(n(a)) = h2^{-K}$ (h objects going into 2^K cells), hence $E(\nu(a)) = 2^{-K}$. We now restate and prove the theorem.

Theorem.
1. If $\alpha_o > \beta_o$, then $\lim_{N \to \infty} E\left(d(\nu_1, \nu_2)\right) = 1$.
2. If $\alpha_o < \beta_o$, then $\lim_{N \to \infty} E\left(d(\nu_1, \nu_2)\right) = 0$.

Proof.
We expand $E\left(d(\nu_1, \nu_2)\right)$ as follows

$$
\begin{aligned}
E\left(d(\nu_1, \nu_2)\right) &= E\left(\frac{1}{2} \sum_{a \in \{0,1\}^K} |\nu_1(a) - \nu_2(a)|\right) \\
&= \frac{1}{2h} \sum_{a \in \{0,1\}^K} E\left(|n_1(a) - n_2(a)|\right) \\
&= \frac{2^K}{2h} E(|n_1 - n_2|)
\end{aligned}
$$

where n_1 and n_2 denote $n_1(0 \cdots 0)$ and $n_2(0 \cdots 0)$, respectively, and the last step follows from the fact that the statistics of $n_1(a)$ and $n_2(a)$ is independent of a. Therefore, to prove the theorem, we evaluate $E(|n_1 - n_2|)$ for large N.

1. Assume $\alpha_o > \beta_o$. Let n denote $n(0 \cdots 0)$, and consider $\Pr(n = 0)$. For n to be zero, all 2^{N-K} strings x of N bits starting with K 0's must *not* be in the environment e. Hence

$$\Pr(n = 0) = (1 - \frac{h}{2^N})(1 - \frac{h}{2^N - 1}) \cdots (1 - \frac{h}{2^N - 2^{N-K} + 1})$$

where the first term is the probability that $0 \cdots 00 \notin e$, the second term is the probability that $0\cdots 01 \notin e$ given that $0\cdots 00 \notin e$, and so on.

$$\geq \left(1 - \frac{h}{2^N - 2^{N-K}}\right)^{2^{N-K}}$$

$$= \left(1 - h2^{-N}(1 - 2^{-K})^{-1}\right)^{2^{N-K}}$$

$$\geq (1 - 2h2^{-N})^{2^{N-K}}$$

$$\geq 1 - 2h2^{-N}2^{N-K}$$

$$= 1 - 2h2^{-K}$$

Hence, $\Pr(n_1 = 0) = \Pr(n_2 = 0) = \Pr(n = 0) \geq 1 - 2h2^{-K}$. However, $E(n_1) = E(n_2) = h2^{-K}$. Therefore,

$$E(|n_1 - n_2|) = \sum_{i=0}^{h}\sum_{j=0}^{h} \Pr(n_1 = i, n_2 = j)|i - j|$$

$$= \sum_{i=0}^{h}\sum_{j=0}^{h} \Pr(n_1 = i)\Pr(n_2 = j)|i - j|$$

$$\geq \sum_{j=0}^{h} \Pr(n_1 = 0)\Pr(n_2 = j)j$$

$$+ \sum_{i=0}^{h} \Pr(n_1 = i)\Pr(n_2 = 0)i$$

which follows by throwing away all the terms where neither i nor j is zero (the term where both i an j are zero appears twice for convenience, but this term is zero anyway).

$$= \Pr(n_1 = 0)E(n_2) + \Pr(n_2 = 0)E(n_1)$$

$$\geq 2(1 - 2h2^{-K})h2^{-K}$$

Substituting this estimate in the expression for $E(d(\nu_1, \nu_2))$, we get

$$E(d(\nu_1, \nu_2)) = \frac{2^K}{2h}E(|n_1 - n_2|)$$

$$\geq \frac{2^K}{2h} \times 2(1 - 2h2^{-K})h2^{-K}$$

$$= 1 - 2h2^{-K}$$

$$= 1 - 2 \times 2^{(\beta-\alpha)N}$$

Since $\alpha_o > \beta_o$ by assumption, this lower bound goes to 1 as N goes to infinity. Since 1 is also an upper bound for $d(\nu_1, \nu_2)$ (and hence an upper bound for the expected value $E(d(\nu_1, \nu_2))$), $\lim_{N\to\infty} E(d(\nu_1, \nu_2))$ must be 1.

2. Assume $\alpha_o < \beta_o$. Consider

$$
\begin{aligned}
E(|n_1 - n_2|) &= E\left(|(n_1 - h2^{-K}) - (n_2 - h2^{-K})|\right) \\
&\leq E(|n_1 - h2^{-K}| + |n_2 - h2^{-K}|) \\
&= E(|n_1 - h2^{-K}|) + E(|n_2 - h2^{-K}|) \\
&= 2E(|n - h2^{-K}|)
\end{aligned}
$$

To evaluate $E(|n - h2^{-K}|)$, we estimate the variance of n and use the fact that $E(|n - h2^{-K}|) \leq \sqrt{\text{var}(n)}$ (recall that $h2^{-K} = E(n)$). Since $\text{var}(n) = E(n^2) - (E(n))^2$, we need an estimate for $E(n^2)$. We write $n = \sum_{\mathbf{a} \in \{0,1\}^{N-K}} \delta_{\mathbf{a}}$, where

$$
\delta_{\mathbf{a}} = \begin{cases} 1, & \text{if } 0\cdots0\mathbf{a} \in e; \\ 0, & \text{otherwise.} \end{cases}
$$

In this notation, $E(n^2)$ can be written as

$$
\begin{aligned}
E(n^2) &= E\left(\sum_{\mathbf{a} \in \{0,1\}^{N-K}} \sum_{\mathbf{b} \in \{0,1\}^{N-K}} \delta_{\mathbf{a}} \delta_{\mathbf{b}} \right) \\
&= \sum_{\mathbf{a} \in \{0,1\}^{N-K}} \sum_{\mathbf{b} \in \{0,1\}^{N-K}} E(\delta_{\mathbf{a}} \delta_{\mathbf{b}})
\end{aligned}
$$

For the 'diagonal' terms $(\mathbf{a} = \mathbf{b})$,

$$
\begin{aligned}
E(\delta_{\mathbf{a}} \delta_{\mathbf{a}}) &= \Pr(\delta_{\mathbf{a}} = 1) \\
&= h2^{-N}
\end{aligned}
$$

There are 2^{N-K} such diagonal terms, hence a total contribution of $2^{N-K} \times h2^{-N} = h2^{-K}$ to the sum. For the 'off-diagonal' terms $(\mathbf{a} \neq \mathbf{b})$,

$$
\begin{aligned}
E(\delta_{\mathbf{a}} \delta_{\mathbf{b}}) &= \Pr(\delta_{\mathbf{a}} = 1, \delta_{\mathbf{b}} = 1) \\
&= \Pr(\delta_{\mathbf{a}} = 1)\Pr(\delta_{\mathbf{b}} = 1 | \delta_{\mathbf{a}} = 1) \\
&= \frac{h}{2^N} \times \frac{h-1}{2^N - 1}
\end{aligned}
$$

There are $2^{N-K}(2^{N-K} - 1)$ such off-diagonal terms, hence a total contribution of $2^{N-K}(2^{N-K} - 1) \times \frac{h(h-1)}{2^N(2^N - 1)} \leq (h2^{-K})^2 \frac{2^N}{2^N - 1}$ to the sum. Putting the contributions

8

from the diagonal and off-diagonal terms together, we get

$$E(n^2) \leq h2^{-K} + (h2^{-K})^2 \frac{2^N}{2^N - 1}$$

$$\mathrm{var}(n) = E(n^2) - (E(n))^2$$

$$\leq \left(h2^{-K} + (h2^{-K})^2 \frac{2^N}{2^N - 1} \right) - \left(h2^{-K} \right)^2$$

$$= h2^{-K} + (h2^{-K})^2 \frac{1}{2^N - 1}$$

$$= h2^{-K} \left(1 + \frac{h2^{-K}}{2^N - 1} \right)$$

$$\leq 2h2^{-K}$$

The last step follows since $h2^{-K}$ is much smaller than $2^N - 1$. Therefore, $E(|n - h2^{-K}|) \leq \sqrt{\mathrm{var}(n)} \leq \left(2h2^{-K} \right)^{\frac{1}{2}}$. Substituting this estimate in the expression for $E(d(\nu_1, \nu_2))$, we get

$$E(d(\nu_1, \nu_2)) = \frac{2^K}{2h} E(|n_1 - n_2|)$$

$$\leq \frac{2^K}{2h} \times 2E(|n - h2^{-K}|)$$

$$\leq \frac{2^K}{2h} \times 2 \times \left(2h2^{-K} \right)^{\frac{1}{2}}$$

$$= \left(2\frac{2^K}{h} \right)^{\frac{1}{2}}$$

$$= \sqrt{2} \times 2^{\frac{1}{2}(\alpha - \beta)N}$$

Since $\alpha_o < \beta_o$ by assumption, this upper bound goes to 0 as N goes to infinity. Since 0 is also a lower bound for $d(\nu_1, \nu_2)$ (and hence a lower bound for the expected value $E(d(\nu_1, \nu_2))$), $\lim_{N \to \infty} E(d(\nu_1, \nu_2))$ must be 0. ∎

REFERENCES

[1] Y. Abu-Mostafa, "Neural networks for computing?," *AIP Conference Proceedings # 151, Neural Networks for Computing*, J. Denker (ed.), pp. 1-6, 1986.

[2] Z. Kohavi, *Switching and Finite Automata Theory*, McGraw-Hill, 1978.

[3] Y. Abu-Mostafa, "The complexity of information extraction," *IEEE Trans. on Information Theory*, vol. IT-32, pp. 513-525, July 1986.

[4] Y. Abu-Mostafa, "Complexity in neural systems," in *Analog VLSI and Neural Systems* by C. Mead, Addison-Wesley, 1988.

Stochastic Learning Networks and their Electronic Implementation

Joshua Alspector*, Robert B. Allen, Victor Hu†, and Srinagesh Satyanarayana‡
Bell Communications Research, Morristown, NJ 07960

We describe a family of learning algorithms that operate on a recurrent, symmetrically connected, neuromorphic network that, like the Boltzmann machine, settles in the presence of noise. These networks learn by modifying synaptic connection strengths on the basis of correlations seen locally by each synapse. We describe a version of the supervised learning algorithm for a network with analog activation functions. We also demonstrate unsupervised competitive learning with this approach, where weight saturation and decay play an important role, and describe preliminary experiments in reinforcement learning, where noise is used in the search procedure. We identify the above described phenomena as elements that can unify learning techniques at a physical microscopic level.

These algorithms were chosen for ease of implementation in vlsi. We have designed a CMOS test chip in 2 micron rules that can speed up the learning about a millionfold over an equivalent simulation on a VAX 11/780. The speedup is due to parallel analog computation for summing and multiplying weights and activations, and the use of physical processes for generating random noise. The components of the test chip are a noise amplifier, a neuron amplifier, and a 300 transistor adaptive synapse, each of which is separately testable. These components are also integrated into a 6 neuron and 15 synapse network. Finally, we point out techniques for reducing the area of the electronic correlational synapse both in technology and design and show how the algorithms we study can be implemented naturally in electronic systems.

1. INTRODUCTION

There has been significant progress, in recent years, in modeling brain function as the collective behavior of highly interconnected networks of simple model neurons. This paper focuses on the issue of learning in these networks especially with regard to their implementation in an electronic system. Learning phenomena that have been studied include associative memory[1], supervised learning by error correction[2] and by stochastic search[3], competitive learning[4] [5] reinforcement learning[6], and other forms of unsupervised learning[7]. From the point of view of neural plausibility as well as electronic implementation, we particularly like learning algorithms that change synaptic connection strengths asynchronously and are based only on information available locally at the synapse. This is illustrated in Fig. 1, where a model synapse uses only the correlations of the neurons it connects and perhaps some weak global evaluation signal not specific to individual neurons to decide how to adjust its conductance.

* Address for correspondence: J. Alspector, Bell Communications Research, 2E-378, 435 South St., Morristown, NJ 07960 / (201) 829-4342 / josh@bellcore.com

† Permanent address: University of California, Berkeley, EE Department, Cory Hall, Berkeley, CA 94720

‡ Permanent address: Columbia University, EE Department, S.W. Mudd Bldg., New York, NY 10027

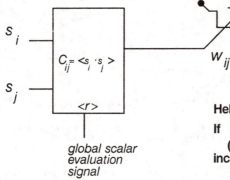

$$s_i$$

$$C_{ij} = <s_i \cdot s_j>$$

$$s_j$$

$$w_{ij}$$

$$<r>$$

*global scalar
evaluation
signal*

Hebb–type learning rule:

If C_{ij} **increases,**
(perhaps in the presence of r **)**
increment w_{ij}

Fig. 1. A local correlational synapse.

We believe that a stochastic search procedure is most compatible with this viewpoint. Statistical procedures based on noise form the communication pathways by which global optimization can take place based only on the interaction of neurons. Search is a necessary part of any learning procedure as the network attempts to find a connection strength matrix that solves a particular problem. Some learning procedures attack the search directly by gradient following through error correction[8] [9] but electronic implementation requires specifying which neurons are input, hidden and output in advance and necessitates global control of the error correction[2] procedure in a way that requires specific connectivity and synchrony at the neural level. There is also the question of how such procedures would work with unsupervised methods and whether they might get stuck in local minima. Stochastic processes can also do gradient following but they are better at avoiding minima, are compatible with asynchronous updates and local weight adjustments, and, as we show in this paper, can generalize well to less supervised learning.

The phenomena we studied are 1) analog activation, 2) noise, 3) semi-local Hebbian synaptic modification, and 4) weight decay and saturation. These techniques were applied to problems in supervised, unsupervised, and reinforcement learning. The goal of the study was to see if these diverse learning styles can be unified at the microscopic level with a small set of physically plausible and electronically implementable phenomena. The hope is to point the way for powerful electronic learning systems in the future by elucidating the conditions and the types of circuits that may be necessary. It may also be true that the conditions for electronic learning may have some bearing on the general principles of biological learning.

2. LOCAL LEARNING AND STOCHASTIC SEARCH

2.1 Supervised Learning in Recurrent Networks with Analog Activations

We have previously shown[10] how the supervised learning procedure of the Boltzmann machine[3] can be implemented in an electronic system. This system works on a recurrent, symmetrically connected network which can be characterized as settling to a minimum in its Liapunov function[1][11]. While this architecture may stretch our criterion of neural plausibility, it does provide for stability and analyzability. The feedback connectivity provides a way for a supervised learning procedure to propagate information back through the network as the stochastic search proceeds. More plausible would be a randomly connected network where symmetry is a statistical approximation and inhibition damps oscillations, but symmetry is more efficient and well matched to our choice of learning rule and search procedure.

We have extended our electronic model of the Boltzmann machine to include analog activations. Fig. 2 shows the model of the neuron we used and its *tanh* or *sigmoid* transfer function. The net input consists of the usual weighted sum of activations from other neurons but, in the case of Boltzmann machine learning, these are added to a noise signal chosen from a variety of distributions so that the neuron performs the physical computation:

$$activation = f(net_i) = f(\Sigma w_{ij}s_j + noise) = \tanh(gain * net_i)$$

Instead of counting the number of on-on and off-off cooccurrences of neurons which a synapse connects, the correlation rule now defines the value of a cooccurrence as:

$$C_{ij} = f_i * f_j$$

where f_i is the activation of neuron i which is a real value from -1 to 1. Note that this rule effectively counts both on-on and off-off cooccurrences in the high gain limit. In this limit, for Gaussian noise, the cumulative probability distribution for the neuron to have activation +1 (on) is close to sigmoidal. The effect of noise "jitter" is illustrated at the bottom of the figure. The weight change rule is still:

$$\text{if } C_{ij}{}^+ > C_{ij}{}^- \text{ then increment } w_{ij} \text{ .... else decrement}$$

where the plus phase clamps the output neurons in their desired states while the minus phase allows them to run free.

As mentioned, we have studied a variety of noise distributions other than those based on the Boltzmann distribution. The 2-2-1 XOR problem was selected as a test case since it has been shown[10] to be easily caught in local minima. The gain was manipulated in conditions with no noise or with noise sampled from one of three distributions. The Gaussian distribution is closest to true electronic thermal noise such as used in our implementation, but we also considered a cut-off uniform distribution and a Cauchy distribution with long noise tails for comparison. The inset to Fig. 3 shows a histogram of samples from the noise distributions used. The noise was multiplied by the temperature to 'jitter' the transfer function. Hence, the jitter decreased as the annealing schedule proceeded.

12

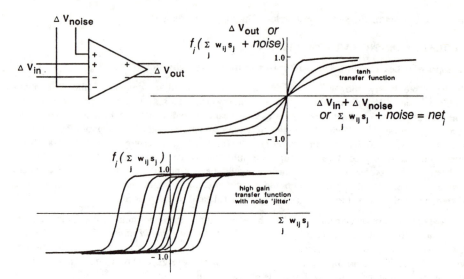

Fig. 2. Electronic analog neuron.

Fig. 3 shows average performance across 100 runs for the last 100 patterns of 2000 training pattern presentations. It can be seen that reducing the gain from a sharp step can improve learning in a small region of gain, even without noise. There seems to be an optimal gain level. However, the addition of noise for any distribution can substantially improve learning at all levels of gain.

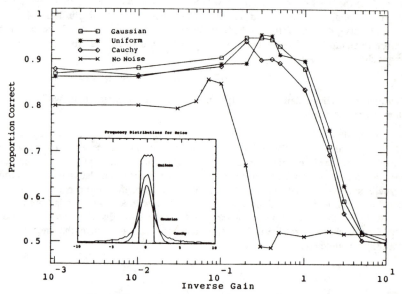

Fig. 3. Proportion correct vs. inverse gain.

2.2 Stochastic Competitive Learning

We have studied how competitive learning[4][5] can be accomplished with stochastic local units. After the presentation of the input pattern, the network is annealed and the weight is increased between the winning cluster unit and the input units which are on. As shown in Fig. 4 this approach was applied to the dipole problem of Rumelhart and Zipser. A 4×4 pixel array input layer connects to a 2 unit competitive layer with recurrent inhibitory connections that are not adjusted. The inhibitory connections provide the competition by means of a winner-take-all process as the network settles. The input patterns are dipoles — only two input units are turned on at each pattern presentation and they must be physically adjacent, either vertically or horizontally. In this way, the network learns about the connectedness of the space and eventually divides it into two equal spatial regions with each of the cluster units responding only to dipoles from one of the halves. Rumelhart and Zipser renormalized the weights after each pattern and picked the winning unit as the one with the highest activation. Instead of explicit normalization of the weights, we include a decay term proportional to the weight. The weights between the input layer and cluster layer are incremented for on-on correlations, but here there are no alternating phases so that even this gross synchrony is not necessary. Indeed, if small time constants are introduced to the weight updates, no external timing should be needed.

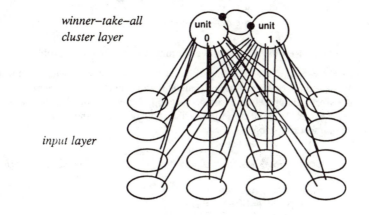

Fig. 4. Competitive learning network for the dipole problem.

Fig. 5 shows the results of several runs. A 1 at the position of an input unit means that unit 1 of the cluster layer has the larger weight leading to it from that position. A + between two units means the dipole from these two units excites unit 1. A 0 and – means that unit 0 is the winner in the complementary case. Note that adjacent 1's should always have a + between them since both weights to unit 1 are stronger. If, however, there is a 1 next to a 0, then there is a tension in the dipole and a competition for dominance in the cluster layer. We define a figure of merit called "surface tension" which is the number of such dipoles in dispute. The smaller the number, the better. Note in Runs A and B, the number is reduced to 4, the minimum possible value, after 2000 pattern presentations. The space is divided vertically and horizontally, respectively. Run C has adopted a less favorable diagonal division with a surface tension of 6.

```
                    Number of dipole pattern presentations

              0          200        800       1400       2000

        0-0-0-0    1+0-0+1    1+1+1+1    1+1+1+1    1+1+1+1
        - - - -    + + + +    + + + -    + + + +    + + + +
        0-0-0-0    1+1+1+1    1+1+1-0    1+1+1+1    1+1+1+1
Run A   - - - -    + + - -    + + - -    + - + -    - - - +
        0-0-0-0    1+1-0-0    1-0-0-0    0-0-0-0    0-0-0-0
        - - - -    + - - -    - - - -    - - - -    - - - -
        0-0-0-0    0-0-0-0    0-0-0-0    0-0-0-0    0-0-0-0


        0-0-0-0    0-0-0-0    0-0-0+1    0-0-0-1    0-0-1+1
        - - - -    - - - +    - - + +    - - - +    - - + +
        0-0-0-0    0-0-0+1    0-0-1+1    0-0-1+1    0-0-1+1
Run B   - - - -    - - - +    - - + +    - - + +    - - + +
        0-0-0-0    1-0-1+1    0-0-1+1    0-0-1+1    0-0-1+1
        - - - -    + - + +    - - + +    - - + +    - - + +
        0-0-0-0    1+0+1+1    0-0+1+1    0-0+1+1    0-0+1+1


        0-0-0-0    0+1+1+1    0+1+1+1    1+1+1+1    1+1+1+1
        - - - -    - + + +    - + + +    + + + +    - + + +
        0-0-0-0    0-1+1+1    0+1+1+1    0+1+1+1    0-0+1+1
Run C   - - - -    - + + +    - + + +    - - + +    - - + +
        0-0-0-0    0-1+1+1    0-0-0-0    0-0-0-0    0-0-0-1
        - - - -    - - - -    - - - -    - - - -    - - - +
        0-0-0-0    0-0-0-0    0-0-0-0    0-0-0-0    0-0-0-1
```

Fig. 5. Results of competitive learning runs on the dipole problem.

Table 1 shows the result of several competitive algorithms compared when averaged over 100 such runs. The deterministic algorithm of Rumelhart and Zipser gives an average surface tension of 4.6 while the stochastic procedure is almost as good. Note that noise is essential in helping the competitive layer settle. Without noise the surface tension is 9.8, showing that the winner-take-all procedure is not working properly.

Competitive learning algorithm	"surface tension"
Stochastic net with decay	
– anneal: T=30→ T=1.0	4.8
– no anneal: 70 @ T=1.0	9.8
Stochastic net with renormalization	5.6
Deterministic, winner–take–all (Rumelhart & Zipser)	4.6

Table 1. Performance of competitive learning algorithms across 100 runs.

We also tried a procedure where, instead of decay, weights were renormalized. The model is that each neuron can support a maximum amount of weight leading into it. Biologically, this might be the area that other neurons can form synapses on, so that one synapse cannot increase its strength except at the expense of some of the others. Electronically, this can be implemented as current emanating from a fixed current source per neuron. As shown in Table 1, this works nearly as well as decay. Moreover, preliminary results show that renormalization is especially effective when more then two cluster units are employed.

Both of the stochastic algorithms, which can be implemented in an electronic synapse in nearly the same way as the supervised learning algorithm, divide the space just as the deterministic normalization procedure[4] does. This suggests that our chip can do both styles of learning, supervised if one includes both phases and unsupervised if only the procedure of the minus phase is used.

2.3 Reinforcement Learning

We have tried several approaches to reinforcement learning using the synaptic model of Fig. 1 where the evaluation signal is a scalar value available globally that represents how well the system performed on each trial. We applied this model to an xor problem with only one output unit. The reinforcement was $r = 1$ for the correct output and $r = -1$ otherwise. To the network, this was similar to supervised learning since for a single unit, the output state is fully specified by a scalar value. A major difference, however, is that we do not clamp the output unit in the desired state in order to compare plus and minus phases. This feature of supervised learning has the effect of adjusting weights to follow a gradient to the desired state. In the reinforcement learning described here, there is no plus phase. This has a satisfying aspect in that no overall synchrony is necessary to compare phases, but is also much slower at converging to a solution because the network has to search the solution space without the guidance of a teacher clamping the output units. This situation becomes much worse when there is more than one output unit. In that case, the probability of reinforcement goes down exponentially with the number of outputs. To test multiple outputs, we chose the simple replication problem whereby the output simply has to replicate the input. We chose the number of hidden units equal to the input (or output).

In the absence of a teacher to clamp the outputs, the network has to find the answer by chance, guided only by a "critic" which rates its effort as "better" or "worse". This means the units must somehow search the space. We use the same stochastic units as in the supervised or unsupervised techniques, but now it is important to have the noise or the annealing temperature set to a proper level. If it is too high, the reinforcement received is random rather than directed by the weights in the network. If it is too low, the available states searched become too small and the probability of finding the right solution decreases. We tuned our annealing schedule by looking at a volatility measure defined at each neuron which is simply the fraction of the time the neuron activation is above zero. We then adjust the final anneal temperature so that this number is neither 0 or 1 (noise too low) nor 0.5 (noise too high). We used both a fixed annealing schedule for all neurons and a unit-specific schedule where the noise was proportional to the sum of weight magnitudes into the unit. A characteristic of reinforcement learning is that the percent correct initially increases but then decreases and often oscillates widely. To avoid this, we added a factor of $(1 - <r>)$ multiplying the final temperature. This helped to stabilize the learning.

In keeping with our simple model of the synapse, we chose a weight adjustment technique that consisted of correlating the states of the connected neurons with the global reinforcement signal. Each synapse measured the quantity $R = r s_i s_j$ for each pattern presented. If $R > 0$, then w_{ij} is incremented and it is decremented if $R < 0$. We later refined this procedure by insisting that the reinforcement be greater than a recent average so that $R = (r - <r>) s_i s_j$. This type of procedure appears in previous work in a number of forms.[12] [13] For $r = \pm 1$ only, this "excess reinforcement" is the same as our previous algorithm but differs if we make a comparison between short term and long term averages or use a graded reinforcement such as the negative of the sum squared error. Following a suggestion by G. Hinton, we also investigated a more complex technique whereby each synapse must store a time average of three quantities: $<r>$, $<s_i s_j>$, and $<r s_i s_j>$. The definition now is $R = <r s_i s_j> - <r><s_i s_j>$ and the rule is the same as before. Statistically, this is the same as "excess reinforcement" if the latter is averaged over trials. For the results reported below the values were collected across 10 pattern presentations. A variation, which employed a continuous moving average, gave similar results.

Table 2 summarizes the performance on the xor and the replication task of these reinforcement learning techniques. As the table shows a variety of increasingly sophisticated weight adjustment rules were explored; nevertheless we were unable to obtain good results with the techniques described for more than 5 output units. In the third column, a small threshold had to be exceeded prior to weight adjustment. In the fourth column, unit-specific temperatures dependent on the sum of weights, were employed. The last column in the table refers to frequency dependent learning where we trained on a single pattern until the network produced a correct answer and then moved on to another pattern. This final procedure is one of several possible techniques related to 'shaping' in operant learning theory in which difficult patterns are presented more often to the network.

network	t=1	time-averaged	$+ \varepsilon = 0.1$	$+T-\Sigma W$	+freq
xor					
2-4-1	(0.60) 0.64	(0.70) 0.88	(0.76) 0.88	(0.92) 0.99	(0.98) 1.00
2-2-1	(0.58) 0.57	(0.69) 0.74	(0.96) 1.00	(0.85) 1.00	(0.78) 0.88
eplication					
2-2-2	(0.94) 0.94	(0.46) 0.46	(0.91) 0.97	(0.87) 0.99	(0.97) 1.00
3-3-3	(0.15) 0.21	(0.31) 0.33	(0.31) 0.62	(0.37) 0.37	(0.97) 1.00
4-4-4	-	-	-	-	(0.75) 1.00
5-5-5	-	-	-	-	(0.13) 0.87
6-6-6	-	-	-	-	(0.02) 0.03

Table 2. Proportion correct performance of reinforcement learning
after (2K) and 10K patterns.

Our experiments, while incomplete, hint that reinforcement learning can also be implemented by the same type of local-global synapse that characterize the other learning paradigms. Noise is also necessary here for the random search procedure.

2.4 Summary of Study of Fundamental Learning Parameters

In summary, we see that the use of noise and our model of a local correlational synapse with a non-specific global evaluation signal are two important features in all the learning paradigms. Graded activation is somewhat less important. Weight decay seems to be quite important although saturation can substitute for it in unsupervised learning. Most interesting from our point of view is that all these phenomena are electronically implementable and therefore physically plausible. Hopefully this means they are also related to true neural phenomena and therefore provide a basis for unifying the various approaches of learning at a microscopic level.

3. ELECTRONIC IMPLEMENTATION

3.1 The Supervised Learning Chip

We have completed the design of the chip previously proposed.[10] Its physical style of computation speeds up learning a millionfold over a computer simulation. Fig. 6 shows a block diagram of the neuron. It is a double differential amplifier. One branch forms a sum of the inputs from the differential outputs of all other neurons with connections to it. The other adds noise from the noise amplifier. This first stage has low gain to preserve dynamic range at the summing nodes. The second stage has high gain and converts to a single ended output. This is fed to a switching arrangement whereby either this output state or some externally applied desired state is fed into the final set of inverter stages which provide for more gain and guaranteed digital complementarity.

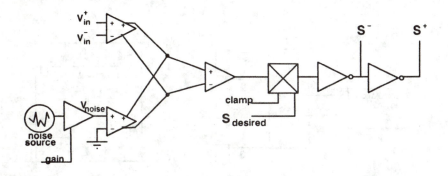

Fig. 6. Block diagram of neuron.

The noise amplifier is shown schematically in Fig. 7. Thermal noise, with an rms level of tens of microvolts, from the channel of an FET is fed into a 3 stage amplifier. Each stage provides a potential gain of 100 over the noise bandwidth. Low pass feedback in each stage stabilizes the DC output as well as controls gain and bandwidth by means of an externally controlled variable resistance for tuning the annealing cycle.

Fig. 8 shows a block diagram of the synapse. The weight is stored in 5 flip-flops as a sign and magnitude binary number. These flip-flops control the conductance from the outputs of neuron i to the inputs of neuron j and vice-versa as shown in the figure. The conductance of the FETs are in the ratio 1:2:4:8 to correspond to the value of the binary number while the sign bit determines whether the true or complementary lines connect. The flip-flops are arranged in a counter which is controlled by the correlation logic. If the plus phase correlations are greater than the minus phase, then the counter is incremented by a single unit. If less, it is decremented.

18

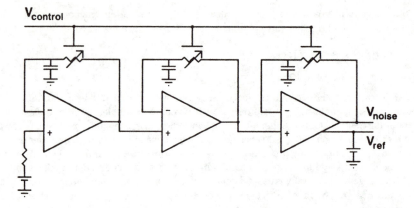

Fig. 7. Block diagram of noise amplifier.

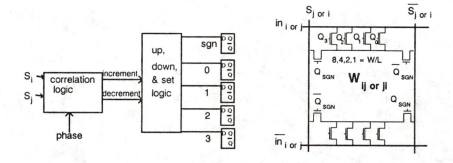

Fig. 8. Block diagram of synapse.

Fig. 9 shows the layout of a test chip. A 6 neuron, 15 synapse network may be seen in the lower left corner. Each neuron has attached to it a noise amplifier to assure that the noise is uncorrelated. The network occupies an area about 2.5 mm on a side in 2 micron design rules. Each 300 transistor synapse occupies 400 by 600 microns. In contrast, a biological synapse occupies only about one square micron. The real miracle of biological learning is in the synapse where plasticity operates on a molecular level, not in the neuron. We can't hope to compete using transistors, however small, especially in the digital domain. Aside from this small network, the rest of the chip is occupied with test structures of the various components.

3.2 Analog Synapse

Analog circuit techniques can reduce the size of the synapse and increase its functionality. Several recent papers[14] [15] have shown how to make a voltage controlled resistor in MOS technology. The voltage controlling the conductance representing the synaptic weight can be obtained by an analog charge integrator from the correlated activation of the neurons which the synapse in question connects. A charge integrator with a "leaky capacitor" has a time constant which can be used to make comparisons as a continuous time average over the last several trials, thereby adding temporal information. One can envision this time constant as being adaptive as well. The charge integrator directly implements the analog Hebb-type[16] correlation rules of section 2.

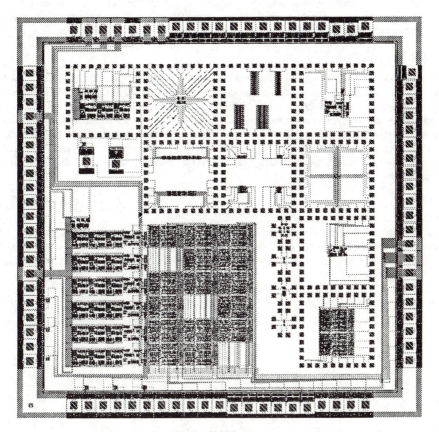

Fig. 9. Chip layout.

3.3 Technological Improvements for Electronic Neural Networks

It is still necessary to store the voltage which controls the analog conductance and we propose the EPROM[17] or EEPROM device for this. Such a device can hold the value of the weight in the same way that flip-flops do in the digital implementation of the synapse[10]. The process which creates this device has two polysilicon layers which are useful for making high valued capacitances in analog circuitry. In addition, the second polysilicon layer could be used to make CCD devices for charge storage and transport. Coupled with the charge storage on a floating gate[18], this forms a compact, low power representation for weight values that approach biological values. Another useful addition would be a high valued stable resistive layer[19]. One could thereby avoid space-wasting long-channel MOSFETs which are currently the only reasonable way to achieve high resistance in MOS technology. Lastly, the addition of a diffusion step or two creates a Bi-CMOS process which adds high quality bipolar transistors useful in analog design. Furthermore, one gets the logarithmic dependence of voltage on current in bipolar technology in a natural, robust way, that is not subject to the variations inherent in using MOSFETs in the subthreshold region. This is especially useful in compressing the dynamic range in sensory processing[20].

4. CONCLUSION

We have shown how a simple adaptive synapse which measures correlations can account for a variety of learning styles in stochastic networks. By embellishing the standard CMOS process and using analog design techniques, a technology suitable for implementing such a synapse electronically can be developed. Noise is an important element in our formulation of learning. It can help a network settle, interpolate between discrete values of conductance during learning, and search a large solution space. Weight decay ("forgetting") and saturation are also important for stability. These phenomena not only unify diverse learning styles but are electronically implementable.

ACKNOWLEDGMENT:
This work has been influenced by many researchers. We would especially like to thank Andy Barto and Geoffrey Hinton for valuable discussions on reinforcement learning, Yannis Tsividis for contributing many ideas in analog circuit design, and Joel Gannett for timely releases of his vlsi verification software.

References

1. J.J. Hopfield, "Neural networks and physical systems with emergent collective computational abilities", Proc. Natl. Acad. Sci. USA **79**, 2554-2558 (1982).

2. D.E. Rumelhart, G.E. Hinton, and R.J. Williams, "Learning internal representations by error propagation", in *Parallel Distributed Processing: Explorations in the Microstructure of Cognition. Vol. 1: Foundations,* edited by D.E. Rumelhart and J.L. McClelland, (MIT Press, Cambridge, MA, 1986), p. 318.

3. D.H. Ackley, G.E. Hinton, and T.J. Sejnowski, "A learning algorithm for Boltzmann machines", Cognitive Science **9**, 147-169 (1985).

4. D.E. Rumelhart and D. Zipser, "Feature discovery by competitive learning", Cognitive Science **9**, 75-112 (1985).

5. S. Grossberg, "Adaptive pattern classification and universal recoding: Part I. Parallel development and coding of neural feature detectors.", Biological Cybernetics **23**, 121-134 (1976).

6. A.G. Barto, R.S. Sutton, and C.W. Anderson, "Neuronlike adaptive elements that can solve difficult learning control problems", IEEE Trans. Sys. Man Cyber. **13**, 835 (1983).

7. B.A. Pearlmutter and G.E. Hinton, "G-Maximization: An unsupervised learning procedure for discovering regularities", in *Neural Networks for Computing,* edited by J.S. Denker, AIP Conference Proceedings 151, American Inst. of Physics, New York (1986), p.333.

8. F. Rosenblatt, *Principles of Neurodynamics: Perceptrons and the Theory of Brain Mechanisms* (Spartan Books, Washington, D.C., 1961).

9. G. Widrow and M.E. Hoff, "Adaptive switching circuits", Inst. of Radio Engineers, Western Electric Show and Convention, Convention Record, Part 4, 96-104 (1960).

10. J. Alspector and R.B. Allen, "A neuromorphic vlsi learning system", in *Advanced Research in VLSI: Proceedings of the 1987 Stanford Conference.* edited by P. Losleben (MIT Press, Cambridge, MA, 1987), pp. 313-349.

11. M.A. Cohen and S. Grossberg, "Absolute stability of global pattern formation and parallel memory storage by competitive neural networks", Trans. IEEE **13**, 815, (1983).

12. B. Widrow, N.K. Gupta, and S. Maitra, "Punish/Reward: Learning with a critic in adaptive threshold systems", IEEE Trans. on Sys. Man & Cyber., SMC-3, 455 (1973).

13. R.S. Sutton, "Temporal credit assignment in reinforcement learning", unpublished doctoral dissertation, U. Mass. Amherst, technical report COINS 84-02 (1984).

14. Z. Czarnul, "Design of voltage-controlled linear transconductance elements with a matched pair of FET transistors", IEEE Trans. Circ. Sys. **33**, 1012, (1986).

15. M. Banu and Y. Tsividis, "Floating voltage-controlled resistors in CMOS technology", Electron. Lett. **18**, 678-679 (1982).

16. D.O. Hebb, *The Organization of Behavior* (Wiley, NY, 1949).

17. D. Frohman-Bentchkowsky, "FAMOS - a new semiconductor charge storage device", Solid-State Electronics **17**, 517 (1974).

18. J.P. Sage, K. Thompson, and R.S. Withers, "An artificial neural network integrated circuit based on MNOS/CCD principles", in *Neural Networks for Computing,* edited by J.S. Denker, AIP Conference Proceedings 151, American Inst. of Physics, New York (1986), p.381.

19. A.P. Thakoor, J.L. Lamb, A. Moopenn, and J. Lambe, "Binary synaptic connections based on memory switching in a-Si:H", in *Neural Networks for Computing,* edited by J.S. Denker, AIP Conference Proceedings 151, American Inst. of Physics, New York (1986), p.426.

20. M.A. Sivilotti, M.A. Mahowald, and C.A. Mead, "Real-Time visual computations using analog CMOS processing arrays", in *Advanced Research in VLSI: Proceedings of the 1987 Stanford Conference.* edited by P. Losleben (MIT Press, Cambridge, MA, 1987), pp. 295-312.

LEARNING ON A GENERAL NETWORK

Amir F. Atiya

Department of Electrical Engineering

California Institute of Technology

Ca 91125

Abstract

This paper generalizes the backpropagation method to a general network containing feedback connections. The network model considered consists of interconnected groups of neurons, where each group could be fully interconnected (it could have feedback connections, with possibly asymmetric weights), but no loops between the groups are allowed. A stochastic descent algorithm is applied, under a certain inequality constraint on each intra-group weight matrix which ensures for the network to possess a unique equilibrium state for every input.

Introduction

It has been shown in the last few years that large networks of interconnected "neuron"-like elements are quite suitable for performing a variety of computational and pattern recognition tasks. One of the well-known neural network models is the backpropagation model [1]-[4]. It is an elegant way for teaching a layered feedforward network by a set of given input/output examples. Neural network models having feedback connections, on the other hand, have also been devised (for example the Hopfield network [5]), and are shown to be quite successful in performing some computational tasks. It is important, though, to have a method for learning by examples for a feedback network, since this is a general way of design, and thus one can avoid using an ad hoc design method for each different computational task. The existence of feedback is expected to improve the computational abilities of a given network. This is because in feedback networks the state iterates until a stable state is reached. Thus processing is performed on several steps or recursions. This, in general allows more processing abilities than the "single step" feedforward case (note also the fact that a feedforward network is a special case of a feedback network). Therefore, in this work we consider the problem of developing a general learning algorithm for feedback networks.

In developing a learning algorithm for feedback networks, one has to pay attention to the following (see Fig. 1 for an example of a configuration of a feedback network). The state of the network evolves in time until it goes to equilibrium, or possibly other types of behavior such as periodic or chaotic motion could occur. However, we are interested in having a steady and and fixed output for every input applied to the network. Therefore, we have the following two important requirements for the network. Beginning in any initial condition, the state should ultimately go to equilibrium. The other requirement is that we have to have a unique equilibrium state. It is in fact that equilibrium state that determines the final output. The objective of the learning algorithm is to adjust the parameters (weights) of the network in small steps, so as to move the unique equilibrium state in a way that will result finally in an output as close as possible to the required one (for each given input). The existence of more than one equilibrium state for a given input causes the following problems. In some iterations one might be updating the weights so as to move one of the equilibrium states in a sought direction, while in other iterations (especially with different input examples) a different equilibrium state is moved. Another important point is that when implementing the network (after the completion of learning), for a fixed input there can be more than one possible output. Independently, other work appeared recently on training a feedback network [6],[7],[8]. Learning algorithms were developed, but solving the problem of ensuring a unique equilibrium was not considered. This problem is addressed in this paper and an appropriate network and a learning algorithm are proposed.

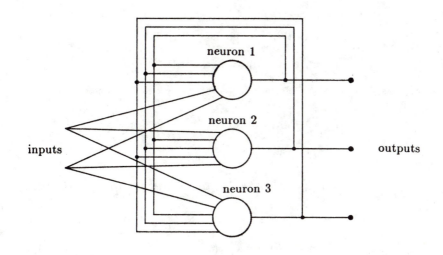

Fig. 1
A recurrent network

The Feedback Network

Consider a group of n neurons which could be fully inter-connected (see Fig. 1 for an example). The weight matrix $\mathbf{W}$ can be asymmetric (as opposed to the Hopfield network). The inputs are also weighted before entering into the network (let $\mathbf{V}$ be the weight matrix). Let $\mathbf{x}$ and $\mathbf{y}$ be the input and output vectors respectively. In our model $\mathbf{y}$ is governed by the following set of differential equations, proposed by Hopfield [5]:

$$\tau \frac{d\mathbf{u}}{dt} = \mathbf{W}\mathbf{f}(\mathbf{u}) - \mathbf{u} + \mathbf{V}\mathbf{x}, \qquad \mathbf{y} = \mathbf{f}(\mathbf{u}) \qquad (1)$$

where $\mathbf{f}(\mathbf{u}) = (f(u_1), ..., f(u_n))^T$, T denotes the transpose operator, f is a bounded and differentiable function, and τ is a positive constant.

For a given input, we would like the network after a short transient period to give a steady and fixed output, no matter what the initial network state was. This means that beginning any initial condition, the state is to be attracted towards a unique equilibrium. This leads to looking for a condition on the matrix $\mathbf{W}$.

Theorem: A network (not necessarily symmetric) satisfying

$$\sum_i \sum_j w_{ij}^2 < 1/\max(f')^2,$$

exhibits no other behavior except going to a unique equilibrium for a given input.

Proof : Let $\mathbf{u}_1(t)$ and $\mathbf{u}_2(t)$ be two solutions of (1). Let

$$J(t) = \|\mathbf{u}_1(t) - \mathbf{u}_2(t)\|^2$$

where $\| \; \|$ is the two-norm. Differentiating J with respect to time, one obtains

$$\frac{dJ(t)}{dt} = 2(\mathbf{u}_1(t) - \mathbf{u}_2(t))^T \left(\frac{d\mathbf{u}_1(t)}{dt} - \frac{d\mathbf{u}_2(t)}{dt} \right).$$

Using (1) , the expression becomes

$$\frac{dJ(t)}{dt} = -\frac{2}{\tau}\|\mathbf{u}_1(t) - \mathbf{u}_2(t)\|^2 + \frac{2}{\tau}(\mathbf{u}_1(t) - \mathbf{u}_2(t))^T \mathbf{W} \big[\mathbf{f}(\mathbf{u}_1(t)) - \mathbf{f}(\mathbf{u}_2(t)) \big].$$

Using Schwarz's Inequality, we obtain

$$\frac{dJ(t)}{dt} \leq -\frac{2}{\tau}\|\mathbf{u}_1(t) - \mathbf{u}_2(t)\|^2 + \frac{2}{\tau}\|\mathbf{u}_1(t) - \mathbf{u}_2(t)\| \cdot \| \mathbf{W} \big[\mathbf{f}(\mathbf{u}_1(t)) - \mathbf{f}(\mathbf{u}_2(t)) \big] \|.$$

Again, by Schwarz's Inequality,

$$\mathbf{w}_i \big[\mathbf{f}(\mathbf{u}_1(t)) - \mathbf{f}(\mathbf{u}_2(t)) \big] \leq \|\mathbf{w}_i\| \cdot \| \mathbf{f}(\mathbf{u}_1(t)) - \mathbf{f}(\mathbf{u}_2(t)) \|, \qquad i = 1, ..., n \qquad (2)$$

where $\mathbf{w}_i$ denotes the ith row of $\mathbf{W}$. Using the mean value theorem, we get

$$\| \mathbf{f}(\mathbf{u}_1(t)) - \mathbf{f}(\mathbf{u}_2(t)) \| \leq (\max|f'|)\|\mathbf{u}_1(t) - \mathbf{u}_2(t)\|. \qquad (3)$$

Using (2),(3), and the expression for $J(t)$, we get

$$\frac{dJ(t)}{dt} \leq -\alpha J(t) \qquad (4)$$

where

$$\alpha = \frac{2}{\tau} - \frac{2}{\tau}(\max|f'|)\sqrt{\sum_i \sum_j w_{ij}^2}.$$

By hypothesis of the Theorem, α is strictly positive. Multiplying both sides of (4) by $exp(\alpha t)$, the inequality

$$\frac{d}{dt}(J(t)e^{\alpha t}) \leq 0$$

results, from which we obtain

$$J(t) \leq J(0)e^{-\alpha t}.$$

From that and from the fact that J is non-negative, it follows that $J(t)$ goes to zero as $t \to \infty$. Therefore, any two solutions corresponding to any two initial conditions ultimately approach each other. To show that this asymptotic solution is in fact an equilibrium, one simply takes $\mathbf{u}_2(t) = \mathbf{u}_1(t + T)$, where T is a constant, and applies the above argument (that $J(t) \to 0$ as $t \to \infty$), and hence $\mathbf{u}_1(t + T) \to \mathbf{u}_1(t)$ as $t \to \infty$ for any T, and this completes the proof.

For example, if the function f is of the following widely used sigmoid-shaped form,

$$f(u) = \frac{1}{1 + e^{-u}},$$

then the sum of the square of the weights should be less than 16. Note that for any function f, scaling does not have an effect on the overall results. We have to work in our updating scheme subject to the constraint given in the Theorem. In many cases where a large network is necessary, this constraint might be too restrictive. Therefore we propose a general network, which is explained in the next Section.

The General Network

We propose the following network (for an example refer to Fig. 2). The neurons are partitioned into several groups. Within each group there are no restrictions on the connections and therefore the group could be fully interconnected (i.e. it could have feedback connections). The groups are connected to each other, but in a way that there are no loops. The inputs to the whole network can be connected to the inputs of any of the groups (each input can have several connections to several groups). The outputs of the whole network are taken to be the outputs (or part of the outputs) of a certain group, say group f. The constraint given in the Theorem is applied on each intra-group weight matrix separately. Let $(\mathbf{q}^a, \mathbf{s}^a)$, $a = 1, ..., N$ be the input/output vector pairs of the function to be implemented. We would like to minimize the sum of the square error, given by

$$E = \sum_{a=1}^{N} e_a$$

where

$$e_a = \sum_{i=1}^{M}(y_i^f - s_i^a)^2,$$

and $\mathbf{y}^f$ is the output vector of group f upon giving input $\mathbf{q}^a$, and M is the dimension of vector $\mathbf{s}^a$. The learning process is performed by feeding the input examples $\mathbf{q}^a$ sequentially to the network, each time updating the weights in an attempt to minimize the error.

26

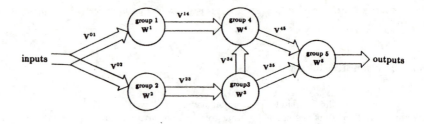

Fig. 2
An example of a general network
(each group represents a recurrent network)

Now, consider a single group l. Let $\mathbf{W}^l$ be the intra-group weight matrix of group l, $\mathbf{V}^{rl}$ be the matrix of weights between the outputs of group r and the inputs of group l, and $\mathbf{y}^l$ be the output vector of group l. Let the respective elements be w_{ij}^l, v_{ij}^{rl}, and y_i^l. Furthermore, let n_l be the number of neurons of group l. Assume that the time constant τ is sufficiently small so as to allow the network to settle quickly to the equilibrium state, which is given by the solution of the equation

$$\mathbf{y}^l = \mathbf{f}(\mathbf{W}^l \mathbf{y}^l + \sum_{r \in A_l} \mathbf{V}^{rl} \mathbf{y}^r). \qquad (5)$$

where A_l is the set of the indices of the groups whose outputs are connected to the inputs of group l. We would like each iteration to update the weight matrices $\mathbf{W}^l$ and $\mathbf{V}^{rl}$ so as to move the equilibrium in a direction to decrease the error. We need therefore to know the change in the error produced by a small change in the weight matrices. Let $\frac{\partial e_a}{\partial \mathbf{W}^l}$, and $\frac{\partial e_a}{\partial \mathbf{V}^{rl}}$ denote the matrices whose $(i,j)^{th}$ element are $\frac{\partial e_a}{\partial w_{ij}^l}$, and $\frac{\partial e_a}{\partial v_{ij}^l}$ respectively. Let $\frac{\partial e_a}{\partial \mathbf{y}^l}$ be the column vector whose i^{th} element is $\frac{\partial e_a}{\partial y_i^l}$. We obtain the following relations:

$$\frac{\partial e_a}{\partial \mathbf{W}^l} = \left[\mathbf{\Lambda}^l - (\mathbf{W}^l)^T\right]^{-1} \frac{\partial e_a}{\partial \mathbf{y}^l}(\mathbf{y}^l)^T,$$

$$\frac{\partial e_a}{\partial \mathbf{V}^{rl}} = \left[\mathbf{\Lambda}^l - (\mathbf{W}^l)^T\right]^{-1} \frac{\partial e_a}{\partial \mathbf{y}^l}(\mathbf{y}^r)^T,$$

where $\mathbf{\Lambda}^l$ is the diagonal matrix whose i^{th} diagonal element is $1/f'(\sum_k w_{ik}^l y_k^l + \sum_r \sum_k v_{ik}^{rl} y_k^r)$ for a derivation refer to Appendix). The vector $\frac{\partial e_a}{\partial \mathbf{y}^l}$ associated with group l can be obtained in terms of the vectors $\frac{\partial e_a}{\partial \mathbf{y}^j}$, $j \epsilon B_l$, where B_l is the set of the indices of the groups whose inputs are connected to the outputs of group l. We get (refer to Appendix)

$$\frac{\partial e_a}{\partial \mathbf{y}^l} = \sum_{j \epsilon B_l} (\mathbf{V}^{lj})^T \left[\mathbf{\Lambda}^j - (\mathbf{W}^j)^T\right]^{-1} \frac{\partial e_a}{\partial \mathbf{y}^j}. \qquad (6)$$

The matrix $\mathbf{\Lambda}^l - (\mathbf{W}^l)^T$ for any group l can never be singular, so we will not face any problem in the updating process. To prove that, let $\mathbf{z}$ be a vector satisfying

$$\left[\mathbf{\Lambda}^l - (\mathbf{W}^l)^T\right]\mathbf{z} = 0.$$

We can write

$$z_i/\max|f'| \le \sum_k w_{ki}^l z_k, \qquad i = 1, ..., n_l$$

where z_i is the ith element of $\mathbf{z}$. Using Schwarz's Inequality, we obtain

$$|z_i|/\max|f'| \le \sqrt{\sum_k z_k^2}\sqrt{\sum_k (w_{ki}^l)^2}, \qquad i = 1, ..., n_l$$

Squaring both sides and adding the inequalities for $i = 1, ..., n_l$, we get

$$\sum_i z_i^2 \le \max(f')^2 \left(\sum_k z_k^2\right)\sum_i\sum_k (w_{ki}^l)^2. \qquad (7)$$

Since the condition

$$\sum_i\sum_k (w_{ik}^l)^2 < 1/\max(f')^2),$$

is enforced, it follows that (7) cannot be satisfied unless $\mathbf{z}$ is the zero vector. Thus, the matrix $\mathbf{\Lambda}^l - (\mathbf{W}^l)^T$ cannot be singular.

For each iteration we begin by updating the weights of group f (the group containing the final outputs). For that group $\frac{\partial e_a}{\partial \mathbf{y}}$ equals simply $2(y_1^f - s_1, ..., y_M^f - s_M, 0, ..., 0)^T)$. Then we move backwards to the groups connected to that group and obtain their corresponding $\frac{\partial e_a}{\partial \mathbf{y}}$ vectors using (6), update the weights, and proceed in the same manner until we complete updating all the groups. Updating the weights is performed using the following stochastic descent algorithm for each group,

$$\Delta\mathbf{W} = -\alpha_1 \frac{\partial e_a}{\partial \mathbf{W}} + \alpha_2 e_a \mathbf{R},$$

$$\Delta\mathbf{V} = -\alpha_3 \frac{\partial e_a}{\partial \mathbf{V}} + \alpha_4 e_a \mathbf{R},$$

where $\mathbf{R}$ is a noise matrix whose elements are characterized by independent zero-mean unity-variance Gaussian densities, and the α's are parameters. The purpose of adding noise is to allow escaping local minima if one gets stuck in any of them. Note that the control parameter is taken to be e_a. Hence the variance of the added noise tends to decrease the more we approach the ideal zero-error solution. This makes sense because for a large error, i.e. for an unsatisfactory solution, it pays more to add noise to the weight matrices in order to escape local minima. On the other hand, if the error is small, then we are possibly near the global minimum or to an acceptable solution, and hence we do not want too much noise in order not to be thrown out of that basin. Note that once we reach the ideal zero-error solution the added noise as well as the gradient of e_a become zero for all a and hence the increments of the weight matrices become zero. If after a certain iteration $\mathbf{W}$ happens to violate the constraint $\sum_{ij} w_{ij}^2 \le constant < 1/max(f')^2$, then its elements are scaled so as to project it back onto the surface of the hypershere.

Implementation Example

A pattern recognition example is considered. Fig. 3 shows a set of two-dimensional training patterns from three classes. It is required to design a neural network recognizer with three output neurons. Each of the neurons should be on if a sample of the corresponding class is presented, and off otherwise, i.e. we would like to design a "winner-take-all" network. A single-layer three neuron feedback network is implemented. We obtained 3.3% error. Performing the same experiment on a feedforward single-layer network with three neurons, we obtained 20% error. For satisfactory results, a feedforward network should be two-layer. With one neuron in the first layer and three in the second layer, we got 36.7% error. Finally, with two neurons in the first layer and three in the second layer, we got a match with the feedback case, with 3.3% error.

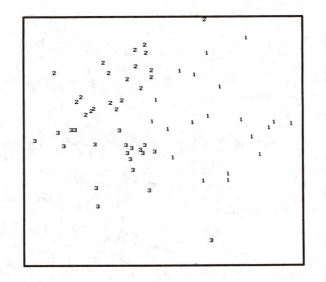

Fig. 3
A pattern recognition example

Conclusion

A way to extend the backpropagation method to feedback networks has been proposed. A condition on the weight matrix is obtained, to insure having only one fixed point, so as to prevent having more than one possible output for a fixed input. A general structure for networks is presented, in which the network consists of a number of feedback groups connected to each other in a feedforward manner. A stochastic descent rule is used to update the weights. The method is applied to a pattern recognition example. With a single-layer feedback network it obtained good results. On the other hand, the feedforward backpropagation method achieved good resuls only for the case of more than one layer, hence also with a larger number of neurons than the feedback case.

Acknowledgement

The author would like to gratefully acknowledge Dr. Y. Abu-Mostafa for the useful discussions. This work is supported by Air Force Office of Scientific Research under Grant AFOSR-86-0296.

Appendix

Differentiating (5), one obtains

$$\frac{\partial y_j^l}{\partial w_{kp}^l} = f'(z_j^l)(\sum_m w_{jm}^l \frac{\partial y_m^l}{\partial w_{kp}^l} + y_p^l \delta_{jk}), \qquad k, p = 1, ..., n_l$$

where

$$\delta_{jk} = \begin{cases} 1 & \text{if } j = k \\ 0 & \text{otherwise,} \end{cases}$$

and

$$z_j^l = \sum_m w_{jm}^l y_m^l + \sum_{r \in A_l} \sum_m v_{jm}^{rl} y_m^r.$$

We can write

$$\frac{\partial \mathbf{y}^l}{\partial w_{kp}^l} = (\mathbf{\Lambda}^l - \mathbf{W}^l)^{-1} \mathbf{b}^{kp} \qquad (A-1)$$

where $\mathbf{b}^{kp}$ is the n_l-dimensional vector whose i^{th} component is given by

$$b_i^{kp} = \begin{cases} y_p^l & \text{if } i = k \\ 0 & \text{otherwise.} \end{cases}$$

By the chain rule,

$$\frac{\partial e_a}{\partial w_{kp}^l} = \sum_j \frac{\partial e_a}{\partial y_j^l} \frac{\partial y_j^l}{\partial w_{kp}^l},$$

which, upon substituting from $(A-1)$, can be put in the form $y_p^l \mathbf{g}_k^T \frac{\partial e_a}{\partial \mathbf{y}^l}$, where $\mathbf{g}_k$ is the kth column of $(\mathbf{\Lambda}^l - \mathbf{W}^l)^{-1}$. Finally, we obtain the required expression, which is

$$\frac{\partial e_a}{\partial \mathbf{W}^l} = \left[\mathbf{\Lambda}^l - (\mathbf{W}^l)^T\right]^{-1} \frac{\partial e_a}{\partial \mathbf{y}^l} (\mathbf{y}^l)^T.$$

Regarding $\frac{\partial e_a}{\partial \mathbf{V}^{rl}}$, it is obtained by differentiating (5) with respect to v_{kp}^{rl}. We get similarly

$$\frac{\partial \mathbf{y}^l}{\partial v_{kp}^{rl}} = (\mathbf{\Lambda}^l - \mathbf{W}^l)^{-1} \mathbf{c}^{kp}$$

where $\mathbf{c}^{kp}$ is the n_l-dimensional vector whose i^{th} component is given by

$$c_i^{kp} = \begin{cases} y_p^r & \text{if } i = k \\ 0 & \text{otherwise.} \end{cases}$$

A derivation very similar to the case of $\frac{\partial e_a}{\partial \mathbf{W}^l}$ results in the following required expression:

$$\frac{\partial e_a}{\partial \mathbf{V}^{rl}} = \left[\mathbf{\Lambda}^l - \left(\mathbf{W}^l\right)^T\right]^{-1} \frac{\partial e_a}{\partial \mathbf{y}^l} (\mathbf{y}^r)^T.$$

Now, finally consider $\frac{\partial e_a}{\partial \mathbf{y}^l}$. Let $\frac{\partial \mathbf{y}^j}{\partial \mathbf{y}^l}$, $j \epsilon B_l$ be the matrix whose $(k,p)^{th}$ element is $\frac{\partial y_k^j}{\partial y_p^l}$. The elements of $\frac{\partial \mathbf{y}^j}{\partial \mathbf{y}^l}$ can be obtained by differentiating the equation for the fixed point for group j, as follows,

$$\frac{\partial y_k^j}{\partial y_p^l} = f'(z_k^j)\left(v_{kp}^{lj} + \sum_m w_{km}^j \frac{\partial y_m^j}{\partial y_p^l}\right).$$

Hence,

$$\frac{\partial \mathbf{y}^j}{\partial \mathbf{y}^l} = \left(\mathbf{\Lambda}^j - \mathbf{W}^j\right)^{-1} \mathbf{V}^{lj}. \qquad (A-2)$$

Using the chain rule, one can write

$$\frac{\partial e_a}{\partial \mathbf{y}^l} = \sum_{j \epsilon B_l} \left(\frac{\partial \mathbf{y}^j}{\partial \mathbf{y}^l}\right)^T \frac{\partial e_a}{\partial \mathbf{y}^j}.$$

We substitute from $(A-2)$ into the previous equation to complete the derivation by obtaining

$$\frac{\partial e_a}{\partial \mathbf{y}^l} = \sum_{j \epsilon B_l} \left(\mathbf{V}^{lj}\right)^T \left[\left(\mathbf{\Lambda}^j - \left(\mathbf{W}^j\right)^T\right)\right]^{-1} \frac{\partial e_a}{\partial \mathbf{y}^j}.$$

References

[1] P. Werbos, "Beyond regression: New tools for prediction and analysis in behavioral sciences", Harvard University dissertation, 1974.

[2] D. Parker, "Learning logic", MIT Tech Report TR-47, Center for Computational Research in Economics and Management Science, 1985.

[3] Y. Le Cun, "A learning scheme for asymmetric threshold network", Proceedings of Cognitiva, Paris, June 1985.

[4] D. Rumelhart, G. Hinton, and R. Williams, "Learning internal representations by error propagation", in D. Rumelhart, J. McLelland and the PDP research group (Eds.), *Parallel distributed processing: Explorations in the microstructure of cognition*, Vol. 1, MIT Press, Cambridge, MA, 1986.

[5] J. Hopfield, "Neurons with graded response have collective computational properties like those of two-state neurons", Proc. Natl. Acad. Sci. USA, May 1984.

[6] L. Almeida, "A learning rule for asynchronous perceptrons with feedback in a combinatorial environment", Proc. of the First Int. Annual Conf. on Neural Networks, San Diego, June 1987.

[7] R. Rohwer, and B. Forrest, "Training time-dependence in neural networks", Proc. of the First Int. Annual Conf. on Neural Networks, San Diego, June 1987.

[8] F. Pineda, "Generalization of back-propagation to recurrent neural networks", Phys. Rev. Lett., vol. 59, no. 19, 9 Nov. 1987.

AN ARTIFICIAL NEURAL NETWORK FOR SPATIO-TEMPORAL BIPOLAR PATTERNS: APPLICATION TO PHONEME CLASSIFICATION

Toshiteru Homma
Les E. Atlas
Robert J. Marks II

Interactive Systems Design Laboratory
Department of Electrical Engineering, FT-10
University of Washington
Seattle, Washington 98195

ABSTRACT

An artificial neural network is developed to recognize spatio-temporal bipolar patterns associatively. The function of a formal neuron is generalized by replacing multiplication with convolution, weights with transfer functions, and thresholding with nonlinear transform following adaptation. The Hebbian learning rule and the delta learning rule are generalized accordingly, resulting in the learning of weights and delays. The neural network which was first developed for spatial patterns was thus generalized for spatio-temporal patterns. It was tested using a set of bipolar input patterns derived from speech signals, showing robust classification of 30 model phonemes.

1. INTRODUCTION

Learning spatio-temporal (or dynamic) patterns is of prominent importance in biological systems and in artificial neural network systems as well. In biological systems, it relates to such issues as classical and operant conditioning, temporal coordination of sensorimotor systems and temporal reasoning. In artificial systems, it addresses such real-world tasks as robot control, speech recognition, dynamic image processing, moving target detection by sonars or radars, EEG diagnosis, and seismic signal processing.

Most of the processing elements used in neural network models for practical applications have been the formal neuron[1] or its variations. These elements lack a memory flexible to temporal patterns, thus limiting most of the neural network models previously proposed to problems of spatial (or static) patterns. Some past solutions have been to convert the dynamic problems to static ones using buffer (or storage) neurons, or using a layered network with/without feedback.

We propose in this paper to use a "dynamic formal neuron" as a processing element for learning dynamic patterns. The operation of the dynamic neuron is a temporal generalization of the formal neuron. As shown in the paper, the generalization is straightforward when the activation part of neuron operation is expressed in the frequency domain. Many of the existing learning rules for static patterns can be easily generalized for dynamic patterns accordingly. We show some examples of applying these neural networks to classifying 30 model phonemes.

2. FORMAL NEURON AND DYNAMIC FORMAL NEURON

The formal neuron is schematically drawn in Fig. 1(a), where

Input	$\vec{x} = [x_1\ x_2\ \cdots\ x_L]^T$	
Activation	$y_i,\ i = 1,2,\ldots,N$	
Output	$z_i,\ i = 1,2,\ldots,N$	
Transmittance	$\vec{w} = [w_{i1}\ w_{i2}\ \cdots\ w_{iL}]^T$	
Node operator	η where $\eta(\cdot)$ is a nonlinear memoryless transform	
Neuron operation	$z_i = \eta(\vec{w}_i^T \vec{x})$	(2.1)

Note that a threshold can be implicitly included as a transmittance from a constant input.

In its original form of formal neuron, $x_i \in \{0,1\}$ and $\eta(\cdot)$ is a unit step function $u(\cdot)$. A variation of it is a bipolar formal neuron where $x_i \in \{-1,1\}$ and $\eta(\cdot)$ is the sign function $sgn(\cdot)$. When the inputs and output are converted to frequency of spikes, it may be expressed as $x_i \in \mathbf{R}$ and $\eta(\cdot)$ is a rectifying function $r(\cdot)$. Other node operators such as a sigmoidal function may be used.

We generalize the notion of formal neuron so that the input and output are functions of time. In doing so, weights are replaced with transfer functions, multiplication with convolution, and the node operator with a nonlinear transform following adaptation as often observed in biological systems.

Fig. 1(b) shows a schematic diagram of a dynamic formal neuron where

Input	$\vec{x}(t) = [x_1(t)\ x_2(t)\ \cdots\ x_L(t)]^T$	
Activation	$y_i(t),\ i = 1,2,\ldots,N$	
Output	$z_i(t),\ i = 1,2,\ldots,N$	
Transfer function	$\vec{w}(t) = [w_{i1}(t)\ w_{i2}(t)\ \cdots\ w_{iL}(t)]^T$	
Adaptation	$a_i(t)$	
Node operator	η where $\eta(\cdot)$ is a nonlinear memoryless transform	
Neuron operation	$z_i(t) = \eta(a_i(-t) * \vec{w}_i(t)^T * \vec{x}(t))$	(2.2)

For convenience, we denote $*$ as correlation instead of convolution. Note that convolving a(t) with b(t) is equivalent to correlating a(-t) with b(t).

If the Fourier transforms $\vec{x}(f) = F\{\vec{x}(t)\}$, $\vec{w}_i(f) = F\{\vec{w}_i(t)\}$, $y_i(f) = F\{y_i(t)\}$, and $a_i(f) = F\{a_i(t)\}$ exist, then

$$y_i(f) = a_i(f)\,[\vec{w}_i(f)^{CT}\ \vec{x}(f)] \tag{2.3}$$

where $\vec{w}_i(f)^{CT}$ is the conjugate transpose of $\vec{w}_i(t)$.

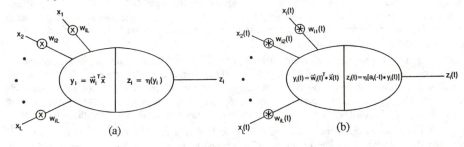

Fig. 1. Formal Neuron and Dynamic Formal Neuron.

3. LEARNING FOR FORMAL NEURON AND DYNAMIC FORMAL NEURON

A number of learning rules for formal neurons has been proposed in the past. In the following paragraphs, we formulate a learning problem and describe two of the existing learning rules, namely, Hebbian learning and delta learning, as examples.

Present to the neural network M pairs of input and desired output samples $\{\vec{x}^{(k)}, \vec{d}^{(k)}\}$, $k = 1, 2, \ldots, M$, in order. Let $\underline{W}^{(k)} = [\vec{w}_1^{(k)} \ \vec{w}_2^{(k)} \ \cdots \ \vec{w}_N^{(k)}]^T$ where $\vec{w}_i^{(k)}$ is the transmittance vector at the k-th step of learning. Likewise, let

$$\underline{X}^{(k)} = [\vec{x}^{(1)} \ \vec{x}^{(2)} \ \cdots \ \vec{x}^{(k)}], \quad \underline{Y}^{(k)} = [\vec{y}^{(1)} \ \vec{y}^{(2)} \ \cdots \ \vec{y}^{(k)}],$$

$$\underline{Z}^{(k)} = [\vec{z}^{(1)} \ \vec{z}^{(2)} \ \cdots \ \vec{z}^{(k)}], \quad \text{and} \quad \underline{D}^{(k)} = [\vec{d}^{(1)} \ \vec{d}^{(2)} \ \cdots \ \vec{d}^{(k)}],$$

where

$$\vec{y}^{(k)} = \underline{W}^{(k)} \vec{x}^{(k)}, \quad \vec{z}^{(k)} = \underline{\eta}(\vec{y}^{(k)}) , \quad \text{and} \quad \underline{\eta}(\vec{y}) = [\eta(y_1) \ \eta(y_2) \ \cdots \ \eta(y_N)]^T.$$

The Hebbian learning rule [2] is described as follows*:

$$\underline{W}^{(k)} = \underline{W}^{(k-1)} + \alpha \vec{d}^{(k)} \vec{x}^{(k)T}$$
(3.1)

The delta learning (or LMS learning) rule[3, 4] is described as follows:

$$\underline{W}^{(k)} = \underline{W}^{(k-1)} - \alpha \{\underline{W}^{(k-1)} \vec{x}^{(k)} - \vec{d}^{(k)}\} \vec{x}^{(k)T}$$
(3.2)

The learning rules described in the previous section are generalized for the dynamic formal neuron by replacing multiplication with correlation. First, the problem is reformulated and then the generalized rules are described as follows.

Present to the neural network M pairs of time-varing input and output samples $\{\vec{x}^{(k)}(t), \vec{d}^{(k)}(t)\}$, $k = 1, 2, \ldots, M$, in order. Let $\underline{W}^{(k)}(t) = [\vec{w}_1(t)^{(k)}(t) \ \vec{w}_2^{(k)}(t) \cdots \vec{w}_N^{(k)}(t)]^T$ where $\vec{w}_i^{(k)}(t)$ is the vector whose elements $w_{ij}^{(k)}(t)$ are transfer functions connecting the input j to the neuron i at the k-th step of learning. The Hebbian learning rule for the dynamic neuron is then

$$\underline{W}^{(k)}(t) = \underline{W}^{(k-1)}(t) + \alpha(-t) * \vec{d}^{(k)}(t) * \vec{x}^{(k)}(t)^T .$$
(3.3)

The delta learning rule for dynamic neuron is then

$$\underline{W}^{(k)}(t) = \underline{W}^{(k-1)}(t) - \alpha(-t) * \{\underline{W}^{(k-1)}(t) * \vec{x}^{(k)}(t) - \vec{d}^{(k)}(t)\} * \vec{x}^{(k)}(t)^T .$$
(3.4)

This generalization procedure can be applied to other learning rules in some linear discriminant systems[5] , the self-organizing mapping system by Kohonen[6] , the perceptron [7] , the back-propagation model[3] , etc. When a system includes a nonlinear operation, more careful analysis is necesssay as pointed out in the Discussion section.

4. DELTA LEARNING, PSEUDO INVERSE AND REGULARIZATION

This section reviews the relation of the delta learning rule to the pseudo-inverse and the technique known as regularization.[4, 6, 8, 9, 10]

Consider a minimization problem as described below: Find $\underline{W}$ which minimizes

$$R = \sum_k \|\vec{y}^{(k)} - \vec{d}^{(k)}\|_2^2 = (\vec{y}^{(k)} - \vec{d}^{(k)})^T (\vec{y}^{(k)} - \vec{d}^{(k)})$$
(4.1)

subject to $\vec{y}^{(k)} = \underline{W} \vec{x}^{(k)}$.

A solution by the delta rule is, using a gradient descent method,

$$\underline{W}^{(k)} = \underline{W}^{(k-1)} - \alpha \frac{\partial}{\partial \underline{W}} R^{(k)}$$
(4.2)

---

* This interpretation assumes a strong supervising signal at the output while learning.

where $R^{(k)} = \| \vec{y}^{(k)} - \vec{d}^{(k)} \|_2^2$. The minimum norm solution to the problem, $\underline{W}^*$, is unique and can be expressed as

$$\underline{W}^* = \underline{D}\,\underline{X}^t \tag{4.3}$$

where $\underline{X}^t$ is the Moore-Penrose pseudo-inverse of $\underline{X}$, i.e.,

$$\underline{X}^t = \lim_{\sigma \to 0}(\underline{X}^T\underline{X} + \sigma^2\underline{I})^{-1}\underline{X}^T = \lim_{\sigma \to 0}\underline{X}^T(\underline{X}\,\underline{X}^T + \sigma^2\underline{I})^{-1}. \tag{4.4}$$

On the condition that $0 < \alpha < \dfrac{2}{\lambda_{max}}$ where λ_{max} is the maximum eigenvalue of $\underline{X}^T\underline{X}$, $\vec{x}^{(k)}$ and $\vec{d}^{(k)}$ are independent, and $\underline{W}^{(k)}$ is uncorrelated with $\vec{x}^{(k)}$,

$$E\{\underline{W}^*\} = E\{\underline{W}^{(\infty)}\} \tag{4.5}$$

where $E\{x\}$ denotes the expected value of x. One way to make use of this relation is to calculate $\underline{W}^*$ for known standard data and refine it by (4.2), thereby saving time in the early stage of learning.

However, this solution results in an ill-conditioned $\underline{W}$ often in practice. When the problem is ill-posed as such, the technique known as regularization can alleviate the ill-conditioning of $\underline{W}$. The problem is reformulated by finding $\underline{W}$ which minimizes

$$R(\sigma) = \sum_k \|\vec{y}^{(k)} - \vec{d}^{(k)}\|_2^2 + \sigma^2 \sum_k \|\vec{w}_k\|_2^2 \tag{4.6}$$

subject to $\vec{y}^{(k)} = \underline{W}\vec{x}^{(k)}$ where $\underline{W} = [\vec{w}_1\vec{w}_2 \cdots \vec{w}_N]^T$.
This reformulation regularizes (4.3) to

$$\underline{W}(\sigma) = \underline{D}\,\underline{X}^T(\underline{X}\,\underline{X}^T + \sigma^2\underline{I})^{-1} \tag{4.7}$$

which is statistically equivalent to $\underline{W}^{(\infty)}$ when the input has an additive noise of variance σ^2 uncorrelated with $\vec{x}^{(k)}$. Interestingly, the leaky LMS algorithm[11] leads to a statistically equivalent solution

$$\underline{W}^{(k)} = \beta\underline{W}^{(k-1)} - \alpha\{\underline{W}^{(k-1)}\vec{x}^{(k)} - \vec{d}^{(k)}\}\vec{x}^{(k)T} \tag{4.8}$$

where $0 < \beta < 1$ and $0 < \alpha < \dfrac{2}{\lambda_{max}}$. These solutions are related as

$$E\{\underline{W}(\sigma)\} = E\{\underline{W}^{(\infty)}\} \tag{4.9}$$

if $\sigma^2 = \dfrac{1-\beta}{\alpha}$ when $\underline{W}^{(k)}$ is uncorrelated with $\vec{x}^{(k)}$.[11]

Equation (4.8) can be generalized for a network using dynamic formal neurons, resulting in a equation similar to (3.4). Making use of (4.9), (4.7) can be generalized for a dynamic neuron network as

$$\underline{W}(t\,;\sigma) = F^{-1}\{\underline{D}(f)\underline{X}(f)^{CT}(\underline{X}(f)\underline{X}(f)^{CT} + \sigma^2\underline{I})^{-1}\} \tag{4.10}$$

where F^{-1} denotes the inverse Fourier transform.

5. SYNTHESIS OF BIPOLAR PHONEME PATTERNS

This section illustrates the scheme used to synthesize bipolar phoneme patterns and to form prototype and test patterns.

The fundamental and first three formant frequencies, along with their bandwidths, of phonemes provided by Klatt[12] were taken as parameters to synthesize 30 prototype phoneme patterns. The phonemes were labeled as shown in Table 1. An array of L (=100) input neurons covered the range of 100 to 4000 Hz. Each neuron had a bipolar state which was +1 only when one of the frequency bands in the phoneme presented to the network was within the critical band of the neuron and -1 otherwise. The center frequencies (f_c) of critical bands were obtained by dividing the 100 to 4000 Hz range into a log scale by L. The critical bandwidth was a constant 100 Hz up to the center frequency f_c = 500 Hz and $0.2f_c$ Hz when f_c >500 Hz.[13]

The parameters shown in Table 1 were used to construct 30 prototype phoneme patterns. For θ, it was constructed as a combination of t and θ. F_1, F_2 ,F_3 were the first, second, and third formants, and B_1, B_2, and B_3. were corresponding bandwidths. The fundamental frequency F_0 = 130 Hz with B_0 = 10 Hz was added when the phoneme was voiced. For plosives, there was a stop before formant traces start. The resulting bipolar patterns are shown in Fig.2. Each pattern had length of 5 time units, composed by linearly interpolating the frequencies when the formant frequency was gliding.

A sequence of phonemes converted from a continuous pronunciation of digits, {o, zero, one, two, three, four, five, six, seven, eight, nine }, was translated into a bipolar pattern, adding two time units of transition between two consequtive phonemes by interpolating the frequency and bandwidth parameters linearly. A flip noise was added to the test pattern and created a noisy test pattern. The sign at every point in the original clean test pattern was flipped with the probability 0.2. These test patterns are shown in Fig. 3.

Table 1. Labels of Phonemes

Label	Phoneme
1	[i^y]
2	[I^ə]
3	[e^y]
4	[ε^ə]
5	[æ^ə]
6	[a]
7	[ɔ^ə]
8	[ʌ]
9	[o^w]
10	[ʊ^ə]
11	[u^w]
12	[ɚ]
13	[a^y]
14	[a^w]
15	[o^y]
16	[w]
17	[y]
18	[r]
19	[l]
20	[f]
21	[v]
22	[θ]
23	[ð]
24	[s]
25	[z]
26	[p]
27	[t]
28	[d]
29	[k]
30	[n]

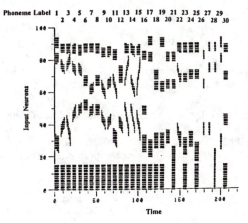

Fig. 2. Prototype Phoneme Patterns. (Thirty phoneme patterns are shown in sequence with intervals of two time units.)

6. SIMULATION OF SPATIO-TEMPORAL FILTERS FOR PHONEME CLASSIFICATION

The network system described below was simulated and used to classify the prototype phoneme patterns in the test patterns shown in the previoius section. It is an example of generalizing a scheme developed for static patterns[13] to that for dynamic patterns. Its operation is in two stages. The first stage operation is a spatio-temporal filter bank:

36

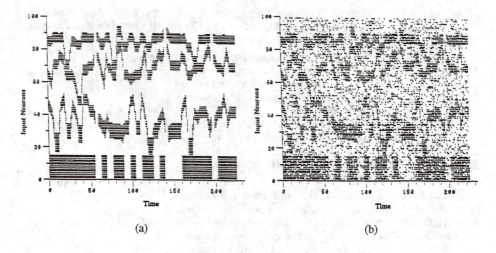

Fig. 3. Test Patterns. (a) Clean Test Pattern. (b) Noisy Test Pattern.

$$\vec{y}(t) = \underline{W}(t) * \vec{x}(t) \text{, and } \vec{z}(t) = \underline{r}(a(-t)\vec{y}(t)) \text{ .} \tag{6.1}$$

The second stage operation is the "winner-take-all" lateral inhibition:

$$\vec{o}(t) = \vec{z}(t) \text{, and } \vec{o}(t+\Delta) = \underline{r}(\underline{A}(-t) * \vec{o}(t) - \vec{h}), \tag{6.2}$$

and

$$\underline{A}(t) = (1 + \frac{1}{5N})\underline{I}\delta(t) - \frac{1}{5N}\vec{1}\vec{1}^T \sum_{n=0}^{4} \delta(t-n\Delta). \tag{6.3}$$

where $\vec{h}$ is a constant threshold vector with elements $h_i = h$ and $\delta(\cdot)$ is the Kronecker delta function. This operation is repeated a sufficient number of times, N_o.[13,14] The output is $\vec{o}(t + N_o \cdot \Delta)$.

Two models based on different learning rules were simulated with parameters shown below.

Model 1 (Spatio-temporal Matched Filter Bank)
Let $\alpha(t) = \delta(t)$, $\vec{d}^{(k)} = \vec{e}_k$ in (3.3) where $\vec{e}_k$ is a unit vector with its elements $e_{ki} = \delta(k-i)$.

$$\underline{W}(t) = \underline{X}(t)^T. \tag{6.4}$$

$$h=200 \text{, and } a(t) = \sum_{n=0}^{4} \frac{1}{5}\delta(t-n\Delta).$$

Model 2 (Spatio-temporal Pseudo-inverse Filter)
Let $\underline{D} = \underline{I}$ in (4.10). Using the alternative expression in (4.4),

$$\underline{W}(t) = F^{-1}\{(\underline{X}(f)^{CT}\underline{X}(f) + \sigma^2\underline{I})^{-1}X^{CT}\}. \tag{6.5}$$

$$h = 0.05 \text{, } \sigma^2 = 1000.0 \text{, and } a(t) = \delta(t).$$

This minimizes

$$R(\sigma,f) = \sum_k \|\vec{y}^{(k)}(f) - \vec{d}^{(k)}(f)\|_2^2 + \sigma^2 \sum_k \|\vec{w}_k(f)\|_2^2 \quad \text{for } all \ f \text{ .} \tag{6.6}$$

Because the time and frequency were finite and discrete in simulation, the result of the inverse discrete Fourier transform in (6.5) may be aliased. To alleviate the aliasing, the transfer functions in the prototype matrix $\underline{X}(t)$ were padded with zeros, thereby doubling the lengths. Further zero-padding the transfer functions did not seem to change teh result significantly.

The results are shown in Fig. 4(a)-(d). The arrows indicate the ideal response positions at the end of a phoneme. When the program was run with different thresholds and adaptation function $a(t)$, the result was not very sensitive to the threshold value, but was, nevertheless affected by the choice of the adaptation function. The maximum number of iterations for the lateral inhibition network to converge was observed: for the experiments shown in Fig. 4(a) - (d), the numbers were 44, 69, 29, and 47, respectively. Model 1 missed one phoneme and falsely responded once in the clean test pattern. It missed three and had one false response in the noisy test pattern. Model 2 correctly recognized all phonemes in the clean test pattern, and false-alarmed once in the noisy test pattern.

7. DISCUSSION

The notion of convolution or correlation used in the models presented is popular in engineering disciplines and has been applied extensively to designing filters, control systems, etc. Such operations also occur in biological systems and have been applied to modeling neural networks.[15,16] Thus the concept of dynamic formal neuron may be helpful for the improvement of artificial neural network models as well as the understanding of biological systems. A portion of the system described by Tank and Hopfield [17] is similar to the matched filter bank model simulated in this paper.

The matched filter bank model (Model 1) performs well when all phonemes (as above) are of the same duration. Otherwise, it would perform poorly unless the lengths were forced to a maximum length by padding the input and transfer functions with -1's during calculation. The pseudo-inverse filter model, on the other hand, should not suffer from this problem. However, this aspect of the model (Model 2) has not yet been explicitly simulated.

Given a spatio-temporal pattern of size $L \times K$, i.e., L spatial elements and K temporal elements, the number of calculations required to process the first stage of filtering by both models is the same as that by a static formal neuron netwoîk in which each neuron is connected to the $L \times K$ input elements. In both cases, $L \times K$ multiplications and additions are necessary to calculate one output value. In the case of bipolar patterns, the mutiplication used for calculation of activation can be replaced by sign-bit check and addition. A future investigation is to use recursive filters or analog filters as transfer functions for faster and more efficient calculation. There are various schemes to obtain optimal recursive or analog filters.[18,19] Besides the lateral inhibition scheme used in the models, there are a number of alternative procedures to realize a "winner-take-all" network in analog or digital fashion.[15,20,21]

As pointed out in the previous section, the Fourier transform in (6.5) requires a precaution concerning the resulting length of transfer functions. Calculating the recursive correlation equation (3.4) also needs such preprocessing as windowing or truncation.[22]

The generalization of static neural networks to dynamic ones along with their learning rules is straignhtforward as shown if the neuron operation and the learning rule are linear. Generalizing a system whose neuron operation and/or learning rule are nonlinear requires more careful analysis and remains for future work. The system described by Watrous and Shastri[16] is an example of generalizing a backpropagation model. Their result showed a good potential of the model and a need for more rigorous analysis of the model. Generalizing a system with recurrent connections is another task to be pursued. In a system with a certain analytical nonlinearity, the signals are expressed by Volterra functionals, for example. A practical learning system can then be constructed if higher kernels are neglected. For example, a cubic function can be used instead of a sigmoidal function.

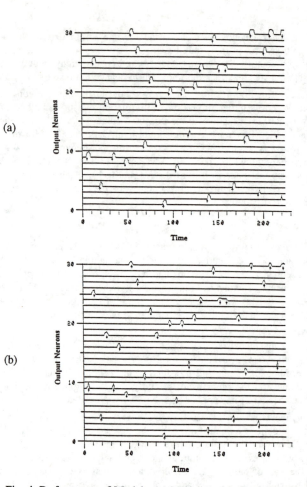

Fig. 4. Performance of Models. (a) Model 1 with Clean Test Pattern. (b) Model 2 with Clean Test Pattern. (c) Model 1 with Noisy Test Pattern. (d) Model 2 with Noisy Test Pattern. Arrows indicate the ideal response positions at the end of phoneme.

8. CONCLUSION

The formal neuron was generalized to the dynamic formal neuron to recognize spatio-temporal patterns. It is shown that existing learning rules can be generalized for dynamic formal neurons.

An artificial neural network using dynamic formal neurons was applied to classifying 30 model phonemes with bipolar patterns created by using parameters of formant frequencies and their bandwidths. The model operates in two stages: in the first stage, it calculates the correlation between the input and prototype patterns stored in the transfer function matrix, and, in the second stage, a lateral inhibition network selects the output of the phoneme pattern close to the input pattern.

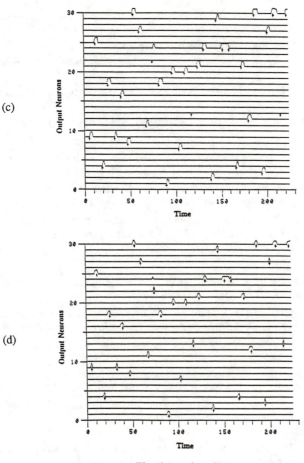

(c)

(d)

Fig. 4 (continued.)

Two models with different transfer functions were tested. Model 1 was a matched filter bank model and Model 2 was a pseudo-inverse filter model. A sequence of phoneme patterns corresponding to continuous pronunciation of digits was used as a test pattern. For the test pattern, Model 1 missed to recognize one phoneme and responded falsely once while Model 2 correctly recognized all the 32 phonemes in the test pattern. When the flip noise which flips the sign of the pattern with the probability 0.2, Model 1 missed three phonemes and falsely responded once while Model 2 recognized all the phonemes and false-alarmed once. Both models detected the phonems at the correct position within the continuous stream.

References

1. W. S. McCulloch and W. Pitts, "A logical calculus of the ideas imminent in nervous activity," *Bulletin of Mathematical Biophysics*, vol. 5, pp. 115-133, 1943.

2. D. O. Hebb, *The Organization of Behavior*, Wiley, New York, 1949.

3. D. E. Rumelhart, G. E. Hinton, and R. J. Williams, "Learning internal representations by error propagation," in *Parallel Distributed Processing, Vol. 1*, MIT, Cambridge, 1986.

4. B. Widrow and M. E. Hoff, "Adaptive switching circuits," *Institute of Radio Engineers, Western Electronics Show and Convention*, vol. Convention Record Part 4, pp. 96-104, 1960.

5. R. O. Duda and P. E. Hart, *Pattern Classification and Scene Analysis*, Chapter 5, Wiley, New York, 1973.

6. T. Kohonen, *Self-organization and Associative Memory*, Springer-Verlag, Berlin, 1984.

7. F. Rosenblatt, *Principles of Neurodynamics*, Spartan Books, Washington, 1962.

8. J. M. Varah, "A practical examination of some numerical methods for linear discrete ill-posed problems," *SIAM Review*, vol. 21, no. 1, pp. 100-111, 1979.

9. C. Koch, J. Marroquin, and A. Yuille, "Analog neural networks in early vision," *Proceedings of the National Academy of Sciences, USA*, vol. 83, pp. 4263-4267, 1986.

10. G. O. Stone, "An analysis of the delta rule and the learning of statistical associations," in *Parallel Distributed Processing., Vol. 1*, MIT, Cambridge, 1986.

11. B. Widrow and S. D. Stearns, *Adaptive Signal Processing*, Prentice-Hall, Englewood Cliffs, 1985.

12. D. H. Klatt, "Software for a cascade/parallel formant synthesizer," *Journal of Acoustical Society of America*, vol. 67, no. 3, pp. 971-995, 1980.

13. L. E. Atlas, T. Homma, and R. J. Marks II, "A neural network for vowel classification," *Proceedings International Conference on Acoustics, Speech, and Signal Processing*, 1987.

14. R. P. Lippman, "An introduction to computing with neural nets," *IEEE ASSP Magazine*, April, 1987.

15. S. Amari and M. A. Arbib, "Competition and cooperation in neural nets," in *Systems Neuroscience*, ed. J. Metzler, pp. 119-165, Academic Press, New York, 1977.

16. R. L. Watrous and L. Shastri, "Learning acoustic features from speech data using connectionist networks," *Proceedings of The Ninth Annual Conference of The Cognitive Science Society*, pp. 518-530, 1987.

17. D. Tank and J. J. Hopfield, "Concentrating information in time: analog neural networks with applications to speech recognition problems," *Proceedings of International Conference on Neural Netoworks*, San Diego, 1987.

18. J. R. Treichler, C. R. Johnson,Jr., and M. G. Larimore, *Theory and Design of Adaptive Filters*, Chapter 5, Wiley, New York, 1987.

19. M Schetzen, *The Volterra and Wiener Theories of Nonlinear Systems*, Chapter 16, Wiley, New York, 1980.

20. S. Grossberg, "Associative and competitive principles of learning," in *Competition and Cooperation in Neural Nets*, ed. M. A. Arbib, pp. 295-341, Springer-Verlag, New York, 1982.

21. R. J. Marks II, L. E. Atlas, J. J. Choi, S. Oh, K. F. Cheung, and D. C. Park, "A performance analysis of associative memories with nonlinearities in the correlation domain," (submitted to Applied Optics), 1987.

22. D. E. Dudgeon and R. M. Mersereau, *Multidimensional Digital Signal Processing*, pp. 230-234, Prentice-Hall, Englewood Cliffs, 1984.

ON PROPERTIES OF NETWORKS
OF NEURON-LIKE ELEMENTS

Pierre Baldi[*] and Santosh S. Venkatesh[†]

15 December 1987

Abstract

The complexity and computational capacity of multi-layered, feedforward neural networks is examined. Neural networks for special purpose (structured) functions are examined from the perspective of circuit complexity. Known results in complexity theory are applied to the special instance of neural network circuits, and in particular, classes of functions that can be implemented in shallow circuits characterised. Some conclusions are also drawn about learning complexity, and some open problems raised. The dual problem of determining the computational capacity of a class of multi-layered networks with dynamics regulated by an algebraic Hamiltonian is considered. Formal results are presented on the storage capacities of programmed higher-order structures, and a tradeoff between ease of programming and capacity is shown. A precise determination is made of the static fixed point structure of random higher-order constructs, and phase-transitions (0-1 laws) are shown.

1 INTRODUCTION

In this article we consider two aspects of computation with neural networks. Firstly we consider the problem of the complexity of the network required to compute classes of specified (structured) functions. We give a brief overview of basic known complexity theorems for readers familiar with neural network models but less familiar with circuit complexity theories. We argue that there is considerable computational and physiological justification for the thesis that shallow circuits (i.e., networks with relatively few layers) are computationally more efficient. We hence concentrate on structured (as opposed to random) problems that can be computed in shallow (constant depth) circuits with a relatively few number (polynomial) of elements, and demonstrate classes of structured problems that are amenable to such low cost solutions. We discuss an allied problem—the complexity of learning—and close with some open problems and a discussion of the observed limitations of the theoretical approach.

We next turn to a rigourous classification of how *much* a network of *given* structure can do; i.e., the computational capacity of a given construct. (This is, in

[*]Department of Mathematics, University of California (San Diego), La Jolla, CA 92093
[†]Moore School of Electrical Engineering, University of Pennsylvania, Philadelphia, PA 19104

a sense, the mirror image of the problem considered above, where we were seeking to design a minimal structure to perform a given task.) In this article we restrict ourselves to the analysis of higher-order neural structures obtained from polynomial threshold rules. We demonstrate that these higher-order networks are a special class of layered neural network, and present formal results on storage capacities for these constructs. Specifically, for the case of programmed interactions we demonstrate that the storage capacity is of the order of n^d where d is the interaction order. For the case of random interactions, a type of phase transition is observed in the distribution of fixed points as a function of attraction depth.

2 COMPLEXITY

There exist two broad classes of constraints on computations.

1. *Physical constraints*: These are related to the hardware in which the computation is embedded, and include among others time constants, energy limitations, volumes and geometrical relations in 3D space, and bandwidth capacities.

2. *Logical constraints*: These can be further subdivided into

   - Computability constraints—for instance, there exist unsolvable problems, i.e., functions such as the halting problem which are not computable in an absolute sense.
   - Complexity constraints—usually giving upper and/or lower bounds on the amount of resources such as the time, or the number of gates required to compute a given function. As an instance, the assertion "There exists an exponential time algorithm for the Traveling Salesman Problem," provides a computational upper bound.

If we view brains as computational devices, it is not unreasonable to think that in the course of the evolutionary process, nature may have been faced several times by problems related to physical and perhaps to a minor degree logical constraints on computations. If this is the case, then complexity theory in a broad sense could contribute in the future to our understanding of parallel computations and architectural issues both in natural and synthetic neural systems.

A simple theory of parallel processing at the macro level (where the elements are processors) can be developed based on the ratio of the time spent on communications between processors [7] for different classes of problems and different processor architecture and interconnections. However, this approach does not seem to work for parallel processing at the level of circuits, especially if calculations and communications are intricately entangled.

Recent neural or connectionist models are based on a common structure, that of highly interconnected networks of linear (or polynomial) threshold (or with sigmoid input-output function) units with adjustable interconnection weights. We shall therefore review the complexity theory of such circuits. In doing so, it will be sometimes helpful to contrast it with the similar theory based on Boolean (AND, OR, NOT) gates. The presentation will be rather informal and technical complements can easily be found in the references.

Consider a circuit as being on a cyclic oriented graph connecting n Boolean inputs to one Boolean output. The nodes of the graph correspond to the gates (the n input units, the "hidden" units, and the output unit) of the circuit. The *size* of the circuit is the total number of gates and the *depth* is the length of the longest path connecting one input to the output. For a layered, feed-forward circuit, the *width* is the average number of computational units in the hidden (or interior) layers of elements. The first obvious thing when comparing Boolean and threshold logic is that they are equivalent in the sense that any Boolean function can be implemented using either logic. In fact, any such function can be computed in a circuit of depth two and exponential size. Simple counting arguments show that the fraction of functions requiring a circuit of exponential size approaches one as $n \to \infty$ in both cases, i.e., a random function will in general require an exponential size circuit. (Paradoxically, it is very difficult to *construct* a family of functions for which we can prove that an exponential circuit is necessary.) Yet, threshold logic is more powerful than Boolean logic. A Boolean gate can compute only one function whereas a threshold gate can compute to the order of $2^{\alpha n^2}$ functions by varying the weights with $1/2 \le \alpha \le 1$ (see [19] for the lower bound; the upper bound is a classical hyperplane counting argument, see for instance [20,30]). It would hence appear plausible that there exist wide classes of problems which can be computed by threshold logic with circuits substantially smaller than those required by Boolean logic. An important result which separates threshold and Boolean logic from this point of view has been demonstrated by Yao [31] (see [10,24] for an elegant proof). The result is that in order to compute a function such as parity in a circuit of constant depth k, at least $\exp(cn^{1/2k})$ Boolean gates with unbounded fanin are required. As we shall demonstrate shortly, a circuit of depth two and linear size is sufficient for the computation of such functions using threshold logic.

It is not unusual to hear discussions about the tradeoffs between the depth and the width of a circuit. We believe that one of the main constributions of complexity analysis is to show that this tradeoff is in some sense minimal and that in fact there exists a very strong bias in favor of shallow (i.e., constant depth) circuits. There are multiple reasons for this. In general, for a fixed size, the number of different functions computable by a circuit of small depth exceeds the number of those computable by a deeper circuit. That is, if one had no a priori knowledge regarding the function to be computed and was given hidden units, then the optimal strategy would be to choose a circuit of depth two with the m units in a single layer. In addition, if we view computations as propagating in a feedforward mode from the inputs to the output unit, then shallow circuits compute faster. And the deeper a circuit, the more difficult become the issues of time delays, synchronisation, and precision on the computations. Finally, it should be noticed that given overall responses of a few hundred milliseconds and given the known time scales for synaptic integration, biological circuitry must be shallow, at least within a "module" and this is corroborated by anatomical data. The relative slowness of neurons and their shallow circuit architecture are to be taken together with the "analog factor" and "entropy factor" [1] to understand the necessary high-connectivity requirements of neural systems.

From the previous analysis emerges an important class of circuits in threshold logic characterised by polynomial size and shallow depth. We have seen that, in general, a random function cannot be computed by such circuits. However, many interesting functions—the *structured problems*—are far from random, and it is then natural to ask what is the class of functions computable by such circuits? While a complete characterisation is probably difficult, there are several sub-classes of structural functions which are known to be computable in shallow poly-size circuits.

The *symmetric* functions, i.e., functions which are invariant under any permutation of the n input variables, are an important class of structured problems that can be implemented in shallow polynomial size circuits. In fact, *any symmetric function can be computed by a threshold circuit of depth two and linear size*; (n hidden units and one output unit are always sufficient). We demonstrate the validity of this assertion by the following instructive construction. We consider n binary inputs, each taking on values -1 and 1 only, and threshold gates as units. Now array the 2^n possible inputs in $n + 1$ rows with the elements in each row being permuted versions of each other (i.e., n-tuples in a row all have the same number of +1's) and with the rows going monotonically from zero +1's to n +1's. Any given symmetric Boolean function clearly assumes the same value for all elements (Boolean n-tuples) in a row, so that contiguous rows where the function assumes the value +1 form bands. (There are at most $n/2$ bands—the worst case occuring for the parity function.) The symmetric function can now be computed with $2B$ threshold gates in a single hidden layer with the topmost "neuron" being activated only if the number of +1's in the input exceeds the number of +1's in the lower edge of the lowest band, and proceeding systematically, the lowest "neuron" being activated only if the number of +1's in the input exceeds the number of +1's in the upper edge of the highest band. An input string will be within a band if and only if an odd number of hidden neurons are activated starting contiguously from the top of the hidden layer, and conversely. Hence, a single output unit can compute the given symmetric function.

It is easy to see that arithmetic operations on binary strings can be performed with polysize small depth circuits. Reif [23] has shown that for a fixed degree of precision, any analytic function such as polynomials, exponentials, and trigonometric functions can be approximated with small and shallow threshold circuits. Finally, in many situations one is interested in the value of a function only for a vanishingly small (i.e., polynomial) fraction of the total number of possible inputs 2^n. These functions can be implemented by polysize shallow circuits and one can relate the size and depths of the circuit to the cardinal of the interesting inputs.

So far we only have been concerned with the complexity of threshold circuits. We now turn to the complexity of learning, i.e., the problem of finding the weights required to implement a given function. Consider the problem of repeating m points in $\mathbb{R}^\ell$ coloured in two colours, using k hyperplanes so that any region contains only monochromatic points. If ℓ and k are fixed the problem can be solved in polynomial time. If either ℓ or k goes to infinity, the problem becomes NP-complete [?]. As a result, it is not difficult to see that the general learning problem is NP-complete (see also [12] for a different proof and [21] for a proof of the fact it is already NP-complete in the case of one single threshold gate).

Some remarks on the limitations of the complexity approach are *a propos* at this juncture:

1. While a variety of structured Boolean functions can be implemented at relatively low cost with networks of linear threshold gates (McCulloch-Pitts neurons), the extension to different input-output functions and the continuous domain is not always straightforward.

2. Even restricting ourselves to networks of relatively simple Boolean devices such as the linear threshold gate, in many instances, only relatively weak bounds are available for computational cost and complexity.

3. Time is probably the single most important ingredient which is completely absent from these threshold units and their interconnections [17,14]; there are, in addition, non-biological aspects of connectionist models [8].

4. Finally, complexity results (where available) are often asymptotic in nature and may not be meaningful in the range corresponding to a particular application.

We shall end this section with a few open questions and speculations. One problem has to do with the time it takes to learn. Learning is often seen as a very slow process both in artificial models (cf. back propagation, for instance) and biological systems (cf. human acquisition of complex skills). However, if we follow the standards of complexity theory, in order to be effective over a wide variety of scales, a single learning algorithm should be polynomial time. We can therefore ask what is learnable by examples in polynomial time by polynomial size shallow threshold circuits? The status of back propagation type of algorithms with respect to this question is not very clear.

The existence of many tasks which are easily executed by biological organisms and for which no satisfactory computer program has been found so far leads to the question of the specificity of learning algorithms, i.e., whether there exists a complexity class of problems or functions for which a "program" can be found only by learning from examples as opposed to by traditional programming. There is some circumstantial evidence against such conjecture. As pointed out by Valiant [25], cryptography can be seen in some sense as the opposite of learning. The conjectures existence of one way function, i.e., functions which can be constructed in polynomial time but cannot be invested (from examples) in polynomial time suggests that learning algorithms may have strict limitations. In addition, for most of the artificial applications seen so far, the programs obtained through learning do not outperform the best already known software, though there may be many other reasons for that. However, even if such a complexity class does not exist, learning algorithm may still be very important because of their inexpensiveness and generality. The work of Valiant [26,13] on polynomial time learning of Boolean formulas in his "distribution free model" explores some additional limitations of what can be learned by examples without including any additional knowledge.

Learning may therefore turn out to be a powerful, inexpensive but limited family of algorithms that need to be incorporated as "sub-routines" of more global programs, the structure of which may be harder to find. Should evolution be regarded as an "exponential" time learning process complemented by the "polynomial" time type of learning occurring in the lifetime of organisms?

3 CAPACITY

In the previous section the focus of our investigation was on the structure and cost of minimal networks that would compute specified Boolean functions. We now consider the dual question: What is the computational capacity of a threshold network of given structure? As with the issues on complexity, it turns out that for fairly general networks, the capacity results favour shallow (but perhaps broad) circuits [29]. In this discourse, however, we shall restrict ourselves to a specified class of higher-order networks, and to problems of associative memory. We will just quote the principal rigourous results here, and present the involved proofs elsewhere [4].

We consider systems of n densely interacting threshold units each of which yields an instantaneous state -1 or +1. (This corresponds in the literature to a system of n Ising spins, or alternatively, a system of n neural states.) The state space is hence the set of vertices of the hypercube. We will in this discussion also restrict our attention throughout to *symmetric interaction systems* wherein the interconnections between threshold elements is bidirectional.

Let $\mathcal{I}_d$ be the family of all subsets of cardinality $d+1$ of the set $\{1, 2, \ldots, n\}$. Clearly $|\mathcal{I}_d| = \binom{n}{d+1}$. For any subset I of $\{1, 2, \ldots, n\}$, and for every state $\mathbf{u} = \{u_1, u_2, \ldots, u_n\} \in \mathbb{B}^n \stackrel{\text{def}}{=} \{-1, 1\}^n$, set $u_I = \prod_{i \in I} u_i$.

Definition 1 A *homogeneous algebraic threshold network* of degree d is a network of n threshold elements with interactions specified by a set of $\binom{n}{d+1}$ real coefficients w_I indexed by I in $\mathcal{I}_d$, and the evolution rule

$$u_i^+ = \text{sgn}\left(\sum_{I \in \mathcal{I}_d : i \in I} w_I u_{I \setminus \{i\}}\right) \tag{1}$$

These systems can be readily seen to be natural generalisations to higher-order of the familiar case $d = 1$ of *linear threshold networks*. The added degrees of freedom in the interaction coefficients can potentially result in enhanced flexibility and programming capability over the linear case as has been noted independently by several authors recently [2,3,4,5,22,27]. Note that each d-wise product $u_{I \setminus i}$ is just the parity of the corresponding d inputs, and by our earlier discussion, this can be computed with d hidden units in one layer followed by a single threshold unit. Thus the higher-order network can be realised by a network of depth three, where the first hidden layer has $d\binom{n}{d}$ units, the second hidden layer has $\binom{n}{d}$ units, and there are n output units which feedback into the n input units. Note that the weights from the input to the first hidden layer, and the first hidden layer to the second are fixed (computing the various d-wise products), and the weights from the second hidden layer to the output are the coefficients w_I which are free parameters.

These systems can be identified either with long range interactions for higher-order spin glasses at zero temperature, or higher-order neural networks. Starting from an arbitrary configuration or state, the system evolves asynchronously by a sequence of single "spin" flips involving spins which are misaligned with the instantaneous "molecular field." The dynamics of these symmetric higher-order systems are regulated analogous to the linear system by higher-order extensions of the classical quadratic Hamiltonian. We define the *homogeneous algebraic Hamiltonian* of degree d by

$$H_d(\mathbf{u}) = - \sum_{I \in \mathcal{I}_d} w_I u_I . \tag{2}$$

The algebraic Hamiltonians are functionals akin in behaviour to the classical quadratic Hamiltonian as has been previously demonstrated [5].

Proposition 1 The functional H_d is non-increasing under the evolution rule 1.

In the terminology of spin glasses, the state trajectories of these higher-order networks can be seen to be following essentially a zero-temperature Monte Carlo (or Glauber) dynamics. Because of the monotonicity of the algebraic Hamiltonians given by equation 2 under the asynchronous evolution rule 1, the system always reaches a stable state (fixed point) where the relation 1 is satisfied for each of the n spins or neural states. The fixed points are hence the arbiters of system dynamics, and determine the computational capacity of the system.

System behaviour and applications are somewhat different depending on whether the interactions are random or programmed. The case of random interactions lends itself to natural extensions of spin glass formulations, while programmed interactions yield applications of higher-order extensions of neural network models. We consider the two cases in turn.

3.1 PROGRAMMED INTERACTIONS

Here we query whether given sets of binary n-vectors can be stored as fixed points by a suitable selection of interaction coefficients. If such sets of *prescribed* vectors can be stored as stable states for some suitable choice of interaction coefficients, then proposition 1 will ensure that the chosen vectors are at the bottom of "energy wells" in the state space with each vector exercising a region of attraction around it—all characterestics of a physical associative memory. In such a situation the dynamical evolution of the network can be interpreted in terms of computations: error-correction, nearest neighbour search and associative memory. Of importance here is the maximum number of states that can be stored as fixed points for an appropriate choice of algebraic threshold network. This represents the *maximal information storage capacity* of such higher-order neural networks.

Let d represent the degree of the algebraic threshold network. Let $\mathbf{u}^{(1)}, \ldots, \mathbf{u}^{(m)}$ be the m-set of vectors which we require to store as fixed points in a suitable algebraic threshold network. We will henceforth refer to these prescribed vectors as *memories.* We define the *storage capacity* of an algebraic threshold network of degree d to be the maximal number m of arbitrarily chosen memories which can be stored with high probability for appropriate choices of coefficients in the network.

Theorem 1 The maximal (algorithm independent) storage capacity of a homogeneous algebraic threshold network of degree d is less than or equal to $2\binom{n}{d}$.

Generalised Sum of Outer-Products Rule: The classical Hebbian rule for the linear case $d = 1$ (cf. [11] and quoted references) can be naturally extended to networks of higher-order. The coefficients w_I, $I \in \mathcal{I}_d$ are constructed as the sum of generalised Kronecker outer-products,

$$w_I = \sum_{\alpha=1}^{m} u_I^{(\alpha)} .$$

Theorem 2 The storage capacity of the outer-product algorithm applied to a homogeneous algebraic threshold network of degree d is less than or equal to $n^d/2(d+1)\log n$ (also cf. [15,27]).

Generalised Spectral Rule: For $d = 1$ the spectral rule amounts to iteratively projecting states orthogonally onto the linear space generated by $\mathbf{u}^{(1)}, \ldots, \mathbf{u}^{(m)}$, and then taking the closest point on the hypercube to this projection (cf. [27,28]). This approach can be extended to higher-orders as we now describe.

Let $\mathbf{W}$ denote the $n \times N_{(n,d)}$ matrix of coefficients w_I arranged lexicographically; i.e.,

$$\mathbf{W} = \begin{bmatrix} w_{1,1,2,\ldots,d-1,d} & w_{1,2,3,\ldots,d,d+1} & \cdots & w_{1,n-d+1,n-d+2,\ldots,n-1,n} \\ w_{2,1,2,\ldots,d-1,d} & w_{2,2,3,\ldots,d,d+1} & \cdots & w_{2,n-d+1,n-d+2,\ldots,n-1,n} \\ \vdots & \vdots & \vdots & \vdots \\ w_{n,1,2,\ldots,d-1,d} & w_{n,2,3,\ldots,d,d+1} & \cdots & w_{n,n-d+1,n-d+2,\ldots,n-1,n} \end{bmatrix}$$

Note that the symmetry and the "zero-diagonal" nature of the interactions have been relaxed to increase capacity. Let $\mathbf{U}$ be the $n \times m$ matrix of memories. Form the extended $N_{(n,d)} \times m$ binary matrix $^1\mathbf{U} = [^1\mathbf{u}^{(1)} \cdots {}^1\mathbf{u}^{(m)}]$, where

$$^1\mathbf{u}^{(\alpha)} = \begin{bmatrix} u_{1,2,\ldots,d-1,d}^{(\alpha)} \\ u_{1,2,\ldots,d-1,d+1}^{(\alpha)} \\ \vdots \\ u_{n-d+1,n-d+2,\ldots,n-1,n}^{(\alpha)} \end{bmatrix}$$

Let $\Lambda = \mathbf{dg}[\lambda^{(1)} \cdots \lambda^{(m)}]$ be a $m \times m$ diagonal matrix with positive diagonal terms. A generalisation of the spectral algorithm for choosing coefficients yields

$$\mathbf{W} = \mathbf{U}\Lambda^1\mathbf{U}^\dagger$$

where $^1\mathbf{U}^\dagger$ is the pseudo-inverse of $^1\mathbf{U}$.

Theorem 3 The storage capacity of the generalised spectral algorithm is at best $\binom{n}{d}$.

3.2 RANDOM INTERACTIONS

We consider homogeneous algebraic threshold networks whose weights w_I are i.i.d., $\mathcal{N}(0,1)$ random variables. This is a natural generalisation to higher-order of Ising spin glasses with Gaussian interactions. We will show an asymptotic estimate for the number of fixed points of the structure. Asymptotic results for the usual case $d = 1$ of linear threshold networks with Gaussian interactions have been reported in the literature [6,9,16].

For $i = 1, \ldots, n$ set

$$S_n^i = u_i \sum_{I \in \mathcal{I}_d : i \in I} w_I u_{I \setminus i} .$$

For each n the random variables S_n^i, $i = 1, \ldots, n$ are identically distributed, jointly Gaussian variables with zero mean, and variance $\sigma_n^2 = \binom{n-1}{d}$.

Definition 2 For any given $\beta \geq 0$, a state $\mathbf{u} \in \mathbb{B}^n$ is β-*strongly stable* iff $S_n^i \geq \beta \sigma_n$, for each $i = 1, \ldots, n$.

The case $\beta = 0$ reverts to the usual case of fixed points. The parameter β is essentially a measure of how deep the well of attraction surrounding the fixed point is. The following proposition asserts that a 0-1 law ("phase transition") governs the expected number of fixed points which have wells of attraction above a certain depth. Let $F_d(\beta)$ be the expected number of β-strongly stable states.

Theorem 4 Corresponding to each fixed interaction order d there exists a positive constant β_d^* such that as $n \to \infty$,

$$F_d^\beta \sim \begin{cases} k_d(\beta)\, 2^{n c_d(\beta)} & \text{if } \beta < \beta_d^* \\ k_d(\beta_d^*) & \text{if } \beta = \beta_d^* \\ 0 & \text{if } \beta > \beta_d^* , \end{cases}$$

where $k_d(\beta) > 0$, and $0 \leq c_d(\beta) < 1$ are parameters depending solely on β and the interaction order d.

4 CONCLUSION

In fine, it appears possible to design shallow, polynomial size threshold circuits to compute a wide class of structured problems. The thesis that shallow circuits compute more efficiently than deep circuits is borne out. For the particular case of higher-order networks, all the garnered results appear to point in the same direction: *For neural networks of fixed degree d, the maximal number of programmable states is essentially of the order of n^d.* The total number of fixed points, however, appear to be exponential in number (at least for the random interaction case) though almost all of them have constant attraction depths.

References

[1] Y. S. Abu-Mostafa, "Number of synapses per neuron," in *Analog VLSI and Neural Systems*, ed. C. Mead, Addison Wesley, 1987.

[2] P. Baldi, *II. Some Contributions to the Theory of Neural Networks*. Ph.D. Thesis, California Insitute of Technology, June 1986.

[3] P. Baldi and S. S. Venkatesh, "Number of stable points for spin glasses and neural networks of higher orders," *Phys. Rev. Lett.*, vol. 58, pp. 913–916, 1987.

[4] P. Baldi and S. S. Venkatesh, "Fixed points of algebraic threshold networks," in preparation.

[5] H. H. Chen, et al, "Higher order correlation model of associative memory," in *Neural Networks for Computing*. New York: AIP Conf. Proc., vol. 151, 1986.

[6] S. F. Edwards and F. Tanaka, "Analytical theory of the ground state properties of a spin glass: I. ising spin glass," *Jnl. Phys. F*, vol. 10, pp. 2769–2778, 1980.

[7] G. C. Fox and S. W. Otto, "Concurrent Computations and the Theory of Complex Systems," *Caltech Concurrent Computation Program*, March 1986.

[8] F. H. Grick and C. Asanuma, "Certain aspects of the anatomy and physiology of the cerebral cortex," in *Parallel Distributed Processing*, vol. 2, eds. D. E. Rumelhart and J. L. McCelland, pp. 333–371, MIT Press, 1986.

[9] D. J. Gross and M. Mezard, "The simplest spin glass," *Nucl. Phys.*, vol. B240, pp. 431–452, 1984.

[10] J. Hasted, "Almost optimal lower bounds for small depth circuits," *Proc. 18-th ACM STOC*, pp. 6–20, 1986.

[11] J. J. Hopfield, "Neural networks and physical sytems with emergent collective computational abilities," *Proc. Natl. Acad. Sci. USA*, vol. 79, pp. 2554–2558, 1982.

[12] J. S. Judd, "Complexity of connectionist learning with various node functions," *Dept. of Computer and Information Science Technical Report*, vol. 87–60, Univ. of Massachussetts, Amherst, 1987.

[13] M. Kearns, M. Li, L. Pitt, and L. Valiant, "On the learnability of Boolean formulae," *Proc. 19-th ACM STOC*, 1987.

[14] C. Koch, T. Poggio, and V. Torre, "Retinal ganglion cells: A functional interpretation of dendritic morphology," *Phil. Trans. R. Soc. London*, vol. B 288, pp. 227–264, 1982.

[15] R. J. McEliece, E. C. Posner, E. R. Rodemich, and S. S. Venkatesh, "The capacity of the Hopfield associative memory," *IEEE Trans. Inform. Theory*, vol. IT–33, pp. 461–482, 1987.

[16] R. J. McEliece and E. C. Posner, "The number of stable points of an infinite-range spin glass memory," *JPL Telecomm. and Data Acquisition Progress Report*, vol. 42–83, pp. 209–215, 1985.

[17] C. A. Mead (ed.), *Analog VLSI and Neural Systems*, Addison Wesley, 1987.

[18] N. Megiddo, "On the complexity of polyhedral separability," to appear in *Jnl. Discrete and Computational Geometry*, 1987.

[19] S. Muroga, "Lower bounds on the number of threshold functions," *IEEE Trans. Elec. Comp.*, vol. 15, pp. 805–806, 1966.

[20] S. Muroga, *Threshold Logic and its Applications*, Wiley Interscience, 1971.

[21] V. N. Peled and B. Simeone, "Polynomial-time algorithms for regular set-covering and threshold synthesis," *Discr. Appl. Math.*, vol. 12, pp. 57–69, 1985.

[22] D. Psaltis and C. H. Park, "Nonlinear discriminant functions and associative memories," in *Neural Networks for Computing*. New York: AIP Conf. Proc., vol. 151, 1986.

[23] J. Reif, "On threshold circuits and polynomial computation," preprint.

[24] R. Smolenski, "Algebraic methods in the theory of lower bounds for Boolean circuit complexity," *Proc. 19-th ACM STOC*, 1987.

[25] L. G. Valiant, "A theory of the learnable," *Comm. ACM*, vol. 27, pp. 1134–1142, 1984.

[26] L. G. Valiant, "Deductive learning," *Phil. Trans. R. Soc. London*, vol. A 312, pp. 441–446, 1984.

[27] S. S. Venkatesh, *Linear Maps with Point Rules: Applications to Pattern Classification and Associative Memory*. Ph.D. Thesis, California Institute of Technology, Aug. 1986.

[28] S. S. Venkatesh and D. Psaltis, "Linear and logarithmic capacities in associative neural networks," to appear *IEEE Trans. Inform. Theory*.

[29] S. S. Venkatesh, D. Psaltis, and J. Yu, private communication.

[30] R. O. Winder, "Bounds on threshold gate realisability," *IRE Trans. Elec. Comp.*, vol. EC–12, pp. 561–564, 1963.

[31] A. C. C. Yao, "Separating the poly-time hierarchy by oracles," *Proc. 26-th IEEE FOCS*, pp. 1–10, 1985.

Supervised Learning of Probability Distributions by Neural Networks

Eric B. Baum

Jet Propulsion Laboratory, Pasadena CA 91109

Frank Wilczek[†]

Department of Physics,Harvard University,Cambridge MA 02138

Abstract:

We propose that the back propagation algorithm for supervised learning can be generalized, put on a satisfactory conceptual footing, and very likely made more efficient by defining the values of the output and input neurons as probabilities and varying the synaptic weights in the gradient direction of the log likelihood, rather than the 'error'.

In the past thirty years many researchers have studied the question of supervised learning in 'neural'-like networks. Recently a learning algorithm called 'back propagation'[1−4] or the 'generalized delta-rule' has been applied to numerous problems including the mapping of text to phonemes[5], the diagnosis of illnesses[6] and the classification of sonar targets[7]. In these applications, it would often be natural to consider imperfect, or probabilistic information. We believe that by considering supervised learning from this slightly larger perspective, one can not only place back propaga-

---

[†] Permanent address: Institute for Theoretical Physics, University of California, Santa Barbara CA 93106

tion on a more rigorous and general basis, relating it to other well studied pattern recognition algorithms, but very likely improve its performance as well.

The problem of supervised learning is to model some mapping between input vectors and output vectors presented to us by some real world phenomena. To be specific, consider the question of medical diagnosis. The input vector corresponds to the symptoms of the patient; the i-th component is defined to be 1 if symptom i is present and 0 if symptom i is absent. The output vector corresponds to the illnesses, so that its j-th component is 1 if the j-th illness is present and 0 otherwise. Given a data base consisting of a number of diagnosed cases, the goal is to construct (learn) a mapping which accounts for these examples and can be applied to diagnose new patients in a reliable way. One could hope, for instance, that such a learning algorithm might yield an expert system to simulate the performance of doctors. Little expert advice would be required for its design, which is advantageous both because experts' time is valuable and because experts often have extraodinary difficulty in describing how they make decisions.

A feedforward neural network implements such a mapping between input vectors and output vectors. Such a network has a set of input nodes, one or several layers of intermediate nodes, and a layer of output nodes. The nodes are connected in a forward directed manner, so that the output of a node may be connected to the inputs of nodes in subsequent layers, but closed loops do not occur. See figure 1. The output of each node is assumed to be a bounded semilinear function of its inputs. That is, if v_j denotes the output of the j-th node and w_{ij} denotes the weight associated with the connection of the output of the j-th node to the input of the i-th, then the i-th neuron takes value $v_i = g(\sum_j w_{ij} v_j)$, where g is a bounded, differentiable function called the activation function. $g(x) = 1/(1 + e^{-x})$, called the logistic function, is frequently used. Given a fixed set of weights $\{w_{ij}\}$, we set the input node values to equal some input vector, compute the value of the nodes layer by layer until we compute the output nodes, and so generate an output vector.

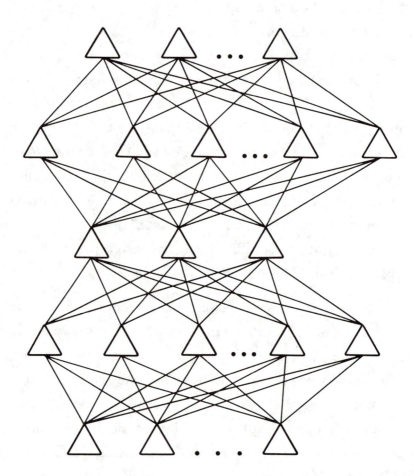

Figure 1: A 5 layer network. Note bottleneck at layer 3.

Such networks have been studied because of analogies to neu-robiology, because it may be easy to fabricate them in hardware, and because learning algorithms such as the Perceptron learning algorithm[8], Widrow- Hoff[9], and backpropagation have been able to choose weights w_{ij} that solve interesting problems.

Given a set of input vectors s_i^μ, together with associated target values t_j^μ, back propagation attempts to adjust the weights so as to minimize the error E in achieving these target values, defined as

$$E = \sum_\mu E_\mu = \sum_{\mu,j} (t_j^\mu - o_j^\mu)^2 \qquad (1)$$

where o_j^μ is the output of the j-th node when s^μ is presented as input. Back propagation starts with randomly chosen w_{ij} and then varies in the gradient direction of E until a local minimum is obtained. Although only a locally optimal set of weights is ob-tained, in a number of experiments the neural net so generated has performed surprisingly well not only on the training set but on subsequent data.[4-6] This performance is probably the main reason for widespread interest in backpropagation.

It seems to us natural, in the context of the medical diagnosis problem, the other real world problems to which backpropagation has been applied, and indeed in any mapping problem where one desires to generalize from a limited and noisy set of examples, to interpret the output vector in probabilistic terms. Such an inter-pretation is standard in the literature on pattern classification.[10] Indeed, the examples might even be probabilistic themselves. That is to say it might not be certain whether symptom i was present in case μ or not.

Let s_i^μ represent the probability symptom i is present in case μ, and let t_j^μ represent the probability disease j ocurred in case

μ. Consider for the moment the case where the t_j^μ are 1 or 0, so that the cases are in fact fully diagnosed. Let $f_j(\hat{s}, \hat{\theta})$ be our prediction of the probability of disease j given input vector $\hat{s}$, where $\hat{\theta}$ is some set of parameters determined by our learning algorithm. In the neural network case, the $\hat{\theta}$ are the connection weights and $f_j(\hat{s}^\mu, \{w_{ij}\}) = o_j^\mu$.

Now lacking a priori knowledge of good $\hat{\theta}$, the best one can do is to choose the parameters $\hat{\theta}$ to maximize the likelihood that the given set of examples should have occurred.[10] The formula for this likelihood, p, is immediate:

$$p = \prod_\mu \left[\prod_{\{j|t_j^\mu=1\}} f_j(\hat{s}^\mu, \hat{\theta}) \prod_{\{j|t_j^\mu=0\}} (1 - f_j(\hat{s}^\mu, \hat{\theta})) \right] \qquad (2)$$

or

$$log(p) = \sum_\mu \left[\sum_{\{j|t_j^\mu=1\}} log(f_j(\hat{s}^\mu, \hat{\theta})) + \sum_{\{j|t_j^\mu=0\}} log(1 - f_j(\hat{s}^\mu, \hat{\theta})) \right]$$
$$(3)$$

The extension of equation (2), and thus equation (3) to the case where the $\hat{t}$ are probabilities, taking values in $[0, 1]$, is straightforward[*1] and yields

$$log(p) = \sum_{\mu,j} \left[t_j^\mu log(f_j(\hat{s}^\mu, \hat{\theta})) + (1 - t_j^\mu)log(1 - f_j((\hat{s}^\mu, \hat{\theta})) \right] \quad (4)$$

Expressions of this sort often arise in physics and information theory and are generally interpreted as an entropy.[11]

We may now vary the $\{\hat{\theta}\}$ in the gradient direction of the entropy. The back propagation algorithm generalizes immediately from minimizing 'Error' or 'Energy' to maximizing entropy or log likelihood, or indeed any other function of the outputs and the inputs[12]. Of course it remains true that the gradient can be computed by back propagation with essentially the same number of computations as are required to compute the output of the network.

A backpropagation algorithm based on log-likelihood is not only more intuitively appealing than one based on an ad-hoc definition of error, but will make quite different and more accurate predictions as well. Consider e.g. training the net on an example which it already understands fairly well. Say $t_j^o = 0$, and $f_j(s^o) = \epsilon$. Now, from eqn(1) $\partial E/\partial f_j = 2\epsilon$, so using 'Error' as a

---

[*1] We may see this by constructing an equivalent larger set of examples with the $\hat{t}$ taking only values 0 or 1 with the appropriate frequency. Thus assume the t_j^μ are rational numbers with denominator d_j^μ and numerator n_j^μ and let $p = \prod_{\mu,j} d_j^\mu$. What we mean by the set of examples $\{t^\mu : \mu = 1, ..., M\}$ can be represented by considering a set of $N = Mp$ examples $\{\tilde{t}_j^\nu\}$ where for each μ, $\tilde{t}_j^\nu = 0$ for $p(\mu - 1) < \nu \leq p\mu$ and $1 \leq \nu mod(d_j^\mu) \leq (d_j^\mu - n_j^\mu)$, and $\tilde{t}_j^\nu = 1$ otherwise. Now applying equation (3) gives equation (4), up to an overall normalization.

criterion the net learns very little from this example, whereas, using eqn(3), $\partial log(p)/\partial f_j = 1/(1 - \epsilon)$, so the net continues to learn and can in fact converge to predict probabilities near 1. Indeed because backpropagation using the standard 'Error' measure can not converge to generate outputs of 1 or 0, it has been customary in the literature[4] to round the target values so that a target of 1 would be presented in the learning algorithm as some ad hoc number such as .8, whereas a target of 0 would be presented as .2.

In the context of our general discussion it is natural to ask whether using a feedforward network and varying the weights is in fact the most effective alternative. Anderson and Abrahams[13] have discussed this issue from a Bayesian viewpoint. From this point of view, fitting output to input using normal distributions and varying the means and covariance matrix may seem to be more logical.

Feedforward networks do however have several advantages for complex problems. Experience with neural networks has shown the importance of including hidden units wherein the network can form an internal representation of the world. If one simply uses normal distributions, any hidden variables included will simply integrate out in calculating an output. It will thus be necessary to include at least third order correlations to implement useful hidden variables. Unfortunately, the number of possible third order correlations is very large, so that there may be practical obstacles to such an approach. Indeed it is well known folklore in curve fitting and pattern classification that the number of parameters must be small compared to the size of the data set if any generalization to future cases is expected.[10]

In feedforward nets the question takes a different form. There can be bottlenecks to information flow. Specifically, if the net is constructed with an intermediate layer which is not bypassed by any connections (i.e. there are no connections from layers preceding to layers subsequent), and if furthermore the activation functions are chosen so that the values of each of the intermediate nodes tend towards either 1 or 0^{*2}, then this layer serves as a bottleneck to information flow. No matter how many input nodes, output nodes, or free parameters there are in the net, the output will be constrained to take on no more than 2^I different patterns, where I is the number of nodes in the bottleneck layer. Thus if I is small, some sort of 'generalization' must occur even if the number of weights is large. One plausible reason for the success of back propagation in adequately solving tasks, in spite of the fact that it finds only local minima, is its ability to vary a large number of parameters. This freedom may allow back propagation to escape from many putative traps and to find an acceptable solution.

A good expert system, say for medical diagnosis, should not only give a diagnosis based on the available information, but should be able to suggest, in questionable cases, which lab tests might be performed to clarify matters. Actually back propagation inherently has such a capability. Back propagation involves calculation of $\partial log(p)/\partial w_{ij}$. This information allows one to compute immediately $\partial log(p)/\partial s_j$. Those input nodes for which this partial derivative is large correspond to important experiments.

In conclusion, we propose that back propagation can be generalized, put on a satisfactory conceptual footing, and very likely made more efficient, by defining the values of the output and in-

---

*2 Alternatively when necessary this can be enforced by adding an energy term to the log-likelihood to constrain the parameter variation so that the neuronal values are near either 1 or 0.

put neurons as probabilities, and replacing the 'Error' by the log-likelihood.

Acknowledgement: E. B. Baum was supported in part by DARPA through arrangement with NASA and by NSF grant DMB-840649, 802. F. Wilczek was supported in part by NSF grant PHY82-17853

References

(1)Werbos,P,"Beyond Regression: New Tools for Prediction and Analysis in the Behavioral Sciences", Harvard University Dissertation (1974)

(2)Parker D. B.,"Learning Logic",MIT Tech Report TR-47, Center for Computationl Research in Economics and Management Science, MIT, 1985

(3)Le Cun, Y., Proceedings of Cognitiva 85,p599-604, Paris (1985)

(4)Rumelhart, D. E., Hinton, G. E., Williams, G. E., "Learning Internal Representations by Error Propagation", in "Parallel Distributed Processing", vol 1, eds. Rumelhart, D. E., McClelland, J. L., MIT Press, Cambridge MA,(1986)

(5)Sejnowski, T. J., Rosenberg, C. R., Complex Systems, v 1, pp 145-168 (1987)

(6)LeCun, Y., Address at 1987 Snowbird Conference on Neural Networks

(7)Gorman, P., Sejnowski, T. J.,"Learned Classification of Sonar Targets Using a Massively Parallel Network", in "Workshop on Neural Network Devices and Applications", JPLD-4406, (1987) pp224-237

(8)Rosenblatt, F.,"Principles of Neurodynamics: Perceptrons and the theory of brain mechanisms", Spartan Books, Washington DC (1962)

(9) Widrow, B., Hoff, M. E., 1960 IRE WESCON Conv. Record, Part 4, 96-104 (1960)

(10) Duda, R. O., Hart, P. E., "Pattern Classification and Scene Analysis", John Wiley and Sons, N.Y., (1973)

(11) Guiasu, S., "Information Theory with Applications", McGraw Hill, NY, (1977)

(12) Baum, E.B., "Generalizing Back Propagation to Computation", in "Neural Networks for Computing", AIP Conf. Proc. 151, Snowbird UT (1986)pp47-53

(13) Anderson, C.H., Abrahams, E., "The Bayes Connection", Proceedings of the IEEE International Conference on Neural Networks, San Diego,(1987)

Centric Models of the Orientation Map in Primary Visual Cortex

William Baxter
Department of Computer Science, S.U.N.Y. at Buffalo, NY 14620

Bruce Dow
Department of Physiology, S.U.N.Y. at Buffalo, NY 14620

Abstract

In the visual cortex of the monkey the horizontal organization of the preferred orientations of orientation-selective cells follows two opposing rules: 1) neighbors tend to have similar orientation preferences, and 2) many different orientations are observed in a local region. Several orientation models which satisfy these constraints are found to differ in the spacing and the topological index of their singularities. Using the rate of orientation change as a measure, the models are compared to published experimental results.

Introduction

It has been known for some years that there exist orientation-sensitive neurons in the visual cortex of cats and monkeys[1,2]. These cells react to highly specific patterns of light occurring in narrowly circumscribed regions of the visual field, i.e., the cell's receptive field. The best patterns for such cells are typically not diffuse levels of illumination, but elongated bars or edges oriented at specific angles. An individual cell responds maximally to a bar at a particular orientation, called the preferred orientation. Its response declines as the bar or edge is rotated away from this preferred orientation.

Orientation-sensitive cells have a highly regular organization in primary cortex[3]. Vertically, as an electrode proceeds into the depth of the cortex, the column of tissue contains cells that tend to have the same preferred orientation, at least in the upper layers. Horizontally, as an electrode progresses across the cortical surface, the preferred orientations change in a smooth, regular manner, so that the recorded orientations appear to rotate with distance. It is this horizontal structure we are concerned with, hereafter referred to as the orientation map. An orientation map is defined as a two-dimensional surface in which every point has associated with it a preferred orientation ranging from $0° \cdots 180°$. In discrete versions, such as the array of cells in the cortex or the discrete simulations in this paper, the orientation map will be considered to be a sampled version of the underlying continuous surface. The investigations of this paper are confined to the upper layers of macaque striate cortex.

Detailed knowledge of the two-dimensional layout of the orientation map has implications for the architecture, development, and function of the visual cortex. The organization of orientation-sensitive cells reflects, to some degree, the organization of intracortical connections in striate cortex. Plausible orientation maps can be generated by models with lateral connections that are uniformly exhibited by all cells in the layer[4,5], or by models which presume no specific intracortical connections, only appropriate patterns of afferent input[6]. In this paper, we examine models in which intracortical connections produce the orientation map but the orientation-controlling circuitry is not displayed by all cells. Rather, it derives from localized "centers" which are distributed across the cortical surface with uniform spacing[7,8,9].

The orientation map also represents a deformation in the retinotopy of primary visual cortex. Since the early sixties it has been known that V1 reflects a topographic map of the retina and hence the visual field[10]. There is some global distortion of this mapping[11,12,13], but generally spatial relations between points in the visual field are maintained on the cortical surface. This well-known phenomenon is only accurate for a medium-grain description of V1, however. At a finer cellular level there is considerable scattering of receptive fields at a given cortical location[14]. The notion of the hypercolumn[3] proposes that such scattering permits each region of the visual field to be analyzed by a population of cells consisting of all the necessary orientations and with inputs from both eyes. A quantitative description of the orientation map will allow prediction of the distances between iso-orientation zones of a particular orientation, and suggest how much cortical machinery is being brought to bear on the analysis of a given feature at a given location in the visual field.

Models of the Orientation Map

Hubel and Wiesel's Parallel Stripe Model

The classic model of the orientation map is the parallel stripe model first published by Hubel and Wiesel in 1972[15]. This model has been reproduced several times in their publications[3,16,17] and appears in many textbooks. The model consists of a series of parallel slabs, one slab for each orientation, which are postulated to be orthogonal to the ocular dominance stripes. The model predicts that a microelectrode advancing tangentially (i.e., horizontally) through the tissue should encounter steadily changing orientations. The rate of change, which is also called the orientation drift rate[18], is determined by the angle of the electrode with respect to the array of orientation stripes.

The parallel stripe model does not account for several phenomena reported in long tangential penetrations through striate cortex in macaque monkeys[17,19]. First, as pointed out by Swindale[20], the model predicts that some penetrations will have flat or very low orientation drift rates over lateral distances of hundreds of micrometers. This is because an electrode advancing horizontally and perpendicular to the ocular dominance stripes (and therefore parallel to the orientation stripes) would be expected to remain within a single orientation column over a considerable distance with its orientation drift rate equal to zero. Such results have never been observed. Second, reversals in the direction of the orientation drift, from clockwise to counterclockwise or vice versa, are commonly seen, yet this phenomenon is not addressed by the parallel stripe model. Wavy stripes in the ocular dominace system[21] do not by themselves introduce reversals. Third, there should be a negative correlation between the orientation drift rate and the ocularity "drift rate". That is, when orientation is changing rapidly, the electrode should be confined to a single ocular dominance stripe (low ocularity drift rate), whereas when ocularity is changing rapidly the electrode should be confined to a single orientation stripe (low orientation drift rate). This is clearly not evident in the recent studies of Livingstone and Hubel[17] (see especially their figs. 3b, 21 & 23), where both orientation and ocularity often have high drift rates in the same electrode track, i.e., they show a positive correlation. Anatomical studies with 2-deoxyglucose also fail to show that the orientation and ocular dominance column systems are orthogonal[22].

Centric Models and the Topological Index

Another model, proposed by Braitenberg and Braitenberg in 1979[7], has the orientations arrayed radially around centers like spokes in a wheel The centers are spaced at distances of about 0.5mm. This model produces reversals and also the sinusoidal progressions frequently encountered in horizontal penetrations. However this approach suggests other possibilities, in fact an entire class of centric models. The organizing centers form discontinuities in the otherwise smooth field of orientations. Different topological types of discontinuity are possible, characterized by their topological index[23]. The topological index is a parameter computed by taking a path around a discontinuity and recording the rotation of the field elements (figure 1). The value of the index indicates the amount of rotation; the sign indicates the direction of rotation. An index of 1 signifies that the orientations rotate through 360°; an index of ½ signifies a 180° rotation. A positive index indicates that the orientations rotate in the same sense as a path taken around the singularity; a negative index indicates the reverse rotation.

Topological singularities are stable under orthogonal transformations, so that if the field elements are each rotated 90° the index of the singularity remains unchanged. Thus a +1 singularity may have orientations radiating out from it like spokes from a wheel, or it may be at the center of a series of concentric circles. Only four types of discontinuities are considered here, +1, -1, +½, -½, since these are the most stable, i.e., their neighborhoods are characterized by smooth change.

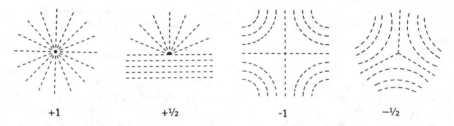

+1   +½   -1   -½

figure 1. Topological singularities. A positive index indicates that the orientations rotate in the same direction as a path taken around the singularity; a negative index indicates the reverse rotation. Orientations rotate through 360° around ±1 centers, 180° around ±½ centers.

Cytochrome Oxidase Puffs

At topological singularities the change in orientation is discontinuous, which violates the structure of a smoothly changing orientation map; modellers try to minimize discontinuities in their models in order to satisfy the smoothness constraint. Interestingly, in the upper layers of striate cortex of monkeys, zones with little or no orientation selectivity have been discovered. These zones are notable for their high cytochrome oxidase reactivity[24] and have been referred to as cytochrome oxidase puffs, dots, spots, patches or blobs[17,25,26,27]. We will refer to them as puffs. If the organizing centers of centric models are located in the cytochrome oxidase puffs then the discontinuities in the orientation map are effectively eliminated (but see below). Braitenberg has indicated[28] that the +1 centers of his model should correspond to the puffs. Dow and Bauer proposed a model[8] with +1 and -1 centers in alternating puffs. Gotz proposed a similar model[9] with alternating +½ and -½ centers in the puffs. The last two models manage to eliminate all discontinuities from the interpuff zones, but they assume a perfect rectangular lattice of cytochrome oxidase puffs.

A Set of Centric Models

There are two parameters for the models considered here. (1) Whether the positive singularities are placed in every puff or in alternate puffs; and (2) whether the singularities are ±1's or ±½'s. This gives four centric models (figure 2):

E1 : +1 centers in puffs, -1 centers in the interpuff zones.
A1 : both +1 and -1 centers in the puffs, interdigitated in a checkerboard fashion.
E½ : +½ centers in the puffs, −½ centers in the interpuff zones.
A½ : both +½ and −½ centers in the puffs, as in A1.

The E1 model corresponds to the Braitenberg model transposed to a rectangular array rather than an hexagonal one, in accordance with the observed organization of the cytochrome oxidase regions[27]. In fact, the rectangular version of the Braitenberg model is pictured in figure 49 of[27]. The A1 model was originally proposed by Dow and Bauer[8] and is also pictured in an article by Mitchison[29]. The A½ model was proposed by Gotz[9]. It should be noted that the E1 and A1 models are the same model rotated and scaled a bit; the E½ and A½ have the same relationship.

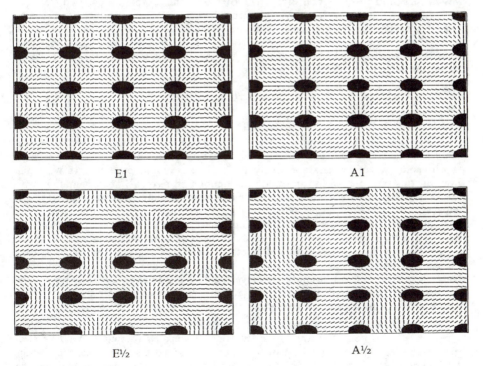

E1   A1

E½   A½

figure 2. The four centric models. Dark ellipses represent cytochrome oxidase puffs. Dots in interpuff zones of E1 & E½ indicate singularities at those points.

Simulations

Simulated horizontal electrode recordings were made in the four models to compare their orientation drift rates with those of published recordings. In the computer simulations (figure 2) the interpuff distances were chosen to correspond to histological measurements[27]. Puff centers are separated by 500μ along their long axes, 350μ along the short axes. The density of the arrays was chosen to approximate the sampling frequency observed in Hubel and Wiesel's horizontal electrode recording experiments[19], about 20 cells per millimeter. Therefore the cell density of the simulation arrays was about six times that shown in the figure.

All of the models produce simulated electrode data that qualitatively resemble the published recording results, e.g., they contain reversals, and runs of constantly changing orientations. The orientation drift rate and number of reversals vary in the different models.

The models of figure 2 are shown in perfectly rectangular arrays. Some important characteristics of the models, such as the absence of discontinuites in interpuff zones, are dependent on this regularity. However, the real arrangement of cytochrome oxidase puffs is somewhat irregular, as in Horton's figure 3[27]. A small set of puffs from the parafoveal region of Horton's figure was enlarged and each of the centric models was embedded in this irregular array. The E1 model and a typical simulated electrode track are shown in figure 3. Several problems are encountered when models developed in a regular lattice are implemented in the irregular lattice of the real system; the models have appreciably different properties. The -1 singularities in E1's interpuff zones have been reduced to $-\frac{1}{2}$'s; the A1 and A$\frac{1}{2}$ models now have some interpuff discontinuities where before they had none.

Quantitative Comparisons

Measurement of the Orientation Drift Rate

There are two sets of centric models in the computer simulations: a set in the perfectly rectangular array (figure 2) and a set in the irregular puff array (as in figure 3). At this point we can generate as many tracks in the simulation arrays as we wish. How can this information be compared to the published records? The orientation drift rate, or slope, is one basis for distinguishing between models. In real electrode tracks however, the data are rather noisy, perhaps from the measuring process or from inherent unevenness of the orientation map. The typical approach is to fit a straight line and use the slope of this line. Reversals in the tracks require that lines be fit piecewise, the approach used by Hubel and Wiesel[19]. Because of the unevenness of the data it is not always clear what constitutes a reversal. Livingstone and Hubel[17] report that the track in their figure 5 has only two reversals in 5 millimeters. Yet there seem to be numerous microreversals between the 1st and 3rd millimeter of their track. At what point is a change in slope considered a true reversal rather than just noise?

The approach used here was to use a local slope measure and ignore the problem of reversals - this permitted the fast calculation of slope by computer. A single electrode track, usually several millimeters long, was assigned a single slope, the average of the derivative taken at each point of the track. Since these are discrete samples, the local derivative must be approximated by taking measurements over a small neighborhood. How large should this neighborhood be? If too small it will be susceptible to noise in the orientation measures, if too large it will "flatten out" true reversals. Slope

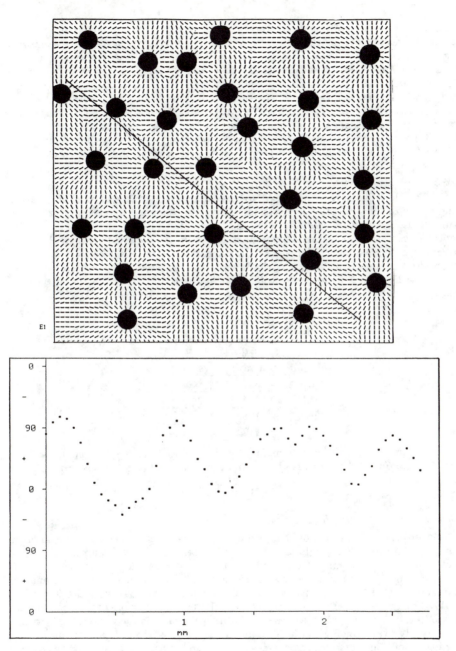

figure 3. A centric model in a realistic puff array (from[27]). A simulated electrode track and resulting data are shown. Only the E1 model is shown here, but other models were similarly embedded in this array.

measures using neighborhoods of several sizes were applied to six published horizontal electrode tracks from the foveal and parafoveal upper layers of macaque striate cortex: figures 5,6,7 from[17], figure 16 from[3], figure 1 from[30]. A neighborhood of 0.1mm, which attempts to fit a line between virtually every pair of points, gave abnormally high slopes. Larger neighborhoods tended to give lower slopes, especially to those tracks which contained reversals. The smallest window that gave consistent measures for all six tracks was 0.2mm; therefore this window was chosen for comparisons between published data and the centric models. This measure gave an average slope of 285 degrees per millimeter in the six published samples of track data, compared to Hubel & Wiesel's measure of 281 deg/mm for the penetrations in their 1974 paper[19].

Slope measures of the centric models

The slope measure was applied to several thousand tracks at random locations and angles in the simulation arrays, and a slope was computed for each simulated electrode track. Average slopes of the models are shown in Table I. Generally, models with ±1 centers have higher slopes than those with ±½ centers; models with centers in every puff have higher slopes than the alternate puff models. Thus E1 showed the highest orientation drift rate, A½ the lowest, with A1 and E½ having intermediate rates. The E1 model, in both the rectangular and irregular arrays, produced the most realistic slope values.

TABLE I   Average slopes of the centric models

	Rectangular array	Irregular array
E1	312	289
A1	216	216
E½	198	202
A½	117	144

Numbers are in degrees/mm. Slope measure (window = 0.2mm) applied to 6 published records yielded an average slope of 285 degrees/mm.

Discussion

Constraints on the Orientation Map

Our original definition of the orientation map permits each cell to have an orientation preference whose angle is completely independent of its neighbors. But this is much too general. Looking at the results of tangential electrode penetrations, there are two striking constraints in the data. The first of these is reflected in the smoothness of the graphs. Orientation changes in a regular manner as the electrode moves horizontally through the upper layers: neighboring cells have similar orientation preferences. Discontinuities do occur but are rare. The other constraint is the fact that the orientation is always changing with distance, although the rate of change may vary. Sequences of constant orientation are very rare and when they do occur they never carry on for any appreciable distance. This is one of the major reasons why the parallel stripe model is untenable. The two major constraints on the orientation map may be put informally as follows:

1. The smoothness constraint : neighboring points have similar orientation preferences.

2. The heterogeneity constraint : all orientations should be represented within a small region of the cortical surface.

This second constraint is a bit stronger than the data imply. The experimental results only show that the orientations change regularly with distance, not that all orientations must be present within a region. But this constraint is important with respect to visual processing and the notion of hypercolumns[3].

These are opposing constraints: the first tends to minimize the slope, or orientation drift rate, while the second tends to maximize this rate. Thus the organization of the orientation map is analogous to physical systems that exhibit "frustration", that is, the elements must satisfy conflicting constraints[31]. One of the properties of such systems is that there are many near-optimal solutions, no one of which is significantly better than the others. As a result, there are many plausible orientation maps: any map that satisfies these two constraints will generate qualitatively plausible simulated electrode tracks. This points out the need for quantitative comparisons between models and experimental results.

Centric models and the two constraints

What are some possible mechanisms of the constraints that generate the orientation map? Smoothness is a local property and could be attributed to the workings of individual cells. It seems to be a fundamental property of cortex that adjacent cells respond to similar stimuli. The heterogeneity requirement operates at a slightly larger scale, that of a hypercolumn rather than a minicolumn. While the first constraint may be modeled as a property of individual cells, the second constraint is distributed over a region of cells. How can such a collection of cells insure that its members cycle through all the required orientations? The topological singularities discussed earlier, by definition, include all orientations within a restricted region. By distributing these centers across the surface of the cortex, the heterogeneity constraint may be satisfied. In fact, the amount of orientation drift rate is a function of the density of this distribution (i.e., more centers per unit area give higher drift rates).

It has been noted that the E1 and the A1 organizations are the same topological model, but on different scales; the low drift rates of the A1 model may be increased by increasing the density of the +1 centers to that of the E1 model. The same relationship holds for the E½ and A½ models. It is also possible to obtain realistic orientation drift rates by increasing the density of +½ centers, or by mixing +1's and +½'s. However, these alternatives increase the number of interpuff singularities. And given the possible combinations of centers, it may be more than coincidental that a set of +1 centers at just the spacing of the cytochrome oxidase regions results in realistic orientation drift rates.

Cortical Architecture and Types of Circuitry

Thus far, we have not addressed the issue of how the preferred orientations are generated. The mechanism is presently unknown, but attempts to depict it have traditionally been of a geometric nature, alluding to the dendritic morphology[1,8,28,32]. More recently, computer simulations have shown that orientation-sensitive units may be obtained from asymmetries in the receptive fields of afferents[6], or developed using simple Hebbian rules for altering synaptic weights[5]. That is, given appropriate network parameters, orientation tuning arises an as inherent property of some neural networks. Centric models propose a quite different approach in which an originally untuned cell is "programmed" by a center located at some distance to respond to a specific orientation. So, for an individual cell, does orientation develop locally, or is it "imposed from without"? Both of these mechanisms may be in effect, acting synergistically to produce the final orientation map. The map may spontaneously form on the embryonic cortex, but with cells that are nonspecific and broadly tuned. The organization imposed by the centers could have two effects on this incipient map. First, the additional influence from centers could "tighten up" the tuning curves, making the cells more specific. Second, the spacing of the centers specifies a distinct and uniform scale for the heterogeneity of the map. An unsupervised developing orientation map could have broad expanses of iso-orientation zones mixed with regions of rapidly changing orientations. The spacing of the puffs, hence the architecture of the cortex, insures that there is an appropriate variety of feature sensitive cells at each location. This has implications for cortical functioning: given the distances of lateral connectivity, for a cell of a given orientation, we can estimate how many other iso-orientation zones of that same orientation the cell may be communicating with. For a given orientation, the E1 model has twice as many iso-orientation zones per unit area as A1.

Ever since the discovery of orientation-specific cells in visual cortex there have been attempts to relate the distribution of cell selectivities to architectural features of the cortex. Hubel and Wiesel originally suggested that the orientation slabs followed the organization of the ocular dominance slabs[15]. The Braitenbergs suggested in their original model[7] that the centers might be identified with the giant cells of Meynert. Later centric models have identified the centers with the cytochrome oxidase regions, again relating the orientation map to the ocular dominance array, since the puffs themselves are closely related to this array.

While biologists have habitually related form to function, workers in machine vision have traditionally relied on general-purpose architectures to implement a variety of algorithms related to the processing of visual information[33]. More recently, many computer scientists designing artificial vision systems have turned their attention towards connectionist systems and neural networks. There is great interest in how the sensitivities to different features and how the selectivities to different values of those features may be embedded in the system architecture[34,35,36]. Linsker has proposed (this volume) that the development of feature spaces is a natural concomitance of layered networks, providing a generic organizing principle for networks. Our work deals with more specific cortical architectonics, but we are convinced that the study of the cortical layout of feature maps will provide important insights for the design of artificial systems.

References

1. D. Hubel & T. Wiesel, *J. Physiol.* (*Lond.*) **160,** 106 (1962).
2. D. Hubel & T. Wiesel, *J. Physiol.* (*Lond.*) **195,** 225 (1968).
3. D. Hubel & T. Wiesel, *Proc. Roy. Soc. Lond. B* **198,** 1 (1977).
4. N. Swindale, *Proc. Roy. Soc. Lond. B* **215,** 211 (1982).
5. R.Linsker, *Proc. Natl. Acad. Sci. USA* **83,** 8779 (1986).
6. R. Soodak, *Proc. Natl. Acad. Sci. USA* **84,** 3936 (1987).

7. V. Braitenberg & C. Braitenberg, *Biol. Cyber.* **33,** 179 (1979).
8. B. Dow & R. Bauer, *Biol. Cyber.* **49,** 189 (1984).
9. K. Gotz, *Biol. Cyber.* **56,** 107 (1987).
10. P. Daniel & D. Whitteridge, *J. Physiol. (Lond.)* **159,** 302 (1961).
11. B. Dow, R. Vautin & R. Bauer, *J. Neurosci.* **5,** 890 (1985).
12. R.B. Tootell, M.S. Silverman, E. Switkes & R. DeValois, *Science* **218,** 902 (1982).
13. D.C. Van Essen, W.T. Newsome & J.H. Maunsell, *Vision Research* **24,** 429 (1984).
14. D. Hubel & T. Wiesel, *J. Comp. Neurol.* **158,** 295 (1974).
15. D. Hubel & T. Wiesel, *J. Comp. Neurol.* **146,** 421 (1972).
16. D. Hubel, *Nature* **299,** 515 (1982).
17. M. Livingstone & D. Hubel, *J. Neurosci.* **4,** 309 (1984).
18. R. Bauer, B. Dow, A. Snyder & R. Vautin, *Exp. Brain Res.* **50,** 133 (1983).
19. D. Hubel & T. Wiesel, *J. Comp. Neurol.* **158,** 267 (1974).
20. N. Swindale, in *Models of the Visual Cortex*, D. Rose & V. Dobson, eds., (Wiley, 1985), p. 452.
21. S. LeVay, D. Hubel, & T. Wiesel, *J. Comp. Neurol.* **159,** 559 (1975).
22. D. Hubel, T. Wiesel & M. Stryker, *J. Comp. Neurol.* **177,** 361 (1978).
23. T. Elsdale & F. Wasoff, *Wilhelm Roux's Archives* **180,** 121 (1976).
24. M.T. Wong-Riley, *Brain Res.* **162,** 201 (1979).
25. A. Humphrey & A. Hendrickson, *J. Neurosci.* **3,** 345 (1983).
26. E. Carroll & M. Wong-Riley, *J. Comp. Neurol.* **222,** 1 (1984).
27. J. Horton, *Proc. Roy. Soc. Lond. B* **304,** 199 (1984).
28. V. Braitenberg, in *Models of the Visual Cortex*, p. 479.
29. G. Mitchison, in *Models of the Visual Cortex*, p. 443.
30. C. Michael, *Vision Research* **25** 415 (1985).
31. S. Kirkpatrick, M. Gelatt & M. Vecchi, *Science* **220,** 671 (1983).
32. S. Tieman & H. Hirsch, in *Models of the Visual Cortex*, p. 432.
33. D. Ballard & C. Brown *Computer Vision* (Prentice-Hall, N.J., 1982).
34. D. Ballard, G. Hinton, & T. Sejnowski, *Nature* **306,** 21 (1983).
35. D. Ballard, *Behav. and Brain Sci.* **9,** 67 (1986).
36. D. Walters, *Proc. First Int. Conf. on Neural Networks* (June 1987).

ANALYSIS AND COMPARISON OF DIFFERENT LEARNING ALGORITHMS FOR PATTERN ASSOCIATION PROBLEMS

J. Bernasconi
Brown Boveri Research Center
CH-5405 Baden, Switzerland

ABSTRACT

We investigate the behavior of different learning algorithms for networks of neuron-like units. As test cases we use simple pattern association problems, such as the XOR-problem and symmetry detection problems. The algorithms considered are either versions of the Boltzmann machine learning rule or based on the backpropagation of errors. We also propose and analyze a generalized delta rule for linear threshold units. We find that the performance of a given learning algorithm depends strongly on the type of units used. In particular, we observe that networks with ±1 units quite generally exhibit a significantly better learning behavior than the corresponding 0,1 versions. We also demonstrate that an adaption of the weight-structure to the symmetries of the problem can lead to a drastic increase in learning speed.

INTRODUCTION

In the past few years, a number of learning procedures for neural network models with hidden units have been proposed[1,2]. They can all be considered as strategies to minimize a suitably chosen error measure. Most of these strategies represent local optimization procedures (e.g. gradient descent) and therefore suffer from all the problems with local minima or cycles. The corresponding learning rates, moreover, are usually very slow.

The performance of a given learning scheme may depend critically on a number of parameters and implementation details. General analytical results concerning these dependences, however, are practically non-existent. As a first step, we have therefore attempted to study empirically the influence of some factors that could have a significant effect on the learning behavior of neural network systems.

Our preliminary investigations are restricted to very small networks and to a few simple examples. Nevertheless, we have made some interesting observations which appear to be rather general and which can thus be expected to remain valid also for much larger and more complex systems.

NEURAL NETWORK MODELS FOR PATTERN ASSOCIATION

An artificial neural network consists of a set of interconnected units (formal neurons). The state of the i-th unit is described by a variable S_i which can be discrete (e.g. $S_i = 0,1$ or $S_i = \pm 1$) or continuous (e.g. $0 \leq S_i \leq 1$ or $-1 \leq S_i \leq +1$), and each connection $j \rightarrow i$ carries a weight W_{ij} which can be positive, zero, or negative.

The dynamics of the network is determined by a local update rule,

$$S_i(t+1) = f(\sum_j W_{ij} S_j(t)) \quad , \tag{1}$$

where f is a nonlinear activation function, specifically a threshold function in the case of discrete units and a sigmoid-type function, e.g.

$$f(x) = 1/(1+e^{-x}) \tag{2}$$

or

$$f(x) = (1-e^{-x})/(1+e^{-x}) \quad , \tag{3}$$

respectively, in the case of continuous units. The individual units can be given different thresholds by introducing an extra unit which always has a value of 1.

If the network is supposed to perform a pattern association task, it is convenient to divide its units into input units, output units, and hidden units. Learning then consists in adjusting the weights in such a way that, for a given input pattern, the network relaxes (under the prescribed dynamics) to a state in which the output units represent the desired output pattern.

Neural networks learn from examples (input/output pairs) which are presented many times, and a typical learning procedure can be viewed as a strategy to minimize a suitably defined error function F. In most cases, this strategy is a (stochastic) gradient descent method: To a clamped input pattern, randomly chosen from the learning examples, the network produces an output pattern $\{O_i\}$. This is compared with the desired output, say $\{T_i\}$, and the error $F(\{O_i\}, \{T_i\})$ is calculated. Subsequently, each weight is changed by an amount proportional to the respective gradient of F,

$$\Delta W_{ij} = -\eta \frac{\partial F}{\partial W_{ij}} \quad , \tag{4}$$

and the procedure is repeated for a new learning example until F is minimized to a satisfactory level.

In our investigations, we shall consider two different types of learning schemes. The first is a deterministic version of the Boltzmann machine learning rule[1] and has been proposed by Yann Le Cun[2]. It applies to networks with symmetric weights, $W_{ij} = W_{ji}$, so that an energy

$$E(\underline{S}) = - \sum_{(i,j)} W_{ij} S_i S_j \tag{5}$$

can be associated with each state $\underline{S} = \{S_i\}$. If $\underline{X}$ refers to the network state when only the input units are clamped and $\underline{Y}$ to the state when both the input and output units are clamped, the error function is defined as

$$F = E(\underline{Y}) - E(\underline{X}) \quad , \tag{6}$$

and the gradients are simply given by

$$-\frac{\partial F}{\partial W_{ij}} = Y_i Y_j - X_i X_j \quad . \tag{7}$$

The second scheme, called <u>backpropagation</u> or <u>generalized delta rule</u>[1,3], probably represents the most widely used learning algorithm. In its original form, it applies to networks with feedforward connections only, and it uses gradient descent to minimize the mean squared error of the output signal,

$$F = \frac{1}{2} \sum_i (T_i - O_i)^2 \quad . \tag{8}$$

For a weight W_{ij} from an (input or hidden) unit j to an output unit i, we simply have

$$-\frac{\partial F}{\partial W_{ij}} = (T_i - O_i)f'(\sum_k W_{ik} S_k)S_j \quad , \tag{9}$$

where f' is the derivative of the nonlinear activation function introduced in Eq. (1), and for weights which do not connect to an output unit, the gradients can successively be determined by applying the chain rule of differentiation.

In the case of discrete units, f is a threshold function, so that the backpropagation algorithm described above cannot be applied. We remark, however, that the perceptron learning rule[4],

$$\Delta W_{ij} = \varepsilon(T_i - O_i)S_j \quad , \tag{10}$$

is nothing else than Eq. (9) with f' replaced by a constant ε. Therefore, we propose that a <u>generalized delta rule for linear threshold units</u> can be obtained if f' is replaced by a constant ε in all the backpropagation expressions for $\partial F/\partial W_{ij}$. This generalization of the perceptron rule is, of course, not unique. In layered networks, e.g., the value of the constant which replaces f' need not be the same for the different layers.

ANALYSIS OF LEARNING ALGORITHMS

The proposed learning algorithms suffer from all the problems of gradient descent on a complicated landscape. If we use small weight changes, learning becomes prohibitively slow, while large weight changes inevitably lead to oscillations which prevent the algorithm from converging to a good solution. The error surface, moreover, may contain many local minima, so that gradient descent is not guaranteed to find a global minimum.

There are several ways to improve a stochastic gradient descent procedure. The weight changes may, e.g., be accumulated over a number of learning examples before the weights are actually changed. Another often used method consists in smoothing the weight changes by overrelaxation,

$$\Delta W_{ij}(k+1) = -\eta \frac{\partial F}{\partial W_{ij}} + \alpha \Delta W_{ij}(k) \quad , \qquad (11)$$

where $\Delta W_{ij}(k)$ refers to the weight change after the presentation of the k-th learning example (or group of learning examples, respectively). The use of a weight decay term,

$$\Delta W_{ij} = -\eta \frac{\partial F}{\partial W_{ij}} - \beta W_{ij} \quad , \qquad (12)$$

prevents the algorithm from generating very large weights which may create such high barriers that a solution cannot be found in reasonable time.

Such smoothing methods suppress the occurrence of oscillations, at least to a certain extent, and thus allow us to use higher learning rates. They cannot prevent, however, that the algorithm may become trapped in bad local minimum. An obvious way to deal with the problem of local minima is to restart the algorithm with different initial weights or, equivalently, to randomize the weights with a certain probability p during the learning procedure. More sophisticated approaches involve, e.g., the use of hill-climbing methods.

The properties of the error-surface over the weight space not only depend on the choice of the error function F, but also on the network architecture, on the type of units used, and on possible restrictions concerning the values which the weights are allowed to assume.

The performance of a learning algorithm thus depends on many factors and parameters. These dependences are conveniently analyzed in terms of the behavior of an appropriately defined learning curve. For our small examples, where the learning set always consists of all input/output cases, we have chosen to represent the performance of a learning procedure by the <u>fraction of networks that are "perfect" after the presentation of N input patterns</u>. (Perfect networks are networks which for every input pattern produce the correct output). Such learning curves give us much more detailed information about the behavior of the system than, e.g., averaged quantities like the mean learning time.

RESULTS

In the following, we shall present and discuss some representative results of our empirical study. All learning curves refer to a set of 100 networks that have been exposed to the same learning procedure, where we have varied the initial weights, or the sequence

76

of learning examples, or both. With one exception (Figure 4), the sequences of learning examples are always random.

A prototype pattern association problem is the exclusive-or (XOR) problem. Corresponding networks have two input units and one output unit. Let us first consider an XOR-network with only one hidden unit, but in which the input units also have direct connections to the output unit. The weights are symmetric, and we use the deterministic version of the Boltzmann learning rule (see Eqs. (5) to (7)). Figure 1 shows results for the case of tabula rasa initial conditions, i.e. the initial weights are all set equal to zero. If the weights are changed after every learning example, about 2/3 of the networks learn the problem with less than 25 presentations per pattern (which corresponds to a total number of 4 × 25 = 100 presentations). The remaining networks (about 1/3), however, never learn to solve the XOR-problem, no matter how many input/output cases are presented. This can be understood by analyzing the corresponding evolution-tree in weight-space which contains an attractor consisting of 14 "non-perfect" weight-configurations. The probability to become trapped by this attractor is exactly 1/3. If the weight changes are accumulated over 4 learning examples, no such attractor

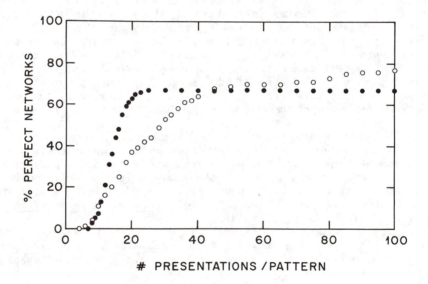

 # PRESENTATIONS / PATTERN

Fig. 1. Learning curves for an XOR-network with one hidden unit (deterministic Boltzmann learning, discrete ±1 units, initial weights zero). Full circles: weights changed after every learning example; open circles: weight changes accumulated over 4 learning examples.

seems to exist (see Fig. 1), but for some networks learning at least
takes an extremely long time. The same saturation effect is observed
with random initial weights (uniformly distributed between -1 and
+1), see Fig. 2.

Figure 2 also exhibits the difference in learning behavior
between networks with ± 1 units and such with 0,1 units. In both
cases, weight randomization leads to a considerably improved learning behavior. A weight decay term, by the way, has the same effect.
The most striking observation, however, is that ± 1 networks learn
much faster than 0,1 networks (the respective average learning times
differ by about a factor of 5). In this connection, we should mention
that $\eta = 0.1$ is about optimal for 0,1 units and that for ± 1 networks
the learning behavior is practically independent of the value of η.
It therefore seems that ± 1 units lead to a much more well-behaved
error-surface than 0,1 units. One can argue, of course, that a
discrete 0,1 model can always be translated into a ± 1 model, but
this would lead to an energy function which has a considerably more
complicated weight dependence than Eq. (5).

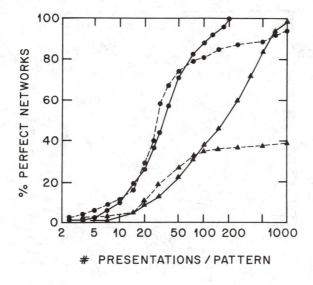

Fig. 2. Learning curves for an XOR-network with one hidden unit
(deterministic Boltzmann learning, initial weights random, weight
changes accumulated over 5 learning examples). Circles: discrete ± 1
units, $\eta = 1$; triangles: discrete 0,1 units, $\eta = 0.1$; broken curves:
without weight randomization; solid curves: with weight randomization (p = 0.025).

Figures 3 and 4 refer to a feedforward XOR-network with 3 hidden units, and to backpropagation or generalized delta rule learning. In all cases we have included an overrelaxation (or momentum) term with $\alpha = 0.9$ (see Eq. (11)). For the networks with continuous units we have used the activation functions given by Eqs. (2) and (3), respectively, and a network was considered "perfect" if for all input/output cases the error was smaller than 0.1 in the 0,1 case, or smaller than 0.2 in the ±1 case, respectively.

In Figure 3, the weights have been changed after every learning example, and all curves refer to an optimal choice of the only remaining parameter, ε or η, respectively. For discrete as well as for continuous units, the ±1 networks again perform much better than their 0,1 counterparts. In the continuous case, the average learning times differ by about a factor of 7, and in the discrete case the discrepancy is even more pronounced. In addition, we observe that in ±1 networks learning with the generalized delta rule for discrete units is about twice as fast as with the backpropagation algorithm for continuous units.

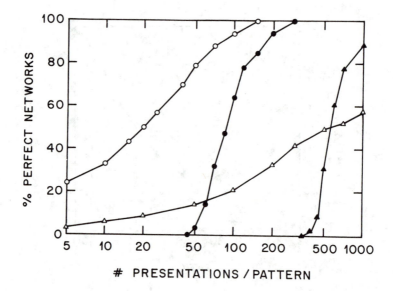

Fig. 3. Learning curves for an XOR-network with three hidden units (backpropagation/generalized delta rule, initial weights random, weights changed after every learning example). Open circles: discrete ±1 units, $\varepsilon = 0.05$; open triangles: discrete 0,1 units, $\varepsilon = 0.025$; full circles: continuous ±1 units, $\eta = 0.125$; full triangles; continuous 0,1 units, $\eta = 0.25$.

In Figure 4, the weight changes are accumulated over all 4 input/output cases, and only networks with continuous units are considered. Also in this case, the ±1 units lead to an improved learning behavior (the optimal η-values are about 2.5 and 5.0, respectively). They not only lead to significantly smaller learning times, but ±1 networks also appear to be less sensitive with respect to a variation of η than the corresponding 0,1 versions.

The better performance of the ±1 models with continuous units can partly be attributed to the steeper slope of the chosen activation function, Eq. (3). A comparison with activation functions that have the same slope, however, shows that the networks with ±1 units still perform significantly better than those with 0,1 units. If the weights are updated after every learning example, e.g., the reduction in learning time remains as large as a factor of 5. In the case of backpropagation learning, the main reason for the better performance of ±1 units thus seems to be related to the fact that the algorithm does not modify weights which emerge from a unit with value zero. Similar observations have been made by Stornetta and Huberman,[5] who further find that the discrepancies become even more pronounced if the network size is increased.

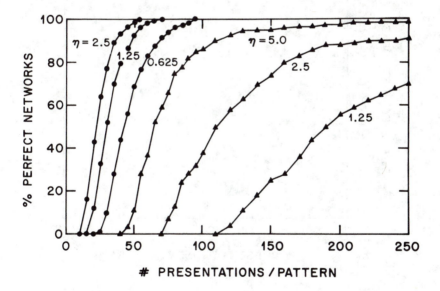

Fig. 4. Learning curves for an XOR-network with three hidden units (backpropagation, initial weights random, weight changes accumulated over all 4 input/output cases). Circles: continuous ±1 units; triangles: continuous 0,1 units.

In Figure 5, finally, we present results for a network that learns to detect mirror symmetry in the input pattern. The network consists of one output, one hidden, and four input units which are also directly connected to the output unit. We use the deterministic version of Boltzmann learning and change the weights after every presentation of a learning pattern. If the weights are allowed to assume arbitrary values, learning is rather slow and on average requires almost 700 presentations per pattern. We have observed, however, that the algorithm preferably seems to converge to solutions in which geometrically symmetric weights are opposite in sign and almost equal in magnitude (see also Ref. 3). This means that the symmetric input patterns are automatically treated as equivalent, as their net input to the hidden as well as to the output unit is zero. We have therefore investigated what happens if the weights are forced to be antisymmetric from the beginning. (The learning procedure, of course, has to be adjusted such that it preserves this antisymmetry). Figure 5 shows that such a problem-adapted weight-structure leads to a dramatic decrease in learning time.

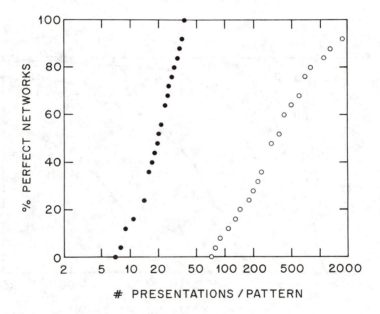

Fig. 5. Learning curves for a symmetry detection network with 4 input units and one hidden unit (deterministic Boltzmann learning, $\eta = 1$, discrete ±1 units, initial weights random, weights changed after every learning example). Full circles: symmetry-adapted weights; open circles: arbitrary weights, weight randomization (p = 0.015).

CONCLUSIONS

The main results of our empirical study can be summarized as follows:
- Networks with ±1 units quite generally exhibit a significantly faster learning than the corresponding 0,1 versions.
- In addition, ±1 networks are often less sensitive to parameter variations than 0,1 networks.
- An adaptation of the weight-structure to the symmetries of the problem can lead to a drastic improvement of the learning behavior.

Our qualitative interpretations seem to indicate that the observed effects should not be restricted to the small examples considered in this paper. It would be very valuable, however, to have corresponding analytical results.

REFERENCES

1. "Parallel Distributed Processing: Explorations in the Microstructure of Cognition", vol. 1: "Foundations", ed. by D.E. Rumelhart and J.L. McClelland (MIT Press, Cambridge), 1986, Chapters 7 & 8.
2. Y. le Cun, in "Disordered Systems and Biological Organization", ed. by E. Bienenstock, F. Fogelman Soulié, and G. Weisbuch (Springer, Berlin), 1986, pp. 233-240.
3. D.E. Rumelhart, G.E. Hinton, and R.J. Williams, Nature 323, 533 (1986).
4. M.L. Minsky and S. Papert, "Perceptrons" (MIT Press, Cambridge), 1969.
5. W.S. Stornetta and B.A. Huberman, IEEE Conference on "Neural Networks", San Diego, California, 21-24 June 1987.

SIMULATIONS SUGGEST INFORMATION PROCESSING ROLES FOR THE DIVERSE CURRENTS IN HIPPOCAMPAL NEURONS

Lyle J. Borg-Graham
Harvard-MIT Division of Health Sciences and Technology and
Center for Biological Information Processing,
Massachusetts Institute of Technology, Cambridge, Massachusetts 02139

ABSTRACT

A computer model of the hippocampal pyramidal cell (HPC) is described which integrates data from a variety of sources in order to develop a consistent description for this cell type. The model presently includes descriptions of eleven non-linear somatic currents of the HPC, and the electrotonic structure of the neuron is modelled with a soma/short-cable approximation. Model simulations qualitatively or quantitatively reproduce a wide range of somatic electrical behavior in HPCs, and demonstrate possible roles for the various currents in information processing.

1 The Computational Properties of Neurons

There are several substrates for neuronal computation, including connectivity, synapses, morphometrics of dendritic trees, linear parameters of cell membrane, as well as non-linear, time-varying membrane conductances, also referred to as currents or channels. In the classical description of neuronal function, the contribution of membrane channels is constrained to that of generating the action potential, setting firing threshold, and establishing the relationship between (steady-state) stimulus intensity and firing frequency. However, it is becoming clear that the role of these channels may be much more complex, resulting in a variety of novel "computational operators" that reflect the information processing occurring in the biological neural net.

2   Modelling Hippocampal Neurons

Over the past decade a wide variety of non-linear ion channels, have been described for many excitable cells, in particular several kinds of neurons. One such neuron is the hippocampal pyramidal cell (HPC). HPC channels are marked by their wide range of temporal, voltage-dependent, and chemical-dependent characteristics, which results in very complex behavior or responses of these stereotypical cortical integrating cells. For example, some HPC channels are activated (opened) transiently and quickly, thus primarily affecting the action potential shape. Other channels have longer kinetics, modulating the response of HPCs over hundreds of milliseconds. The measurement these channels is hampered by various technical constraints, including the small size and extended electrotonic structure of HPCs and the diverse preparations used in experiments. Modelling the electrical behavior of HPCs with computer simulations is one method of integrating data from a variety of sources in order to develop a consistent description for this cell type.

In the model referred to here putative mechanisms for voltage-dependent and calcium-dependent channel gating have been used to generate simulations of the somatic electrical behavior of HPCs, and to suggest mechanisms for information processing at the single cell level. The model has also been used to suggest experimental protocols designed to test the validity of simulation results. Model simulations qualitatively or quantitatively reproduce a wide range of somatic electrical behavior in HPCs, and explicitly demonstrate possible functional roles for the various currents [1].

The model presently includes descriptions of eleven non-linear somatic currents, including three putative Na^+ currents – $I_{Na-trig}$, I_{Na-rep}, and $I_{Na-tail}$; six K^+ currents that have been reported in the literature – I_{DR} (Delayed Rectifier), I_A, I_C, I_{AHP} (After-hyperpolarization), I_M, and I_Q; and two Ca^{2+} currents, also reported previously – I_{Ca} and I_{CaS}.

The electrotonic structure of the HPC is modelled with a soma/short-cable approximation, and the dendrites are assumed to be linear. While the conditions for reducing the dendritic tree to a single cable are not met for HPC (the so-called Rall conditions [3]), the Z_{in} of the cable is close to that of the tree. In addition, although HPC dendrites have non-linear membrane, it assumed that as a first approximation the contribution of currents from this membrane may be ignored in the somatic response to somatic stimulus. Likewise, the model structure assumes that axon-soma current under these conditions can be lumped into the soma circuit.

In part this paper will address the following question: if neural nets are realizable using elements that have simple integrative all-or-nothing responses, connected to each other with regenerative conductors, then what is the function for all the channels observed experimentally in real neurons? The results of this HPC model study suggest some purpose for these complexities, and in this paper we shall investigate some of the possible roles of non-linear channels in neuronal information processing. However, given the speculative nature of many of the currents that we have presented in the model, it is important to view results based on the interaction of the many model elements as preliminary.

3   Defining Neural Information Coding is the First Step in Describing Biological Computations

Determination of computational properties of neurons requires *a priori* assumptions as to how information is encoded in neuronal output. The classical description assumes that information is encoded as spike frequency. However, a single output variable, proportional to firing frequency, ignores other potentially information-rich degrees of freedom, including:

- Relative phase of concurrent inputs.

- Frequency modulation during single bursts.

- Cessation of firing due to intrinsic mechanisms.

- Spike shape.

Note that these variables apply to patterns of repetitive firing[1]. The relative phase of different inputs to a single cell is very important at low firing rates, but becomes less so as firing frequency approaches the time constant of the postsynaptic membrane or some other rate-limiting process in the synaptic transduction (e.g. neurotransmitter release or post synaptic channel activation/deactivation kinetics). Frequency modulation during bursts/spike trains may be important in the interaction of a given axon's output with other inputs at the target neuron. Cessation of firing due to mechanisms intrinsic to the cell (as opposed to the end of input) may be

---

[1] Single spikes may be considered as degenerate cases of repetitive firing responses.

important, for example, in that cell's transmission function. Finally, modulation of spike shape may have several consequences, which will be discussed later.

4   Physiological Modulation of HPC Currents

In order for modulation of HPC currents to be considered as potential information processing mechanisms *in vivo*, it is necessary to identify physiological modulators. For several of the currents described here such factors have been identified. For example, there is evidence that I_M is inhibited by muscarinic (physiologically, cholinergic) agonists [2], that I_A is inhibited by acetylcholine [6], and that I_{AHP} is inhibited by noradrenaline [5]. In fact, the list of neurotransmitters which are active non-synaptically is growing rapidly. It remains to be seen whether there are as yet undiscovered mechanisms for modulating other HPC currents, for example the three Na^+ currents proposed in the present model. Some possible consequences of such mechanisms will be discussed later.

5   HPC Currents and Information Processing

The role of a given channel on the HPC electrical response depends on its temporal characteristics as a function of voltage, intracellular messengers, and other variables. This is complicated by the fact that the opening and closing of channels is equivalent to varying *conductances*, allowing both linear and non-linear operations (e.g. [4] and [7]). In particular, a current which is activated/deactivated over a period of hundreds of milliseconds will, to a first approximation, act by slowly changing the time constant of the membrane. At the other extreme, currents which activate/deactivate with sub-millisecond time constants act by changing the trajectory of the membrane voltage in complicated ways. The classic example of this is the role of Na^+ currents underlying the action potential.

To investigate how the different HPC currents may contribute to the information processing of this neuron, we have looked at how each current shapes the HPC response to a simple repertoire of inputs. At this stage in our research the inputs have been very basic – short somatic current steps that evoke single spikes, long lasting somatic current steps that evoke spike trains, and current steps at the distal end of the dendritic cable. By examining the response to these inputs the functional roles of the HPC

Current	Spike Shape	Spike Threshold	τ_m/Frequency-Intensity
$I_{Na-trig}$	+	+++	−
I_{Na-rep}	+	++	+++
I_{Ca}	− (++)	− (+)	+ (+++)
I_{DR}	++	+	++
I_A	+	++	++
I_C	+	−	+++
I_{AHP}	−	++	+++
I_M	−	+	+

Table 1: Putative functional roles of HPC somatic currents. Entries in parentheses indicate secondary role, e.g. Ca^{2+} activation of K^+ current.

currents can be tentatively grouped into three (non-exclusive) categories:

- Modulation of spike shape.

- Modulation of firing threshold, both for single and repetitive spikes.

- Modulation of semi-steady-state membrane time constant.

- Modulation of repetitive firing, specifically the relationship between strength of tonic input and frequency of initial burst and later "steady state" spike train.

Table 1 summarizes speculative roles for some of the HPC currents as suggested by the simulations. Note that while all four of the listed characteristics are interrelated, the last two are particularly so and are lumped together in Table 1.

5.1 Possible Roles for Modulation of FI Characteristic

Again, it has been traditionally assumed that neural information is encoded by (steady-state) frequency modulation, e.g. the number of spikes per second over some time period encodes the output information of a neuron. For example, muscle fiber contraction is approximately proportional to the spike frequency of its motor neuron [2]. If the physiological inhibition of a specific

---

[2] In fact, where action potential propagation is a stereotyped phenomena, such as in long axons, then the timing of spikes is the only parameter that may be modulated.

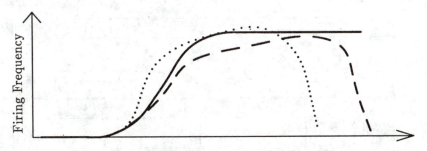

Stimulus Intensity (Constant Current)

Figure 1: Classical relation between total neuronal input (typically tonic current stimulus) and spike firing frequency [solid line] and (qualitative) biological relationships [dashed and dotted lines]. The dotted line applies when I_{Na-rep} is blocked.

current changes the FI characteristic, this allows one way to modulate that neuron's information processing by various agents.

Figure 1 contrasts the classical input-output relation of a neuron and more biological input-output relations. The relationships have several features which can be potentially modulated either physiologically or pathologically, including saturation, threshold, and shape of the curves. Note in particular the cessation of output with increased stimulation, as the depolarizing stimulus prevents the resetting of the transient inward currents.

For the HPC, simulations show (Figure 2 and Figure 3) that blocking the putative I_{Na-rep} has the effect of causing the cell to "latch-up" in response to tonic stimulus that would otherwise elicit stable spike trains. Both depolarizing currents and repolarizing currents play a role here. First, spike upstroke is mediated by both I_{Na-rep} and the lower threshold $I_{Na-trig}$; at high stimuli repolarization between spikes does not get low enough to reset $I_{Na-trig}$. Second, spikes due to only one of these Na^+ currents are weaker and as a result do not activate the repolarizing K^+ currents as much as normal because a) reduced time at depolarized levels activates the voltage-dependent K^+ currents less and b) less Ca^{2+} influx with smaller spikes reduces the Ca^{2+}-dependent activation of some K^+ currents. The net result is that repolarization between spikes is weaker and, again, does not reset $I_{Na-trig}$.

Although the current being modulated here (I_{Na-rep}) is theoretical, the

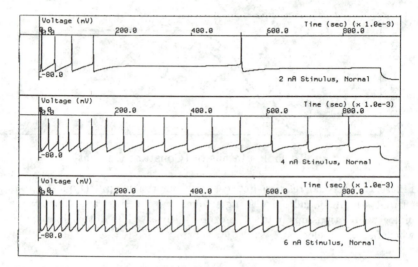

Figure 2: Simulation of repetitive firing in response to constant current injection into the soma. In this series, with the "normal" cell, a stimulus of about 8 nA (not shown) will cause to cell to fire a short burst and then cease firing.

possibility of selective blocking of I_{Na-rep} allows a mechanism for shifting the saturation of the neuron's response to the left and, as can be seen by comparing Figures 2 and 3, making the FI curve steeper over the response range.

5.2 Possible Roles for Modulation of Spike Threshold

The somatic firing threshold determines the minimal input for eliciting a spike, and in effect change the sensitivity of a cell. As a simple example, blocking $I_{Na-trig}$ in the HPC model raises threshold by about 10 millivolts. This could cause the cell to ignore input patterns that would otherwise generate action potentials.

There are two aspects of the firing "threshold" for a cell – static and dynamic. Thus, the *rate* at which the soma membrane approaches threshold is important along with the magnitude of that threshold. In general the threshold level rises with a slower depolarization for several reasons, including partial inactivation of inward currents (e.g. $I_{Na-trig}$) and partial activation of outward currents (e.g. I_A [8]) at subthreshold levels.

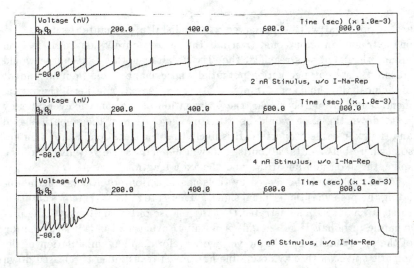

Figure 3: Blocking one of the putative Na^+ currents (I_{Na-rep}) causes the HPC repetitive firing response to fail at lower stimulus than "normal". This corresponds to the leftward shift in the saturation of the response curve shown in Figure 1.

Thus it is possible, for example, that I_A helps to distinguish tonic dendritic distal synaptic input from proximal input. For input that eventually will supply the same depolarizing current at the soma, dendritic input will have a slower onset due to the cable properties of the dendrites. This slow onset could allow I_A to delay the onset of the spike or spikes. A similar depolarizing current applied more proximally would have a faster onset. Sub-threshold activation of I_A on the depolarizing phase would then be insufficient to delay the spike.

5.3   Possible Roles for Modulation of Somatic Spike Shape

How important is the *shape* of an individual spike generated at the soma? First, we can assume that spike shape, in particular spike width, is unimportant at the soma spike-generating membrane – once the soma fires, it fires. However, the effect of the spike beyond the soma may or may not depend on the spike shape, and this is dependent on both the degree which spike propagation is linear and on the properties of the pre-synaptic membrane.

Axon transmission is both a linear and non-linear phenomena, and the shorter the axon's electrotonic length, the more the shape of the somatic action potential will be preserved at the distal pre-synaptic terminal. At one extreme, an axon could transmit the spike a purely non-linear fashion – once threshold was reached, the classic "all-or-nothing" response would transmit a stereotyped action potential whose shape would be independent of the post-threshold soma response. At the other extreme, i.e. if the axonal membrane were purely linear, the propagation of the somatic event at any point down the axon would be a linear convolution of the somatic signal and the axon cable properties. It is likely that the situation in the brain lies somewhere between these limits, and will depend on the wavelength of the spike, the axon non-linearities and the axon length.

What role could be served by the somatic action potential shape modulating the pre-synaptic terminal signal? There are at least three possibilities. First, it has been demonstrated that the release of transmitter at some pre-synaptic terminals is not an "all-or-nothing" event, and in fact is a function of the pre-synaptic membrane voltage waveform. Thus, modulation of the somatic spike width may determine how much transmitter is released down the line, providing a mechanism for changing the effective strength of the spike as seen by the target neuron. Modulation of somatic spike width could be equivalent to a modulation of the "loudness" of a given neuron's message.

Second, pyramidal cell axons often project collateral branches back to the originating soma, forming axo-somatic synapses which result in a feedback loop. In this case, modulation of the somatic spike could affect this feedback in complicated ways, particularly since the collaterals are typically short.

Finally, somatic spike shape may also play a role in the transmission of spikes at axonal branch points. For example, consider a axonal branch point with an impedance mismatch and two daughter branches, one thin and one thick. Here a spike that is too narrow may not be able to depolarize the thick branch sufficiently for transmission of the spike down that branch, with the spike propagating only down the thin branch. Conversely, a wider spike may be passed by both branches. Modulation of the somatic spike shape could then be used to direct how a cell's output is broadcast, some times allowing transmission to all the destinations of an HPC , and at other times inhibiting transmission to a limited set of the target neurons.

For HPCs much evidence has been obtained which implicate the roles of various HPC currents on modulating somatic spike shape, for example the Ca^{2+}-dependent K^+ current I_C [9]. Simulations which demonstrate the effect of I_C on the shape of individual action potentials are shown in Figure 4.

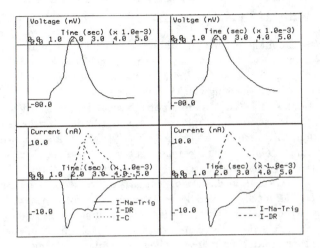

Figure 4: Role of I_C during repolarization of spike. In the simulation on the left, I_C is the largest repolarizing current. In the simulation on the right, blocking I_C results in an wider spike.

6 The Assumption of Somatic Vs. Non-Somatic Currents

In this research the *somatic* response of the HPC has been modelled under the assumption that the data on HPC currents reflect activity of channels localized at the soma. However, it must be considered that all channel proteins, regardless of their final functional destination, are manufactured at the soma. Some of the so-called somatic channels may therefore be vestiges of channels intended for dendritic, axonal, or pre-synaptic membrane. For example, if the spike-shaping channels are intended to be expressed for pre-synaptic membrane, then modulation of these channels by endogenous factors (e.g. ACh) takes place at target neuron. This may seem disadvantageous if a factor is to act selectively on some afferent tract. On the other hand, in the dendritic field of a given neuron it is possible only some afferents have certain channels, thus allowing selective response to modulating agents. These possibilities further expand the potential roles of membrane channels for computation.

7 Other Possible Roles of Currents for Modulating HPC Response

There are many other potential ways that HPC currents may modulate the HPC response. For example, the relationship between intracellular Ca^{2+} and the Ca^{2+}-dependent K^+ currents, I_C and I_{AHP}, may indicate possible information processing mechanisms.

Intracellular Ca^{2+} is an important second messenger for several intracellular processes, for example muscular contraction, but excessive $[Ca^{2+}]_{in}$ is noxious. There are at least three negative feedback mechanisms for limiting the flow of Ca^{2+} : voltage-dependent inactivation of Ca^{2+} currents; reduction of E_{Ca} (and thus the Ca^{2+} driving force) with Ca^{2+} influx; and the just mentioned Ca^{2+}-mediation of repolarizing currents. A possible information processing mechanism could be by modulation of I_{AHP}, which plays an important role in limiting repetitive firing[3]. Simulations suggest that blocking this current causes I_C to step in and eventually limit further repetitive firing, though after many more spikes in a train. Blocking both these currents may allow other mechanisms to control repetitive firing, perhaps ones that operate independently of $[Ca^{2+}]_{in}$. Conceivably, this could put the neuron into quite a different operating region.

8 Populations of Neurons Vs. Single Cells: Implications for Graded Modulation of HPC Currents

In this paper we have considered the all-or-nothing contribution of the various channels, i.e. the entire population of a given channel type is either activated normally or all the channels are disabled/blocked. This description may be oversimplified in two ways. First, it is possible that a blocking mechanism for a given channel may have a graded effect. For example, it is possible that cholinergic input is not homogeneous over the soma membrane, or that at a given time only a portion of these afferents are activated. In either case it is possible that only a portion of the cholinergic receptors are bound, thus inhibiting a portion of channels. Second, the result of channel inhibition by neuromodulatory projections must consider both single cell

---

[3] The slowing down of the spike trains in Figure 2 and Figure 3 is mainly due to the buildup of $[Ca^{2+}]_{in}$, which progressively activates more I_{AHP}.

response and population response, the size of the population depending on the neuro-architecture of a cortical region and the afferents. For example, activation of a cholinergic tract which terminates in a localized hippocampal region may effect thousands of HPCs. Assuming that the I_M of individual HPCs in the region may be either turned on or off completely with some probability, the behavior of the *population* will be that of a graded response of I_M inhibition. This graded response will in turn depend on the strength of the cholinergic tract activity.

The key point is that the information processing properties of isolated neurons may be reflected in the behavior of a population, and vica-versa. While it is likely that removal of a single pyramidal cell from the hippocampus will have zero functional effect, no neuron is an island. Understanding the central nervous system begins with the spectrum of behavior in its functional units, which may range from single channels, to specific areas of a dendritic tree, to the single cell, to cortical or nuclear subfields, on up through the main subsystems of CNS.

References

[1] L. Borg-Graham. *Modelling the Somatic Electrical Behavior of Hippocampal Pyramidal Neurons.* Master's thesis, Massachusetts Institute of Technology, 1987.

[2] J. Halliwell and P. Adams. Voltage clamp analysis of muscarinic excitation in hippocampal neurons. *Brain Research*, 250:71–92, 1982.

[3] J. J. B. Jack, D. Noble, and R. W. Tsien. *Electric Current Flow In Excitable Cells.* Clarendon Press, Oxford, 1983.

[4] C. Koch and T. Poggio. Biophysics of computation: neurons, synapses and membranes. *C.B.I.P. Paper*, (008), 1984. Center for Biological Information Processing, MIT.

[5] D. Madison and R. Nicoll. Noradrenaline blocks accommodation of pyramidal cell discharge in the hippocampus. *Nature*, 299:, Oct 1982.

[6] Y. Nakajuma, S. Nakajima, R. Leonard, and K. Yamaguchi. Actetylcholine inhibits a-current in dissociated cultured hippocampal neurons. *Biophysical Journal*, 49:575a, 1986.

[7] T. Poggio and V. Torre. *Theoretical Approaches to Complex Systems, Lecture Notes in Biomathematics*, pages 28–38. Volume 21, Springer Verlag, Berlin, 1978. A New Approach to Synaptic Interaction.

[8] J. Storm. A-current and ca-dependent transient outward current control the initial repetitive firing in hippocampal neurons. *Biophysical Journal*, 49:369a, 1986.

[9] J. Storm. Mechanisms of action potential repolarization and a fast after-hyperpolarization in rat hippocampal pyramidal cells. *Journal of Physiology*, 1986.

OPTIMAL NEURAL SPIKE CLASSIFICATION

Amir F. Atiya(*) and James M. Bower(**)
(*) Dept. of Electrical Engineering
(**) Division of Biology
California Institute of Technology
Ca 91125

Abstract

Being able to record the electrical activities of a number of neurons simultaneously is likely to be important in the study of the functional organization of networks of real neurons. Using one extracellular microelectrode to record from several neurons is one approach to studying the response properties of sets of adjacent and therefore likely related neurons. However, to do this, it is necessary to correctly classify the signals generated by these different neurons. This paper considers this problem of classifying the signals in such an extracellular recording, based upon their shapes, and specifically considers the classification of signals in the case when spikes overlap temporally.

Introduction

How single neurons in a network of neurons interact when processing information is likely to be a fundamental question central to understanding how real neural networks compute. In the mammalian nervous system we know that spatially adjacent neurons are, in general, more likely to interact, as well as receive common inputs. Thus neurobiologists are interested in devising techniques that allow adjacent groups of neurons to be sampled simultaneously. Unfortunately, the small scale of real neural networks makes inserting one recording electrode per cell impractical. Therefore, one is forced to use single electrodes designed to sample neural signals evoked by several cells at once. While this approach provides the multi-neuron recordings being sought, it also presents a rather serious waveform classification problem because the actual temporal sequence of action potentials in each individual neuron must be deciphered. This paper describes a method for classifying the activities of several individual neurons recorded simultaneously using a single electrode.

Description of the Problem

Over the last two decades considerable attention[1-8] has been devoted to the problem of classification of action potentials in multi-neuron recordings. These action potentials (also referred to as "spikes") are the extracellularly recorded signal produced by a single neuron when it is passing information to other neurons (Fig. 1). Fortunately, spikes recorded from the same cell are more or less similar in shape, while spikes coming from different neurons usually have somewhat different shapes, depending on the neuron type, electrode characteristics, the distance between the electrode and the neuron, and the intervening medium. Fig. 1 illustrates some representative variations in spike shapes. It is our objective to detect and classify different spikes based on their shapes. However, relying entirely on the shape of the spikes presents difficulties. For example spikes from different neurons can overlap temporally producing novel waveforms (see Fig. 2 for an example of an overlap). To deal with these overlaps, one has first to detect the occurrence of an overlap, and then estimate the constituent spikes. Unfortunately, only a few of the available spike separation algorithms consider these events, even though they are potentially very important in understanding neural networks. Those few tend to rely on heuristic rules and subtractive methods to resolve overlap cases. No currently published method we are aware of attempts to use knowledge of the likelihood of overlap events for detecting them, which is at the basis of the method we will describe.

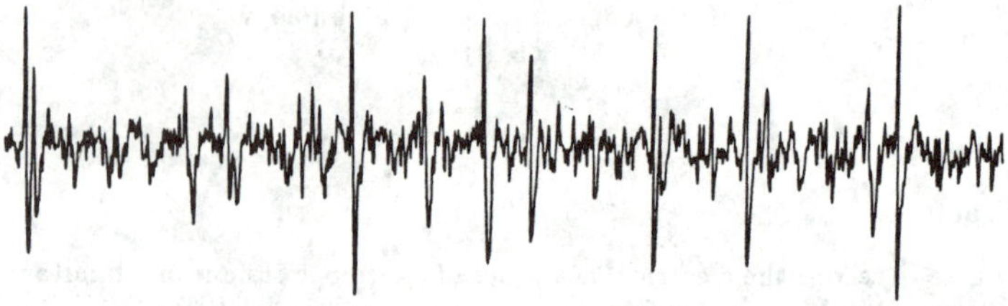

Fig. 1

An example of a multi-neuron recording

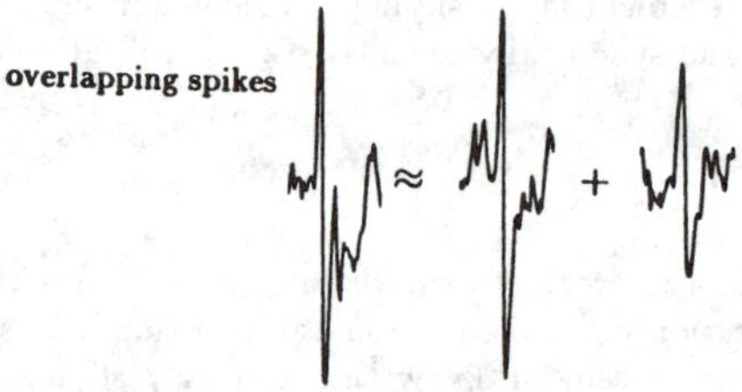

Fig. 2

An example of a temporal overlap of action potentials

General Approach

The first step in classifying neural waveforms is obviously to identify the typical spike shapes occurring in a particular recording. To do this we have applied a learning algorithm on the beginning portion of the recording, which in an unsupervised fashion (i.e. without the intervention of a human operator) estimates the shapes. After the learning stage we have the classification stage, which is applied on the remaining portion of the recording. A new classification method is proposed, which gives minimum probability of error, even in case of the occurrence of overlapping spikes. Both the learning and the classification algorithms require a preprocessing step to detect the position of the spike candidate in the data record.

Detection: For the first task of detection most researchers use a simple level detecting algorithm, that signals a spike when recorded voltage levels cross a certain voltage threshold. However, variations in recording position due to natural brain movements during recording (e.g. respiration) can cause changes in relative height of the positive to the negative peak. Thus, a level detector (using either a positive or a negative threshold) can miss some spikes. Alternatively, we have chosen to detect an event by sliding a window of fixed length until a time when the peak to peak value within the window exceeds a certain threshold.

Learning: Learning is performed on the beginning portion of the sampled data using the Isodata clustering algorithm[9]. The task is to estimate the number of neurons n whose spikes are represented in the waveform and learn the different shapes of the spikes of the various neurons. For that purpose we apply the clustering algorithm choosing only one feature from the spike, the peak to peak value which we have found to be quite an effective feature. Note that using the peak to peak value in the learning stage does not necessitate using it for classification (one might need additional or different features, especially for tackling the case of spike overlap).

The Optimal Classification Rule: Once we have identified the number of different events present, the classification stage is concerned with estimating the identities of the spikes in the recording, based on the typical spike shapes obtained in the learning stage. In our classification scheme we make use of the information given by the shape of the detected spike as well as the firing rates of the different neurons. Although the shape plays in general the most important role in the classification, the rates become a more significant factor when dealing with overlapping events. This is because in general overlap is considerably less frequent than single spikes. The shape information is given by a set of features extracted from the waveform. Let $\mathbf{x}$ be the feature vector of the detected spike (e.g. the samples of the spike waveform). Let $N_1, ..., N_n$ represent the different neurons. The detection algorithm tells us only that at least one spike occurred in the narrow interval $(t - T_1, t + T_2)$ ($=$ say I) where t is the instant of the peak of the detected spike, T_1 and T_2 are constants chosen subjectively according to the smallest possible time separation between two consecutive spikes, identifiable as two separate (nonoverlapping) spikes. By definition, if more than one spike occurs in the interval I, then we have an overlap. As a matter of convention, the instant of the occurrence of a spike is taken to be that of the spike peak. For simplicity, we will consider the case of two possibly overlapping spikes, though the method can be extended easily to more. The classification rule which results in minimum probability of error is the one which chooses the neuron (or pair of neurons in case of overlap) which has the maximum likelihood. We have therefore to compare the P_i's and the P_{lj}'s, defined as

$$P_i = P(N_i \text{ fired in } I | \mathbf{x}, A), \qquad i = 1, ..., n$$

$$P_{lj} = P(N_l \text{ and } N_j \text{ fired in } I | \mathbf{x}, A), \qquad l, j = 1, ..., n, \quad j < l$$

where A represents the event that one or two spikes occurred in the interval I. In other words P_i the probability that what has been detected is a single spike from neuron i, whereas P_{lj} is the probability that we have two overlapping spikes from neurons l and j (note that spikes from the same neuron never overlap). Henceforth we will use f to denote probability density. For the purpose of abbreviation let $B_i(t)$ mean "neuron N_i fired at t". The classification problem can be reduced to comparing the following likelihood functions:

$$L_i = f(B_i(t)) \int_{t-T_1}^{t+T_2} f(\mathbf{x}|B_i(t_1)) dt_1, \qquad i = 1, ..., n \qquad (1a)$$

$$L_{lj} = f(B_j(t)) f(B_l(t)) \int_{t-T_1}^{t+T_2} \int_{t-T_1}^{t+T_2} f(\mathbf{x}|B_l(t_1), B_j(t_2)) dt_1 dt_2, \qquad l, j = 1, ..., n, \quad j < l \qquad (1b)$$

(for a derivation refer to Appendix). Let f_i be the density of the inter-spike interval and τ_i be the most recent firing instant of neuron N_i. If we are given the fact that neuron N_i has been idle for at least a period of duration $t - \tau_i$, we get

$$f(B_i(t)) = \frac{f_i(t - \tau_i)}{\int_{t-\tau_i}^{\infty} f_i(\tau) d\tau}. \qquad (2)$$

A disadvantage of using (2) is that the available f_i's and τ_i's are only estimates, which depend on the previous classification results, Further, for reliable estimation of the densities f_i, one needs a large number of spikes and therefore a long learning period since we are estimating a whole function. Therefore, we have not used this form, but instead have used the following two schemes. In the first one, we ignore the knowledge about the previous firing pattern except for the estimated firing rates $\lambda_1, ..., \lambda_n$ of the different neurons $N_1, ..., N_n$ respectively. Then the probability of a spike coming from neuron N_i in an interval of duration dt is simply $\lambda_i dt$. Hence

$$f(B_i(t)) = \lambda_i. \qquad (3)$$

In the second scheme we do not use any previous knowledge except for the total firing rate (of all neurons), say α. Then

$$f(B_i(t)) = \frac{\alpha}{n}. \qquad (4)$$

Although the second scheme does not use as much of the information about the firing pattern as the first scheme does, it has the advantage of obtaining and using a more reliable estimate of the firing rate, because in general the overall firing rate changes less with time than the individual rates and because the estimate of α does not depend on previous classification results. However, it is useful mostly when the firing rates of the different neurons do not vary much, otherwise the firt scheme is preferred.

In real recording situations, sometimes one encounters voltage signals which are much different than any of the previously learned typical spike shapes or their pairwise overlaps. This can happen for example due to a falsely detected noise event, a spike from a class not encountered in the learning stage, or to the overlap of three or more spikes. To cope with these cases we use the reject option. This means that we refuse to classify the detected spike because of the unlikeliness of the assumed event A. The reject option is therefore employed whenever $P(A|\mathbf{x})$ is smaller than a certain threshold. We know that

$$P(A|\mathbf{x}) = f(A, \mathbf{x})/[f(A, \mathbf{x}) + f(A^c, \mathbf{x})]$$

where A^c is the complement of the event A. The density $f(A^c, \mathbf{x})$ can be approximated as uniform (over the possible values of $\mathbf{x}$) because a large variety of cases are covered by the event A^c. It follows that one can just compare $f(A, \mathbf{x})$ to a threshold. Hence the decision strategy becomes finally: Reject if the sum of the likelihood functions is less than a threshold. Otherwise choose the neuron (or pair of neurons) corresponding to the largest likelihood functions. Note that the sum of the likelihood functions equals $f(A, \mathbf{x})$ (refer to Appendix).

Now, let us evaluate the integrals in (1). Overlapping spikes are assumed to add linearly. Since we intend to handle the overlap case, we have to use a set of features x_m which obeys the following. Given the features of two of the waveforms, then one can compute those of their overlap. A good such candidate is the set of the samples of the spike (or possibly also just part of the samples). The added noise, partly thermal noise from the electrode and partly due to firings from distant neurons, can usually be approximated as white Gaussian. Let the variance be σ^2. The integrals in the likelihood functions can be approximated as summations (note in fact that we have samples available, not a continuous waveform). Let $\mathbf{y}^i$ represent the typical feature vector (template) associated with neuron N_i, with the mth component being y_m^i. Then

$$f(\mathbf{x}|B_i(k_1)) = \frac{1}{(2\pi)^{M/2}\sigma^M} exp\left[-\frac{1}{2\sigma^2}\sum_{m=1}^{M}(x_m - y_{m-k_1}^i)^2\right]$$

$$f(\mathbf{x}|B_l(k_1), B_j(k_2)) = \frac{1}{(2\pi)^{M/2}\sigma^M} exp\left[-\frac{1}{2\sigma^2}\sum_{m=1}^{M}(x_m - y_{m-k_1}^l - y_{m-k_2}^j)^2\right]$$

where x_m is the mth component of $\mathbf{x}$, and M is the dimension of $\mathbf{x}$. This leads to the following likelihood functions

$$L'_i = f(B_i(k)) \sum_{k_1=-M_1}^{M_2} exp\Big[-\frac{1}{2\sigma^2}\sum_{m=1}^{M}(x_m - y^i_{m-k_1})^2\Big]$$

$$L'_{lj} = f(B_l(k))f(B_j(k)) \sum_{k_1=-M_1}^{M_2}\sum_{k_2=-M_1}^{M_2} exp\Big[-\frac{1}{2\sigma^2}\sum_{m=1}^{M}(x_m - y^l_{m-k_1} - y^j_{m-k_2})^2\Big]$$

where k is the spike instant, and the interval from $-M_1$ to M_2 corresponds to the interval I defined at the beginning of the Section.

Implementation

The techniques we have just described were tested in the following way. For the first experiment we identified two spike classes in a recording from the rat cerebellum. A signal is created, composed of a number of spikes from the two classes at random instants, plus noise. To make the situation as realistic as possible, the added noise is taken from idle periods (i.e. non-spiking) of a real recording. The reason for using such an artificially generated signal is to be able to know the class identities of the spikes, in order to test our approach quantitatively. We implement the detection and classification techniques on the obtained signal, with various values of noise amplitude. In our case the ratio of the peak to peak values of the templates turns out to be 1.375. Also, the spike rate of one of the classes is twice that of the other class. Fig.3a shows the results with applying the first scheme (i.e. using Eq. 3). The overall percentage correct classification for all spikes (solid curve) and the percentage correct classification for overlapping spikes (dashed curve) are plotted versus the standard deviation of the noise σ normalized with respect to the peak h of the large template. Notice that the overall classification accuracy is near 100% for σ/h less than 0.15, which is actually the range of noise amplitudes we mostly encountered in our work with real recordings. Observe also the good results for classifying overlapping events. We have applied also the second scheme (i.e. using Eq. 4) and obtained similar results. We wish to mention that the thresholds for detection and for the reject option are set up so as to obtain no more than 3% falsely detected spikes.

A similar experiment is performed with three waveforms (three classes), where two of the waveforms are the same as those used in the first experiment. The third is the average of the first two. All the three neurons have the same spike rate (i.e. $\lambda_1 = \lambda_2 = \lambda_3$). Hence both classification schemes are equivalent in this case. Fig. 3b shows the overall as well as the sub-category of overlap classification results. One observes that the results are worse than those for the two-class case. This is because the spacings between the templates are in general smaller. Notice also that the accuracy in resolving overlapping events is now tangibly less than the overall accuracy. However, one can say that the results are acceptable in the range of σ/h less than 0.1. The following experiment is also performed using the same data. We would like to investigate the importance of the information given by the (overall) firing rate on the problem of classifying overlapping events. In our method the summation in the likelihood functions for single spikes is multiplied by α/n, while that for overlapping spikes is multiplied by $(\alpha/n)^2$. Usually α/n is considerably less than one. Hence we have a factor which gives less weight for overlapping events. Now, consider the case of ignoring completely the information given by the firing rate and relying solely on shape information. We assume that overlapping spikes from any two given classes represent "new" class of waveforms and that each of these overlap classes has the same rate as that of a single-spike class. In that case we can obtain expressions for the likelihood functions as consisting just the summations, i.e. free of the rate

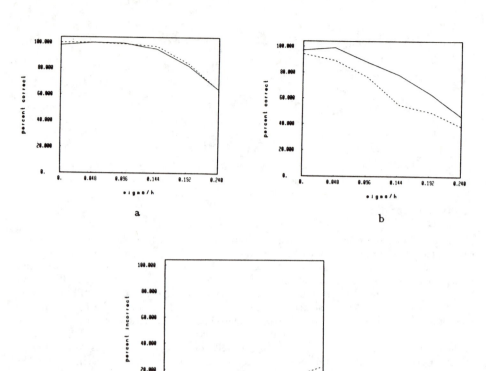

a

b

c

Fig. 3
a) Overall (solid curve) and overlap (dashed curve)
classification accuracy for a two class case
b) Overall (solid curve) and overlap (dashed curve)
classification accuracy for a three class case
c)Percent of incorrect classification of single spikes as overlap
solid curve: scheme utilzing the spike rate
dashed curve: scheme not utilizing the spike rate

factor α/n (refer to Appendix). An experiment is performed using that scheme (on the same three class data). One observes that the method classifies a number of single spikes wrongly as overlaps, much more than our original scheme does (see Fig. 3c), especially for the large noise case. On the other hand, the number of overlaps which are classified wrongly as single spikes is near zero for both schemes.

Finally, in the last experiment the techniques are implemented on real recordings from the rat cerebellum. The recorded signal is band-pass-filtered in the frequency range 300 Hz - 10 KHz, then sampled with a rate of 20KHz. For classification, we take 20 samples per spike as features. Fig. 4 shows the results of the proposed method, using the first scheme (Eq. 3). The number of neurons whose spikes are represented in the waveform is estimated to be four. The detection threshold is set up so that spikes which are too small are disregarded, because they come from several neurons far away from the electrode and are hard to distinguish. Notice the overlap of classes 1 and 2, which was detected. We used the second scheme also on the same portion and it gave similar results as those of the first scheme (only one of the spikes is classified differently). Overall, the discrepancies between classifications done by the proposed method and an experienced human observer were found to be small.

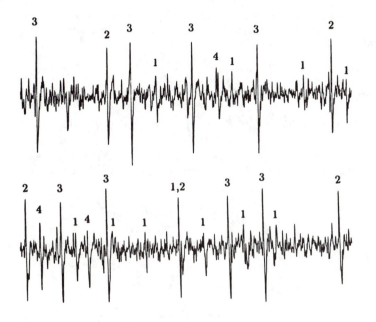

Fig. 4

Classification results for a recording from the rat cerebellum

Conclusion

Many researchers have considered the problem of spike classification in multi-neuron recordings, but only few have tackled the case of spike overlap, which could occur frequently, particularly if the group of neurons under study is stimulated. In this work we propose a method for spike classification, which can also aid in detecting and classifying overlapping spikes. By taking into account the statistical properties of the discharges of the neurons sampled, this method minimizes the probability of classification error. The application of the method to artificial as well as real recordings confirm its effectiveness.

Appendix

Consider first P_{lj}. We can write

$$P_{lj} = \int_{t-T_1}^{t+T_2} \int_{t-T_1}^{t+T_2} f(B_l(t_1), B_j(t_2)|\mathbf{x}, A) \, dt_1 \, dt_2.$$

We can also obtain

$$P_{lj} = \int_{t-T_1}^{t+T_2} \int_{t-T_1}^{t+T_2} \frac{f(\mathbf{x}, A | B_l(t_1), B_j(t_2))}{f(\mathbf{x}, A)} f(B_l(t_1), B_j(t_2)) dt_1 dt_2.$$

Now, consider the two events $B_l(t_1)$ and $B_j(t_2)$. In the absense of any information about their dependence, we assume that they are independent. We get

$$f(B_l(t_1), B_j(t_2)) = f(B_l(t_1)) f(B_j(t_2)).$$

Within the interval I, both $f(B_l(t_1))$ and $f(B_j(t_2))$ hardly vary because the duration of I is very small compared to a typical inter-spike interval. Therefore we get the following approximation:

$$f(B_l(t_1)) \approx f(B_l(t))$$
$$f(B_j(t_2)) \approx f(B_j(t)).$$

The expression for P_{lj} becomes

$$P_{lj} \approx \frac{f(B_l(t)) f(B_j(t))}{f(\mathbf{x}, A)} \int_{t-T_1}^{t+T_2} \int_{t-T_1}^{t-T_2} f(\mathbf{x} | B_l(t_1), B_j(t_2)) dt_1 dt_2.$$

Notice that the term A was omitted from the argument of the density inside the integral, because the occurrence of two spikes at t_1 and $t_2 \epsilon I$ implies the occurrence of A. A similar derivation for P_i results in

$$P_i \approx \frac{f(B_i(t))}{f(\mathbf{x}, A)} \int_{t-T_1}^{t+T_2} f(\mathbf{x} | B_i(t_1)) dt_1.$$

The term $f(\mathbf{x}, A)$ is common to all the P_{lj}'s and the P_i's. Hence one can simply compare the following likelihood functions:

$$L_i = f(B_i(t)) \int_{t-T_1}^{t+T_2} f(\mathbf{x} | B_i(t_1)) dt_1$$

$$L_{lj} = f(B_l(t)) f(B_j(t)) \int_{t-T_1}^{t+T_2} \int_{t-T_1}^{t+T_2} f(\mathbf{x} | B_l(t_1), B_j(t_2)) dt_1 dt_2.$$

Acknowledgement

Our thanks to Dr. Yaser Abu-Mostafa for his assistance with this work. This project was supported by the Caltech Program of Advanced Technology (sponsored by Aerojet, GM, GTE, and TRW), and the Joseph Drown Foundation.

References

[1] M. Abeles and M. Goldstein, *Proc. IEEE*, 65, pp.762-773, 1977.
[2] G. Dinning and A. Sanderson, *IEEE Trans. Bio − Med. Eng.*, BME-28, pp. 804-812, 1981.
[3] E. D'Hollander and G. Orban, *IEEE Trans. Bio-Med. Eng.*, BME-26, pp. 279-284, 1979.
[4] D. Mishelevich, *IEEE Trans. Bio-Med. Eng.*, BME-17, pp. 147-150, 1970.
[5] V. Prochazka and H. Kornhuber, *Electroenceph. clin. Neurophysiol.*, 32, pp. 91-93, 1973.
[6] W. Roberts, *Biol. Cybernet.*, 35, pp. 73-80, 1979.
[7] W. Roberts and D. Hartline, *Brain Res.*, 94, pp. 141-149, 1975.
[8] E. Schmidt, *J. Neurosci. Methods*, 12, pp. 95-111, 1984.
[9] R. Duda and P. Hart, *Pattern Classification and Scene Analysis*, John Wiley, 1973.

NEURAL NETWORKS FOR TEMPLATE MATCHING: APPLICATION TO REAL-TIME CLASSIFICATION OF THE ACTION POTENTIALS OF REAL NEURONS

Yiu-fai Wong[†], Jashojiban Banik[†] and James M. Bower[‡]
†Division of Engineering and Applied Science
‡Division of Biology
California Institute of Technology
Pasadena, CA 91125

ABSTRACT

Much experimental study of real neural networks relies on the proper classification of extracellulary sampled neural signals (i.e. action potentials) recorded from the brains of experimental animals. In most neurophysiology laboratories this classification task is simplified by limiting investigations to single, electrically well-isolated neurons recorded one at a time. However, for those interested in sampling the activities of many single neurons simultaneously, waveform classification becomes a serious concern. In this paper we describe and constrast three approaches to this problem each designed not only to recognize isolated neural events, but also to separately classify temporally overlapping events in real time. First we present two formulations of waveform classification using a neural network template matching approach. These two formulations are then compared to a simple template matching implementation. Analysis with real neural signals reveals that simple template matching is a better solution to this problem than either neural network approach.

INTRODUCTION

For many years, neurobiologists have been studying the nervous system by using single electrodes to serially sample the electrical activity of single neurons in the brain. However, as physiologists and theorists have become more aware of the complex, nonlinear dynamics of these networks, it has become apparent that serial sampling strategies may not provide all the information necessary to understand functional organization. In addition, it will likely be necessary to develop new techniques which sample the activities of multiple neurons simultaneously[1]. Over the last several years, we have developed two different methods to acquire multineuron data. Our initial design involved the placement of many tiny microelectrodes individually in a tightly packed pseudo-floating configuration within the brain[2]. More recently we have been developing a more sophisticated approach which utilizes recent advances in silicon technology to fabricate multi-ported silicon based electrodes (Fig. 1). Using these electrodes we expect to be able to readily record the activity patterns of larger number of neurons.

As research in multi-single neuron recording techniques continue, it has become very clear that whatever technique is used to acquire neural signals from many brain locations, the technical difficulties associated with sampling, data compressing, storing, analyzing and interpreting these signals largely dwarf the development of the sampling device itself. In this report we specifically consider the need to assure that neural action potentials (also known as "spikes") on each of many parallel recording channels are correctly classified, which is just one aspect of the problem of post-processing multi-single neuron data. With more traditional single electrode/single neuron recordings, this task usually involves passing analog signals through a Schmidt trigger whose output indicates the occurence of an event to a computer, at the same time as it triggers an oscilloscope sweep of the analog data. The experimenter visually monitors the oscilloscope to verify the accuracy of the discrimination as a well-discriminated signal from a single neuron will overlap on successive oscilloscope traces (Fig. 1c). Obviously this approach is impractical when large numbers of channels are recorded at the same time. Instead, it is necessary to automate this classification procedure. In this paper we will describe and contrast three approaches we have developed to do this.

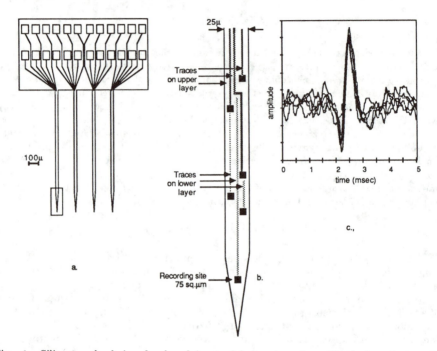

Fig. 1. Silicon probe being developed in our lababoratory for multi-single unit recording in cerebellar cortex. a) a complete probe; b) surface view of one recording tip; c) several superimposed neuronal action potentials recorded from such a silicon electrode in cerebellar cortex.

While our principal design objective is the assurance that neural waveforms are adequately discriminated on multiple channels, technically the overall objective of this research project is to sample from as many single neurons as possible. Therefore, it is a natural extention of our effort to develop a neural waveform classification scheme robust enough to allow us to distinguish activities arising from more than one neuron per recording site. To do this, however, we now not only have to determine that a particular signal is neural in origin, but also from which of several possible neurons it arose (see Fig. 2a). While in general signals from different neurons have different waveforms aiding in the classification, neurons recorded on the same channel firing simultaneously or nearly simultaneously will produce novel combination waveforms (Fig. 2b) which also need to be classified. It is this last complication which particularly bedevils previous efforts to classify neural signals (For review see 5, also see 3-4). In summary, then, our objective was to design a circuit that would:

1. **distinguish different waveforms** even though neuronal discharges tend to be quite similar in shape (Fig. 2a);
2. **recognize the same waveform** even though unavoidable movements such as animal respiration often result in periodic changes in the amplitude of a recorded signal by moving the brain relative to the tip of the electrode;
3. be considerably **robust to recording noise** which variably corrupts all neural recordings (Fig. 2);
4. **resolve overlapping waveforms**, which are likely to be particularly interesting events from a neurobiological point of view;
5. provide **real-time performance** allowing the experimenter to detect problems with discrimination and monitor the progress of the experiment;
6. be **implementable in hardware** due to the need to classify neural signals on many channels simultaneously. Simply duplicating a software-based algorithm for each channel will not work, but rather, multiple, small, independent, and programmable hardware devices need to be constructed.

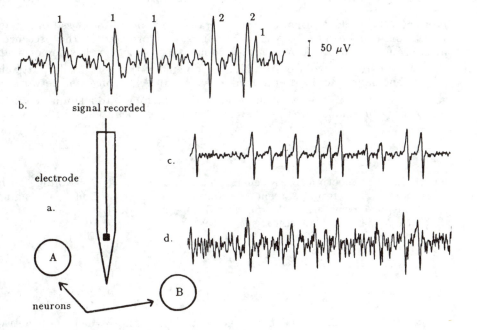

Fig. 2. a) Schematic diagram of an electrode recording from two neuronal cell bodies b) An actual multi-neuron recording. Note the similarities in the two waveforms and the overlapping event. c) and d) Synthesized data with different noise levels for testing classification algorithms (c: 0.3 NSR ; d: 1.1 NSR).

METHODS

The problem of detecting and classifying multiple neural signals on single voltage records involves two steps. First, the waveforms that are present in a particular signal must be identified and the templates be generated; second, these waveforms must be detected and classified in ongoing data records. To accomplish the first step we have modified the principal component analysis procedure described by Abeles and Goldstein[3] to automatically extract templates of the distinct waveforms found in an initial sample of the digitized analog data. This will not be discussed further as it is the means of accomplishing the second step which concerns us here. Specifically, in this paper we compare three new approaches to ongoing waveform classification which deal explicitly with overlapping spikes and variably meet other design criteria outlined above. These approaches consist of a modified template matching scheme, and two applied neural network implementations. We will first consider the neural network approaches. On a point of nomenclature, to avoid confusion in what follows, the real neurons whose signals we want to classify will be referred to as "neurons" while computing elements in the applied neural networks will be called "Hopons."

Neural Network Approach — Overall, the problem of classifying neural waveforms can best be seen as an optimization problem in the presence of noise. Much recent work on neural-type network algorithms has demonstrated that these networks work quite well on problems of this sort[6-8]. In particular, in a recent paper Hopfield and Tank describe an A/D converter network and suggest how to map the problem of template matching into a similar context[8]. The energy functional for the network they propose has the form:

$$E = \frac{-1}{2} \sum_i \sum_j T_{ij} V_i V_j - \sum_i V_i I_i \qquad (1)$$

where T_{ij} = connectivity between Hopon i and Hopon j, V_i = voltage output of Hopon i, I_i = input current to Hopon i and each Hopon has a sigmoid input-output characteristic $V = g(u) = 1/(1 + exp(-au))$.

If the equation of motion is set to be:

$$du_i/dt = -\partial E/\partial V = \sum_j T_{ij} V_j + I_i \qquad (1a)$$

then we see that $dE/dt = -\left(\sum_j T_{ij} V_j + I_i\right) dV/dt = -(du/dt)(dV/dt) = -g'(u)(du/dt)^2 \leq 0$. Hence E will go to to a minimum which, in a network constructed as described below, will correspond to a proposed solution to a particular waveform classification problem.

Template Matching using a Hopfield-type Neural Net — We have taken the following approach to template matching using a neural network. For simplicity, we initially restricted the classification problem to one involving two waveforms and have accordingly constructed a neural network made up of two groups of Hopons, each concerned with discriminating one or the other waveform. The classification procedure works as follows: first, a Schmidt trigger is used to detect the presence of a voltage on the signal channel above a set threshold. When this threshold is crossed, implying the presence of a possible neural signal, 2 msecs of data around the crossing are stored in a buffer (40 samples at 20 KHz). Note that biophysical limitations assure that a single real neuron cannot discharge more than once in this time period, so only one waveform of a particular type can occur in this data sample. Also, action potentials are of the order of 1 msec in duration, so the 2 msec window will include the full signal for single or overlapped waveforms. In the next step (explained later) the data values are correlated and passed into a Hopfield network designed to minimize the mean-square error between the actual data and the linear combination of different delays of the templates. Each Hopon in the set of Hopons concerned with one waveform represents a particular temporal delay in the occurrence of that waveform in the buffer. To express the network in terms of an energy function formulation: Let $x(t)$ = input waveform amplitude in the tth time bin, $s_j(t)$ = amplitude of the jth template, V_{jk} denote if $s_j(t-k)(j^{th}$ template delayed by k time bins)is present in the input waveform. Then the appropriate energy function is:

$$E = \frac{1}{2}\sum_t \left(x(t) - \sum_{j,k} V_{jk} s_j(t-k) \right)^2$$

$$- \frac{1}{2}\sum_{t,j,k} V_{jk}(V_{jk} - 1)s_j^2(t-k) \qquad (2)$$

$$+ \gamma \sum_{j,k_1 < k_2} V_{jk_1} V_{jk_2}$$

The first term is designed to minimize the mean-square error and specifies the best match. Since $V \in [0,1]$, the second term is minimized only when each V_{jk} assumes values 0 or 1. It also sets the diagonal elements T_{ij} to 0. The third term creates mutual inhibition among the processing nodes evaluating the same neuronal signal, which as described above can only occur once per sample.

Expanding and simplifying expression (2), the connection matrix is:

$$T_{(j_1,k_1),(j_2,k_2)} = \begin{cases} -\sum_t s_{j_1}(t-k_1)s_{j_2}(t-k_2) - \gamma\delta_{j_1 j_2} \\ 0 \qquad\qquad\qquad\qquad\qquad\qquad\quad \text{if } j_1 = j_2, k_1 = k_2 \end{cases} \qquad (3a)$$

and the input current

$$I_{jk} = \sum_t x(t)s_j(t-k) - \frac{1}{2}\sum_t s_j^2(t-k) \qquad (3b)$$

As it can be seen, the inputs are the correlations between the actual data and the various delays of the templates subtracting a constant term.

Modified Hopfield Network — As documented in more detail in Fig. 3-4, the above full Hopfield-type network works well for temporally isolated spikes at moderate noise levels, but for overlapping spikes it has a local minima problem. This is more severe with more than two waveforms in the network.

Further, we need to build our network in hardware and the full Hopfield network is difficult to implement with current technology (see below). For these reasons, we developed a modified neural network approach which significantly reduces the necessary hardware complexity and also has improved performance. To understand how this works, let us look at the information contained in the quantities T_{ij} and I_{ij} (eq. 3a and 3b) and make some use of them. These quantities have to be calculated at a pre-processing stage before being loaded into the Hopfield network. If after calculating these quantities, we can quickly rule out a large number of possible template combinations, then we can significantly reduce the size of the problem and thus use a much smaller (and hence more efficient) neural network to find the optimal solution. To make the derivation simple, we define slightly modified versions of T_{ij} and I_{ij} (eq. 4a and 4b) for two-template case.

$$T_{ij} = \sum_t s_1(t-i)s_2(t-j) \tag{4a}$$

$$I_{ij} = \sum_t x(t)\left[\frac{1}{2}s_1(t-i) + \frac{1}{2}s_2(t-j)\right] - \frac{1}{2}\sum_t s_1^2(t-i) - \frac{1}{2}\sum_t s_2^2(t-j) \tag{4b}$$

In the case of overlaping spikes the T_{ij}'s are the cross-correlations between $s_1(t)$ and $s_2(t)$ with different delays and I_{ij}'s are the cross-correlations between input $x(t)$ and weighted combination of $s_1(t)$ and $s_2(t)$. Now if $x(t) = s_1(t-i) + s_2(t-j)$ (i.e. the overlap of the first template with i time bin delay and the second template with j time bin delay), then $\Delta_{ij} = |T_{ij} - I_{ij}| = 0$. However in the presence of noise, Δ_{ij} will not be identically zero, but will equal to the noise, and if $\Delta_{ij} > \Delta T_{ij}$ (where $\Delta T_{ij} = |T_{ij} - T_{i'j'}|$ for $i \neq i'$ and $j \neq j'$) this simple algorithm may make unacceptable errors. A solution to this problem for overlapping spikes will be described below, but now let us consider the problem of classifying non-overlapping spikes. In this case, we can compare the input cross-correlation with the auto-correlations (eq. 4c and 4d).

$$T_i' = \sum_t s_1^2(t-i); \quad T_i'' = \sum_t s_2^2(t-i) \tag{4c}$$

$$I_i' = \sum_t x(t)s_1(t-i); \quad I_i'' = \sum_t x(t)s_2(t-i) \tag{4d}$$

So for non-overlapping cases, if $x(t) = s_1(t-i)$, then $\Delta_i' = |T_i' - I_i'| = 0$. If $x(t) = s_2(t-i)$, then $\Delta_i'' = |T_i'' - I_i''| = 0$.

In the absence of noise, then the minimum of Δ_{ij}, Δ_i' and Δ_i'' represents the correct classification. However, in the presence of noise, none of these quantities will be identically zero, but will equal the noise in the input $x(t)$ which will give rise to unacceptible errors. Our solution to this noise related problem is to choose a few minima (three have chosen in our case) instead of one. For each minimum there is either a known corresponding linear combination of templates for overlapping cases or a simple template for non-overlapping cases. A three neuron Hopfield-type network is then programmed so that each neuron corresponds to each of the cases. The input $x(t)$ is fed to this tiny network to resolve whatever confusion remains after the first step of "cross-correlation" comparisons. (Note: Simple template matching as described below can also be used in the place of the tiny Hopfield type network.)

Simple Template Matching — To evaluate the performances of these neural network approaches, we decided to implement a simple template matching scheme, which we will now describe. However, as documented below, this approach turned out to be the most accurate and require the least complex hardware of any of the three approaches. The first step is, again, to fill a buffer with data based on the detection of a possible neural signal. Then we calculate the difference between the recorded waveform and all possible combinations of the two previously identified templates. Formally, this consists of calculating the distances between the input $x(m)$ and all possible cases generated by all the combinations of the two templates.

$$d_{ij} = \sum_t |x(t) - \{s_1(t-i) + s_2(t-j)\}|$$

$$d'_i = \sum_t |x(t) - s_1(t-i)|; \quad d''_i = \sum_t |x(t) - s_2(t-i)|$$

$$d_{min} = min(d_{ij}, d'_i, d''_i)$$

d_{min} gives the best fit of all possible combinations of templates to the actual voltage signal.

TESTING PROCEDURES

To compare the performance of each of the three approaches, we devised a common set of test data using the following procedures. First, we used the principal component method of Abeles and Goldstein[3] to generate two templates from a digitized analog record of neural activity recorded in the cerebellum of the rat. The two actual spike waveform templates we decided to use had a peak-to-peak ratio of 1.375. From a second set of analog recordings made from a site in the cerebellum in which no action potential events were evident, we determined the spectral characteristics of the recording noise. These two components derived from real neural recordings were then digitally combined, the objective being to construct realistic records, while also knowing absolutely what the correct solution to the template matching problem was for each occurring spike. As shown in Fig. 2c and 2d, data sets corresponding to different noise to signal ratios were constructed. We also carried out simulations with the amplitudes of the templates themselves varied in the synthesized records to simulate waveform changes due to brain movements often seen in real recordings. In addition to two waveform test sets, we also constructed three waveform sets by generating a third template that was the average of the first two templates. To further quantify the comparisons of the three diffferent approaches described above we considered non-overlapping and overlapping spikes separately. To quantify the performance of the three different approaches, two standards for classification were devised. In the first and hardest case, to be judged a correct classification, the precise order and timing of two waveforms had to be reconstructed. In the second and looser scheme, classification was judged correct if the order of two waveforms was correct but timing was allowed to vary by ± 100 μsecs(i.e. ± 2 time bins) which for most neurobiological applications is probably sufficient resolution. Figs. 3-4 compare the performance results for the three approaches to waveform classification implemented as digital simulations.

PERFORMANCE COMPARISON

Two templates – non-overlapping waveforms: As shown in Fig. 3a, at low noise-to-signal ratios (NSRs below .2) each of the three approaches were comparable in performance reaching close to 100% accuracy for each criterion. As the ratio was increased, however the neural network implementations did less and less well with respect to the simple template matching algorithm with the full Hopfield type network doing considerably worse than the modified network. In the range of NSR most often found in real data (.2 - .4) simple template matching performed considerably better than either of the neural network approaches. Also it is to be noted that simple template matching gives an estimate of the goodness of fit betwwen the waveform and the closest template which could be used to identify events that should not be classified (e.g. signals due to noise).

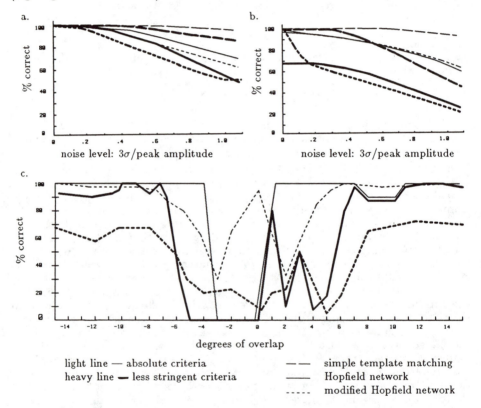

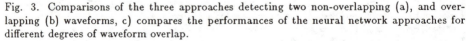

Fig. 3. Comparisons of the three approaches detecting two non-overlapping (a), and overlapping (b) waveforms, c) compares the performances of the neural network approaches for different degrees of waveform overlap.

Two templates – overlapping waveforms: Fig. 3b and 3c compare performances when waveforms overlapped. In Fig. 3b the serious local minima problem encountered in the full neural network is demonstrated as is the improved performance of the modified network. Again, overall performance in physiological ranges of noise is clearly best for simple template matching. When the noise level is low, the modified approach is the better of the two neural networks due to the reliability of the correlation number which reflects the resemblence between the input data and the template. When the noise level is high, errors in the correlation numbers may exclude the right combination from the smaller network. In this case its performance is actually a little worse than the larger Hopfield network. Fig. 3c documents in detail which degrees of overlap produce the most trouble for the neural network approaches at average NSR levels found in real neural data. It can be seen that for the neural networks, the most serious problem is encountered when the delays between the two waveforms are small enough that the resulting waveform looks like the larger waveform with some perturbation.

Three templates – overlapping and non-overlapping: In Fig. 4 are shown the comparisons between the full Hopfield network approach and the simple template matching approach. For nonoverlapping waveforms, the performance of these two approaches is much more comparable than for the two waveform case (Fig. 4a), although simple template matching is still the optimal method. In the overlapping waveform condition, however, the neural network approach fails badly (Fig. 4b and 4c). For this particular application and implementation, the neural network approach does not scale well.

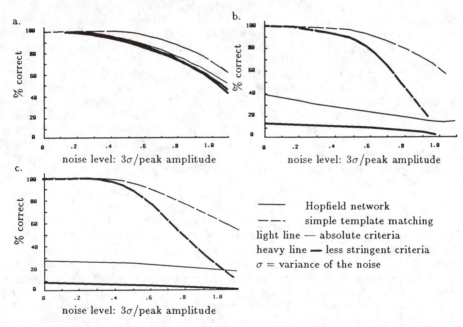

Fig. 4. Comparisons of performance for three waveforms. a) nonoverlapping waveforms; b) two waveforms overlapping; c) three waveforms overlapping.

HARDWARE COMPARISONS

As described earlier, an important design requirement for this work was the ability to detect neural signals in analog records in real-time originating from many simultaneously active sampling electrodes. Because it is not feasible to run the algorithms in a computer in real time for all the channels simultaneously, it is necessary to design and build dedicated hardware for each channel. To do this, we have decided to design VLSI implementations of our circuitry. In this regard, it is well recognized that large modifiable neural networks need very elaborate hardware implementations. Let us consider, for example, implementing hardwares for a two-template case for comparisons. Let $n = $ no. of neurons per template (one neuron for each delay of the template), $m = $ no. of iterations to reach the stable state (in simulating the discretized differential equation, with step size $= 0.05$), $l = $ no. of samples in a template $t_j(m)$. Then, the number of connections in the full Hopfield network will be $4n^2$. The total no. of synaptic calculations $= 4mn^2$. So, for two templates and $n = 16, m = 100, 4mn^2 = 102,400$. Thus building the full Hopfield-type network digitally requires a system too large to be put in a single VLSI chip which will work in real time. If we want to build an analog system, we need to have many $(O(4n^2))$ easily modifiable synapses. As yet this technology is not available for nets of this size. The modified Hopfield-type network on the other hand is less technically demanding. To do the preprocessing to obtain the minimum values we have to do about $n^2 = 256$ additions to find all possible $I_{ij}s$ and require 256 subtractions and comparisons to find three minima. The costs associated with doing input cross-correlations are the same as for the full neural network (i.e. $2nl = 768(l = 24)$ multiplications). The saving with the modified approach is that the network used is small and fast (120 multiplications and 120 additions to construct the modifiable synapses, no. of synaptic calculations $= 90$ with $m = 10, n = 3$).

In contrast to the neural networks, simple template matching is simple indeed. For example, it must perform about $n^2l + n^2 = 10,496$ additions and $n^2 = 256$ comparisons to find the minimum d_{ij}. Additions are considerably less costly in time and hardware than multiplications. In fact, because this method needs only addition operations, our preliminary design work suggests it can be built on a single chip and will be able to do the two-template classification in as little as 20 microseconds. This actually raises the possibility that with switching and buffering one chip might be able to service more than one channel in essentially real time.

CONCLUSIONS

Template matching using a full Hopfield-type neural network is found to be robust to noise and changes in signal waveform for the two neural waveform classification problem. However, for a three-waveform case, the network does not perform well. Further, the network requires many modifiable connections and therefore results in an elaborate hardware implementation. The overall performance of the modified neural network approach is better than the full Hopfield network approach. The computation has been reduced largly and the hardware requirements are considerably less demanding demonstrating the value of designing a specific network to a specified problem. However, even the modified neural network performs less well than a simple template-matching algorithm which also has the simplest hardware implementation. Using the simple template matching algorithm, our simulations suggest it will be possible to build a two or three waveform classifier on a single VLSI chip using CMOS technology that works in real time with excellent error characteristics. Further, such a chip will be able to accurately classify variably overlapping neural signals.

REFERENCES

[1] G. L. Gerstein, M. J. Bloom, I. E. Espinosa, S. Evanczuk & M. R. Turner, IEEE Trans. Sys. Cyb. Man., <u>SMC-13</u>, 668(1983).
[2] J. M. Bower & R. Llinas, Soc. Neurosci. Abst., <u>9</u>, 607(1983).
[3] M. Abeles & M. H. Goldstein, Proc. IEEE, <u>65</u>, 762(1977).
[4] W. M. Roberts & D. K. Hartline, Brain Res., <u>94</u>, 141(1976).
[5] E. M. Schmidt, J. of Neurosci. Methods, <u>12</u>, 95(1984).
[6] J. J. Hopfield, Proc. Natl. Acad. Sci.(USA), <u>81</u>, 3088(1984).
[7] J. J. Hopfield & D. W. Tank, Biol. Cybern., <u>52</u>, 141(1985).
[8] D. W. Tank & J. J. Hopfield, IEEE Trans. Circuits Syst., <u>CAS-33</u>, 533(1986).

ACKNOWLEDGEMENTS

We would like to acknowledge the contribution of Dr. Mark Nelson to the intellectual development of these projects and the able assistance of Herb Adams, Mike Walshe and John Powers in designing and constructing support equipment. This work was supported by NIH grant NS22205, the Whitaker Foundation and the Joseph Drown Foundation.

A Computer Simulation of Olfactory Cortex With Functional Implications for Storage and Retrieval of Olfactory Information

Matthew A. Wilson and James M. Bower
Computation and Neural Systems Program
Division of Biology, California Institute of Technology, Pasadena, CA 91125

ABSTRACT

Based on anatomical and physiological data, we have developed a computer simulation of piriform (olfactory) cortex which is capable of reproducing spatial and temporal patterns of actual cortical activity under a variety of conditions. Using a simple Hebb-type learning rule in conjunction with the cortical dynamics which emerge from the anatomical and physiological organization of the model, the simulations are capable of establishing cortical representations for different input patterns. The basis of these representations lies in the interaction of sparsely distributed, highly divergent/convergent interconnections between modeled neurons. We have shown that different representations can be stored with minimal interference, and that following learning these representations are resistant to input degradation, allowing reconstruction of a representation following only a partial presentation of an original training stimulus. Further, we have demonstrated that the degree of overlap of cortical representations for different stimuli can also be modulated. For instance similar input patterns can be induced to generate distinct cortical representations (discrimination), while dissimilar inputs can be induced to generate overlapping representations (accommodation). Both features are presumably important in classifying olfactory stimuli.

INTRODUCTION

Piriform cortex is a primary olfactory cerebral cortical structure which receives second order input from the olfactory receptors via the olfactory bulb (Fig. 1). It is believed to play a significant role in the classification and storage of olfactory information[1,2,3]. For several years we have been using computer simulations as a tool for studying information processing within this cortex[4,5]. While we are ultimately interested in higher order functional questions, our first modeling objective was to construct a computer simulation which contained sufficient neurobiological detail to reproduce experimentally obtained cortical activity patterns. We believe this first step is crucial both to establish correspondences between the model and the cortex, and to assure that the model is capable of generating output that can be compared to data from actual physiological experiments. In the current case, having demonstrated that the behavior of the simulation at least approximates that of the actual cortex[4] (Fig. 3), we are now using the model to explore the types of processing which could be carried out by this cortical structure. In particular, in this paper we will describe the ability of the simulated cortex to store and recall cortical activity patterns generated by stimulus various conditions. We believe this approach can be used to provide experimentally testable hypotheses concerning the functional organization of this cortex which would have been difficult to deduce solely from neurophysiological or neuroanatomical data.

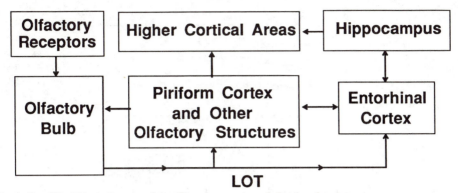

Fig. 1. Simplified block diagram of the olfactory system and closely related structures.

MODEL DESCRIPTION

This model is largely instructed by the neurobiology of piriform cortex[3]. Axonal conduction velocities, time delays, and the general properties of neuronal integration and the major intrinsic neuronal connections approximate those currently described in the actual cortex. However, the simulation reduces both the number and complexity of the simulated neurons (see below). As additional information concerning the these or other important features of the cortex is obtained it will be incorporated in the model. Bracketed numbers in the text refer to the relevent mathematical expressions found in the appendix.

Neurons. The model contains three distinct populations of intrinsic cortical neurons, and a fourth set of cells which simulate cortical input from the olfactory bulb (Fig. 2). The intrinsic neurons consist of an excitatory population of pyramidal neurons (which are the principle neuronal type in this cortex), and two populations of inhibitory interneurons. In these simulations each population is modeled as 100 neurons arranged in a 10x10 array (the actual piriform cortex of the rat contains on the order of 10^6 neurons). The output of each modeled cell type consists of an all-or-none action potential which is generated when the membrane potential of the cell crosses a threshold [2.3]. This output reaches other neurons after a delay which is a function of the velocity of the fiber which connects them and the cortical distance from the originating neuron to each target neuron [2.0, 2.4]. When an action potential arrives at a destination cell it triggers a conductance change in a particular ionic channel type in that cell which has a characteristic time course, amplitude, and waveform [2.0, 2.1]. The effect of this conductance change on the transmembrane potential is to drive it towards the equilibrium potential of that channel. Na^+, Cl^-, and K^+ channels are included in the model. These channels are differentially activated by activity in synapses associated with different cell types (see below).

116

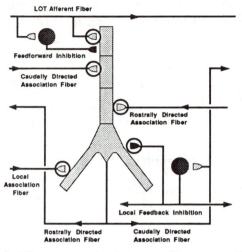

LOT Afferent Fiber

Feedforward Inhibition

Caudally Directed Association Fiber

Rostrally Directed Association Fiber

Local Association Fiber

Local Feedback Inhibition

Rostrally Directed Association Fiber

Caudally Directed Association Fiber

Fig. 2. Schematic diagram of piriform cortex showing an excitatory pyramidal cell and two inhibitory interneurons with their local interactions. Circles indicate sites of synaptic modifiability.

Connection Patterns. In the olfactory system, olfactory receptors project to the olfactory bulb which, in turn, projects directly to the piriform cortex and other olfactory structures (Fig. 1). The input to the piriform cortex from the olfactory bulb is delivered via a fiber bundle known as the lateral olfactory tract (LOT). This fiber tract appears to make sparse, non-topographic, excitatory connections with pyramidal and feedforward inhibitory neurons across the extent of the cortex[3,6]. In the model this input is simulated as 100 independent cells each of which make random connections (p=0.05) with pyramidal and feedforward inhibitory neurons (Fig. 1 and 2).

In addition to the input connections from the olfactory bulb, there is also an extensive set of connections between the neurons intrinsic to the cortex (Fig. 2). For example, the association fiber system arises from pyramidal cells and makes sparse, distributed excitatory connections with other pyramidal cells all across the cortex[7,8,9]. In the model these connections are randomly distributed with 0.05 probability. In the model and in the actual cortex, pyramidal cells also make excitatory connections with nearby feedforward and feedback inhibitory cells. These interneurons, in turn, make reciprocal inhibitory connections with the group of nearby pyramidal cells. The primary effect of the feedback inhibitory neurons is to inhibit pyramidal cell firing through a Cl^- mediated current shunting mechanism[10,11,12]. Feedforward interneurons inhibit pyramidal cells via a long latency, long duration, K^+ mediated hyperpolarizing potential[12,13]. Pyramidal cell axons also constitute the primary output of both the model and the actual piriform cortex[7,14].

Synaptic Properties and Modification Rules. In the model, each synaptic connection has an associated weight which determines the peak amplitude of the conductance change induced in the postsynaptic cell following presynaptic activity [2.0]. To study learning in the model, synaptic weights associated with some of the fiber systems are modifiable in an activity-dependent fashion (Fig. 2). The basic modification rule in each case is Hebb-like; i.e. change in synaptic strength is proportional to presynaptic activity multiplied by the offset of the postsynaptic membrane potential from a baseline potential. This baseline potential is set slightly more positive than the Cl⁻ equilibrium potential associated with the shunting feedback inhibition. This means that synapses activated while a destination cell is in a depolarized or excited state are strengthened, while those activated during a period of inhibition are weakened. In the model, synapses which follow this rule include the association fiber connections between excitatory pyramidal neurons as well as the connections between inhibitory neurons and pyramidal neurons. Whether these synapses are modifiable in this way in the actual cortex is a subject of active research in our lab. However, the model does mimic the actual synaptic properties associated with the input pathway (LOT) which we have shown to undergo a transient increase in synaptic strength following activation which is independent of postsynaptic potential[15]. This increase is not permanent and the synaptic strength subsequently returns to its baseline value.

Generation of Physiological Responses. Neurons in the model are represented as first-order "leaky" integrators with multiple, time-varying inputs [1.0]. During simulation runs, membrane potentials and currents as well as the time of occurence of action potentials are stored for comparison with actual data. An explicit compartmental model (5 compartments) of the pyramidal cells is used to generate the spatial current distributions used for calculation of field potentials (evoked potentials, EEGs) [3.0, 4.0].

Stimulus Characteristics. To compare the responses of the model to those of the actual cortex, we mimicked actual experimental stimulation protocols in the simulated cortex and contrasted the resulting intracellular and extracellular records. For example, shock stimuli applied to the LOT are often used to elicit characteristic cortical evoked potentials *in vivo*[16,17,18]. In the model we simulated this stimulus paradigm by simultaneously activating all 100 input fibers. Another measure of cortical activity used most successfully by Freeman and colleagues involves recording EEG activity from piriform cortex in behaving animals[19,20]. These odor-like responses were generated in the model through steady, random stimulation of the input fibers.

To study learning in the model, once physiological measures were established, it was required that we use more refined stimulation procedures. In the absence of any specific information about actual input activity patterns along the LOT, we constructed each stimulus out of a randomly selected set of 10 out of the 100 input

118

fibers. Each stimulus episode consisted of a burst of activity in this subset of fibers with a duration of 10 msec at 25 msec intervals to simulate the 40 Hz periodicity of the actual olfactory bulb input. This pattern of activity was repeated in trials of 200 msec duration which roughly corresponds to the theta rhythm periodicity of bulbar activity and respiration[21,22]. Each trial was then presented 5 times for a total exposure time of 1 second (cortical time). During this period the Hebb-type learning rule could be used to modify the connection weights in an activity-dependent fashion.

Output Measure for Learning. Given that the sole output of the cortex is in the form of action potentials generated by the pyramidal cells, the output measure of the model was taken to be the vector of spike frequency for all pyramidal neurons over a 200 msec trial, with each element of the vector corresponding to the firing frequency of a single pyramidal cell. Figures 5 through 8 show the 10 by 10 array of pyramidal cells. The size of the box placed at each cell position represents the magnitude of the spike frequency for that cell. To evaluate learning effects, overlap comparisons between response pairs were made by taking the normalized dot product of their response vectors and expressing that value as a percent overlap (Fig. 4).

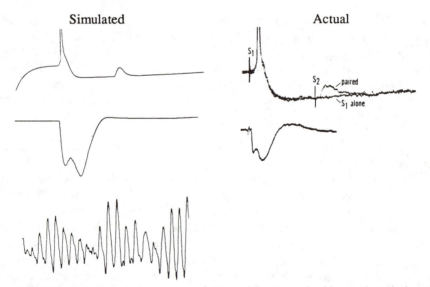

Fig. 3. Simulated physiological responses of the model compared with actual cortical responses. Upper: Simulated intracellular response of a single cell to paired stimulation of the input system (LOT) (left) compared with actual response (right) (Haberly & Bower,'84). Middle: Simulated extracellular response recorded at the cortical surface to stimulation of the LOT (left), compared with actual response (right) (Haberly,'73b). Lower: Stimulated EEG response recorted at the cortical surface to odor-like input (left), for actual EEG see Freeman 1978.

Computational Requirements. All simulations were carried out on a Sun Microsystems 3/260 model microcomputer equipped with 8 Mbytes of memory and a floating point accelerator. Average time for a 200 msec simulation was 3 cpu minutes.

RESULTS

Physiological Responses

As described above, our initial modeling objective was to accurately simulate a wide range of activity patterns recorded, by ourselves and others, in piriform cortex using various physiological procedures. Comparisons between actual and simulated records for several types of response are shown in figure 3. In general, the model replicated known physiological responses quite well (Wilson et al in preparation describes, in detail, the analysis of the physiological results). For example in response to shock stimulation of the input pathway (LOT), the model reproduces the principle characteristics of both the intracellular and location-dependent extracellular waveforms recorded in the actual cortex[9,17,18] (Fig. 3).

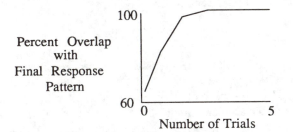

Fig. 4. Convergence of the cortical response during training with a single stimulus with synaptic modification.

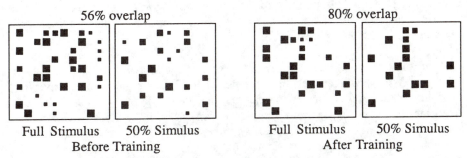

Fig. 5. Reconstruction of cortical response patterns with partially degraded stimuli. Left: Response, before training, to the full stimulus (left) and to the same stimulus with 50% of the input fibers inactivated (right). There is a 44% degradation in the response. Right: Response after training, to the full stimulus (left), and to the same stimulus with 50% of the input fibers inactivated (right). As a result of training, the degradation is now only 20%.

Trained on A   Trained on B   Retains A Response

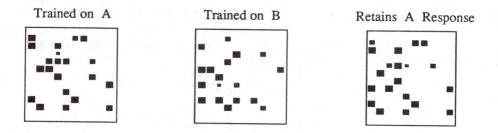

Fig. 6. Storage of multiple patterns. Left: Response to stimulus A after training. Middle: Response to stimulus B after training on A followed by training on B. Right: Response to stimulus A after training on A followed by training on B. When compared with the original response (left) there is an 85% congruence.

Further, in response to odor-like stimulation the model exhibits 40 Hz oscillations which are characteristic of the EEG activity in olfactory cortex in awake, behaving animals[19]. Although beyond the scope of the present paper, the simulation also duplicates epileptiform[9] and damped oscillatory[16] type activity seen in the cortex under special stimulus or pharmacological conditions[4].

Learning

Having simulated characteristic physiological responses, we wished to explore the capabilities of the model to store and recall information. Learning in this case is defined as the development of a consistent representation in the activity of the cortex for a particular input pattern with repeated stimulation and synaptic modification. Figure 4 shows how the network converges, with training, on a representation for a stimulus. Having demonstrated that, we studied three properties of learned responses - the reconstruction of trained cortical response patterns with partially degraded stimuli, the simultaneous storage of separate stimulus response patterns, and the modulation of cortical response patterns independent of relative stimulus characteristics.

Reconstruction of Learned Cortical Response Patterns with Partially Degraded Stimuli. We were interested in knowing what effect training would have on the sensitivity of cortical responses to fluctuations in the input signal. First we presented the model with a random stimulus A for one trial (without synaptic modification). On the next trial the model was presented with a degraded version of A in which half of the original 10 input fibers were inactivated. Comparison of the responses to these two stimuli in the naive cortex showed a 44% variation. Next, the model was trained on the full stimulus A for 1 second (with synaptic modification). Again, half of the input was removed and the model was presented with the degraded stimulus for 1 trial (without synaptic modification). In this case the dif-

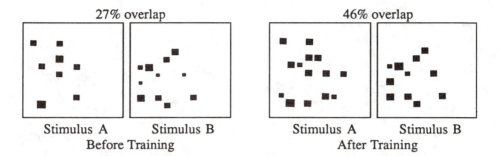

Fig. 7. Results of merging cortical response patterns for dissimilar stimuli. Left: Response to stimulus A and stimulus B before training. Stimuli A and B do not activate any input fibers in common but still have a 27% overlap in cortical response patterns. Right: Response to stimulus A and stimulus B after training in the presence of a common modulatory input E1. The overlap in cortical response patterns is now 46%.

ference between cortical responses was only 20% (Fig. 5) showing that training increased the robustness of the response to degradation of the stimulus.

Storage of Two Patterns. The model was first trained on a random stimulus A for 1 second. The response vector for this case was saved. Then, continuing with the weights obtained during this training, the model was trained on a new non-overlapping (i.e. different input fibers activated) stimulus B. Both stimulus A and stimulus B alone activated roughly 25% of the cortical pyramidal neurons with 25% overlap between the two responses. Following the second training period we assessed the amount of interference in recalling A introduced by training with B by presenting stimulus A again for a single trial (without synaptic modification). The variation between the response to A following additional training with B and the initially saved reponse to A alone was less than 15% (Fig. 6) demonstrating that learning B did not substantially interfere with the ability to recall A.

Modulation of Cortical Response Patterns. It has been previously demonstrated that the stimulus evoked response of olfactory cortex can be modulated by factors not directly tied to stimulus qualities, such as the behavioral state of the animal[1,20,23]. Accordingly we were interested in knowing whether the representations stored in the model could be modulated by the influence of such a "state" input.

One potential role of a "state" input might be to merge the cortical response patterns for dissimilar stimuli; an effect we refer to as accomodation. To test this in the model, we presented it with a random input stimulus A for 1 trial. It was then presented with a random input stimulus B (non-overlapping input fibers). The amount of overlap in the cortical responses for these untrained cases was 27%. Next, the model was trained for 1 second on stimulus A in the presence of an additional random "state" stimulus E1 (activity in a set of 10 input fibers distinct

122

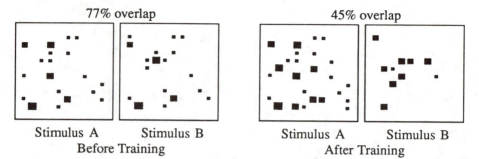

77% overlap   45% overlap

Stimulus A   Stimulus B   Stimulus A   Stimulus B
       Before Training   After Training

Fig. 8. Results of differentiating cortical response patterns for similar stimuli. Left: Response to stimulus A and stimulus B before training. Stimuli A and B activate 75% of their input fibers in common and have a 77% overlap in cortical response patterns. Right: Response to stimulus A and stimulus B after training A in the presence of modulatory input E1 and training B with a different modulatory input E2. The overlap in cortical response patterns is now 45%.

from both A and B). The model was then trained on stimulus B in the presence of the same "state" stimulus E1. After training, the model was presented with stimulus A alone for 1 trial and stimulus B alone for 1 trial. Results showed that now, even without the coincident E1 input, the amount of overlap between A and B responses was found to have increased to 46% (Fig 7). The role of E1 in this case was to provide a common stimulus component during learning which reinforced shared components of the responses to input stimuli A and B.

To test the ability of a state stimulus to induce differentiation of cortical response patterns for similar stimuli, we presented the model with a random input stimulus A for 1 trial, followed by 1 trial of a random input stimulus B (75% of the input fibers overlapping). The amount of overlap in the cortical responses for these untrained cases was 77%. Next, the model was trained for a period of 1 second on stimulus A in the presence of an additional random "state" stimulus E1 (a set of 10 input fibers not overlapping either A or B). It was then trained on input stimulus B in the presence of a different random "state" stimulus E2 (10 input fibers not overlapping either A, B, or E1) After this training the model was presented with stimulus A alone for 1 trial and stimulus B alone for 1 trial. The amount of overlap was found to have decreased to 45% (Fig 8). In this situation E1 and E2 provided a differential signal during learning which reinforced distinct components of the responses to input stimuli A and B.

DISCUSSION

Physiological Responses. Detailed discussion of the mechanisms underlying the simulated patterns of physiological activity in the cortex is beyond the scope of the current paper. However, the model has been of value in suggesting roles for specific features of the cortex in generating physiologically recorded activity. For example, while actual input to the cortex from the olfactory bulb is modulated into 40 Hz bursts[24], continuous stimulation of the model allowed us to demonstrate the model's capability for intrinsic periodic activity independent of the complementary pattern of stimulation from the olfactory bulb. While a similar ability has also been demonstrated by models of Freeman[25], by studying this oscillating property in the model we were able to associate these oscillatory characteristics with specific interactions of local and distant network properties (e.g. inhibitory and excitatory time constants and trans-cortical axonal conduction velocities). This result suggests underlying mechanisms for these oscillatory patterns which may be somewhat different than those previously proposed.

Learning. The main subject of this paper is the examination of the learning capabilities of the cortical model. In this model, the apparently sparse, highly distributed pattern of connectivity characteristic of piriform cortex is fundamental to the way in which the model learns. Essentially, the highly distributed pattern of connections allows the model to develop stimulus-specific cortical response patterns by extracting correlations from randomly distributed input and association fiber activity. These correlations are, in effect, stored in the synaptic weights of the association fiber and local inhibitory connections.

The model has also demonstrated robustness of a learned cortical response against degradation of the input signal. A key to this property is the action of sparsely distributed association fibers which provide reinforcment for previously established patterns of cortical activity. This property arises from the modification of synaptic weights due to correlations in activity between intra-cortical association fibers. As a result of this modification the activity of a subset of pyramidal neurons driven by a degraded input drives the remaining neurons in the response.

In general, in the model, similar stimuli will map onto similar cortical responses and dissimilar stimuli will map onto dissimilar cortical responses. However, a presumably important function of the cortex is not simply to store sensory information, but to represent incoming stimuli as a function of the absolute stimulus qualities and the context in which the stimulus occurs. The fact that many of the structures that piriform cortex projects to (and receives projections from) may be involved in multimodal "state" generation[14] is circumstantial evidence that such modulation may occur. We have demonstrated in the model that such a modulatory input can modify the representations generated by pairs of stimuli so as to push the representations of like stimuli apart and pull the representations of dissimilar stimuli together. It should be pointed out that this modulatory input was not an "instructive" signal which explicitly directed the course of the representation, but rather a "state" signal which did not require *a priori* knowledge of the representational structure. In the model, this modulatory phenomenon is a simple consequence of the degree of overlap in the combined (odor stimulus + modulator) stimulus. Both cases approached approximately 50% overlap in cortical responses reflecting the approximately 50% overlap in the combined stimuli for both cases.

Of interest was the use of the model's reconstructive capabilities to maintain the modulated response to each input stimulus even in the absence of the modulatory input.

CAVEATS AND CONCLUSIONS

Our approach to studying this system involves using computer simulation to investigate mechanisms of information processing which could be implemented given what is known about biological constraints. The significance of results presented here lies primarily in the finding that the structure of the model and the parameter settings which were appropriate for the reproduction of physiological responses were also appropriate for the proper convergence of a simple, biologically plausible learning rule under various conditions. Of course, the model we have developed is only an approximation to the actual cortex limited by our knowledge of its organization and the computing power available. For example, the actual piriform cortex of the rat contains on the order of 10^6 cells (compared with 10^2 in the simulations) with a sparsity of connection on the order of p=0.001 (compared with p=0.05 in the simulations). Our continuing research effort will include explorations of the scaling properties of the network.

Other assumptions made in the context of the current model include the assumption that the representation of information in piriform cortex is in the form of spatial distributions of rate-coded outputs. Information contained in the spatio-temporal patterns of activity was not analyzed, although preliminary observation suggests that this may be of significance. In fact, the dynamics of the model itself suggest that temporally encoded information in the input at various time scales may be resolvable by the cortex. Additionally, the output of the cortex was assumed to have spatial uniformity, i.e. no differential weighting of information was made on the basis of spatial location in the cortex. But again, observation of the dynamics of the model, as well as the details of known anatomical distribution patterns for axonal connections, indicate that this is a major oversimplification. Preliminary evidence from the model would indicate that some form of hierarchical structuring of information along rostral/caudal lines may occur. For example it may be that cells found in progressively more rostral locations would have increasingly non-specific odor responses.

Further investigations of learning within the model will explore each of these issues more fully, with attempts to correlate simulated findings with actual recordings from awake, behaving animals. At the same time, new data pertaining to the structure of the cortex will be incorporated into the model as it emerges.

ACKNOWLEDGEMENTS

We wish to thank Dr. Lewis Haberly and Dr. Joshua Chover for their roles in the development and continued support of the modeling effort. We also wish to thank Dave Bilitch for his technical assistance. This work was supported by NIH grant NS22205, NSF grant EET-8700064, the Lockheed Corporation, and a fellowship from the ARCS foundation.

APPENDIX

$$\frac{dV_i}{dt} = \frac{1}{c_m}\left[\sum_{k=1}^{n_{types}} I_{ik}(t) + \frac{E_r - V_i(t)}{r_l}\right]$$ (1.0)

Somatic Integration

$$I_{ik}(t) = [E_k - V_i(t)]g_{ik}(t)$$ (1.1)

n_{types} = number of input types
$V_i(t)$ = membrane potential of i th cell
$I_{ik}(t)$ = current into cell i due to input type k
E_k = equilibrium potential associated with input type k

E_r = resting potential
r_l = membrane leakage resistance
c_m = membrane capacitance
$g_{ik}(t)$ = conductance due to input type k in cell i

$$g_{ik}(t) = \sum_{j=1}^{n_{cells}} \int_{\lambda=0}^{\lambda=d_k} F_k(\lambda)\, A_{ijk}\, W_{ij}\, S_j(t - \lambda - \frac{L_{ij}}{v_k} - \epsilon_k)\, d\lambda$$ (2.0)

$$F_k(t) = \frac{t}{\tau}e^{(1-\frac{t}{\tau})}\left[(1 - U(t-\tau)) + U(t-\tau)\cos\left[\frac{\pi}{2}\frac{(t-\tau)}{(d_k-\tau)}\right]\right] \quad, \quad \tau = \gamma d_k$$ (2.1)

Spike Propagation

and Synaptic Input

$$A_{ijk} = (1 - \rho_k^{min})e^{-L_{ij}\rho_k} + \rho_k^{min}$$ (2.2)

$$S_j(t) = \begin{cases} 1 & V_j(t) > T_j, \ S_j(\lambda) = 0 \ \text{ for } \ \lambda = t..t - \Delta t_r \\ 0 & \text{otherwise} \end{cases}$$ (2.3)

$$L_{ij} = |i - j|\Delta x$$ (2.4)

n_{cells} = number of cells in the simulation
Δx = distance between adjacent cells
d_k = duration of conductance change due to input type k
v_k = velocity of signals for input type k
ϵ_k = latency for input type k
ρ_k = spatial attenuation factor for input type k
ρ_k^{min} = minimum spatial attenuation for input type k
Δt_r = refractory period

T_j = threshold for cell j
L_{ij} = distance from cell i to cell j
A_{ijk} = distribution of synaptic density for input type k
W_{ij} = synaptic weight from cell j to cell i
$g_{ik}(t)$ = conductance due to input type k in cell i
$F_k(t)$ = conductance waveform for input type k
$S_j(t)$ = spike output of cell j at time t
$U(t)$ = unit step function

Field Potentials

$$V_{ep}^j(t) = \frac{R_e}{4\pi}\sum_{i=1}^{n_{cells}}\sum_{n=1}^{n_{sgs}} \frac{I_m^n(t)}{\left[(z_{elec} - z_n)^2 + (x_j - x_i)^2\right]^{\frac{1}{2}}}$$ (3.0)

n_{cells} = number of cells in the simulation
n_{segs} = number of segments in the compartmental model
$V_{ep}^{jj}(t)$ = approximate extracellular field potential at cell j
$I_m^{in}(t)$ = membrane current for segment n in cell i

z_{elec} = depth of recording site
z_n = depth of segment n
x_j = x location of the j th cell
R_e = extracellular resistance per unit length

$$\frac{dV_n}{dt} = \frac{1}{c_m^n}\left[I_a^{n-}(t) + I_a^{n+}(t) + \frac{E_r - V_n(t)}{r_m^n} + \sum_{c=1}^{n_{chan}} [E_c - V_n(t)]g_{nc}(t)\right]$$ (4.0)

Dendritic Model

$$I_a^{n-}(t) = \frac{V_{n-1}(t) - V_n(t)}{r_a^{n-1} + r_a^n} \quad, \quad I_a^{n+}(t) = \frac{V_{n+1}(t) - V_n(t)}{r_a^{n+1} + r_a^n}$$ (4.1)

126

$$I_m^n(t) = I_a^{n-}(t) + I_a^{n+}(t) \qquad (4.2)$$

$$r_a^n = \frac{1}{2}\left[R_e l_n + R_i \frac{l_n}{\pi\left[\frac{d_n}{2}\right]^2}\right] \quad , \quad r_m^n = \frac{R_m}{\pi\, l_n d_n} \quad , \quad c_m^n = C_m \pi\, l_n d_n \quad (4.3)$$

n_{chan} = number of different channels per segment
$V_n(t)$ = membrane potential of nth segment
c_m^n = membrane capacitance for segment n
r_a^n = axial resistance for segment n
r_m^n = membrane resistance for segment n
$g_{nc}(t)$ = conductance of channel c in segment n
E_c = equilibrium potential associated with channel c
$I_a^{n\pm}(t)$ = axial current between segment $n\pm1$ and n

$I_m^n(t)$ = membrane current for segment n
l_n = length of segment n
d_n = diameter of segment n
R_m = membrane resistivity
R_i = intracellular resistivity per unit length
R_e = extracellular resistance per unit length
C_m = capacitance per unit surface area

REFERENCES

1. W. J. Freeman, J. Neurophysiol., 23, 111 (1960).
2. T. Tanabe, M. Iino, and S. F. Takagi, J. Neurophysiol., 38,1284 (1975).
3. L. B. Haberly, Chemical Senses, 10, 219 (1985).
4. M. Wilson, J. M. Bower, J. Chover, and L. B. Haberly, Soc. Neuro. Abs., 11, 317 (1986).
5. M. Wilson and J. M. Bower, Soc. Neurosci. Abs., 12, 310 (1987).
6. M. Devor, J. Comp. Neur., 166, 31 (1976).
7. L. B. Haberly and J. L. Price, J. Comp. Neurol., 178, 711 (1978a).
8. L. B. Haberly and S. Presto, J. Comp. Neur., 248, 464 (1986).
9. L. B. Haberly and J. M. Bower, J. Neurophysiol., 51, 90 (1984).
10. M. A. Biedenbach and C. F. Stevens, J. Neurophysiol., 32, 193 (1969).
11. M. A. Biedenbach and C. F. Stevens, J. Neurophysiol., 32, 204 (1969).
12. M. Satou, K. Mori, Y. Tazawa, and S. F. Takagi, J. Neurophysiol., 48, 1157 (1982).
13. G. F. Tseng and L. B. Haberly, Soc. Neurosci. Abs. 12, 667 (1986).
14. L. B. Luskin and J. L. Price, J. Comp. Neur., 216, 264 (1983).
15. J. M. Bower and L. B. Haberly,L.B., Proc. Natl. Acad. Sci. USA, 83, 1115 (1985).
16. W. J. Freeman, J. Neurophysiol., 31, 1 (1968).
17. L. B. Haberly, J. Neurophysiol., 36, 762 (1973).
18. L. B. Haberly, J. Neurophysiol., 36, 775 (1973).
19. W. J. Freeman, Electroenceph. and Clin. Neurophysiol., 44, 586 (1978).
20. W.J. Freeman and W. Schneider, Psychophysiology, 19, 44 (1982).
21. F. Macrides and S. L. Chorover, Science, 175, 84 (1972).
22. F. Macrides, H. B. Eigenbaum, and W. B. Forbes, J. Neurosci., 2, 12, 1705 (1982).
23. P. D. MacLean, N. H. Horwitz, and F. Robinson, Yale J. Biol. Med., 25, 159 (1952).
24. E. D. Adrian, Electroenceph. and Clin. Neurophysiol., 2, 377 (1950).
25. W. J. Freeman, Exp. Neurol., 10, 525 (1964).

Neural Network Implementation Approaches
for the
Connection Machine

Nathan H. Brown, Jr.

MRJ/Perkin Elmer, 10467 White Granite Dr. (Suite 304), Oakton, Va. 22124

ABSTRACT

The SIMD parallelism of the Connection Machine (CM) allows the construction of neural network simulations by the use of simple data and control structures. Two approaches are described which allow parallel computation of a model's nonlinear functions, parallel modification of a model's weights, and parallel propagation of a model's activation and error. Each approach also allows a model's interconnect structure to be physically dynamic. A Hopfield model is implemented with each approach at six sizes over the same number of CM processors to provide a performance comparison.

INTRODUCTION

Simulations of neural network models on digital computers perform various computations by applying linear or nonlinear functions, defined in a program, to weighted sums of integer or real numbers retrieved and stored by array reference. The numerical values are model dependent parameters like time averaged spiking frequency (activation), synaptic efficacy (weight), the error in error back propagation models, and computational temperature in thermodynamic models. The interconnect structure of a particular model is implied by indexing relationships between arrays defined in a program. On the Connection Machine (CM), these relationships are expressed in hardware processors interconnected by a 16-dimensional hypercube communication network. Mappings are constructed to define higher dimensional interconnectivity between processors on top of the fundamental geometry of the communication network. Parallel transfers are defined over these mappings. These mappings may be dynamic. CM parallel operations transform array indexing from a temporal succession of references to memory to a single temporal reference to spatially distributed processors.

Two alternative approaches to implementing neural network simulations on the CM are described. Both approaches use "data parallelism"[1] provided by the *Lisp virtual machine. Data and control structures associated with each approach and performance data for a Hopfield model implemented with each approach are presented.

DATA STRUCTURES

The functional components of a neural network model implemented in *Lisp are stored in a uniform parallel variable (pvar) data structure on the CM. The data structure may be viewed as columns of pvars. Columns are given to all CM virtual processors. Each CM physical processor may support 16 virtual processors. In the first approach described, CM processors are used to represent the edge set of a models graph structure. In the second approach described, each processor can represent a unit, an outgoing link, or an incoming link in a model's structure. Movement of activation (or error) through a model's interconnect structure is simulated by moving numeric values over the CM's hypercube. Many such movements can result from the execution of a single CM macroinstruction. The CM transparently handles message buffering and collision resolution. However, some care is required on the part of the user to insure that message traffic is distributed over enough processors so that messages don't stack up at certain processors, forcing the CM to sequentially handle large numbers of buffered messages. Each approach requires serial transfers of model parameters and states over the communication channel between the host and the CM at certain times in a simulation.

The first approach, "the edge list approach," distributes the edge list of a network graph to the CM, one edge per CM processor. Interconnect weights for each edge are stored in the memory of the processors. An array on the host machine stores the current activation for all units. This approach may be considered to represent abstract synapses on the CM. The interconnect structure of a model is described by product sets on an ordered pair of identification (id) numbers, rid and sid. The rid is the id of units receiving activation and sid the id of units sending activation. Each id is a unique integer. In a hierarchical network, the ids of input units are never in the set of rids and the ids of output units are never in the set of sids. Various set relations (e.g. inverse, reflexive, symmetric, etc.) defined over id ranges can be used as a high level representation of a network's interconnect structure. These relations can be translated into pvar columns. The limits to the interconnect complexity of a simulated model are the virtual processor memory limits of the CM configuration used and the stack space required by functions used to compute the weighted sums of activation. Fig. 1 shows a R^3 -> R^2 -> R^4 interconnect structure and its edge list representation on the CM.

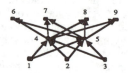

CM PROCESSOR 0 1 2 3 4 5 6 7 8 9 10 11 12 13
RACT (a_i) 4 4 4 5 5 5 6 6 7 7 8 8 9 9

WEIGHT (w_{ij})

SACT (a_j) 1 2 3 1 2 3 4 5 4 5 4 5 4 5

Fig. 1. Edge List Representation of a R^3-> R^2 -> R^4 Interconnect Structure

This representation can use as few as six pvars for a model with Hebbian adaptation: rid (i), sid (j), interconnect weight (w_{ij}), ract (a_i), sact (a_j), and learn rate (η). Error back propagation requires the addition of: error (e_i), old interconnect weight $(w_{ij}(t-1))$, and the momentum term (α). The receiver and sender unit identification pvars are described above. The interconnect weight pvar stores the weight for the interconnect. The activation pvar, sact, stores the current activation, a_j, transfered to the unit specified by rid from the unit specified by sid. The activation pvar, ract, stores the current weighted activation $a_j w_{ij}$. The error pvar stores the error for the unit specified by the sid. A variety of proclaims (e.g. integer, floating point, boolean, and field) exist in *Lisp to define the type and size of pvars. Proclaims conserve memory and speed up execution. Using a small number of pvars limits the amount of memory used in each CM processor so that maximum virtualization of the hardware processors can be realized. Any neural model can be specified in this fashion. Sigma-pi models require multiple input activation pvars be specified. Some edges may have a different number of input activation pvars than others. To maintain the uniform data structure of this approach a tag pvar has to be used to determine which input activation pvars are in use on a particular edge.

The edge list approach allows the structure of a simulated model to "physically" change because edges may be added (up to the virtual processor limit), or deleted at any time without affecting the operation of the control structure. Edges may also be placed in any processor because the subselection (on rid or sid) operation performed before a particular update operation insures that all processors (edges) with the desired units are selected for the update.

The second simulation approach, "the composite approach," uses a more complicated data structure where units, incoming links, and outgoing links are represented. Update routines for this approach use parallel segmented scans to form the weighted sum of input activation. Parallel segmented scans allow a MIMD like computation of the weighted sums for many units at once. Pvar columns have unique values for unit, incoming link, and outgoing link representations. The data structures for input units, hidden units, and output units are composed of sets of the three pvar column types. Fig. 2 shows the representation for the same model as in Fig. 1 implemented with the composite approach.

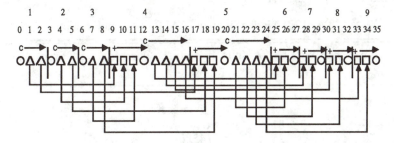

Fig. 2. Composite Representation of a R^3 -> R^2 -> R^4 Interconnect Structure

In Fig. 2, CM processors acting as units, outgoing links, and incoming links are represented respectively by circles, triangles, and squares. CM cube address pointers used to direct the parallel transfer of activation are shown by arrows below the structure. These pointers define the model interconnect mapping. Multiple sets of these pointers may be stored in seperate pvars. Segmented scans are represented by operation-arrow icons above the structure. A basic composite approach pvar set for a model with Hebbian adaptation is: forward B, forward A, forward transfer address,

interconnect weight (w_{ij}), act-1 (a_i), act-2 (a_j), threshold, learn rate (η), current unit id (i), attached unit id (j), level, and column type. Back progagation of error requires the addition of: backward B, backward A, backward transfer address, error (e_j), previous

interconnect weight ($w_{ij}(t-1)$), and the momentum term (α). The forward and backward boolean pvars control the segmented scanning operations over unit constructs. Pvar A of each type controls the plus scanning and pvar B of each type controls the copy scanning. The forward transfer pvar stores cube addresses for forward (ascending cube address) parallel transfer of activation. The backward transfer pvar stores cube addresses for backward (descending cube address) parallel transfer of error. The interconnect weight, activation, and error pvars have the same functions as in the edge list approach. The current unit id stores the current unit's id number. The attached unit id stores the id number of an attached unit. This is the edge list of the network's graph. The contents of these pvars only have meaning in link pvar columns. The level pvar stores the level of a unit in a hierarchical network. The type pvar stores a unique arbitrary tag for the pvar column type. These last three pvars are used to subselect processor ranges to reduce the number of processors involved in an operation.

Again, edges and units can be added or deleted. Processor memories for deleted units are zeroed out. A new structure can be placed in any unused processors. The level, column type, current unit id, and attached unit id values must be consistent with the desired model interconnectivity.

The number of CM virtual processors required to represent a given model on the CM differs for each approach. Given N units and N(N-1) non-zero interconnects (e.g. a symmetric model), the edge list approach simply distributes N(N-1) edges to N(N-1) CM virtual processors. The composite approach requires two virtual processors for each interconnect and one virtual processor for each unit or N+2 N(N-1) CM virtual processors total. The difference between the number of processors required by the two approaches is N^2. Table I shows the processor and CM virtualization requirements for each approach over a range of model sizes.

TABLE I Model Sizes and CM Processors Required

Run No.	Grid Size	Number of Units	Edge List N(N-1)	Quart CM Virt. Procs.	Virt. LeveL
1	8^2	64	4032	8192	0
2	9^2	81	6480	8192	0
3	11^2	121	14520	16384	0
4	13^2	169	28392	32768	2
5	16^2	256	65280	65536	4
6	19^2	361	129960	131072	8

Run No.	Grid Size	Number of Units	Composite N+2N(N-1)	Quart CM Virt. Procs.	Virt. LeveL
7	8^2	64	8128	8192	0
8	9^2	81	13041	16384	0
9	11^2	121	29161	32768	2
10	13^2	169	56953	65536	4
11	16^2	256	130816	131072	8
12	19^2	361	260281	262144	16

CONTROL STRUCTURES

The control code for neural network simulations (in *Lisp or C*) is stored and executed sequentially on a host computer (e.g. Symbolics 36xx and VAX 86xx) connected to the CM by a high speed communication line. Neural network simulations executed in *Lisp use a small subset of the total instruction set: processor selection reset (*all), processor selection (*when), parallel content assignment (*!!), global summation (*sum), parallel multiplication (*!!), parallel summation (+!!), parallel exponentiation (exp!!), the parallel global memory references (*pset) and (pref!!), and the parallel segmented scans (copy!! and +!!). Selecting CM processors puts them in a "list of active processors" (loop) where their contents may be arithmetically manipulated in parallel. Copies of the list of active processors may be made and used at any time. A subset of the processors in the loop may be "subselected" at any time, reducing the loop contents. The processor selection reset clears the current selected set by setting all processors as selected. Parallel content assignment allows pvars in the currently selected processor set to be assinged allowed values in one step. Global summation executes a tree reduction sum across the CM processors by grid or cube address for particular pvars. Parallel multiplications and additions multiply and add pvars for all selected CM processors in one step. The parallel exponential applies the function, e^x, to the contents of a specified pvar, x, over all selected processors. Parallel segmented scans apply two functions, copy!! and +!!, to subsets of CM processors by scanning across grid or cube addresses. Scanning may be forward or backward (i.e. by ascending or descending cube address order, respectively).

Figs. 3 and 4 show the edge list approach kernels required for Hebbian learning for a $R^2 \rightarrow R^2$ model. The loop construct in Fig. 3 drives the activation update

$$a_i(t+1)=F[\Sigma w_{ij}(t+1)a_j(t)] \qquad (1)$$

operation. The usual loop to compute each weighted sum for a particular unit has been replaced by four parallel operations: a selection reset (*all), a subselection of all the processors for which the particular unit is a receiver of activation (*when (=!! rid (!! (1+ u)))), a parallel multiplication (*!! weight sact), and a tree reduction sum (*sum ...). Activation is spread for a particular unit, to all others it is connected to, by: storing the newly computed activation in an array on the host, then subselecting the processors where the particular unit is a sender of activation (*when (=!! sid (!! (1+ u)))), and broadcasting the array value on the host to those processors.

```
(dotimes (u 4)
     (*all (*when (=!! rid (!! (1+ u)))
          (setf (aref activation u)
                    (some-nonlinearity (*sum (*!! weight sact))))
          (*set ract (!! (aref activation u)))
     (*all (*when (=!! sid (!! (1+ u)))
          (*set sact (!! (aref activation u)))))))
```

Fig. 3. Activation Update Kernel for the Edge Lst Approach.

Fig. 4 shows the Hebbian weight update kernel

$$w_{ij}(t+1) = \eta a_i(t+1)a_j(t+1). \tag{2}$$

```
(*all
   (*set weight
      (*!! learn-rate ract sact))))
```

Fig. 4. Hebbian Weight Modification Kernel for the Edge List Approach

The edge list activation update kernel is essentially serial because the steps involved can only be applied to one unit at a time. The weight modification is parallel. For error back propagation a seperate loop for computing the errors for the units on each layer of a model is required. Activation update and error back propagation also require transfers to and from arrays on the host on every iteration step incurring a concomitant overhead.

Other common computations used for neural networks can be computed in parallel using the edge list approach. Fig. 5 shows the code kernel for parallel computation of Lyapunov engergy equations

$$E = -1/2\Sigma^N_{i \neq j} w_{ij}a_ia_j + \Sigma^N_{i=1} I_ia_i \tag{3}$$

where i=1 to number of units (N).

```
(+ (* -.5 (*sum (*!! weight ract sact))) (*sum (*!! input sact)))
```

Fig. 5. Kernel for Computation of the Lyapunov Energy Equation

Although an input pvar, input, is defined for all edges, it is only non-zero for those edges associated with input units. Fig. 6 shows the pvar structure for parallel computation of a Hopfield weight prescription, with segmented scanning to produce the weights in one step,

$$w_{ij} = \Sigma^S_{r=1}(2a^r_i-1)(2a^r_j-1) \tag{4}$$

where $w_{ii}=0$, $w_{ij}=w_{ji}$, and r=1 to the number of patterns, S, to be stored.

```
seg      t   n            n    t   n           n
ract    v1_1 v2_1 ... vS_1   v1_1 v2_1 ... vS_1 ...
sact    v1_2 v2_2 ... vS_2   v1_3 v2_3 ... vS_3 ...
weight            w12                 w13
```

Fig. 6. Pvar Structure for Parallel Computation of Hopfield Weight Prescription

Fig. 7 shows the *Lisp kernel used on the pvar structure in Fig. 6.

```
(set weight
   (scan '+!! (*!! (-!! (*!! ract (!! 2)) (!! 1)) (-!! (*!! sact (!! 2)) (!! 1))))
      :segment-pvar seg :include-self t)
```

Fig. 7. Parallel Computation of Hopfield Weight Prescription

The inefficiencies of the edge list activation update are solved by the updating method used in the composite approach. Fig. 8 shows the *Lisp kernel for activation update using the composite approach. Fig. 9 shows the *Lisp kernel for the Hebbian learning operation in the composite approach.

```
(*all
    (*when (=!! level (!! 1))
        (*set act (scan!! act-1 'copy!! :segment-pvar forwardb :include-self t))
        (*set act (*!! act-1 weight))
        (*when (=!! type (!! 2)) (*pset :overwrite act-1 act-1 ftransfer)))
    (*when (=!! level (!! 2))
        (*set act (scan!! act-1 '+!! :segment-pvar forwarda :include-self t))
        (*when (=!! type (!! 1)) (some-nonlinearity!! act-1))))
```

Fig. 8. Activation Update Kernel for the Composite Approach

```
(*all
    (*set act-1 (scan!! act-1 'copy!! :segment-pvar forwardb
                                        :include-self t))
    (*when (=!! type (!! 2))
        (*set act-2 (pref!! act-1 btransfer)))
    (*set weight
        (+!! weight
            (*!! learn-rate act-1 act-2)))))
```

Fig. 9. Hebbian Weight Update Kernel for the Composite Approach

It is immediately obvious that no looping is invloved. Any number of interconnects may be updated by the proper subselection. However, the more subselection is used the less efficient the computation becomes because less processors are invloved.

COMPLEXITY ANALYSIS

The performance results presented in the next section can be largely anticipated from an analysis of the space and time requirements of the CM implementation approaches. For simplicity I use a $R^n \to R^n$ model with Hebbian adaptation. The oder of magnitude requirements for activation and weight updating are compared for both CM implementation approaches and a basic serial matrix arithmetic approach.

For the given model, the space requirements on a conventional serial machine are $2n+n^2$ locations or $O(n^2)$. The growth of the space requirement is dominated by the nxn weight matrix defining the system interconnect structure. The edge list approach uses six pvars for each processor and uses nxn processors for the mapping, or $6n^2$ locations or $O(n^2)$. The composite approach uses 11 pvars. There are 2n processors for units and $2n^2$ processors for interconnects in the given model. The composite approach uses $11(2n+2n^2)$ locations or $O(n^2)$. The CM implementations take up roughly the same space as the serial implementation, but the space for the serial implementation is composed of passive memory whereas the space for the CM implementations is composed of interconnected processors with memory.

The time analysis for the approaches compares the time order of magnitudes to compute the activation update (1) and the Hebbian weight update (2). On a serial machine, the n weighted sums computed for the activation update require n^2 multiplications and $n(n-1)$ additions. There are $2n^2-n$ operations or time order of magnitude $O(n^2)$. The time order of magnitude for the weight matrix update is $O(n^2)$ since there are n^2 weight matrix elements.

The edge list approach forms n weighted sums by performing a parallel product of all of the weights and activations in the model, (*!! weight sact), and then a tree reduction sum, (*sum ...), of the products for the n units (see Fig. 4). There are $1+n(n\log_2 n)$ operations or time order of magnitude $O(n^2)$. This is the same order of magnitude as obtained on a serial machine. Further, the performance of the activation update is a function of the number of interconnects to be processed.

The composite approach forms n weighted sums in nine steps (see Fig. 8): five selection operations; the segmented copy scan before the parallel multiplication; the parallel multiplication; the parallel transfer of the products; and the segmented plus scan, which forms the n sums in one step. This gives the composite activation update a time order of magnitude $O(1)$. Performance is independent of the number of interconnects processed. The next section shows that this is not quite true.

The n^2 weights in the model can be updated in three parallel steps using the edge list approach (see Fig. 4). The n^2 weights in the model can be updated in eight parallel steps using the composite approach (see Fig. 9). In either case, the weight update operation has a time order of magnitude $O(1)$.

The time complexity results obtained for the composite approach apply to computation of the Lyaponov energy equation (3) and the Hopfield weighting prescription (4), given that pvar structures which can be scanned (see Figs. 1 and 6) are used. The same operations performed serially are time order of magnitude $O(n^2)$.

The above operations all incur a one time overhead cost for generating the addresses in the pointer pvars, used for parallel transfers, and arranging the values in segments for scanning. What the above analysis shows is that time complexity is traded for space complexity. The goal of CM programming is to use as many processors as possible at every step.

PERFORMANCE COMPARISON

Simulations of a Hopfield spin-glass model[2] were run for six different model sizes over the same number (16,384) of physical CM processors to provide a performance comparison between implementation approaches. The Hopfield network was chosen for the performance comparison because of its simple and well known convergence dynamics and because it uses a small set of pvars which allows a wide range of network sizes (degrees of virtualization) to be run. Twelve treaments are run. Six with the edge list approach and six with the composite approach. Table 3-1 shows the model sizes run for each treatment. Each treatment was run at the virtualization level just necessary to accomodate the number of processors required for each simulation.

Two exemplar patterns are stored. Five test patterns are matched against the two exemplars. Two test patterns have their centers removed, two have a row and column removed, and one is a random pattern. Each exemplar was hand picked and tested to insure that it did not produce cross-talk. The number of rows and columns in the exemplars and patterns increase as the size of the networks for the treatments increases.

Since the performance of the CM is at issue, rather than the performance of the network model used, a simple model and a simple pattern set were chosen to minimize consideration of the influence of model dynamics on performance.

Performance is presented by plotting execution speed versus model size. Size is measured by the number of interconnects in a model. The execution speed metric is interconnects updated per second, $N*(N-1)/t$, where N is the number of units in a model and t is the time used to update the activations for all of the units in a model. All of the units were updated three times for each pattern . Convergence was determined by the output activation remaining stable over the final two updates. The value of t for a treatment is the average of 15 samples of t. Fig. 10 shows the activation update cycle time for both approaches. Fig. 11 shows the interconnect update speed plots for both approaches. The edge list approach is plotted in black. The composite approach is plotted in white. The performance shown excludes overhead for interpretation of the *Lisp instructions. The model size categories for each plot correspond to the model sizes and levels of CM virtualization shown in Table I.

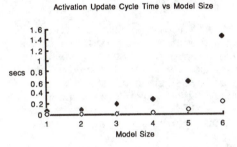

Fig. 10. Activation Update Cycle Times

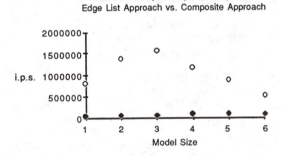

Fig. 11. Edge List Interconnect Update Speeds

Fig. 11 shows an order of magnitude performance difference between the approaches and a roll off in performance for each approach as a function of the number of virtual processors supported by each physical processor. The performance turn around is at 4x virtualization for the edge list approach and 2x virtualization for the composite approach.

CONCLUSIONS

Representing the interconnect structure of neural network models with mappings defined over the set of fine grain processors provided by the CM architecture provides good performance for a modest programming effort utilizing only a small subset of the instructions provided by *Lisp. Further, the performance will continue to scale up linearly as long as not more than 2x virtualization is required. While the complexity analysis of the composite activation update suggests that its performance should be independent of the number of interconnects to be processed, the performance results show that the performance is indirectly dependent on the number of interconnects to be processed because the level of virtualization required (after the physical processors are exhausted) is dependent on the number of interconnects to be processed and virtualization decreases performance linearly. The complexity analysis of the edge list activation update shows that its performance should be roughly the same as serial implementations on comparable machines. The results suggest that the composite approach is to be prefered over the edge list approach but not be used at a virtualization level higher than 2x.

The mechanism of the composite activation update suggest that hierarchical networks simulated in this fashion will compare in performance to single layer networks because the parallel transfers provide a type of pipeline for activation for synchronously updated hierarchical networks while providing simultaneous activation transfers for asynchronously updated single layer networks. Researchers at Thinking Machines Corporation and the M.I.T. AI Laboratory in Cambridge Mass. use a similar approach for an implementation of NETtalk. Their approach overlaps the weights of connected units and simultaneously pipelines activation forward and error backward.[3]

Performance better than that presented can be gained by translation of the control code from interpreted *Lisp to PARIS and use of the CM2. In addition to not being interpreted, PARIS allows explicit control over important registers that aren't accessable through *Lisp. The CM2 will offer a number of new features which will enhance performance of neural network simulations: a *Lisp compiler, larger processor memory (64K), and floating point processors. The complier and floating point processors will increase execution speeds while the larger processor memories will provide a larger number of virtual processors at the performance turn around points allowing higher performance through higher CM utilization.

REFERENCES

1. "Introduction to Data Level Parallelism," Thinking Machines Technical Report 86.14, (April 1986).

2. Hopfield, J. J., "Neural networks and physical systems with emergent collective computational abilities," Proc. Natl. Acad. Sci., Vol. 79, (April 1982), pp. 2554-2558.

3. Blelloch, G. and Rosenberg, C. Network Learning on the Connection Machine, M.I.T. Technical Report, 1987.

On the Power of Neural Networks for Solving Hard Problems

Jehoshua Bruck
Joseph W. Goodman
Information Systems Laboratory
Department of Electrical Engineering
Stanford University
Stanford, CA 94305

Abstract

This paper deals with a neural network model in which each neuron performs a threshold logic function. An important property of the model is that it always converges to a stable state when operating in a serial mode [2,5]. This property is the basis of the potential applications of the model such as associative memory devices and combinatorial optimization [3,6].

One of the motivations for use of the model for solving hard combinatorial problems is the fact that it can be implemented by optical devices and thus operate at a higher speed than conventional electronics.

The main theme in this work is to investigate the power of the model for solving NP-hard problems [4,8], and to understand the relation between speed of operation and the size of a neural network. In particular, it will be shown that for any NP-hard problem the existence of a polynomial size network that solves it implies that NP=co-NP. Also, for Traveling Salesman Problem (TSP), even a polynomial size network that gets an ϵ-approximate solution does not exist unless P=NP.

The above results are of great practical interest, because right now it is possible to build neural networks which will operate fast but are limited in the number of neurons.

1 Background

The neural network model is a discrete time system that can be represented by a weighted and undirected graph. There is a weight attached to each edge of the graph and a threshold value attached to each node (neuron) of the graph.

The *order* of the network is the number of nodes in the corresponding graph. Let N be a neural network of order n; then N is uniquely defined by (W, T) where:

- W is an $n \times n$ symmetric matrix, W_{ij} is equal to the weight attached to edge (i, j).

- T is a vector of dimension n, T_i denotes the threshold attached to node i.

Every node (neuron) can be in one of two possible states, either 1 or -1. The state of node i at time t is denoted by $V_i(t)$. The *state* of the neural network at time t is the vector $V(t)$.

The next state of a node is computed by:

$$V_i(t+1) = sgn(H_i(t)) = \begin{cases} 1 & \text{if } H_i(t) \geq 0 \\ -1 & \text{otherwise} \end{cases} \tag{1}$$

where

$$H_i(t) = \sum_{j=1}^{n} W_{ji} V_j(t) - T_i$$

The next state of the network, i.e. $V(t+1)$, is computed from the current state by performing the evaluation (1) at a subset of the nodes of the network, to be denoted by S. The modes of operation are determined by the method by which the set S is selected in each time interval. If the computation is performed at a single node in any time interval, i.e. $\mid S \mid = 1$, then we will say that the network is operating in a *serial* mode; if $\mid S \mid = n$ then we will say that that the network is operating in a *fully parallel* mode. All the other cases, i.e. $1 < \mid S \mid < n$ will be called *parallel* modes of operation. The set S can be chosen at random or according to some deterministic rule.

A state $V(t)$ is called *stable* iff $V(t) = sgn(WV(t) - T)$, i.e. there is no change in the state of the network no matter what the mode of operation is. One of the most important properties of the model is the fact that it always converges to a stable state while operating in a serial mode. The main idea in the proof of the convergence property is to define a so called *energy function* and to show that this energy function is nondecreasing when the state of the network changes. The energy function is:

$$E(t) = V^T(t)WV(t) - 2V^T(t)T \tag{2}$$

An important note is that originally the energy function was defined such that it is nonincreasing [5]; we changed it such that it will comply with some known graph problems (e.g. Min Cut).

A neural network will always get to a stable state which corresponds to a local maximum in the energy function. This suggests the use of the network as a device for performing a local search algorithm for finding a maximal value of the energy function [6]. Thus, the network will perform a local search by operating in a random and serial mode. It is also known [2,9] that maximization of E associated with a given network N in which $T = 0$ is equivalent to finding the Minimum Cut in N. Actually, many hard problems can be formulated as maximization of a quadratic form (e.g. TSP [6]) and thus can be mapped to a neural network.

2   The Main Results

The set of stable states is the set of possible final solutions that one will get using the above approach. These final solutions correspond to local maxima of the energy function but do not necessarily correspond to global optima of the corresponding problem. The main question is: suppose we allow the network to operate for a very long time until it converges; can we do better than just getting some local optimum? i.e., is it possible to design a network which will always find the exact solution (or some guaranteed approximation) of the problem?

Definition: Let X be an instance of problem. Then $| X |$ denotes the size of X, that is, the number of bits required to represent X. For example, for X being an instance of TSP, $| X |$ is the number of bits needed to represent the matrix of the distances between cities.

Definition: Let N be a neural network. Then $| N |$ denotes the size of the network N. Namely, the number of bits needed to represent W and T.

Let us start by defining the desired setup for using the neural network as a model for solving hard problems.

Consider an optimization problem L, we would like to have for every instance X of L a neural network N_X with the following properties:

- Every local maximum of the energy function associated with N_X corresponds to a global optimum of X.

- The network N_X is small, that is, $| N_X |$ is bounded by some polynomial in $| X |$.

Moreover, we would like to have an algorithm, to be denoted by A_L, which given an instance $X \in L$, generates the description for N_X in polynomial (in $| X |$) time.

Now, we will define the desired setup for using the neural network as a model for finding approximate solutions for hard problems.

Definition: Let E_{glo} be the global maximum of the energy function. Let E_{loc}

be a local maximum of the energy function. We will say that a local maximum is an ϵ-approximate of the global iff:

$$\frac{E_{glo} - E_{loc}}{E_{glo}} \leq \epsilon$$

The setup for finding approximate solutions is similar to the one for finding exact solutions. For $\epsilon \geq 0$ being some fixed number. We would like to have a network N_{X_ϵ} in which every local maximum is an ϵ-approximate of the global and that the global corresponds to an optimum of X. The network N_{X_ϵ} should be small, namely, $\mid N_{X_\epsilon} \mid$ should be bounded by a polynomial in $\mid X \mid$. Also, we would like to have an algorithm A_{L_ϵ}, such that, given an instance $X \in L$, it generates the description for N_{X_ϵ} in polynomial (in $\mid X \mid$) time.

Note that in both the exact case and the approximate case we do not put any restriction on the time it takes the network to converge to a solution (it can be exponential).

At this point the reader should convince himself that the above description is what he imagined as the setup for using the neural network model for solving hard problems, because that is what the following definition is about.

Definition: We will say that a neural network for solving (or finding an ϵ-approximation of) a problem L exists if the algorithm A_L (or A_{L_ϵ}) which generates the description of N_X (or N_{X_ϵ}) exists.

The main results in the paper are summarized by the following two propositions. The first one deals with exact solutions of NP-hard problems while the second deals with approximate solutions to TSP.

Proposition 1 *Let L be an NP-hard problem. Then the existence of a neural network for solving L implies that NP = co-NP.*

Proposition 2 *Let $\epsilon \geq 0$ be some fixed number. The existence of a neural network for finding an ϵ-approximate solution to TSP implies that P=NP.*

Both (P=NP) and (NP=co-NP) are believed to be false statements, hence, we can not use the model in the way we imagine.

The key observation for proving the above propositions is the fact that a single iteration in a neural network takes time which is bounded by a polynomial in the size of the instance of the corresponding problem. The proofs of the above two propositions follow directly from known results in complexity theory and should not be considered as new results in complexity theory.

3 The Proofs

Proof of Proposition 1: The proof follows from the definition of the classes NP and co-NP, and Lemma 1. The definitions and the lemma appear in Chapters 15 and 16 in [8] and also in Chapters 2 and 7 in [4].

Lemma 1 *If the complement of an NP-complete problem is in NP, then NP=co-NP.*

Let L be an NP-hard problem. Suppose there exists a neural network that solves L. Let $\hat{L}$ be an NP-complete problem. By definition, $\hat{L}$ can be polynomialy reduced to L. Thus, for every instance $X \in \hat{L}$, we have a neural network such that from any of its global maxima we can efficiently recognize whether X is a 'yes' or a 'no' instance of $\hat{L}$.

We claim that we have a nondeterministic polynomial time algorithm to decide that a given instance $X \in \hat{L}$ is a 'no' instance. Here is how we do it: for $X \in \hat{L}$ we construct the neural network that solves it by using the reduction to L. We then check every state of the network to see if it is a local maximum (that is done in polynomial time). In case it is a local maximum, we check if the instance is a 'yes' or a 'no' instance (this is also done in polynomial time).

Thus, we have a nondeterministic polynomial time algorithm to recognize any 'no' instance of $\hat{L}$. Thus, the complement of the problem $\hat{L}$ is in NP. But $\hat{L}$ is an NP-complete problem, hence, from Lemma 1 it follows that NP=co-NP. $\square$

Proof of Proposition 2: The result is a corollary of the results in [7], the reader can refer to it for a more complete presentation.

The proof uses the fact that the Restricted Hamiltonian Circuit (RHC) is an NP-complete problem.

Definiton of RHC: Given a graph $G = (V, E)$ and a Hamiltonian path in G. The question is whether there is a Hamiltonian circuit in G?

It is proven in [7] that RHC is NP-complete.

Suppose there exists a polynomial size neural network for finding an ϵ-approximate solution to TSP. Then it can be shown that an instance $X \in RHC$ can be reduced to an instance $\hat{X} \in TSP$, such that in the network $N_{\hat{X}_\epsilon}$ the following holds: if the Hamiltonian path that is given in X corresponds to a local maximum in $N_{\hat{X}_\epsilon}$ then X is a 'no' instance; else, if it does not correspond to a local maximum in $N_{\hat{X}_\epsilon}$ then X is a 'yes' instance. Note that we can check for locality in polynomial time.

Hence, the existence of $N_{\hat{X}_\epsilon}$ for all $\hat{X} \in TSP$ implies that we have a polynomial time algorithm for RHC. $\square$

4 Concluding Remarks

1. In Proposition 1 we let $| W |$ and $| T |$ be arbitrary but bounded by a polynomial in the size of a given instance of a problem. If we assume that $| W |$ and $| T |$ are fixed for all instances then a similar result to Proposition 1 can be proved without using complexity theory; this result appears in [1].

2. The network which corresponds to TSP, as suggested in [6], can not solve the TSP with guaranteed quality. However, one should note that all the analysis in this paper is a worst case type of analysis. So, it might be that there exist networks that have good behavior on the average.

3. Proposition 1 is general to all NP-hard problems while Proposition 2 is specific to TSP. Both propositions hold for any type of networks in which an iteration takes polynomial time.

4. Clearly, every network has an algorithm which is equivalent to it, but an algorithm does not necessarily have a corresponding network. Thus, if we do not know of an algorithmic solution to a problem we also will not be able to find a network which solves the problem. If one believes that the neural network model is a good model (e.g. it is amenable to implementation with optics), one should develop techniques to program the network to perform an algorithm that is known to have some guaranteed good behavior.

Acknowledgement: Support of the U.S. Air Force Office of Scientific Research is gratefully acknowledged.

References

[1] Y. Abu Mostafa, *Neural Networks for Computing?* in Neural Networks for Computing, edited by J. Denker (AIP Conference Proceedings no. 151, 1986).

[2] J. Bruck and J. Sanz, *A Study on Neural Networks*, IBM Tech Rep, RJ 5403, 1986. To appear in International Journal of Intelligent Systems, 1988.

[3] J. Bruck and J. W. Goodman, *A Generalized Convergence Theorem for Neural Networks and its Applications in Combinatorial Optimization*, IEEE First ICNN, San-Diego, June 1987.

[4] M. R. Garey and D. S. Johnson, *Computers and Intractability: A Guide to the Theory of NP-Completeness*, W. H. Freeman and Company, 1979.

[5] J. J. Hopfield, *Neural Networks and Physical Systems with Emergent Collective Computational Abilities*, Proc. Nat. Acad. Sci. . USA, Vol. 79, pp. 2554-2558, 1982.

[6] J. J. Hopfield and D. W. Tank, *Neural Computations of Decisions in Optimization Problems*, Biol. Cybern. 52, pp. 141-152, 1985.

[7] C. H. Papadimitriou and K. Steiglitz, *On the Complexity of Local Search for the Traveling Salesman Problem*, SIAM J. on Comp., Vol. 6, No. 1, pp. 76-83, 1977.

[8] C. H. Papadimitriou and K. Steiglitz, *Combinatorial Optimization: Algorithms and Complexity*, Prentice-Hall, Inc., 1982.

[9] J. C. Picard and H. D. Ratliff, *Minimum Cuts and Related Problems*, Networks, Vol 5, pp. 357-370, 1974.

SPEECH RECOGNITION EXPERIMENTS
WITH PERCEPTRONS

D. J. Burr
Bell Communications Research
Morristown, NJ 07960

ABSTRACT

Artificial neural networks (ANNs) are capable of accurate recognition of simple speech vocabularies such as isolated digits [1]. This paper looks at two more difficult vocabularies, the alphabetic E-set and a set of polysyllabic words. The E-set is difficult because it contains weak discriminants and polysyllables are difficult because of timing variation. Polysyllabic word recognition is aided by a time pre-alignment technique based on dynamic programming and E-set recognition is improved by focusing attention. Recognition accuracies are better than 98% for both vocabularies when implemented with a single layer perceptron.

INTRODUCTION

Artificial neural networks perform well on simple pattern recognition tasks. On speaker trained spoken digits a layered network performs as accurately as a conventional nearest neighbor classifier trained on the same tokens [1]. Spoken digits are easy to recognize since they are for the most part monosyllabic and are distinguished by strong vowels.

It is reasonable to ask whether artificial neural networks can also solve more difficult speech recognition problems. Polysyllabic recognition is difficult because multi-syllable words exhibit large timing variation. Another difficult vocabulary, the alphabetic E-set, consists of the words B, C, D, E, G, P, T, V, and Z. This vocabulary is hard since the distinguishing sounds are short in duration and low in energy.

We show that a simple one-layer perceptron [7] can solve both problems very well if a good input representation is used and sufficient examples are given. We examine two spectral representations — a smoothed FFT (fast Fourier transform) and an LPC (linear prediction coefficient) spectrum. A time stabilization technique is described which pre-aligns speech templates based on peaks in the energy contour. Finally, by focusing attention of the artificial neural network to the beginning of the word, recognition accuracy of the E-set can be consistently increased.

A layered neural network, a relative of the earlier perceptron [7], can be trained by a simple gradient descent process [8]. Layered networks have been applied successfully to speech recognition [1], handwriting recognition [2], and to speech synthesis [11]. A variation of a layered network [3] uses feedback to model causal constraints, which can be useful in learning speech and language. Hidden neurons within a layered network are the building blocks that are used to form solutions to specific problems. The number of hidden units required is related to the problem [1,2]. Though a single hidden layer can form any mapping [12], no more than two layers are needed for disjunctive normal form [4]. The second layer may be useful in providing more stable learning and representation in the presence of noise. Though neural nets have been shown to perform as well as conventional techniques[1,5], neural nets may do better when classes have outliers [5].

PERCEPTRONS

A simple perceptron contains one input layer and one output layer of neurons directly connected to each other (no hidden neurons). This is often called a one-layer system, referring to the single layer of weights connecting input to output. Figure 1. shows a one-layer perceptron configured to sense speech patterns on a two-dimensional grid. The input consists of a 64-point spectrum at each of twenty time slices. Each of the 1280 inputs is connected to each of the output neurons, though only a sampling of connections are shown. There is one output neuron corresponding to each pattern class. Neurons have standard linear-weighted inputs with logistic activation.

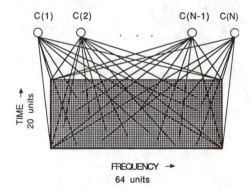

Figure 1. A single layer perceptron sensing a time-frequency array of sample data. Each output neuron $C(i)$ $(1 \leq i \leq N)$ corresponds to a pattern class and is full connected to the input array (for clarity only a few connections are shown).

An input word is fit to the grid region by applying an automatic endpoint detection algorithm. The algorithm is a variation of one proposed by Rabiner and Sambur [9] which employs a double threshold successive approximation method. Endpoints are determined by first detecting threshold crossings of energy and then of zero crossing rate. In practice a level crossing other than zero is used to prevent endpoints from being triggered by background sounds.

INPUT REPRESENTATIONS

Two different input representations were used in this study. The first is a Fourier representation smoothed in both time and frequency. Speech is sampled at 10 KHz and Hamming windowed at a number of sample points. A 128-point FFT spectrum is computed to produce a template of 64 spectral samples at each of twenty time frames. The template is smoothed twice with a time window of length three and a frequency window of length eight.

For comparison purposes an LPC spectrum is computed using a tenth order model on 300-sample Hamming windows. Analysis is performed using the autocorrelation method with Durbin recursion [6]. The resulting spectrum is smoothed over three time frames.

Sample spectra for the utterance "neural-nets" is shown in Figure 2. Notice the relative smoothness of the LPC spectrum which directly models spectral peaks.

FFT   LPC

Figure 2. FFT and LPC time-frequency plots for the utterance "neural nets". Time is toward the left, and frequency, toward the right.

DYNAMIC TIME ALIGNMENT

Conventional speech recognition systems often employ a time normalization technique based on dynamic programming [10]. It is used to warp the time scales of two utterances to obtain optimal alignment between their spectral frames. We employ a variation of dynamic programming which aligns energy contours rather than spectra. A reference energy template is chosen for each pattern class, and incoming patterns are warped onto it. Figure 3 shows five utterances of "neural-nets" both before and after time alignment. Notice the improved alignment of energy peaks.

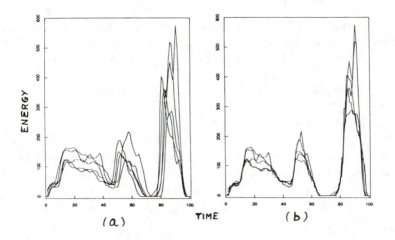

Figure 3. (a) Superimposed energy plots of five different utterances of "neural nets". (b). Same utterances after dynamic time alignment.

POLYSYLLABLE RECOGNITION

Twenty polysyllabic words containing three to five syllables were chosen, and five tokens of each were recorded by a single male speaker. A variable number of tokens were used to train a simple perceptron to study the effect of training set size on performance. Two performance measures were used: classification accuracy, and an RMS error measure. Training tokens were permuted to obtain additional experimental data points.

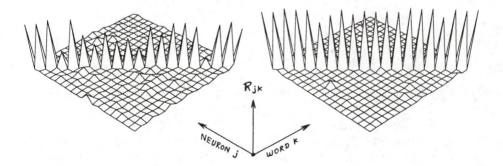

Figure 4. Output responses of a perceptron trained with one token per class (left) and four tokens per class (right).

Figure 4 shows two representative perspective plots of the output of a perceptron trained on one and four tokens respectively per class. Plots show network response (z-coordinate) as a function of output node (left axis) and test word index (right axis). Note that more training tokens produce a more ideal map — a map should have ones along the diagonal and zeroes everywhere else.

Table 1 shows the results of these experiments for three different representations: (1) FFT, (2) LPC and (3) time aligned LPC. This table lists classification accuracy as a function of number of training tokens and input representation. The perceptron learned to classify the unseen patterns perfectly for all cases except the FFT with a single training pattern.

Table 1. Polysyllabic Word Recognition Accuracy				
Number Training Tokens	1	2	3	4
FFT	98.7%	100%	100%	100%
LPC	100%	100%	100%	100%
Time Aligned LPC	100%	100%	100%	100%
Permuted Trials	400	300	200	100

A different performance measure, the RMS error, evaluates the degree to which the trained network output responses R_{jk} approximate the ideal targets T_{jk}. The measure is evaluated over the N non-trained tokens and M output nodes of the network. T_{jk} equals 1 for $j=k$ and 0 for $j \neq k$.

$$RMS\ Error = \frac{\left[\sum_{j=1}^{M} \sum_{k=1}^{N} (T_{jk} - R_{jk})^2 \right]^{1/2}}{MN}$$

Figure 5 shows plots of RMS error as a function of input representation and training patterns. Note that the FFT representation produced the highest error, LPC was about 40% less, and time-aligned LPC only marginally better than non-aligned LPC. In a situation where many choices must be made (i.e. vocabularies much larger than 20 words) LPC is the preferred choice, and time alignment could be useful to disambiguate similar words. Increased number of training tokens results in improved performance in all cases.

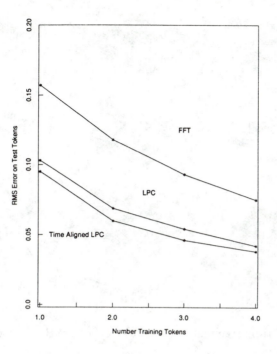

Figure 5. RMS error versus number of training tokens for various input representations.

E-SET VOCABULARY

The E-Set vocabulary consists of the nine E-words of the English alphabet — B, C, D, E, G, P, T, V, Z. Twenty tokens of each of the nine classes were recorded by a single male speaker. To maximize the sizes of training and test sets, half were used for training and the other half for testing. Ten permutations produced a total of **900** separate recognition trials.

Figure 6 shows typical LPC templates for the nine classes. Notice the double formant ridge due to the "E" sound, which is common to all tokens. Another characteristic feature is the F0 ridge — the upward fold on the left of all tokens which characterizes voicing or pitched sound.

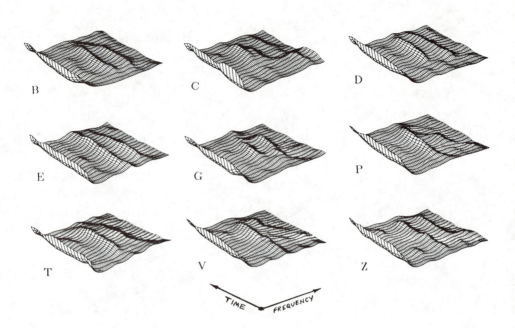

Figure 6. LPC time-frequency plots for representative tokens of the E-set words.

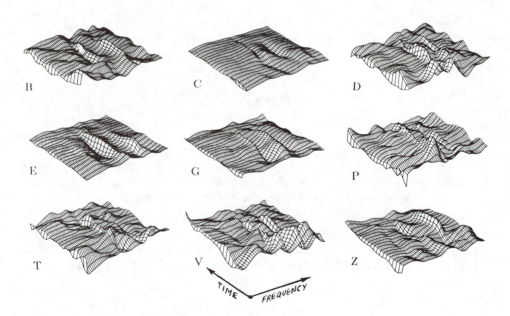

Figure 7. Time-frequency plots of weight values connected to each output neuron "B" through "Z" in a trained perceptron.

Figure 7 shows similar plots illustrating the weights learned by the network when trained on ten tokens of each class. These are plotted like spectra, since one weight is associated with each spectral sample. Note that the patterns have some formant structure. A recognition accuracy of 91.4% included perfect scores for classes C, E, and G.

Notice that weights along the F0 contour are mostly small and some are slightly negative. This is a response to the voiced "E" sound common to all classes. The network has learned to discount "voicing" as a discriminator for this vocabulary.

Notice also the strong "hilly" terrain near the beginning of most templates. This shows where the network has decided to focus much of its discriminating power. Note in particular the hill-valley pair at the beginning of "P" and "T". These are near to formants F2/F3 and could conceivably be formant onset detectors. Note the complicated detector pattern for the "V" sound.

The classes that are easy to discriminate (C, E, G) produce relatively flat and uninteresting weight spaces. A highly convoluted weight space must therefore be correlated with difficulty in discrimination. It makes little sense however that the network should be working hard in the late time ("E" sound) portion of the utterance. Perhaps additional training might reduce this activity, since the network would eventually find little consistent difference there.

A second experiment was conducted to help the network to focus attention. The first k frames of each input token were averaged to produce an average spectrum. These average spectra were then used in a simple nearest neighbor recognizer scheme. Recognition accuracy was measured as a function of k. The highest performance was for k=8, indicating that the first 40% of the word contained most of the "action".

	B	C	D	E	G	P	T	V	Z
B	98	0	1	0	0	0	0	1	0
C	0	100	0	0	0	0	0	0	0
D	0	0	98	0	0	2	0	0	0
E	0	0	0	100	0	0	0	0	0
G	0	0	0	0	100	0	0	0	0
P	0	0	3	0	0	93	4	0	0
T	0	0	0	0	0	0	100	0	0
V	2	0	0	0	0	2	0	96	0
Z	0	0	0	0	0	0	0	1	99

Figure 8. Confusion matrix of the E-set focused on the first 40% of each word.

All words were resampled to concentrate 20 time frames into the first 40% of the word. LPC spectra were recomputed using a 16th order model and the network was trained on the new templates. Performance increased from 91.4% to 98.2%. There were only 16 classification errors out of the 900 recognition tests. The confusion matrix is shown in Figure 8. Learning times for all experiments consisted of about ten passes through the training set. When weights were primed with average spectral values rather than random values, learning time decreased slightly.

CONCLUSIONS

Artificial neural networks are capable of high performance in pattern recognition applications, matching or exceeding that of conventional classifiers. We have shown that for difficult speech problems such as time alignment and weak discriminability, artificial neural networks perform at high accuracy exceeding 98%. One-layer perceptrons learn these difficult tasks almost effortlessly — not in spite of their simplicity, but because of it.

REFERENCES

1. D. J. Burr, "A Neural Network Digit Recognizer", Proceedings of IEEE Conference on Systems, Man, and Cybernetics, Atlanta, GA, October, 1986, pp. 1621-1625.

2. D. J. Burr, "Experiments with a Connectionist Text Reader," IEEE International Conference on Neural Networks, San Diego, CA, June, 1987.

3. M. I. Jordan, "Serial Order: A Parallel Distributed Processing Approach," ICS Report 8604, UCSD Institute for Cognitive Science, La Jolla, CA, May 1986.

4. S. J. Hanson, and D. J. Burr, "What Connectionist Models Learn: Toward a Theory of Representation in Multi-Layered Neural Networks," submitted for publication.

5. W. Y. Huang and R. P. Lippmann, "Comparisons Between Neural Net and Conventional Classifiers," IEEE International Conference on Neural Networks, San Diego, CA, June 21-23, 1987.

6. J. D. Markel and A. H. Gray, Jr., Linear Prediction of Speech, Springer-Verlag, New York, 1976.

7. M. L. Minsky and S. Papert, Perceptrons, MIT Press, Cambridge, Mass., 1969.

8. D. E. Rumelhart, G. E. Hinton, and R. J. Williams, "Learning Internal Representations by Error Propagation," in Parallel Distributed Processing, Vol. 1, D. E. Rumelhart and J. L. McClelland, eds., MIT Press, 1986, pp. 318-362.

9. L. R. Rabiner and M. R. Sambur, "An Algorithm for Determining the End-points of Isolated Utterances," BSTJ, Vol. 54, 297-315, Feb. 1975.

10. H. Sakoe and S. Chiba, "Dynamic Programming Optimization for Spoken Word Recognition," IEEE Trans. Acoust., Speech, Signal Processing, Vol. ASSP-26, No. 1, 43-49, Feb. 1978.

11. T. J. Sejnowski and C. R. Rosenberg, "NETtalk: A Parallel Network that Learns to Read Aloud," Technical Report JHU/EECS-86/01, Johns Hopkins University Electrical Engineering and Computer Science, 1986.

12. A. Wieland and R. Leighton, "Geometric Analysis of Neural Network Capabilities," IEEE International Conference on Neural Networks, San Deigo, CA, June 21-24, 1987.

PRESYNAPTIC NEURAL INFORMATION PROCESSING

L. R. Carley
Department of Electrical and Computer Engineering
Carnegie Mellon University, Pittsburgh PA 15213

ABSTRACT

The potential for presynaptic information processing within the arbor of a single axon will be discussed in this paper. Current knowledge about the activity dependence of the firing threshold, the conditions required for conduction failure, and the similarity of nodes along a single axon will be reviewed. An electronic circuit model for a site of low conduction safety in an axon will be presented. In response to single frequency stimulation the electronic circuit acts as a lowpass filter.

I. INTRODUCTION

The axon is often modeled as a wire which imposes a fixed delay on a propagating signal. Using this model, neural information processing is performed by synaptically summing weighted contributions of the outputs from other neurons. However, substantial information processing may be performed in by the axon itself. Numerous researchers have observed periodic conduction failures at normal physiological impulse activity rates (e.g., in cat[1], in frog[2], and in man[3]). The oscillatory nature of these conduction failures is a result of the dependence of the firing threshold on past impulse conduction activity.

The simplest view of axonal (presynaptic) information processing is as a switch: the axon will either conduct an impulse or not. The state of the switch depends on how past impulse activity modulates the firing threshold, which will result in conduction failure if firing threshold is bigger than the incoming impulse strength. In this way, the connectivity of a synaptic neural network could be modulated by past impulse activity at sites of conduction failure within the network. More sophisticated presynaptic neural information processing is possible when the axon has more than one terminus, implying the existence of branch points within the axon. Section II will present a general description of potential for presynaptic information processing.

The after–effects of previous activity are able to vary the connectivity of the axonal arbor at sites of low conduction safety according to the temporal pattern of the impulse train at each site (Raymond and Lettvin, 1978; Raymond, 1979). In order to understand the information processing potential of presynaptic networks it is necessary to study the after–effects of activity on the firing threshold. Each impulse is normally followed by a brief refractory period (about 10 ms in frog sciatic nerve) of increased threshold and a longer superexcitable period (about 1 s in frog sciatic nerve) during which the threshold is actually below its resting level. During prolonged periods of activity, there is a gradual increase in firing threshold which can persist long (> 1 hour in frog nerve) after cessation of impulse activity (Raymond and Lettvin, 1978). In section III, the methods used to measure the firing threshold and the after—effects of activity will be presented.

In addition to understanding how impulse activity modulates sites of low conduction safety, it is important to explore possible constraints on the distribution of sites of low conduction safety within the axon's arbor. Section IV presents results from a study of the distribution of the after—effects of activity along an axon.

Section V presents an electronic circuit model for a region of low conduction safety within an axonal arbor. It has been designed to have a firing threshold that depends on the past activity in a manner similar to the activity dependence measured for frog sciatic nerve.

II. PRESYNAPTIC SIGNAL PROCESSING

Conduction failure has been observed in many different organisms, including man, at normal physiological activity rates.[1,2,3] The after—effects of activity can "modulate" conduction failures at a site of low conduction safety. One common place where the conduction safety is low is at branch points where an impedance mismatch occurs in the axon.

In order to clarify the meaning of presynaptic information processing, a simple example is in order. Parnas reported that in crayfish a single axon separately activates the medial (DEAM) and lateral (DEAL) branches of the deep abdominal extensor muscles.[4,5] At low stimulus frequencies (below 40—50 Hz) impulses travel down both branches; however, each impulse evokes much smaller contractions in DEAL than in DEAM resulting in contraction of DEAM without significant contraction of DEAL. At higher stimulus frequencies conduction in the branch leading to DEAM fails and DEAL contracts without DEAM contracting. Both DEAL and DEAM can be stimulated separately by stimulus patterns more complicated than a single frequency.

The theory of "fallible trees", which has been discussed by Lettvin, McCulloch and Pitts, Raymond, and Waxman and Grossman among others, suggests that one axon which branches many times forms an information processing element with one input and many outputs. Thus, the after—effects of previous activity are able to vary the connectivity of the axonal arbor at regions of low conduction safety according to the temporal pattern of the impulse train in each branch. The transfer function of the fallible tree is determined by the distribution of sites of low conduction safety and the distribution of superexcitability and depressibility at those sites. Thus, a single axon with 1000 terminals can potentially be in 2^{1000} different states as a function of the locations of sites of conduction failure within the axonal arbor. And, each site of low conduction safety is modulated by the past impulse activity at that site.

Fallible trees have a number of interesting properties. They can be used to cause different input frequencies to excite different axonal terminals. Also, fallible trees, starting at rest, will preserve timing information in the input signal; i.e., starting from rest, all branches will respond to the first impulse.

III. AFTER–EFFECTS OF ACTIVITY

In this section, the firing threshold will be defined and an experimental method for its measurement will be described. In addition, the after-effects of activity will be characterized and typical results of the characterization process will be given.

The following method was used to measure the firing threshold. Whole nerves were placed in the experimental setup (shown in figure 1). The whole nerve fiber was stimulated with a gross electrode. The response from a single axon was recorded using a suction microelectrode. Firing threshold was measured by applying test stimuli through the gross stimulating electrode and looking for a response in the suction microelectrode.

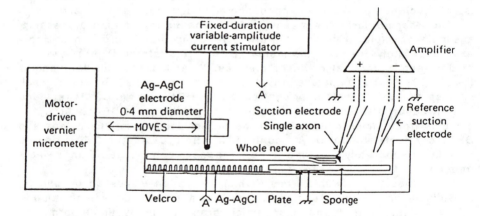

Figure 1. Drawing of the experimental recording chamber.

Threshold Hunting, a process for choosing the test stimulus strength, was used to characterize the axons.[6] It uses the following paradigm. A test stimulus which fails to elicit a conducting impulse causes a small increase the strength of subsequent test stimuli. A test stimulus which elicits an impulse causes a small decrease in the strength of subsequent test stimuli. Conditioning Stimuli, ones large enough to guarantee firing an impulse, can be interspersed between test stimuli in order to achieve a controlled overall activity rate. Rapid variations in threshold following one or more conditioning impulses can be measured by slowly increasing the time delay between the conditioning stimuli and the test stimulus. Several phases follow each impulse. First, there is a refractory period of short duration (about 10ms in frog nerve) during which another impulse cannot be initiated. Following the refractory period the axon actually becomes more excitable than at rest for a period (ranging from 200ms to 1s in frog nerve, see figure 2). The superexcitable period is measured by applying a conditioning stimulus and then delaying by a gradually increasing time delay and applying a test stimulus (see figure 3). There is only a slight increase in the peak of the superexcitable period following multiple impulses.[7] The superexcitability of an axon was characterized by the % decrease of the threshold from its resting level at the peak of the superexcitable period.

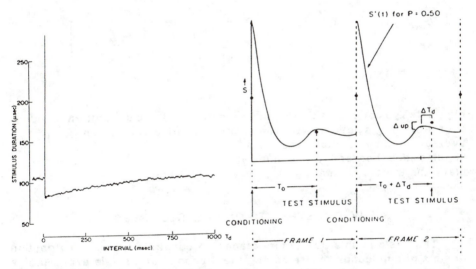

Figure 2. Typical superexcitable period in axon from frog sciatic nerve.

Figure 3. Stimulus pattern used for measuring superexcitability.

During a period of repetitive impulse conduction, the firing threshold may gradually increase. After the period of increased impulse activity ends, the threshold gradually recovers from its maximum over the course of several minutes or more with complete return of the threshold to its resting level taking as long as an hour or two (in frog nerve) depending on the extent of the preceding impulse activity. The depressibility of an axon can be characterized by the initial upward slope of the depression and the time

158

constant of the recovery phase (see figure 4). The pattern of conditioning and test stimuli used to generate the curve in figure 4 is shown in figure 5.

Depression may be correlated with microanatomical changes which occur in the glial cells in the nodal region during periods of increased activity.[8] During periods of repetitive stimulation the size and number of extracellular paranodal intramyelinic vacuoles increases causing changes in the paranodal geometry.

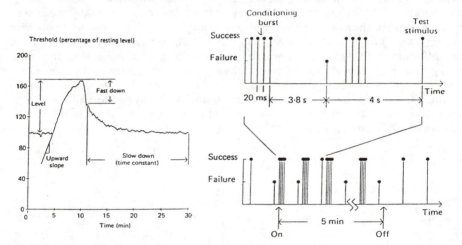

Figure 4. Typical depression in an axon from frog sciatic nerve. The average activity rate was 4 impulses/sec between the 5 min mark and the 10 min mark.

Figure 5. Stimulus pattern used for measuring depression.

IV. CONSTRAINTS ON FALLIBLE TREES

The basic fallible tree theory places no constraints on the distribution of sites of conduction failure among the branches of a single axon. In this section one possible constraint on the distribution of sites of conduction failure will be presented. Experiments have been performed in an attempt to determine if the extremely wide variations in superexcitability and depressibility found between nodes from different axons in a single nerve[9] (particularly for depressibility) also occur between nodes from the same axon.

A study of the distribution of the after—effects of activity along an unbranching length of frog sciatic nerve found only small variations in the after—effects along a single axon.[10] Both superexcitability and depressibility were extremely consistent for nodes from along a single unbranching length of axon (see figures 6 and 7). This suggests that there may be a cell—wide regulatory system that maintains the depressibility and superexcitability at comparable levels throughout the extent of the axon. Thus, portions of a fallible tree which have the same axon diameter would be expected to have the same superexcitability and depressibility.

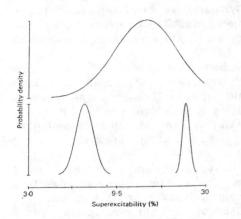
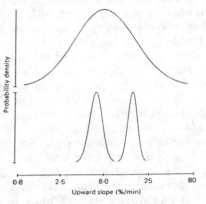

Figure 6. PDF of Superexcitability. The upper trace represents the PDF of the entire population of nodes studied and the two lower traces represent the separate populations of nodes from two different axons.

Figure 7. PDF of Depressibility. The upper trace represents the PDF of the entire population of nodes studied and the two lower traces represent the separate populations of nodes from two different axons.

This study did not examine axons which branched, therefore it cannot be concluded that superexcitability and depressibility must remain constant throughout a fallible tree. For example, it is quite likely that the cell actually regulates quantities like pump-site density, not depressibility. In that case, daughter branches of smaller diameter might be expected to show consistently higher depressibility. Further research is needed to determine how the activity dependence of the threshold scales with axon diameter along a single axon before the consistency of the after-effects along an unbranching axon can be used as a constraint on presynaptic information processing networks.

V. ELECTRICAL AXON CIRCUIT

This section presents a simple electronic circuit which has been designed to have a firing threshold that depends on the past states of the output in a manner similar to the activity dependence measured for frog sciatic nerve. In response to constant frequency stimuli, the circuit acts as a lowpass filter whose corner frequency depends on the coefficients which determine the after–effects of activity.

Figure 8 shows the circuit diagram for a switched capacitor circuit which approximates the after–effects of activity found in the frog sciatic nerve. The circuit employs a two phase nonoverlapping clock, e for the even clock and o for the odd clock, typical of switched capacitor circuits. It incorporates a basic model for superexcitability and depressibility. V_{TH} represents the resting threshold of the axon. On each clock cycle the V_{IN} is compared with $V_{TH}+V_D-V_S$.

The two capacitors and three switches at the bottom of figure 8 model the change in threshold caused by superexcitability. Note that each impulse resets the comparator's minus input to $(1-\alpha_s)V_{TH}$, which decays back to V_{TH} on subsequent clock cycles with a time constant inversely proportional to β_S. This is a slight deviation from the actual physiological situation in which multiple conditioning impulses will generate slightly more superexcitability than a single impulse.[7]

The two capacitors and two switches at the upper left of figure 8 model the depressibility of the axon. The current source represents a fixed increment in the firing threshold with every past impulse. The depression voltage decays back to 0 on subsequent clock cycles with a time constant inversely proportional to β_D.

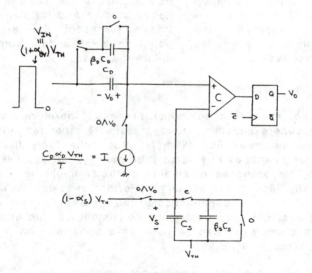

Figure 8. Circuit diagram for electrical circuit analog of nerve threshold.

The electrical circuit exhibits response patterns similar to those of neurons that are conducting intermittently (see figure 9). During bursts of conduction, the depression voltage increases linearly until the comparator

fails to fire. The electrical axon then fails to fire until the depression
voltage decays back to $(1+\alpha_{OV})V_{TH}$. The connectivity between the input
and output of the axon is defined to be the average fraction of impulses
which are conducted. In terms of connectivity, the electrical axon model
acts as a lowpass filter (see figure 10).

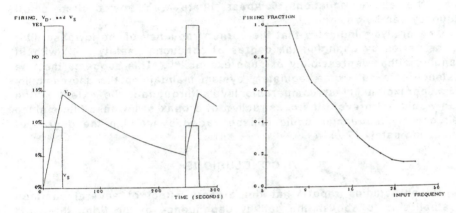

Figure 9. Typical waveforms for
intermittent conduction. The
upper trace indicates whether
impulses are conducted or not.
V_D and V_S are the depression
voltage and the superexcitable
voltage respectively.

Figure 10. Frequency response of
electrical axon model. The
connectivity is reflected by the
fraction of impulses which are
conducted out of a sequence of
100,000 stimuli where the
frequency is in stimuli/second.

For a fixed stimulus frequency, the average fraction of impulses
which are conducted by the electrical model can be predicted analytically.
The expressions can be greatly simplified by making the assumption that
V_D increases and decreases in a linear fashion. Under that assumption, in
terms of the variables indicated on the schematic diagram,

$$P(firing) = \frac{\alpha_{OV}(1-(1-\beta_D)^M)}{\alpha_{OV}(1-(1-\beta_D)^M)+\alpha_D}$$

where M is the number of clock cycles between input stimuli, which is
inversely proportional to the input frequency. The frequency at which only
half of the impulses are conducted is defined as the corner frequency of
the lowpass filter. The corner frequency is

$$f(P = 0.5) = \frac{1}{M} = \frac{log(1 - \beta_D)}{log(1 - \frac{\alpha_D}{\alpha_{OV}})}$$

Using the above equations, lowpass filters with any desired cutoff frequency can be designed.

The analysis indicates that the corner frequency of the lowpass filter can be varied by changing the degree of conduction safety (α_{OV}) without changing either depressibility or superexcitability. This suggests that the existence of a cell—wide regulatory system maintaining the depressibility and superexcitability at comparable levels throughout the extent of the axon would not prevent the construction of a bank of lowpass filters since their corner frequencies could still be varied by varying the degree of conduction safety (α_{OV}).

VI. CONCLUSIONS

Recent studies report that the primary effect of several common anesthetics is to abolish the activity dependence of the firing threshold without interfering with impulse conduction.[11] This suggests that presynaptic processing may play an important role in human consciousness. This paper has explored some of the basic ideas of presynaptic information processing, especially the after—effects of activity and their modulation of impulse conduction at sites of low conduction safety. A switched capacitor circuit which simulates the activity dependent conduction block that occurs in axons has been designed and simulated. Simulation results are very similar to the intermittent conduction patterns measured experimentally in frog axons. One potential information processing possibility for the arbor of a single axon, suggested by the analysis of the electronic circuit, is to act as a filterbank; every terminal could act as a lowpass filter with a different corner frequency.

BIBLIOGRAPHY

[1] Barron D. H. and B. H. C. Matthews, Intermittent conduction in the spinal chord. *J. Physiol.* 85, p. 73—103 (1935).

[2] Fuortes M. G. F., Action of strychnine on the "intermittent conduction" of impulses along dorsal columns of the spinal chord of frogs. *J. Physiol.* 112, p.42 (1950).

[3] Culp W. and J. Ochoa, *Nerves and Muscles as Abnormal Impulse Generators.* (Oxford University Press, London, 1980).

[4] Grossman Y., I. Parnas, and M. E. Spira, Ionic mechanisms involved in differential conduction of action potentials at high frequency in a branching axon. *J. Physiol.* 295, p.307−322 (1978).

[5] Parnas I., Differential block at high frequency of branches of a single axon innervating two muscles. *J. Physiol.* 35, p. 903−914, 1972.

[6] Carley, L.R. and S.A. Raymond, Threshold Measurement: Applications to Excitable Membranes of Nerve and Muscle. *J. Neurosci. Meth.* 9, p. 309−333 (1983).

[7] Raymond S. A. and J. Y. Lettvin, After−effects of activity in peripheral axons as a clue to nervous coding. In *Physiology and Pathobiology of Axons*, S. G. Waxman (ed.), (Raven Press, New York, 1978), p. 203−225.

[8] Wurtz C. C. and M. H. Ellisman, Alternations in the ultrastructure of peripheral nodes of Ranvier associated with repetitive action potential propagation. *J. Neurosci.* 6(11), 3133−3143 (1986).

[9] Raymond S. A., Effects of nerve impulses on threshold of frog sciatic nerve fibers. *J. Physiol.* 290, 273−303 (1979).

[10] Carley, L.R. and S.A. Raymond, Comparison of the after−effects of impulse conduction on threshold at nodes of Ranvier along single frog Sciatic axons. *J. Physiol.* 386, p. 503−527 (1987).

[11] Raymond S. A. and J. G. Thalhammer, Endogenous activity−dependent mechanisms for reducing hyperexcitability of axons: Effects of anesthetics and CO_2. In *Inactivation of Hypersensistive Neurons*, N. Chalazonitis and M. Gola, (eds.), (Alan R. Liss Inc., New York, 1987), p. 331−343.

MATHEMATICAL ANALYSIS OF LEARNING BEHAVIOR OF NEURONAL MODELS

BY
JOHN Y. CHEUNG
MASSOUD OMIDVAR

SCHOOL OF ELECTRICAL ENGINEERING AND COMPUTER SCIENCE
UNIVERSITY OF OKLAHOMA
NORMAN, OK 73019

Presented to the IEEE Conference on "Neural Information Processing Systems–Natural and Synthetic," Denver, November 8–12, 1987, and to be published in the Collection of Papers from the IEEE Conference on NIPS.

Please address all further correspondence to:

John Y. Cheung
School of EECS
202 W. Boyd, CEC 219
Norman, OK 73019
(405)325-4721

November, 1987

MATHEMATICAL ANALYSIS OF LEARNING BEHAVIOR OF NEURONAL MODELS

John Y. Cheung and Massoud Omidvar
School of Electrical Engineering
and Computer Science

ABSTRACT

In this paper, we wish to analyze the convergence behavior of a number of neuronal plasticity models. Recent neurophysiological research suggests that the neuronal behavior is adaptive. In particular, memory stored within a neuron is associated with the synaptic weights which are varied or adjusted to achieve learning. A number of adaptive neuronal models have been proposed in the literature. Three specific models will be analyzed in this paper, specifically the Hebb model, the Sutton–Barto model, and the most recent trace model. In this paper we will examine the conditions for convergence, the position of convergence and the rate at convergence, of these models as they applied to classical conditioning. Simulation results are also presented to verify the analysis.

INTRODUCTION

A number of static models to describe the behavior of a neuron have been in use in the past decades. More recently, research in neurophysiology suggests that a static view may be insufficient. Rather, the parameters within a neuron tend to vary with past history to achieve learning. It was suggested that by altering the internal parameters, neurons may adapt themselves to repetitive input stimuli and become conditioned. Learning thus occurs when the neurons are conditioned. To describe this behavior of neuronal plasticity, a number of models have been proposed. The earliest one may have been postulated by Hebb and more recently by Sutton and Barto [1]. We will also introduce a new model, the most recent trace (or MRT) model in this paper. The primary objective of this paper, however, is to analyze the convergence behavior of these models during adaptation.

The general neuronal model used in this paper is shown in Figure 1. There are a number of neuronal inputs $x_i(t), i = 1, \ldots, N$. Each input is scaled by the corresponding synaptic weights $w_i(t), i = 1, \ldots, N$. The weighted inputs are arithmetically summed.

$$y(t) = \sum_{i=1}^{N} x_i(t) w_i(t) - \Theta(t) \qquad (1)$$

where $\Theta(t)$ is taken to be zero.

Neuronal inputs are assumed to take on numerical values ranging from zero to one inclusively. Synaptic weights are allowed to take on any reasonable values for the purpose of this paper though in reality, the weights may very well be bounded. Since the relative magnitude of the weights and the neuronal inputs are not well defined at this point, we will not put a bound on the magnitude of the weights also. The neuronal output is normally the result of a sigmoidal transformation. For simplicity, we will approximate this operation by a linear transformation.

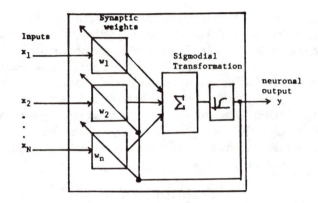

Figure 1. A general neuronal model.

For convergence analysis, we will assume that there are only two neuronal inputs in the traditional classical conditioning environment for simplicity. Of course, the analysis techniques can be extended to any number of inputs. In classical conditioning, the two inputs are the conditioned stimulus $x_c(t)$ and the unconditioned stimulus $x_u(t)$.

THE SUTTON–BARTO MODEL

More recently, Sutton and Barto [1] have proposed an adaptive model based on both the signal trace $\overline{x}_i(t)$ and the output trace $\overline{y}(t)$ as given below:

$$w_i(t + 1) = w_i(t) + c\overline{x}_i(t)(y(t)) - \overline{y}(t) \tag{2a}$$

$$\overline{y}(t + 1) = \beta\overline{y}(t) + (1 - \beta)y(t) \tag{2b}$$

$$\overline{x}_i(t + 1) = \alpha\overline{x}_i(t) + x_i(t) \tag{2c}$$

where both α and β are positive constants.

Condition of Convergence

In order to simplify the analysis, we will choose $\alpha = 0$ and $\beta = 0$, i.e.:

$$\bar{x}_i(t) = x_i(t-1)$$

and

$$\bar{y}(t) = y(t-1)$$

In other words, (2a) becomes:

$$w_i(t+1) = w_i(t) + cx_i(t)(y(t) - y(t-1)) \tag{3}$$

The above assumption only serves to simplify the analysis and will not affect the convergence conditions because the boundedness of $\bar{x}_i(t)$ and $\bar{y}(t)$ only depends on that for $x_i(t)$ and $y(t-1)$ respectively.

As in the previous section, we recognize that (3) is a recurrence relation so convergence can be checked by the ratio test. It is also possible to rewrite (3) in matrix format. Due to the recursion of the neuronal output in the equation, we will include the neuronal output $y(t)$ in the parameter vector also:

$$\begin{pmatrix} w_1(t+1) \\ w_2(t+1) \\ y(t) \end{pmatrix} = \begin{pmatrix} 1 + cx_1^2(t) & cx_1(t)x_2(t) & -cx_1(t) \\ cx_1(t)x_2(t) & 1 + cx_2^2(t) & -cx_2(t) \\ x_1(t) & x_2(t) & 0 \end{pmatrix} \begin{pmatrix} w_1(t) \\ w_2(t) \\ y(t-1) \end{pmatrix} \tag{4}$$

or

$$W^{(S-B)}(t+1) = A^{(S-B)} W^{(S-B)}(t)$$

To show convergence, we need to set the magnitude of the determinant of $A^{(S-B)}$ to be less than unity.

$$|A^{(S-B)}| = c(x_1^2(t) + x_2^2(t)) \tag{5}$$

Hence, the condition for convergence is:

$$c < \frac{1}{x_1^2(t) + x_2^2(t)} \tag{6}$$

From (6), we can see that the adaptation constant must be chosen to be less than the reciprocal of the Euclidean sum of energies of all the inputs. The same techniques can be extended to any number of inputs. This can be proved merely by following the same procedures outlined above.

Position At Convergence

Having proved convergence of the Sutton–Barto model equations of neuronal plasticity, we want to find out next at what location the system remains when converged. We have seen earlier that at convergence, the weights cease to change and so does the neuronal output. We will denote this converged position as $(W^{(S-B)})^* \equiv W^{(S-B)}(\infty)$. In other words:

$$(W^{(S-B)})^* = A^{(S-B)}(W^{(S-B)})^* \tag{7}$$

Since any arbitrary parameter vector can always be decomposed into a weighted sum of the eigenvectors, i.e.

$$W^{(S-B)}(0) = \alpha_1 V_1 + \alpha_2 V_2 + \alpha_3 V_3 \tag{8}$$

The constants α_1, α_2, and α_3 can easily be found by inverting $A^{(S-B)}$. The eigenvalues of $A^{(S-B)}$ can be shown to be 1, 1, and $c(x_1^2 + x_2^2)$. When c is within the region of convergence, the magnitude of the third eigenvalue is less than unity. That means that at convergence, there will be no contribution from the third eigenvector. Hence,

$$(W^{(S-B)})^* = \lim_{t \to \infty} W^{(S-B)}(t) = \alpha_1 V_1 + \alpha_2 V_2 \tag{9}$$

From (9), we can predict precisely what the converged position would be given only with the initial conditions.

Rate of Convergence

We have seen that when c is carefully chosen, the Sutton–Barto model will converge and we have also derived an expression for the converged position. Next we want to find out how fast convergence can be attained. The rate of convergence is a measure of how fast the initial parameter approaches the optimal position. The asymptotic rate of convergence is[2]:

$$R_\infty(A^{(S-B)}) = -\log S(A^{(S-B)}) \tag{10}$$

where $S(A^{(S-B)})$ is the spectral radius and is equalled to $c(x_1^2 + x_2^2)$ in this case. This completes the convergence analysis on the Sutton–Barto model of neuronal plasticity.

THE MRT MODEL OF NEURONAL PLASTICITY

The most recent trace (MRT) model of neuronal plasticity [3] developed by the authors can be considered as a cross between the Sutton–Barto model and the Klopf's model [4]. The adaptation of the synaptic weights can be expressed as follows:

$$w_i(t+1) = w_i(t) + cw_i(t)x_i(t)(y(t) - y(t-1)) \tag{11}$$

A comparison of (11) and the Sutton-Barto model in (3) shows that the second term on the right hand side contains an extra factor, $w_i(t)$, which is used to speed up the convergence as shown later. The output trace has been replaced by $y(t-1)$, the most recent output, hence the name, the most recent trace model. The input trace is also replaced by the most recent input.

Condition of Convergence

We can now proceed to analyze the condition of convergence for the MRT model. Due to the presence of the $w_i(t)$ factor in the second term in (31), the ratio test cannot be applied here. To analyze the convergence behavior further, let us rewrite (11) in matrix format:

$$
\begin{pmatrix} w_1(t+1) \\ w_2(t+1) \\ y(t) \end{pmatrix} = \begin{pmatrix} 1 & 0 & 0 \\ 0 & 1 & 0 \\ x_1(t) & x_2(t) & 0 \end{pmatrix} \begin{pmatrix} w_1(t) \\ w_2(t) \\ y(t-1) \end{pmatrix}
$$

$$
+ c(w_1(t)\ w_2(t)\ y(t-1)) \begin{pmatrix} x_1(t) \\ x_2(t) \\ -1 \end{pmatrix} \begin{pmatrix} x_1(t) & 0 & 0 \\ 0 & x_2(t) & 0 \\ 0 & 0 & 0 \end{pmatrix} \begin{pmatrix} w_1(t) \\ w_2(t) \\ y(t-1) \end{pmatrix} \quad (12)
$$

or

$$
W^{(MRT)}(t+1) = A^{(MRT)}W^{(MRT)}(t) + c\,(W^{(MRT)}(t))^T BCW^{(MRT)}(t)
$$

The superscript T denotes the matrix transpose operation. The above equation is quadratic in $W^{(MRT)}(t)$. Complete convergence analysis of this equation is extremely difficult.

In order to understand the convergence behavior of (12), we note that the dominant term that determines convergence mainly relates to the second quadratic term. Hence for convergence analysis only, we will ignore the first term:

$$
W^{(MRT)}(t+1) \approx c(W^{(MRT)}(t))^T BCW^{(MRT)}(t) \quad (13)
$$

We can readily see from above that the primary convergence factor is $B^T C$. Since C is only dependent on $x_i(t)$, convergence can be obtained if the duration of the synaptic inputs being active is bounded. It can be shown that the condition of convergence is bounded by:

$$
c < \frac{1}{(x_1^2 w_1(\infty) + x_2^2 w_2(\infty))} \quad (14)
$$

170

We can readily see that the adaptation constant c can be chosen according to (14) to ensure convergence for $t < T$.

SIMULATIONS

To verify the theoretical analysis of these three adaptive neuronal models based on classical conditioning, these models have been simulated on the IBM 3081 mainframe using the FORTRAN language in single precision. Several test scenarios have been designed to compare the analytical predictions with actual simulation results.

To verify the conditions for convergence, we will vary the value of the adaptation constant c. The conditioned and unconditioned stimuli were set to unity and the value of c varies between 0.1 to 1.0. For the Sutton–Barto model the simulation given in Fig. 2 shows that convergence is obtained for $c < 0.5$ as expected from theoretical analysis. For the MRT model, simulation results given in Fig. 3 shows that convergence is obtained for $c < 0.7$, also as expected from theoretical analysis. The theoretical location at convergence for the Sutton and Barto model is also shown in Figure 2. It is readily seen that the simulation results confirm the theoretical expectations.

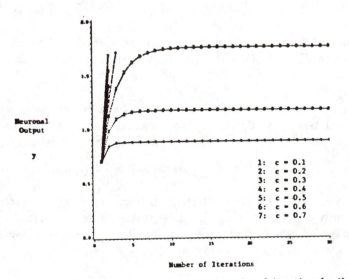

Figure 2. Plots of neuronal outputs versus the number of iterations for the Sutton–Barto model with different values of adaptation constant c.

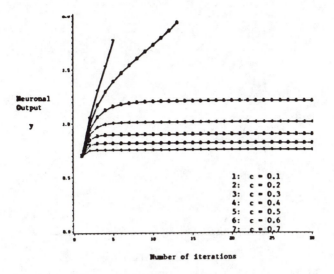

Figure 3. Plots of neuronal outputs versus the number of iterations
for the MRT model with different values of adaptation
constant c.

To illustrate the rate of convergence, we will plot the trajectory of the deviation in synaptic weights from the optimal values in the logarithmic scale since this error is logarithmic as found earlier. The slope of the line yields the rate of convergence. The trajectory for the Sutton–Barto Model is given in Figure 4 while that for the MRT model is given in Figure 5. It is clear from Figure 4 that the trajectory in the logarithmic form is a straight line. The slope $\hat{R}_n(A^{(S-B)})$ can readily be calculated. The curve for the MRT model given in Figure 5 is also a straight line but with a much larger slope showing faster convergence.

SUMMARY

In this paper, we have sought to discover analytically the convergence behavior of three adaptive neuronal models. From the analysis, we see that the Hebb model does not converge at all. With constant active inputs, the output will grow exponentially. In spite of this lack of convergence the Hebb model is still a workable model realizing that the divergent behavior would be curtailed by the sigmoidal transformation to yield realistic outputs. The

172

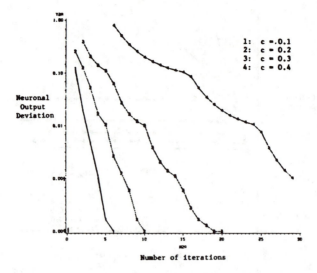

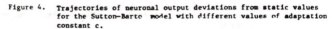

Number of iterations

Figure 4. Trajectories of neuronal output deviations from static values
for the Sutton-Barto model with different values of adaptation
constant c.

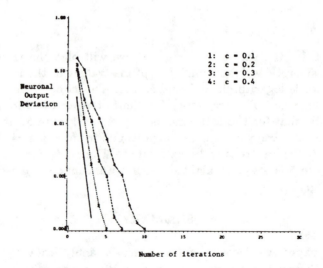

Number of iterations

Figure 5. Trajectories of neuronal output deviations from static
values for the MRT model with different values of
adaptation constant c.

analysis on the Sutton and Barto model shows that this model will converge when the adaptation constant c is carefully chosen. The bounds for c is also found for this model. Due to the structure of this model, both the location at convergence and the rate of convergence are also found. We have also introduced a new model of neuronal plasticity called the most recent trace (MRT) model. Certain similarities exist between the MRT model and the Sutton–Barto model and also between the MRT model and the Klopf model. Analysis shows that the update equations for the synaptic weights are quadratic resulting in polynomial rate of convergence. Simulation results also show that much faster convergence rate can be obtained with the MRT model.

REFERENCES

1. Sutton, R.S. and A.G. Barto, Psychological Review, vol. 88, p. 135, (1981).
2. Hageman, L. A. and D.M. Young. Applied Interactive Methods. (Academic Press, Inc. 1981).
3. Omidvar, Massoud. Analysis of Neuronal Plasticity. Doctoral dissertation, School of Electrical Engineering and Computer Science, University of Oklahoma, 1987.
4. Klopf, A.H. Proceedings of the American Institute of Physics Conference #151 on Neural Networks for Computing, p. 265–270, (1986).

A Neural Network Classifier Based on Coding Theory

Tzi-Dar Chiueh and Rodney Goodman
California Institute of Technology, Pasadena, California 91125

ABSTRACT

The new neural network classifier we propose transforms the classification problem into the coding theory problem of decoding a noisy codeword. An input vector in the feature space is transformed into an internal representation which is a codeword in the code space, and then error correction decoded in this space to classify the input feature vector to its class. Two classes of codes which give high performance are the Hadamard matrix code and the maximal length sequence code. We show that the number of classes stored in an N-neuron system is linear in N and significantly more than that obtainable by using the Hopfield type memory as a classifier.

I. INTRODUCTION

Associative recall using neural networks has recently received a great deal of attention. Hopfield in his papers [1,2] describes a mechanism which iterates through a feedback loop and stabilizes at the memory element that is nearest the input, provided that not many memory vectors are stored in the machine. He has also shown that the number of memories that can be stored in an N-neuron system is about 0.15N for N between 30 and 100. McEliece et al. in their work [3] showed that for synchronous operation of the Hopfield memory about $N/(2\log N)$ data vectors can be stored reliably when N is large. Abu-Mostafa [4] has predicted that the upper bound for the number of data vectors in an N-neuron Hopfield machine is N. We believe that one should be able to devise a machine with M, the number of data vectors, *linear* in N and larger than the 0.15N achieved by the Hopfield method.

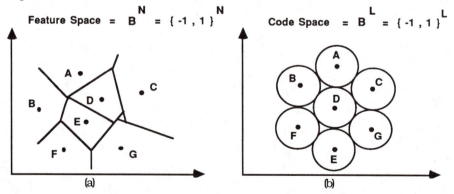

Figure 1 (a) Classification problems versus (b) Error control decoding problems

In this paper we are specifically concerned with the problem of classification as in pattern recognition. We propose a new method of building a neural network classifier, based on the well established techniques of error control coding. Consider a typical classification problem (Fig. 1(a)), in which one is given *a priori* a set of classes, $C^{(\alpha)}$, $\alpha = 1, \ldots , M$. Associated with each class is a feature vector which labels the class (the *exemplar* of the class), i.e. it is the most representative point in the class region. The input is classified into the class with the nearest exemplar to the input. Hence for each class there is a region in the N-dimensional binary feature space $\mathbf{B}^N \equiv \{1,-1\}^N$, in which every vector will be classified to the corresponding class.

A similar problem is that of decoding a codeword in an error correcting code as shown in Fig. 1(b). In this case codewords are constructed by design and are usually at least d_{min} apart. The received corrupted codeword is the input to the decoder, which then finds the nearest codeword to the input. In principle then, if the distance between codewords is greater than $2t+1$, it is possible to decode (or classify) a noisy codeword (feature vector) into the correct codeword (exemplar) provided that the Hamming distance between the noisy codeword and the correct codeword is no more than t. Note that there is no guarantee that the exemplars are uniformly distributed in $\mathbf{B}^N$, consequently the *attraction radius* (the maximum number of errors that can occur in any given feature vector such that the vector can still be correctly classified) will depend on the *minimum* distance between exemplars.

Many solutions to the minimum Hamming distance classification have been proposed, the one commonly used is derived from the idea of matched filters in communication theory. Lippmann [5] proposed a two-stage neural network that solves this classification problem by first correlating the input with all exemplars and then picking the maximum by a "winner-take-all" circuit or a network composed of two-input comparators. In Figure 2, $f_1, f_2, ..., f_N$ are the N input bits, and $s_1, s_2, ... s_M$ are the matching scores(similarity) of $\mathbf{f}$ with the M exemplars. The second block picks the maximum of $s_1, s_2, ..., s_M$ and produces the index of the exemplar with the largest score. The main disadvantage of such a classifier is the complexity of the maximum-picking circuit, for example a "winner-take-all" net needs connection weights of large dynamic range and graded-response neurons, whilst the comparator maximum net demands M-1 comparators organized in $\log_2 M$ stages.

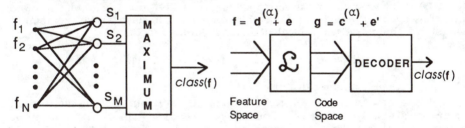

Fig. 2 A matched filter type classifier   Fig. 3 Structure of the proposed classifier

Our main idea is thus to transform every vector in the feature space to a vector in some code space in such a way that every exemplar corresponds to a codeword in that code. The code should preferably (but not necessarily) have the property that codewords are uniformly distributed in the code space, that is, the Hamming distance between every pair of codewords is the same. With this transformation, we turn the problem of classification into the coding problem of decoding a noisy codeword. We then do error correction decoding on the vector in the code space to obtain the index of the noisy codeword and hence classify the original feature vector, as shown in Figure 3.

This paper develops the construction of such a classification machine as follows. First we consider the problem of transforming the input vectors from the feature space to the code space. We describe two hetero-associative memories for doing this, the first method uses an outer product matrix technique similar to that of Hopfield's, and the second method generates its matrix by the pseudoinverse technique[6,7]. Given that we have transformed the problem of associative recall, or classification, into the problem of decoding a noisy codeword, we next consider suitable codes for our machine. We require the codewords in this code to have the property of orthogonality or *pseudo*-orthogonality, that is, the ratio of the cross-correlation to the auto-correlation of the codewords is small. We show two classes of such good codes for this particular decoding problem i.e. the Hadamard matrix codes, and the maximal length sequence codes[8]. We next formulate the complete decoding algorithm, and describe the overall structure of the classifier in terms of a two layer neural network. The first layer performs the mapping operation on the input, and the second one decodes its output to produce the index of the class to which the input belongs.

The second part of the paper is concerned with the performance of the classifier. We first analyze the performance of this new classifier by finding the relation between the maximum number of classes that can be stored and the classification error rate. We show (when using a transform based on the outer product method) that for negligible misclassification rate and large N, a not very tight lower bound on M, the number of stored classes, is 0.22N. We then present comprehensive simulation results that confirm and exceed our theoretical expectations. The simulation results compare our method with the Hopfield model for both the outer product and pseudo-inverse method, and for both the analog and hard limited connection matrices. In all cases our classifier exceeds the performance of the Hopfield memory in terms of the number of classes that can be reliably recovered.

II. TRANSFORM TECHNIQUES

Our objective is to build a machine that can discriminate among input vectors and classify each one of them into the appropriate class. Suppose $\mathbf{d}^{(\alpha)} \in \mathbf{B}^N$ is the exemplar of the corresponding class $C^{(\alpha)}$, $\alpha = 1, 2, \dots, M$. Given the input $\mathbf{f}$, we want the machine to be able to identify the class whose exemplar is closest to $\mathbf{f}$, that is, we want to calculate the following function,

$$class(\mathbf{f}) = \alpha \qquad iff \quad |\mathbf{f} - \mathbf{d}^{(\alpha)}| < |\mathbf{f} - \mathbf{d}^{(\beta)}| \quad \forall \, \alpha \neq \beta$$

where $|\ |$ denotes Hamming distance in $\mathbf{B}^N$.

We approach the problem by seeking a transform $\mathcal{L}$ that maps each exemplar $\mathbf{d}^{(\alpha)}$ in $\mathbf{B}^N$ to the corresponding codeword $\mathbf{w}^{(\alpha)}$ in $\mathbf{B}^L$. And an input feature vector $\mathbf{f} = \mathbf{d}^{(\gamma)} + \mathbf{e}$ is thus mapped to a noisy codeword $\mathbf{g} = \mathbf{w}^{(\gamma)} + \mathbf{e}'$ where $\mathbf{e}$ is the error added to the exemplar, and $\mathbf{e}'$ is the corresponding error pattern in the code space. We then do error correction decoding on $\mathbf{g}$ to get the index of the corresponding codeword. Note that $\mathbf{e}'$ may not have the same Hamming weight as $\mathbf{e}$, that is, the transformation $\mathcal{L}$ may either generate more errors or eliminate errors that are present in the original input feature vector. We require $\mathcal{L}$ to satisfy the following equation,

$$\mathcal{L}\mathbf{d}^{(\alpha)} = \mathbf{w}^{(\alpha)} \qquad \alpha = 0, 1, \dots, M-1$$

and $\mathcal{L}$ will be implemented using a single-layer feedforward network.

Thus we first construct a matrix according to the sets of $\mathbf{d}^{(\alpha)}$'s and $\mathbf{w}^{(\alpha)}$'s, call it $\mathbf{T}$, and define $\mathcal{L}$ as

$$\mathcal{L} \equiv sgn \circ \mathbf{T}$$

where sgn is the threshold operator that maps a vector in $\mathbf{R}^L$ to $\mathbf{B}^L$ and $\mathbf{R}$ is the field of real numbers.

Let $\mathbf{D}$ be an N x M matrix whose αth column is $\mathbf{d}^{(\alpha)}$ and $\mathbf{W}$ be an L x M matrix whose βth column is $\mathbf{w}^{(\beta)}$. The two possible methods of constructing the matrix for $\mathcal{L}$ are as follows:

Scheme A (outer product method) [3,6] : In this scheme the matrix $\mathbf{T}$ is defined as the sum of outer products of all exemplar-codeword pairs, i.e.

$$T^{(A)}_{ij} = \sum_{\alpha=0}^{M-1} w_i(\alpha) \cdot d_j(\alpha)$$

or equivalently,

$$\mathbf{T}^{(A)} = \mathbf{W} \mathbf{D}^t$$

Scheme B (pseudo-inverse method) [6,7] : We want to find a matrix $\mathbf{T}^{(B)}$ satisfying the following equation,

$$\mathbf{T}^{(B)} \mathbf{D} = \mathbf{W}$$

In general $\mathbf{D}$ is not a square matrix, moreover $\mathbf{D}$ may be singular, so $\mathbf{D}^{-1}$ may not exist. To circumvent this difficulty, we calculate the *pseudo-inverse* (denoted $\mathbf{D}^\dagger$) of the matrix $\mathbf{D}$ instead of its real inverse, let $\mathbf{D}^\dagger \equiv (\mathbf{D}^t\mathbf{D})^{-1}\mathbf{D}^t$. $\mathbf{T}^{(B)}$ can be formulated as,

$$\mathbf{T}^{(B)} = \mathbf{W} \mathbf{D}^\dagger = \mathbf{W} (\mathbf{D}^t \mathbf{D})^{-1}\mathbf{D}^t$$

III. CODES

The codes we are looking for should preferably have the property that its codewords be distributed uniformly in $\mathbf{B}^L$, that is, the distance between each two codewords must be the same and as large as possible. We thus seek classes of *equidistant* codes. Two such classes are the Hadamard matrix codes, and the maximal length sequence codes.

First define the word *pseudo-orthogonal* .

Definition : Let $\mathbf{w}^{(\alpha)} = (w_0(\alpha), w_1(\alpha), \ldots\ldots, w_{L-1}(\alpha)) \in \mathbf{B}^L$ be the αth codeword of code C, where $\alpha = 1, 2, \ldots, M$. Code C is said to be *pseudo-orthogonal* iff

$$(\mathbf{w}^{(\alpha)}, \mathbf{w}^{(\beta)}) = \sum_{i=0}^{L-1} w_i(\alpha) \, w_i(\beta)$$

$$= \begin{cases} L & \alpha = \beta \\ \epsilon & \alpha \neq \beta \end{cases} \qquad \text{where } \epsilon \ll L$$

where (,) denotes inner product of two vectors.

Hadamard Matrices: An orthogonal code of length L whose L codewords are rows or columns of an L x L Hadamard matrix. In this case $\epsilon = 0$ and the distance between any two codewords is L/2. It is conjectured that there exist such codes for all L which are multiples of 4, thus providing a large class of codes[8].

Maximal Length Sequence Codes: There exists a family of maximal length sequence (also called pseudo-random or PN sequence) codes[8], generated by shift registers, that satisfy pseudo-orthogonality with $\epsilon = -1$. Suppose $g(x)$ is a primitive polynomial over $GF(2)$ of degree D, and let $L = 2^D - 1$, and if

$$f(x) = 1/g(x) = \sum_{k=0}^{\infty} c_k \cdot x^k$$

then $c_0, c_1, \ldots$ is a periodic sequence of period L (since $g(x) \mid x^L - 1$). If code C is made up of the L cyclic shifts of

$$\mathbf{c} = (1 - 2c_0, 1 - 2c_1, \ldots, 1 - 2c_{L-1})$$

then code C satisfies pseudo-orthogonality with $\epsilon = -1$. One then easily sees that the minimum distance of this code is $(L - 1)/2$ which gives a correcting power of approximately $L/4$ errors for large L.

IV. OVERALL CLASSIFIER STRUCTURE

We shall now describe the overall classifier structure, essentially it consists of the mapping $\mathcal{L}$ followed by the error correction decoder for the maximal length sequence code or Hadamard matrix code. The decoder operates by correlating the input vector with every codeword and then thresholding the result at $(L + \epsilon)/2$. The rationale of this algorithm is as follows, since the distance between every two codewords in this code is exactly $(L - \epsilon)/2$ bits, the decoder should be able to correct any error pattern with less than $(L - \epsilon)/4$ errors if the threshold is set halfway between L and ϵ i.e. $(L + \epsilon)/2$.

Suppose the input vector to the decoder is $\mathbf{g} = \mathbf{w}^{(\alpha)} + \mathbf{e}$ and $\mathbf{e}$ has Hamming weight s (i.e. s nonzero components) then we have

$$(\mathbf{g}, \mathbf{w}^{(\alpha)}) = L - 2s$$
$$(\mathbf{g}, \mathbf{w}^{(\beta)}) \leq 2s + \epsilon \qquad \text{where} \quad \beta \neq \alpha$$

From the above equation, if $\mathbf{g}$ is less than $(L - \epsilon)/4$ errors away from $\mathbf{w}^{(\alpha)}$ (i.e. $s < (L - \epsilon)/4$) then $(\mathbf{g}, \mathbf{w}^{(\alpha)})$ will be more than $(L + \epsilon)/2$ and $(\mathbf{g}, \mathbf{w}^{(\beta)})$ will be less than $(L + \epsilon)/2$, for all $\beta \neq \alpha$. As a result, we arrive at the following decoding algorithm,

$$decode(\mathbf{g}) = sgn(\mathbf{W}^t \mathbf{g} - ((L + \epsilon)/2)\mathbf{j})$$

where $\mathbf{j} = [1\ 1\ \ldots\ 1]^t$, which is an M x 1 vector.

In the case when $\epsilon = -1$ and less than $(L+1)/4$ errors in the input, the output will be a vector in $\mathbf{B}^M \equiv \{1, -1\}^M$ with only *one* component positive (+1), the index of which is the index of the class that the input vector belongs. However if there are more than $(L+1)/4$ errors, the output can be either the all negative(-1) vector (decoder failure) or another vector with one positive component(decoder error).

The function *class* can now be defined as the composition of $\mathcal{L}$ and *decode*, the overall structure of the new classifier is depicted in Figure 4. It can be viewed as a two-layer neural network with L hidden units and M output neurons. The first layer is for mapping the input feature vector to a noisy codeword in the code space (the "internal representation") while the second one decodes the first's output and produces the index of the class to which the input belongs.

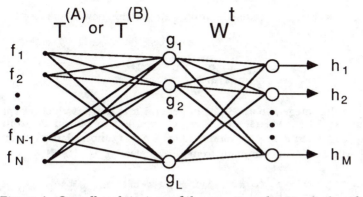

Figure 4   Overall architecture of the new neural network classifier

V. PERFORMANCE ANALYSIS

From the previous section, we know that our classifier will make an error only if the transformed vector in the code space, which is the input to the decoder, has no less than $(L - \epsilon)/4$ errors. We now proceed to find the error rate for this classifier in the case when the input is one of the exemplars (i.e. no error), say $\mathbf{f} = \mathbf{d}^{(\beta)}$ and an outer product connection matrix for $\mathcal{L}$. Following the approach of McEliece et. al.[3], we have

$$(\mathcal{L}\, \mathbf{d}^{(\beta)})_i = sgn\, (\, \sum_{j=0}^{N-1} \sum_{\alpha=0}^{M-1} w_i^{(\alpha)}\, d_j^{(\alpha)}\, d_j^{(\beta)}\,)$$

$$= sgn(\, N\, w_i^{(\beta)} + \sum_{\substack{j=0 \\ }}^{N-1} \sum_{\substack{\alpha=0 \\ \alpha \neq \beta}}^{M-1} w_i^{(\alpha)}\, d_j^{(\alpha)}\, d_j^{(\beta)}\,)$$

Assume without loss of generality that $w_i^{(\beta)} = -1$, and if

$$X \equiv \sum_{\substack{j=0 \\ }}^{N-1} \sum_{\substack{\alpha=0 \\ \alpha \neq \beta}}^{M-1} w_i^{(\alpha)}\, d_j^{(\alpha)}\, d_j^{(\beta)} \geq N$$

then

$$(\mathcal{L}\, \mathbf{d}^{(\beta)})_i \neq w_i^{(\beta)}$$

Notice that we assumed all $\mathbf{d}^{(\alpha)}$'s are random, namely each component of any $\mathbf{d}^{(\alpha)}$ is the outcome of a Bernoulli trial, accordingly, X is the sum of $N(M-1)$ independent identically distributed random variables with mean 0 and variance 1. In the asymptotic case, when N and M are both very large, X can be approximated by a normal distribution with mean 0, variance NM. Thus

$$p \equiv Pr\{\, (\mathcal{L}\, \mathbf{d}^{(\beta)})_i \neq w_i^{(\beta)}\, \}$$

$$\simeq Q(\sqrt{N/M})$$

$$\text{where}\quad Q(x) = \frac{1}{\sqrt{2\pi}} \int_x^\infty e^{\,t^2/2}\ dt$$

Next we calculate the misclassification rate of the new classifier as follows (assuming $\epsilon \ll L$),

$$P_e = \sum_{k=\lfloor L/4 \rfloor}^{L} \binom{L}{k} p^k (1-p)^{L-k}$$

where $\lfloor \ \rfloor$ is the integer floor. Since in general it is not possible to express the summation explicitly, we use the Chernoff method to bound P_e from above. Multiplying each term in the summation by a number larger than unity ($e^{t(k-L/4)}$ with $t > 0$) and summing from $k = 0$ instead of $k = \lfloor L/4 \rfloor$,

$$P_e < \sum_{k=0}^{L} \binom{L}{k} p^k (1-p)^{L-k} e^{t(k-L/4)} = e^{-Lt/4} (1-p+pe^t)^L$$

Differentiating the RHS of the above equation w.r.t. t and set it to 0, we find the optimal t_0 as $e^{t_0} = (1-p)/3p$. The condition that $t_0 > 0$ implies that $p < 1/4$, and since we are dealing with the case where p is small, it is automatically satisfied. Substituting the optimal t_0, we obtain

$$P_e < c^L \cdot p^{L/4} \cdot (1-p)^{3L/4} \qquad \text{where } c = 4/(3^{3/4}) = 1.7547654$$

From the expression for P_e, we can estimate M, the number of classes that can be classified with negligible misclassification rate, in the following way, suppose $P_e = \delta$ where $\delta \ll 1$ and $p \ll 1$, then

$$\delta^{4/L} < c^4 \cdot p \cdot (1-p)^3 \quad \Rightarrow \quad p = Q(\sqrt{N/M}) > c^{-4} \cdot (1-p)^{-3} \cdot \delta^{4/L}$$

For small x we have $Q^{-1}(x) \sim \sqrt{2 \log(1/x)}$ and since δ is a fixed value, as L approaches infinity, we have

$$M > \frac{N}{8 \log c} = \frac{N}{4.5}$$

From the above lower bound for M, one easily see that this new machine is able to classify a constant times N classes, which is better than the number of memory items a Hopfield model can store i.e. $N/(2\log N)$. Although the analysis is done assuming N approaches infinity, the simulation results in the next section show that when N is moderately large (e.g. 63) the above lower bound applies.

VI. SIMULATION RESULTS AND A CHARACTER RECOGNITION EXAMPLE

We have simulated both the Hopfield model and our new machine(using maximal length sequence codes) for L = N = 31, 63 and for the following four cases respectively.
(i) connection matrix generated by <u>outer product method</u>
(ii) connection matrix generated by <u>pseudo-inverse method</u>
(iii) connection matrix generated by <u>outer product method</u>, the components of the connection matrix are <u>hard limited</u>.
(iv) connection matrix generated by <u>pseudo-inverse method</u>, the components of the connection matrix are <u>hard limited</u>.

For each case and each choice of N, the program fixes M and the number of errors in the input vector, then randomly generates 50 sets of M exemplars and computes the connection matrix for each machine. For each machine it randomly picks an exemplar and adds noise to it by randomly complementing the specified number of bits to generate 20 trial input vectors, it then simulates the machine and checks whether or not the input is classified to the nearest class and reports the percentage of success for each machine.

The simulation results are shown in Figure 5, in each graph the horizontal axis is M and the vertical axis is the attraction radius. The data we show are obtained by collecting only those cases when the success rate is more than 98%, that is for fixed M what is the largest attraction radius (number of bits in error of the input vector) that has a success rate of more than 98%. Here we use the attraction radius of -1 to denote that for this particular M, with the input being an exemplar, the success rate is less than 98% in that machine.

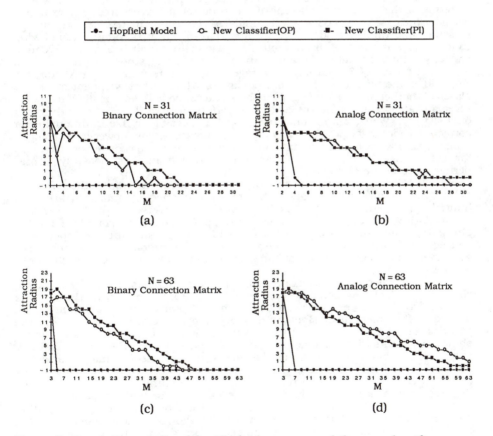

Figure 5 Simulation results of the Hopfield memory and the new classifier

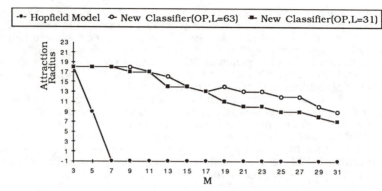

Figure 6 Performance of the new classifier using codes of different lengths

In all cases our classifier exceeds the performance of the Hopfield model in terms of the number of classes that can be reliably recovered. For example, consider the case of N = 63 and a hard limited connection matrix for both the new classifier and the Hopfield model, we find that for an attraction radius of zero, that is, no error in the input vector, the Hopfield model has a classification capacity of approximately 5, while our new model can store 47. Also, for an attraction radius of 8, that is, an average of N/8 errors in the input vector, the Hopfield model can reliably store 4 classes while our new model stores 27 classes. Another simulation (Fig. 6) using a shorter code (L = 31 instead of L = 63) reveals that by shortening the code, the performance of the classifier degrades only slightly. We therefore conjecture that it is possible to use traditional error correcting codes (e.g. BCH code) as internal representations, however, by going to a higher rate code, one is trading minimum distance of the code (error tolerance) for complexity (number of hidden units), which implies possibly poorer performance of the classifier.

We also notice that the superiority of the pseudoinverse method over the outer product method appears only when the connection matrices are hard limited. The reason for this is that the pseudoinverse method is best for decorrelating the dependency among exemplars, yet the exemplars in this simulation are generated randomly and are presumably independent, consequently one can not see the advantage of pseudoinverse method. For correlated exemplars, we expect the pseudoinverse method to be clearly better (see next example).

Next we present an example of applying this classifier to recognizing characters. Each character is represented by a 9 x 7 pixel array, the input is generated by flipping every pixel with 0.1 and 0.2 probability. The input is then passed to five machines: Hopfield memory, the new classifier with either pseudoinverse method or outer product method, and L = 7 or L = 31. Figure 7 and 8 show the results of all 5 machines for 0.1 and 0.2 pixel flipping probability respectively, a blank output means that the classifier refuses to make a decision. First note that the L = 7 case is not necessarily worse than the L = 31 case, this confirms the earlier conjecture that fewer hidden units (shorter code) only degrades performance slightly. Also one easily sees that the pseudoinverse method is better than the outer product method because of the correlation between exemplars. Both methods outperform the Hopfield memory since the latter mixes exemplars that are to be remembered and produces a blend of exemplars rather than the exemplars themselves, accordingly it cannot classify the input without mistakes.

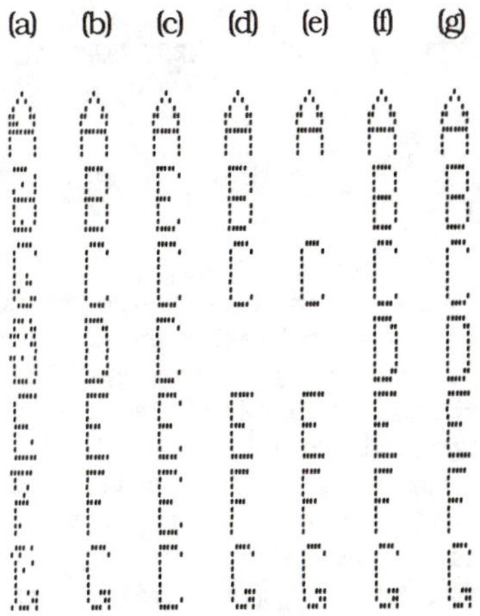

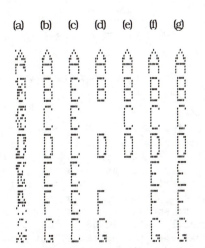

Figure 7 The character recognition example with 10% pixel reverse probability (a) input (b) correct output (c) Hopfield Model (d)-(g) new classifier (d) OP, L = 7 (e)OP, L = 31 (f) PI, L = 7 (g) PI, L = 31

Figure 8 The character recognition example with 20% pixel reverse probability (a) input (b) correct output (c) Hopfield Model (d)-(g) new classifier (d) OP, L = 7 (e)OP, L = 31 (f) PI, L = 7 (g) PI, L = 31

VII. CONCLUSION

In this paper we have presented a new neural network classifier design based on coding theory techniques. The classifier uses codewords from an error correcting code as its internal representations. Two classes of codes which give high performance are the Hadamard matrix codes and the maximal length sequence codes. In performance terms we have shown that the new machine is significantly better than using the Hopfield model as a classifier. We should also note that when comparing the new classifier with the Hopfield model, the increased performance of the new classifier does not entail extra complexity, since it needs only L + M hard limiter neurons and L(N + M) connection weights versus N neurons and N^2 weights in a Hopfield memory.

In conclusion we believe that our model forms the basis of a fast, practical method of classification with an efficiency greater than other previous neural network techniques.

REFERENCES

[1] J. J. Hopfield, *Proc. Nat. Acad. Sci. USA* , Vol. 79, pp. 2554-2558 (1982).

[2] J. J. Hopfield, *Proc. Nat. Acad. Sci. USA* , Vol. 81, pp. 3088-3092 (1984).

[3] R. J. McEliece, et. al, *IEEE Tran. on Information Theory* , Vol. IT-33, pp. 461-482 (1987).

[4] Y. S. Abu-Mostafa and J. St. Jacques, *IEEE Tran. on Information Theory* , Vol. IT-31, pp. 461-464 (1985).

[5] R. Lippmann, *IEEE ASSP Magazine* , Vol. 4, No. 2, pp. 4-22 (April 1987).

[6] T. Kohonen, *Associative Memory - A System-Theoretical Approach* (Springer-Verlag, Berlin Heidelberg, 1977).

[7] S. S. Venkatesh,*Linear Map with Point Rules* , Ph. D Thesis, Caltech, 1987.

[8] E. R. Berlekamp, *Algebraic Coding Theory* , Aegean Park Press, 1984.

THE CAPACITY OF THE KANERVA ASSOCIATIVE MEMORY IS EXPONENTIAL

P. A. Chou[1]
Stanford University, Stanford, CA 94305

ABSTRACT

The capacity of an associative memory is defined as the maximum number of words that can be stored and retrieved reliably by an address within a given sphere of attraction. It is shown by sphere packing arguments that as the address length increases, the capacity of any associative memory is limited to an exponential growth rate of $1 - h_2(\delta)$, where $h_2(\delta)$ is the binary entropy function in bits, and δ is the radius of the sphere of attraction. This exponential growth in capacity can actually be achieved by the Kanerva associative memory, if its parameters are optimally set. Formulas for these optimal values are provided. The exponential growth in capacity for the Kanerva associative memory contrasts sharply with the sub-linear growth in capacity for the Hopfield associative memory.

ASSOCIATIVE MEMORY AND ITS CAPACITY

Our model of an associative memory is the following. Let (X, Y) be an (address, datum) pair, where X is a vector of n ±1s and Y is a vector of m ±1s, and let $(X^{(1)}, Y^{(1)}), \ldots, (X^{(M)}, Y^{(M)})$, be M (address, datum) pairs stored in an associative memory. If the associative memory is presented at the input with an address X that is close to some stored address $X^{(j)}$, then it should produce at the output a word Y that is close to the corresponding contents $Y^{(j)}$. To be specific, let us say that an associative memory can *correct fraction δ errors* if an X within Hamming distance $n\delta$ of $X^{(j)}$ retrieves Y equal to $Y^{(j)}$. The Hamming sphere around each $X^{(j)}$ will be called the sphere of attraction, and δ will be called the radius of attraction.

One notion of the capacity of this associative memory is the maximum number of words that it can store while correcting fraction δ errors. Unfortunately, this notion of capacity is ill-defined, because it depends on exactly which (address, datum) pairs have been stored. Clearly, no associative memory can correct fraction δ errors for *every* sequence of stored (address, datum) pairs. Consider, for example, a sequence in which several different words are written to the same address. No memory can reliably retrieve the contents of the overwritten words. At the other extreme, any associative memory can store an unlimited number of words and retrieve them all reliably, if their contents are identical.

A useful definition of capacity must lie somewhere between these two extremes. In this paper, we are interested in the largest M such that for *most* sequences of addresses $X^{(1)}, \ldots, X^{(M)}$ and *most* sequences of data $Y^{(1)}, \ldots, Y^{(M)}$, the memory can correct fraction δ errors. We define

[1]This work was supported by the National Science Foundation under NSF grant IST-8509860 and by an IBM Doctoral Fellowship.

'*most* sequences' in a probabilistic sense, as some set of sequences with total probability greater than say, .99. When all sequences are equiprobable, this reduces to the deterministic version: 99% of all sequences.

In practice it is too difficult to compute the capacity of a given associative memory with inputs of length n and outputs of length m. Fortunately, though, it is easier to compute the asymptotic rate at which M increases, as n and m increase, for a given family of associative memories. This is the approach taken by McEliece et al. [1] towards the capacity of the Hopfield associative memory. We take the same approach towards the capacity of the Kanerva associative memory, and towards the capacities of associative memories in general. In the next section we provide an upper bound on the rate of growth of the capacity of any associative memory fitting our general model. It is shown by sphere packing arguments that capacity is limited to an exponential rate of growth of $1 - h_2(\delta)$, where $h_2(\delta)$ is the binary entropy function in bits, and δ is the radius of attraction. In a later section it will turn out that this exponential growth in capacity can actually be achieved by the Kanerva associative memory, if its parameters are optimally set. This exponential growth in capacity for the Kanerva associative memory contrasts sharply with the sub-linear growth in capacity for the Hopfield associative memory [1].

A UNIVERSAL UPPER BOUND ON CAPACITY

Recall that our definition of the capacity of an associative memory is the largest M such that for *most* sequences of addresses $X^{(1)}, \ldots, X^{(M)}$ and *most* sequences of data $Y^{(1)}, \ldots, Y^{(M)}$, the memory can correct fraction δ errors. Clearly, an upper bound to this capacity is the largest M for which there exists *some* sequence of addresses $X^{(1)}, \ldots, X^{(M)}$ such that for *most* sequences of data $Y^{(1)}, \ldots, Y^{(M)}$, the memory can correct fraction δ errors. We now derive an expression for this upper bound.

Let δ be the radius of attraction and let $D_H(X^{(j)}, d)$ be the sphere of attraction, *i.e.*, the set of all Xs at most Hamming distance $d = \lfloor n\delta \rfloor$ from $X^{(j)}$. Since by assumption the memory corrects fraction δ errors, every address $X \in D_H(X^{(j)}, d)$ retrieves the word $Y^{(j)}$. The size of $D_H(X^{(j)}, d)$ is easily shown to be independent of $X^{(j)}$ and equal to $\nu_{n,d} = \sum_{k=0}^{d} \binom{n}{k}$, where $\binom{n}{k}$ is the binomial coefficient $n!/k!(n-k)!$. Thus out of a total of 2^n n-bit addresses, at least $\nu_{n,d}$ addresses retrieve $Y^{(1)}$, at least $\nu_{n,d}$ addresses retrieve $Y^{(2)}$, at least $\nu_{n,d}$ addresses retrieve $Y^{(3)}$, and so forth. It follows that the total number of distinct $Y^{(j)}$s can be at most $2^n/\nu_{n,d}$. Now, from Stirling's formula it can be shown that if $d \leq n/2$, then $\nu_{n,d} = 2^{nh_2(d/n) + \mathcal{O}(\log n)}$, where $h_2(\delta) = -\delta \log_2 \delta - (1-\delta) \log_2(1-\delta)$ is the binary entropy function in bits, and $\mathcal{O}(\log n)$ is some function whose magnitude grows more slowly than a constant times $\log n$. Thus the total number of distinct $Y^{(j)}$s can be at most $2^{n(1 - h_2(\delta)) + \mathcal{O}(\log n)}$. Since any set containing '*most* sequences' of M m-bit words will contain a large number of distinct words (if m is

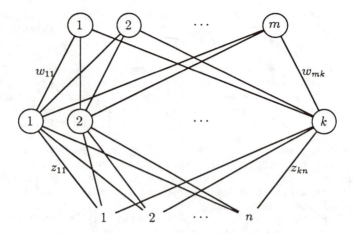

Figure 1: Neural net representation of the Kanerva associative memory. Signals propagate from the bottom (input) to the top (output). Each arc multiplies the signal by its weight; each node adds the incoming signals and then thresholds.

sufficiently large --- see [2] for details), it follows that

$$M \le 2^{n(1-h_2(\delta))+\mathcal{O}(\log n)}. \tag{1}$$

In general a function $f(n)$ is said to be $\mathcal{O}(g(n))$ if $f(n)/g(n)$ is bounded, *i.e.*, if there exists a constant α such that $|f(n)| \le \alpha|g(n)|$ for all n. Thus (1) says that there exists a constant α such that $M \le 2^{n(1-h_2(\delta))+\alpha \log n}$. It should be emphasized that since α is unknown, this bound has no meaning for fixed n. However, it indicates that asymptotically in n, the maximum exponential rate of growth of M is $1 - h_2(\delta)$.

Intuitively, only a sequence of addresses $X^{(1)}, \ldots, X^{(M)}$ that optimally pack the address space $\{-1, +1\}^n$ can hope to achieve this upper bound. Remarkably, *most* such sequences are optimal in this sense, when n is large. The Kanerva associative memory can take advantage of this fact.

THE KANERVA ASSOCIATIVE MEMORY

The Kanerva associative memory [3,4] can be regarded as a two-layer neural network, as shown in Figure 1, where the first layer is a preprocessor and the second layer is the usual Hopfield style array. The preprocessor essentially encodes each n-bit input address into a very large k-bit internal representation, $k \gg n$, whose size will be permitted to grow exponentially in n. It does not seem surprising, then, that the capacity of the Kanerva associative memory can grow exponentially in n, for it is known that the capacity of the Hopfield array grows almost linearly in k, assuming the coordinates of the k-vector are drawn at random by independent flips of a fair coin [1].

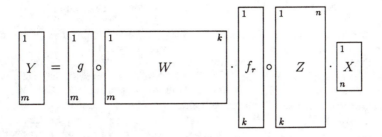

Figure 2: Matrix representation of the Kanerva associative memory. Signals propagate from the right (input) to the left (output). Dimensions are shown in the box corners. Circles stand for functional composition; dots stand for matrix multiplication.

In this situation, however, such an assumption is ridiculous: Since the k-bit internal representation is a function of the n-bit input address, it can contain at most n bits of information, whereas independent flips of a fair coin contain k bits of information. Kanerva's primary contribution is therefore the specification of the preprocessor, that is, the specification of how to map each n-bit input address into a very large k-bit internal representation.

The operation of the preprocessor is easily described. Consider the matrix representation shown in Figure 2. The matrix Z is randomly populated with ± 1s. This randomness assumption is required to ease the analysis. The function f_r is 1 in the ith coordinate if the ith row of Z is within Hamming distance r of X, and is 0 otherwise. This is accomplished by thresholding the ith input against $n - 2r$. The parameters r and k are two essential parameters in the Kanerva associative memory. If r and k are set correctly, then the number of 1s in the representation $f_r(ZX)$ will be very small in comparison to the number of 0s. Hence $f_r(ZX)$ can be considered to be a sparse internal representation of X.

The second stage of the memory operates in the usual way, except on the internal representation of X. That is, $Y = g(Wf_r(ZX))$, where

$$W = \sum_{j=1}^{M} Y^{(j)}[f_r(ZX^{(j)})]^t, \qquad (2)$$

and g is the threshold function whose ith coordinate is $+1$ if the ith input is greater than 0 and -1 is the ith input is less than 0. The ith column of W can be regarded as a memory location whose address is the ith row of Z. Every X within Hamming distance r of the ith row of Z accesses this location. Hence r is known as the *access radius*, and k is the *number of memory locations*.

The approach taken in this paper is to fix the linear rate ρ at which r grows with n, and to fix the exponential rate κ at which k grows with n. It turns out that the capacity then grows at a fixed exponential rate $C_{\rho,\kappa}(\delta)$, depending on ρ, κ, and δ. These exponential rates are sufficient to overcome the standard loose but simple polynomial bounds on the errors due to combinatorial approximations.

THE CAPACITY OF THE KANERVA ASSOCIATIVE MEMORY

Fix $0 \leq \kappa \leq 1$, $0 \leq \rho \leq 1/2$, and $0 \leq \delta \leq \min\{2\rho, 1/2\}$. Let n be the input address length, and let m be the output word length. It is assumed that m is at most polynomial in n, i.e., $m = \exp\{\mathcal{O}(\log n)\}$. Let $r = \lfloor \rho n \rfloor$ be the access radius, let $k = 2^{\lfloor \kappa n \rfloor}$ be the number of memory locations, and let $d = \lfloor \delta n \rfloor$ be the radius of attraction. Let M_n be the number of stored words. The components of the n-vectors $X^{(1)}, \ldots, X^{(M_n)}$, the m-vectors $Y^{(1)}, \ldots, Y^{(M_n)}$, and the $k \times n$ matrix Z are assumed to be IID equiprobable ± 1 random variables. Finally, given an n-vector X, let $Y = g(W f_r(ZX))$ where $W = \sum_{j=1}^{M_n} Y^{(j)} [f_r(ZX^{(j)})]^t$.

Define the quantity

$$C_{\rho,\kappa}(\delta) = \begin{cases} 2\delta + 2(1-\delta)h(\frac{\rho - \delta/2}{1-\delta}) + \kappa - 2h(\rho) & \text{if } \kappa \leq \kappa_0(\rho) \\ C_{\rho,\kappa_0(\rho)}(\delta) & \text{if } \kappa > \kappa_0(\rho) \end{cases} \quad , \tag{3}$$

where

$$\kappa_0(\rho) = 2h(\rho) - 2\gamma - 2(1-\gamma)h(\frac{\rho - \gamma/2}{1-\gamma}) + 1 - h(\gamma) \tag{4}$$

and

$$\gamma = \tfrac{3}{4} - \sqrt{\tfrac{9}{16} - 2\rho(1-\rho)}.$$

Theorem: If

$$M_n \leq 2^{nC_{\rho,\kappa}(\delta) + \mathcal{O}(\log n)}$$

then for all $\epsilon > 0$, all sufficiently large n, all $j \in \{1, \ldots, M_n\}$, and all $X \in D_H(X^{(j)}, d)$,

$$P\{Y \neq Y^{(j)}\} < \epsilon.$$

Proof: See [2].

Interpretation: If the exponential growth rate of the number of stored words M_n is asymptotically less than $C_{\rho,\kappa}(\delta)$, then for every sufficiently large address length n, there is some realization of the $n \times 2^{n\kappa}$ preprocessor matrix Z such that the associative memory can correct fraction δ errors for *most* sequences of M_n (address, datum) pairs. Thus $C_{\rho,\kappa}(\delta)$ is a lower bound on the exponential growth rate of the capacity of the Kanerva associative memory with access radius $n\rho$ and number of memory locations $2^{n\kappa}$.

Figure 3 shows $C_{\rho,\kappa}(\delta)$ as a function of the radius of attraction δ, for $\kappa = \kappa_0(\rho)$ and $\rho = 0.1$, 0.2, 0.3, 0.4 and 0.45. For any fixed access radius ρ, $C_{\rho,\kappa_0(\rho)}(\delta)$ decreases as δ increases. This reflects the fact that fewer (address, datum) pairs can be stored if a greater fraction of errors must be corrected. As ρ increases, $C_{\rho,\kappa_0(\rho)}(\delta)$ begins at a lower point but falls off less steeply. In a moment we shall see that ρ can be adjusted to provide the optimal performance for a given δ.

Not shown in Figure 3 is the behavior of $C_{\rho,\kappa}(\delta)$ as a function of κ. However, the behavior is simple. For $\kappa > \kappa_0(\rho)$, $C_{\rho,\kappa}(\delta)$ remains unchanged, while for $\kappa \leq \kappa_0(\rho)$, $C_{\rho,\kappa}(\delta)$ is simply shifted down by the difference $\kappa_0(\rho) - \kappa$. This establishes the conditions under which the Kanerva associative memory is robust against random component failures. Although increasing the number of memory locations beyond $2^{n\kappa_0(\rho)}$ does not increase the capacity, it does increase robustness. Random

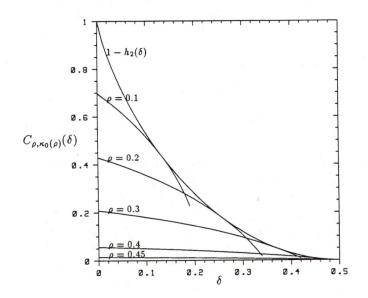

Figure 3: Graphs of $C_{\rho,\kappa_0(\rho)}(\delta)$ as defined by (3). The upper envelope is $1 - h_2(\delta)$.

component failures will not affect the capacity until so many components
have failed that the number of surviving memory locations is less than
$2^{n\kappa_0(\rho)}$.

Perhaps the most important curve exhibited in Figure 3 is the
sphere packing upper bound $1 - h_2(\delta)$, which is achieved for a particular
ρ by $\delta = \frac{3}{4} - \sqrt{\frac{9}{16} - 2\rho(1-\rho)}$. Equivalently, the upper bound is achieved
for a particular δ by ρ equal to

$$\rho_0(\delta) = \frac{1}{2} - \sqrt{\frac{1}{4} - \frac{3}{4}\delta(1 - \frac{2}{3}\delta)}. \tag{5}$$

Thus (4) and (5) specify the optimal values of the parameters κ and ρ,
respectively. These functions are shown in Figure 4. With these
optimal values, (3) simplifies to

$$C_{\rho,\kappa}(\delta) = 1 - h(\delta),$$

the sphere packing bound.

It can also be seen that for $\delta = 0$ in (3), the exponential growth
rate of the capacity is asymptotically equal to κ, which is the
exponential growth rate of the number of memory locations, k_n. That is,
$M_n = 2^{n\kappa + \mathcal{O}(\log n)} = k_n \cdot 2^{\mathcal{O}(\log n)}$. Kanerva [3] and Keeler [5] have argued
that the capacity at $\delta = 0$ is proportional to the number of memory
locations, i.e., $M_n = k_n \cdot \beta$, for some constant β. Thus our results are
consistent with those of Kanerva and Keeler, provided the 'polynomial'
$2^{\mathcal{O}(\log n)}$ can be proved to be a constant. However, the usual statement of
their result, $M = k \cdot \beta$, that the capacity is simply proportional to the
number of memory locations, is false, since in light of the universal

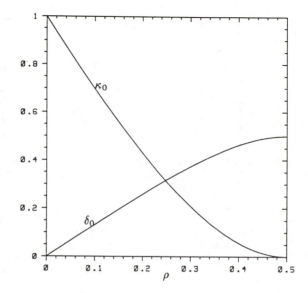

Figure 4: Graphs of $\kappa_0(\rho)$ and $\delta_0(\rho)$, the inverse of $\rho_0(\delta)$, as defined by (4) and (5).

upper bound, it is impossible for the capacity to grow without bound, with no dependence on the dimension n. In our formulation, this difficulty does not arise because we have explicitly related the number of memory locations to the input dimension: $k_n = 2^{n\kappa}$. In fact, our formulation provides explicit, coherent relationships between all of the following variables: the capacity M, the number of memory locations k, the input and output dimensions n and m, the radius of attraction δ, and the access radius ρ. We are therefore able to generalize the results of [3,5] to the case $\delta > 0$, and provide explicit expressions for the asymptotically optimal values of ρ and κ as well.

CONCLUSION

We described a fairly general model of associative memory and selected a useful definition of its capacity. A universal upper bound on the growth of the capacity of such an associative memory was shown by a sphere packing argument to be exponential with rate $1 - h_2(\delta)$, where $h_2(\delta)$ is the binary entropy function and δ is the radius of attraction. We reviewed the operation of the Kanerva associative memory, and stated a lower bound on the exponential growth rate of its capacity. This lower bound meets the universal upper bound for optimal values of the memory parameters ρ and κ. We provided explicit formulas for these optimal values. Previous results for $\delta = 0$ stating that the capacity of the Kanerva associative memory is proportional to the number of memory locations cannot be strictly true. Our formulation corrects the problem and generalizes those results to the case $\delta > 0$.

REFERENCES

1. R.J. McEliece, E.C. Posner, E.R. Rodemich, and S.S. Venkatesh, ''The capacity of the Hopfield associative memory,'' *IEEE Transactions on Information Theory*, submitted.
2. P.A. Chou, ''The capacity of the Kanerva associative memory,'' *IEEE Transactions on Information Theory*, submitted.
3. P. Kanerva, ''Self-propagating search: a unified theory of memory,'' Tech. Rep. CSLI-84-7, Stanford Center for the Study of Language and Information, Stanford, CA, March 1984.
4. P. Kanerva, ''Parallel structures in human and computer memory,'' in *Neural Networks for Computing*, (J.S. Denker, ed.), New York: American Institute of Physics, 1986.
5. J.D. Keeler, ''Comparison between sparsely distributed memory and Hopfield-type neural network models,'' Tech. Rep. RIACS TR 86.31, NASA Research Institute for Advanced Computer Science, Mountain View, CA, Dec. 1986.

PHASE TRANSITIONS IN NEURAL NETWORKS

Joshua Chover
University of Wisconsin, Madison, WI 53706

ABSTRACT

Various simulations of cortical subnetworks have evidenced
something like phase transitions with respect to key parameters.
We demonstrate that such transitions must indeed exist in analogous
infinite array models. For related finite array models classical
phase transitions (which describe steady-state behavior) may not
exist, but there can be distinct qualitative changes in
("metastable") transient behavior as key system parameters pass
through critical values.

INTRODUCTION

Suppose that one stimulates a neural network – actual or
simulated – and in some manner records the subsequent firing
activity of cells. Suppose further that one repeats the experiment
for different values of some parameter (p) of the system; and that
one finds a "critical value" (p_c) of the parameter, such that
(say) for values $p > p_c$ the activity tends to be much higher than
it is for values $p < p_c$. Then, by analogy with statistical
mechanics (where, e.g., p may be temperature, with critical
values for boiling and freezing) one can say that the neural
network undergoes a "phase transition" at p_c. <u>Intra</u>cellular phase
transitions, parametrized by membrane potential, are well known.
Here we consider <u>inter</u>cellular phase transitions. These have been
evidenced in several detailed cortical simulations: e.g., of the
piriform cortex[1] and of the hippocampus[2]. In the piriform case,
the parameter p represented the frequency of high amplitude
spontaneous EPSPs received by a typical pyramidal cell; in the
hippocampal case, the parameter was the ratio of inhibitory to
excitatory cells in the system.

By what mechanisms could approach to, and retreat from, a
critical value of some parameter be brought about? An intriguing
conjecture is that neuromodulators can play such a role in certain
networks; temporarily raising or depressing synaptic efficacies[3].
What possible interesting consequences could approach to
criticality have for system performance. <u>Good</u> effects could be
these: for a network with plasticity, heightened firing response
to a stimulus can mean faster changes in synaptic efficacies, which
would bring about faster memory storage. More and longer activity
could also mean faster access to memory. A <u>bad</u> effect of

near-criticality – depending on other parameters – can be wild, epileptiform activity.

Phase transitions as they might relate to neural networks have been studied by many authors[4]. Here, for clarity, we look at a particular category of network models – abstracted from the piriform cortex setting referred to above – and show the following:

a) For "elementary" reasons, phase transition would have to exist if there were infinitely many cells; and the near-subcritical state involves prolonged cellular firing activity in response to an initial stimulation.

b) Such prolonged firing activity takes place for analogous large finite cellular arrays – as evidenced also by computer simulations.

What we shall be examining is space-time patterns which describe the mid-term transient activity of (Markovian) systems that tend to silence (with high probability) in the long run. (There is no reference to energy functions, nor to long-run stable firing rates – as such rates would be zero in most of our cases.)

In the following models time will proceed in discrete steps. (In the more complicated settings these will be short in comparison to other time constants, so that the effect of quantization becomes smaller.) The parameter p will be the probability that at any given time a given cell will experience a certain amount of excitatory "spontaneous firing" input: by itself this amount will be insufficient to cause the cell to fire, but in conjunction with sufficiently many excitatory inputs from other cells it can assist in reaching firing threshold. (Other related parameters such as average firing threshold value and average efficacy value give similar results.) In all the models there is a refractory period after a cell fires, during which it cannot fire again; and there may be local (shunt type) inhibition by a firing cell on near neighbors as well as on itself – but there is no long-distance inhibition. We look first at limiting cases where there are infinitely many cells and — classically – phase transition appears in a sharp form.

A "SIMPLE" MODEL

We consider an infinite linear array of similar cells which obey the following rules, pictured in Fig. 1A:

(i) If cell k fires at time n, then it must be silent at time n+1;

(ii) if cell k is silent at time n but both of its neighbors k-1 and k+1 do fire at time n, then cell k fires at time n+1;

(iii) if cell k is silent at time n and just one of its neighbors (k-1 or k+1) fires at time n, then cell k will fire at time n+1 with probability p and not fire with probability 1-p, independently of similar decisions at other cells and at other times.

194

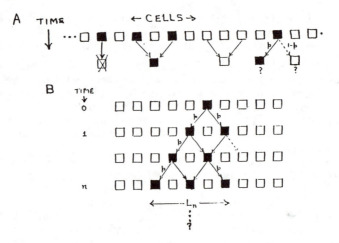

Fig. 1. "Simple model". A: firing rules; cells are represented
horizontally, time proceeds downwards; filled squares
denote firing. B: sample development.

Thus, effectively, signal propagation speed here is one cell
per unit time, and a cell's firing threshold value is 2 (EPSP
units). If we stimulate <u>one</u> cell to fire at time n=0, will its
influence necessarily die out or can it go on forever? (See
Fig. 1B.) For an answer we note that in this simple case the
firing pattern (if any) at time n must be an alternating stretch
of firing/silent cells of some length, call it L_n. Moreover,

$L_{n+1} = L_n+2$ with probability p^2 (when there are sponteneous
firing assists on both ends of the stretch), or $L_{n+1} = L_n-2$ with

probability $(1-p)^2$ (when there is no assist at either end of the
stretch), or $L_{n+1} = L_n$ with probability $2p(1-p)$ (when there is
an assist at just one end of the stretch).

Starting with any finite alternating stretch L_0, the

successive values L_n constitute a "random walk" among the

nonnegative integers. Intuition and simple analysis[5] lead to the
same conclusion: if the probability for L_n to decrease $((1-p)^2)$

is greater than that for it to increase (p^2) – i.e. if the average
step taken by the random walk is negative – then ultimately L_n
will reach 0 and the firing response dies out. Contrariwise, if

$p^2 > (1-p)^2$ then the L_n can drift to even higher values with
positive probability. In Fig. 2A we sketch the probability for
ultimate die-out as a function of p; and in Fig. 2B, the average
time until die out. Figs. 2A and B show a classic example of phase
transition ($p_c = 1/2$) for this infinite array.

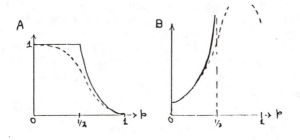

Fig. 2. Critical behavior. A: probability of ultimate die out (or
        of reaching other traps, in finite array case).
        B: average time until die-out (or for reaching other
        traps). Solid curves refer to an infinite array; dashed,
        to finite arrays.

MORE COMPLEX MODELS

For an infinite linear array of cells, as sketched in Fig. 3 ,
we describe now a much more general (and hopefully more realistic)
set of rules:
        (i') A cell cannot fire, nor receive excitatory inputs, at
time n if it has fired at any time during the preceding m_R time
units (refraction and feedback inhibition).
        (ii') Each cell x has a local "inhibitory neighborhood"
consisting of a number (j) of cells to its immediate right and
left. The given cell x cannot fire or receive excitatory inputs
at time n if any other cell y in its inhibitory neighborhood
has fired at any time between t and $t+m_I$ units preceding n,
where t is the time it would take for a message to travel from y
to x at a speed of v_I cells per unit time. (This rule
represents local shunt-type inhibition.)
        (iii') Each cell x has an "excitatory neighborhood"
consisting of a number (e) of cells to the immediate right and left
of its inhibitory neighborhood. If a cell y in that neighborhood
fires at a certain time, that firing causes a unit impulse to
travel to cell x at a speed of v_E cells per unit time. The
impulse is received at x subject to rules (i') and (ii').

(iv') All cells share a "firing threshold" value θ and an "integration time constant." s (s < θ). In addition each cell, at each time n and independently of other times and other cells, can receive a random amount X_n of "spontaneous excitatory input".

The variable X_n can have a general distribution; however, for simplicity we suppose here that it assumes only one of two values: b or 0, with probabilities p and 1-p respectively. (We suppose that b < θ, so that the spontaneous "assist" itself is insufficient for firing.) The above quantities enter into the following <u>firing rule</u>: a cell will fire at time n if it is not prevented by rules (i') and (ii') and if the total number of inputs from other cells, received during the integration "window" lasting between times n-s+1 and n inclusive, plus the assist X_n,

equals or exceeds the threshold θ.

(The propagation speeds v_I and V_E and the neighborhoods are here given left-right symmetry merely for ease in exposition.)

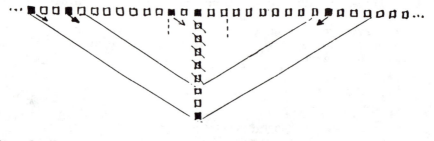

Fig. 3. Message travel in complex model: see text rules (i')-(iv').

Will such a model display phase transition at some critical value of the spontaneous firing frequency p ? The dependence of responses upon the initial conditions and upon the various parameters is intricate and will affect the answer. We briefly discuss here conditions under which the answer is again yes.

(1) For a given configuration of parameters and a given initial stimulation (of a stretch of contiguous cells) we compare the development of the model's firing response first to that of an auxiliary "more active" system: Suppose that L_n now denotes the distance at time n between the left- and right-most cells which are either firing or in refractory mode. Because no cell can fire without influence from others and because such influence travels at a given speed, there is a maximal amount (D) whereby L_{n+1} can exceed L_n. There is also a maximum probability Q(p) - which

depends on the spontaneous firing parameter p – that $L_{n+1} \geq L_n$
(whatever n). We can compare L_n with a random walk "A_n"
defined so that $A_{n+1} = A_n + D$ with probability $Q(p)$ and
$A_{n+1} = A_n - 1$ with probability $1 - Q(p)$. At each transition, A_n is
more likely to increase than L_n. Hence L_n is more likely to die
out than A_n. In the many cases where $Q(p)$ tends to zero as p
does, the average step size of A_n (viz., $DQ(p) + (-1)(1 - Q(b))$)
will become negative for p below a "critical" value p_a. Thus,
as in the "simple" model above, the probability of ultimate die-out
for the A_n, hence also for the L_n of the complex model, will be
1 when $0 \leq p < p_a$.

(2) There will be a phase transition for the complex model if
its probability of die out – given the same parameters and initial
stimulation is in (1) – becomes less than 1 for some p values
with $p_a < p < 1$. Comparison of the complex process with a simpler
"less active" process is difficult in general. However, there are
parameter configurations which ultimately can channel all or part
of the firing activity into a (space-time) sublattice analgous to
that in Fig. 1. Fig. 4 illustrates such a case. For p
sufficiently large there is positive probability that the activity
will not die out, just as in the "simple" model.

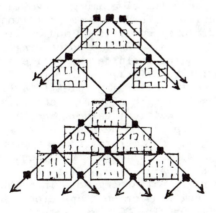

Fig. 4. Activity on a sublattice. (Parameter values: j=2, e=6,
M_R=2, M_I=1, V_R=V_I=1, θ=3, s=2, and b=1.) Rectangular
areas indicate refraction/inhibition; diagonal lines,
excitatory influence.

LARGE FINITE ARRAYS

Consider now a large finite array of N cells, again as
sketched in Fig. 3 ; and operating according to rules similar to
(i')-(iv') above, with suitable modifications near the edges.
Appropriately encoded, its activity can be described by a (huge)
Markov transition matrix, and – depending on the initial
stimulation – must tend[5] to one of a set of steady-state
distributions over firing patterns. For example, (α) if N is
odd and the rules are those for Fig. 1, then extinction is the
unique steady state, for <u>any</u> $p < 1$ (since the L_n form a random
walk with "reflecting" upper barrier). But, (β) if N is even
and the cells are arranged in a ring, then, for any p with
$0 < p < 1$, both extinction and an alternate flip-flop firing
pattern of period 2 are "traps" for the system – with relative long
run probabilities determined by the initial state. See the dashed
line in Fig. 2A for the extinction probability in the (β) case,
and in Fig. 2B for the expected time until hitting a trap in the
(α) case $(p<\frac{1}{2})$ and the (β) case.

What qualitative properties related to phase transition and
critical p values carry over from the infinite to the finite
array case? The (α) example above shows that <u>long term</u> activity
may now be the same for all $0 < p < 1$ but that parameter
intervals can exist whose key feature is a <u>particularly large
expected time</u> before the system hits a trap. (Again, the critical
region can depend upon the initial stimulation.) Prior to being
trapped the system spends its time among many states in a kind of
"metastable" equilibrium. (We have some preliminary theoretical
results on this conditional equilibrium and on its relation to the
infinite array case. See also Ref. 6 concerning time scales for
which certain corresponding infinite and finite stochastic automata
systems display similar behavior.)

Simulation of models satisfying rules (i')-(iv') does indeed
display large changes in length of firing activity corresponding to
parameter changes near a critical value. See Fig. 5 for a typical
example: As a function of p, the expected time until the system
is trapped (for the given parameters) rises approximately linearly
in the interval $.05<p<.12$, with most runs resulting in extinction
– as is the case in Fig. 5A at time $n=115$ (for $p=.10$). But for
$p>.15$ a relatively rigid patterning sets in which leads with high
probability to very long runs or to traps other than extinction –
as is the case in Fig. 5B $(p=.20)$ where the run is arbitrarity
truncated at $n=525$. (The patterning is highly influenced by the
large size of the excitatory neighborhoods.)

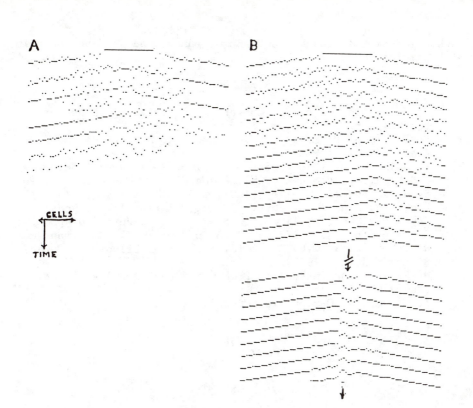

Fig. 5. Space time firing patterns for one configuration of basic parameters. (There are 200 cells; j=2, e=178, M_R=10, M_I=9, V_R=V_I=7, θ=25, s=2, and b=12; 50 are stimulated initially.) A: p=.10. B: p=.20.

CONCLUSION

Mechanisms such as neuromodulators, which can (temporarily) bring spontaneous firing levels – or synaptic efficacies, or average firing thresholds, or other similar parameters – to near-critical values, can thereby induce large amplification of response activity to selected stimuli. The repertoire of such responses is an important aspect of the system's function.

[Acknowledgement: Thanks to C. Bezuidenhout and J. Kane for help with simulations.]

REFERENCES

1. M. Wilson, J. Bower, J. Chover, L. Haberly, 16th Neurosci. Soc. Mtg. Abstr. 370.11 (1986).
2. R. D. Traub, R. Miles, R.K.S. Wong, 16th Neurosci. Soc. Mtg. Abstr. 196.12 (1986).
3. A. Selverston, this conference, also, Model Neural Networks and Behavior, Plenum (1985); E. Marder, S. Hooper, J. Eisen, Synaptic Function, Wiley (1987) p.305.
4. E.g.: W. Kinzel, Z. Phys. B58, p. 231 (1985); A. Noest. Phys. Rev. Let. 57(1), p. 90 (1986); R. Durrett (to appear); G. Carpenter, J. Diff. Eqns. 23, p.335 (1977); G. Ermentraut, S. Cohen, Biol. Cyb. 34, p.137 (1979); H. Wilson, S. Cowan, Biophys. J. 12 (1972).
5. W. Feller, An Introd. to Prob. Th'y. and Appl'ns. I. Wiley (1968) Ch. 14, 15.
6. T. Cox and A. Graven (to appear).

NEW HARDWARE FOR MASSIVE NEURAL NETWORKS

D. D. Coon and A. G. U. Perera
Applied Technology Laboratory
University of Pittsburgh
Pittsburgh, PA 15260.

ABSTRACT

Transient phenomena associated with forward biased silicon $p^+ - n - n^+$ structures at 4.2K show remarkable similarities with biological neurons. The devices play a role similar to the two-terminal switching elements in Hodgkin-Huxley equivalent circuit diagrams. The devices provide simpler and more realistic neuron emulation than transistors or op-amps. They have such low power and current requirements that they could be used in massive neural networks. Some observed properties of simple circuits containing the devices include action potentials, refractory periods, threshold behavior, excitation, inhibition, summation over synaptic inputs, synaptic weights, temporal integration, memory, network connectivity modification based on experience, pacemaker activity, firing thresholds, coupling to sensors with graded signal outputs and the dependence of firing rate on input current. Transfer functions for simple artificial neurons with spiketrain inputs and spiketrain outputs have been measured and correlated with input coupling.

INTRODUCTION

Here we discuss the simulation of neuron phenomena by electronic processes in silicon from the point of view of hardware for new approaches to electronic processing of information which parallel the means by which information is processed in intelligent organisms. Development of this hardware basis is pursued through exploratory work on circuits which exhibit some basic features of biological neural networks. Fig. 1 shows the basic circuit used to obtain spiketrain outputs. A distinguishing feature of this hardware basis is the spontaneous generation of action potentials as a device physics feature.

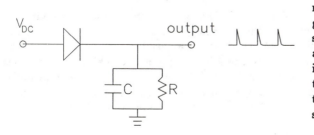

Figure 1: Spontaneous, neuronlike spiketrain generating circuit. The spikes are nearly equal in amplitude so that information is contained in the frequency and temporal pattern of the spiketrain generation.

TWO-TERMINAL SWITCHING ELEMENTS

The use of transistor based circuitry[1] is avoided because transistor electrical characteristics are not similar to neuron characteristics. The use of devices with fundamentally non-neuronlike character increases the complexity of artificial neural networks. Complexity would be an important drawback for massive neural networks and most neural networks in nature achieve their remarkable performance through their massive size. In addition, transistors have three terminals whereas the switching elements of Hodgkin-Huxley equivalent circuits have two terminals. Motivated in part by Hodgkin-Huxley equivalent circuit diagrams, we employ two-terminal $p^+ - n - n^+$ devices which execute transient switching between low conductance and high conductance states. (See Fig. 2) We call these devices injection mode devices (IMDs). In the "OFF-STATE", a typical current through the devices is $\sim 100\,\mathrm{fA/mm^2}$, and in the "ON-STATE" a typical current is $\sim 10\,\mathrm{mA/mm^2}$. Hence this device is an extremely good switch with a ON/OFF ratio of 10^{11}. As in real neurons[2], the current in the device is a function of voltage and time, not only voltage. The devices require cryogenic cooling but this results in an advantageously low quiescent power drain of $< 1\,\mathrm{nanowatt/cm^2}$ of chip area and the very low leakage currents mentioned above. In addition, the highly unique ability of the neural networks described here to operate in a cryogenic environment is an important advantage for infrared image processing at the focal plane (see Fig. 3 and further discussion below). Vision systems begin processing at the focal plane and there are many benefits to be gained from the vision system approach to IR image processing.

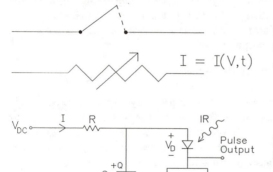

$$I = I(V,t)$$

Figure 2: Switching element in Hodgkin-Huxley equivalent circuits.

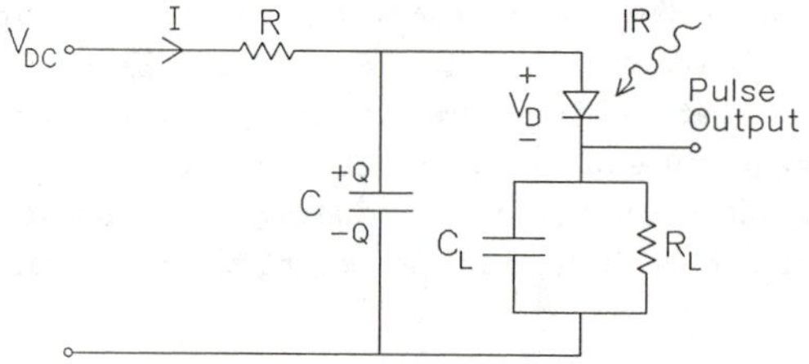

Figure 3: Single stage conversion of infrared intensity to spiketrain frequency with a neuron-like semiconductor device. No pre-amplifiers are necessary.

Coding of graded input signals (see Fig. 4) such as photocurrents into action potential spike trains with millimeter scale devices has been experimentally demonstrated[3] with currents from 1 μA down to about 1 picoampere with coding noise referred to input of < 10 femtoamperes. Coding of much smaller current levels should be possible with smaller devices. Figure 5 clearly shows the threshold behavior of the IMD. For devices studied to date, a transition from action potential output to graded signal output is observed for input currents of the order of 0.5 picoamperes[13].

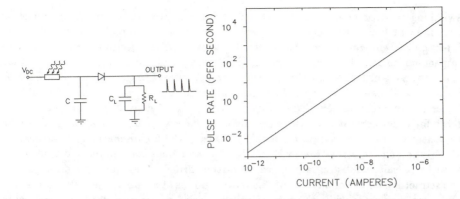

Figure 4: Coding of NIR-VISIBLE-UV intensity into firing frequency of a spiketrain and the experimentally determined firing rate vs. the input current for one device. Note that the dynamic range is about 10^7.

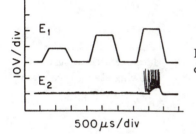

Figure 5: Illustration of the threshold firing of the device in response to input step functions.

This transition is remarkably well described in von Neumann's discussion[5,6] of the mixed character of neural elements which he relates to the concept of subliminal stimulation levels which are too low to produce the stereotypical all-or-nothing response. Neural network modelers frequently adopt viewpoints which ignore this interesting mixed character. The von Neumann viewpoint links the mixed character to concepts of nonlinear dynamics in a way which is not apparent in recent neural network modeling literature. The scaling down of IMD size should result in even lower current requirements for all-or-nothing response.

DEVICE PHYSICS

Recently, neuronlike action potential transients in IMDs have been the subject of considerable research[3,4,7,8,9,10,11,12,13]. In the simple circuits of Fig. 1, the IMD gives rise to a spontaneous neuronlike spiketrain output. Between pulses, the IMD is polarized in the sense that it is in a low conductance state with a substantial voltage occurring across it, even though it is forward biased. The low conductance has been attributed to small interfacial work functions due to band offsets at the n^+-n and p^+-n interfaces[8].

Low temperatures inhibit thermionic injection of electrons and holes into the n-region from the n^+-layer and p^+-layer impurity bands[14]. Pulses are caused by

204

switching to depolarized states with low diode potential drops and large injection currents which are believed to be triggered by the slow buildup of a small thermionic injection current from the n^+-layer into the n-region. The injection current can cause impact ionization of n-region donor impurities resulting in an increasingly positive space charge which further enhances the injection current to the point where the IMD abruptly switches to the low conductance state with large injection current. Switching times are typically under 100ns. Charging of the load capacitance C_L cuts off the large injection current and resets the diode to its low conductance state. The load capacitor C_L then discharges through R_L. During the C_L discharging time constant $R_L C_L$ the voltage across the IMD itself is low and therefore the bias voltage would have to be raised substantially to cause further firing. Thus, $R_L C_L$ is analogous to the refractory period of a neuron. The output pulses of an IMD generally have about the same amplitude while the rate of pulsing varies over a wide range depending on the bias voltage and the presence of electromagnetic radiation.[7,8,10]

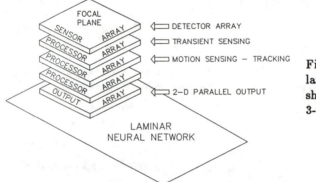

Figure 6: Illustrative laminar architecture showing stacked wafers in 3-dimensions.

REAL TIME PARALLEL ASYNCHRONOUS PROCESSING

The devices described here could form the hardware basis for a parallel asynchronous processor in much the same way that transistors form the basis for digital computers. The devices could be used to construct networks which could perform real time signal processing. Pulse propagation through silicon chips (parallel firethrough, see Fig. 7) as opposed to the lateral planar propagation in conventional integrated circuits has been proposed.[15] This would permit the use of laminar, stacked wafer architectures. See Fig. 6.

Such architectures would eliminate the serial processing limitations of standard processors which utilize multiplexing and charge transfer. There are additional advantages in terms of elimination of pre-amplifiers and reduction in power consumption. The approach would utilize the *low power, low noise* devices[10] described here to perform input signal-to-frequency conversion in every processing channel.

POWER CONSUMPTION FOR A BRAIN SCALE SYSTEM

The low power and low current requirements together with the electronic simplicity (lower parts-count as compared with transistor and op-amp approaches) and

INPUTS

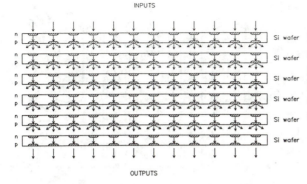

Si wafer
Si wafer
Si wafer
Si wafer
Si wafer
Si wafer

OUTPUTS

Figure 7: Schematic illustration of the signal flow pattern through a real time parallel asynchronous processor consisting of stacked silicon wafers.

the natural emulation of neuron features means that the approach described here would be especially advantageous for very large neural networks, e.g. systems comparable to supercomputers in which power dissipation and system complexity are important considerations. The *power consumption* of large scale analog[16] and digital[17] systems is always a major concern. For example, the power consumption of the CRAY XMP-48 is of the order of 300 kilowatts. For the devices described here, the power consumption is very low. For these devices, we have observed quiescent power drains of about $1 \, \mathrm{nW/cm^2}$ and pulse power consumption of about $500 \, \mathrm{nJ/pulse/cm^2}$. We estimate that a system with 10^{11} active $10 \, \mu m \times 10 \, \mu m$ elements (comparable to the number of neurons in the brain[18]) all firing with an average pulse rate of 1 KHz (corresponding to a high neuronal firing rate[5]) would consume about 50 watts. The quiescent power drain for this system would be 0.1 milliwatts. Thus, power (P) requirements for such an artificial neural network with the size scale (10^{11} pulse generating elements) of the human brain and a range of activity between zero and the maximum conceivable sustained activity for neurons in the brain would be 0.1 milliwatts $< P <$ 50 watts for 10 micron technology. For comparison, we note that von Neumann's estimate for the power dissipation of the brain is of order 10 to 25 watts.[5,6] Fabrication of a 10^{11} element $10 \, \mu m$ artificial neural network would require processing of about 1500 four inch wafers.

NETWORK CONNECTIVITY

For a network with coupling between many IMD's[3] we have shown[4] that

$$C_i R_i \frac{dV_i}{dt} + V_i = \sum_{j=1}^{N} T_{ij} F_j(V_j) + R_i I_i \tag{1}$$

where V_i is the voltage across the diode and the input capacitance C_i of the i-th network node, R_i represents a leakage resistance in parallel with C_i, and I_i represents an external current input to the i-th diode. i,j=1,2,3,..... label different network nodes and T_{ij} incoporates coupling between network elements. Equation 1 has the same form as equations which occur in the Hopfield model[20,21,22,23] for neural networks. Sejnowski has also discussed similar equations in connection with skeleton filters in

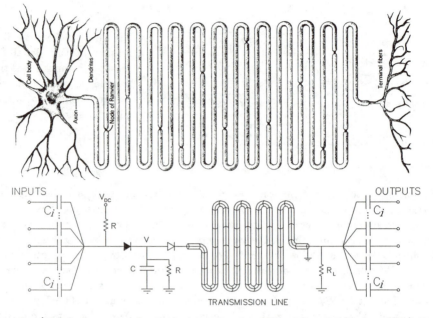

Figure 8: a) Main features of a typical neuron from Kandel and Schwartz.[19] b) Our artificial neuron, which shows the summation over synaptic inputs and fan-out.

the brain.[24,25] Nonlinear threshold behavior of IMD's enters through $F(V)$ as it does in the neural network models.

In Fig. 8-b a range of input capacitances is possible. This range of capacitances is related to the range of possible synaptic weights. The circuit in Fig. 8 accomplishes pulse height discrimination and each pulse can contribute to the charge stored on the central node capacitance C. The charge added to C during each input pulse is linearly related to the input capacitance except at extreme limits. The range of input capacitances for a particular experiment was $.002 \, \mu F$ to $.2 \, \mu F$ which differ by a factor of about 100. The effect of various input capacitance values (synaptic weights) on input-output firing rates is shown in Fig. 9. Also the Fig. 8-b shows many capacitive inputs/outputs to/from a single IMD. i.e. fan-in and fan-out. For pulses which arrive at different inputs at about the same time, the effect of the pulses is additive. The time within which inputs are summed is just the stored charge lifetime. Summation over many inputs is an important feature of neural information processing.

EXCITATION, INHIBITION, MEMORY

Both excitatory and inhibitory input circuits are shown in Fig. 10. Input pulses cause the accumulation of charge on C in excitatory circuits and the depletion of charge on C in inhibitory circuits. Charge associated with input spiketrains is integrated/stored on C. The temporally integrated charge is depleted by the firing of the IMD. Thus, the storage time is related to the firing rate. After an input spiketrain raises the potential across C to a value above the firing threshold, the resulting IMD

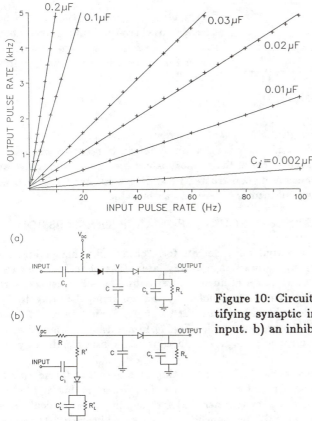

Figure 9: Output pulse rate vs. the input pulse rate for different input capacitance values C_i values

Figure 10: Circuits which incorporate rectifying synaptic inputs. a) an excitatory input. b) an inhibitory input.

output spiketrain codes the input information. The output firing rate is linearly related to the input firing rate times the synaptic coupling strength (linearly related to C_i). See Fig. 9. If the input ceases, then the potential across C relaxes back to a value just below the firing threshold. When not firing, the IMD has a high impedance. If there is negligible leakage of charge from C, then V can remain near V_T (threshold voltage) for a long time and a new input signal will quickly take the IMD over the firing threshold. See Fig. 11. We have observed stored charge lifetimes of 56 days and longer times may be acheivable. The lifetime of charge stored on C can be reduced by adding a resistance in parallel with C.

From the discussion of integration, we see that long term storage of charge on C is equivalent to long term memory. The memory can be read by seeing if a new input pulse or spiketrain produces a prompt output pulse or spiketrain. The read signal input channel in Fig. 8-b can be the same as or different from the channel which resulted in the charge storage. In either case memory would produce a change in the pattern of connectivity if the circuit was imbedded in a neural network. Changes in patterns of connectivity are similar to Hebb's rule considerations[26] in which memory is associated with increases in the strength (weight) of synaptic couplings. Frequently,

208

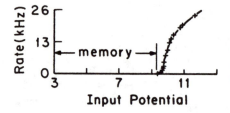

Figure 11: Firing rate vs. the bias voltage. The region where the firing is negligible is associated with memory. The state of the memory is associated with the proximity to the firing threshold.

the increase in synaptic weights is modeled by increased conductance whereas in the circuits in Figs. 10(a) and 8-b memory is achieved by integration and charge storage. Note that for these particular circuits, the memory is not eraseable although volatile (short term) memory can easily be constructed by adding a resistor in parallel with C. Thus, a continuous range of memory lifetimes can be achieved.

2-D PARALLEL ASYNCHRONOUS CHIP-TO-CHIP TRANSMISSION

For many IMD's the output pulse heights for a circuit like that in Fig. 1 are >3 volts. Thus, output from the first stage or any later stage of the network could easily be transmitted to other parts of an overall system. Two-dimensional arrays of devices on different chips could be coupled by indium bump bonding to form the laminar architecture described above. Planar technology could be used for local lateral interconnections in the processor. (See Fig. 7) In addition to transmission of electrical pulses, optical transmission is possible because the pulses can directly drive LED's.

Emerging GaAs-on-Si technology is interesting as a means of fabricating two dimensional emitter arrays. Optical transmission is not necessary but it might be useful (A) for processed image data transfer, (B) for coupling to an optical processor, or (C) to provide 2-D optical interconnects between chips bearing 2-D arrays of $p^+ - n - n^+$ diodes. Note that with optical interconnects between chips, the circuits employed here would be internal receivers. The p-i-n diodes employed in the present work would be well suited to the receiver role. An interesting possibility would entail the use optical interconnects between chips to achieve local, lateral interaction. This would be accomplished by having each optical emitter in a 2-D array broadcast locally to multiple receivers rather than to a single receiver. Similarly, each receiver would have a receptive field extending over multiple transmitters. It is also possible that an optical element could be placed in the gap between parallel transmitter and receiver planes to structure, control or alter 2-D patterns of interconnection. This would be an alternative to a planar technology approach to lateral interconnection. If the optical elements were active then the system would constitute a hybrid optical/electronic processor, whereas if passive optical elements were employed, we would regard the system as an optoelectronic processor. In either case, we picture the processing functions of temporal integration, spatial summation over inputs, coding and pulse generation as residing on-chip.

ACKNOWLEDGEMENTS

The work was supported in part by U.S. DOE under contract #DE-ACO2-80ER10667 and NSF under grant # ECS-8603075.

References

[1] L. D. Harmon, Kybernetik **1**, 89 (1961).

[2] A. L. Hodgkin and A. F. Huxley, J. Physiol **117**, 500 (1952).

[3] D. D. Coon and A. G. U. Perera, Int. J. Electronics **63**, 61 (1987).

[4] K. M. S. V. Bandara, D. D. Coon and R. P. G. Karunasiri, *Infrared Transient Sensing*, to be published.

[5] J. von Neumann, *The Computer and the Brain*, Yale University Press, New Haven and London, 1958.

[6] J. von Neumann, *Collected Works*, Pergamon Press, New York, 1961.

[7] D. D. Coon and A. G. U. Perera, Int. J. Infrared and Millimeter Waves **7**, 1571 (1986).

[8] D. D. Coon and S. D. Gunapala, J. Appl. Phys **57**, 5525 (1985).

[9] D. D. Coon, S. N. Ma and A. G. U. Perera, Phys. Rev. Let. **58**, 1139 (1987).

[10] D. D. Coon and A. G. U. Perera, Applied Physics Letters **51**, 1711 (1987).

[11] D. D. Coon and A. G. U. Perera, Solid-State Electronics **29**, 929 (1986).

[12] D. D. Coon and A. G. U. Perera, Applied Physics Letters **51**, 1086 (1987).

[13] K. M. S. V. Bandara, D.D. Coon and R. P. G. Karunasiri, Appl. Phys. Lett **51**, 961 (1987).

[14] Y. N. Yang, D. D. Coon and P. F. Shepard, Applied Physics Letters **45**, 752 (1984).

[15] D. D. Coon and A. G. U. Perera, Int. J. IR and Millimeter Waves **8**, 1037 (1987).

[16] M. A. Sivilotti, M. R. Emerling and C. A. Mead, *VLSI Architectures for Implementation of Neural Networks*, **Neural Networks for Computing**, A.I.P., 1986, pp. 408–413.

[17] R. W. Keyes, Proc. IEEE **63**, 740 (1975).

[18] E. R. Kandel and J. H. Schwartz, *Principles of Neural Science*, Elsevier, New York, 1985.

[19] E. R. Kandel and J. H. Schwartz, *Principles of Neural Science*, Elsevier, New York, 1985, page 15, Reproduced by permission of Elsevier Science Publishing Co., N.Y..

[20] J. J. Hopfield, Proc. Natl. Acad. Sci. U.S.A **81**, 3088 (1984).

[21] J. J. Hopfield and D. W. Tank, Biol. Cybern **52**, 141 (1985).

[22] J. J. Hopfield and D. W. Tank, Science **233**, 625 (1986).

[23] D. W. Tank and J. J. Hopfield, IEEE. Circuits Syst. **CAS-33**, 533 (1986).

[24] T. J. Sejnowski, J. Math. Biology **4**, 303 (1977).

[25] T. J. Sejnowski, *Skeleton Filters in the Brain*, Lawrence Erlbaum, New Jersey, 1981, pp. 189–212, edited by G. E. Hinton and J. A. Anderson.

[26] J. L. McClelland, D. E. Rumelhart and the PDP research group, *Parallel Distributed Processing*, The MIT Press, Cambridge, Massachusetts, 1986, two volumes.

HIGH DENSITY ASSOCIATIVE MEMORIES[1]

Amir Dembo
Information Systems Laboratory, Stanford University
Stanford, CA 94305

Ofer Zeitouni
Laboratory for Information and Decision Systems
MIT, Cambridge, MA 02139

ABSTRACT

A class of high density associative memories is constructed, starting from a description of desired properties those should exhibit. These properties include high capacity, controllable basins of attraction and fast speed of convergence. Fortunately enough, the resulting memory is implementable by an artificial Neural Net.

INTRODUCTION

Most of the work on associative memories has been structure oriented; i.e., given a Neural architecture, efforts were directed towards the analysis of the resulting network. Issues like capacity, basins of attractions, etc. were the main objects to be analyzed cf., e.g. [1], [2], [3], [4] and references there, among others.

In this paper, we take a different approach; we start by explicitly stating the desired properties of the network, in terms of capacity, etc. Those requirements are given in terms of axioms (c.f. below). Then, we bring a synthesis method which enables one to design an architecture which will yield the desired performance. Surprisingly enough, it turns out that one gets rather easily the following properties:

(a) High capacity (unlimited in the continuous state-space case, bounded only by sphere-packing bounds in the discrete state case).

(b) Guaranteed basins of attractions in terms of the natural metric of the state space.

(c) High speed of convergence in the guaranteed basins of attraction.

Moreover, it turns out that the architecture suggested below is the only one which satisfies all our axioms ("desired properties")!

Our approach is based on defining a potential and following a descent algorithm (e.g., a gradient algorithm). The main design task is to construct such a potential (and, to a lesser extent, an implementation of the descent algorithm via a Neural network). In doing so, it turns out that, for reasons described below, it is useful to regard each desired memory location as a "particle" in the state space. It is natural to require now the following requirement from a

[1] An expanded version of this work has been submitted to Phys. Rev. A. This work was carried out at the Center for Neural Science, Brown University.

memory:

(P1) The potential should be linear w.r.t. adding partic les in the sense that the potential of two particles should be the sum of the potentials induced by the individual particles (i.e., we do not allow interparticles interaction).

(P2) Particle locations are the only possible sites of stable memory locations.

(P3) The system should be invariant to translations and rotations of the coordinates.

We note that the last requirement is made only for the sake of simplicity. It is not essential and may be dropped without affecting the results.

In the sequel, we construct a potential which satisfies the above requirements. We refer the reader to [5] for details of the proofs, etc.

Acknowledgements. We would like to thank Prof. L.N. Cooper and C.M. Bachmann for many fruitful discussions. In particular, section 2 is part of a joint work with them ([6]).

2. HIGH DENSITY STORAGE MODEL

In what follows we present a particular case of a method for the construction of a high storage density neural memory. We define a function with an arbitrary number of minima that lie at preassigned points and define an appropriate relaxation procedure. The general case in presented in [5].

Let $\bar{x}_1, \ldots, \bar{x}_m$ be a set of m arbitrary distinct memories in R^N. The "energy" function we will use is:

$$\xi = -\frac{1}{L} \sum_{i=1}^{m} Q_i |\bar{\mu} - \bar{x}_i|^{-L} \tag{1}$$

where we assume throughout that $N \geq 3$, $L \geq (N - 2)$, and $Q_i > 0$ and use $|\ldots|$ to denote the Euclidean distance. Note that for $L = 1$, $N=3$, ξ is the electrostatic potential induced by negative fixed particles with charges $-Q_i$. This "energy" function possesses global minima at $\bar{x}_1, \ldots, \bar{x}_m$ (where $\xi(\bar{x}_i) = -\infty$) and has no local minima except at these points. A rigorous proof is presented in [5] together with the complete characterization of functions having this property.

As a relaxation procedure, we can choose any dynamical system for which ξ is strictly decreasing, uniformly in compacts. In this instance, the theory of dynamical systems guarantees that for almost any initial data, the trajectory of the system converges to one of the desired points $\bar{x}^1, \ldots, \bar{x}^m$. However, to give concrete results and to further exploit the resemblance to electrostatic, consider the relaxation:

$$\dot{\bar{\mu}} = \bar{E}_{\bar{\mu}} = -\sum_{i=1}^{m} Q_i |\bar{\mu} - \bar{x}_i|^{-(L+2)} (\bar{\mu} - \bar{x}_i) \tag{2}$$

where for N=3, L=1, equation (2) describes the motion of a positive test particle in the electrostatic field $\bar{E}_{\bar{\mu}}$ generated by the negative fixed charges $-Q_1, \ldots, -Q_m$ at $\bar{x}_1, \ldots, \bar{x}_m$.

Since the field $\bar{E}_{\bar{\mu}}$ is just minus the gradient of ξ, it is clear that along trajectories of (2), $d\xi/dt \leq 0$, with equality only at the fixed points of (2), which are exactly the stationary points of ξ.

Therefore, using (2) as the relaxation procedure, we can conclude that entering at any $\bar{\mu}(0)$, the system converges to a stationary point of ξ. The space of inputs is partitioned into m domains of attraction, each one corresponding to a different memory, and the boundaries (a set of measure zero), on which $\bar{\mu}(0)$ will converge to a saddle point of ξ.

We can now explain why $\xi_{\bar{\mu}}$ has no spurious local minima, at least for L=1, N=3, using elementary physical arguments. Suppose ξ has a spurious local minima at $\bar{y} \neq \bar{x}_1, \ldots, \bar{x}_m$, then in a small neighborhood of $\bar{y}$ which does not include any of the $\bar{x}_i$, the field $\bar{E}_{\bar{\mu}}$ points towards $\bar{y}$. Thus, on any closed surface in that neighborhood, the integral of the normal inward component of $\bar{E}_{\bar{\mu}}$ is positive. However, this integral is just the total charge included inside the surface, which is zero. Thus we arrive at a contradiction, so $\bar{y}$ can not be a local minimum.

We now have a relaxation procedure, such that almost any $\bar{\mu}(0)$ is attracted by one of the $\bar{x}_i$, but we have not yet specified the shapes of the basins of attraction. By varying the charges Q_i, we can enlarge one basin of attraction at the expense of the others (and vice versa).

Even when all of the Q_i are equal, the position of the $\bar{x}_i$ might cause $\bar{\mu}(0)$ not to converge to the closest memory, as emphasized in the example in fig. 1. However, let $r = \min_{1 \leq i \neq j \leq m} |\bar{x}_i - \bar{x}_j|$ be the minimal distance between any two memories; then if $|\bar{\mu}(0) - \bar{x}_i| \leq \frac{r}{(1 + 3^{1/k})}$ it can be shown that $\bar{\mu}(0)$ will converge to $\bar{x}_i$, (provided that $k = \frac{L+1}{N+1} \geq 1$). Thus, if the memories are densely packed in a hypersphere, by choosing k large enough (i.e. enlarging the parameter L), convergence to the closest memory for any "interesting" input, that is an input $\bar{\mu}(0)$ with a distinct closest memory, is guaranteed. The detailed proof of the above property is given in [5]. It is based on bounding the number of $\bar{x}_j$, $j \neq i$, in a hypersphere of radius $R(R \geq r)$ around $\bar{x}_i$, by $[2R/r + 1]^N$, then bounding the magnitude of the field induced by any $\bar{x}_j$, $j \neq i$, on the boundary of such a hypersphere by $(R - |\bar{\mu}(0) - \bar{x}_i|)^{-(L+1)}$, and finally integrating to show that for $|\bar{\mu}(0) - \bar{x}_i| \leq \frac{\theta r}{(1 + 3^{1/k})}$, with $\theta < 1$, the convergence of $\bar{\mu}(0)$ to $\bar{x}_i$ is within finite time T, which behaves like θ^{L+2} for L $\gg$ 1 and $\theta < 1$ and fixed. Intuitively the reason for

214

this behaviour is the short-range nature of the fields used in equation (2). Because of this, we also expect extremely low convergence rate for inputs $\bar{\mu}(0)$ far away from all of the $\bar{x}_i$.

Figure 1

$R \gg 1$ and $\delta \ll 1$

The radial nature of these fields suggests a way to overcome this difficulty, that is to increase the convergence rate from points very far away, without disturbing all of the aforementioned desirable properties of the model. Assume that we know in advance that all of the $\bar{x}_i$ lie inside some large hypersphere S around the origin. Then, at any point $\bar{\mu}$ outside S, the field $\bar{E}_{\bar{\mu}}$ has a positive projection radially into S. By adding a long-range force to $\bar{E}_{\bar{\mu}}$, effective only outside of S, we can hasten the movement towards S, from points far away, without creating additional minima inside of S. As an example the force ($-\bar{\mu}$ for $\bar{\mu} \notin$ S; 0 for $\bar{\mu} \in$ S) will pull any test input $\bar{\mu}(0)$ to the boundary of S within the small finite time $T \approx 1/|S|$, and from then on the system will behave inside S according to the original field $\bar{E}_{\bar{\mu}}$.

Up to this point, our derivations have been for a continuous system, but from it we can deduce a discrete system. We shall do this mainly for a clearer comparison between our high density memory model and the discrete version of Hopfield's model. Before continuing in that direction, note that our continuous system has unlimited storage capacity unlike Hopfield's continuous system, which like his discrete model, has limited capacity.

For the discrete system, assume that the $\bar{x}_i$ are composed of elements +1 and replace the Euclidean distance in (1) with the normalized Hamming distance $|\bar{\mu}_1 - \bar{\mu}_2| = \frac{1}{N} \sum_{j=1}^{N} |\mu_j^1 - \mu_j^2|$. This places the vectors $\bar{x}_i$ on the unit hypersphere.

The relaxation process for the discrete system will be of the type defined in Hopfield's model in [1]. Choose at random a component to be updated (that is, a neighbor $\bar{\mu}'$ of $\bar{\mu}$ such that $|\bar{\mu}' - \bar{\mu}| = 2/N$), calculate the "energy" difference, $\delta\xi = \xi(\bar{\mu}') - \xi(\bar{\mu})$, and only if $\delta\xi < 0$, change this component, that is:

$$\mu_i \to \mu_i \; \text{sign}(\xi(\bar{\mu}') - \xi(\bar{\mu})), \tag{3}$$

where $\xi(\bar{\mu})$ is the potential energy in (1). Since there is a finite number of possible $\bar{\mu}$ vectors (2^N), convergence in finite time is guaranteed.

This relaxation procedure is rigid since the movement is limited to points with components +1. Therefore, although the local minima of $\xi(\bar{\mu})$ defined in (2) are only at the desired points $\bar{x}_i$, the relaxation may get stuck at some $\bar{\mu}$ which is not a stationary point of $\xi(\bar{\mu})$. However, the short range behaviour of the potential $\xi(\bar{\mu})$, unlike the long-range behavior of the quadratic potential used by Hopfield, gives rise to results similar to those we have quoted for the continuous model (equation (1)).

Specifically, let the stored memories $\bar{x}_1, \ldots, \bar{x}_m$ be separated from one another by having at least ρN different components ($0 < \rho \leq 1/2$ and ρ fixed), and let $\bar{\mu}(0)$ agree up to at least one $\bar{x}_i$ with at most $\theta\rho N$ errors between them ($0 \leq \theta < 1/2$, with θ fixed), then $\bar{\mu}(0)$ converges monotonically to $\bar{x}_i$ by the relaxation procedure given in equation (3).

This result holds independently of m, provided that N is large enough (typically, $N\rho \ln(\frac{1-\theta}{\theta}) \geq 1$) and L is chosen so that $\frac{N}{L} \leq \ln(\frac{1-\theta}{\theta})$

The proof is constructed by bounding the cummulative effect of terms $|\bar{\mu} - \bar{x}_j|^{-L}$, $j \neq i$, to the energy difference $\delta\xi$ and showing that it is dominated by $|\bar{\mu} - \bar{x}_i|^{-L}$. For details, we refer the reader again to [5].

Note the importance of this property: unlike the Hopfield model which is limited to $m \leq N$, the suggested system is optimal in the sense of Information Theory, since for every set of memories $\bar{x}_1, \ldots, \bar{x}_m$ separated from each other by a Hamming distance ρN, up to $1/2$ ρN errors in the input can be corrected, provided that N is large and L properly chosen.

As for the complexity of the system, we note that the nonlinear operation a^{-L}, for $a > 0$ and L integer (which is at the heart of our system computationally) is equivalent to $e^{-L\ln(a)}$ and can be implemented, therefore, by a simple electrical circuit composed of diodes, which have exponential input-output characteristics, and resistors, which can carry out the necessary multiplications (cf. the implementation of section 3).

Further, since both $|\bar{x}_i|$ and $|\bar{\mu}|$ are held fixed in the discrete system, where all states are on the unit hypersphere, $|\bar{\mu} - \bar{x}_i|^2$ is equivalent to the inner product of $\bar{\mu}$ and $\bar{x}_i$, up to a constant.

To conclude, the suggested model involves about $m \cdot N$ multiplications, followed by m nonlinear operations, and then $m \cdot N$ additions. The original model of Hopfield involves N^2 multiplications and additions, and then N nonlinear operations, but is limited to $m \leq N$. Therefore, whenever the Hopfield model is applicable the complexity of both models is comparable.

3. IMPLEMENTATION

We propose below one possible network which implements the discrete time and space version of the model described above. An implementation for the ocntinuous time case, which is even simpler, is also hinted. We point out that the implementation described below is by no means unique, (and maybe even not the simplest one). Moreover, the "neurons" used are artificial neurons which perform various tasks, as follows: There are (N+1) neurons which are delay elements, and m pointwise non-linear functions (which may be interpreted as delay-less, intermediate neurons). There are mN synaptic connections between those two layers of neurons. In addition, as in the Hopfield model, we have at each iteration to specify (either deterministically or stochastically) which coordinate are we updating. To do that, we use an N dimensional "control register" whose content is always a unit vector of $\{0, 1\}^N$ (and the location of the '1' will denote the next coordiante to be changed). This vector may be varied from instant n to n + 1 either by shift ("sequential coordinate update") or at random.

Let Δ_i, $i \leq i \leq N$ be the i-th output of the "control" register, x_i, $1 \leq i \leq N$ and V be the (N+1) neurons inputs and $\tilde{x}_i = x_i(1-2\Delta_i)$ the corresponding outputs (where $\tilde{x}_i$, $x_i \varepsilon \{+1,-1\}$, $\Delta_i \varepsilon \{0,1\}$, but V is a real number), ϕ_j, $1 \leq j \leq m$ be the input of the j-th intermediate neuron $(-1 \leq \phi_j \leq 1)$, $\eta_j = -(1-\phi_j)^{-L}$ be its output, and $W_{ji} = U_i^{(j)}/N$ be the synaptic weight of the ij - th synapsis, where $U_i^{(j)}$ refers here to the i-th element of the j-th memory.

The system's equations are:

$$\tilde{x}_i = x_i (1 - 2\Delta_i) \qquad 1 \leq i \leq N \qquad (4a)$$

$$\phi_j = \sum_{i=1}^{N} W_{ji} \tilde{x}_i \qquad 1 \leq j \leq m \qquad (4b)$$

$$\eta_j = -(1 - \phi_j)^{-L} \qquad 1 \leq j \leq m \qquad (4c)$$

$$\tilde{V} = \sum_{j=1}^{m} \eta_j \qquad (4d)$$

$$S = \frac{1}{2}(1 - sign(\tilde{V} - V)) \qquad (4e)$$

$$x_i \leftarrow x_i + S\tilde{x}_i \qquad 1 \leq i \leq N \qquad (4f)$$

$$V \leftarrow V + S\tilde{V} \qquad (4g)$$

The system is initialized by $x_i = x_i(0)$ (the probe vector), and $V = +\infty$. A block diagram of this sytem appears in Fig. 2. Note that we made use of N + m + 1 neurons and O(Nm) connections.

As for the continuous time case (with memories on the unit sphere) we will get the equations:

$$\dot{x}_i + 2m\tilde{V}x_i = LN \sum_{j=1}^{m} W_{ji}\eta_j, \qquad 1 \le i \le N \quad (5a)$$

$$\phi_j = N \sum_{i=1}^{N} W_{ji}x_i, \quad \delta = \sum_{i=1}^{N} x_i^2, \qquad 1 \le j \le m \quad (5b)$$

$$\eta_j = (1 + \delta - 2\phi_j)^{-\left(\frac{L}{2}+1\right)}, \qquad 1 \le j \le m \quad (5c)$$

$$\tilde{V} = \sum_{j=1}^{m} \eta_j \qquad\qquad\qquad (5d)$$

with similar interpretation (here there is no 'control' register as all components are updated continuously).

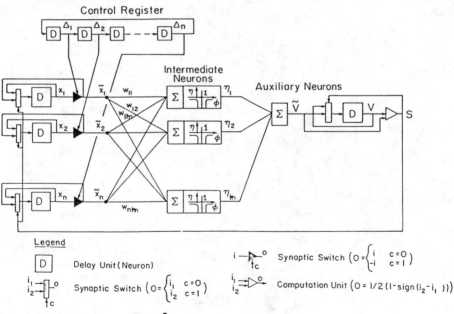

Figure 2   Neural Network Implementation

REFERENCES

1. J.J. Hopfield, "Neural Networks and Physical Systems with Emergent Collective Computational Abilities", Proc. Nat. Acad. Sci. U.S.A., Vol. 79 (1982), pp. 2554-2558.
2. R.J. McEliece, et al., "The Capacity of the Hopfield Associative Memory", IEEE Trans. on Inf. Theory, Vol. IT-33 (1987), pp. 461-482.
3. A. Dembo, "On the Capacity of the Hopfield Memory", submitted, IEEE Trans. on Inf. Theory.
4. Kohonen, T., Self Organization and Associative Memory, Springer, Berlin, 1984.
5. Dembo, A. and Zeitouni, O., General Potential Surfaces and Neural Networks, submitted, Phys. Rev. A.
6. Bachmann, C.M., Cooper, L.N., Dembo, A. and Zeitouni, O., A relazation Model for Memory with high storage density, to appear, Proc. Natl. Ac. Science.

Network Generality, Training Required, and Precision Required

John S. Denker and Ben S. Wittner [1]
AT&T Bell Laboratories
Holmdel, New Jersey 07733

Keep your hand on your wallet.
— Leon Cooper, 1987

Abstract

We show how to estimate (1) the number of functions that can be implemented by a particular network architecture, (2) how much analog precision is needed in the connections in the network, and (3) the number of training examples the network must see before it can be expected to form reliable generalizations.

Generality versus Training Data Required

Consider the following objectives: First, the network should be very powerful and versatile, i.e., it should implement any function (truth table) you like, and secondly, it should learn easily, forming meaningful generalizations from a small number of training examples. Well, it is information-theoretically impossible to create such a network. We will present here a simplified argument; a more complete and sophisticated version can be found in Denker et al. (1987).

It is customary to regard learning as a dynamical process: adjusting the weights (etc.) in a single network. In order to derive the results of this paper, however, we take a different viewpoint, which we call the ensemble viewpoint. Imagine making a very large number of replicas of the network. Each replica has the same architecture as the original, but the weights are set differently in each case. No further adjustment takes place; the "learning process" consists of winnowing the ensemble of replicas, searching for the one(s) that satisfy our requirements.

Training proceeds as follows: We present each item in the training set to every network in the ensemble. That is, we use the abscissa of the training pattern as input to the network, and compare the ordinate of the training pattern to see if it agrees with the actual output of the network. For each network, we keep a score reflecting how many times (and how badly) it disagreed with a training item. Networks with the lowest score are the ones that agree best with the training data. If we had complete confidence in

---

[1] Currently at NYNEX Science and Technology, 500 Westchester Ave., White Plains, NY 10604

the reliability of the training set, we could at each step simply throw away all networks that disagree.

For definiteness, let us consider a typical network architecture, with N_0 input wires and N_l units in each processing layer l, for $l \in \{1 \cdots L\}$. For simplicity we assume $N_L = 1$. We recognize the importance of networks with continuous-valued inputs and outputs, but we will concentrate for now on training (and testing) patterns that are discrete, with $N \equiv N_0$ bits of abscissa and $N_L = 1$ bit of ordinate. This allows us to classify the networks into bins according to what Boolean input-output relation they implement, and simply consider the ensemble of bins.

There are 2^{2^N} possible bins. If the network architecture is completely general and powerful, all 2^{2^N} functions will exist in the ensemble of bins. On average, one expects that each training item will throw away at most half of the bins. Assuming maximal efficiency, if m training items are used, then when $m \gtrsim 2^N$ there will be only one bin remaining, and that must be the unique function that consistently describes all the data. But there are only 2^N possible abscissas using N bits. Therefore a truly general network cannot possibly exhibit meaningful generalization — 100% of the possible data is needed for training.

Now suppose that the network is not completely general, so that even with all possible settings of the weights we can only create functions in 2^{S_0} bins, where $S_0 \ll 2^N$. We call S_0 the initial entropy of the network. A more formal and general definition is given in Denker et al. (1987). Once again, we can use the training data to winnow the ensemble, and when $m \gtrsim S_0$, there will be only one remaining bin. That function will presumably generalize correctly to the remaining $2^N - m$ possible patterns. Certainly that function is the best we can do with the network architecture and the training data we were given.

The usual problem with automatic learning is this: If the network is too general, S_0 will be large, and an inordinate amount of training data will be required. The required amount of data may be simply unavailable, or it may be so large that training would be prohibitively time-consuming. The shows the critical importance of building a network that is not more general than necessary.

Estimating the Entropy

In real engineering situations, it is important to be able to estimate the initial entropy of various proposed designs, since that determines the amount of training data that will be required. Calculating S_0 directly from the definition is prohibitively difficult, but we can use the definition to derive useful approximate expressions. (You wouldn't want to calculate the thermodynamic entropy of a bucket of water directly from the definition, either.)

Suppose that the weights in the network at each connection i were not continuously adjustable real numbers, but rather were specified by a discrete code with b_i bits. Then the total number of bits required to specify the configuration of the network is

$$B = \sum_i b_i \tag{1}$$

Now the total number of functions that could possibly be implemented by such a network architecture would be at most 2^B. The actual number will always be smaller than this, since there are various ways in which different settings of the weights can lead to identical functions (bins). For one thing, for each hidden layer $l \in \{1 \cdots L-1\}$, the numbering of the hidden units can be permuted, and the polarity of the hidden units can be flipped, which means that 2^{S_0} is less than 2^B by a factor (among others) of $\prod_l N_l! \, 2^{N_l}$. In addition, if there is an inordinately large number of bits b_i at each connection, there will be many settings where small changes in the connection will be immaterial. This will make 2^{S_0} smaller by an additional factor. We expect $\partial S_0 / \partial b_i \approx 1$ when b_i is small, and $\partial S_0 / \partial b_i \approx 0$ when b_i is large; we must now figure out where the crossover occurs.

The number of "useful and significant" bits of precision, which we designate b^*, typically scales like the logarithm of number of connections to the unit in question. This can be understood as follows: suppose there are N connections into a given unit, and an input signal to that unit of some size A is observed to be significant (the exact value of A drops out of the present calculation). Then there is no point in having a weight with magnitude much larger than A, nor much smaller than A/N. That is, the dynamic range should be comparable to the number of connections. (This argument is not exact, and it is easy to devise exceptions, but the conclusion remains useful.) If only a fraction $1/S$ of the units in the previous layer are active (nonzero) at a time, the needed dynamic range is reduced. This implies $b^* \approx \log(N/S)$.

Note: our calculation does not involve the dynamics of the learning process. Some numerical methods (including versions of back propagation) commonly require a number of temporary "guard bits" on each weight, as pointed out by Richard Durbin (private communication). Another $\log N$ bits ought to suffice. These bits are not needed after learning is complete, and do not contribute to S_0.

If we combine these ideas and apply them to a network with N units in each layer, fully connected, we arrive at the following expression for the number of different Boolean functions that can be implemented by such a network:

$$2^{S_0} \approx \frac{2^B}{N! \, 2^N} \tag{2}$$

where

$$B \approx L N^2 \log N \tag{3}$$

These results depend on the fact that we are considering only a very restricted type of processing unit: the output is a monotone function of a weighted sum of inputs. Cover

(1965) discussed in considerable depth the capabilities of such units. Valiant (1986) has explored the learning capabilities of various models of computation.

Abu-Mustafa has emphasized the principles of information and entropy and applied them to measuring the properties of the training set. At this conference, formulas similar to equation 3 arose in the work of Baum, Psaltis, and Venkatesh, in the context of calculating the number of different training patterns a network should be *able to* memorize. We originally proposed equation 2 as an estimate of the number of patterns the network would *have to* memorize before it could form a reliable generalization. The basic idea, which has numerous consequences, is to estimate the number of (bins of) networks that can be realized.

References

1. Yasser Abu-Mustafa, these proceedings.

2. Eric Baum, these proceedings.

3. T. M. Cover, "Geometrical and statistical properties of systems of linear inequalities with applications in pattern recognition," *IEEE Trans. Elec. Comp.*, **EC-14**, 326-334, (June 1965)

4. John Denker, Daniel Schwartz, Ben Wittner, Sara Solla, John Hopfield, Richard Howard, and Lawrence Jackel, <u>Complex Systems</u>, in press (1987).

5. Demetri Psaltis, these proceedings.

6. L. G. Valiant, SIAM J. Comput. **15(2)**, 531 (1986), and references therein.

7. Santosh Venkatesh, these proceedings.

'Ensemble' Boltzmann Units have Collective Computational Properties like those of Hopfield and Tank Neurons

Mark Derthick and Joe Tebelskis
Department of Computer Science
Carnegie-Mellon University

1 Introduction

There are three existing connectionist models in which network states are assigned a computational energy. These models—Hopfield nets, Hopfield and Tank nets, and Boltzmann Machines—search for states with minimal energy. Every link in the network can be thought of as imposing a constraint on acceptable states, and each violation adds to the total energy. This is convenient for the designer because constraint satisfaction problems can be mapped easily onto a network. Multiple constraints can be superposed, and those states satisfying the most constraints will have the lowest energy.

Of course there is no free lunch. Constraint satisfaction problems are generally combinatorial and remain so even with a parallel implementation. Indeed, Merrick Furst (personal communication) has shown that an NP-complete problem, graph coloring, can be reduced to deciding whether a connectionist network has a state with an energy of zero (or below). Therefore designing a practical network for solving a problem requires more than simply putting the energy minima in the right places. The topography of the energy space affects the ease with which a network can find good solutions. If the problem has highly interacting constraints, there will be many local minima separated by energy barriers. There are two principal approaches to searching these spaces: monotonic gradient descent, introduced by Hopfield [1] and refined by Hopfield and Tank [2]; and stochastic gradient descent, used by the Boltzmann Machine [3]. While the monotonic methods are not guaranteed to find the optimal solution, they generally find good solutions much faster than the Boltzmann Machine. This paper adds a refinement to the Boltzmann Machine search algorithm analogous to the Hopfield and Tank technique, allowing the user to trade off the speed of search for the quality of the solution.

2 Hopfield nets

A Hopfield net [1] consists of binary-valued units connected by symmetric weighted links. The global energy of the network is defined to be

$$E = -\frac{1}{2}\sum_i \sum_{j \neq i} w_{ij}s_i s_j - \sum_i I_i s_i$$

where s_i is the state of unit i, and w_{ij} is the weight on the link between units i and j.

The search algorithm is: randomly select a unit and probe it until quiescence. During a probe, a unit decides whether to be on or off, determined by the states of its neighbors. When a unit is probed, there are two possible resulting global states. The difference in energy between these states is called the unit's *energy gap*:

$$\Delta_k \equiv E_{s_k=0} - E_{s_k=1} = \sum_i w_{ik}s_i + I_k$$

The decision rule is

$$s_i = \begin{cases} 0 \text{ if } \Delta_i < 0 \\ 1 \text{ otherwise} \end{cases}$$

This rule chooses the state with lower energy. With time, the global energy of the network monotonically decreases. Since there are only a finite number of states, the network must eventually reach quiescence.

3 Boltzmann Machines

A Boltzmann Machine [3] also has binary units and weighted links, and the same energy function is used. Boltzmann Machines also have a learning rule for updating weights, but it is not used in this paper. Here the important difference is in the decision rule, which is stochastic. As in probing a Hopfield unit, the energy gap is determined. It is used to determine a probability of adopting the on state:

$$P(s_i = 1) = \frac{1}{1 + e^{-\Delta_i/T}}$$

where T is the computational temperature. With this rule, energy does not decrease monotonically. The network is more likely to adopt low energy states, but it sometimes goes uphill. The idea is that it can search a number of minima, but spends more time in deeper ones. At low temperatures, the ratio of time spent in the deepest minima is so large that the chances of not being in the global minimum are negligible. It has been proven [4] that after searching long enough, the probabilities of the states are given by the Boltzmann distribution, which is strictly a function of energy and temperature, and is independent of topography:

$$\frac{P_\alpha}{P_\beta} = e^{-(E_\alpha - E_\beta)/T} \tag{1}$$

The approach to equilibrium, where equation 1 holds, is speeded by initially searching at a high temperature and gradually decreasing it. Unfortunately, reaching equilibrium stills takes exponential time. While the Hopfield net settles quickly and is not guaranteed to find the best solution, a Boltzmann Machine can theoretically be run long enough to guarantee that the global optimum is found. Most of the time the uphill moves which allow the network to escape local minima are a waste of time, however. It is a direct consequence of the guaranteed ability to find the best solution that makes finding even approximate solutions slow.

4 Hopfield and Tank networks

In Hopfield and Tank nets [2], the units take on continuous values between zero and one, so the search takes place in the interior of a hypercube rather than only on its vertices. The search algorithm is deterministic gradient descent. By beginning near the center of the space and searching in the direction of steepest descent, it seems likely that the deepest minimum will be found. There is still no guarantee, but good results have been reported for many problems.

The modified energy equation is

$$E = -\frac{1}{2}\sum_i \sum_j w_{ij}s_i s_j + \sum_i \frac{1}{R_i} \int_0^{s_i} g^{-1}(s)ds - \sum_i I_i s_i \qquad (2)$$

R_i is the input resistance to unit i, and $g(u)$ is the sigmoidal unit transfer function $\frac{1}{1+e^{2\lambda u}}$. The second term is zero for extreme values of s_i, and is minimized at $s_i = \frac{1}{2}$.

The Hopfield and Tank model is continuous in time as well as value. Instead of proceeding by discrete probes, the system is described by simultaneous differential equations, one for each unit. Hopfield and Tank show that the following equation of motion results in a monotonic decrease in the value of the energy function:

$$\frac{du_i}{dt} = -u_i/\tau + \sum_j w_{ij}s_j + I_i$$

where $\tau = RC$, C is a constant determining the speed of convergence, $u_i = g^{-1}(s_i)$, and the gain, λ, is analogous to (the inverse of) temperature in a Boltzmann Machine. λ determines how important it is to satisfy the constraints imposed by the links to other units. When λ is low, these constraints are largely ignored and the second term dominates, tending to keep the system near the center of the search space, where there is a single global minimum. At high gains, the minima lie at the corners of the search space, in the same locations as for the Hopfield model and the Boltzmann model. If the system is run at high gain, but the initial state is near the center of the space, the search gradually moves out towards the corners, on the way encountering "continental divides" between watersheds leading to all the various local minima. The initial steepness of the watersheds serves as a heuristic for choosing which minima is

226

likely to be lower. This search heuristic emerges automatically from the architecture, making network design simple. For many problems this single automatic heuristic results in a system comparable to the best knowledge intensive algorithms in which many domain specific heuristics are laboriously hand programmed.

For many problems, Hopfield and Tank nets seem quite sufficient [5, 6]. However for one network we have been using [7] the Hopfield and Tank model invariably settles into poor local minima. The solution has been to use a new model combining the advantages of Boltzmann Machines and Hopfield and Tank networks.

5 'Ensemble' Boltzmann Machines

It seems the Hopfield and Tank model gets its advantage by measuring the actual gradient, giving the steepest direction to move. This is much more informative than picking a random direction and deciding which of the two corners of the space to try, as models using binary units must do. Peter Brown (personal communication) has investigated continuous Boltzmann Machines, in which units stochastically adopt a state between zero and one. The scheme presented here has a similar effect, but the units actually take on discrete states between zero and one. Each ensemble unit can be thought of as an ensemble of identically connected conventional Boltzmann units. To probe the ensemble unit, each of its constituents is probed, and the state of the ensemble unit is the average of its constituents' states. Because this average is over a number of identical independent binary random variables, the ensemble unit's state is binomially distributed.

Figure 1 shows an ensemble unit with three constituents. At infinite temperature, all unit states tend toward $\frac{1}{2}$, and at zero temperature the states go to zero or one unless the energy gap is exactly zero. This is similar to the behavior of a Hopfield and Tank network at low and high gain, respectively. In Ensemble Boltzmann Machines (EBMs) the tendency towards $\frac{1}{2}$ in the absence of constraints from other units results from the shape of the binomial distribution. In contrast, the second term in the energy equation is responsible for this effect in the Hopfield and Tank model.

Although an EBM proceeds in discrete time using probes, over a large number of probes the search tends to proceed in the direction of the gradient. Every time a unit is probed, a move is made along one axis whose length depends on the magnitude of the gradient in that direction. Because probing still contains a degree of stochasticity, EBMs can escape from local minima, and if run long enough are guaranteed to find the global minimum. By varying n, the number of components of each ensemble unit, the system can exhibit any intermediate behavior in the tradeoff between the speed of convergence of Hopfield and Tank networks, and the ability to escape local minima of Boltzmann Machines.

Clearly when $n = 1$ the performance is identical to a conventional Boltzmann Machine, because each unit consists of a single Boltzmann unit. As $n \to \infty$ the

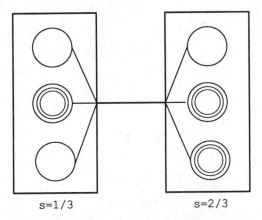

s=1/3   s=2/3

Figure 1: The heavy lines depict an 'Ensemble' Boltzmann Machine with two units. With an ensemble size of three, this network behaves like a conventional Boltzmann Machine consisting of six units (light lines). The state of the ensemble units is the average of the states of its components.

value a unit takes on after probing becomes deterministic. The stable points of the system are then identical to the ones of the Hopfield and Tank model.

To prove this, it suffices to show that at each probe the ensemble Boltzmann unit takes on the state which gives rise to the lowest (Hopfield and Tank) energy. Therefore the energy must monotonically decrease. Further, if the system is not at a global (Hopfield and Tank) energy minimum, there is some unit which can be probed so as to lower the energy.

To show that the state resulting from a probe is the minimum possible, we show first that the derivative of the energy with resepect to the unit's state is zero at the resulting state, and second that the second derivative is positive over the entire range of possible states, zero to one.

Taking the derivative of equation 2 gives

$$\frac{dE}{ds_k} = -\sum_i w_{ik}s_i + \frac{1}{R_i}g^{-1}(s_k) - I_k$$

Now

$$g(u) = \frac{1}{1 + e^{-2\lambda u}}$$

so

$$g^{-1}(u) = \frac{1}{2\lambda}\ln\frac{s}{1-s}$$

Let $T = \frac{1}{2\lambda R}$. The EBM update rule is

$$s_k = \frac{1}{1 + e^{-\Delta_k/T}}$$

Therefore

$$
\frac{dE}{ds_k}\bigg|_{s_k=\frac{1}{1+e^{-\Delta_k/T}}} = -\Delta_k + T \ln \left[\frac{\frac{1}{1+e^{-\Delta_k/T}}}{\frac{e^{-\Delta_k/T}}{1+e^{-\Delta_k/T}}} \right]
$$

$$
= -\Delta_k + T \ln e^{\Delta_k/T}
$$

$$
= -\Delta_k + T(\Delta_k/T)
$$

$$
= 0
$$

and

$$
\frac{d^2E}{ds_k^2} = \frac{1}{2\lambda R} \cdot \frac{1-s_k}{s_k} \cdot \left[\frac{(1-s_k)-(-s_k)}{(1-s_k)^2} \right]
$$

$$
= \frac{1}{2\lambda R s_k (1-s_k)}
$$

$$
> 0 \text{ on } 0 < s_k < 1
$$

In writing a program to simulate an EBM, it would be wasteful to explicitly represent the components of each ensemble unit. Since each component has an identical energy gap, the average of their values is given by the binomial distribution b(n,p) where n is the ensemble size, and p is $\frac{1}{1+e^{-\Delta/T}}$. There are numerical methods for sampling from this distribution in time independent of n [8]. When n is infinite, there is no need to bother with the distribution because the result is just p.

Hopfield and Tank suggest [2] that the Hopfield and Tank model is a mean field approximation to the original Hopfield model. In a mean field approximation, the average value of a variable is used to calculate its effect on other variables, rather than calculating all the individual interactions. Consider a large ensemble of Hopfield nets with two units, A and B. To find the distribution of final states exactly, each B unit must be updated based on the A unit in the same network. The calculation must be repeated for every network in the ensemble. Using a mean field approximation, the average value of all the B units is calculated based on the average value of all the A units. This calculation is no harder than that of the state of a single Hopfield network, yet is potentially more informative since it approximates an average property of a whole ensemble of Hopfield networks. The states of Hopfield and Tank units can be viewed as representing the ensemble average of the states of Hopfield units in this way. Peterson and Anderson [9] demonstrate rigorously that the behavior is a mean field approximation.

In the EBM, it is intuitively clear that a mean field approximation is being made. The network can be thought of as a real ensemble of Boltzmann networks, except with additional connections between the networks so that each Boltzmann unit sees not only its neighbors in the same net, but also sees the average state of the neighboring units in all the nets (see figure 1).

6   Traveling Salesman Problem

The traveling salesman problem illustrates the use of energy-based connectionist networks, and the ease with which they may be designed. Given a list of city locations, the task is to find a tour of minimum length through all the cities and returning to the starting city. To represent a solution to an n city problem in a network, it is convenient to use n columns of n rows of units [2]. If a unit at coordinates (i, j) is on, it indicates that the ith city is the jth to be visited. A valid solution will have n units on, one in every column and one in every row. The requirements can be divided into four constraints: there can be no more than one unit on in a row, no more that one unit on in a column, there must be n units on, and the distances between cities must be minimized. Hopfield and Tank use the following energy function to effect these constraints:

$$
\begin{aligned}
E \;=\; & A/2 \sum_X \sum_i \sum_{j \neq i} s_{Xi} s_{Xj} \;+ \\
& B/2 \sum_i \sum_X \sum_{Y \neq X} s_{Xi} s_{Yi} \;+ \\
& C/2 \left(\sum_X \sum_i s_{Xi} - n \right)^2 \;+ \\
& D/2 \sum_X \sum_{Y \neq X} \sum_i d_{XY} s_{Xi}(s_{Y,i+1} + s_{Y,i-1})
\end{aligned}
\tag{3}
$$

Here units are given two subscripts to indicate their row and column, and the subscripts "wrap around" when outside the range $1 \leq i \leq n$. The first term is implemented with inhibitory links between every pair of units in a row, and is zero only if no two are on. The second term is inhibition within columns. In the third term, n is the number of cities in the tour. When the system reaches a vertex of the search space, this term is zero only if exactly n units are on. This constraint is implemented with inhibitory links between all n^4 pairs of units plus an excitatory input current to all units. In the last term d_{XY} is the distance between cities X and Y. At points in the search space representing valid tours, the summation is numerically equal to the length of the tour.

As long as the constraints ensuring that the solution is a valid tour are stronger than those minimizing distance, the global energy minimum will represent the shortest tour. However *every* valid tour will be a *local* energy minimum. Which tour is chosen will depend on the random initial starting state, and on the random probing order.

7   Empirical Results

The evidence that convinced me EBMs offer improved performance over Hopfield and Tank networks was the ease of tuning them for the Ted Turner problem reported

230

in [7]. However this evidence is entirely subjective; it is impossible to show that no set of parameters exist which would make the Hopfield and Tank model perform well. Instead we have chosen to repeat the traveling salesman problem experiments reported by Hopfield and Tank [2], using the same cities and the same values for the constants in equation 3. The tour involves 10 cities, and the shortest tour is of length 2.72. An average tour has length 4.55. Hopfield and Tank report finding a valid tour in 16 of 20 settlings, and that half of these are one of the two shortest tours.

One advantage of Hopfield and Tank nets over Boltzmann Machines is that they move continuously in the direction of the gradient. EBMs move in discrete jumps whose size is the value of the gradient along a given axis. When the system is far from equilibrium these jumps can be quite large, and the search is inefficient. Although Hopfield and Tank nets can do a whole search at high gain, Boltzmann Machines usually vary the temperature so the system can remain close to equilibrium as the low temperature equilibrium is approached. For this reason our model was more sensitive to the gain parameter than the Hopfield and Tank model, and we used temperatures much higher than $\frac{1}{2\lambda R}$.

As expected, when n is infinite, an EBM produces results similar to those reported by Hopfield and Tank. 85 out of 100 settlings resulted in valid tours, and the average length was 2.73. Table 1 shows how n affects the number of valid tours and the average tour length. As n decreases from infinity, both the average tour length and the number of valid tours increases. (We have no explanation for the anomalously low number of valid tours for $n = 40$.) Both of these effects result from the increased sampling noise in determining the ensemble unit states for lower n. With more noise, the system has an easier time escaping local minima which do not represent valid tours. Yet at the same time the discriminability between the very best tours and moderately good tours decreases, because these smaller energy differences are swamped by the noise.

Rather than stop trials when the network was observed to converge, a constant number of probes, 200 per unit, was made. However we noted that convergence was generally faster for larger values of n. Thus for the traveling salesman problem, large n give faster and better solutions, but a smaller values gives the highest reliability. Depending on the application, a value of either infinity or 50 seems best.

8   Conclusion

'Ensemble' Boltzmann Machines are completely upward compatible with conventional Boltzmann Machines. The above experiment can be taken to show that they perform better at the traveling salesman problem. In addition, at the limit of infinite ensemble size they perform similarly to Hopfield and Tank nets. For TSP and perhaps many other problems, the latter model seems an equally good choice. Perhaps due to the extreme regularity of the architecture, the energy space must be nicely behaved

Ensemble Size	Percent Valid	Average Tour Length
1	93	3.32
40	84	2.92
50	95	2.79
100	89	2.79
1000	90	2.80
infinity	85	2.73

Table 1: Number of valid tours out of 100 trials and average tour length, as a function of ensemble size. An ensemble size of one corresponds to a Boltzmann Machine. Infinity loosely corresponds to a Hopfield and Tank network.

in that the ravine steepness near the center of the space is a good indication of its eventual depth. In this case the ability to escape local minima is not required for good performance.

For the Ted Turner problem, which has a very irregular architecture and many more constraint types, the ability to escape local minima seems essential. Conventional Boltzmann Machines are too noisy, both for efficient search and for debugging. EBMs allow the designer the flexibility to add only as much noise as is necessary. In addition, lower noise can be used for debugging. Even though this may give poorer performance, a more deterministic search is easier for the debugger to understand, allowing the proper fix to be made.

Acknowledgements

We appreciate receiving data and explanations from David Tank, Paul Smolensky, and Erik Sobel. This research has been supported by an ONR Graduate Fellowship, by NSF grant EET-8716324, and by the Defense Advanced Research Projects Agency (DOD), ARPA Order No. 4976 under contract F33615-87-C-1499 and monitored by the:
Avionics Laboratory
Air Force Wright Aeronautical Laboratories
Aeronautical Systems Division (AFSC)
Wright-Patterson AFB, OH 45433-6543

This research was also sponsored by the same agency under contract N00039-87-C-0251 and monitored by the Space and Naval Warfare Systems Command.

References

[1] J. J. Hopfield, "Neural networks and physical systems with emergent collective computational abilities," *Proceedings of the National Academy of Sciences U.S.A.*, vol. 79, pp. 2554–2558, April 1982.

[2] J. Hopfield and D. Tank, "'Neural' computation of decisions in optimization problems," *Biological Cybernetics*, vol. 52, pp. 141–152, 1985.

[3] G. E. Hinton and T. J. Sejnowski, "Learning and relearning in Boltzmann Machines," in *Parallel distributed processing: Explorations in the microstructure of cognition*, Cambridge, MA: Bradford Books, 1986.

[4] S. Geman and D. Geman, "Stochastic relaxation, Gibbs distributions, and the Bayesian restoration of images," *IEEE Transactions on Pattern Analysis and Machine Intelligence*, vol. PAMI-6, pp. 721–741, 1984.

[5] J. L. Marroquin, *Probabilistic Solution of Inverse Problems*. PhD thesis, MIT, September 1985.

[6] J. Hopfield and D. Tank, "Simple 'Neural' optimization networks: an a/d converter, signal decision circuit and a linear programming circuit," *IEEE Transactions on Circuits and Systems*, vol. 33, pp. 533–541, 1986.

[7] M. Derthick, "Counterfactual reasoning with direct models," in *AAAI-87*, Morgan Kaufmann, July 1987.

[8] D. E. Knuth, *The Art of Computer Programming. Second Edition.* Vol. 2, Addison-Wesley, 1981.

[9] C. Peterson and J. R. Anderson, "A mean field theory learning algorithm for neural networks," Tech. Rep. EI-259-87, MCC, August 1987.

HIGH ORDER NEURAL NETWORKS FOR EFFICIENT
ASSOCIATIVE MEMORY DESIGN

I. GUYON*, L. PERSONNAZ*, J. P. NADAL** and G. DREYFUS*

* Ecole Supérieure de Physique et de Chimie Industrielles de la Ville de Paris
Laboratoire d'Electronique
10, rue Vauquelin
75005 Paris (France)

** Ecole Normale Supérieure
Groupe de Physique des Solides
24, rue Lhomond
75005 Paris (France)

ABSTRACT

We propose learning rules for recurrent neural networks with high-order interactions between some or all neurons. The designed networks exhibit the desired associative memory function : perfect storage and retrieval of pieces of information and/or sequences of information of any complexity.

INTRODUCTION

In the field of information processing, an important class of potential applications of neural networks arises from their ability to perform as associative memories. Since the publication of J. Hopfield's seminal paper[1], investigations of the storage and retrieval properties of recurrent networks have led to a deep understanding of their properties. The basic limitations of these networks are the following :
- their storage capacity is of the order of the number of neurons ;
- they are unable to handle structured problems ;
- they are unable to classify non-linearly separable data.

In order to circumvent these limitations, one has to introduce additional non-linearities. This can be done either by using "hidden", non-linear units, or by considering multi-neuron interactions[2]. This paper presents learning rules for networks with multiple interactions, allowing the storage and retrieval, either of static pieces of information (autoassociative memory), or of temporal sequences (associative memory), while preventing an explosive growth of the number of synaptic coefficients.

AUTOASSOCIATIVE MEMORY

The problem that will be addressed in this paragraph is how to design an autoassociative memory with a recurrent (or feedback) neural network when the number p of prototypes is large as compared to the number n of neurons.
We consider a network of n binary neurons, operating in a synchronous mode, with period τ. The state of neuron i at time t is denoted by $\sigma_i(t)$, and the state of the network at time t is represented by a vector $\underline{\sigma}(t)$ whose components are the $\sigma_i(t)$. The dynamics of each neuron is governed by the following relation :

$$\sigma_i(t+\tau) = \text{sgn } v_i(t). \tag{1}$$

In networks with two-neuron interactions only, the potential $v_i(t)$ is a linear function of the state of the network :

$$v_i(t) = \sum_{j=1}^{n} C_{ij} \, \sigma_j(t).$$

For autoassociative memory design, it has been shown[3] that any set of correlated patterns, up to a number of patterns p equal to 2^n, can be made the stable states of the system, provided the synaptic matrix is computed as the orthogonal projection matrix onto the subspace spanned by the stored vectors. However, as p increases, the rank of the family of prototype vectors will increase, and finally reach the value of n. In such a case, the synaptic matrix reduces to the identity matrix, so that all 2^n states are stable and the energy landscape becomes flat. Even if such an extreme case is avoided, the attractivity of the stored states decreases with increasing p, or, in other terms, the number of fixed points which are not the stored patterns increases ; this problem can be alleviated to a large extent by making a useful use of these "spurious" fixed points[4]. Another possible solution consists in "gardening" the state space in order to enlarge the basins of attraction of the fixed points[5]. Anyway, no dramatic improvements are provided by all these solutions since the storage capacity is always O(n).

We now show that the introduction of high-order interactions between neurons, increases the storage capacity proportionally to the number of connections per neuron. The dynamical behaviour of neuron i is still governed by (1). We consider two and three-neuron interactions, extension to higher order are straightforward.
The potential $v_i(t)$ is now defined as

$$v_i(t) = \sum_j C_{i,j}\, \sigma_j(t) + \sum_{j,l} C_{i,jl}\, \sigma_j(t)\, \sigma_l(t).$$

It is more convenient, for the derivation of learning rules, to write the potential in the matrix form :

$$\underline{v}(t) = C\, \underline{\chi}(t),$$

where $\underline{\chi}(t)$ is an m dimensional vector whose components are taken among the set of the $(n^2+n)/2$ values : $\sigma_1, \dots, \sigma_n, \sigma_1\sigma_2, \dots, \sigma_j\sigma_l, \dots, \sigma_{n-1}\sigma_n$.

 As in the case of the two-neuron interactions model, we want to compute the interaction coefficients so that the prototypes are stable and attractor states.
A condition to store a set of states $\underline{\sigma}^k$ (k=1 to p) is that $\underline{v}^k = \underline{\sigma}^k$ for all k. Among the solutions, the most convenient solution is given by the (n,m) matrix

$$C = \Sigma\, \Gamma^I \tag{2}$$

where Σ is the (n,p) matrix whose columns are the $\underline{\sigma}^k$ and Γ^I is the (p,m) pseudoinverse of the (m,p) matrix Γ whose columns are the $\underline{\chi}^k$. This solution satisfies the above requirements, up to a storage capacity which is related to the dimension m of vectors $\underline{\chi}$. Thus, in a network with three-neuron interactions, the number of patterns that can be stored is $O(n^2)$. Details on these derivations are published in Ref.6.

By using only a subset of the products $\{\sigma_j \sigma_l\}$, the increase in the number of synaptic coefficients can remain within acceptable limits, while the attractivity of the stored patterns is enhanced, even though their number exceeds the number of neurons ; this will be examplified in the simulations presented below.

Finally, it can be noticed that, if vector γ contains all the $\{\sigma_i \sigma_j\}$, i=1,...n, j=1,...n, only, the computation of the vector potential $\underline{v} = C\gamma$ can be performed after the following expression :

$$\underline{v} = \Sigma \{ (\Sigma^T\Sigma)^② \}^I \ (\Sigma^T\underline{\sigma})^②$$

where ② stands for the operation which consists in squaring all the matrix coefficients. Hence, the computation of the synaptic coefficients is avoided, memory and computing time are saved if the simulations are performed on a conventional computer. This formulation is also meaningful for optical implementations, the function ② being easily performed in optics[7].

In order to illustrate the capabilities of the learning rule, we have performed numerical simulations which show the increase of the size of the basins of attraction when second-order interactions, in addition to the first-order ones, are used. The simulations were carried out as follows. The number of neurons n being fixed, the amount of second-order interactions was chosen ; p prototype patterns were picked randomly, their components being ±1 with probability 0.5 ; the second-order interactions were chosen randomly. The synaptic matrix was computed from relation (2). The neural network was forced into an initial state lying at an initial Hamming distance H_i from one of the prototypes $\underline{\sigma}^k$; it was subsequently left to evolve until it reached a stable state at a distance H_f from $\underline{\sigma}^k$. This procedure was repeated many times for each prototype and the H_f were averaged over all the tests and all the prototypes.

Figures 1a. and 1b. are charts of the mean values of H_f as a function of the number of prototypes, for n = 30 and for various values of m (the dimension of vector χ). These curves allowed us to determine the maximum number of prototype states which can be stored for a given quality of recall. Perfect recall implies $H_f = 0$; when the number of prototypes increases, the error in recall may reach $H_f \approx H_i$: the associative memory is degenerate. The results obtained for $H_i/n = 10\%$ are plotted on Figure 1a. When no high-order interactions were used, H_f reached H_i for $p/n \approx 1$, as expected ; conversely, virtually no error in recall occured up to $p/n \approx 2$ when all second-order interactions were taken into account (m=465). Figure 1b shows the same quantities for H_i=20% ; since the initial states were more distant from the prototypes, the errors in recall were more severe.

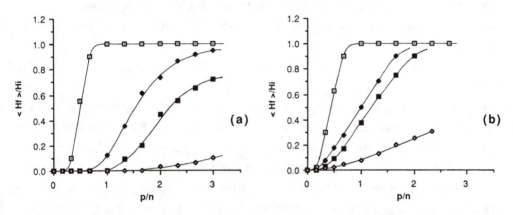

Fig. 1. Improvement of the attractivity by addition of three-neuron interactions to the two-neuron interactions. All prototypes are always stored exactly (all curves go through the origin). Each point corresponds to an average over min(p,10) prototypes and 30 tests for each prototype.

□ Projection : m = n = 30 ; ◆ m = 120 ; ■ m = 180 ; ◇ m = 465 (all interactions)

1 a : H_i/n =10% ; 1 b : H_i/n =20%.

TEMPORAL SEQUENCES (ASSOCIATIVE MEMORY)

The previous section was devoted to the storage and retrieval of items of information considered as fixed points of the dynamics of the network (autoassociative memory design). However, since fully connected neural networks are basically dynamical systems, they are natural candidates for storing and retrieving information which is dynamical in nature, i.e., temporal sequences of patterns[8]. In this section, we propose a general solution to the problem of storing and retrieving sequences of arbitrary complexity, in recurrent networks with parallel dynamics.

Sequences consist in sets of transitions between states $\underline{\sigma}^k \rightarrow \underline{\sigma}^{k+1}$, k=1,..., p. A sufficient condition to store these sets of transitions is that $\underline{v}^k = \underline{\sigma}^{k+1}$ for all k. In the case of a linear potential $\underline{v} = C \, \underline{\sigma}$, the storage prescription proposed in ref.3 can be used : $C = \Sigma^+ \Sigma^I$,

where Σ is a matrix whose columns are the $\underline{\sigma}^k$ and Σ^+ is the matrix whose columns are the successors $\underline{\sigma}^{k+1}$ of $\underline{\sigma}^k$. If p is larger than n, one can use high-order interactions, which leads to introduce a non-linear potential $\underline{v} = C \, \underline{\gamma}$, with $\underline{\gamma}$ as previously defined. We proposed in ref.10 the following storage prescription :

$$C = \Sigma^+ \Gamma^I \tag{3}$$

The two above prescriptions are only valid for storing simple sequences, where no patterns occur twice (or more). Suppose that one pattern occurs twice ; when the network reaches this *bifurcation point,* it is unable to make a decision according the deterministic dynamics described in (1), since the knowledge of the present state is not sufficient. Thus, complex sequences require to keep, at each time step of the dynamics, a non-zero memory span.

The vector potential $\underline{v} = C\underline{\gamma}$ must involve the states at time t and t-τ, which leads to define the vector $\underline{\gamma}$ as a concatenation of vectors $\underline{\sigma}(t)$, $\underline{\sigma}(t-\tau)$, $\underline{\sigma}(t) \otimes \underline{\sigma}(t)$, $\underline{\sigma}(t) \otimes \underline{\sigma}(t-\tau)$, or a suitable subset thereof. The subsequent vector $\underline{\sigma}(t+\tau)$ is still determined by relation (1). In this form, the problem is a generalization of the storage of patterns with high order interactions, as described above. The storage of sequences can be still processed by relation (3).

The solution presented above has the following features :
i)   Sequences with bifurcation points can be stored and retrieved.
ii) The dimension of the synaptic matrix is at most $(n, 2(n^2+n))$, and at least $(n, 2n)$ in the linear case, so that at most $2n(n^2+n)$ and at least $2n^2$ synapses are required.

iii) The storage capacity is 0(m), where m is the dimension of the vector γ.

iv) Retrieval of a sequence requires initializing the network with *two* states in succession.

The example of <u>Figure 2</u> illustrates the retrieval performances of the latter learning rule. We have limited vector γ to $\underline{\sigma}(t) \otimes \underline{\sigma}(t-\tau)$. In a network of n=48 neurons, a large number of poems have been stored, with a total of p=424 elementary transitions. Each state is consists in the 6 bit codes of 8 letters.

ALOUETTE	
JE TE	**JE NE**
PLUMERAI	**OLVMERAI**
ALOUETTE	AQFUETTE
GENTILLE	JEHKILLE
ALOUETTE	SLOUETTE
ALOUETTE	ALOUETTE
JE TE	JE TE
PLUMERAI	PLUMERAI
...	...

Fig. 2. One of the stored poems is shown in the first column. The network is initialized with two states (the first two lines of the second column). After a few steps, the network reaches the nearest stored sequence.

LOCAL LEARNING

Finally, it should be mentioned that all the synaptic matrices introduced in this paper can be computed by iterative, local learning rules.

For autoassociative memory, it has been shown analytically[9] that the procedure :

$$C_{ij}(k) = C_{ij}(k-1) + (1/n)\,(\sigma_i^k - v_i^k)\,\sigma_j^k \qquad \text{with } C_{ij}(0) = 0,$$

which is a Widrow-Hoff type learning rule, yields the projection matrix, when the number of presentations of the prototypes $\{\underline{\sigma}^k\}$ goes to infinity, if the latter are linearly independent.

A derivation along the same lines shows that, by repeated presentations of the prototype transitions, the learning rules :

$$C_{ij}(k) = C_{ij}(k-1) + (1/n)\ (\ \sigma_i{}^k - v_i{}^k\ )\ \gamma_j{}^k \qquad \text{with } C_{ij}(0) = 0$$

$$C_{ij}(k) = C_{ij}(k-1) + (1/n)\ (\ \sigma_i{}^{k+1} - v_i{}^k\ )\ \gamma_j{}^k \quad \text{with } C_{ij}(0) = 0$$

lead to the exact solutions (relations (2) and (3) respectively), if the vectors $\underline{\gamma}^k$ are linearly independent.

GENERALIZATION TASKS

Apart from storing and retrieving static pieces of information or sequences, neural networks can be used to solve problems in which there exists a structure or regularity in the sample patterns (for example presence of clumps, parity, symmetry...) that the network must discover. Feed-forward networks with multiple layers of first-order neurons can be trained with back-propagation algorithms for these purposes; however, one-layer feed-forward networks with multi-neuron interactions provide an interesting alternative. For instance, a proper choice of vector $\underline{\gamma}$ (second-order terms only) with the above learning rule yields a perfectly straightforward solution to the exclusive-OR problem. Maxwell et al. have shown that a suitable high-order neuron is able to exhibit the "ad hoc network solution" for the contiguity problem[11].

CONCLUSION

The use of neural networks with high-order interactions has long been advocated as a natural way to overcome the various limitations of the Hopfield model. However, no procedure guaranteed to store any set of information as fixed points or as temporal sequences had been proposed. The purpose of the present paper is to present briefly such storage prescriptions and show some illustrations of the use of these methods. Full derivations and extensions will be published in more detailed papers.

REFERENCES

1. J. J. Hopfield, Proc. Natl. Acad. Sci. (USA) $\underline{79}$, 2554 (1982).
2. P. Peretto and J. J. Niez, Biol. Cybern. $\underline{54}$, 53 (1986).
   P. Baldi and S. S. Venkatesh, Phys. Rev. Lett. $\underline{58}$, 913 (1987).
   For more references see ref.6.
3. L. Personnaz, I. Guyon, G. Dreyfus, J. Phys. Lett. $\underline{46}$, 359 (1985).
   L. Personnaz, I. Guyon, G. Dreyfus, Phys. Rev. A $\underline{34}$, 4217 (1986).
4. I. Guyon, L. Personnaz, G. Dreyfus, in "Neural Computers", R. Eckmiller and C. von der Malsburg eds (Springer, 1988).
5. E. Gardner, Europhys. Lett. $\underline{4}$, 481 (1987).
   G. Pöppel and U.Krey, Europhys. Lett., $\underline{4}$, 979 (1987).
6. L. Personnaz, I. Guyon, G. Dreyfus, Europhys. Lett. $\underline{4}$, 863 (1987).
7. D. Psaltis and C. H. Park, in "Neural Networks for Computing", J. S. Denker ed., (A.I.P. Conference Proceedings 151, 1986).
8. P. Peretto, J. J. Niez, in "Disordered Systems and Biological Organization", E. Bienenstock, F. Fogelman, G. Weisbush eds (Springer, Berlin 1986).
   S. Dehaene, J. P. Changeux, J. P. Nadal, PNAS (USA) $\underline{84}$, 2727 (1987).
   D. Kleinfeld, H. Sompolinsky, preprint 1987.
   J. Keeler, to appear in J. Cog. Sci.
   For more references see ref. 9.
9. I. Guyon, L. Personnaz, J.P. Nadal and G. Dreyfus, submitted for publication.
10. S. Diederich, M. Opper, Phys. Rev. Lett. $\underline{58}$, 949 (1987).
11. T. Maxwell, C. Lee Giles, Y. C. Lee, Proceedings of ICNN-87, San Diego, 1987.

THE SIGMOID NONLINEARITY IN PREPYRIFORM CORTEX

Frank H. Eeckman
University of California, Berkeley, CA 94720

ABSTRACT

We report a study on the relationship between EEG amplitude values and unit spike output in the prepyriform cortex of awake and motivated rats. This relationship takes the form of a sigmoid curve, that describes normalized pulse-output for normalized wave input. The curve is fitted using nonlinear regression and is described by its slope and maximum value.

Measurements were made for both excitatory and inhibitory neurons in the cortex. These neurons are known to form a monosynaptic negative feedback loop. Both classes of cells can be described by the same parameters.

The sigmoid curve is asymmetric in that the region of maximal slope is displaced toward the excitatory side. The data are compatible with Freeman's model of prepyriform burst generation. Other analogies with existing neural nets are being discussed, and the implications for signal processing are reviewed. In particular the relationship of sigmoid slope to efficiency of neural computation is examined.

INTRODUCTION

The olfactory cortex of mammals generates repeated nearly sinusoidal bursts of electrical activity (EEG) in the 30 to 60 Hz. range[1]. These bursts ride on top of a slower (1 to 2 Hz.), high amplitude wave related to respiration. Each burst begins shortly after inspiration and terminates during expiration. They are generated locally in the cortex. Similar bursts occur in the olfactory bulb (OB) and there is a high degree of correlation between the activity in the two structures[1].

The two main cell types in the olfactory cortex are the superficial pyramidal cell (type A), an excitatory neuron receiving direct input from the OB, and the cortical granule cell (type B), an inhibitory interneuron. These cell groups are monosynaptically connected in a negative feedback loop[2].

Superficial pyramidal cells are mutually excitatory[3, 4, 5] as well as being excitatory to the granule cells. The granule cells are inhibitory to the pyramidal cells as well as to each other[3, 4, 6].

In this paper we focus on the analysis of amplitude dependent properties: How is the output of a cellmass (pulses) related to the synaptic potentials (ie. waves)? The concurrent recording of multi-unit spikes and EEG allows us to study these phenomena in the olfactory cortex.

The anatomy of the olfactory system has been extensively studied beginning with the work of S. Ramon y Cajal [7]. The regular geometry and the simple three-layered architecture makes these structures ideally suitable for EEG recording [4, 8]. The EEG generators in the various olfactory regions have been identified and their synaptic connectivities have been extensively studied[9, 10, 5, 4, 11, 6].

The EEG is the scalar sum of synaptic currents in the underlying cortex. It can be recorded using low impedance (< .5 Mohm) cortical or depth electrodes. Multi-unit signals are recorded in the appropriate cell layers using high impedance (> .5 Mohm) electrodes and appropriate high pass filtering.

Here we derive a function that relates waves (EEG) to pulses in the olfactory cortex of the rat. This function has a sigmoidal shape. The derivative of this curve gives us the gain curve for wave-to-pulse conversion. This is the forward gain for neurons embedded in the cortical cellmass. The product of the forward gain values of both sets of neurons (excitatory and inhibitory) gives us the feedback gain values. These ultimately determine the dynamics of the system under study.

MATERIALS AND METHODS

A total of twenty-nine rats were entered in this study. In each rat a linear array of 6 100 micron stainless steel electrodes was chronically implanted in the prepyriform (olfactory) cortex. The tips of the electrodes were electrolytically sharpened to produce a tip impedance on the order of .5 to 1 megaohm. The electrodes were implanted laterally in the midcortex, using stereotaxic coordinates. Their position was verified electrophysiologically using a stimulating electrode in the olfactory tract. This procedure has been described earlier by Freeman [12]. At the end of the recording session a small iron deposit was made to help in histological verification. Every electrode position was verified in this manner.

Each rat was recorded from over a two week period following implantation. All animals were awake and attentive. No stimulation (electrical or olfactory) was used. The background environment for recording was the animal's home cage placed in the same room during all sessions.

For the present study two channels of data were recorded concurrently. Channel 1 carried the EEG signal, filtered between 10 and 300 Hz. and digitized at 1 ms intervals. Channel 2 carried standard pulses 5 V, 1.2 ms wide, that were obtained by passing the multi-unit signal (filtered between 300 Hz. and 3kHz.) through a window discriminator.

These two time-series were stored on disk for off-line processing using a Perkin-Elmer 3220 computer. All routines were written in FORTRAN. They were tested on data files containing standard sine-wave and pulse signals.

DATA PROCESSING

The procedures for obtaining a two-dimensional conditional pulse probability table have been described earlier [4]. This table gives us the probability of occurrence of a spike conditional on both time and normalized EEG amplitude value.

By counting the number of pulses at a fixed time-delay, where the EEG is maximal in amplitude, and plotting them versus the normalized EEG amplitudes, one obtains a sigmoidal function: The Pulse probability Sigmoid Curve (PSC) [13, 14]. This function is normalized by dividing it by the average pulse level in the record. It is smoothed by passing it through a digital 1:1:1 filter and fitted by nonlinear regression.

The equations are:

$$Q = Q_{max} (1 - \exp [- (e^v - 1) / Q_{max}]) \quad \text{for } v > - u_0 \qquad (1)$$
$$Q = -1 \qquad\qquad\qquad\qquad\qquad\qquad \text{for } v < - u_0$$

where u_0 is the steady state voltage, and $Q = (p-p_0)/p_0$.

and $Q_{max} = (p_{max}-p_0)/p_0$.

p_0 is the background pulse count, p_{max} is the maximal pulse count.

These equations rely on one parameter only. The derivation and justification for these equations were discussed in an earlier paper by Freeman [13].

244

RESULTS

Data were obtained from all animals. They express normalized pulse counts, a dimensionless value as a function of normalized EEG values, expressed as a Z-score (ie. ranging from - 3 sd. to + 3 sd., with mean of 0.0). The true mean for the EEG after filtering is very close to 0.0 mV and the distribution of amplitude values is very nearly Gaussian.

The recording convention was such that high EEG-values (ie. > 0.0 to + 3.0 sd.) corresponded to surface-negative waves. These in turn occur with activity at the apical dendrites of the cells of interest. Low EEG values (ie. from - 3.0 sd. to < 0.0) corresponded to surface-positive voltage values, representing inhibition of the cells.

The data were smoothed and fitted with equation (1). This yielded a Q_{max} value for every data file. There were on average 5 data files per animal. Of these 5, an average of 3.7 per animal could be fitted succesfully with our technique. In 25 % of the traces, each representing a different electrode pair, no correlations between spikes and the EEG were found.

Besides Q_{max} we also calculated Q' the maximum derivative of the PSC, representing the maximal gain.

There were 108 traces in all. In the first 61 cases the Q_{max} value described the wave-to-pulse conversion for a class of cells whose maximum firing probability is in phase with the EEG. These cells were labelled type A cells [2]. These traces correspond to the excitatory pyramidal cells. The mean for Q_{max} in that group was 14.6, with a standard deviation of 1.84. The range was 10.5 to 17.8.

In the remaining 47 traces the Q_{max} described the wave-to-pulse conversion for class B cells. Class B is a label for those cells whose maximal firing probability lags the EEG maximum by approximately 1/4 cycle. The mean for Q_{max} in that group was 14.3, with a standard deviation of 2.05. The range in this group was 11.0 to 18.8.

The overall mean for Q_{max} was 14.4 with a standard deviation of 1.94. There is no difference in Q_{max} between both groups as measured by the Student t-test. The nonparametric Wilcoxon rank-sum test also found no difference between the groups (p = 0.558 for the t-test; p = 0.729 for the Wilcoxon).

Assuming that the two groups have Q_{max} values that are normally distributed (in group A, mean = 14.6, median = 14.6; in group B, mean = 14.3, median = 14.1), and that they have equal variances (st. deviation group A is 1.84; st. deviation group B is 2.05) but different means, we estimated the power of the t-test to detect that difference in means.

A difference of 3 points between the Q_{max}'s of the respective groups was considered to be physiologically significant. Given these assumptions the power of the t-test to detect a 3 point difference was greater than .999 at the alpha .05 level for a two sided test. We thus feel reasonably confident that there is no difference between the Q_{max} values of both groups.

The first derivative of the PSC gives us the gain for wave-to-pulse conversion[4]. The maximum value for this first derivative was labelled Q'. The location at which the maximum Q' occurs was labelled V_{max}. V_{max} is expressed in units of standard deviation of EEG amplitudes.

The mean for Q' in group A was 5.7, with a standard deviation of .67, in group B it was 5.6 with standard deviation of .73. Since Q' depends on Q_{max}, the same statistics apply to both: there was no significant difference between the two groups for slope maxima.

Figure 1. Distribution of Qmax values

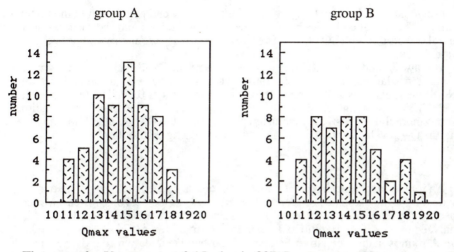

The mean for V_{max} was at 2.15 sd. +/- .307. In every case V_{max} was on the excitatory side from 0.00, ie. at a positive value of EEG Z-scores. All values were greater than 1.00. A similar phenomenon has been reported in the olfactory bulb 4, 14, 15.

Figure 2. Examples of sigmoid fits.

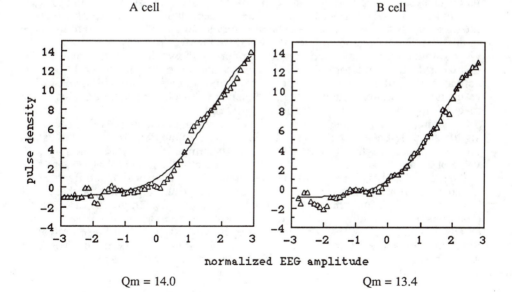

COMPARISON WITH DATA FROM THE OB

Previously we derived Q_{max} values for the mitral cell population in the olfactory bulb[14]. The mitral cells are the output neurons of the bulb and their axons form the lateral olfactory tract (LOT). The LOT is the main input to the pyramidal cells (type A) in the cortex.

For awake and motivated rats (N = 10) the mean Q_{max} value was 6.34 and the standard deviation was 1.46. The range was 4.41- 9.53. For anesthetized animals (N= 8) the mean was 2.36 and the standard deviation was 0.89. The range was 1.15- 3.62. There was a significant difference between anesthetized and awake animals. Furthermore there is a significant difference between the Q_{max} value for cortical cells and the Q_{max} value for bulbar cells (non - overlapping distributions).

DISCUSSION

An important characteristic of a feedback loop is its feedback gain. There is ample evidence for the existence of feedback at all levels in the nervous system. Moreover specific feedback loops between populations of neurons have been described and analyzed in the olfactory bulb and the prepyriform cortex 3, 9, 4.

A monosynaptic negative feedback loop has been shown to exist in the PPC, between the pyramidal cells and inhibitory cells, called granule cells 3, 2, 6, 16. Time series analysis of concurrent pulse and EEG recordings agrees with this idea.

The pyramidal cells are in the forward limb of the loop: they excite the granule cells. They are also mutually excitatory 2,4,16. The granule cells are in the feedback limb: they inhibit the pyramidal cells. Evidence for mutual inhibition (granule to granule) in the PPC also exists 17, 6.

The analysis of cell firings versus EEG amplitude at selected time-lags allows one to derive a function (the PSC) that relates synaptic potentials to output in a neural feedback system. The first derivative of this curve gives an estimate of the forward gain at that stage of the loop. The procedure has been applied to various structures in the olfactory system 4, 13, 15, 14. The olfactory system lends itself well to this type of analysis due to its geometry, topology and well known anatomy.

Examination of the experimental gain curves shows that the maximal gain is displaced to the excitatory side. This means that not only will the cells become activated by excitatory input, but their mutual interaction strength will increase. The result is an oscillatory burst of high frequency (30- 60 Hz.) activity. This is the mechanism behind bursting in the olfactory EEG 4, 13.

In comparison with the data from the olfactory bulb one notices that there is a significant difference in the slope and the maximum of the PSC. In cortex the values are substantially higher, however the V_{max} is similar. C. Gray 15 found a mean value of 2.14 +/- 0.41 for V_{max} in the olfactory bulb of the rabbit (N= 6). Our value in the present study is 2.15 +/- .31. The difference is not statistically significant.

There are important aspects of nonlinear coupling of the sigmoid type that are of interest in cortical functioning. A sigmoid interaction between groups of elements ("neurons") is a prominent feature in many artificial neural nets. S. Grossberg has extensively studied the many desirable properties of sigmoids in these networks. Sigmoids can be used to contrast-enhance certain features in the stimulus. Together with a thresholding operation a sigmoid rule can effectively quench noise. Sigmoids can also provide for a built in gain control mechanism 18, 19.

Changing sigmoid slopes have been investigated by J. Hopfield. In his network changing the slope of the sigmoid interaction between the elements affects the number of attractors that the system can go to 20. We have previously remarked upon the similarities between this and the change in sigmoid slope between waking and anesthetized animals 14. Here we present a system with a steep slope (the PPC) in series with a system with a shallow slope (the OB).

Present investigations into similarities between the olfactory bulb and Hopfield networks have been reported 21, 22. Similarities between the cortex and Hopfield-like networks have also been proposed 23.

Spatial amplitude patterns of EEG that correlate with significant odors exist in the bulb 24. A transmission of "wave-packets" from the bulb to the cortex is known to occur 25. It has been shown through cofrequency and phase analysis that the bulb can drive the cortex 25, 26. It thus seeems likely that spatial patterns may also exist in the cortex. A steeper sigmoid, if the analogy with neural networks is correct, would allow the cortex to further classify input patterns coming from the olfactory bulb.

In this view the bulb could form an initial classifier as well as a scratch-pad memory for olfactory events. The cortex could then be the second classifier, as well as the more permanent memory.

These are at present speculations that may turn out to be premature. They nevertheless are important in guiding experiments as well as in modelling. Theoretical studies will have to inform us of the likelihood of this kind of processing.

REFERENCES

1 S.L. Bressler and W.J. Freeman, Electroencephalogr. Clin. Neurophysiol. **50** : 19 (1980).
2 W.J. Freeman, J. Neurophysiol. **31**: 1 (1968).
3 W.J. Freeman, Exptl. Neurol. **10**: 525 (1964).
4 W.J. Freeman, Mass Action in the Nervous System. (Academic Press, N.Y., 1975), Chapter 3.
5 L.B. Haberly and G.M. Shepherd, Neurophys. **36**: 789 (1973).
6 L.B. Haberly and J.M. Bower, J. Neurophysiol. **51**: 90 (1984).
7 S. Ramon y Cajal, Histologie du Systeme Nerveux de l'Homme et des Vertebres. (Ed. Maloine, Paris, 1911) .
8 W.J. Freeman, Biol. Cybernetics. **35**: 21 (1979).
9 W. Rall and G.M. Shepherd, J. Neurophysiol. **31**: 884 (1968).
10 G.M. Shepherd, Physiol. Rev. **52**: 864 (1972).
11 L.B. Haberly and J.L. Price, J. Comp. Neurol. **178**; 711 (1978).
12 W.J. Freeman, Exptl. Neurol. **6**: 70 (1962).
13 W.J. Freeman, Biol. Cybernetics. **33**: 237 (1979).
14 F.H. Eeckman and W.J. Freeman, AIP Proc. **151**: 135 (1986).
15 C.M. Gray, Ph.D. thesis, Baylor College of Medicine (Houston,1986)
16 L.B. Haberly, Chemical Senses, **10**: 219 (1985).
17 M. Satou et al., J. Neurophysiol. **48**: 1157 (1982).
18 S. Grossberg, Studies in Applied Mathematics, Vol LII, 3 (MIT Press, 1973) p 213.
19 S. Grossberg, SIAM-AMS Proc. **13**: 107 (1981).
20 J.J Hopfield, Proc. Natl. Acad. Sci. USA **81**: 3088 (1984).
21 W.A. Baird, Physica **22D**: 150 (1986).
22 W.A. Baird, AIP Proceedings **151**: 29 (1986).
23 M. Wilson and J. Bower, Neurosci. Abstr. **387.10** (1987).

24 K.A. Grajski and W.J. Freeman, AIP Proc. **151**: 188 (1986).
25 S.L. Bressler, Brain Res. **409**: 285 (1986).
26 S.L. Bressler, Brain Res. **409**: 294 (1986).

HIERARCHICAL LEARNING CONTROL -

AN APPROACH WITH NEURON-LIKE ASSOCIATIVE MEMORIES

E. Ersü
ISRA Systemtechnik GmbH, Schöfferstr. 15, D-6100 Darmstadt, FRG

H. Tolle
TH Darmstadt, Institut für Regelungstechnik,
Schloßgraben 1, D-6100 Darmstadt, FRG

ABSTRACT

Advances in brain theory need two complementary approaches:
Analytical investigations by in situ measurements and as well synthetic modelling supported by computer simulations to generate
suggestive hypothesis on purposeful structures in the neural
tissue. In this paper research of the second line is described:
Starting from a neurophysiologically inspired model of stimulus-
response (S-R) and/or associative memorization and a psychologically motivated ministructure for basic control tasks, pre-conditions
and conditions are studied for cooperation of such units in a
hierarchical organisation, as can be assumed to be the general
layout of macrostructures in the brain.

I. INTRODUCTION

Theoretic modelling in brain theory is a highly speculative
subject. However, it is necessary since it seems very unlikely to
get a clear picture of this very complicated device by just analyzing the available measurements on sound and/or damaged brain parts
only. As in general physics, one has to realize, that there are
different levels of modelling: in physics stretching from the ato-
mary level over atom assemblies till up to general behavioural
models like kinematics and mechanics, in brain theory stretching
from chemical reactions over electrical spikes and neuronal cell
assembly cooperation till general human behaviour.

The research discussed in this paper is located just above the
direct study of synaptic cooperation of neuronal cell assemblies as
studied e. g. in /Amari 1988/. It takes into account the changes of
synaptic weighting, without simulating the physical details of such
changes, and makes use of a general imitation of learning situation
(stimuli) - response connections for building up trainable basic
control loops, which allow dynamic S-R memorization and which are
themselves elements of some more complex behavioural loops. The
general aim of this work is to make first steps in studying structures, preconditions and conditions for building up purposeful
hierarchies and by this to generate hypothesis on reasons and

meaning behind substructures in the brain like the columnar organization of the cerebral cortex (compare e. g. /Mountcastle 1978/).

The paper is organized as follows: In Chapter II a short description is given of the basic elements for building up hierarchies, the learning control loop LERNAS and on the role of its subelement AMS, some associative memory system inspired by neuronal network considerations. Chapter III starts from certain remarks on substructures in the brain and discusses the cooperation of LERNAS-elements in hierarchies as possible imitations of substructures. Chapter IV specifies the steps taken in this paper in the direction of Chapter III and Chapter V presents the results achieved by computer simulations. Finally an outlook will be given on further investigations.

II. LERNAS AND AMS

Since the formal neuron was introduced by /McCulloch and Pitts 1943/, various kinds of neural network models have been proposed, such as the perceptron by /Rosenblatt 1957/ the neuron equation of /Caianello 1961/, the cerebellar model articulation controller CMAC by /Albus 1972, 1975/ or the associative memory models by /Fukushima 1973/, /Kohonen 1977/ and /Amari 1977/. However, the ability of such systems to store information efficiently and to perform certain pattern recognition jobs is not adequate for survival of living creatures. So they can be only substructures in the overall brain organization; one may call them a microstructure.

Purposeful acting means a goal driven coordination of sensory information and motor actions. Although the human brain is a very complex far end solution of evolution, the authors speculated in 1978 that it might be a hierarchical combination of basic elements, which would perform in an elementary way like the human brain in total, especially since there is a high similarity in the basic needs as well as in the neuronal tissue of human beings and relatively simple creatures. This led to the design of the learning control loop LERNAS in 1981 by one of the authors - /Ersü 1984/ - on the basis of psychological findings. He transformed the statement of /Piaget 1970/, that the complete intelligent action needs three elements: "1) the question, which directs possible search actions, 2) the hypothesis, which anticipates eventual solutions, 3) the control, which selects the solution to be chosen" into the structure shown in Fig. 1, by identifying the "question" with an performance criterion for assessment of possible advantages/disadvantages of certain actions, the "hypothesis" with a predictive model of environment answers and the "control" with a control strategy which selects for known situations the best action, for unknown situations some explorative action (active learning).

In detail, Fig. 1 has to be understood in the following way: The predictive model is built up in a step by step procedure from a characterization of the actual situation at the time instant $k \bullet T_s$

T_s sampling time) and the measured response of the unknown environment at time instant $(k+1)T_s$. The actual situation consists of measurements regarding the stimuli and responses of the environment at time instant $k \cdot T_s$ plus – as far as necessary for a unique characterization – of the situation-stimuli and responses at time instants $(k-1)T_s$, $(k-2)T_s$..., provided by the short term memory. To reduce learning effort, the associative memory system used to store the predictive model has the ability of local generalization, that means making use of the trained response value not only for the corresponding actual situation, but also in similar situations. The assessment module generates on the basis of a given goal – a wanted environment response – with an adequate performance criterion an evaluation of possible actions through testing them with the predictive model, as far as this is already built up and gives meaningful answers. The result is stored in the control strategyAMS together with its quality: real optimal action for the actual situation or only relatively optimal action, if the testing reached the border of the known area in the predictive model of the environment. In the second case, the real action is changed in a sense of curiosity, so that by the action the known area of the predictive model is extended. By this, one reaches more and more the first case, in which the real optimal actions are known. Since the first guess for a good action in the optimization phase is given to the assessment module from the control strategy AMS – not indicated in Fig. 1 to avoid unnecessary complication – finally the planning level gets superfluous and one gets very quick optimal reactions, the checking with the planning level being necessary and helpful only to find out, whether the environment has not changed, possibly. Again the associative memory system used for the control strategy is locally generalizing to reduce the necessary training effort.

The AMS storage elements for the predictive model, and for optimized actions are a refinement and implementation for on-line application of the neuronal network model CMAC from J. Albus – see e. g. /Ersü, Militzer 1982/ –, but it could be any other locally generalizing neural network model and even a storage element based on pure mathematical considerations, as has been shown in /Militzer, Tolle 1986/.

The important property to build up an excellent capability to handle different tasks in an environment known only by some sensory information – the property which qualifies LERNAS as a possible basic structure (a "ministructure") in the nervous system of living creatures – has been proven by its application to the control of a number of technical processes, starting with empty memories for the predictive model and the control strategy storage. Details on this as well as on the mathematical equations describing LERNAS can be found in /Ersü, Mao 1983/, /Ersü, Tolle 1984/ and /Ersü, Militzer 1984/.

It should be mentioned that the concept of an explicit predictive environmental model - as used in LERNAS - is neither the only meaningful description of human job handling nor a necessary part of our basic learning element. It suffices to use a prediction whether a certain action is advantegeous to reach the actual goal or whether this is not the case. More information on such a basic element MINLERNAS, which may be used instead of LERNAS in general (however, with the penalty of some performance degradation) are given in /Ersü, Tolle 1988/.

III. HIERARCHIES

There are a number of reasons to believe, that the brain is built up as a hierarchy of control loops, the higher levels having more and more coordinative functions. A very simple example shows the necessity in certain cases. The legs of a jumping jack can move together, only. If one wants to move them separately, one has to cut the connection, has to build up a separate controller for each leg and a coordinating controller in a hierarchically higher level to restore the possibility of coordinated movements. Actually, one can find such an evolution in the historical development of certain animals. In a more complex sense a multilevel hierarchy exists in the extrapyramidal motor system. Fig. 2 from /Albus 1979/ specifies five levels of hierarchy for motor control. It can be speculated, that hierarchical organizations are not existing in the senso-motoric level only, but also in the levels of general abstractions and thinking. E. g. /Dörner 1974/ supports this idea.

If one assumes out of these indications, that hierarchies are a fundamental element of brain structuring - the details and numbers of hierarchy-levels not being known - one has to look for certain substructures and groupings of substructures in the brain. In this connection one finds as a first subdivision the cortical layers, but then as another more detailed subdivision the columns, cell assemblies heavily connected in the axis vertical to cortical layers and sparsely connected horizontally. /Mountcastle 1978/ defines minicolumns, which comprise in some neural tissue roughly 100 in other neural tissue roughly 250 individual cells. In addition to these minicolumns certain packages of minicolumns, consisting out of several hundreds of the minicolumns, can be located. They are called macrocolumns by /Mountcastle 1978/. Fig. 3 gives some abstraction, how such structures could be interpreted: each minicolumn is considered to be a ministructure of the type LERNAS, a number of LERNAS units - here shown in a ring structure instead of a filled up cylindrical structure - building up a macrocolumn. The signals between the LERNAS elements could be overlapping and cooperating. Minicolumns being elements of macrocolumns of a higher cortical layer - here layer j projecting to layer k - could initiate and/or coordinate this cooperation in a hierarchical sense.

Such a complex system is difficult to simulate. One has to go into this direction in a step by step procedure. In a first step the overlapping or crosstalk between the minicolumns may be suppressed and the number of ministructures LERNAS representing the minicolumns should be reduced heavily. This motivates Fig. 4 as a fundamental blockdiagram for research on cooperation of LERNAS elements.

IV. TOPICS ADDRESSED

From Fig. 4 only the lowest level of coordination (layer 1), that means the coordination of two subprocesses was implemented up to now - right half of Fig. 5. This has two reasons: Firstly, a number of fundamental questions can be posed and discussed with such a formulation already. Secondly, it is difficult to set up meaningful subprocesses and coordination goals for a higher order system.

The problem discussed in the following can be understood as the coordination of two minicolumns as described in Chapter III, but also as the coordination of higher level subtasks, which may be detailed themselves by ministructures and/or systems like Fig. 4. This is indicated in the left half of Fig. 5.

Important questions regarding hierarchies of learning control loops are:

I.   What seem to be meaningful interventions from the coordinator onto the lower level systems?

II. Is parallel learning in both levels possible or requires a meaningful learning strategy that the control of subtasks has to be learned at first before the coordination can be learned?

III. Normally one expects, that the lower level takes care of short term requirements and the upper level of long term strategies. Is that necessary or what happens if the upper level works on nearly the same time horizon as the lower levels?

IV. Furtheron one expects, that the upper level may look after other goals than the lower level, e. g. the lower level tries to suppress disturbances effects since the upper level tries to minimize overall energy consumption. But can such different strategies work without oscillations or destabilization of the system?

Question I can be discussed by some general arguments, for questions II-IV only indications of possible answers can be given from simulation results. This will be postponed to Chapter V.

Fig. 6 shows three possible intervention schemes from the coordinator.

By case a) an intervention into the structure or the parameters of the sublevel (=local) controllers is meant. Since associative mappings like AMS have no parameters being directly responsible for the behaviour of the controller – as would be the case with a parametrized linear or non-linear differential equation being the description of a conventional controller – this does not make sense for the controller built up in LERNAS. However, one could consider the possibility to change parameters or even elements, that means structural terms of the performance criterion, which is responsible for the shaping of the controller. But this would require to learn anew, which takes a too long time span in general.

By case b) a distribution of work load regarding control commands is meant. The possible idea could be, that the coordinator gives control inputs to hold the long range mean value required, since the local controllers take into account fast dynamic fluctuations only. However, this has the disadvantage that the control actions of the upper level have to be included into the inputs to the local controllers, extending the dimension of in-put space of these storage devices, since otherwise the process appears to be highly time variant for the local controllers, which is difficult to handle for LERNAS.

So case c) seems to be the best solution. In this case the coordinator commands the set points of the local controllers, generating by this local subgoals for the lower level controllers. Since this requires no input space extension for the local controllers and is in full agreement with the working conditions of single LERNAS loops, it is a meaningful and effective approach.

Fig. 7 shows the accordingly built up structure in detail. The control strategy of Fig. 1 is divided here in two parts the storage element (the controller C) and the active learning AL. The elements are explicitly characterized for the upper level only. The whole lower level is considered by the coordinator as a single pseudo-process to be controlled (see Fig. 4).

V. SIMULATION RESULTS

For answering questions II and III the very simple non-linear process shown in Fig. 8 – detailing the subprocesses SP1, SP2 and their coupling in Fig. 7 – was used. For the comparison of bottom up and parallel learning suitably fixed PI-controllers were used for bottom up learning instead of LERNAS 1 and LERNAS 2, simulating optimally trained local controllers. Fig. 9a shows the result due to which in the first run a certain time is required for achieving a good set point following through coordinator assistance. However, with the third repetition (4th run) a good performance is reached from the first set point change on already. For parallel learning all (and not only the coordinator AMS-memories) were empty in the beginning. Practically the same performance was achieved as in bottom up training – Fig. 9b –, indicating, that at least in simple problems, as considered here, parallel learning is a real possibility. However - what is not illustrated here - the coordinator sampling time must be sufficiently long, so that the local controllers can reach the defined subgoals at least qualitatively in this time span.

For answering question III, in which respect a higher difference in the time horizon between local controller and coordinator changes the picture, a doubling of the sampling rate for the coordinator was implemented. Fig. 10 give the results. They can be interpreted as follows: Smaller sampling rates allow the coordinator to get more information about the pseudo-sub-processes, the global goal is reached faster. Larger sampling rates lead to a better overall performance when the goal is reached: there is a higher amount of averaging regarding informations about the pseudo-sub-processes.

Up to now in both levels the goal or performance criterion was the minimization of differences between the actual plant output and the requested plant output. The influence of different coordinator goals - question IV - was investigated by simulating a two stage waste water neutralization process. A detailed description of this process set up and the simulation results shall not be given here out of space reasons. It was found that:

- in hierarchical systems satisfactory overall behaviour may be reached by well defined subgoals with clearly different coordinator goals.

- since learning is goal driven, one has to accept that implicit wishes on closed loop behaviour are fulfilled by chance only. Therefore important requirements have to be included in the performance criteria explicitly.

It should be remarked finally, that one has to keep in mind, that simulation results with one single process are indications of possible behaviour only, not excluding that in other cases a fundamentally different behaviour can be met.

VI. OUTLOOK

As has been mentioned already in Chapter III and IV, this work is one of many first steps of investigations regarding hierarchical organization in the brain, its preconditions and possible behaviour.

Subjects of further research should be the self-organizing task distribution between the processing units of each layer, and the formation of interlayer projections in order to build up meta-tasks composed of a sequence of frequently occuring elementary tasks. These investigations will on the other hand show to what extent this kind of higher-learning functions can be achieved by a hierarchy of LERNAS-type structures which model more or less low-level basic learning behaviour.

256

VII. ACKNOWLEDGEMENTS

The work presented has been supported partly by the Stiftung Volkswagenwerk. The detailed evaluations of Chapter IV and V have been performed by Dipl.-Ing. M. Zoll and Dipl.-Ing. S. Gehlen. We are very thankful for this assistance.

VIII. REFERENCES

Albus, J. S.	Theoretical and Experimental Aspects of a Cerebellar Model, Ph.D. Thesis, Univ. of Maryland, 1972
Albus, J. S.	A New Approach to Manipulator Control: The Cerebellar Model Articulation Controller (CMAC), Trans. ASmE series, G, 1975
Albus, J. S.	A Model of the Brain for Robot Control – Part 3: A Comparison of the Brain and Our Model, Byte, 1979
Amari, S. I.	Neural Theory of Association and Concept Formation, Biol. Cybernetics, Vol. 26, 1977
Amari, S. I.	Mathematical Theory of Self-Organization in Neural Nets, in: Organization of Neural Networks, Structures and Models, ed. by von Seelen, Shaw, Leinhos, VHC-Verlagsges. Weinheim, W.-Germany, 1988
Caianello, E. R.	Outline of a Theory of Thought Process and Thinking Machines, Journal of Theoretical Biology, Vol. 1, 1961
Dörner, D.	Problemlösen als Informationsverarbeitung Verlag H. Huber, 1974
Ersü, E.	On the Application of Associative Neural Network Models to Technical Control Problems, in: Localization and Orientation in Biology and Engineering, ed. by Varju, Schnitzler, Springer Verlag Berlin, W.-Germany, 1984
Ersü, E. Mao, X.	Control of pH by Use of a Self-Organizing Concept with Associative Memories, ACI'83, Kopenhagen (Denmark), 1983

Ersü, E.
Militzer, J.

Software Implementation of a Neuron-Like
Associative Memory System for Control
Application, Proceedings of the 2nd IASTED
Conference on Mini- and Microcomputer Appli-
cations, MIMI'82, Davos (Switzerland), 1982

Ersü, E.
Militzer, J.

Real-Time Implementation of an Associative
Memory-Based Learning Control Scheme for Non-
Linear Multivariable Processes, Symposium
"Applications of Multivariable System
Techniques", Plymouth (UK), 1984

Ersü, E.
Tolle, H.

A New Concept for Learning Control Inspired by
Brain Theory, Proceed. 9th IFAC World Congress,
Budapest (Hungary), 1984

Ersü, E.
Tolle, H.

Learning Control Structures with Neuron-Like
Associative Memory Systems, in: Organization of
Neural Networks, Structures and Models, ed. by
von Seelen, Shaw, Leinhos, VCH Verlagsgesell-
schaft Weinheim, W.-Germany, 1988

Fukushima, K.

A Model of Associative Memory in the Brain
Biol. Cybernetics, Vol. 12, 1973

Kohonen, T.

Associative Memory, Springer Verlag Berlin,
W.-Germany, 1977

McCulloch, W. S.
Pitts, W. H.

A Logical Calculus of the Ideas, Immanent in
Nervous Activity, Bull. Math. Biophys. 9, 1943

Militzer, J.
Tolle, H.

Vertiefungen zu einem Teilbereiche der mensch-
lichen Intelligenz imitierenden Regelungsansatz
Tagungsband-DGLR-Jahrestagung, München,
W.-Germany, 1986

Mountcastle, V. B.

An Organizing Principle for Cerebral Function:
The Unit Module and the Distributed System, in:
The Mindful Brain by G. M. Edelman,
V. B. Mountcastle, The MIT-Press, Cambridge,
USA, 1978

Piaget, J.

Psychologie der Intelligenz, Rascher Verlag,
4th printing, 1970

Rosenblatt, F.

The Perceptron: A Perceiving and Recognizing
Automation, Cornell Aeronautical Laboratory,
Report No. 85-460-1, 1957

258

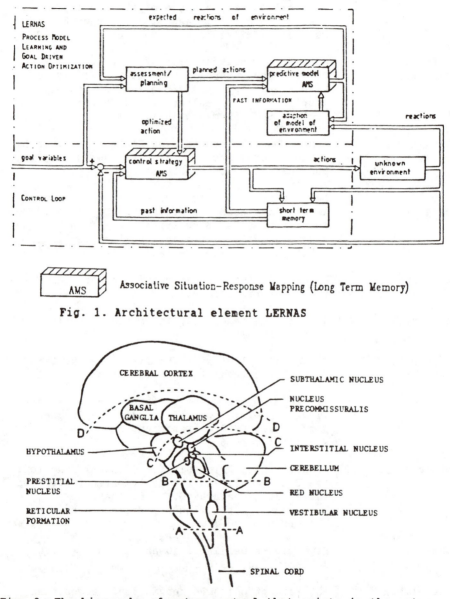

Fig. 1. Architectural element LERNAS

Fig. 2. The hierarchy of motor control that exists in the extra-pyramidal motor system. Basic reflexes remain even if the brain stem is cut at A-A. Coordination of these reflexes for standing is possible if the cut is at B-B. The sequential coordination required for walking requires the area below C-C to be operable. Simple tasks can be executed if the region below D-D is intact. Lengthy tasks and complex goals require the cerebral cortex. (/Albus 1979/)

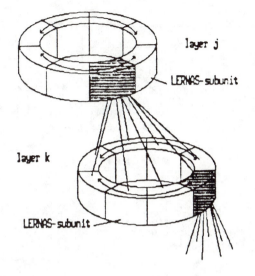

Fig. 3. Generic scetch of macrocolumns - drawn as ring
structures - from different cortical layers with
LERNAS-subunits representing minicolumns

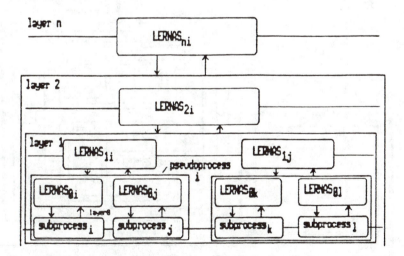

Fig. 4. LERNAS-hierarchy as a simplified research model
for cooperation of columnar structures

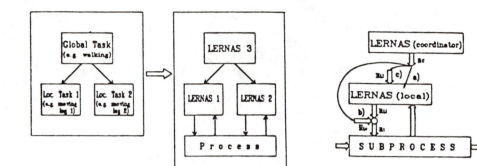

Fig. 5. Hierarchical work/
control distribution

Fig. 6. Methods of
intervention from the
coordinator

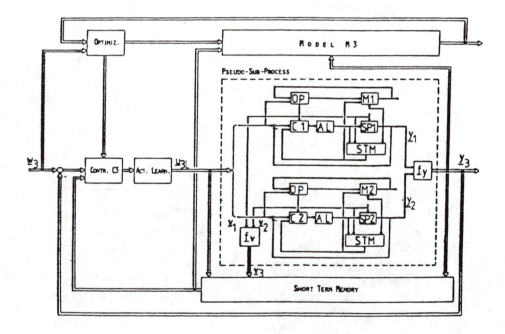

Fig. 7. Implementation of the hierarchical structure

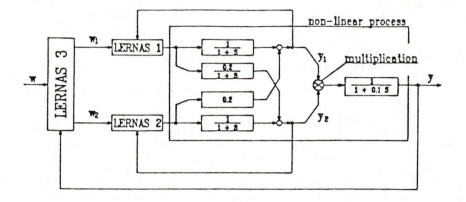

Fig. 8. Hierarchical structure with non-linear
multivariable test-process

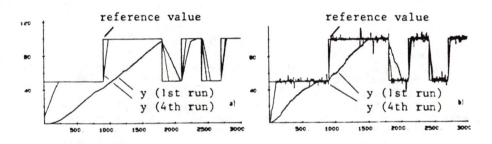

Fig. 9. Learning on coordinator level using already
trained (a) and untrained (b) lower levels
$(T_{coord} = 2 \text{ sec}, T_{loc} = 0.5 \text{ sec})$

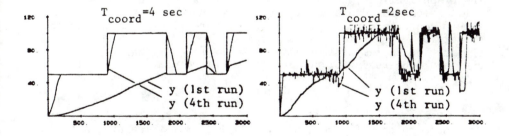

Fig. 10. Coordinator learning behaviour using different
coordinator horizons $(T_{loc} = 0.5 \text{ sec})$

ON TROPISTIC PROCESSING AND ITS APPLICATIONS

Manuel F. Fernández
General Electric Advanced Technology Laboratories
Syracuse, New York 13221

ABSTRACT

The interaction of a set of tropisms is sufficient in many cases to explain the seemingly complex behavioral responses exhibited by varied classes of biological systems to combinations of stimuli. It can be shown that a straightforward generalization of the tropism phenomenon allows the efficient implementation of effective algorithms which appear to respond "intelligently" to changing environmental conditions. Examples of the utilization of tropistic processing techniques will be presented in this paper in applications entailing simulated behavior synthesis, path-planning, pattern analysis (clustering), and engineering design optimization.

INTRODUCTION

The goal of this paper is to present an intuitive overview of a general unsupervised procedure for addressing a variety of system control and cost minimization problems. This procedure is based on the idea of utilizing "stimuli" produced by the environment in which the systems are designed to operate as basis for dynamically providing the necessary system parameter updates.

This is by no means a new idea: countless examples of this approach abound in nature, where innate reactions to specific stimuli ("tropisms" or "taxis" --not to be confused with "instincts") provide organisms with built-in first-order control laws for triggering varied responses [8]. (It is hypothesized that "knowledge" obtained through evolution/adaptation or through learning then refines or suppresses most of these primal reactions).

Several examples of the implicit utilization of this approach can also be found in the literature, in applications ranging from behavior modeling to pattern analysis. We very briefly depict some these applications, underlining a common pattern in their formulation and generalizing it through the use of basic field theory concepts and representations. A more rigorous and detailed exposition --regarding both mathematic and application/implementation aspects-- is presently under preparation and should be ready for publication sometime next year ([6]).

TROPISMS

Tropisms can be defined in general as class-invariant systemic responses to specific sets of stimuli [6]. All time-invariant systems can thus be viewed as tropistic provided that we allow all possible stimuli to form part of our set of inputs. In most tropistic systems, however, response- (or time-) invariance applies only to specific inputs: green plants, for example, twist and grow in the direction of light (phototropism), some birds' flight patterns follow changes in the Earth's magnetic field (magnetotropism), various organisms react to gravitational field variations (geotropism), etc.

Tropism/stimuli interactions can be portrayed in terms of the superposition of scalar (e.g., potential) or vector (e.g., force) fields exhibiting properties paralleling those of the suitably constrained "reactions" we wish to model [1],[6]. The resulting field can then be used as a basis for assessing the intrinsic cost of pursuing any given path of action, and standard techniques (e.g., gradient-following in the case of scalar fields or divergence computation in the case of vector fields) utilized in determining a response*. In addition, the global view of the situation provided by field representations suggest that a basic theory of tropistic behavior can also be formulated in terms of energy expenditure minimization (Euler-Lagrange equations). This formulation would yield integral-based representations (Feynman path integrals [4],[11]) satisfying the observation that tropistic processes typically obey the principle of least action.

Alternatively, fields may also be collapsed into "attractors" (points of a given "mass" or "charge" in cost space) through laws defining the relationships that are to exist among these "attractors" and the other particles traveling through the space. This provides the simplification that when updating dynamically changing situations only the effects caused by the interaction of the attractors with the particles of interest --rather than the whole cost field-- may have to be recalculated.

For example, appropriately positioned point charges exerting on each other an electrostatic force inversely proportional to the square of their distance can be used to represent the effects of a coulombic-type cost potential field. A particle traveling through this field would now be affected by the combination of forces ensuing from the interaction of the attractors' charges with its own. If this particle were then to passively follow the composite of the effects of these forces it would be following the gradient of the cost field (i.e., the vector resulting from the superposition of the forces acting on the particle would point in the direction of steepest change in potential).

Finally, other representations of tropism/stimuli interactions (e.g., Value-Driven Decision Theory approaches) entail associating "profit" functions (usually sigmoidal) with each tropism, modeling the relative desirability of triggering a reaction as a function of the time since it was last activated [9]. These representations are

---

* In order to bring extra insight into tropism/stimuli interactions and simplify their formulation, one may exchange vector and scalar field representations through the utilization of appropriately selected mappings. Some of the most important of such mappings are the gradient operator (particularly so because the gradient of a scalar --potential-- field is proportional to a "force" --vector-- field), the divergence (which may be thought of as performing in vector fields a function analogous to that performed in scalar fields by the gradient), and their combinations (e.g., the Laplacian, a scalar-to-scalar mapping which can be visualized as performing on potential fields the equivalent of a second derivative operation.

- *Model fly as a positive geotropistic point of mass M.*
- *Model fence stakes as negative geotropistic points with masses $m_1, m_2, \ldots, m_N$.*
- *At each update time compute sum of forces acting on frog:*

$$F = k \left(\frac{M}{d_M^2} - \sum_{i=1}^{N} \frac{m_i}{d_{m_i}^2} \right) m,$$

- *Compute frog's heading and acceleration based on the ensuing force; then update frog's position.*

Figure 1: Attractor-based representation of a frog-fence-fly scenario (see [1] for a vector-field representation). The objective is to model a frog's path-planning decision-making process when approaching a fly in the presence of obstacles. (The picket fence is represented by the elliptical outline with an opening in the back, the fly --inside the fenced space-- is represented by a "+" sign, and arrows are used to indicate the direction of a frog's trajectory into and out of fenced area).

particularly amenable to neural-net implementations [6].

TROPISTIC PROCESSING

Tropistic processing entails building into systems tropisms appropriate for the environment in which these systems are expected to operate. This allows taking advantage of environment-produced "stimuli" for providing the required control for the systems' behavior.

The idea of tropistic processing has been utilized with good results in a variety of applications. Arbib et.al., for example, have implicitly utilized tropistic processing to describe a batrachian's reaction to its environment in terms of what may be visualized as magnetic (vector) fields' interactions [1].

Watanabe [12] devised for pattern analysis purposes an interaction of tropisms ("geotropisms") in which pattern "atoms" are attracted to each other, and hence "clustered", subject to a squared-inverse-distance ("feature distance") law similiar to that from gravitational mechanics. It can be seen that if each pattern atom were considered an "organism", its behavior would not be conceptually different from that exhibited by Arbibian frogs: in both cases organisms passively follow the force vectors resulting from the interaction of the environmental stimuli with the organisms' tropisms. It is interesting, though, to note that the "organisms'" behavior will nonetheless appear "intelligent" to the casual observer.

The ability of tropistic processes to emulate seemingly rational behavior is now begining to be explored and utilized in the development of synthetic-psychological models and experiments. Braitenberg, for example, has placed tropisms as the primal building block from which his models for cognition, reason, and emotions evolve [3]**; Barto [2] has suggested the possibility of combining tropisms and associative (reinforced) learning, with aims at enabling the automatic triggering of behavioral responses by previously experienced situations; and Fernández [6] has used CROBOTS [10], a virtual multiprocessor emulator, as laboratory for evaluating the effects of modifying tropistic responses on the basis of their projected future consequences.

Other applications of tropistic processing presently being investigated include path-planning and engineering design optimization [6]. For example, consider an air-reconnaissance mission deep behind enemy lines; as the mission progresses and unexpected SAM sites are discovered, contingency flight paths may be developed in real time simply by modeling each SAM or interdiction site as a mass point towards which the aircraft exhibits negative geotropistic tendencies (i.e., gravitational forces repel it), and modeling the objective as a positive geotropistic point. A path to

---

** Of particular interest within the sole context of Tropistic Processing is Dewdney's [5] commented version of the first chapters of Braitenberg's book [3], in which the "behavior" of mechanically very simple cars, provided with "eyes" and phototropism-supporting connections (including Ledley-type "neurons" [4]), is "analyzed".

266

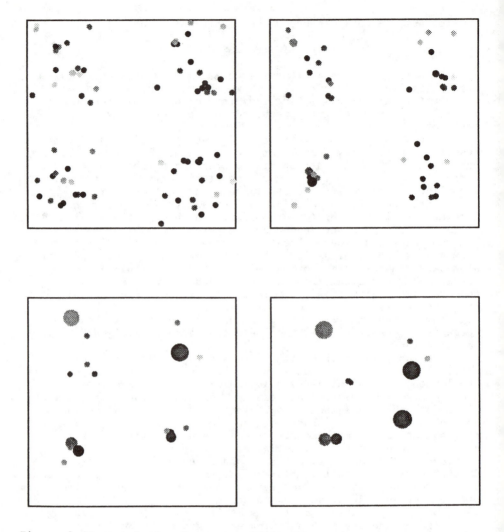

Figure 2 (Geotropistic clustering [12]): The problem being portrayed here is that of clustering dots distributed in [x,y]-space as shown and uniformly in color ([red,blue,green]). The approach followed is that outlined in Figure 1, with the differences that normalized (Mahalanobis) distances are used and when merges occur, conservation of momentum is observed. Tags are also kept --specifying with which dots and in what order merges occur-- to allow drawing cluster boundaries in the original data set. (Efficient implementation of this clustering technique entails using a ring of processors, each of which is assigned the "features" of one or more "dots" and the task of carrying out computations with respect to these features. If the features of each dot are then transmitted through the ring, all the forces imposed on it by the rest will have been determined upon completion of the circuit).

the target will then be automatically drawn by the interaction of the tropisms with the gravitational forces. (Once the mission has been completed, the target and its effects can be eliminated, leaving active only the repulsive forces, which will then "guide" the airplane out of the danger zone).

In engineering design applications such as lens modeling and design, lenses (gradient-index type, for example) can be modeled in terms of photons attempting to reach an objective plane through a three-dimensional scalar field of refraction indices; modeling the process tropistically (in a manner analogous to that of the air-reconnaissance example above) would yield the least-action paths that the individual photons would follow. Similarly, in "surface-of-revolution" fuselage design ("Newton's Problem"), the characteristics of the interaction of forces acting within a sheet of metal foil when external forces (collisions with a fluid's molecules) are applied can be modeled in terms of tropistic reactions which will tend to reconfigure the sheet so as to make it present the least resistance to friction when traversing a fluid.

Additional applications of tropistic processing include target tracking and multisensor fusion (both can be considered instances of "clustering") [6], resource allocation and game theory (both closely related to path-planning) [9], and an assortment of other cost-minimization functions. Overall, however, one of the most important applications of tropistic processing may be in the modeling and understanding of analog processes [6], the imitation of which may in turn lead to the development of effective strategies

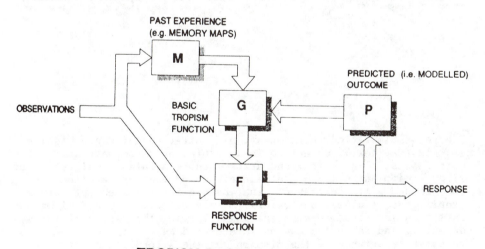

TROPISM-BASED SYSTEM

Figure 3: The combination of tropisms and associative (reinforced) learning can be used to enable the automatic triggering of behavioral responses by previously experienced situations [2]. Also, the modeled projection of the future consequences of a tropistic decision can be utilized in the modification of such decision [6]. (Note analogy to filtering problem in which past history and predicted behavior are used to smooth present observations).

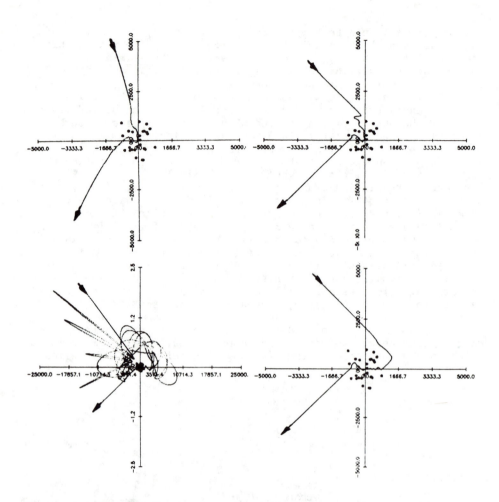

Figure 4: Simplified representation of air-reconnaissance mission
example (see text): objective is at center of coordinate axis, thick
dots represent SAM sites, and arrows denote airplane's direction of
flight (airplane's maximum attainable speed and acceleration are
constrained). All portrayed scenarios are identical except for
tropistic control-law parameters (mainly objective to SAM-sites mass
ratios in the first three scenarios). Varying the masses of the
objective and SAM sites can be interpreted as trading off the
relative importance of the mission vs. the aircraft's safety, and
can produce dramatically differing flight paths, induce chaotic
behavior (bottom-left scenario), or render the system unstable. The
bottom-right scenario portrays the situation in which a tropistic
decision is projected into the future and, if not meeting some
criterion, modified (altering the direction of flight --e.g.,
following an isokline--, re-evaluating the mission's relative
importance --revising masses--, changing the update rate, etc.).

for taking full advantage of parallel architectures [11]***. It is thus expected that the flexibility of tropistic processes to adapt to changing environmental conditions will prove highly valuable to the advancement of areas such as robotics, parallel processing and artificial intelligence, where at the very least they will provide some decision-making capabilities whenever unforeseen circumstances are encountered.

ACKNOWLEDGEMENTS

Special thanks to D. P. Bray for the ideas provided in our many discussions and for the development of the finely detailed simulations that have enabled the visualization of unexpected aspects of our work.

REFERENCES

[1] Arbib, M.A. and House, D.H.: "Depth and Detours: Decision Making in Parallel Systems". IEEE Workshop on Languages for Automation: Cognitive Aspects in Information Processing; pp. 172-180 (1985).

[2] Barto, A.G. (Editor): "Simulation Experiments with Goal-Seeking Adaptive Elements". Avionics Laboratory, Wright-Patterson Air Force Base, OH. Report # AFWAL-TR-84-1022. (1984).

[3] Braitenberg, V.: Vehicles: Experiments in Synthetic Psychology. The MIT Press. (1984).

[4] Cheng, G.C.; Ledley, R.S.; and Ouyang, B.: "Pattern Recognition with Time Interval Modulation Information Coding". IEEE Transactions on Aerospace and Electronic Systems. AES-6, No.2; pp. 221-227 (1970).

[5] Dewdney, A.K.: "Computer Recreations". Scientific American. Vol.256, No.3; pp. 16-26 (1987).

[6] Fernández, M.F.: "Tropistic Processing". To be published (1988).

[7] Feynman, R.P.: Statistical Mechanics: A Set of Lectures. Frontiers in Physics Lecture Note Series (1982).

[8] Hirsch, J.: "Nonadaptive Tropisms and the Evolution of Behavior". Annals of the New York Academy of Sciences. Vol.223; pp. 84-88 (1973).

[9] Lucas, G. and Pugh, G.: "Applications of Value-Driven Automation Methodology for the Control and Coordination of Netted Sensors in Advanced C**3". Report # RADC-TR-80-223. Rome Air Development Center, NY. (1980).

[10] Poindexter, T.: "CROBOTS". Manual, programs, and files (1985). 2903 Winchester Dr., Bloomington, IL., 61701.

[11] Wallqvist, A.; Berne, B.J.; and Pangali, C.: "Exploiting Physical Parallelism Using Supercomputers: Two Examples from Chemical Physics". Computer. Vol.20, No.5; pp. 9-21 (1987).

[12] Watanabe, S.: Pattern Recognition: Human and Mechanical. John Wiley & Sons; pp. 160-168 (1985).

*** Optical Fourier transform operations, for instance, can be modeled in high-granularity machines through a procedure analogous to the gradient-index lens simulation example, with processors representing diffraction-grating "atoms" [6].

Correlational Strength and Computational Algebra of Synaptic Connections Between Neurons

Eberhard E. Fetz
Department of Physiology & Biophysics,
University of Washington, Seattle, WA 98195

ABSTRACT

Intracellular recordings in spinal cord motoneurons and cerebral cortex neurons have provided new evidence on the correlational strength of monosynaptic connections, and the relation between the shapes of postsynaptic potentials and the associated increased firing probability. In these cells, excitatory postsynaptic potentials (EPSPs) produce cross-correlogram peaks which resemble in large part the derivative of the EPSP. Additional synaptic noise broadens the peak, but the peak area -- i.e., the number of above-chance firings triggered per EPSP -- remains proportional to the EPSP amplitude. A typical EPSP of 100 µv triggers about .01 firings per EPSP. The consequences of these data for information processing by polysynaptic connections is discussed. The effects of sequential polysynaptic links can be calculated by convolving the effects of the underlying monosynaptic connections. The net effect of parallel pathways is the sum of the individual contributions.

INTRODUCTION

Interactions between neurons are determined by the strength and distribution of their synaptic connections. The strength of synaptic interactions has been measured directly in the central nervous system by two techniques. Intracellular recording reveals the magnitude and time course of postsynaptic potentials (PSPs) produced by synaptic connections, and cross-correlation of extracellular spike trains measures the effect of the PSP's on the firing probability of the connected cells. The relation between the shape of excitatory postsynaptic potentials (EPSPs) and the shape of the cross-correlogram peak they produce has been empirically investigated in cat motoneurons [2,4,5] and in neocortical cells [10].

RELATION BETWEEN EPSP'S AND CORRELOGRAM PEAKS

Synaptic interactions have been studied most thoroughly in spinal cord motoneurons. Figure 1 illustrates the membrane potential of a rhythmically firing motoneuron, and the effect of EPSPs on its firing. An EPSP occurring sufficiently close to threshold (Θ) will cause the motoneuron to fire and will advance an action potential to its rising edge (top). Mathematical analysis of this threshold-crossing process predicts that an EPSP with shape e(t) will produce a firing probability f(t), which resembles

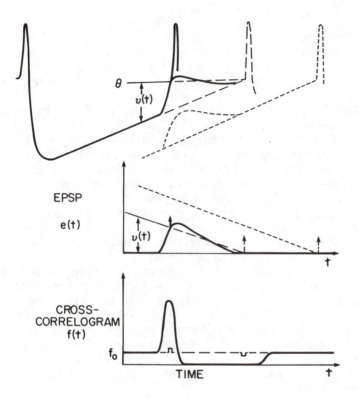

Fig. 1. The relation between EPSP's and motoneuron firing. Top: membrane trajectory of rhythmically firing motoneuron, showing EPSP crossing threshold (Θ) and shortening the normal interspike interval by advancing a spike. V(t) is difference between membrane potential and threshold. Middle: same threshold-crossing process aligned with EPSP, with v(t) plotted as falling trajectory. Intercept (at upward arrow) indicates time of the advanced action potential. Bottom: Cross-correlation histogram predicted by threshold crossings. The peak in the firing rate f(t) above baseline (f_0) is produced by spikes advanced from baseline , as indicated by the changed counts for the illustrated trajectory. Consequently, the area in the peak equals the area of the subsequent trough.

the derivative of the EPSP [4,8]. Specifically, for smooth membrane potential trajectories approaching threshold (the case of no additional synaptic noise):

$$f(t) = f_0 + (f_0/\dot{v})\, de/dt \tag{1}$$

where f_0 is the baseline firing rate of the motoneuron and $\dot{v}$ is the rate of closure between motoneuron membrane potential and threshold. This relation can be derived analytically by tranforming the process to a coordinate system aligned with the EPSP (Fig. 1, middle) and calculating the relative timing of spikes advanced by intercepts of the threshold trajectories with the EPSP [4]. The above relation (1) is also valid for the correlogram trough during the falling phase of the EPSP, as long as $de/dt > -\dot{v}$; if the EPSP falls more rapidly than $-\dot{v}$, the trough is limited at zero firing rate (as illustrated for the correlogram at bottom). The fact that the shape of the correlogram peak above baseline matches the EPSP derivative has been empirically confirmed for large EPSPs in cat motoneurons [4]. This relation implies that the <u>height</u> of the correlogram peak above baseline is proportional to the EPSP <u>rate of rise</u>. The integral of this relationship predicts that the <u>area</u> between the correlogram peak and baseline is proportional to the EPSP <u>amplitude</u>. This linear relation further implies that the effects of simultaneously arriving EPSPs will add linearly.

The presence of additional background synaptic "noise", which is normally produced by randomly occurring synaptic inputs, tends to make the correlogram peak broader than the duration of the EPSP risetime. This broadening is produced by membrane potential fluctuations which cause additional threshold crossings during the decay of the EPSP by trajectories that would have missed the EPSP (e.g., the dashed trajectory in Fig. 1, middle). On the basis of indirect empirical comparisons it has been proposed [6,7] that the broader correlogram peaks can be described by the sum of two linear functions of $e(t)$:

$$f(t) = f_0 + a\, e(t) + b\, de/dt \tag{2}$$

This relation provides a reasonable match when the coefficients (a and b) can be optimized for each case [5,7], but direct empirical comparisons [2,4] indicate that the difference between the correlogram peak and the derivative is typically briefer than the EPSP.

The effect of synaptic noise on the transform between EPSP and correlogram peak has not yet been analytically derived (except for the case of Gaussian noise[1]). However the threshold-crossing process has been simulated by a computer model which adds synaptic noise to the trajectories intercepting the EPSP [1]. The correlograms generated by the simulation match the correlograms measured empirically for small EPSP's in motoneurons [2], confirming the validity of the model.

Although synaptic noise distributes the triggered firings over a wider peak, the area of the correlogram peak, i.e., the number of motoneuron firings produced by an EPSP, is essentially preserved and remains proportional to EPSP amplitude for moderate noise levels. For unitary EPSP's (produced by a single afferent fiber) in cat motoneurons, the number of firings triggered per EPSP (N_p) was linearly related to the amplitude (h) of the EPSP [2]:

$$N_p = (0.1/mv) \cdot h \ (mv) + .003 \tag{3}$$

The fact that the number of triggered spikes increases in proportion to EPSP amplitude has also been confirmed for neocortical neurons [10]; for cells recorded in sensorimotor cortex slices (probably pyramidal cells) the coefficient of h was very similar: 0.07/mv. This means that a typical unitary EPSP with amplitude of 100 µv, raises the probability that the postsynaptic cell fires by less than .01. Moreover, this increase occurs during a specific time interval corresponding to the rise time of the EPSP -- on the order of 1 - 2 msec. The net increase in firing rate of the postsynaptic cell is calculated by the proportional decrease in interspike intervals produced by the triggered spikes [4]. (While the above values are typical, unitary EPSP's range in size from several hundred µv down to undetectable levels of several µv., and have risetimes of .2 - 4 msec.)

Inhibitory connections between cells, mediated by inhibitory postsynaptic potentials (IPSPs), produce a trough in the cross-correlogram. This reduction of firing probability below baseline is followed by a subsequent broad, shallow peak, representing the spikes that have been delayed during the IPSP. Although the effects of inhibitory connections remain to be analyzed more quantitatively, preliminary results indicate that small IPSP's in synaptic noise produce decreases in firing probability that are similar to the increases produced by EPSP's [4,5].

DISYNAPTIC LINKS

The effects of polysynaptic links between neurons can be understood as combinations of the underlying monosynaptic connections. A monosynaptic connection from cell A to cell B would produce a first-order cross-correlation peak $P_1(B|A,t)$, representing the conditional probability that neuron B fires above chance at time t, given a spike in cell A at time t = 0. As noted above, the shape of this first-order correlogram peak is largely proportional to the EPSP derivative (for cells whose interspike interval exceeds the duration of the EPSP). The latency of the peak is the conduction time from A to B (Fig. 2 top left).

In contrast, several types of disynaptic linkages between A and B, mediated by a third neuron C, will produce a second-order correlation peak between A and B. A disynaptic link may be produced by two serial monosynaptic connections, from A to C and from C to B (Fig. 2, bottom left), or by a common synaptic input from C ending on both A and B (Fig. 2, bottom right). In both cases, the second-order correlation between A and B produced by the disynaptic link would be the convolution of the two first-order correlations between the monosynaptically connected cells:

$$P_2(B|A) = P_1(B|C) \otimes P_1(C|A) \tag{4}$$

As indicated by the diagram, the cross-correlogram peak $P_2(B|A,t)$ would be smaller and more dispersed than the peaks of the underlying first-order correlation peaks. For serial connections the peak would appear to the right of the origin, at a latency that is the sum of the two monosynaptic latencies. The peak produced by a common input typically straddles the origin, since its timing reflects the difference between the underlying latencies.

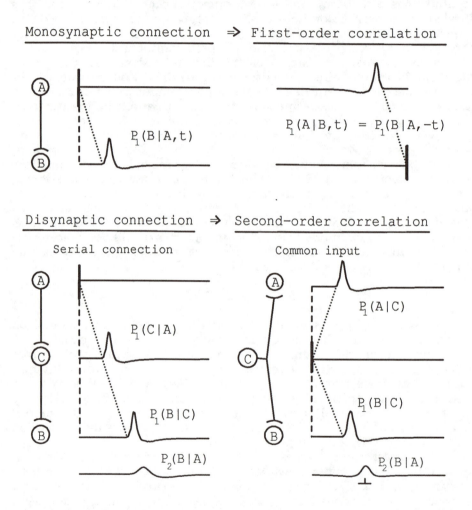

Fig. 2. Correlational effects of monosynaptic and disynaptic links between two neurons. Top: monosynaptic excitatory link from A to B produces an increase in firing probability of B after A (left). As with all correlograms this is the time-inverted probability of increased firing in A relative to B (right). Bottom: Two common disynaptic links between A and B are a serial connection via C (left) and a common input from C. In both cases the effect of the disynaptic link is the convolution of the underlying monosynaptic links.

This relation means that the probability that a spike in cell A will produce a correlated spike in cell B would be the product of the two probabilities for the intervening monosynaptic connections. Given a typical N_p of .01/EPSP, this would reduce the effectiveness of a given disynaptic linkage by two orders of magnitude relative to a monosynaptic connection. However, the net strength of *all* the disynaptic linkages between two given cells is proportional to the number of mediating interneurons {C}, since the effects of parallel pathways add. Thus, the net potency of all the disynaptic linkages between two cells could approach that of a monosynaptic linkage if the number of mediating interneurons were sufficiently large. It should also be noted that some interneurons may fire more than once per EPSP and have a higher probability of being triggered to fire than motoneurons [11].

For completeness, two other possible disynaptic links between A and B involving a third cell C may be considered. One is a serial connection from B to C to A, which is the reverse of the serial connection from A to B. This would produce a $P_2(B|A)$ with peak to the left of the origin. The fourth circuit involves convergent connections from both A and B to C; this is the only combination that would not produce any causal link between A and B.

The effects of still higher-order polysynaptic linkages can be computed similarly, by convolving the effects produced by the sequential connections. For example, trisynaptic linkages between four neurons are equivalent to combinations of disynaptic and monosynaptic connections.

The cross-correlograms between two cells have a certain symmetry, depending on which is the reference cell. The cross-correlation histogram of cell B referenced to A is identical to the time-inverted correlogram of A referenced to B. This is illustrated for the monosynaptic connection in Fig.2, top right, but is true for all correlograms. This symmetry represents the fact that the above-chance probability of B firing after A is the same as the probability of A firing before B:

$$P(B|A, t) = P(A|B, -t) \tag{5}$$

As a consequence, polysynaptic correlational links can be computed as the same convolution integral (Eq. 4), independent of the direction of impulse propagation.

PARALLEL PATHS AND FEEDBACK LOOPS

In addition to the simple combinations of pair-wise connections between neurons illustrated above, additional connections between the same cells may form circuits with various kinds of loops. Recurrent connections can produce feedback loops, whose correlational effects are also calculated by convolving effects of the underlying synaptic links. Parallel feed-forward paths can form multiple pathways between the same cells. These produce correlational effects that are the sum of the effects of the individual underlying connections.

The simplest feedback loop is formed by reciprocal connections between a pair of cells. The effects of excitatory feedback can be computed by successive convolutions of the underlying monosynaptic connections (Fig. 3 top). Note that such a positive feedback loop would be capable of sustaining activity only if the connections were sufficiently potent to ensure postsynaptic firing. Since the probabilities of triggered firings at a single synapse are considerably less than one, reverberating activity can be sustained only if the number of interacting cells is correspondingly increased. Thus, if the probability for a single link is on the order of .01, reverberating activity can be sustained if A and B are similarly interconnected with at least a hundred cells in parallel.

Connections between three neurons may produce various kinds of loops. *Feedforward parallel pathways* are formed when cell A is monosynaptically connected to B and in addition has a serial disynaptic connection through C, as illustrated in Fig. 3 (bottom left); the correlational effects of the two linkages from A to B would sum linearly, as shown for excitatory connections. Again, the effect of a larger set of cells {C} would be additive. *Feedback loops* could be formed with three cells by recurrent connections between any pair; the correlational consequences of the loop again are the convolution of the underlying links. Three cells can form another type loop if both A and B are monosynaptically connected, and simultaneously influenced by a common interneuron C (Fig. 3 bottom right). In this case the expected correlogram between A and B would be the sum of the individual components -- a common input peak around the origin plus a delayed peak produced by the serial connection.

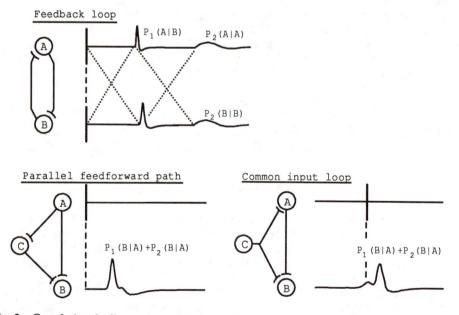

Fig. 3. Correlational effects of parallel connections between two neurons. Top: feedback loop between two neurons A and B produces higher-order effects equivalent to convolution of monosynaptic effects. Bottom: Loops formed by parallel feedforward paths (left) and by a common input concurrent with a monosynaptic link (right) produce additive effects.

CONCLUSIONS

Thus, a simple computational algebra can be used to derive the correlational effects of a given network structure. Effects of sequential connections can be computed by convolution and effects of parallel paths by summation. The inverse problem, of deducing the circuitry from the correlational data is more difficult, since similar correlogram features may be produced by different circuits [9].

The fact that monosynaptic links produce small correlational effects on the order of .01 represents a significant constraint in the mechanisms of information processing in real neural nets. For example, secure propagation of activity through serial polysynaptic linkages requires that the small probability of triggered firing via a given link is compensated by a proportional increase in the number of parallel links. Thus, reliable serial conduction would require hundreds of neurons at each level, with appropriate divergent and convergent connections. It should also be noted that the effect of interneurons can be modulated by changing their activity. The intervening cells need to be active to mediate the correlational effects. As indicated by eq. 1, the size of the correlogram peak is proportional to the firing rate (f_0) of the postsynaptic cell. This allows dynamic modulation of polysynaptic linkages. The greater the number of links, the more susceptible they are to modulation.

Acknowledgements: The author thanks Mr. Garrett Kenyon for stimulating discussions and the cited colleagues for collaborative efforts. This work was supported in part by NIH grants NS 12542 and RR00166.

REFERENCES

1. Bishop, B., Reyes, A.D., and Fetz E.E., Soc. for Neurosci Abst. 11:157 (1985).
2. Cope, T.C., Fetz, E.E., and Matsumura, M., J. Physiol. 390:161-18 (1987).
3. Fetz, E.E. and Cheney, P.D., J. Neurophysiol. 44:751-772 (1980).
4. Fetz, E.E. and Gustafsson, B., J. Physiol. 341:387-410 (1983).
5. Gustafsson, B., and McCrea, D., J. Physiol. 347:431-451 (1984).
6. Kirkwood, P.A., J. Neurosci. Meth. 1:107-132 (1979).
7. Kirkwood, P.A., and Sears, T. J. Physiol. 275:103-134 (1978).
8. Knox, C.K., Biophys. J. 14: 567-582 (1974).
9. Moore, G.P., Segundo, J.P., Perkel, D.H. and Levitan, H., Biophys. J. 10:876-900 (1970).
10. Reyes, A.D., Fetz E.E. and Schwindt, P.C., Soc. for Neurosci Abst. 13:157 (1987).
11. Surmeier, D.J. and Weinberg, R.J., Brain Res. 331:180-184 (1985).

THE HOPFIELD MODEL WITH MULTI-LEVEL NEURONS

Michael Fleisher
Department of Electrical Engineering
Technion - Israel Institute of Technology
Haifa 32000, Israel

ABSTRACT

The Hopfield neural network model for associative memory is generalized. The generalization replaces two state neurons by neurons taking a richer set of values. Two classes of neuron input output relations are developed guaranteeing convergence to stable states. The first is a class of "continuous" relations and the second is a class of allowed quantization rules for the neurons. The information capacity for networks from the second class is found to be of order N^3 bits for a network with N neurons.

A generalization of the sum of outer products learning rule is developed and investigated as well.

I. INTRODUCTION

The ability to perform collective computation in a distributed system of flexible structure without global synchronization is an important engineering objective. Hopfield's neural network [1] is such a model of associative content addressable memory.

An important property of the Hopfield neural network is its guaranteed convergence to stable states (interpreted as the stored memories). In this work we introduce a generalization of the Hopfield model by allowing the outputs of the neurons to take a richer set of values than Hopfield's original binary neurons. Sufficient conditions for preserving the convergence property are developed for the neuron input output relations. Two classes of relations are obtained. The first introduces neurons which simulate multi thres-hold functions, networks with such neurons will be called quantized neural networks (Q.N.N.). The second class introduces continuous neuron input output relations and networks with such neurons will be called continuous neural networks (C.N.N.).

In Section II, we introduce Hopfield's neural network and show its convergence property. C.N.N. are introduced in Section III and a sufficient condition for the neuron input output continuous relations is developed for preserving convergence. In Section IV, Q.N.N. are introduced and their input output rela-tions are analyzed in the same manner as in III. In Section IV we look further at Q.N.N. by using the definition of information capacity for neural networks of [2] to obtain a tight asymptotic estimate of the capacity for a Q.N.N. with N neurons. Section VI is a generalized sum of outer products learning for the Q.N.N. and section VII is the discussion.

II. THE HOPFIELD NEURAL NETWORK

A neural network consists of N pairwise connected neurons. The i 'th neuron can be in one of two states: $X_i = -1$ or $X_i = +1$. The connections are fixed real numbers denoted by W_{ij} (the connection from neuron i to neuron j). Define the state vector $\underline{X}$ to be a binary vector whose i 'th component corresponds to the state of the i 'th neuron. Randomly and asynchronously, each neuron examines its input and decides its next output in the following manner. Let t_i be the threshold voltage of the i 'th neuron. If the weighted sum of the present other $N-1$ neuron outputs (which compose the i 'th neuron input) is greater or equal to t_i, the next X_i (X_i^+) is $+1$, if not, X_i^+ is -1. This action is given in (1).

$$X_i^+ = sgn \ [\ \sum_{j=1}^{N} W_{ij}X_j - t_i \ ] \tag{1}$$

We give the following theorem

Theorem 1 (of [1])

The network described with symmetric ($W_{ij} = W_{ji}$) zero diagonal ($W_{ii} = 0$) connection matrix $\mathbf{W}$ has the convergence property.

Proof

Define the quantity

$$E(\underline{X}) = -\frac{1}{2} \sum_{i}^{N} \sum_{j=1}^{N} W_{ij}X_iX_j + \sum_{i=1}^{N} t_iX_i \tag{2}$$

We show that $E(\underline{X})$ can only decrease as a result of the action of the network. Suppose that X_k changed to $X_k^+ = X_k + \Delta X_k$, the resulting change in E is given by

$$\Delta E = -\Delta X_k \ (\ \sum_{j=1}^{N} W_{kj}X_j - t_k) \tag{3}$$

(Eq. (3) is correct because of the restrictions on $\mathbf{W}$). The term in brackets is exactly the argument of the sgn function in (1) and therefore the signs of ΔX_k and the term in brackets is the same (or $\Delta X_k = 0$) and we get $\Delta E \leq 0$. Combining this with the fact that $E(\underline{X})$ is bounded shows that eventually the network will remain in a local minimum of $E(\underline{X})$. This completes the proof.

The technique used in the proof of Theorem 1 is an important tool in analyzing neural networks. A network with a particular underlying $E(\underline{X})$ function can be used to solve optimization problems with $E(\underline{X})$ as the object of optimization. Thus we see another use of neural networks.

III. THE C.N.N.

We ask ourselves the following question: How can we change the sgn function in (1) without affecting the convergence property? The new action rule for the i 'th neuron is

$$X_i^+ = f_i [\sum_{j=1}^{N} W_{ij} X_j] \tag{4}$$

Our attention is focused on possible choices for $f_i(\cdot)$. The following theorem gives a part of the answer.

Theorem 2

The network described by (4) (with symmetric zero diagonal $\mathbf{W}$) has the convergence property if $f_i(\cdot)$ are strictly increasing and bounded.

Proof

Define

$$E(\underline{X}) = -\frac{1}{2} \sum_{i}^{N} \sum_{j}^{N} W_{ij} X_i X_j + \sum_{i=1}^{N} \int_{0}^{X_i} f_i^{-1}(u) du \tag{5}$$

We show as before that $E(\underline{X})$ can only decrease and since E is bounded (because of the boundedness of f_i's) the theorem is proved.

Using $g_i(X_i) = \int_{0}^{X_i} f_i^{-1}(u) du$ we have

$$\Delta E = -\Delta X_k [\sum_{i=1}^{N} W_{kj} X_j - \frac{g_k(X_k + \Delta X_k) - g(X_k)}{\Delta X_k}] \tag{6}$$

Using the intermediate value theorem we get

$$\Delta E = -\Delta X_k [\sum_{j=1}^{N} W_{kj} X_j - g_k'(C)] = -\Delta X_k [f_k^{-1}(X_k + \Delta X_k) - f_k^{-1}(C)] \tag{7}$$

where C is a point between X_k and $X_k + \Delta X_k$. Now, if $\Delta X_k > 0$ we have $C \leq X_k + \Delta X_k = > f_k^{-1}(C) \leq f_k^{-1}(X_k + \Delta X_k)$ and the term in brackets is greater or equal to zero $=> \Delta E \leq 0$. A similar argument holds for $\Delta X_k < 0$ (of course $\Delta X_k = 0 => \Delta E = 0$). This completes the proof.

Some remarks:

(a) Strictly increasing bounded neuron relations are not the whole class of relations conserving the convergence property. This is seen immediately from the fact that Hopfield's original model (1) is not in this class.

(b) The $E(\underline{X})$ in the C.N.N. coincides with Hopfield's continuous neural network [3]. The difference between the two networks lies in the updating scheme. In our C.N.N. the neurons update their outputs at the moments they examine their inputs while in [3] the updating is in the form of a set of differential equations featuring the time evolution of the network outputs.

(c) The boundedness requirement of the neuron relations results from the boundedness of $E(\underline{X})$. It is possible to impose further restrictions on W resulting in unbounded neuron relations but keeping $E(\underline{X})$ bounded (from below). This was done in [4] where the neurons exhibit linear relations.

IV. THE Q.N.N.

We develop the class of quantization rules for the neurons, keeping the convergence property. Denote the set of possible neuron outputs by $Y_o < Y_1 < ... < Y_n$ and the set of threshold values by $t_1 < t_2 < \cdots < t_n$ the action of the neurons is given by

$$X_i^+ = Y_l \quad \text{if} \quad t_l < \sum_{j=1}^{N} W_{ij} X_j \leq t_{l+1} \quad l = 0,...,n \tag{8}$$

and $t_o = -\infty, t_{n+1} = +\infty$.

The following theorem gives a class of quantization rules with the convergence property.

Theorem 3

Any quantization rule for the neurons which is an increasing step function that is

$$Y_0 < Y_1 < \cdots Y_n \; ; t_1 < \cdots < t_n \qquad (9)$$

Yields a network with the convergence property (with a $\mathbf{W}$ symmetric and zero diagonal).

Proof

We proceed to prove.

Define

$$E(\underline{X}) = -\frac{1}{2} \sum_{i}^{N} \sum_{j=1}^{N} W_{ij} X_i X_j + \sum_{i=1}^{N} tG(X_i) + \sum_{i=1}^{N} dX_i \qquad (10)$$

where $G(X)$ is a piecewise linear convex U function defined by the relation

$$t \frac{G(Y_l) - G(Y_{l-1})}{Y_l - Y_{l-1}} + d = t_l \quad l = 1, \ldots, n \qquad (11)$$

As before we show $\Delta E \leq 0$. Suppose a change occurred in X_k such that $X_k = Y_{i-1}, X_k^+ = Y_i$. We then have

$$\Delta E = -\Delta X_k \left[\sum_{j=1}^{N} W_{kj} X_j - t \frac{G(X_k^+) - G(X_k)}{\Delta X_k} - d \right] = -\Delta X_k \left[\sum_{j=1}^{N} W_{kj} X_j - t_k \right] \leq 0$$

$$(12)$$

A similar argument follows when $X_k = Y_i, X_k^+ = Y_{i-1} < X_k$. Any bigger change in X_k (from Y_i to Y_j with $|i-j| > 1$) yields the same result since it can be viewed as a sequence of $|i-j|$ changes from Y_i to Y_j each resulting in $\Delta E \leq 0$. The proof is completed by noting that $\Delta X_k = 0 => \Delta E = 0$ and $E(\underline{X})$ is bounded.

Corollary

Hopfield's original model is a special case of (9).

V. INFORMATION CAPACITY OF THE Q.N.N.

We use the definition of [2] for the information capacity of the Q.N.N.

Definition 1

The information capacity of the Q.N.N. (bits) is the $\log$ (Base 2) of the number of distinguishable networks of N neurons. Two networks are distinguishable if observing the state transitions of the neurons yields different observations. For Hopfield's original model it was shown in [2] that the capacity C of a network of N neurons is bounded by $C \leq \log (2^{(N-1)^2})^N = 0(N^3)b$. It was also shown that $C \geq \Omega(N^3)b$ and thus is exactly of the order $N^3 b$. It is obvious that in our case (which contains the original model) we must have $C \geq \Omega(N^3)b$ as well (since the lower bound cannot decrease in this richer case). It is shown in the Appendix that the number of multi threshold functions of $N-1$ variables with $n+1$ output levels is at most $(n+1)^{N^2+N+1}$ since we have N neurons there will be $((n+1)^{N^2+N+1})^N$ distinguishable networks and thus

$$C \leq \log ((n+1)^{N^2+N+1})^N = 0(N^3)b \qquad (14)$$

or as before, C is exactly of $0(N^3)b$. In fact, the rise in C is probably a factor of $0(\log_2 n)$ as can be seen from the upper bound.

VI. "OUTER PRODUCT" LEARNING RULE

For Hopfield's original network with two state neurons (taking the values ± 1) a natural and extensively investigated [ ],[ ],[ ] learning rule is the so called sum of outer products construction.

$$W_{ij} = \frac{1}{N} \sum_{l=1}^{K} X_i^l X_j^l \qquad (15)$$

where $\underline{X}^1, \ldots, \underline{X}^K$ are the desired stable states of the network. A well-known result for (15) is that the asymptotic capacity K of the network is

$$K = \frac{N-1}{4\log N} + 1 \tag{16}$$

In this section we introduce a natural generalization of (15) and prove a similar result for the asymptotic capacity. We first limit the possible quantization rules to:

$$X_i = F(u_i) = \begin{cases} Y_o & t_1 > u_i \geq t_o \\ \cdot \\ \cdot \\ \cdot \\ Y_n & t_{n+1} > u_i \geq t_n \end{cases} \tag{17}$$

with $Y_o < \cdots < Y_n$

$$t_j = \frac{1}{2} \left[Y_j + Y_{j-1} \right] \quad j=1, \cdots n$$

$$t_o = -\infty \; ; \; t_{n+1} = \infty$$

with

    (a) $n+1$ is even
    (b) $\forall \; i \quad Y_i \neq 0$
    (c) $Y_i = -Y_{n-i} \quad i=0, \ldots, n$

Next we state that the desired stable vectors $\underline{X}^1, \cdots \underline{X}^K$ are such that each component is picked independently at random from $\{ Y_o, \cdots Y_M \}$ with equal probability. Thus, the $K \cdot N$ components of the $\underline{X}$'s are zero mean i.i.D random variables. Our modified learning rule is

$$W_{ij} = \frac{1}{N} \sum_{l=1}^{K} X_i^l \cdot \left[\frac{1}{X_j^l} \right] \tag{18}$$

Note that for $X_i \in \{+1, -1\}$ (18) is identical to (16).

Define

$$\widetilde{\Delta Y} \overset{\Delta}{=} \min_{i \neq j} |Y_i - Y_j|$$

$$A = \max_{i,j} \frac{|Y_i|^2}{|Y_j|}$$

We state that

PROPOSITION:

The asymptotic capacity of the above network is given by

$$K = \frac{N}{\dfrac{16A^2}{(\widetilde{\Delta Y})^2} \log N} \tag{19}$$

PROOF:

Define

$$P(K, N) = P_r \left\{ \begin{array}{l} K \text{ vectors chosen randomly as described} \\ \text{are stable states with the } W \text{ of ( )} \end{array} \right\}$$

$$P(K, N) = 1 - P_r(\bigcup A_{ij}) \geq 1 - \sum_{i,j} P_r(A_{ij}) \quad \begin{array}{l} i = 1, \ldots, N \\ j = 1, \ldots, K \end{array} \tag{20}$$

where A_{ij} is the event that the i th component of j th vector is in error. We concentrate on the event A_{11} W.L.G.

The input u_1 when $\underline{X}'$ is presented is given by

$$u_1 = \sum_{j=1}^{N} W_{1j} X_j^1 = X_1^1 + \frac{K-1}{N} X_1^1 + \frac{1}{N} \sum_{l=2}^{K} \sum_{j=2}^{N} X_1^l \frac{X_j^1}{X_j^l} \tag{21}$$

The first term is mapped by (17) into itself and corresponds to the desired signal.

The last term is a sum of $(K-1)(N-1)$ i.i.D zero mean random variables and corresponds to noise.

The middle term $\dfrac{K-1}{N} X_1^1$ is disposed of by assuming $\dfrac{K-1}{N} \underset{N \to \infty}{\to} 0$. (With a zero diagonal

choice of W (using (18) with $i \neq j$) this term does not appear).

$P_r(\mathbf{A}_{11}) = P_r \{$ noise gets us out of range $\}$

Denoting the noise by I we have

$$P_r(\mathbf{A}_{11}) \le P_r(|I| > \frac{\widetilde{\Delta Y}}{2}) \le 2\exp\left\{ -\frac{\frac{1}{2}(\widetilde{\Delta Y})^2 N^2}{(K-1)(N-1)4A^2} \right\} \tag{22}$$

where the first inequality is from the definition of $\widetilde{\Delta Y}$ and the second uses the lemma of [6] p. 58. We thus

get

$$P(K,N) \ge 1 - K \cdot N \cdot 2\exp\left\{ -\frac{(\widetilde{\Delta Y})^2 N^2}{8(K-1)(N-1)A^2} \right\} \tag{23}$$

substituting (19) and taking $N \to \infty$ we get $P(K,N) \to 1$ and this completes the proof.

VII. DISCUSSION

Two classes of generalization of the Hopfield neural network model were presented. We give some remarks:

(a) Any combination of neurons from the two classes will have the convergence property as well.

(b) Our definition of the information capacity for the C.N.N. is useless since a full observation of the possible state transitions of the network is impossible.

APPENDIX

We prove the following theorem.

Theorem

An upper bound on the number of multi threshold functions with N inputs and M points in the domain (out of $(n+1)^N$ possible points) C_N^M is the solution of the recurrence relation

$$C_N^M = C_N^{M-1} + n \cdot C_{N-1}^{M-1} \tag{A.1}$$

Proof

Let us look on the N dimensional weight space $\underline{W}$. Each input point $\underline{X}$ divides the weight space into $n+1$ regions by n parallel hyperplanes $\sum\limits_{i=1}^{N} W_i X_i = t_k$ $k=1,\ldots,n$. We keep adding points in such a way that the new n hyperplanes corresponding to each added point partition the $\underline{W}$ space into as many regions as possible. Assume $M-1$ points have made C_N^{M-1} regions and we add the M 'th point. Each hyperplane (out of n) is divided into at most C_{N-1}^{M-1} regions (being itself an $N-1$ dimensional space divided by $(M-1)n$ hyperlines). We thus have after passing the n hyperplanes:

$$C_N^M = C_N^{M-1} + n \cdot C_{N-1}^{M-1}$$

is $C_N^M = (n+1) \sum\limits_{i=o}^{N-1} \begin{bmatrix} M-1 \\ i \end{bmatrix} n^i$ and the theorem is proved.

The solution of the recurrence in the case $M = (n+1)^N$ (all possible points) we have a bound on the number of multithreshold functions of N variables equal to

$$C_n^{(n+1)^N} = (n+1) \sum\limits_{i=1}^{N-1} \begin{bmatrix} (n+1)^N - 1 \\ i \end{bmatrix} n^i \leq (n+1)^{N^2+N+1}$$

and the result used is established.

LIST OF REFERENCES

[1] Hopfield J. J., "Neural networks and physical systems with emergent collective computational abilities", Proc. Nat. Acad. Sci. USA, Vol. 79 (1982), pp. 2554-2558.

[2] Abu-Mostafa Y.S. and Jacques J. St., "Information capacity of the Hopfield model", IEEE Trans. on Info. Theory, Vol. IT-31 (1985, pp. 461-464.

[3] Hopfield J. J., "Neurons with graded response have collective computational properties like those of two state neurons", Proc. Nat. Acad. Sci. USA, Vol. 81 (1984).

[4] Fleisher M., "Fast processing of autoregressive signals by a neural network", to be presented at IEEE Conference, Israel 1987.

[5] Levin, E., Private communication.

[6] Petrov, "Sums of independent random variables".

CYCLES: A Simulation Tool for Studying Cyclic Neural Networks

Michael T. Gately
Texas Instruments Incorporated, Dallas, TX 75265

ABSTRACT

A computer program has been designed and implemented to allow a researcher to analyze the oscillatory behavior of simulated neural networks with cyclic connectivity. The computer program, implemented on the Texas Instruments Explorer/Odyssey system, and the results of numerous experiments are discussed.

The program, CYCLES, allows a user to construct, operate, and inspect neural networks containing cyclic connection paths with the aid of a powerful graphics-based interface. Numerous cycles have been studied, including cycles with one or more activation points, non-interruptible cycles, cycles with variable path lengths, and interacting cycles. The final class, interacting cycles, is important due to its ability to implement time-dependent goal processing in neural networks.

INTRODUCTION

Neural networks are capable of many types of computation. However, the majority of researchers are currently limiting their studies to various forms of mapping systems; such as content addressable memories, expert system engines, and artificial retinas. Typically, these systems have one layer of fully connected neurons or several layers of neurons with limited (forward direction only) connectivity. I have defined a new neural network topology; a two-dimensional lattice of neurons connected in such a way that circular paths are possible.

The neural networks defined can be viewed as a grid of neurons with one edge containing input neurons and the opposite edge containing output neurons [Figure 1]. Within the grid, any neuron can be connected to any other. Thus from one point of view, this is a multi-layered system with full connectivity. I view the weights of the connections as being the long term memory (LTM) of the system and the propagation of information through the grid as being it's short term memory (STM).

The topology of connectivity between neurons can take on any number of patterns. Using the mammalian brain as a guide, I have limited the amount of connectivity to something much less then total. In addition to making analysis of such systems less complex, limiting the connectivity to some small percentage of the total number of neurons reduces the amount of memory used in computer simulations. In general, the connectivity can be purely random, or can form any of a number of patterns that are repeated across the grid of neurons.

The program CYCLES allows the user to quickly describe the shape of the neural network grid, the source of input data, the destination of the output data, the pattern of connectivity. Once constructed, the network can be "run." during which time the STM may be viewed graphically.

291

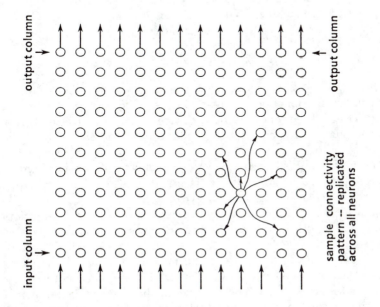

Figure 1. COMPONENTS OF A CYCLES NEURAL NETWORK

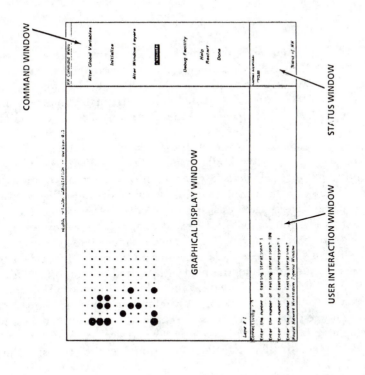

Figure 2. NEURAL NETWORK WORKSTATION INTERFACE

IMPLEMENTATION

CYCLES was implemented on a TI Explorer/Odyssey computer system with 8MB of RAM and 128MB of Virtual Memory. The program was written in Common LISP. The program was started in July of 1986, put aside for a while, and finished in March of 1987. Since that time, numerous small enhancements have been made – and the system has been used to test various theories of cyclic neural networks.

The code was integrated into the Neural Network Workstation (NNW), an interface to various neural network algorithms. The NNW utilizes the window interface of the Explorer LISP machine to present a consistent command input and graphical output to a variety of neural network algorithms [Figure 2].

The backpropagation-like neurons are collected together into a large three-dimensional array. The implementation actually allows the use of multiple two-dimensional grids; to date, however, I have studied only single-grid systems.

Each neuron in a CYCLES simulation consists of a list of information; the value of the neuron, the time that the neuron last fired, a temporary value used during the computation of the new value, and a list of the neurons connectivity. The connectivity list stores the location of a related neuron and the strength of the connection between the two neurons. Because the system is implemented in arrays and lists, large systems tend to be very slow. However, most of my analysis has taken place on very small systems (< 80 neurons) and for this size the speed is acceptable.

To help gauge the speed of CYCLES, a single grid system containing 100 neurons takes 0.8 seconds and 1235 cons cells (memory cells) to complete one update within the LISP machine. If the graphics interface is disabled, a test requiring 100 updates takes a total of 10.56 seconds.

TYPES OF CYCLES

As mentioned above, several types of cycles have been observed. Each of these can be used for different applications. Figure 3 shows some of these cycles.

1. SIMPLE cycles are those that have one or more points of activation traveling across a set number of neurons in a particular order. The path length can be any size.

2. NON-INTERRUPTABLE cycles are those that have sufficiently strong connectivity strengths that random flows of activation which interact with the cycle will not upset or vary the original cycle.

3. VARIABLE PATH LENGTH cycles can, based upon external information, change their path length. There must be one or more neurons that are always a part of the path.

4. INTERACTING cycles typically have one neuron in common. Each cycle must have at least one other neuron involved at the junction point in order to keep the cycles separate. This type of cycle has been shown to implement a complex form of a clock where the product of the two (or more) path lengths are the fundamental frequency.

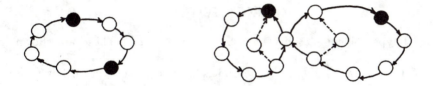

Figure 3. Types of Cycles [Simple and Interacting]

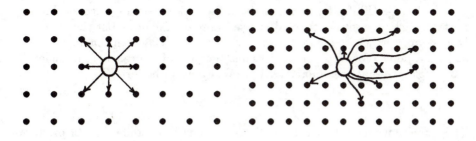

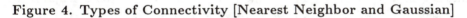

Figure 4. Types of Connectivity [Nearest Neighbor and Gaussian]

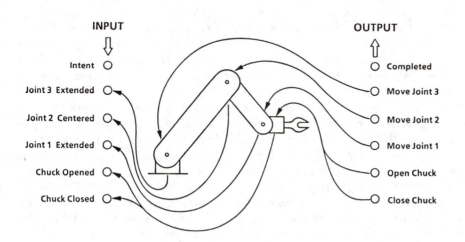

Figure 5. Robot Arm used in Example

CONNECTIVITY

Several types of connectivity have been investigated. These are shown in Figure 4.

1. In TOTAL connectivity, every neuron is connected to every other neuron. This particular pattern produces very complex interactions with no apparent stability.
2. With RANDOM connectivity, each neuron is connected to a random number of other neurons. These other neurons can be anywhere in the grid.
3. A very useful type of connectivity is to have a PATTERN. The patterns can be of any shape, typically having one neuron feed its nearest neighbors.
4. Finally, the GAUSSIAN pattern has been used with the most success. In this pattern, each neuron is connected to a set number of nodes – but the selection is random. Further, the distribution of nodes is in a Gaussian shape, centered around a point "forward" of itself. Thus the flow of information, in general, moves forward, but the connectivity allows cycles to be formed.

ALGORITHM

The algorithm currently being used in the system is a standard inner product equation with a sigmoidal threshold function. Each time a neuron's weight is to be calculated, the value of each contributing neuron on the connectivity list is multiplied by the strength of the connection and summed. This sum is passed through a sigmoidal thresholding function. The value of the neuron is changed to be the result of this threshold function. As you can see, the system updates neurons in an ordered fashion, thus certain interactions will not be observed. Since timing information is saved in the neurons, asynchrony could be simulated.

Initially, the weights of the connections are set randomly. A number of interesting cycles have been observed as a result of this randomness. However, several experiments have required specific weights. To accommodate this, an interface to the weight matrix is used. The user can create any set of connection strengths desired.

I have experimented with several learning algorithms–that is, algorithms that change the connection weights. The first mechanism was a simple Hebbian rule that states that if two neurons both fire, and there is a connection between them, then strengthen the strength of that connection. A second algorithm I experimented with used a pain/pleasure indicator to strengthen or weaken weights.

An algorithm that is currently under development actually presets the weights from a grammar of activity required of the network. Thus, the user can describe a process that must be controlled by a network using a simple grammar. This description is then "compiled" into a set of weights that contain cycles to indicate time-independent components of the activity.

USAGE

Even without a biological background, it is easy to see that the processing power of the human brain is far more than present associative memories. Our repertoire of capabilities includes, among other things: memory of a time line, creativity, numerous types of biological clocks, and the ability to create and execute complex plans. The CYCLES algorithm has been shown to be capable of executing complex, time-variable plans.

A plan can be defined as a sequence of actions that must be performed in some preset order. Under this definition, the execution of a plan would be very straightforward. However, when individual actions within the plan take an indeterminate length of time, it is necessary to construct an execution engine capable of dealing with unexpected time delays. Such a system must also be able to abort the processing of a plan based on new data.

With careful programming of connection weights, I have been able to use CYCLES to execute time-variable plans. The particular example I have chosen is for a robot arm to change its tool. In this activity, once the controller receives the signal that the motion required, a series of actions take place that result in the tool being changed.

As input to this system I have used a number of sensors that may be found in a robot; extension sensors in 2-D joints and pressure sensors in articulators. The outputs of this network are pulses that I have defined to activate motors on the robot arm. Figure 5 shows how this system could be implemented. Figure 6 indicates the steps required to perform the task. Simple time delays, such as found with binding motors and misplaced objects are accommodated with the built in time-independence.

The small cycles that occur within the neural network can be thought of as short term memory. The cycle acts as a place holder – keeping track of the system's current place in a series of tasks. This type of pausing is necessary in many "real" activities such as simple process control or the analysis of time varying data.

IMPLICATIONS

The success of CYCLES to simple process control activities such as robot arm control implies that there is a whole new area of applications for neural networks beyond present associative memories. The exploitation of the flow of activation as a form of short term memory provides us with a technique for dealing with many of the "other" type of computations which humans perform.

The future of the CYCLES algorithm will take two directions. First, the completion of a grammar and compiler for encoding process control tasks into a network. Second, other learning algorithms will be investigated which are capable of adding and removing connections and altering the strengths of connections based upon an abstract pain/pleasure indicator.

The robot gets a signal
to begin the tool change
process. A cycle is started
that outputs a signal to
the chuck motor.

When sensor indicates
that the chuck is open,
the first cycle is stopped
and a second begins
activating the motor
in the first joint.

When the first joint is
fully extended, the joint
sensor sends a signal
that stops that cycle, and
begins one that outputs
a signal to the second
joint.

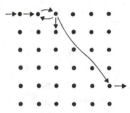
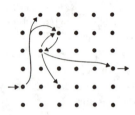
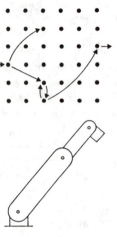

When the joint indicator
indicates that the joint is
centered, it changes the
flow of activation to cause
a cycle that activates the
third joint.

Next, the chuck is closed
around the new tool bit.

The last signal ends the
sequence of cycles and
sends the completed
signal.

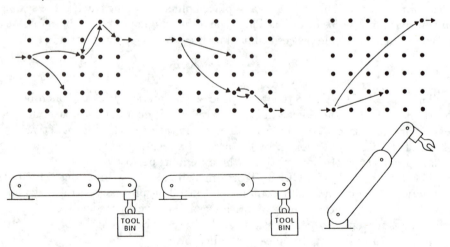

Figure 6. Example use of CYCLES to control a Robot Arm

TEMPORAL PATTERNS OF ACTIVITY IN NEURAL NETWORKS

Paolo Gaudiano
Dept. of Aerospace Engineering Sciences,
University of Colorado, Boulder CO 80309, USA

January 5, 1988

Abstract

Patterns of activity over real neural structures are known to exhibit time–dependent behavior. It would seem that the brain may be capable of utilizing temporal behavior of activity in neural networks as a way of performing functions which cannot otherwise be easily implemented. These might include the origination of sequential behavior and the recognition of time–dependent stimuli. A model is presented here which uses neuronal populations with recurrent feedback connections in an attempt to observe and describe the resulting time-dependent behavior. Shortcomings and problems inherent to this model are discussed. Current models by other researchers are reviewed and their similarities and differences discussed.

METHODS / PRELIMINARY RESULTS

In previous papers,[2,3] computer models were presented that simulate a net consisting of two spatially organized populations of realistic neurons. The populations are richly interconnected and are shown to exhibit internally sustained activity. It was shown that if the neurons have response times significantly shorter than the typical unit time characteristic of the input patterns (usually 1 msec), the populations will exhibit time–dependent behavior. This will typically result in the net falling into a limit cycle. By a limit cycle, it is meant that the population falls into activity patterns during which all of the active cells fire in a cyclic, periodic fashion. Although the period of firing of the individual cells may be different, after a fixed time the overall population activity will repeat in a cyclic, periodic fashion. For populations organized in 7x7 grids, the limit cycle will usually start 20-200 msec after the input is turned off, and its period will be in the order of 20-100 msec.

The point of interest is that if the net is allowed to undergo synaptic modifications by means of a modified hebbian learning rule while being presented with a specific spatial pattern (i.e., cells at specific spatial locations within the net are externally stimulated), subsequent presentations of the same pattern with different temporal characteristics will cause the population to recall patterns which are spatially identical (the same cells will be active) but which have different temporal qualities. In other words, the net can fall into a different limit cycle. These limit cycles seem to behave as attractors in that similar input patterns will result in the same limit cycle, and hence each distinct limit cycle appears to have a basin of attraction. Hence a net which can only learn a small number of spatially distinct patterns can recall the patterns in a number of temporal modes. If it were possible to quantitatively discriminate between such temporal modes, it would seem reasonable to speculate that different limit cycles could correspond to different memory traces. This would significantly increase estimates on the capacity of memory storage in the net.

It has also been shown that a net being presented with a given pattern will fall and stay into a limit cycle until another pattern is presented which will cause the system to fall into a different basin of attraction. If no other patterns are presented, the net will remain in the same limit cycle indefinitely. Furthermore, the net will fall into the same limit cycle independently of the duration of the input stimulus, so long as the input stimulus is presented for a long enough time to raise the population activity level beyond a minimum necessary to achieve self-sustained activity. Hence, if we suppose that the net "recognizes" the input when it falls into the corresponding limit cycle, it follows that the net will recognize a string of input patterns regardless of the duration of each input pattern, so long as each input is presented long enough for the net to fall into the appropriate limit cycle. In particular, our system is capable of falling into a limit cycle within some tens of milliseconds. This can be fast enough to encode, for example, a string of phonemes as would typically be found in continuous speech. It may be possible, for instance, to create a model similar to Rumelhart and McClelland's 1981 model on word recognition by appropriately connecting multiple layers of these networks. If the response time of the cells were increased in higher layers, it may be possible to have the lowest level respond to stimuli quickly enough to distinguish phonemes (or some sub–phonemic basic linguistic unit), then have populations from this first level feed into a slower, word–recognizing population layer, and so on. Such a model may be able to perform word recognition from an input consisting of continuous phoneme strings even when the phonemes may vary in duration of presentation.

SHORTCOMINGS

Unfortunately, it was noticed a short time ago that a consistent mistake had been made in the process of obtaining the above–mentioned results. Namely, in the process of decreasing the response time of the cells I accidentally reached a response time below the time step used in the numerical approximation that updates the state of each cell during a simulation. The equations that describe the state of each cell depend on the state of the cell at the previous time step as well as on the input at the present time. These equations are of first order in time, and an explicit discrete approximation is used in the model. Unfortunately it is a known fact that care must be taken in selecting the size of the time step in order to obtain reliable results. It is infact the case that by reducing the time step to a level below the response time of the cells the dynamics of the system varied significantly. It is questionable whether it would be possible to adjust some of the population parameters within reson to obtain the same results with a smaller step size, but the following points should be taken into account: 1) other researchers have created similar models that show such cyclic behavior (see for example Silverman, Shaw and Pearson[7]). 2) biological data exists which would indicate the existance of cyclic or periodic bahvior in real neural systems (see for instance Baird[1]).

As I just recently completed a series of studies at this university, I will not be able to perform a detailed examination of the system described here, but instead I will more than likely create new models on different research equipment which will be geared more specifically towards the study of temporal behavior in neural networks.

OTHER MODELS

It should be noted that in the past few years some researchers have begun investigating the possibility of neural networks that can exhibit time–dependent behavior, and I would like to report on some of the available results as they relate to the topic of temporal patterns. Baird[1] reports findings from the rabbit's olfctory bulb which indicate the existance of phase–locked oscillatory states corresponding to olfactory stimuli presented to the subjects. He outlines an elegant model which attributes pattern recognition abilities to competing instabilities in the dynamic activity of neural structures. He further speculates that inhomogeneous connectivity in the bulb can be selectively modified to achieve input–sensitive oscillatory states.

Silverman, Shaw and Pearson[7] have developed a model based on a biologically–inspired idealized neural structure, which they call the trion. This unit represents a localized group of neurons with a discrete firing period. It was found that small ensembles of trions with symmetric connections can exhibit quasi–stable periodic firing patterns which do not require pacemakers or external driving. Their results are inspired by existing physiological data and are consistent with other works.

Kleinfeld[6], and Sompolinsky and Kanter[8] independently developed neural network models that can generate and recognize sequential or cyclic patterns. Both models rely on what could be summarized as the recirculation of information through time–delayed channels.

Very similar results are presented by Jordan[4] who extends a typical connectionist or PDP model to include state and plan units with recurrent connections and feedback from output units through hidden units. He employs supervised learning with fuzzy constraints to induce learning of sequences in the system.

From a slightly different approach, Tank and Hopfield[9] make use of patterned sets of delays which effectively compress information in time. They develop a model which recognizes patterns by falling into local minima of a state–space energy function. They suggest that a systematic selection of delay functions can be done which will allow for time distortions that would be likely to occur in the input.

Finally, a somewhat different approach is taken by Homma, Atlas and Marks[5], who generalize a network for spatial pattern recognition to one that performs spatio–temporal patterns by extending classical principles from spatial networks to dynamic networks. In particular, they replace multiplication with convolution, weights with transfer functions, and thresholding with non linear transforms. Hebbian and Delta learning rules are similarly generalized. The resulting models are able to perform temporal pattern recognition.

The above is only a partial list of some of the relevant work in this field, and there are probably various other results I am not aware of.

DISCUSSION

All of the above results indicate the importance of temporal patterns in neural networks. The need is apparent for further formal models which can successfully quantify temporal behavior in neural networks. Several questions must be answered to further clarify the role and meaning of temporal patterns in neural nets. For instance, there is an apparent difference between a model that performs sequential tasks and one that performs recognition of dynamic patterns. It seems that appropriate selection of delay mechanisms will be necessary to account for many types of temporal pattern recognition. The question of scaling must also be explored: mechanism are known to exist in the brain which can cause delays ranging from the millisecond–range (e.g. variations in synaptic cleft size) to the tenth of a second range (e.g. axonal transmission times). On the other hand, the brain is capable of recognizing sequences of stimuli that can be much longer than the typical neural event, such as for instance being able to remember a song in its entirety. These and other questions could lead to interesting new aspects of brain function which are presently unclear.

References

[1] Baird, B., "Nonlinear Dynamics of Pattern Formation and Pattern Recognition in the Rabbit Olfactory Bulb". Physica 22D, 150-175. 1986.

[2] Gaudiano, P., "Computer Models of Neural Networks". Unpublished Master's Thesis. University of Colorado. 1987.

[3] Gaudiano, P., MacGregor, R.J., "Dynamic Activity and Memory Traces in Computer-Simulated Recurrently-Connected Neural Networks". Proceedings of the First International Conference on Neural Networks. 2:177-185. 1987.

[4] Jordan, M.I., "Attractor Dynamics and Parallelism in a Connectionist Sequential Machine". Proceedings of the Eighth Annual Conference of the Cognitive Sciences Society. 1986.

[5] Homma, T., Atlas, L.E., Marks, R.J.II, "An Artificial Neural Network for Spatio-Temporal Bipolar Patterns: Application to Phoneme Classification". To appear in proceedings of Neural Information Processing Systems Conference (AIP). 1987.

[6] Kleinfeld, D., "Sequential State Generation by Model Neural Networks". Proc. Natl. Acad. Sci. USA. 83: 9469-9473. 1986.

[7] Silverman, D.J., Shaw, G.L., Pearson, J.C. "Associative Recall Properties of the Trion Model of Cortical Organization". Biol. Cybern. 53:259-271. 1986.

[8] Sompolinsky, H., Kanter, I. "Temporal Association in Asymmetric Neural Networks". Phys. Rev. Let. 57:2861-2864. 1986.

[9] Tank, D.W., Hopfield, J.J. "Neural Computation by Concentrating Information in Time". Proc. Natl. Acad. Sci. USA. 84:1896-1900. 1987.

ENCODING GEOMETRIC INVARIANCES IN
HIGHER-ORDER NEURAL NETWORKS

C.L. Giles
Air Force Office of Scientific Research, Bolling AFB, DC 20332

R.D. Griffin
Naval Research Laboratory, Washington, DC 20375-5000

T. Maxwell
Sachs-Freeman Associates, Landover, MD 20785

ABSTRACT

We describe a method of constructing higher-order neural
networks that respond invariantly under geometric transformations on
the input space. By requiring each unit to satisfy a set of
constraints on the interconnection weights, a particular structure is
imposed on the network. A network built using such an architecture
maintains its invariant performance independent of the values the
weights assume, of the learning rules used, and of the form of the
nonlinearities in the network. The invariance exhibited by a first-
order network is usually of a trivial sort, e.g., responding only to
the average input in the case of translation invariance, whereas
higher-order networks can perform useful functions and still exhibit
the invariance. We derive the weight constraints for translation,
rotation, scale, and several combinations of these transformations,
and report results of simulation studies.

INTRODUCTION

A persistent difficulty for pattern recognition systems is the
requirement that patterns or objects be recognized independent of
irrelevant parameters or distortions such as orientation (position,
rotation, aspect), scale or size, background or context, doppler
shift, time of occurrence, or signal duration. The remarkable
performance of humans and other animals on this problem in the visual
and auditory realms is often taken for granted, until one tries to
build a machine with similar performance. Though many methods have
been developed for dealing with these problems,[1] we have classified
them into two categories: 1) preprocessing or transformation
(inherent) approaches, and 2) case-specific or "brute force"
(learned) approaches. Common transformation techniques include:
Fourier, Hough, and related transforms; moments; and Fourier
descriptors of the input signal. In these approaches the signal is
usually transformed so that the subsequent processing ignores
arbitrary parameters such as scale, translation, etc. In addition,
these techniques are usually computationally expensive and are
sensitive to noise in the input signal. The "brute force" approach
is exemplified by training a device, such as a perceptron, to
classify a pattern independent of it's position by presenting the

training pattern at all possible positions. MADALINE machines[2] have been shown to perform well using such techniques. Often, this type of invariance is pattern specific, does not easily generalize to other patterns, and depends on the type of learning algorithm employed. Furthermore, a great deal of time and energy is spent on learning the invariance, rather than on learning the signal. We describe a method that has the advantage of inherent invariance but uses a higher-order neural network approach that must learn only the desired signal. Higher-order units have been shown to have unique computational strengths and are quite amenable to the encoding of a priori knowledge.[3-7]

MATHEMATICAL DEVELOPMENT

Our approach is similar to the group invariance approach,[8,10] although we make no appeal to group theory to obtain our results. We begin by selecting a transformation on the input space, then require the output of the unit to be invariant to the transformation. The resulting equations yield constraints on the interconnection weights, and thus imply a particular form or structure for the network architecture.

For the i-th unit y_i of order M defined on a discrete input space, let the output be given by

$$y_i[W_i^M(\mathbf{x}), p(\mathbf{x})] = f(\ w_i^0 + \Sigma\ w_i^1(\mathbf{x}_1)\ p(\mathbf{x}_1)$$

$$+ \Sigma\Sigma\ w_i^2(\mathbf{x}_1, \mathbf{x}_2)\ p(\mathbf{x}_1)\ p(\mathbf{x}_2) + \ldots$$

$$+ \Sigma\ldots\Sigma\ w_i^M(\mathbf{x}_1, \ldots \mathbf{x}_M)\ p(\mathbf{x}_1)..p(\mathbf{x}_M)\ ), \qquad (1)$$

where $p(\mathbf{x})$ is the input pattern or signal function (sometimes called a pixel) evaluated at position vector $\mathbf{x}$, $w_i^m(\mathbf{x}_1, \ldots \mathbf{x}_m)$ is the weight of order m connecting the outputs of units at $\mathbf{x}_1$, $\mathbf{x}_2$, .. $\mathbf{x}_m$ to the i-th unit, i.e., it correlates m values, $f(u)$ is some threshold or sigmoid output function, and the summations extend over the input space. $W_i^M(\mathbf{x})$ represents the entire set of weights associated with the i-th unit. These units are equivalent to the sigma-pi units[a] defined by Rumelhart, Hinton, and Williams.[7] Systems built from these units suffer from a combinatorial explosion of terms, hence are more complicated to build and train. To reduce the severity of this problem, one can limit the range of the interconnection weights or the number of orders, or impose various other constraints. We find that, in addition to the advantages of inherent invariance, imposing an invariance constraint on Eq. (1) reduces the number of allowed

---

[a]The sigma-pi neural networks are multi-layer networks with higher-order terms in any layer. As such, most of the neural networks described here can be considered as a special case of the sigma-pi units. However, the sigma-pi units as originally formulated did not have invariant weight terms, though it is quite simple to incorporate such invariances in these units.

weights, thus simplifying the architecture and shortening the training time.

We now define what we mean by invariance. The output of a unit is invariant with respect to the transformation **T** on the input pattern if[9]

$$T[y_i(W_i^M, p(x))] = y(W_i^M, T[p(x)]) = y(W_i^M, p(x)) \qquad (2)$$

An example of the class of invariant response defined by Eq. (2) would be invariant detection of an object in the receptive field of a panning or zooming camera. An example of a different class would be invariant detection of an object that is moving within the field of a fixed camera. One can think of this latter case as consisting of a fixed field of "noise" plus a moving field that contains only the object of interest. If the detection system does not respond to the fixed field, then this latter case is included in Eq. (2).

To illustrate our method we derive the weight constraints for one-dimensional translation invariance. We will first switch to a continuous formulation, however, for reasons of simplicity and generality, and because it is easier to grasp the physical significance of the results, although any numerical simulation requires a discrete formulation and has significant implications for the implementation of our results. Instead of an index i, we now keep track of our units with the continuous variable **u**. With these changes Eq. (2) now becomes

$$y[u; W^M(x), p(x)] = f(w^0 + \int dx_1 \ w^1(u; x_1) \ p(x_1) + \ldots$$

$$+ \int \ldots \int dx_1 \ldots dx_M \ w^M(u; x_1, \ldots x_M) \ p(x_1) \ldots p(x_M) \ ), \qquad (3)$$

The limits on the integrals are defined by the problem and are crucial in what follows. Let **T** be a translation of the input pattern by $-x_0$, so that

$$T[p(x)] = p(x+x_0) \qquad (4)$$

where x_0 is the translation of the input pattern. Then, from eq (2),

$$Ty[u; W^M(x), p(x)] = y[u; W^M(x), p(x+x_0)] = y[u; W^M(x), p(x)] \qquad (5)$$

Since p(x) is arbitrary we must impose term-by-term equality in the argument of the threshold function; i.e.,

$$\int dx_1 \ w^1(u; x_1) \ p(x_1) = \int dx_1 \ w^1(u; x_1) \ p(x_1+x_0), \qquad (5a)$$

$$\int \int dx_1 \ dx_2 \ w^2(u; x_1, x_2) \ p(x_1) \ p(x_2) =$$

$$\int \int dx_1 \ dx_2 \ w^2(u; x_1, x_2) \ p(x_1+x_0) \ p(x_2+x_0), \qquad (5b)$$

etc.

Making the substitutions $x_1 \to x_1-x_0$, $x_2 \to x_2-x_0$, etc, we find that

$$\int dx_1 \; w^1(u;x_1) \; p(x_1) = \int dx_1 \; w^1(u;x_1-x_0) \; p(x_1), \tag{6a}$$

$$\int\int dx_1 \; dx_2 \; w^2(u;x_1,x_2) \; p(x_1) \; p(x_2) =$$

$$\int\int dx_1 \; dx_2 \; w^2(u;x_1-x_0,x_2-x_0) \; p(x_1) \; p(x_2), \tag{6b}$$

etc.

Note that the limits of the integrals on the right hand side must be adjusted to satisfy the change-of-variables. If the limits on the integrals are infinite or if one imposes some sort of periodic boundary condition, the limits of the integrals on both sides of the equation can be set equal. We will assume in the remainder of this paper that these conditions can be met; normally this means the limits of the integrals extend to infinity. (In an implementation, it is usually impractical or even impossible to satisfy these requirements, but our simulation results indicate that these networks perform satisfactorily even though the regions of integration are not identical. This question must be addressed for each class of transformation; it is an integral part of the implementation design.) Since the functions $p(x)$ are arbitrary and the regions of integration are the same, the weight functions must be equal. This imposes a constraint on the functional form of the weight functions or, in the discrete implementation, limits the allowed connections and thus the number of weights. In the case of translation invariance, the constraint on the functional form of the weight functions requires that

$$w^1(u;x_1) = w^1(u;x_1-x_0), \tag{7a}$$

$$w^2(u;x_1,x_2) = w^2(u;x_1-x_0,x_2-x_0), \tag{7b}$$

etc.

These equations imply that the first order weight is independent of input position, and depends only on the output position u. The second order weight is a function only of vector differences,[10] i.e.,

$$w^1(u;x_1) = w^1(u), \tag{8a}$$

$$w^2(u;x_1,x_2) = w^2(u;x_1-x_2). \tag{8b}$$

For a discrete implementation with N input units (pixels) fully connected to an output unit, this requirement reduces the number of second-order weights from order N^2 to order N, i.e., only weights for differences of indexes are needed rather than all unique pair combinations. Of course, this advantage is multiplied as the number of fully-connected output units increases.

FURTHER EXAMPLES

We have applied these techniques to several other transformations of interest. For the case of transformation of scale define the scale operator S such that

$$Sp(\mathbf{x}) = a^n p(a\mathbf{x}) \qquad (9)$$

where a is the scale factor, and $\mathbf{x}$ is a vector of dimension n. The factor a^n is used for normalization purposes, so that a given figure always contains the same "energy" regardless of its scale. Application of the same procedure to this transformation leads to the following constraints on the weights:

$$w^1(u;x_1/a) = w^1(u;x_1), \qquad (10a)$$

$$w^2(u;x_1/a,x_2/a) = w^2(u;x_1,x_2), \qquad (10b)$$

$$w^3(u;x_1/a,x_2/a,x_3/a) = w^3(u;x_1,x_2,x_3), \text{ etc.} \qquad (10c)$$

Consider a two-dimensional problem viewed in polar coordinates (r,t). A set of solutions to these constraints is

$$w^1(u;r_1,t_1) = w^1(u;t_1), \qquad (11a)$$

$$w^2(u;r_1,r_2;t_1,t_2) = w^2(u;r_1/r_2;t_1,t_2), \qquad (11b)$$

$$w^3(u;r_1,r_2,r_3;t_1,t_2,t_3) = w^3(u;(r_1-r_2)/r_3;t_1,t_2,t_3). \qquad (11c)$$

Note that with increasing order comes increasing freedom in the selection of the functional form of the weights. Any solution that satisfies the constraint may be used. This gives the designer additional freedom to limit the connection complexity, or to encode special behavior into the net architecture. An example of this is given later when we discuss combining translation and scale invariance in the same network.

Now consider a change of scale for a two-dimensional system in rectangular coordinates, and consider only the second-order weights. A set of solutions to the weight constraint is:

$$w^2(u;x_1,y_1;x_2,y_2) = w^2(u;x_1/y_1;x_2/y_2), \qquad (12a)$$

$$w^2(u;x_1,y_1;x_2,y_2) = w^2(u;x_1/x_2;y_1/y_2), \qquad (12b)$$

$$w^2(u;x_1,y_1;x_2,y_2) = w^2(u;(x_1-x_2)/(y_1-y_2)), \text{ etc.} \qquad (12c)$$

We have done a simulation using the form of Eq. (12b). The simulation was done using a small input space (8x8) and one output unit. A simple least-mean-square (back-propagation) algorithm was used for training the network. When taught to distinguish the letters T and C at one scale, it distinguished them at changes of scale of up to 4X with about 15 percent maximum degradation in the output strength. These results are quite encouraging because no special effort was required to make the system work, and no corrections or modifications were made to account for the boundary condition requirements as discussed near Eq. (6). This and other simulations are discussed further later.

As a third example of a geometric transformation, consider the case of rotation about the origin for a two-dimensional space in polar coordinates. One can readily show that the weight constraints are satisfied if

$$w^1(u;r_1,t_1) = w^1(u;r_1),$$ (13a)

$$w^2(u;r_1,r_2;t_1,t_2) = w^2(u;r_1,r_2;t_1-t_2), \text{ etc.}$$ (13b)

These results are reminiscent of the results for translation invariance. This is not uncommon: seemingly different problems often have similar constraint requirements if the proper change of variable is made. This can be used to advantage when implementing such networks but we will not discuss it further here.

An interesting case arises when one considers combinations of invariances, e.g., scale and translation. This raises the question of the effect of the order of the transformations, i.e., is scale followed by translation equivalent to translation followed by scale? The obvious answer is no, yet for certain cases the order is unimportant. Consider first the case of change-of-scale by a, followed by a translation x_0; the constraints on the weights up to second order are:

$$w^1(u;x_1) = w^1(u;(x_1-x_0)/a),$$ (14a)

$$w^2(u;x_1,x_2) = w^2(u;(x_1-x_0)/a,(x_2-x_0)/a),$$ (14b)

and for translation followed by scale the constraints are:

$$w^1(u;x_1) = w^1(u;(x_1/a)-x_0), \text{ and}$$ (15a)

$$w^2(u;x_1,x_2) = w^2(u;(x_1/a)-x_0,(x_2/a)-x_0).$$ (15b)

Consider only the second-order weights for the two-dimensional case. Choose rectangular coordinate variables (x,y) so that the translation is given by (x_0,y_0). Then

$$w^2(u;x_1,y_1;x_2,y_2) =$$
$$w^2(u;(x_1/a)-x_0,(y_1/a)-y_0;(x_2/a)-x_0,(y_2/a)-y_0),$$ (16a)

or

$$w^2(u;x_1,y_1;x_2,y_2) =$$
$$w^2(u;(x_1-x_0)/a,(y_1-y_0)/a;(x_2-x_0)/a,(y_2-y_0)/a).$$ (16b)

If we take as our solution

$$w^2(u;x_1,y_1;x_2,y_2) = w^2(u;(x_1-x_2)/(y_1-y_2)),$$ (17)

then w^2 is invariant to scale and translation, and the order is unimportant. With higher-order weights one can be even more adventurous.

As a final example consider the case of a change of scale by a factor a and rotation about the origin by an amount t_0 for a two-dimensional system in polar coordinates. (Note that the order of transformation makes no difference.) The weight constraints up to second order are:

$$w^1(u;r_1,t_1) = w^1(u;r_1/a,t_1-t_0), \text{ and}$$ (18a)

$$w^2(r_1,t_1;r_2,t_2) = w^2(u;r_1/a,t_1-t_0;r_2/a,t_2-t_0). \qquad (18b)$$

The first-order constraint requires that w^1 be independent of the input variables, but for the second-order term one can obtain a more useful solution:

$$w^2(u;r_1,t_1;r_2,t_2) = w^2(u;r_1/r_2;t_1-t_2). \qquad (19)$$

This implies that with second-order weights, one can construct a unit that is insensitive to changes in scale and rotation of the input space. How useful it is depends upon the application.

SIMULATION RESULTS

We have constructed several higher-order neural networks that demonstrated invariant response to transformations of scale and of translation of the input patterns. The systems were small, consisting of less than 100 input units, were constructed from second-and first-order units, and contained only one, two, or three layers. We used a back-propagation algorithm modified for the higher-order (sigma-pi) units. The simulation studies are still in the early stages, so the performance of the networks has not been thoroughly investigated. It seems safe to say, however, that there is much to be gained by a thorough study of these systems. For example, we have demonstrated that a small system of second-order units trained to distinguish the letters T and C at one scale can continue to distinguish them over changes in scale of factors of at least four without retraining and with satisfactory performance. Similar performance has been obtained for the case of translation invariance.

Even at this stage, some interesting facets of this approach are becoming clear: 1) Even with the constraints imposed by the invariance, it is usually necessary to limit the range of connections in order to restrict the complexity of the network. This is often cited as a problem with higher-order networks, but we take the view that one can learn a great deal more about the nature of a problem by examining it at this level rather than by simply training a network that has a general-purpose architecture. 2) The higher-order networks seem to solve problems in an elegant and simple manner. However, unless one is careful in the design of the network, it performs worse than a simpler conventional network when there is noise in the input field. 3) Learning is often "quicker" than in a conventional approach, although this is highly dependent on the specific problem and implementation design. It seems that a tradeoff can be made: either faster learning but less noise robustness, or slower learning with more robust performance.

DISCUSSION

We have shown a simple way to encode geometric invariances into neural networks (instead of training them), though to be useful the networks must be constructed of higher-order units. The invariant encoding is achieved by restricting the allowable network

architectures and is <u>independent</u> of learning rules and the form of
the sigmoid or threshold functions. The invariance encoding is
normally for an entire layer, although it can be on an individual
unit basis. It is easy to build one or more invariant layers into a
multi-layer net, and different layers can satisfy different
invariance requirements. This is useful for operating on internal
features or representations in an invariant manner. For learning in
such a net, a multi-layered learning rule such as generalized back-
propagation[7] must be used. In our simulations we have used a
generalized back-propagation learning rule to train a two-layer
system consisting of a second-order, translation-invariant input
layer and a first-order output layer. Note that we have not shown
that one can not encode invariances into layered first-order
networks, but the analysis in this paper implies that such invariance
would be dependent on the form of the sigmoid function.

When invariances are encoded into higher-order neural networks,
the number of interconnections required is usually reduced by orders
of powers of N where N is the size of the input. For example, a
fully connected, first-order, single-layer net with a single output
unit would have order N interconnections; a similar second-order net,
order N^2. If this second-order net (or layer) is made shift
invariant, the order is reduced to N. The number of multiplies and
adds is still of order N^2.

We have limited our discussion in this paper to geometric
invariances, but there seems to be no reason why temporal or other
invariances could not be encoded in a similar manner.

REFERENCES

1.   D.H. Ballard and C.M. Brown, <u>Computer Vision</u> (Prentice-Hall,
     Englewood Cliffs, NJ, 1982).

2.   B. Widrow, IEEE First Intl. Conf. on Neural Networks, 87TH0191-
     7, Vol. 1, p. 143, San Diego, CA, June 1987.

3.   J.A. Feldman, Biological Cybernetics **46**, 27 (1982).

4.   C.L. Giles and T. Maxwell, Appl. Optics **26**, 4972 (1987).

5.   G.E. Hinton, Proc. 7th Intl. Joint Conf. on Artificial
     Intelligence, ed. A. Drina, 683 (1981).

6.   Y.C. Lee, G. Doolen, H.H. Chen, G.Z. Sun, T. Maxwell, H.Y. Lee,
     C.L. Giles, Physica **22D**, 276 (1986).

7.   D.E. Rumelhart, G.E. Hinton, and R.J. Williams, <u>Parallel
     Distributed Processing</u>, Vol. 1, Ch. 8, D.E. Rumelhart and J.L.
     McClelland, eds., (MIT Press, Cambridge, 1986).

8. T. Maxwell, C.L. Giles, Y.C. Lee, and H.H. Chen, Proc. IEEE Intl. Conf. on Systems, Man, and Cybernetics, 86CH2364-8, p. 627, Atlanta, GA, October 1986.

9. W. Pitts and W.S. McCulloch, Bull. Math. Biophys. 9, 127 (1947).

10. M. Minsky and S. Papert, Perceptrons (MIT Press, Cambridge, Mass., 1969).

PROBABILISTIC CHARACTERIZATION OF

NEURAL MODEL COMPUTATIONS

Richard M. Golden [†]

University of Pittsburgh, Pittsburgh, Pa. 15260

ABSTRACT

Information retrieval in a neural network is viewed as a procedure in which the network computes a "most probable" or MAP estimate of the unknown information. This viewpoint allows the class of probability distributions, P, the neural network can acquire to be explicitly specified. Learning algorithms for the neural network which search for the "most probable" member of P can then be designed. Statistical tests which decide if the "true" or environmental probability distribution is in P can also be developed. Example applications of the theory to the highly nonlinear back-propagation learning algorithm, and the networks of Hopfield and Anderson are discussed.

INTRODUCTION

A connectionist system is a network of simple neuron-like computing elements which can store and retrieve information, and most importantly make generalizations. Using terminology suggested by Rumelhart & McClelland [1], the computing elements of a connectionist system are called *units,* and each unit is associated with a real number indicating its *activity level.* The activity level of a given unit in the system can also influence the activity level of another unit. The degree of influence between two such units is often characterized by a parameter of the system known as a *connection strength.* During the *information retrieval process* some subset of the units in the system are activated, and these units in turn activate neighboring units via the inter-unit connection strengths. The activation levels of the neighboring units are then interpreted as

---

[†] Correspondence should be addressed to the author at the Department of Psychology, Stanford University, Stanford, California, 94305, USA.

the retrieved information. During the *learning process,* the values of the inter-unit connection strengths in the system are slightly modified each time the units in the system become activated by incoming information.

DERIVATION OF THE SUBJECTIVE PF

Smolensky [2] demonstrated how the class of possible probability distri-butions that could be represented by a Harmony theory neural network model can be derived from basic principles. Using a simple variation of the arguments made by Smolensky, a procedure for deriving the class of probability distribu-tions associated with *any* connectionist system whose information retrieval dynamics can be summarized by an additive energy function is briefly sketched. A rigorous presentation of this proof may be found in Golden [3].

Let a sample space, S_p, be a subset of the activation pattern state space, S_d, for a particular neural network model. For notational convenience, define the term *probability function* (pf) to indicate a function that assigns numbers between zero and one to the elements of S_p. For discrete random variables, the pf is a probability mass function. For continuous random variables, the pf is a probability density function. Let a particular stationary stochastic environment be represented by the scalar-valued pf, $p_e(X)$, where X is a particular activation pattern. The pf, $p_e(X)$, indicates the *relative frequency* of occurrence of activa-tion pattern X in the network model's environment. A second pf defined with respect to sample space S_p also must be introduced. This probability function, $p_s(X)$, is called the network's subjective pf. The pf $p_s(X)$ is interpreted as the *network's belief* that X will occur in the network's environment.

The subjective pf may be derived by making the assumption that the information retrieval dynamical system, D_s, is optimal. That is, it is assumed that D_s is an algorithm designed to transform a less probable state X into a more probable state X^* where the probability of a state is defined by the subjective pf $p_s(X;A)$, and where the elements of A are the connection strengths among the units. Or in traditional engineering terminology, it is assumed that D_s is a MAP (maximum a posteriori) estimation algorithm. The second assumption is that an *energy* function, $V(X)$, that is minimized by the system during the information retrieval process can be found with an *additivity* property. The additivity pro-perty says that if the neural network were partitioned into two physically unconnected subnetworks, then V(X) can be rewritten as $V_1(X_1) + V_2(X_2)$ where V_1 is the energy function minimized by the first subnetwork and V_2 is the energy function minimized by the second subnetwork. The third assumption is that V(X) provides a sufficient amount of information to specify the probability of activation pattern X. That is, $p_s(X) = G(V(X))$ where G is some continuous function. And the final assumption (following Smolensky [2]) is that statistical and physical independence are equivalent.

To derive $p_s(X)$, it is necessary to characterize G more specifically. Note that if probabilities are assigned to activation patterns such that physically independent substates of the system are also statistically independent, then the additivity property of V(X) forces G to be an exponential function since the only continuous function that maps addition into multiplication is the exponential [4]. After normalization and the assignment of unity to an irrelevant free parameter [2], the unique subjective pf for a network model that minimizes V(X) during the information retrieval process is:

$$p_s(X;A) = Z^{-1} exp[-V(X;A)] \tag{1}$$

$$Z = \int exp[-V(X;A)]dX \tag{2}$$

provided that $Z < C < \infty$. Note that the integral in (2) is taken over S_p. Also note that the pf, p_s, and sample space, S_p, specify a Markov Random Field since (1) is a Gibbs distribution [5].

Example 1: Subjective pfs for associative back-propagation networks

The information retrieval equation for an associative back-propagation [6] network can be written in the form $O=\Phi[I;A]$ where the elements of the vector O are the activity levels for the output units and the elements of the vector I are the activity levels for the input units. The parameter vector A specifies the values of the "connection strengths" among the units in the system. The function Φ specifies the *architecture* of the network.

A natural additive energy function for the information retrieval dynamics of the least squares associative back-propagation algorithm is:

$$V(O) = |O - \Phi(I;A)|^2. \tag{3}$$

If S_p is defined to be a real vector space such that $O \in S_p$, then direct substitution of V(O) for $V_d(X;A)$ into (1) and (2) yields a multivariate Gaussian density function with mean $\Phi(I;A)$ and covariance matrix equal to the identity matrix multiplied by 1/2. This multivariate Gaussian density function is $p_s(O|I;A)$. That is, with respect to $p_s(O|I;A)$, information retrieval in an associative back-propagation network involves retrieving the "most probable" output vector, O, for a given input vector, I.

Example 2: Subjective pfs for Hopfield and BSB networks.

The Hopfield [7] and BSB model [8,9] neural network models minimize the following energy function during information retrieval:

$$V(X) = -X^T M X \tag{4}$$

where the elements of X are the activation levels of the units in the system, and the elements of M are the connection strengths among the units. Thus, the subjective pf for these networks is:

$$p_s(\mathbf{X}) = Z^{-1} \, exp \, [\mathbf{X}^T \mathbf{M} \mathbf{X}] \quad where \quad Z = \sum exp \, [\mathbf{X}^T \mathbf{M} \mathbf{X}] \qquad (5)$$

where the summation is taken over S_p.

APPLICATIONS OF THE THEORY

If the subjective pf for a given connectionist system is known, then traditional analyses from the theory of statistical inference are immediately applicable. In this section some examples of how these analyses can aid in the design and analysis of neural networks are provided.

Evaluating Learning Algorithms

Learning in a neural network model involves searching for a set of connection strengths or parameters that obtain a global minimum of a *learning* energy function. The theory proposed here explicitly shows how an optimal learning energy function can be constructed using the model's subjective pf and the environmental pf. In particular, optimal learning is defined as searching for the *most probable* connection strengths, given some set of observations (samples) drawn from the environmental pf. Given some mild restrictions upon the form of the a priori pf associated with the connection strengths, and for a sufficiently large set of observations, estimating the most probable connection strengths (MAP estimation) is equivalent to maximum likelihood estimation [10]

A well-known result [11] is that if the parameters of the subjective pf are represented by the parameter vector $\mathbf{A}$, then the maximum likelihood estimate of $\mathbf{A}$ is obtained by finding the $\mathbf{A}^*$ that minimizes the function:

$$E(\mathbf{A}) = - <LOG\,[p_s(\mathbf{X};\mathbf{A})]> \tag{6}$$

where $< >$ is the expectation operator taken with respect to the environmental pf. Also note that (6) is the Kullback-Leibler [12] distance measure plus an irrelevant constant. Asymptotically, $E(\mathbf{A})$ is the logarithm of the probability of $\mathbf{A}$ given some set of observations drawn from the environmental pf.

Equation (6) is an important equation since it can aid in the evaluation and design of optimal learning algorithms. Substitution of the multivariate Gaussian associated with (3) into (6) shows that the back-propagation algorithm is doing gradient descent upon the function in (6). On the other hand, substitution of (5) into (6) shows that the Hebbian and Widrow-Hoff learning rules proposed for the Hopfield and BSB model networks are not doing gradient descent upon (6).

Evaluating Network Architectures

The global minimum of (6) occurs if and only if the subjective and environmental pfs are equivalent [12]. Thus, one crucial issue is whether *any* set of connection strengths exists such that the neural network's subjective pf can be made equivalent to a given environmental pf. If no such set of connection strengths exists, the subjective pf, p_s, is defined to be *misspecified*. White [11] and Lancaster [13] have introduced a statistical test designed to reject the null hypothesis that the subjective pf, p_s, is not misspecified. Golden [3] suggests a version of this test that is suitable for subjective pfs with many parameters.

REFERENCES

1. D. E. Rumelhart, J. L. McClelland, and the PDP Research Group, Parallel distributed processing: Explorations in the microstructure of cognition, *1*, (MIT Press, Cambridge, 1986).
2. P. Smolensky, In D. E. Rumelhart, J. L. McClelland and the PDP Research Group (Eds.), Parallel distributed processing: Explorations in the microstructure of cognition, *1*, (MIT Press, Cambridge, 1986), pp. 194-281.

3. R. M. Golden, A unified framework for connectionist systems. Unpublished manuscript.

4. C. Goffman, Introduction to real analysis. (Harper and Row, N. Y., 1966), p. 65.

5. J. L. Marroquin, Probabilistic solution of inverse problems. A.I. Memo 860, MIT Press (1985).

6. D. E. Rumelhart, G. E. Hinton, & R. J. Williams, In D. E. Rumelhart, J. L. McClelland, and the PDP Research Group (Eds.), Parallel distributed processing: Explorations in the microstructure of cognition, *1,* (MIT Press, Cambridge, 1986), pp. 318-362.

7. J. J. Hopfield, Proceedings of the National Academy of Sciences, USA, *79,* 2554-2558 (1982).

8. J. A. Anderson, R. M. Golden, & G. L. Murphy, In H. Szu (Ed.), Optical and Hybrid Computing, SPIE, *634,* 260-276 (1986).

9. R. M. Golden, Journal of Mathematical Psychology, *30,* 73-80 (1986).

10. H. L. Van Trees, Detection, estimation, and modulation theory. (Wiley, N. Y., 1968).

11. H. White, Econometrica, *50,* 1-25 (1982).

12. S. Kullback & R. A. Leibler, Annals of Mathematical Statistics, *22,* 79-86 (1951).

13. T. Lancaster, Econometrica, *52,* 1051-1053 (1984).

ACKNOWLEDGEMENTS

This research was supported in part by the Mellon foundation while the author was an Andrew Mellon Fellow in the Psychology Department at the University of Pittsburgh, and partly by the Office of Naval Research under Contract No. N-0014-86-K-0107 to Walter Schneider. This manuscript was revised while the author was an NIH postdoctoral scholar at Stanford University. This research was also supported in part by grants from the Office of Naval Research (Contract No. N00014-87-K-0671), and the System Development Foundation to David Rumelhart. I am very grateful to Dean C. Mumme for comments, criticisms, and helpful discussions concerning an earlier version of this manuscript. I would also like to thank David B. Cooper of Brown University for his suggestion that many neural network models might be viewed within a unified statistical framework.

PARTITIONING OF SENSORY DATA BY A CORTICAL NETWORK[1]

Richard Granger, José Ambros-Ingerson, Howard Henry, Gary Lynch
Center for the Neurobiology of Learning and Memory
University of California
Irvine, CA. 91717

SUMMARY

To process sensory data, sensory brain areas must preserve information about both the similarities and differences among learned cues: without the latter, acuity would be lost, whereas without the former, degraded versions of a cue would be erroneously thought to be distinct cues, and would not be recognized. We have constructed a model of piriform cortex incorporating a large number of biophysical, anatomical and physiological parameters, such as two-step excitatory firing thresholds, necessary and sufficient conditions for long-term potentiation (LTP) of synapses, three distinct types of inhibitory currents (short IPSPs, long hyperpolarizing currents (LHP) and long cell-specific afterhyperpolarization (AHP)), sparse connectivity between bulb and layer-II cortex, caudally-flowing excitatory collateral fibers, nonlinear dendritic summation, etc. We have tested the model for its ability to learn similarity- and difference-preserving encodings of incoming sensory cues; the biological characteristics of the model enable it to produce multiple encodings of each input cue in such a way that different readouts of the cell firing activity of the model preserve both similarity and difference information.

In particular, probabilistic quantal transmitter-release properties of piriform synapses give rise to probabilistic postsynaptic voltage levels which, in combination with the activity of local patches of inhibitory interneurons in layer II, differentially select bursting vs. single-pulsing layer-II cells. Time-locked firing to the theta rhythm (Larson and Lynch, 1986) enables distinct spatial patterns to be read out against a relatively quiescent background firing rate. Training trials using the physiological rules for induction of LTP yield stable layer-II-cell spatial firing patterns for learned cues. Multiple simulated olfactory input patterns (i.e., those that share many chemical features) will give rise to strongly-overlapping bulb firing patterns, activating many shared lateral olfactory tract (LOT) axons innervating layer Ia of piriform cortex, which in turn yields highly overlapping layer-II-cell excitatory potentials, enabling this spatial layer-II-cell encoding to preserve the overlap (similarity) among similar inputs. At the same time, those synapses that are enhanced by the learning process cause stronger cell firing, yielding strong, cell-specific afterhyperpolarizing (AHP) currents. Local inhibitory interneurons effectively select alternate cells to fire once strongly-firing cells have undergone AHP. These alternate cells then activate their caudally-flowing recurrent collaterals, activating distinct populations of synapses in caudal layer Ib. Potentiation of these synapses in combination with those of still-active LOT axons selectively enhance the response of caudal cells that tend to accentuate the differences among even very-similar cues.

Empirical tests of the computer simulation have shown that, after training, the initial spatial layer II cell firing responses to similar cues enhance the similarity of the cues, such that the overlap in response is equal to or greater than the overlap in

---

[1]This research was supported in part by the Office of Naval Research under grants N00014-84-K-0391 and N00014-87-K-0838 and by the National Science Foundation under grant IST-85-12419.

input cell firing (in the bulb): e.g., two cues that overlap by 65% give rise to response patterns that overlap by 80% or more. Reciprocally, later cell firing patterns (after AHP), increasingly enhance the differences among even very-similar patterns, so that cues with 90% input overlap give rise to output responses that overlap by less than 10%. This difference-enhancing response can be measured with respect to its acuity; since 90% input overlaps are reduced to near zero response overlaps, it enables the structure to distinguish between even very-similar cues. On the other hand, the similarity-enhancing response is properly viewed as a *partitioning* mechanism, mapping quite-distinct input cues onto nearly-identical response patterns (or category indicators). We therefore use a statistical metric for the information value of categorizations to measure the value of partitionings produced by the piriform simulation network.

INTRODUCTION

The three primary dimensions along which network processing models vary are their learning rules, their performance rules and their architectural structures. In practice, performance rules are much the same across different models, usually being some variant of a 'weighted-sum' rule (in which a unit's output is calculated as some function of the sum of its inputs multiplied by their 'synaptic' weights). Performance rules are usually either 'static' rules (calculating unit outputs and halting) or 'settling' rules (iteratively calculating outputs until a convergent solution is reached). Most learning rules are either variants of a 'correlation' rule, loosely based on Hebb's (1949) postulate; or a 'delta' rule, e.g., the perceptron rule (Rosenblatt, 1962), the adaline rule (Widrow and Hoff, 1960) or the generalized delta or 'backpropagation' rule (Parker, 1985; Rumelhart et al., 1986). Finally, architectures vary by and large with learning rules: e.g., multi-layered feedforward nets require a generalized delta rule for convergence; bidirectional connections usually imply a variant of a Hebbian or correlation rule, etc.

Architectures and learning and performance rules are typically arrived at for reasons of their convenient computational properties and analytical tractability. These rules are sometimes based in part on some results borrowed from neurobiology: e.g., 'units' in some network models are intended to correspond loosely to neurons, and 'weights' loosely to synapses; the notions of parallelism and distributed processing are based on metaphors derived from neural processes.

An open question is how much of the rest of the rich literature of neurobiological results should or could profitably be incorporated into a network model. From the point of view of constructing mechanisms to perform certain pre-specified computatonal functions (e.g., correlation, optimization), there are varying answers to this question. However, the goal of understanding brain circuit function introduces a fundamental problem: there are no known, pre-specified functions of any given cortical structures. We have constructed and studied a physiologically- and anatomically-accurate model of a particular brain structure, olfactory cortex, that is strictly based on biological data, with the goal of elucidating the local function of this circuit from its performance in a 'bottom-up' fashion. We measure our progress by the accuracy with which the model corresponds to known data, and predicts novel physiological results (see, e.g., Lynch and Granger, 1988; Lynch et al., 1988).

Our initial analysis of the circuit reveals a mechanism consisting of a learning rule that is notably simple and restricted compared to most network models, a relatively novel architecture with some unusual properties, and a performance rule that is extraordinarily complex compared to typical network-model performance rules. Taken together, these rules, derived directly from the known biology of the olfactory cortex, generate a coherent mechanism that has interesting computational properties. This paper describes the learning and performance rules and the architecture of the model; the relevant physiology and anatomy underlying these rules and structures, respectively; and an analysis of the coherent mechanism that results.

LEARNING RULES DERIVED FROM LONG-TERM POTENTIATION

Long-term potentiation (LTP) of synapses is a phenomenon in which a brief series of biochemical events gives rise to an enhancement of synaptic efficacy that is extraordinarily long-lasting (Bliss and Lømo, 1973; Lynch and Baudry, 1984; Staubli and Lynch, 1987); it is therefore a candidate mechanism underlying certain forms of learning, in which few training trials are required for long-lasting memory. The physiological characteristics of LTP form the basis for a straightforward network learning rule.

It is known that simultaneous pre- and post-synaptic activity (i.e., intense depolarization) result in LTP (e.g., Wigstrøm et al., 1986). Since excitatory cells are embedded in a meshwork of inhibitory interneurons, the requisite induction of adequate levels of pre- and postsynaptic activity is achieved by stimulation of large numbers of afferents for prolonged periods, by voltage clamping the postsynaptic cell, or by chemically blocking the activity of inhibitory interneurons. In the intact animal, however, the question of how simultaneous pre- and postsynaptic activity might be induced has been an open question. Recent work (Larson and Lynch, 1986) has shown that when hippocampal afferents are subjected to patterned stimulation with particular temporal and frequency parameters, inhibition is naturally eliminated within a specific time window, and LTP can arise as a result. Figure 1 shows that LTP naturally occurs using short (3-4 pulse) bursts of high-frequency (100Hz) stimulation with a 200ms interburst interval; only the second of a pair of two such bursts causes potentiation. This occurs because the normal short inhibitory currents (IPSPs), which prevent the first burst from depolarizing the postsynaptic cell sufficiently to produce LTP, are maximally refractory at 200ms after being stimulated, and therefore, although the second burst arrives against a hyperpolarized background resulting from the long hyperpolarizing currents (LHP) initiated by the first burst, the second burst does not initiate its own IPSPs, since they are then refractory. The studies leading to these conclusions were performed in *in vitro* hippocampal slices; LTP induced by this patterned stimulation technique in intact animals shows no measurable decrement prior to the time at which recording arrangements deteriorate: more than a month in some cases (see Staubli and Lynch, 1987).

PERFORMANCE RULES DERIVED FROM
OLFACTORY PHYSIOLOGY AND BEHAVIOR

From the above data we may infer that LTP itself depends on simultaneous pre- and postsynaptic activity, as Hebb postulated, but that a sufficient degree of the latter occurs only under particular conditions. Those conditions (patterned stimulation) suggest the beginnings of a performance rule for the network. Drawing this out requires a review of the inhibitory currents active in hippocampus and in piriform cortex. Three classes of such currents are known to be present: short IPSPs, long LHPs and extremely long, cell-specific afterhyperpolarization, or AHP (see Figure 2). Short IPSPs arise from both feedforward and feedback activation of inhibitory interneurons which in turn synapse on excitatory cells (e.g., layer II cells, which are primary excitatory cells in piriform). IPSPs develop more slowly than excitatory postsynaptic potentials (EPSPs) but quickly shunt the EPSP, thus reversing the depolarization that arises from EPSPs, and bringing the cell voltage down below its original resting potential. IPSPs last approximately 50–100ms, and then enter a refractory period during which they cannot be reactivated from about 100–300ms after they have been once activated. Longer hyperpolarization (LHP) is presumably dependent on a distinct type of inhibitory interneuron or inhibitory receptor, and arises in much the same way; however, these cells are apparently not refractory once activated. LHP lasts for 300-500ms.

Taken together, IPSPs and LHP constitute a form of high-pass frequency filter: 200ms after an input burst, a subsequent input will arrive against a background of hyperpolarization due to LHP, yet this input will not initiate its own IPSP due to the refractory period. If the input is a single pulse, its EPSP will fail to trigger the postsynaptic cell, since it will not be able to overcome the LHP-induced hyperpolarized potential of the cell. Yet if the input is a high-frequency burst, the pulses comprising the burst will give rise to different behavior. Ordinarily, the first EPSP would have been driven back to resting potential by its accompanying IPSP, before the second pulse in the burst could arrive. But when the IPSP is absent, the first EPSP is not driven rapidly down to resting potential, and the second pulse sums with it, raising the voltage of the postsynaptic cell and allowing voltage-dependent channels to open, thereby further depolarizing the cell, and causing it to spike (Figure 3). Hence these high-frequency bursts fire the cell, while single pulses or lower-frequency bursts would not do so. When these cells fire, then active synapses can be potentiated.

The third inhibitory mechanism, AHP, is a current that causes an excitatory cell to become refractory after it has fired strongly or rapidly. This mechanism is therefore specific to those cells that have fired, unlike the first two mechanisms. AHP can prevent a cell from firing again for as long as 1000ms (1 second).

It has long been observed that EEG waves in the hippocampi of learning animals are dominated by the theta rhythm, i.e., activity occuring at about 4-8Hz. This is now seen to correspond to the optimal rate for firing postsynaptic cells and for enhancing synapses via LTP; i.e., this rhythmic aspect of the performance rules of these networks is suggested by the physiology of LTP. The resulting activation patterns may take the following form: relatively synchronized cell firing occurring approximately once every 200ms, i.e., spatial patterns of induced activity occurring at the rate of one new spatial cell-firing pattern every 200ms. The cells most strongly participating in any one firing pattern will not participate in subsequent patterns (at least the next 4-5 patterns, i.e., 800-1000ms), due to AHP. This raises the interesting possibility that different spatial patterns (at different times) may be conveying different information about their inputs. In summary, postsynaptic cells fire in pulses or bursts depending on the synaptically-weighted sums of their active axonal inputs; this firing is synchronized across the cells in a structure, giving rise to a spatial pattern of activity across these cells; once cells fire they will not fire again in subsequent patterns; each pattern (occuring at the theta rhythm, i.e., approximately once every 200ms) will therefore consist of extremely different spatial patterns of cell activity. Hence the 'output' of such a network is a sequence of spatial patterns.

In an animal engaged in an olfactory discrimination learning task, the theta rhythm dominates the animals behavior: the animals literally sniff at theta. We have been able to sustitute direct stimulation (in theta-burst mode) of the lateral olfactory tract (LOT), which is the input to the olfactory cortex, for odors: these 'electrical odors' are learned and discriminated by the animals, either from other electrical odors (via different stimulating electrodes) or from real odors. Furthermore, behavioral learning in this paradigm is accompanied by LTP of piriform synapses (Roman et al., 1987). This experimental paradigm thus provides us with a known set of behaviorally-relevant inputs to the olfactory cortex that give rise to synaptic potentiation that apparently underlies the learning of the stimuli.

ARCHITECTURE OF OLFACTORY CORTEX

Nasal receptor cells respond differentially to different chemicals; these cells topographically innervate the olfactory bulb, which is arranged such that combinations of specific spatial 'patches' of bulb characteristically respond to specific odors. Bulb also receives a number of centrifugal afferents from brain, most of which terminate on the inhibitory granule cells. The excitatory mitral cells in bulb send out axons that form the lateral olfactory tract (LOT), which constitutes the only major input to olfactory (piriform) cortex. This cortex in turn has some feedback connections to bulb via the anterior olfactory nucleus.

Figure 4 illustrates the anatomy of the superficial layers of olfactory cortex: the LOT axons flow across layer Ia, synapsing with the dendrites of piriform layer-II cells. Those cells in turn give rise to collateral axon outputs which flow, in layer Ib, parallel and subjacent to the LOT, in a predominantly rostral-to-caudal direction, eventually terminating in entorhinal cortex. Layer Ia is very sparsely connected; the probability of synapses between LOT axons and layer-II cell dendrites is less than 0.10 (Lynch, 1986), and decreases caudally. Layer Ib (where collaterals synapse with dendrites) is also sparse, but its density increases caudally, as the number of collaterals increases; the overall connectivity density on layer-II-cell dendrites is approximately constant throughout most of piriform. Layer II also contains, in addition to the principal excitatory cells (modified stellates), inhibitory interneurons which synapse on excitatory cells within a specified radius, forming a 'patchwork' of cells affected by a particular inhibitory cell; the spheres of influence of inhibitory cells almost certainly overlap somewhat. There are approximately 50,000 LOT axons, 500,000 piriform layer II cells, and a much smaller number of inhibitory cells that divide layer II roughly into functional patches. (See Price, 1973; Luskin and Price, 1983; Krettek and Price, 1977; Price and Slotnick, 1983; Haberly and Price, 1977, 1978a, 1978b).

The layer II cell collateral axons flow through layer III for a distance before rising up to layer Ib (Haberly, 1985); taken in combination with the predominantly caudal directionality of these collaterals, this means that rostral piriform will be dominated by LOT inputs. Extreme caudal piriform (and all of lateral entorhinal cortex) is dominated by collaterals from more rostral cells; moving from rostral to caudal piriform, cells increasingly can be thought of as 'hybrid cells': cells receiving inputs from both the bulb (via the LOT) and from rostral piriform (via collateral axons). The architectural characteristics of rostral piriform is therefore quite different from that of caudal piriform, and differential analysis must be performed of rostral cells vs. hybrid cells, as will be seen later in the paper.

SIMULATION AND FORMAL ANALYSIS: INTRODUCTION

We have conducted several simulations of olfactory cortex incorporating many of the physiological features discussed earlier. Two hundred layer II cells are used with 100 input (LOT) lines and 200 collateral axons; both the LOT and collateral axons flow caudally. LOT axons connect with rostral dendrites with a probability of 0.2, which decreases linearly to 0.05 by the caudal end of the model. The connectivity is arranged randomly, subject to the constraint that the number of contacts for axons and dendrites is fixed within certain narrow boundaries (in the most severe case, each axon forms 20 synapses and each dendrite receives 20 contacts). The resulting matrix is thus hypergeometric in both dimensions. There are 20 simulated inhibitory interneurons, such that the layer II cells are arranged in 20 overlapping patches, each within the influence of one such inhibitory cell. Inhibition rules are approximately as discussed above; i.e., the short IPSP is longer than an EPSP but only one fifth the length of the LHP; cell-specific AHP in turn is twice as long as LHP.

Synaptic activity in the model is probabilistic and quantal: for any presynaptic activation, there is a fixed probability that the synapse will allow a certain amount of conductance to be contributed to the postsynaptic cell. Long-term potentiation was represented by a 40% increase in contact strength, as well as an increase in the probability of conductance being transmitted. These effects would be expected to arise, *in situ*, from modifying existing synapses as well as adding new ones (Lynch, 1986), two results obtained in electron microscopic studies (Lee et al., 1980). Only excitatory cell synapses are subject to LTP. LTP occurred when a cell was activated twice at a simulated 200ms interval: the first input 'primes' the synapse so that a subsequent burst input can drive it past a threshold value; following from the physiological results, previously potentiated synapses were much less different from "naive" synapses when driven at high frequency (see Lynch et al., 1988). The simulation used theta burst activation (i.e., bursts of pulses with the bursts occurring at 5Hz) of inputs during learning, and operated according to these synchronized fixed time steps, as discussed above.

The network was trained on sets of "odors", each of which was represented as a group of active LOT lines, as in the "electric odor" experiments already described. Usually three or four "components" were used in an odor, with each component consisting of a group of contiguous LOT lines. We assumed that the bulb normalized the output signal to about 20% of all LOT fibers. In some cases, more specific bulb rules were used and in particular inhibition was assumed to be greatest in areas surrounding an active bulb "patch".

The network exhibited several interesting behaviors. Learning, as expected, increased the robustness of the response to specific vectors; thus adding or subtracting LOT lines from a previously learned input did not, within limits, greatly change the response. The model, like most network simulations, dealt reasonably well with degraded or noisy known signals. An unexpected result developed after the network had learned a succession of cues. In experiments of this type, the simulation would begin to generate two quite distinct output signals within a given sampling episode; that is, a single previously learned cue would generate two successive responses in successive 'sniffs' presented to an "experienced" network. The first of these response patterns proved to be common to several signals while the second was specific to each learned signal. The common signal was found to occur when the network had learned 3–5 inputs which had substantial overlap in their components (e.g., four odors that shared ≈70% of their components). It appeared then that the network had begun to produce "category" or "clustering" responses, on the first sniff of a simulated odor, and "individual" or "differentiation" responses on subsequent sniffs of that same odor. When presented with a novel cue which contained elements shared with other, previously learned signals, the network produced the cluster response but no subsequent individual or specific output signal. Four to five cluster response patterns and 20 − 25 individual responses were produced in the network without distortion.

In retrospect, it was clear that the model accomplished two necessary and in some senses opposing operations: 1) it detected similarities in the members of a cue category or cluster, and, 2) it nonetheless distinguished between cues that were quite similar. Its first response was to the similarity-based category and its second to the specific signal.

ANALYSIS OF CATEGORIZATION IN ROSTRAL PIRIFORM

Assume that a set of input cues (or 'simulated odors') $X^\alpha, X^\beta \ldots X^\zeta$ differ from each other in the firing of d_X LOT input lines; similarly, inputs $Y^\alpha, Y^\beta \ldots Y^\zeta$ differ in d_Y lines, but that inputs from the sets X and Y differ from each other in $D_{X,Y} >> d$ lines, such that the Xs and the Ys form distinct natural categories. Then the performance of the network should give rise to output (layer II cell) firing patterns that are very similar among members of either category, but different for members of different categories; i.e., there should be a single spatial pattern of response for members of X, with little variation in response across members, and there should be a distinct spatial pattern of response for members of Y.

Considering a matrix constructed by uniform selection of neurons, each with a hypergeometric distribution for its synapses, as an approximation of the bidimensional hypergeometric matrix described above, the following results can be derived. The expected value of $\hat{d}$, the Hamming distance between responses for two input cues differing by $2d$ LOT lines (input Hamming distance of d) is: •

$$E(\hat{d}) = \sum_{k=1}^{N_o} \left[\sum_{\substack{i \geq \theta \\ j < \theta}} S_i I(i,j) + \sum_{\substack{i < \theta \\ j \geq \theta}} S_i I(i,j) \right]$$

where N_o is the number of postsynaptic cells, each S_i is the probability that a cell will have precisely i active contacts from one of the two cues, and $I(i,j)$ is the probability that the number of contacts on the cell will increase (or decrease) from i to j with the change in d LOT lines; i.e., changing from the first cue to the second. Hence, the first term denotes the probability of a cell decreasing its number of active contacts from above to below some threshold, θ, such that that cell fired in response to one cue but not the other (and therefore is one of the cells that will contribute to the difference between responses to the two cues). Reciprocally, the second term is the probability that the cell increases its number of active synapses such that it is now over the threshold; this cell also will contribute to the difference in response. We restrict our analysis for now to rostral piriform, in which there are assumed to be few if any collateral axons. We will return to this issue in the next subsection.

The value for each S_a, the probability of a active contacts on a cell, is a hypergeometric function, since there are a fixed number of contacts anatomically between LOT and (rostral) piriform cells:

$$S_a = p(a \text{ active synapses}) = \frac{\binom{A}{a}\binom{N-A}{n-a}}{\binom{N}{n}}$$

where N is the number of LOT lines, A is the number of active (firing) LOT lines, n is the number of synapses per dendrite formed by the LOT, and a is the number of active such synapses. The formula can be read by noting that the first binomial indicates the number of ways of choosing a active synapses on the dendrite from the A active incoming LOT lines; for each of these, the next expression calculates the number of ways in which the remaining $n - a$ (inactive) synapses on the dendrite are chosen from the $N - A$ inactive incoming LOT lines; the probability of active synapses on a dendrite depends on the sparseness of the matrix (i.e., the probability of connection between any given LOT line and dendrite); the solution must be normalized by the number of ways in which n synapses on a dendrite can be chosen from N incoming LOT lines.

The probability of a cell changing its number of contacts from a to $\hat{a}$ is:

$$I(a,\hat{a}) = \sum_{\substack{g-l= \\ a-\hat{a}}} \left[\frac{\binom{a}{l}\binom{A-a}{d-l}}{\binom{A}{d}} \frac{\binom{n-a}{g}\binom{N-A-(n-a)}{d-g}}{\binom{N-A}{d}} \right]$$

where N, n, A, and a are as above, l is the "loss" or reduction in the number of active synapses, and g is the gain or increase. Hence the left expression is the probability of losing l active synapses by changing d LOT lines, and the right-hand expression is the probability of gaining g active synapses. The product of the expressions are summed over all the ways of choosing l and g such that the net change $g - l$ is the desired difference $a - \hat{a}$.

If training on each cue induces only fractional LTP, then over trials, synapses contacted by any overlapping parts of the input cues should become stronger than those contacted only by unique parts of the cue. Comparing two cues from within a category, vs. two cues from between categories, there may be the same number of active synapses lost across the two cues in either case, but the expected *strength* of the synapses lost in the former case (within category) should be significantly lower than in the latter case (across categories). Hence, for a given threshold, the difference $\hat{d}$ between output firing patterns will be smaller for two within-category cues than for cues from two different categories.

It is important to note that clustering is an operation that is quite distinct from stimulus generalization. Observing that an object is a car does not occur because of a comparison with a specific, previously learned car. Instead the category "car" emerges from the learning of many different cars and may be based on a "prototype" that has no necessary correspondence with a specific, real object. The same could be said of the network. It did not produce a categorical response when one cue had been learned and second similar stimulus was presented. Category or cluster responses, as noted, required the learning of several exemplars of a similarity-based cluster. It is the process of extracting commonalities from the environment that defines clustering, not the simple noting of similarities between two cues.

An essential question in clustering concerns the location of the boundaries of a given group; i.e., what degree of similarity must a set of cues possess to be grouped together? This issue has been discussed from any number of theoretical positions (e.g., information theory); all these analyses incorporate the point that the breadth of a category must reflect the overall homogeneity or heterogeneity of the environment. In a world where things are quite similar, useful categories will necessarily be composed of objects with much in common. Suppose, for instance, that subjects were presented with a set of four distinct coffee cups of different colors, and asked later to recall the objects. The subjects might respond by listing the cups as a blue, red, yellow and green coffee cup, reflecting a relatively specific level of description in the hierarchy of objects that are coffee cups. In contrast, if presented with four different objects, a blue coffee cup, a drinking glass, a silver fork and a plastic spoon, the cup would be much more likely to be recalled as simply a cup, or a coffee cup, and rarely as a blue coffee cup; the specificity of encoding chosen depends on the overall heterogeneity of the environment. The categories formed by the simulation were quite appropriate when judged by an information theoretic measure, but how well it does across a wide range of possible worlds has not been addressed.

ANALYSIS OF PROBLEMS ARISING FROM CAUDAL AXON FLOW

The anatomical feature of directed flow of collateral axons gives rise to an immediate problem in principle. In essence, the more rostral cells that fire in response to an input, the more active inputs there are from these cells to the caudal cells, via collateral axons, such that the probability of caudal cell firing increases precipitously with probability of rostral cell firing. Conversely, reducing the number of rostral cells from firing, either by reducing the number of active input LOT axons or by raising the layer II cell firing threshold, prevents sufficient input to the caudal cells to enable their probability of firing to be much above zero.

This problem can be stated formally, by making assumptions about the detailed nature of the connectivity of LOT and collateral axons in layer I as these axons proceed from rostral to caudal piriform. The probability of contact between LOT axons and layer-II-cell dendrites decreases caudally, as the number of collateral axons is increasing, given their rostral to caudal flow tendency. This situation is depicted in Figure 4. Assuming that probability of LOT contact tends to go to zero, we may adopt a labelling scheme for axons and synaptic contacts, as in the diagram, in which some combination of LOT axons (x_k) and collateral axons (h_m) contact any particular layer II cell dendrite (h_n), each of which is itself the source of an additional collateral axon flowing to cells more caudal than itself. Then the cell firing function for layer II cell h_n is:

$$h_n = H\left(\sum_{m<n} h_m w_{nm} + \sum_{k \geq n} x_k w_{nk} - \theta\right)$$

where the x_k denote LOT axon activity of those axons still with nonzero probability of contact for layer II cell h_n, the h_m denote activity of layer II cells rostral of h_n, θ is the cell firing threshold, w_{nm} is the synaptic strength between axon m and dendrite n, and H is the Heaviside step function, equal to 1 or 0 according to whether its argument is positive or negative. If we assume instead that probability of cell firing is a graded function rather than a step function, we may eliminate the H step function and calculate the firing of the cell (h_n) from its inputs $(h_{n,net})$ via the logistic:

$$h_{n,net} = \sum_{m<n} h_m w_{nm} + \sum_{k \geq n} x_k w_{nk}$$

$$h_n = \frac{1}{1 + e^{-\left(k h_{n,net} + \theta_n\right)}}$$

Then we may expand the expression for firing of cell h_n as follows:

$$h_n = \left[1 + e^{-\left(\sum_{m<n} h_m w_{nm} + \sum_{k \geq n} x_k w_{nk} + \theta\right)}\right]^{-1}$$

By assuming a fixed firing threshold, and varying the number of active input LOT lines, the probability of cell firing can be examined. Numerical simulation of the above expressions across a range of LOT spatial activation patterns demonstrates that probability of cell firing remains near zero until a critical number of LOT lines are active, at which point the probability flips to close to 100% (Figure 5). This means that, for any given firing threshold, given fewer than a certain amount of LOT input, practically no piriform cells will fire, whereas a slight increase in the number of active LOT lines will mean that practically all piriform cells should fire.

This excruciating dependence of cell firing on amount of LOT input indicates that normalization of the size of the LOT input alone will be insufficient to stabilize the size of the layer II response; even slight variation of LOT activity in either direction has extreme consequences. A number of solutions are possible; in particular, the known local anatomy and physiology of layer II inhibitory interneurons provides a mechanism for controlling the amount of layer II response. As discussed, inhibitory interneurons give rise to both feedforward (activated by LOT input) and feedback (activated by collateral axons) activity; the influence of any particuar interneuron is limited anatomically to a relatively small radius around itself within layer II, and the influence of multiple interneurons probably overlap to some extent. Nonetheless, the 'sphere of influence' of a particular inhibitory interneuron can be viewed as a local patch in layer II, within which the number of active excitatory cells is in large measure controlled by the activity of the inhibitory cell in that patch. If a number of excitatory cells are firing with varying depolarization levels within a patch in layer II, activation of the inhibitory cells by the excitatory cells will tend to weaken those excitatory cells that are less depolarized than the most strongly-firing cell within the patch, leading to a competition in which only those cells firing most strongly within a patch will burst, and these cells will, via the interneuron, suppress multiple firing of other cells within the patch. Thus the patch takes on some of the characteristics of a 'winner-take-all' network (Feldman, 1982): only the most strongly firing cells will be able to overcome inhibition sufficiently to burst, some additional cells will pulse once and then be overwhelmed by inhibition, and the rest of the cells in the patch will be silent, even though that patch may be receiving a large amount of excitatory input via LOT and collateral axon activity in layer I.

EMERGENT CATEGORIZATION BEHAVIOR IN THE MODEL

The probabilistic quantal transmitter-release properties of piriform synapses described above give rise to probabilistic levels of postsynaptic depolarization. This inherent randomness of cell firing, in combination with activity of local inhibitory patches in layer II, selects different sets of bursting and pulsing cells on different trials if no synaptic enhancement has taken place. The time-locked firing to the theta rhythm enables distinct spatial patterns of firing to be read out against a relatively quiescent background firing rate. Synaptic LTP enhances the conductances and alters the probabilistic nature of communication between a given axon and dendrite, which tends to overcome the randomness of the cell firing patterns in untrained cells, yielding a stable spatial pattern that will reliably appear in response to the same input in the future, and in fact will appear even in response to degraded or noisy versions of the input pattern. Furthermore, subsequent input patterns that differ in only minor respects from a learned LOT input pattern will contact many of the already-potentiated synapses from the original pattern, thereby tending to give rise to a very similar (and stable) output firing pattern. Thus as multiple cues sharing many overlapping LOT lines are learned, the layer II cell responses to each of these cues will strongly resemble the responses to the others. Hence, the response(s) behave as though simply labelling a *category* of very-similar cues; sufficiently different cues will give rise to quite-different category responses.

EMERGENT DIFFERENTIATION BEHAVIOR IN THE MODEL

Potentiated synapses cause stronger depolarization and firing of those cells participating in a 'category' response to a learned cue. This increased depolarization causes strong, cell-specific afterhyperpolarization (AHP), effectively putting those cells into a relatively long-lasting ($\approx$ 1sec) refractory period that prevents them from firing in response to the next few sampling sniffs of the cue. Then the inhibitory 'winner-take-all' behavior within patches effectively selects alternate cells to fire, once these strongly-firing (learned) cells have undergone AHP. These alternates will be selected with some randomness, given the probabilistic release characteristics discussed above, since these cells will tend not to have potentiated synapses. These alternate cells then activate their caudally-flowing recurrent collaterals, activating distinct populations of synapses in caudal layer Ib. Potentiation of these synapses in combination with those of still-active LOT axons tends to 'recruit' stable subpopulations of caudal cells that are distinct for each simulated odor. They are distinct for each odor because first rostral cells are selected from the population of unpotentiated or weakly-potentiated cells (after the strongly potentiated cells have been removed via AHP); hence they will at first tend to be selected randomly. Then, of the caudal cells that receive some activation from the weakening caudal LOT lines, those that also receive collateral innervation from these semi-randomly selected rostrals will be those that will tend to fire most strongly, and hence to be potentiated.

The probability of a cell participating in the rostral semi-randomly selected groups for more than one odor (e.g., for two similar odors) is lower than the probability of cells being recruited by these two odors initially, since the population are those that receive not enough input from the LOT to have been recruited as a category cell and potentiated, yet receive enough input to fire as an alternate cell. The probability of any caudal cell then being recruited for more than one odor by these rostral cell collaterals in combination with weakening caudal LOT lines is similarly low. The product of these two probabilities is of course lower still. Hence, the probability that any particular caudal cell potentiated as part of this process will participate in response to more than one odor is very low.

This means that, when sampling (sniffing), the first pattern of cell firing will indicate similarity among learned odors, causing AHP of those patterns; thus later sniffs will generate patterns of firing that tend to be quite different for different odors, even when those odors are very similar. Empirical tests of the simulation have shown that odors consisting of 90%-overlapping LOT firing patterns will give rise to overlaps of between 85% and 95% in their initial layer II spatial firing patterns, whereas these same cues give rise to layer II patterns that overlap by less than 20% on 2nd and 3rd sniffs. The spatio-temporal pattern of layer II firing over multiple samples thus can be taken as a strong differentiating mechanism for even very-similar cues, while the initial sniff-response for those cues will nonetheless give rise to a spatial firing pattern that indicates the similarity of sets of learned cues, and therefore their 'category membership' in the clustering sense.

CLUSTERING

Incremental clustering of cues into similarity-based categories is a more subtle process than might be thought and while it is clear that the piriform simulation performs this function, we do not know how optimal its performance is in an information-theoretic sense, relative to some measure of the value or cost of information in the encoding. Building a categorical scheme is a non-monotonic, combinatorial problem: that is, each new item to be learned can have disproportionate effects on the existing scheme, and the number of potential categories (clusters) climbs factorially with the number of items to be categorized. Algorithmic solutions to problems of this type are computationally very expensive. Calculation of an ideal categorization scheme (with respect to particular cost measures in a performance task), using a hill-climbing algorithm derived from an information-theoretic measure of category value, applied to a problem involving 22 simulated odors, required more than 4 hours on a 68020-based processor. The simulation network reached the same answer as the game-theoretic program, but did so in seconds. It is worth mentioning again that the simulation did so while simultaneously learning unique encodings for the cues, as described above, which is itself a nontrivial task.

Humans, on at least some tasks, may carry out clustering by building initial clusters and then merging or splitting them as more cues are presented. Thus far, the networks do not pass through successive categorization schema. However, experiments on human categorization have almost exclusively involved situations in which all cues were presented in rapid succession and category membership is taught explicitly, rather than developed independently by the subject. Hence, it is not clear from the experimental literature whether or not stable clusters develop in this way from stimuli presented at widely spaced intervals with no category membership information given, which is the problem corresponding to that given the network (and that is likely common in nature). It will be of interest to test categorizing skills of rats learning successive olfactory discriminations over several days. Using appropriately selected stimuli, it should be possible to determine if stable clusters are constructed and whether merging and splitting occurs over trials.

Any useful clustering device must utilize information about the heterogenity of the stimulus world in setting the heterogeneity of individual categories. Heterogeneity of categories refers to the degree of similarity that is used to determine if cues are to be grouped together or not. Several network parameters will influence category size and we are exploring how these influence the individuation function; one particularly interesting possibility involves a shifting threshold function, an idea used with great success by Cooper in his work on visual cortex. The problems presented to the simulation thus far involve a totally naive system, one that has had no "developmental" history. We are currently exploring a model in which early experiences are not learned by the network but instead set parameters for later ("adult") learning episodes. The idea is that early experience determines the heterogenity of the stimulus world and imprints this on the network, not by specific changes in synaptic strengths, but in a more general fashion.

CONCLUSIONS

Neurons have a nearly bewildering array of biophysical, chemical, electrophysiological and anatomical properties that control their behavior; an open question in neural network research is which of these properties need be incorporated into networks in order to simulate brain circuit function. The simulation described here incorporates an extreme amount of biological data, and in fact has given rise to novel physiological questions, which we have tested experimentally with results that are counterintuitive and previously unsuspected in the existing physiological literature (see, e.g., Lynch and Granger, 1988; Lynch et al., 1988). Incorporation of this mass of physiological parameters into the simulation gives rise to a coherent architecture and learning and performance rules, when interpreted in terms of computational function of the network, which generates a robust capability to encode multiple levels of information about learned stimuli. The coherence of the data in the model is useful in two ways: to provide a framework for understanding the purposes and interactions of many apparently-disparate biological properties of neurons, and to aid in the design of novel artificial network architectures inspired by biology, which may have useful computational functions.

It is instructive to note that neurons are capable of many possible biophysical functions, yet early results from chronic recording of cells from olfactory cortex *in animals actively engaged in learning many novel odors in an olfactory discrimination task* clearly shows a particular operating mode of this cortical structure when it is actively in use by the animal (Larson et al., unpublished data). The rats in this task are very familiar with the testing paradigm and exhibit very raid learning, with no difficulty in acquiring large numbers of discriminations. Sampling, detection and responding occur in fractions of a second, indicating that the utilization of recognition memories in the olfactory system can be a rapid operation; it is not surprising, then, that the odor-coded units so far encountered in our physiological experiments have rapid and stereotyped responses. Given the dense innervation of the olfactory bulb by the brain, it is possible that the type of spatial encoding that appears to be responsible for the preliminary results of these chronic experiments would not appear in animals that were not engaged in active sampling or were confronted with unfamiliar problems. That is, the operation of the olfactory cortex might be as dependent upon the behavioral 'state' and behavioral history of the rat as upon the actual odors presented to it. It will be of interest to compare the results from well-trained freely-moving animals with those obtained using more restrictive testing conditions.

The temporal properties of synaptic currents and afterpotentials, results from simulations and chronic recording studies, taken together, suggest two useful caveats for biological models:

- Cell firing in cortical structures (e.g., piriform, hippocampus and possibly neocortex) is linked to particular rhythms (theta in the case of piriform and hippocampus) during real learning behavior, and thus it is likely that the 'coding language' of these structures involves spatial cell firing patterns within a brief time window. This stands in contrast to other methods such as frequency coding that appears in other structures (such as peripheral sensory structures, e.g., retina and cochlea; see, e.g., Sivilotti et al., 1987).

- Temporal sequences of spatial patterns may encode different types of information, such as hierarchical encodings of perceptions, in contrast with views in which either asynchronous 'cycling' activity occurs or a system yields a single punctate output and then halts.

In particular, simulation of piriform gives rise to temporal sequences of spatial patterns of synchronized cell firing in layer II, and the patterns change over time: the physiology and anatomy of the structure cause successive 'sniffs' of the same olfactory stimulus to give rise to a sequence of spatial patterns, each of which encodes successively more specific information about the stimulus, beginning with its similarity to other previously-learned stimuli, and ending with a unique encoding of its characteristics. It is possible that both the early similarity-based 'cluster' information and the late unique encodings are used, for different purposes, by brain structures that receive these signals as output from piriform.

ACKNOWLEDGEMENTS

Much of the theoretical underpinning of this work depends critically on data generated by John Larson; we are grateful for his insightful advice and help. This work has benefited from discussions with Michel Baudry, Mark Gluck, and Ursula Staubli. José Ambros-Ingerson is supported by a fellowship from Hewlett-Packard, México, administered by UC MEXUS.

REFERENCES

Bliss, T.V.P. and Lømo, T. (1973). Long-lasting potentiation of synaptic transmission in the dentate area of the anesthetized rabbit following stimulation of the perforant path. *J.Physiol.Lond.* 232:357–374.

Feldman, J.A. (1982). Dynamic connections in neural networks. *Biological Cybernetics* 46:27–39.

Haberly, L.B. (1985). Neuronal circuitry in olfactory cortex: Anatomy and functional implications. *Chemical Senses* 10:219–238.

Haberly, L.B. and J.L. Price (1977). The axonal projection patterns of the mitral and tufted cells of the olfactory bulb in the rat. *Brain Res* 129:152–157.

Haberly, L.B. and J.L. Price (1978a). Association and commissural fiber systems of the olfactory cortex of the rat. I. Systems originating in the piriform cortex and adjacent areas. *J. Comp. Neurol.* 178:711–740.

Haberly, L.B. and J.L. Price (1978b). Association and commissural fiber systems of the olfactory cortex of the rat. II. Systems originating in the olfactory peduncle. *J. Comp. Neurol.* 181:781–808.

Hebb, D.O. (1949). *The Organization of Behavior.* New York: Wiley.

Krettek, J.E. and J.L. Price (1977). Projections from the amygdaloid complex and adjacent olfactory structures to the entorhinal cortex and to the subiculum in the rat and cat. *J Comp Neurol* 172:723–752.

Larson, J. and G. Lynch (1986). Synaptic potentiation in hippocampus by patterned stimulation involves two events. *Science* 232:985–988.

Lee, K., Schottler, F., Oliver, M. and Lynch, G. (1980). Brief bursts of high-frequency stimulation produce two types of structural change in rat hippocampus. *J.Neurophysiol.* 44:247–258.

Lynch, G. and Baudry, M. (1984). The biochemistry of memory: a new and specific hypothesis. *Science* 224:1057–1063.

Lynch, G. (1986). Synapses, circuits, and the beginnings of memory. Cambridge, Mass: MIT Press.

Lynch, G., Larson, J., Staubli, U., and Baudry, M. (1987). New perspectives on the physiology, chemistry and pharmacology of memory. *Drug Devel.Res.* 10:295–315.

Lynch, G., Granger, R., Levy, W. and Larson, J. (1988). Some possible functions of simple cortical networks suggested by computer modeling. In: *Neural Models of Plasticity: Theoretical and Empirical Approaches,* Byrne, J. and Berry, W.O. (Eds.), (in press).

Lynch, G. and Granger, R. (1988). Simulation and analysis of a cortical network. *The Psychology of Learning and Motivation,* Vol.22 (in press).

Luskin, M.B. and J.L. Price (1983). The laminar distribution of intracortical fibers orginating in the olfactory cortex of the rat. *J Comp Neurol* 216:292–302.

Parker, D.B. (1985). Learning-logic. MIT TR-47, Massachusetts Institute of Technology, Center for Computational Research in Economics and Management Science, Cambridge, Mass.

Price, J.L. (1973). An autoradiographic study of complementary laminar patterns of termination of afferent fibers to the olfactory cortex. *J.Comp.Neur.* 150:87–108.

Price, J.L. and B.M. Slotnick (1983). Dual olfactory representation in the rat thalamus: An anatomical and electrophysiological study. *J Comp Neurol* 215:63–77.

Roman, F., Staubli, U. and Lynch, G. (1987). Evidence for synaptic potentiation in a cortical network during learning. *Brain Res.* 418:221–226.

Rosenblatt, F. (1962). Principles of neurodynamics. New York: Spartan.

Rumelhart, D., Hinton, G. and Williams, R. (1986). Learning Internal Representations by Error Propagation. In D.Rumelhart and J.McClelland (Eds.), *Parallel*

Distributed Processing, Cambridge: MIT Press.

Sivilotti, M.A., Mahowald, M.A. and Mead, C.A. (1987). Real-time visual computations using analog CMOS processing arrays. In: Advanced Research in VLSI (Ed. Paul Losleben), MIT Press, Cambridge.

Staubli, U. and Lynch, G. (1987). Stable hippocampal long-term potentiation elicited by "theta" pattern stimulation. *Brain Res.* (in press).

Widrow, G. and Hoff, M.E. (1960). Adaptive Switching Circuits. *Institute of Radio Engineers, Western Electronic Convention Record, Part 4*, pp.96–104.

Wigstrøm, H., B. Gustaffson, Y.Y. Huang and W.C. Abraham (1986). Hippocampal long-term potentiation is induced by pairing single afferent volleys with intracellularly injected depolarizing current pulses. *Acta Physiol Scand* 126:317–319.

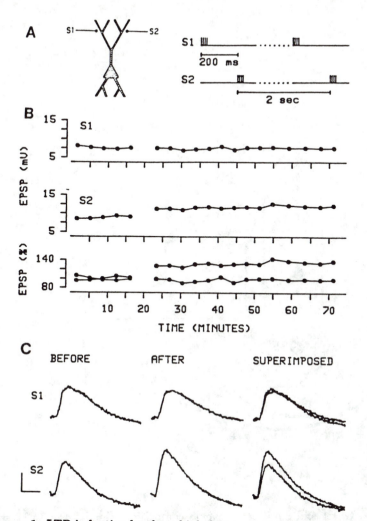

Figure 1. LTP induction by short high-frequency bursts involves sequential "priming" and "consolidation" events.

A) S1 and S2 represent separate groups of Shaffer/commissural fibers converging on a single CA1 pyramidal neuron. The stimulation pattern employed consisted of pairs of bursts (each 4 pulses at 100Hz) given to S1 and S2 respectively, with a 200ms delay between them. The pairs were repeated 10 times at 2 sec intervals.

B) Only the synapses activated by the delayed burst (S2) showed LTP. The top panel shows measurements of amplitudes of intracellular EPSPs evoked by single pulses to S1 before and after patterned stimulation (given at 20 min into the experiment). The middle panel shows the amplitude of EPSPs evoked by S2. Bottom panel shows EPSP amplitudes for both pathways expressed as a percentage of their respective sizes before burst stimulation.

C) Shown are records of EPSPs evoked by S1 and S2 five min. before and 40 min. after patterned burst stimulation. Calibration bar: 5mV, 5msec. (From Larson and Lynch, 1986).

334

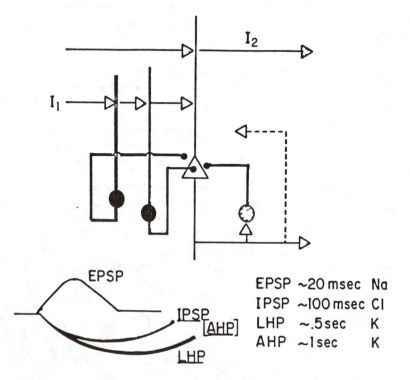

Figure 2. Onset and duration of events comprising stimulation of a layer II cell in piriform cortex. Axonal stimulation via the lateral olfactory tract (LOT) activates feedforward EPSPs with rapid onset and short duration ($\approx$20msec) and two types of feedforward inhibition: short feedforward IPSPs with slower onset and somewhat longer duration ($\approx$100msec) than the EPSPs, and longer hyperpolarizing potentials (LHP) lasting $\approx$500msec. These two types of inhibition are not specific to firing cells; an additional, very long-lasting ($\approx$1sec) inhibitory afterhyperpolarizing current (AHP) is induced in a cell-specific fashion in those cells with intense firing activity. Finally, feedback EPSPs and IPSPs are induced by activation via recurrent collateral axons from layer II cells.

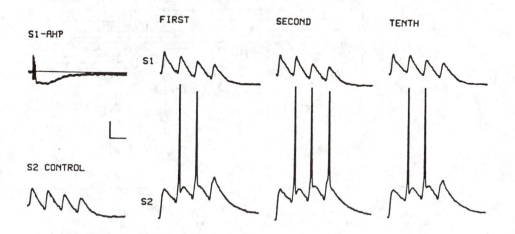

Figure 3. When short, high-frequency bursts are input to cells 200ms after an initial 'priming' event, the broadened EPSPs (see Figure 1) will allow the contributions of the second and subsequent pulses comprising the burst to sum with the depolarization of the first pulse, yielding higher postsynaptic depolarization sufficient to cause the cell to spike. (From Lynch, Larson, Staubli and Baudry, 1987).

336

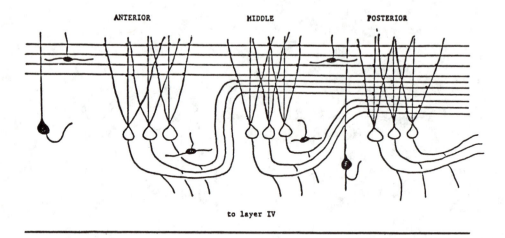

ANTERIOR   MIDDLE   POSTERIOR

to layer IV

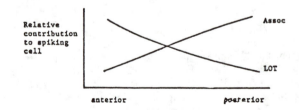

ant.→ post.

→ probability of LOT contact per axon decreases- "n" is constant

→ probability of assoc. contact per axon is constant- "n" increases

Relative
contribution
to spiking
cell

Assoc

LOT

anterior   posterior

Figure 4. Organization of extrinsic and feedback inputs to layer-II cells of piriform cortex. The axons comprising the lateral olfactory tract (LOT), originating from the bulb, innervate distal dendrites, whereas the feedback collateral or associational fibers contact proximal dendrites. Layer II cells in anterior (rostral) piriform are depicted as being dominated by extrinsic (LOT) input, whereas feedback inputs are more prominent on cells in posterior (caudal) piriform.

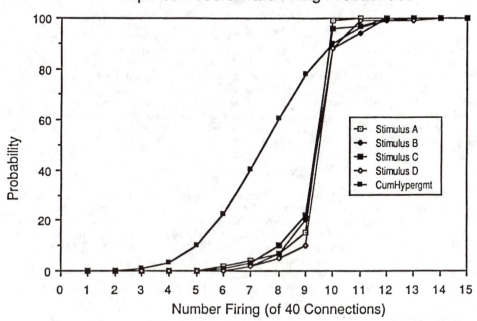

Figure 5. Probability of layer-II-cell firing as a function of number of LOT axons active, in the absence of local inhibitory patches. The hypergeometric function ('CumHypergmt') specifies the probability of layer II cell firing in the absence of caudally-directed feedback collaterals, i.e., assuming that all collaterals are equally probable to travel either rostrally or caudally. In this case, there is a smooth S-shaped function for probability of cell firing with increasing LOT activity, so that adjustment of global firing threshold (e.g., via nonspecific cholinergic inputs affecting all piriform inhibitory interneurons) can effectively normalize piriform layer II cell firing. However, when feedback axons are caudally directed, then probability steepens markedly, becoming a near step function, in which the probability of cell firing is exquisitely sensitive to the number of active inputs, across a range of empirically-tested LOT stimulation patterns (A – D in the figure). In this case, global adjustment of inhibition will fail to adequately normalize layer II cell firing: the probability of cell firing will always be either near zero or near 1.0; i.e., either nearly all cells will fire or almost none will fire. Local inhibitory control of 'patches' of layer II solve this problem (refer to text).

338

The Connectivity Analysis of Simple Association
- or -
How Many Connections Do You Need?

Dan Hammerstrom *
Oregon Graduate Center, Beaverton, OR 97006

ABSTRACT

The efficient realization, using current silicon technology, of Very Large Connection Networks (VLCN) with more than a billion connections requires that these networks exhibit a high degree of communication locality. Real neural networks exhibit significant locality, yet most connectionist/neural network models have little. In this paper, the connectivity requirements of a simple associative network are analyzed using communication theory. Several techniques based on communication theory are presented that improve the robustness of the network in the face of sparse, local interconnect structures. Also discussed are some potential problems when information is distributed too widely.

INTRODUCTION

Connectionist/neural network researchers are learning to program networks that exhibit a broad range of cognitive behavior. Unfortunately, existing computer systems are limited in their ability to emulate such networks efficiently. The cost of emulating a network, whether with special purpose, highly parallel, silicon-based architectures, or with traditional parallel architectures, is *directly* proportional to the number of connections in the network. This number tends to increase geometrically as the number of nodes increases. Even with large, massively parallel architectures, connections take time and silicon area. Many existing neural network models scale poorly in learning time and connections, precluding large implementations.

The connectivity costs of a network are directly related to its locality. A network exhibits *locality of communication* [1] if most of its processing elements connect to other physically adjacent processing elements in any reasonable mapping of the elements onto a planar surface. There is much evidence that real neural networks exhibit locality [2]. In this paper, a technique is presented for analyzing the effects of locality on the process of association. These networks use a complex node similar to the higher-order learning units of Maxwell et al. [3].

NETWORK MODEL

The network model used in this paper is now defined (see Figure 1).

Definition 1: A *recursive neural network*, called a *c-graph* is a graph structure, $\Gamma(V,E,C)$, where:

- There is a set of CNs (network nodes), V, whose outputs can take a range of positive real values, v_i, between 0 and 1. There are N_v nodes in the set.

- There is a set of *codons*, E, that can take a range of positive real values, e_{ij} (for codon j of node i), between 0 and 1. There are N_c codons dedicated to each CN (the output of each codon is only used by its local CN), so there are a total of $N_c N_v$ codons in the network. The fan-in or *order* of a codon is f_c. It is assumed that f_c is the same for each codon, and N_c is the same for each CN.

*This work was supported in part by the Semiconductor Research Corporation contract no. 86-10-097, and jointly by the Office of Naval Research and Air Force Office of Scientific Research, ONR contract no. N00014 87 K 0259.

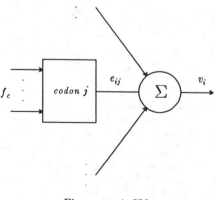

Figure 1 - A CN

- $c_{ijk} \in C$ is a set of *connections* of CNs to codons, $1 \leq i,k \leq N_v$ and $1 \leq j \leq N_c$, c_{ijk} can take two values $\{0,1\}$ indicating the existence of a connection from CN k to codon j of CN i. □

Definition 2: The value of CN i is

$$v_i = F\left[\theta + \sum_{j=1}^{N_c} e_{ij}\right] \tag{1}$$

The function, F, is a continuous non-linear, monotonic function, such as the sigmoid function. □

Definition 3: Define a mapping, $D(i,j,\vec{x}) \rightarrow \vec{y}$, where $\vec{x}$ is an input vector to Γ and $\vec{y}$ is the f_c element input vector of codon j of CN i. That is, $\vec{y}$ has as its elements those elements of x_k of $\vec{x}$ where $c_{ijk} = 1$, $\forall\ k$. □

The D function indicates the subset of $\vec{x}$ seen by codon j of CN i. Different input vectors may map to the same codon vectors, e.g., $D(i,j,\vec{x}) \rightarrow \vec{y}$ and $D(i,j,\vec{z}) \rightarrow \vec{y}$, where $\vec{x} \neq \vec{z}$.

Definition 4: The codon values e_{ij} are determined as follows. Let $\vec{x}(m)$ be input vector m of the M learned input vectors for CN i. For codon e_{ij} of CN i, let T_{ij} be the set of f_c-dimensional vectors such that $\vec{\tau}_{ij}(m) \in T_{ij}$, and $D(i,j,\vec{x}(m)) \rightarrow \vec{\tau}_{ij}(m)$. That is, each vector, $\vec{\tau}_{ij}(m)$ in T_{ij} consists of those subvectors of $\vec{x}(m)$ that are in codon ij's receptive field.

The variable l indexes the $L(i,j)$ vectors of T_{ij}. The number of distinct vectors in T_{ij} may be less than the total number of learned vectors ($L(i,j) \leq M$). Though the $\vec{x}(m)$ are distinct, the subsets, $\vec{\tau}_{ij}(m)$, need not be, since there is a possible many to one mapping of the $\vec{x}$ vectors onto each vector $\vec{\tau}_{ij}$.

Let X^1 be the subset of vectors where $v_i = 1$ (CN i is supposed to output a 1), and X^0 be those vectors where $v_i = 0$, then define

$$n_{ij}^q(l) = size\_of\left\{D(i,j,\vec{x}(m))\ st\ v_i = q\right\} \tag{2}$$

for $q = 0,1$, and $\forall\ m$ that map to this l. That is, $n_{ij}^0(l)$ is the number of $\vec{x}$ vectors that map into $\vec{\tau}_{ij}(l)$ where $v_i=0$ and $n_{ij}^1(l)$ is the number of $\vec{x}$ vectors that map into $\vec{\tau}_{ij}(l)$, where $v_i=1$.

The *compression* of a codon for a vector $\vec{\tau}_{ij}(l)$ then is defined as

$$HC_{ij}(l) = \frac{n_{ij}^1(l)}{n_{ij}^1(l)+n_{ij}^0(l)} \tag{3}$$

($HC_{ij}(l)\equiv 0$ when both n^1, $n^0=0$.) The output of codon ij, e_{ij}, is the maximum-likelyhood decoding

$$e_{ij} = HC_{ij}(l'). \tag{4}$$

Where HC indicates the likelyhood of $v_i=1$ when a vector $\vec{x}$ that maps to l' is input, and l' is that vector $\tau(l')$ where $min[d_h(\tau(l'),\vec{y})]\ \forall\ l$, $D(i,j,\vec{x})\rightarrow\vec{y}$, and $\vec{x}$ is the current input vector. In other words, l' is that vector (of the set of subset learned vectors that codon ij receives) that is closest (using distance measure d_h) to $\vec{y}$ (the subset input vector). $\square$

The output of a codon is the "most-likely" output according to its inputs. For example, when there is no code compression at a codon, $e_{ij}=1$, if the "closest" (in terms of some measure of vector distance, e.g. Hamming distance) subvector in the receptive field of the codon belongs to a learned vector where the CN is to output a 1. The codons described here are very similar to those proposed by Marr [4] and implement nearest-neighbor classification. It is assumed that codon function is determined statically prior to network operation, that is, the desired categories have already been learned.

To measure performance, network capacity is used.

Definition 5: The *input noise*, Ω_I, is the average d_h between an input vector and the closest (minimum d_h) learned vector, where d_h is a measure of the "difference" between two vectors - for bit vectors this can be Hamming distance. The *output noise*, Ω_O, is the average distance between network output and the learned output vector associated with the closest learned input vector. The *information gain*, G_I, is just

$$G_I \equiv -\log\left[\frac{\Omega_I}{\Omega_O}\right] \tag{5}$$

$\square$

Definition 6: The *capacity* of a network is the maximum number of learned vectors such that the information gain, G_I, is strictly positive (>0). $\square$

COMMUNICATION ANALOGY

Consider a single connection network node, or CN. (The remainder of this paper will be restricted to a single CN.) Assume that the CN output value space is restricted to two values, 0 and 1. Therefore, the CN must decide whether the input it sees belongs to the class of "0" codes, those codes for which it remains off, or the class of "1" codes, those codes for which it becomes active. The inputs it sees in its receptive field constitute a subset of the input vectors (the $D(...)$ function) to the network. It is also assumed that the CN is an *ideal* 1-NN (Nearest Neighbor) classifier or feature detector. That is, given a particular set of learned vectors, the CN will classify an arbitrary input according to the class of the nearest (using d_h as a measure of distance) learned vector. This situation is equivalent to the case where a single CN has a single codon whose receptive field size is equivalent to that of the CN.

Imagine a sender who wishes to send one bit of information over a noisy channel. The sender has a probabilistic encoder that choses a code word (learned vector) according to some probability distribution. The receiver knows this code set, though it has no knowledge of which bit is being sent. Noise is added to the code word during its transmission over the channel, which is analogous to applying an input vector to a network's inputs, where the vector lies within some learned vector's region. The "noise" is represented by the distance (d_h) between the input vector and the associated learned vector.

The code word sent over the channel consists of those bits that are seen in the receptive field of the CN being modeled. In the associative mapping of input vectors to output vectors, each CN must respond with the appropriate output (0 or 1) for the associated learned output vector. Therefore, a CN is a decoder that estimates in which class the received code word belongs. This is a classic block encoding problem, where increasing the field size is equivalent to increasing code length. As the receptive field size increases, the performance of the decoder improves in the presence of noise. Using communication theory then, the trade-off between interconnection costs as they relate to field size and the functionality of a node as it relates to the correctness of its decision making process (output errors) can be characterized.

As the receptive field size of a node increases, so does the redundancy of the input, though this is dependent on the particular codes being used for the learned vectors, since there are situations where increasing the field size provides no additional information. There is a point of diminishing returns, where each additional bit provides ever less reduction in output error. Another factor is that interconnection costs increase exponentially with field size. The result of these two trends is a cost performance measure that has a single global maximum value. In other words, given a set of learned vectors and their probabilities, and a set of interconnection costs, a "best" receptive field size can be determined, beyond which, increasing connectivity brings diminishing returns.

SINGLE CODON, WITH NO CODE COMPRESSION

A single neural element with a single codon and with no code compression can be modelled exactly as a communication channel (see Figure 2). Each network node is assumed to have a single codon whose receptive field size is equal to that of the receptive field size of the node.

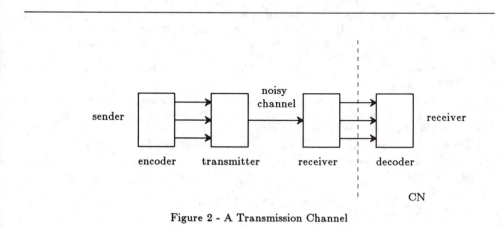

Figure 2 - A Transmission Channel

The operation of the channel is as follows. A bit is input into the channel encoder, which selects a random code of length N and transmits that code over the channel. The receiver then, using nearest neighbor classification, decides if the original message was either a 0 or a 1.

Let M be the number of code words used by the encoder. The rate* then indicates the density of the code space.

Definition 7: The *rate*, R, of a communication channel is

$$R \equiv \frac{\log M}{N} \tag{6}$$

□

The block length, N, corresponds directly to the receptive field size of the codon, i.e., $N=f_c$. The derivations in later sections use a related measure:

Definition 8: The *code utilization*, b, is the number of learned vectors assigned to a particular code or

$$b \equiv \frac{M}{2^N} \tag{7}$$

b can be written in terms of R

$$b = 2^{N(R-1)} \tag{8}$$

As b approaches 1, code compression increases. b is essentially unbounded, since M may be significantly larger than 2^N. □

The decode error (information loss) due to code compression is a random variable that depends on the compression rate and the *a priori* probabilities, therefore, it will be different with different learned vector sets and codons within a set. As the average code utilization for all codons approaches 1, code compression occurs more often and codon decode error is unavoidable.

Let $\overline{x_i}$ be the vector output of the encoder, and the input to the channel, where each element of $\overline{x_i}$ is either a 1 or 0. Let $\overline{y_i}$ be the vector output of the channel, and the input to the decoder, where each element is either a 1 or a 0. The Noisy Channel Coding Theorem is now presented for a general case, where the individual M input codes are to be distinguished. The result is then extended to a CN, where, even though M input codes are used, the CN need only distinguish those codes where it must output a 1 from those where it must output a 0. The theorem is from Gallager (5.6.1)[5]. Random codes are assumed throughout.

Theorem 1: Let a discrete memoryless channel have transition probabilities $P_N(j/k)$ and, for any positive integer N and positive number R, consider the ensemble of (N,R) block codes in which each letter of each code word is independently selected according to the probability assignment $Q(k)$. Then, for each message m, $1 \leq m \leq \lceil e^{NR} \rceil$, and all ρ, $0 \leq \rho \leq 1$, the ensemble average probability of decoding error using maximum-likelyhood decoding satisfies

$$P'_{e,m} \leq \exp \left\{ -N \left[E_0(\rho, Q) - \rho R \right] \right\} \tag{9}$$

where

---

*In the definitions given here and the theorems below, the notation of Gallager[5] is used. Many of the definitions and theorems are also from Gallager.

$$E_0(\rho,Q) = -\ln \sum_{j=0}^{J-1} \left[\sum_{k=0}^{K-1} Q(k)P(j/k)^{\frac{1}{1+\rho}} \right]^{1+\rho} \tag{10}$$

□

These results are now adjusted for our special case.

Theorem 2: For a single CN, the average channel error rate for random code vectors is

$$P_{cdn} \leq 2q(1-q)P'_{e,m} \tag{11}$$

where $q = Q(k) \; \forall \; k$ is the probability of an input vector bit being a 1. □

These results cover a wide range of models. A more easily computable expression can be derived by recognizing some of the restrictions inherent in the CN model. First, assume that all channel code bits are equally likely, that is, $\forall \; k$, $Q(k)=q$, that the error model is the Binary Symmetric Channel (BSC), and that the errors are identically distributed and independent — that is, each bit has the same probability, ϵ, of being in error, independent of the code word and the bit position in the code word.

A simplified version of the above theorem can be derived. Maximizing ρ gives the tightest bounds:

$$P_{cdn} \leq 0.5 \max_{0 \leq \rho \leq 1} P'_e(\rho) \tag{12}$$

where (letting codon input be the block length, $N = f_c$)

$$P'_e(\rho) \leq \exp\left\{ -f_c[E_0(\rho) - \rho R] \right\} \tag{13}$$

The minimum value of this expression is obtained when $\rho=1$ (for $q=0.5$):

$$E_0 = -\log 2 \left[\left(0.5\sqrt{\epsilon} + 0.5\sqrt{1-\epsilon} \right)^2 \right] \tag{14}$$

SINGLE-CODON WITH CODE COMPRESSION

Unfortunately, the implementation complexity of a codon grows exponentially with the size of the codon, which limits its practical size. An alternative is to approximate single codon function of a single CN with many smaller, overlapped codons. The goal is to maintain performance and reduce implementation costs, thus improving the cost/performance of the decoding process. As codons get smaller, the receptive field size becomes smaller relative to the number of CNs in the network. When this happens there is codon compression, or *vector aliasing*, that introduces its own errors into the decoding process due to information loss. Networks can overcome this error by using multiple redundant codons (with overlapping receptive fields) that tend to correct the compression error.

Compression occurs when two code words requiring different decoder output share the same representation (within the receptive field of the codon). The following theorem gives the probability of incorrect codon output with and without compression error.

Theorem 3: For a BSC model where $q=0.5$, the codon receptive field is f_c, the code utilization is b, and the channel bits are selected randomly and independently, the probability of a codon decoding error when $b>1$ is approximately

$$P_{cdn} \leq (1-\epsilon)^{f_c}\bar{p}_c - \left[1 - (1-\epsilon)^{f_c}\right]0.5 \tag{15}$$

where the expected compression error per codon is approximated by

$$\bar{p}_c = 0.5 - \frac{2\sqrt{bq(1-q)}}{b\sqrt{2\pi}} \tag{16}$$

and from equations 13-14, when $b < 1$

$$P_{cdn} \le \exp\left\{-f_c\left[-\log\left[\left[(0.5\sqrt{\epsilon}+0.5\sqrt{1-\epsilon}\,\right]^2\right]-R\right]\right\} \tag{17}$$

Proof is given in Hammerstrom[6] . □

As b grows, $\bar{p}_c$ approaches 0.5 asymptotically. Thus, the performance of a single codon degrades rapidly in the presence of even small amounts of compression.

MULTIPLE CODONS WITH CODE COMPRESSION

The use of multiple small codons is more efficient than a few large codons, but there are some fundamental performance constraints. When a codon is split into two or more smaller codons (and the original receptive field is subdivided accordingly), there are several effects to be considered. First, the error rate of each new codon increases due to a decrease in receptive field size (the codon's block code length). The second effect is that the code utilization, b, will increase for each codon, since the same number of learned vectors is mapped into a smaller receptive field. This change also increases the error rate per codon due to code compression. In fact, as the individual codon receptive fields get smaller, significant code compression occurs. For higher-order input codes, there is an added error that occurs when the order of the individual codons is decreased (since random codes are being assumed, this effect is not considered here). The third effect is the mass action of large numbers of codons. Even though individual codons may be in error, if the majority are correct, then the CN will have correct output. This effect decreases the total error rate.

Assume that each CN has more than one codon, $c > 1$. The union of the receptive fields for these codons is the receptive field for the CN with no no restrictions on the degree of overlap of the various codon receptive fields within or between CNs. For a CN with a large number of codons, the codon overlap will generally be random and uniformly distributed. Also assume that the transmission errors seen by different receptive fields are independent.

Now consider what happens to a codon's compression error rate (ignoring transmission error for the time being) when a codon is replaced by two or more smaller codons covering the same receptive field. This replacement process can continue until there are only 1-codons, which, incidentally, is analogous to most current neural models. For a multiple codon CN, assume that each codon votes a 1 or 0. The summation unit then totals this information and outputs a 1 if the majority of codons vote for a 1, etc.

Theorem 4: The probability of a CN error due to compression error is

$$P_c = \frac{1}{\sqrt{2\pi}} \int\limits_{\frac{c/2-c\bar{p}_c-1/2}{\sqrt{c\bar{p}_c(1-\bar{p}_c)}}}^{\infty} e^{-\frac{1}{2}x^2} \, dy \tag{18}$$

where $\bar{p}_c$ is given in equation 16 and $q = 0.5$.

P_c incorporates the two effects of moving to multiple smaller codons and adding more codons. Using equation 17 gives the total error probability (per bit), P_{CN}:

$$P_{CN} = P_{cdn} + P_c - P_{cdn}P_c \tag{19}$$

Proof is in Hammerstrom[6] . □

For networks that perform association as defined in this paper, the connection weights rapidly approach a single uniform value as the size of the network grows. In information theoretic terms, the information content of those weights approaches zero as the compression increases. Why then do simple non-conjunctive networks (1-codon equivalent) work at all? In the next section I define connectivity cost constraints and show that the answer to the first question is that the general associative structures defined here *do not* scale cost-effectively and more importantly that there are limits to the degree of distribution of information.

CONNECTIVITY COSTS

It is much easier to assess costs if some implementation medium is assumed. I have chosen standard silicon, which is a two dimensional surface where CN's and codons take up surface area according to their receptive field sizes. In addition, there is area devoted to the metal lines that interconnect the CNs. A specific VLSI technology need not be assumed, since the comparisons are relative, thus keeping CNs, codons, and metal in the proper proportions, according to a standard metal width, m_w (which also includes the inter-metal pitch). For the analyses performed here, it is assumed that m_l levels of metal are possible.

In the previous section I established the relationship of network performance, in terms of the transmission error rate, ϵ, and the network capacity, M. In this section I present an implementation cost, which is total silicon area, A. This figure can then be used to derive a cost/performance figure that can be used to compare such factors as codon size and receptive field size. There are two components to the total area: A_{CN}, the area of a CN, and A_{MI}, the area of the metal interconnect between CNs. A_{CN} consists of the silicon area requirements of the codons for all CNs. The metal area for local, intra-CN interconnect is considered to be much smaller than that of the codons themselves and of that of the more global, inter-CN interconnect, and is not considered here. The area per CN is roughly

$$A_{CN} = c f_c m_c \left(\frac{m_w}{m_l}\right)^2 \qquad (20)$$

where m_c is the maximum number of vectors that each codon must distinguish, for $b \geq 1$, $m_c = 2^{f_c}$.

Theorem 5: Assume a rectangular, *unbounded** grid of CNs (all CNs are equi-distant from their four nearest neighbors), where each CN has a bounded receptive field of its n_{CN} nearest CNs, where n_{CN} is the receptive field size for the CN, $n_{CN} = \frac{c f_c}{R}$, where c is the number of codons, and R is the intra-CN redundancy, that is, the ratio of inputs to synapses (e.g., when $R=1$ each CN input is used once at the CN, when $R=2$ each input is used on the average at two sites). The metal area required to support each CN's receptive field is (proof is giving by Hammerstrom[6]):

$$A_{MI} = \left[\frac{n_{CN}{}^3}{16} + \frac{3 n_{CN}{}^{\frac{5}{2}}}{2} + 9 n_{CN}{}^2\right] \left(\frac{m_w}{m_l}\right)^2 \qquad (21)$$

The total area per CN, A, then is

---

*Another implementation strategy is to place all CNs along a diagonal, which gives n^2 area. However, this technique only works for a *bounded* number of CNs and when dendritic computation can be spread over a large area, which limits the range of possible CN implementations. The theorem stated here covers an infinite plane of CNs each with a *bounded* receptive field.

$$A = (A_{MI} + A_{CN}) = O(n_{CN}^3) \qquad (22)$$

☐

Even with the assumption of maximum locality, the total metal interconnect area increases as the *cube* of the per CN receptive field size!

SINGLE CN SIMULATION

What do the bounds tell us about CN connectivity requirements? From simulations, increasing the CN's receptive field size improves the performance (increases capacity), but there is also an increasing cost, which increases faster than the performance! Another observation is that redundancy is quite effective as a means for increasing the effectiveness of a CN with constrained connectivity. (There are some limits to R, since it can reach a point where the intra-CN connectivity approaches that of inter-CN for some situations.) With a fixed n_{CN}, increasing cost-effectiveness (A/m) is possible by increasing both order and redundancy.

In order to verify the derived bounds, I also wrote a discrete event simulation of a CN, where a random set of learned vectors were chosen and the CN's codons were programmed according to the model presented earlier. Learned vectors were chosen randomly and subjected to random noise, ϵ. The CN then attempted to categorize these inputs into two major groups (CN output = 1 and CN output = 0). For the most part the analytic bounds agreed with the simulation, though they tended to be optimistic in slightly underestimating the error. These differences can be easily explained by the simplifying assumptions that were made to make the analytic bounds mathematically tractable.

DISTRIBUTED VS. LOCALIZED

Throughout this paper, it has been tacitly assumed that representations are distributed across a number of CNs, and that any single CN participates in a number of representations. In a *local* representation each CN represents a single concept or feature. It is the distribution of representation that makes the CN's decode job difficult, since it is the cause of the code compression problem.

There has been much debate in the connectionist/neuromodelling community as to the advantages and disadvantages of each approach; the interested reader is referred to Hinton[7], Baum et al. [8], and Ballard[9]. Some of the results derived here are relevant to this debate. As the distribution of representation increases, the compression per CN increases accordingly. It was shown above that the mean error in a codon's response quickly approaches 0.5, independent of the input noise. This result also holds at the CN level. For each individual CN, this error can be offset by adding more codons, but this is expensive and tends to obviate one of the arguments in favor of distributed representations, that is, the multi-use advantage, where fewer CNs are needed because of more complex, redundant encodings. As the degree of distribution increases, the required connectivity and the code compression increases, so the added information that each codon adds to its CN's decoding process goes to zero (equivalent to all weights approaching a uniform value).

SUMMARY AND CONCLUSIONS

In this paper a single CN (node) performance model was developed that was based on Communication Theory. Likewise, an implementation cost model was derived.

The communication model introduced the codon as a higher-order decoding element and showed that for small codons (much less than total CN fan-in, or convergence) code compression, or vector aliasing, within the codon's receptive field is a severe problem for large networks. As code compression increases, the information added by any individual codon to the CN's decoding task rapidly approaches zero.

The cost model showed that for 2-dimensional silicon, the area required for inter-node metal connectivity grows as the cube of a CN's fan-in.

The combination of these two trends indicates that past a certain point, which is highly dependent on the probability structure of the learned vector space, increasing the fan-in of a CN (as is done, for example, when the distribution of representation is increased) yields diminishing returns in terms of total cost-performance. Though the rate of diminishing returns can be decreased by the use of redundant, higher-order connections.

The next step is to apply these techniques to ensembles of nodes (CNs) operating in a competitive learning or feature extraction environment.

REFERENCES

[1]   J. Bailey, "A VLSI Interconnect Structure for Neural Networks," Ph.D. Dissertation, Department of Computer Science/Engineering, OGC. In Preparation.

[2]   V. B. Mountcastle, "An Organizing Principle for Cerebral Function: The Unit Module and the Distributed System," in *The Mindful Brain*, MIT Press, Cambridge, MA, 1977.

[3]   T. Maxwell, C. L. Giles, Y. C. Lee and H. H. Chen, "Transformation Invariance Using High Order Correlations in Neural Net Architectures," *Proceedings International Conf. on Systems, Man, and Cybernetics*, 1986.

[4]   D. Marr, "A Theory for Cerebral Neocortex," *Proc. Roy. Soc. London*, vol. 176(1970), pp. 161-234.

[5]   R. G. Gallager, *Information Theory and Reliable Communication*, John Wiley and Sons, New York, 1968.

[6]   D. Hammerstrom, "A Connectivity Analysis of Recursive, Auto-Associative Connection Networks," Tech. Report CS/E-86-009, Dept. of Computer Science/Engineering, Oregon Graduate Center, Beaverton, Oregon, August 1986.

[7]   G. E. Hinton, "Distributed Representations," Technical Report CMU-CS-84-157, Computer Science Dept., Carnegie-Mellon University, Pittsburgh, PA 15213, 1984.

[8]   E. B. Baum, J. Moody and F. Wilczek, "Internal Representations for Associative Memory," Technical Report NSF-ITP-86-138, Institute for Theoretical Physics, Santa Barbara, CA, 1986.

[9]   D. H. Ballard, "Cortical Connections and Parallel Processing: Structure and Function," Technical Report 133, Computer Science Department, Rochester, NY, January 1985.

Minkowski-r Back-Propagation: Learning in Connectionist Models with Non-Euclidian Error Signals

Stephen José Hanson and David J. Burr

Bell Communications Research
Morristown, New Jersey 07960

Abstract

Many connectionist learning models are implemented using a gradient descent in a least squares error function of the output and teacher signal. The present model generalizes, in particular, back-propagation [1] by using Minkowski-r power metrics. For small $r's$ a "city-block" error metric is approximated and for large $r's$ the "maximum" or "supremum" metric is approached, while for $r=2$ the standard back-propagation model results. An implementation of Minkowski-r back-propagation is described, and several experiments are done which show that different values of r may be desirable for various purposes. Different r values may be appropriate for the reduction of the effects of outliers (noise), modeling the input space with more compact clusters, or modeling the statistics of a particular domain more naturally or in a way that may be more perceptually or psychologically meaningful (e.g. speech or vision).

1. Introduction

The recent resurgence of connectionist models can be traced to their ability to do complex modeling of an input domain. It can be shown that neural-like networks containing a single hidden layer of non-linear activation units can learn to do a *piece-wise linear* partitioning of a feature space [2]. One result of such a partitioning is a complex gradient surface on which decisions about new input stimuli will be made. The generalization, categorization and clustering properties of the network are therefore determined by this mapping of input stimuli to this gradient surface in the output space. This gradient surface is a function of the conditional probability distributions of the output vectors given the input feature vectors as well as a function of the error relating the teacher signal and output.

Presently many of the models have been implemented using least squares error. In this paper we describe a new model of gradient descent back-propagation [1] using Minkowski-r power error metrics. For small r's a "city-block" error measure (r=1) is approximated and for larger r's a "maximum" or supremum error measure is approached, while the standard case of Euclidian back-propagation is a special case with r=2. First we derive the general case and then discuss some of the implications of varying the power in the general metric.

2. Derivation of Minkowski-r Back-propagation

The standard back-propagation is derived by minimizing least squares error as a function of connection weights within a completely connected layered network. The error for the Euclidian case is (for a single input-output pair),

$$E = \frac{1}{2} \sum_i (y_i - \hat{y}_i)^2, \tag{1}$$

where y is the activation of a unit and $\hat{y}$ represents an independent teacher signal. The activation of a unit (y) is typically computed by normalizing the input from other units (x) over the interval $(0,1)$ while compressing the high and low end of this range. A common function used for this normalization is the logistic,

$$y_i = \frac{1}{1 + e^{-x_i}} \tag{2}$$

The input to a unit (x) is found by summing products of the weights and corresponding activations from other units,

$$x_i = \sum_h y_h w_{hi}, \tag{3}$$

where y_h represents units in the fan in of unit i and w_{hi} represents the strength of the connection between unit i and unit h.

A gradient for the Euclidian or standard back-propagation case could be found by finding the partial of the error with respect to each weight, and can be expressed in this three term differential,

$$\frac{\partial E}{\partial w_{hi}} = \frac{\partial E}{\partial y_i} \frac{\partial y_i}{\partial x_i} \frac{\partial x_i}{\partial w_{hi}}$$

(4)

which from the equations before turns out to be,

$$\frac{\partial E}{\partial w_{hi}} = (y_i - \hat{y}_i)y_i(1-y_i)y_h$$

(5)

Generalizing the error for Minkowski-r power metrics (see Figure 1 for the family of curves),

$$E = \frac{1}{r} \sum_i |(y_i - \hat{y}_i)|^r$$

(6)

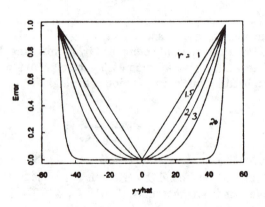

Figure 1: Minkowski-r Family

Using equations 2-4 above with equation 6 we can easily find an expression for the gradient in the general Minkowski-r case,

$$\frac{\partial E}{\partial w_{hi}} = (|y_i - \hat{y}_i|)^{r-1}y_i(1-y_i)y_h sgn(y_i - \hat{y}_i)$$

(7)

This gradient is used in the weight update rule proposed by Rumelhart, Hinton and Williams [1],

$$w_{hi}(n+1) = \alpha \frac{\partial E}{\partial w_{hi}} + w_{hi}(n) \qquad (8)$$

Since the gradient computed for the hidden layer is a function of the gradient for the output, the hidden layer weight updating proceeds in the same way as in the Euclidian case [1], simply substituting this new Minkowski-r gradient.

It is also possible to define a gradient over r such that a minimum in error would be sought. Such a gradient was suggested by White [3, see also 4] for maximum likelihood estimation of r, and can be shown to be,

$$\frac{d\log(E)}{dr} = (1-1/r)(1/r) + (1/r)^2 \log(r) + (1/r)^2 \psi(1/r) + (1/r)^2 |y_i - \hat{y}_i|$$

$$-(1/r)(|y_i - \hat{y}_i|)^r \log(|y_i - \hat{y}_i|) \qquad (9)$$

An approximation of this gradient (using the last term of equation 9) has been implemented and investigated for simple problems and shown to be fairly robust in recovering similar r values. However, it is important that the r update rule changes slower than the weight update rule. In the simulations we ran r was changed once for every 10 times the weight values were changed. This rate might be expected to vary with the problem and rate of convergence. Local minima may be expected in larger problems while seeking an optimal r. It may be more informative for the moment to examine different classes of problems with fixed r and consider the specific rationale for those classes of problems.

3. Variations in r

Various r values may be useful for various aspects of representing information in the feature domain. Changing r basically results in a reweighting of errors from output bits[1]. Small r's give less weight for large deviations and tend to reduce the influence of outlier points in the feature space during learning. In fact, it can be shown that if the distributions of feature vectors are non-gaussian, then the r=2 case

---

1. It is possible to entertain r values that are negative, which would give largest weight to small errors close to zero and smallest weight to very large errors. Values of r less than 1 generally are non-metric, i.e. they violate at least one of the metric axioms. For example, r<0 violates the triangle inequality. For some problems this may make sense and the need for a metric error weighting may be unnecessary. These issues are not explored in this paper.

will *not* be a maximum likelihood estimator of the weights [5]. The city block case, r=1, in fact, arises if the underlying conditional probability distributions are Laplace [5]. More generally, r's less than two will tend to model non-gaussian distributions where the tails of the distributions are more pronounced than in the gaussian. Better estimators can be shown to exist for general noise reduction and have been studied in the area of *robust estimation* procedures [5] of which the Minkowski-r metric is only one possible case to consider.

r<2. It is generally recommended that r=1.5 may be optimal for many noise reduction problems [6]. However, noise reduction may also be expected to vary with the problem and nature of the noise. One example we have looked at involves the recovery of an arbitrary 3 dimensional smooth surface as shown in Figure 2a, after the addition of random noise. This surface was generated from a gaussian curve in the 2 dimensions. Uniform random noise equal to the width (standard deviation) of the surface shape was added point-wise to the surface producing the noise plus surface shape shown in Figure 2b.

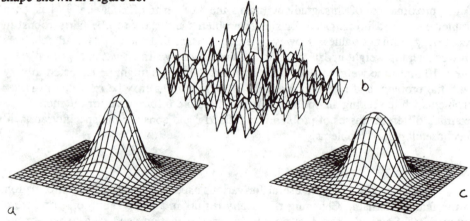

Figure 2: Shape surface (2a), Shape plus noise surface (2b) and recovered Shape surface (2c)

The shape in Figure 2a was used as target points for Minkowski-r back-propagation[2] and recovered with some distortion of the slope of the shape near the peak of the

---

2. All simulation runs, unless otherwise stated, used the same learning rate (.05) and smoothing value (.9) and stopping criterion defined in terms of absolute mean deviation. The number of iterations to meet the stopping criterion varied considerably as r was changed (see below).

surface (see Figure 2c). Next the noise plus shape surface was used as target points for the learning procedure with r=2. The shape shown in Figure 3a was recovered, however, with considerable distortion iaround the base and peak. The value of r was reduced to 1.5 (Figure 3b) and then finally to 1.2 (Figure 3c) before shape distortions were eliminated. Although, the major properties of the shape of the surface were recovered, the scale seems distorted (however, easily restored with renormalization into the 0,1 range).

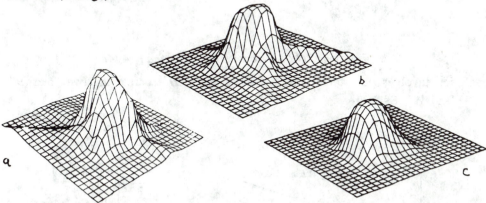

Figure 3: Shape surface recovered with r=2 (3a), r=1.5 (3b) and r=1.2 (3c)

$r>2$. Large r's tend to weight large deviations. When noise is not possible in the feature space (as in an arbitrary boolean problem) or where the token clusters are compact and isolated then simpler (in the sense of the number and placement of partition planes) generalization surfaces may be created with larger r values. For example, in the simple XOR problem, the main effect of increasing r is to pull the decision boundaries closer into the non-zero targets (compare high activation regions in Figure 4a and 4b).

In this particular problem clearly such compression of the target regions does not constitute simpler decision surfaces. However, if more hidden units are used than are needed for pattern class separation, then increasing r during training will tend to reduce the number of cuts in the space to the minimum needed. This seems to be primarily due to the sensitivity of the hyper-plane placement in the feature space to the geometry of the targets.

A more complex case illustrating the same idea comes from an example suggested by Minsky & Papert [7] called "the mesh". This type of pattern recognition problem is also, like XOR, a non-linearly separable problem. An optimal

Figure 4: XOR solved with r=2 (4a) and r=4 (4b)

solution involves only three cuts in feature space to separate the two "meshed" clusters (see Figure 5a).

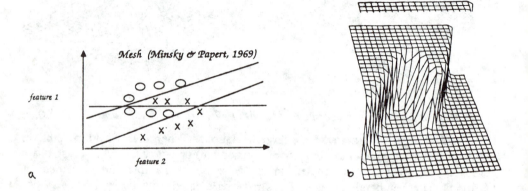

Figure 5: Mesh problem with minimum cut solution (5a) and Performance Surface(5b)

Typical solutions for r=2 in this case tend to use a large number of hidden units to separate the two sets of exemplars (see Figure 5b for a performance surface). For example, in Figure 6a notice that a typical (based on several runs) Euclidian back-prop starting with 16 hidden units has found a solution involving five decision boundaries (lines shown in the plane also representing hidden units) while the r=3 case used primarily three decision boundaries and placed a number of other boundaries redundantly near the center of the meshed region (see Figure 6b) where there is maximum uncertainty about the cluster identification.

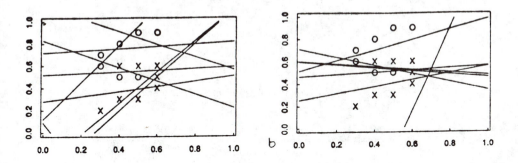

Figure 6: Mesh solved with r=2 (6a) and r=3 (6b)

Speech Recognition. A final case in which large r's may be appropriate is data that has been previously processed with a transformation that produced compact regions requiring separation in the feature space. One example we have looked at involves spoken digit recognition. The first 10 cepstral coefficients of spoken digits ("one" through "ten") were used for input to a network. In this case an advantage is shown for larger r's with smaller training set sizes. Shown in Figure 7 are transfer data for 50 spoken digits replicated in ten different runs per point (bars show standard error of the mean). Transfer shows a training set size effect for both r=2 and r=3, however for the larger r value at smaller training set sizes (10 and 20) note that transfer is enhanced.

We speculate that this may be due to the larger r backprop creating discrimination regions that are better able to capture the compactness of the clusters inherent in a small number of training points.

4. Convergence Properties

It should be generally noted that as r increases, convergence time tends to grow roughly linearly (although this may be problem dependent). Consequently, decreasing r can significantly improve convergence, without much change to the nature of solution. Further, if noise is present decreasing r may reduce it dramatically. Note finally that the gradient for the Minkowski-r back-propagation is nonlinear and therefore more complex for implementing learning procedures.

356

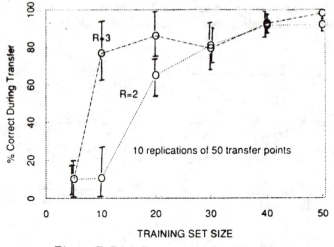

Figure 7: Digit Recognition Set Size Effect

5. Summary and Conclusion

A new procedure which is a variation on the Back-propagation algorithm is derived and simulated in a number of different problem domains. Noise in the target domain may be reduced by using power values less than 2 and the sensitivity of partition planes to the geometry of the problem may be increased with increasing power values. Other types of objective functions should be explored for their potential consequences on network resources and ensuing pattern recognition capabilities.

References

1. Rumelhart D. E., Hinton G. E., Williams R., Learning Internal Representations by error propagation. Nature, 1986.

2. Burr D. J. and Hanson S. J., Knowledge Representation in Connectionist Networks, Bellcore, Technical Report,

3. White, H. Personal Communication, 1987.

4. White, H. Some Asymptotic Results for Learning in Single Hidden Layer Feedforward Network Models, Unpublished Manuscript, 1987.

5. Mosteller, F. & Tukey, J. Robust Estimation Procedures, Addison Wesley, 1980.

6. Tukey, J. Personal Communication, 1987.

7. Minsky, M. & Papert, S., Perceptrons: An Introduction to Computational Geometry, MIT Press, 1969.

LEARNING REPRESENTATIONS BY RECIRCULATION

Geoffrey E. Hinton
Computer Science and Psychology Departments, University of Toronto,
Toronto M5S 1A4, Canada

James L. McClelland
Psychology and Computer Science Departments, Carnegie-Mellon University,
Pittsburgh, PA 15213

ABSTRACT

We describe a new learning procedure for networks that contain groups of non-linear units arranged in a closed loop. The aim of the learning is to discover codes that allow the activity vectors in a "visible" group to be represented by activity vectors in a "hidden" group. One way to test whether a code is an accurate representation is to try to reconstruct the visible vector from the hidden vector. The difference between the original and the reconstructed visible vectors is called the reconstruction error, and the learning procedure aims to minimize this error. The learning procedure has two passes. On the first pass, the original visible vector is passed around the loop, and on the second pass an average of the original vector and the reconstructed vector is passed around the loop. The learning procedure changes each weight by an amount proportional to the product of the "presynaptic" activity and the *difference* in the post-synaptic activity on the two passes. This procedure is much simpler to implement than methods like back-propagation. Simulations in simple networks show that it usually converges rapidly on a good set of codes, and analysis shows that in certain restricted cases it performs gradient descent in the squared reconstruction error.

INTRODUCTION

Supervised gradient-descent learning procedures such as back-propagation[1] have been shown to construct interesting internal representations in "hidden" units that are not part of the input or output of a connectionist network. One criticism of back-propagation is that it requires a teacher to specify the desired output vectors. It is possible to dispense with the teacher in the case of "encoder" networks[2] in which the desired output vector is identical with the input vector (see Fig. 1). The purpose of an encoder network is to learn good "codes" in the intermediate, hidden units. If for, example, there are less hidden units than input units, an encoder network will perform data-compression[3]. It is also possible to introduce other kinds of constraints on the hidden units, so we can view an encoder network as a way of ensuring that the input can be reconstructed from the activity in the hidden units whilst also making

This research was supported by contract N00014-86-K-00167 from the Office of Naval Research and a grant from the Canadian National Science and Engineering Research Council. Geoffrey Hinton is a fellow of the Canadian Institute for Advanced Research. We thank Mike Franzini, Conrad Galland and Geoffrey Goodhill for helpful discussions and help with the simulations.

the hidden units satisfy some other constraint.

A second criticism of back-propagation is that it is neurally implausible (and hard to implement in hardware) because it requires all the connections to be used backwards and it requires the units to use different input-output functions for the forward and backward passes. Recirculation is designed to overcome this second criticism in the special case of encoder networks.

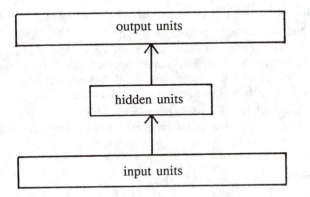

Fig. 1. A diagram of a three layer encoder network that learns good codes using back-propagation. On the forward pass, activity flows from the input units in the bottom layer to the output units in the top layer. On the backward pass, error-derivatives flow from the top layer to the bottom layer.

Instead of using a separate group of units for the input and output we use the very same group of "visible" units, so the input vector is the initial state of this group and the output vector is the state after information has passed around the loop. The difference between the activity of a visible unit before and after sending activity around the loop is the derivative of the squared reconstruction error. So, if the visible units are linear, we can perform gradient descent in the squared error by changing each of a visible unit's incoming weights by an amount proportional to the product of this difference and the activity of the hidden unit from which the connection emanates. So learning the weights from the hidden units to the output units is simple. The harder problem is to learn the weights on connections coming into hidden units because there is no direct specification of the desired states of these units. Back-propagation solves this problem by back-propagating error-derivatives from the output units to generate error-derivatives for the hidden units. Recirculation solves the problem in a quite different way that is easier to implement but much harder to analyse.

THE RECIRCULATION PROCEDURE

We introduce the recirculation procedure by considering a very simple architecture in which there is just one group of hidden units. Each visible unit has a directed connection to every hidden unit, and each hidden unit has a directed connection to every visible unit. The total input received by a unit is

$$x_j = \sum_i y_i w_{ji} - \theta_j \tag{1}$$

where y_i is the state of the ith unit, w_{ji} is the weight on the connection from the ith to the jth unit and θ_j is the threshold of the jth unit. The threshold term can be eliminated by giving every unit an extra input connection whose activity level is fixed at 1. The weight on this special connection is the negative of the threshold, and it can be learned in just the same way as the other weights. This method of implementing thresholds will be assumed throughout the paper.

The functions relating inputs to outputs of visible and hidden units are smooth monotonic functions with bounded derivatives. For hidden units we use the logistic function:

$$y_j = \sigma(x_j) = \frac{1}{1+e^{-x_j}} \tag{2}$$

Other smooth monotonic functions would serve as well. For visible units, our mathematical analysis focuses on the linear case in which the output equals the total input, though in simulations we use the logistic function.

We have already given a verbal description of the learning rule for the hidden-to-visible connections. The weight, w_{ij}, from the jth hidden unit to the ith visible unit is changed as follows:

$$\Delta w_{ij} = \varepsilon y_j(1) [y_i(0) - y_i(2)] \tag{3}$$

where $y_i(0)$ is the state of the ith visible unit at time 0 and $y_i(2)$ is its state at time 2 after activity has passed around the loop once. The rule for the visible-to-hidden connections is identical:

$$\Delta w_{ji} = \varepsilon y_i(2) [y_j(1) - y_j(3)] \tag{4}$$

where $y_j(1)$ is the state of the jth hidden unit at time 1 (on the first pass around the loop) and $y_j(3)$ is its state at time 3 (on the second pass around the loop). Fig. 2 shows the network exploded in time.

In general, this rule for changing the visible-to-hidden connections does not perform steepest descent in the squared reconstruction error, so it behaves differently from back-propagation. This raises two issues: Under what conditions does it work, and under what conditions does it approximate steepest descent?

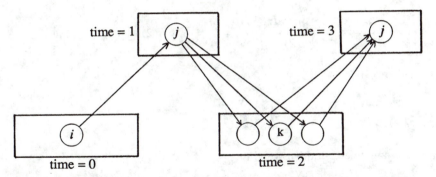

Fig. 2. A diagram showing the states of the visible and hidden units exploded in time. The visible units are at the bottom and the hidden units are at the top. Time goes from left to right.

CONDITIONS UNDER WHICH RECIRCULATION APPROXIMATES GRADIENT DESCENT

For the simple architecture shown in Fig. 2, the recirculation learning procedure changes the visible-to-hidden weights in the direction of steepest descent in the squared reconstruction error provided the following conditions hold:

1. The visible units are linear.
2. The weights are symmetrical (i.e. $w_{ji} = w_{ij}$ for all i,j).
3. The visible units have high regression.

"Regression" means that, after one pass around the loop, instead of setting the activity of a visible unit, i, to be equal to its current total input, $x_i(2)$, as determined by Eq 1, we set its activity to be

$$y_i(2) = \lambda y_i(0) + (1-\lambda) x_i(2) \tag{5}$$

where the regression, λ, is close to 1. Using high regression ensures that the visible units only change state slightly so that when the new visible vector is sent around the loop again on the second pass, it has very similar effects to the first pass. In order to make the learning rule for the hidden units as similar as possible to the rule for the visible units, we also use regression in computing the activity of the hidden units on the second pass

$$y_j(3) = \lambda y_j(1) + (1-\lambda) \sigma(x_j(3)) \tag{6}$$

For a given input vector, the squared reconstruction error, E, is

$$E = \frac{1}{2}\sum_k [y_k(2) - y_k(0)]^2$$

For a hidden unit, j,

$$\frac{\partial E}{\partial y_j(1)} = \sum_k \frac{\partial E}{\partial y_k(2)} \frac{dy_k(2)}{dx_k(2)} \frac{\partial x_k(2)}{\partial y_j(1)} = \sum_k [y_k(2) - y_k(0)] \, y_k'(2) \, w_{kj} \qquad (7)$$

where

$$y_k'(2) = \frac{dy_k(2)}{dx_k(2)}$$

For a visible-to-hidden weight w_{ji}

$$\frac{\partial E}{\partial w_{ji}} = y_j'(1) \, y_i(0) \, \frac{\partial E}{\partial y_j(1)}$$

So, using Eq 7 and the assumption that $w_{kj} = w_{jk}$ for all k,j

$$\frac{\partial E}{\partial w_{ji}} = y_j'(1) \, y_i(0) \, [\sum_k y_k(2) \, y_k'(2) \, w_{jk} - \sum_k y_k(0) \, y_k'(2) \, w_{jk}]$$

The assumption that the visible units are linear (with a gradient of 1) means that for all k, $y_k'(2) = 1$. So using Eq 1 we have

$$\frac{\partial E}{\partial w_{ji}} = y_j'(1) \, y_i(0) \, [x_j(3) - x_j(1)] \qquad (8)$$

Now, with sufficiently high regression, we can assume that the states of units only change slightly with time so that

$$y_j'(1) [x_j(3) - x_j(1)] \approx \sigma(x_j(3)) - \sigma(x_j(1)) = \frac{1}{(1 - \lambda)} [y_j(3) - y_j(1)]$$

and $\quad y_i(0) \approx y_i(2)$

So by substituting in Eq 8 we get

$$\frac{\partial E}{\partial w_{ji}} \approx \frac{1}{(1 - \lambda)} y_i(2) [y_j(3) - y_j(1)] \qquad (9)$$

An interesting property of Eq 9 is that it does not contain a term for the gradient of the input-output function of unit j so recirculation learning can be applied even when unit j uses an *unknown* non-linearity. To do back-propagation it is necessary to know the gradient of the non-linearity, but recirculation *measures* the gradient by measuring the effect of a small difference in input, so the term $y_j(3) - y_j(1)$ implicitly contains the gradient.

A SIMULATION OF RECIRCULATION

From a biological standpoint, the symmetry requirement that $w_{ij} = w_{ji}$ is unrealistic unless it can be shown that this symmetry of the weights can be learned. To investigate what would happen if symmetry was not enforced (and if the visible units used the same non-linearity as the hidden units), we applied the recirculation learning procedure to a network with 4 visible units and 2 hidden units. The visible vectors were 1000, 0100, 0010 and 0001, so the 2 hidden units had to learn 4 different codes to represent these four visible vectors. All the weights and biases in the network were started at small random values uniformly distributed in the range −0.5 to +0.5. We used regression in the hidden units, even though this is not strictly necessary, but we ignored the term $1/(1-\lambda)$ in Eq 9.

Using an ε of 20 and a λ of 0.75 for both the visible and the hidden units, the network learned to produce a reconstruction error of less than 0.1 on every unit in an average of 48 weight updates (with a maximum of 202 in 100 simulations). Each weight update was performed after trying all four training cases and the change was the sum of the four changes prescribed by Eq 3 or 4 as appropriate. The final reconstruction error was measured using a regression of 0, even though high regression was used during the learning. The learning speed is comparable with back-propagation, though a precise comparison is hard because the optimal values of ε are different in the two cases. Also, the fact that we ignored the term $1/(1-\lambda)$ when modifying the visible-to-hidden weights means that recirculation tends to change the visible-to-hidden weights more slowly than the hidden-to-visible weights, and this would also help back-propagation.

It is not immediately obvious why the recirculation learning procedure works when the weights are not constrained to be symmetrical, so we compared the weight changes prescribed by the recirculation procedure with the weight changes that would cause steepest descent in the sum squared reconstruction error (i.e. the weight changes prescribed by back-propagation). As expected, recirculation and back-propagation agree on the weight changes for the hidden-to-visible connections, even though the gradient of the logistic function is not taken into account in weight adjustments under recirculation. (Conrad Galland has observed that this agreement is only slightly affected by using visible units that have the non-linear input-output function shown in Eq 2 because at any stage of the learning, all the visible units tend to have similar slopes for their input-output functions, so the non-linearity scales all the weight changes by approximately the same amount.)

For the visible-to-hidden connections, recirculation initially prescribes weight changes that are only randomly related to the direction of steepest descent, so these changes do not help to improve the performance of the system. As the learning proceeds, however, these changes come to agree with the direction of steepest descent. The crucial observation is that this agreement occurs *after* the hidden-to-visible weights have changed in such a way that they are approximately aligned (symmetrical up to a constant factor) with the visible-to-hidden weights. So it appears that changing the hidden-to-visible weights in the direction of steepest descent creates the conditions that are necessary for the recirculation procedure to cause changes in the visible-to-hidden weights that follow the direction of steepest descent.

It is not hard to see why this happens if we start with random, zero-mean visible-to-hidden weights. If the visible-to-hidden weight w_{ji} is positive, hidden unit j will tend to have a higher than average activity level when the ith visible unit has a higher than average activity. So y_j will tend to be higher than average when the reconstructed value of y_i should be higher than average -- i.e. when the term $[y_i(0)-y_i(2)]$ in Eq 3 is positive. It will also be lower than average when this term is negative. These relationships will be reversed if w_{ji} is negative, so w_{ij} will grow faster when w_{ji} is positive than it will when w_{ji} is negative. Smolensky[4] presents a mathematical analysis that shows why a similar learning procedure creates symmetrical weights in a purely linear system. Williams[5] also analyses a related learning rule for linear systems which he calls the "symmetric error correction" procedure and he shows that it performs principle components analysis. In our simulations of recirculation, the visible-to-hidden weights become aligned with the corresponding hidden-to-visible weights, though the hidden-to-visible weights are generally of larger magnitude.

A PICTURE OF RECIRCULATION

To gain more insight into the conditions under which recirculation learning produces the appropriate changes in the visible-to-hidden weights, we introduce the pictorial representation shown in Fig. 3. The initial visible vector, A, is mapped into the reconstructed vector, C, so the error vector is AC. Using high regression, the visible vector that is sent around the loop on the second pass is P, where the difference vector AP is a small fraction of the error vector AC. If the regression is sufficiently high and all the non-linearities in the system have bounded derivatives and the weights have bounded magnitudes, the difference vectors AP, BQ, and CR will be very small and we can assume that, to first order, the system behaves linearly in these difference vectors. If, for example, we moved P so as to double the length of AP we would also double the length of BQ and CR.

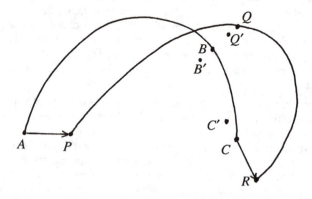

Fig. 3. A diagram showing some vectors (A, P) over the visible units, their "hidden" images (B, Q) over the hidden units, and their "visible" images (C, R) over the visible units. The vectors B' and C' are the hidden and visible images of A after the visible-to-hidden weights have been changed by the learning procedure.

Suppose we change the visible-to-hidden weights in the manner prescribed by Eq 4, using a very small value of ε. Let Q' be the hidden image of P (i.e. the image of P in the hidden units) after the weight changes. To first order, Q' will lie between B and Q on the line BQ. This follows from the observation that Eq 4 has the effect of moving each $y_j(3)$ towards $y_j(1)$ by an amount proportional to their difference. Since B is close to Q, a weight change that moves the hidden image of P from Q to Q' will move the hidden image of A from B to B', where B' lies on the extension of the line BQ as shown in Fig. 3. If the hidden-to-visible weights are not changed, the visible image of A will move from C to C', where C' lies on the extension of the line CR as shown in Fig. 3. So the visible-to-hidden weight changes will reduce the squared reconstruction error provided the vector CR is approximately parallel to the vector AP.

But why should we expect the vector CR to be aligned with the vector AP? In general we should not, except when the visible-to-hidden and hidden-to-visible weights are approximately aligned. The learning in the hidden-to-visible connections has a tendency to cause this alignment. In addition, it is easy to modify the recirculation learning procedure so as to increase the tendency for the learning in the hidden-to-visible connections to cause alignment. Eq 3 has the effect of moving the visible image of A closer to A by an amount proportional to the magnitude of the error vector AC. If we apply the same rule on the next pass around the loop, we move the visible image of P closer to P by an amount proportional to the magnitude of PR. If the vector CR is anti-aligned with the vector AP, the magnitude of AC will exceed the magnitude of PR, so the result of these two movements will be to improve the alignment between AP and CR. We have not yet tested this modified procedure through simulations, however.

This is only an informal argument and much work remains to be done in establishing the precise conditions under which the recirculation learning procedure approximates steepest descent. The informal argument applies equally well to systems that contain longer loops which have several groups of hidden units arranged in series. At each stage in the loop, the same learning procedure can be applied, and the weight changes will approximate gradient descent provided the difference of the two visible vectors that are sent around the loop aligns with the difference of their images. We have not yet done enough simulations to develop a clear picture of the conditions under which the changes in the hidden-to-visible weights produce the required alignment.

USING A HIERARCHY OF CLOSED LOOPS

Instead of using a single loop that contains many hidden layers in series, it is possible to use a more modular system. Each module consists of one "visible" group and one "hidden" group connected in a closed loop, but the visible group for one module is actually composed of the hidden groups of several lower level modules, as shown in Fig. 4. Since the same learning rule is used for both visible and hidden units, there is no problem in applying it to systems in which some units are the visible units of one module and the hidden units of another. Ballard[6] has experimented with back-propagation in this kind of system, and we have run some simulations of recirculation using the architecture shown in Fig. 4. The network learned to encode a set of vectors specified over the bottom layer. After learning, each of the vectors became an attractor and the network was capable of completing a partial vector, even though this involved passing information through several layers.

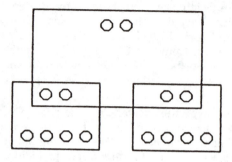

Fig 4. A network in which the hidden units of the bottom two modules are the visible units of the top module.

CONCLUSION

We have described a simple learning procedure that is capable of forming representations in non-linear hidden units whose input-output functions have bounded derivatives. The procedure is easy to implement in hardware, even if the non-linearity is unknown. Given some strong assumptions, the procedure performs gradient descent in the reconstruction error. If the symmetry assumption is violated, the learning procedure still works because the changes in the hidden-to-visible weights produce symmetry. If the assumption about the linearity of the visible units is violated, the procedure still works in the cases we have simulated. For the general case of a loop with many non-linear stages, we have an informal picture of a condition that must hold for the procedure to approximate gradient descent, but we do not have a formal analysis, and we do not have sufficient experience with simulations to give an empirical description of the general conditions under which the learning procedure works.

REFERENCES

1. D. E. Rumelhart, G. E. Hinton and R. J. Williams, *Nature* **323**, 533-536 (1986).

2. D. H. Ackley, G. E. Hinton and T. J. Sejnowski, *Cognitive Science* **9**, 147-169 (1985).

3. G. Cottrell, J. L. Elman and D. Zipser, Proc. Cognitive Science Society, Seattle, WA (1987).

4. P. Smolensky, Technical Report CU-CS-355-87, University of Colorado at Boulder (1986).

5. R. J. Williams, Technical Report 8501, Institute of Cognitive Science, University of California, San Diego (1985).

6. D. H. Ballard, Proc. American Association for Artificial Intelligence, Seattle, WA (1987).

SCHEMA FOR MOTOR CONTROL
UTILIZING A NETWORK MODEL OF THE CEREBELLUM

James C. Houk, Ph.D.
Northwestern University Medical School, Chicago, Illinois
60201

ABSTRACT

This paper outlines a schema for movement control based on two stages of signal processing. The higher stage is a neural network model that treats the cerebellum as an array of adjustable motor pattern generators. This network uses sensory input to preset and to trigger elemental pattern generators and to evaluate their performance. The actual patterned outputs, however, are produced by intrinsic circuitry that includes recurrent loops and is thus capable of self-sustained activity. These patterned outputs are sent as motor commands to local feedback systems called motor servos. The latter control the forces and lengths of individual muscles. Overall control is thus achieved in two stages: (1) an adaptive cerebellar network generates an array of feedforward motor commands and (2) a set of local feedback systems translates these commands into actual movements.

INTRODUCTION

There is considerable evidence that the cerebellum is involved in the adaptive control of movement[1], although the manner in which this control is achieved is not well understood. As a means of probing these cerebellar mechanisms, my colleagues and I have been conducting microelectrode studies of the neural messages that flow through the intermediate division of the cerebellum and onward to limb muscles via the rubrospinal tract. We regard this cerebellorubrospinal pathway as a useful model system for studying general problems of sensorimotor integration and adaptive brain function. A summary of our findings has been published as a book chapter[2].

On the basis of these and other neurophysiological results, I recently hypothesized that the cerebellum functions as an array of adjustable motor pattern generators[3]. The outputs from these pattern generators are assumed to function as motor commands, i.e., as neural control signals that are sent to lower-level motor systems where they produce movements. According to this hypothesis, the cerebellum uses its extensive sensory input to preset the pattern generators, to trigger them to initiate the production of patterned outputs and to evaluate the success or failure of the patterns in controlling a motor behavior. However, sensory input appears not to play a major role in shaping the waveforms of the patterned outputs. Instead, these waveforms seem to be produced by intrinsic circuitry.

The initial purpose of the present paper is to provide some ideas for a neural network model of the cerebellum that might be capable of accounting for adjustable motor pattern generation. Several previous authors have described network models of the cerebellum that, like the present model, are based on the neuroanatomical organization of this brain structure[4,5,6]. While the present model borrows heavily from these previous models, it has some additional features that may explain the unique manner in which the cerebellum processes sensory input to produce motor commands. A second purpose of this paper is to outline how this network model fits within a broader schema for motor control that I have been developing over the past several years[3,7]. Before presenting these ideas, let me first review some basic physiology and anatomy of the cerebellum[1].

SIGNALS AND CIRCUITS IN THE CEREBELLUM

There are three main categories of input fibers to the cerebellum, called mossy fibers, climbing fibers and noradrenergic fibers. As illustrated in Fig. 1, the mossy fiber input shows considerable fan-out via granule cells and parallel fibers. The parallel fibers in turn are arranged to provide a high degree of fan-in to individual Purkinje cells (P). These P cells are the sole output elements of the cortical portion of the cerebellum. Via the parallel fiber input, each P cell is exposed to approximately 200,000 potential messages. In marked contrast, the climbing fiber input to P cells is highly focused. Each climbing fiber branches to only 10 P cells, and each cell receives input from only one climbing fiber. Although less is known about input via noradrenergic fibers, it appears to be diffuse and even more divergent than the mossy fiber input.

Mossy fibers originate from several brain sites transmitting a diversity of information about the external world and the internal state of the body. Some mossy fiber inputs are clearly sensory. They come fairly directly from cutaneous, muscle or vestibular receptors. Others are routed via the cerebral cortex where they represent highly processed visual, auditory or somatosensory information. Yet another category of mossy fiber transmits information about central motor commands (Fig. 1 shows one such pathway, from collaterals of the rubrospinal tract relayed

through the lateral reticular nucleus (L)). The discharge
rates of mossy fibers are modulated over a wide dynamic
range which permits them to transmit detailed parametric
information about the state of the body and its external
environment.

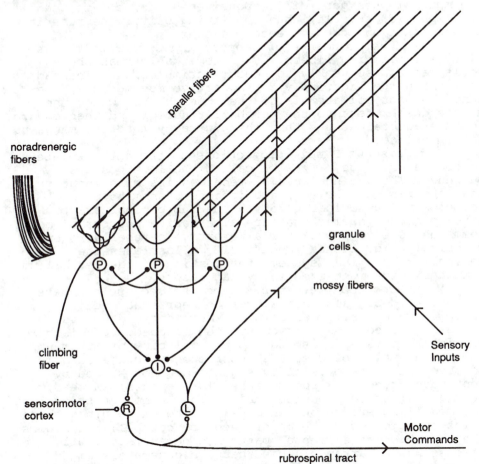

Figure 1: Pathways through the cerebellum. This diagram,
which highlights the cerebellorubrospinal system, also
constitutes a circuit diagram for the model of an
elemental pattern generator.

The sole source of climbing fibers is from cells
located in the inferior olivary nucleus. Olivary neurons
are selectively sensitive to sensory events. These cells
have atypical electrical properties which limit their
discharge to rates less than 10 impulses/sec, and usual
rates are closer to 1 impulse/sec. As a consequence,

individual climbing fibers transmit very little parametric
information about the intensity and duration of a stimulus;
instead, they appear to be specialized to detect simply the
occurrences of sensory events. There are also motor inputs
to this pathway, but they appear to be strictly inhibitory.
The motor inputs gate off responsiveness to self-induced
(or expected) stimuli, thus converting olivary neurons into
detectors of unexpected sensory events.

Given the abundance of sensory input to P cells via
mossy and climbing fibers, it is remarkable that these
cells respond so weakly to sensory stimulation. Instead,
they discharge vigorously during active movements. P cells
send abundant collaterals to their neighbors, while their
main axons project to the cerebellar nuclei and then onward
to several brain sites that in turn relay motor commands to
the spinal cord.

Fig. 1 shows P cell projections to the intermediate
cerebellar nucleus (I), also called the interpositus
nucleus. The red nucleus (R) receives its main input from
the interpositus nucleus, and it then transmits motor
commands to the spinal cord via the rubrospinal tract.
Other premotor nuclei that are alternative sources of motor
commands receive input from alternative cerebellar output
circuits. Fig. 1 thus specifically illustrates the
cerebellorubrospinal system, the portion of the cerebellum
that has been emphasized in my laboratory.

Microelectrode recordings from the red nucleus have
demonstrated signals that appear to represent detailed
velocity commands for distal limb movements. Bursts of
discharge precede each movement, the frequency of discharge
within the burst corresponds to the velocity of movement,
and the duration of the burst corresponds to the duration
of movement. These velocity signals are not shaped by
continuous feedback from peripheral receptors; instead,
they appear to be produced centrally. An important goal of
the modelling effort outlined here is to explain how these
velocity commands might be produced by cerebellar circuits
that function as elemental pattern generators. I will then
discuss how an array of these pattern generators might
serve well in an overall schema of motor control.

ELEMENTAL PATTERN GENERATORS

The motivation for proposing pattern generators rather
than more conventional network designs derives from the
experimental observation that motor commands, once initiated, are not affected, or are only minimally affected, by
alterations in sensory input. This observation indicates
that the temporal features of these motor commands are
produced by self-sustained activity within the neural
network rather than by the time courses of network inputs.

Two features of the intrinsic circuitry of the cerebellum may be particularly instrumental in explaining self-sustained activity. One is a recurrent pathway from cerebellar nuclei that returns back to cerebellar nuclei. In the case of the cerebellorubrospinal system in Fig. 1, the recurrent pathway is from the interpositus nucleus to red nucleus to lateral reticular nucleus and back to interpositus, what I will call the IRL loop. The other feature of intrinsic cerebellar circuitry that may be of critical importance in pattern generation is mutual inhibition between P cells. Fig. 1 shows how mutual inhibition results from the recurrent collaterals of P-cell axons. Inhibitory interneurons called basket and stellate cells (not shown in Fig. 1) provide additional pathways for mutual inhibition. Both the IRL loop and mutual inhibition between P cells constitute positive feedback circuits and, as such, are capable of self-sustained activity.

Self-sustained activity in the form of high-frequency spontaneous discharge has been observed in the IRL loop under conditions in which the inhibitory P-cell input to I cells is blocked [3]. Trace A in Fig. 2 shows this unrestrained discharge schematically, and the other traces illustrate how a motor command might be sculpted out of this tendency toward high-frequency, repetitive discharge.

Trace B shows a brief burst of input presumed to be sent from the sensorimotor cortex to the R cell in Fig. 1. This burst serves as a trigger that initiates repetitive discharge in an IRL loop, and trace D illustrates the discharge of an I cell in the active loop. The intraburst discharge frequency of this cell is presumed to be determined by the summed magnitude of inhibitory input (shown in trace C) from the set of P cells that project to it (Fig. 1 shows only a few P cells from this set). Since the inhibitory input to I was reduced to an appropriate magnitude for controlling this intraburst frequency some time prior to the arrival of the trigger event, this example illustrates a mechanism for presetting the pattern generator. Note that the same reduction of inhibition that presets the intraburst frequency would bring the loop closer to the threshold for repetitive firing, thus serving to enable the triggering operation. The I-cell burst, after continuing for a duration appropriate for the desired motor behavior, is assumed to be terminated by an abrupt increase in inhibitory input from the set of P cells that project to I (trace C).

The time course of bursting discharge illustrated in Fig. 2D would be expected to propagate throughout the IRL loop and be transmitted via the rubrospinal tract to the spinal cord where it could serve as a motor command. Bursts of R-cell discharge similar to this are observed to precede movements in trained monkey subjects [2].

372

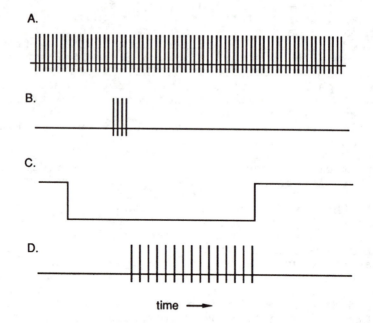

A. Repetitive discharge of I cell in the absence of P-cell inhibition. B. Trigger burst sent to the IRL loop from sensorimotor cortex.

Figure 2: Signals Contributing to Pattern Generation. A. Repetitive discharge of I cell in the absence of P-cell inhibition. B. Trigger burst sent to the IRL loop from sensorimotor cortex. C. Summed inhibition produced by the set of P cells projecting to the I cell. D. Resultant motor pattern in I cell.

The sculpting of a motor command out of a repetitive firing tendency in the IRL loop clearly requires timed transitions in the discharge rates of specific P cells. The present model postulates that the latter result from state transitions in the network of P cells. Bell and Grimm[8] described spontaneous transitions in P-cell firing that occur intermittently, and I have frequently observed them as well. These transitions appear to be produced by intrinsic mechanisms and are difficult to influence with sensory stimulation. The mutual recurrent inhibition between P cells might explain this tendency toward state transitions.

Recurrent inhibition between P cells is mediated by synapses near the cell bodies and primary dendrites of the P cells whereas parallel fiber input extends far out on the dendritic tree. This arrangement may explain why sensory input via parallel fibers does not have a strong, continuous effect on P cell discharge. This sensory input may serve mainly to promote state transitions in the network of P cells, perhaps by modulating the likelihood that a given P cell would participate in a state transition. Once the transition starts, the activity of the P cell may be dominated by the recurrent inhibition close to the cell body.

The mechanism responsible for the adaptive adjustment of these elemental pattern generators may be a change in the synaptic strengths of parallel fiber input to P cells[9]. Such alterations in the efficacy of sensory input would influence the state transitions discussed in the previous paragraph, thus mediating adaptive adjustments in the amplitude and timing of patterned output. Elsewhere I have suggested that this learning process is analogous to operant conditioning and includes both positive and negative reinforcement[3]. Noradrenergic fibers might mediate positive reinforcement, whereas climbing fibers might mediate negative reinforcement. For example, if the network were controlling a limb movement, negative reinforcement might occur when the limb bumps into an object in the work space (climbing fibers fire in response to unexpected somatic events such as this), whereas positive reinforcement might occur whenever the limb successfully acquires the desired target (the noradrenergic fibers to the cerebellum are thought to receive input from reward centers in the brain). Positive reinforcement may be analogous to the associative reward-punishment algorithm described by Barto[10] which would fit with the diffuse projections of noradrenergic fibers. Negative reinforcement might be capable of a higher degree of credit assignment in view of the more focused projections of climbing fibers.

In summary, the previous paragraphs outline some ideas that may be useful in developing a network model of the cerebellum. This particular set of ideas was motivated by a desire to explain the unique manner in which the cerebellum uses sensory input to control patterned output. The model deals explicitly with small circuits within a much larger network. The small circuits are considered elemental pattern generators, whereas the larger network can be considered an array of these pattern generators. The assembly of many elements into an array may give rise to some emergent properties of the network, due to interactions between the elements. However, the highly compartmentalized anatomical structure of the cerebellum fosters the notion of relatively independent elemental pattern generators as hypothesized in the schema for movement control presented in the next section.

SCHEMA FOR MOTOR CONTROL

A major aim in developing the elemental pattern generator model described in the previous section was to explain the intriguing manner in which the cerebellum uses sensory input. Stated succinctly, sensory input is used to preset and to trigger each elemental pattern generator and to evaluate the success of previous output patterns in controlling motor behavior. However, sensory input is not used to shape the waveform of an ongoing output pattern. This means that continuous feedback is not available, at the level of the cerebellum, for any immediate adjustments of motor commands.

Is this kind of behavior actually advantageous in the control of movement? I would propose the affirmative, particularly on the grounds that this strategy seems to have withstood the test of evolution. Elsewhere I have reviewed the global strategies that are used to control several different types of body function[11]. A common theme in each of these physiological control systems is the use of negative feedback only as a low-level strategy, and this coupled with a high-level stage of adaptive feedforward control. It was argued that this particular two-stage control strategy is well suited for utilizing the advantageous features of feedback, feedforward and adaptive control in combination.

The adjustable pattern generator model of the cerebellum outlined in the previous section is a prime example of an adaptive, feedforward controller. In the subsequent paragraphs I will outline how this high-level feedforward controller communicates with low-level feedback systems called motor servos to produce limb movements (Fig. 3).

The array of adjustable pattern generators (PG_n) in the first column of Fig. 3 produce an array of elemental commands that are transmitted via descending fibers to the spinal cord. The connectivity matrix for descending fibers represents the consequences of their branching patterns. Any given fiber is likely to branch to innervate several motor servos. Similarly, each member of the array of motor servos (MS_m) receives convergent input from a large number of pattern generators, and the summed total of this input constitutes its overall motor command.

A motor servo consists of a muscle, its stretch receptors and the spinal reflex pathways back to the same muscle[12]. These reflex pathways constitute negative feedback loops that interact with the motor command to control the discharge of the motor neuron pool innervating the particular muscle. Negative feedback from the muscle receptors functions to maintain the stiffness of the muscle relatively constant, thus providing a spring-like interface between the body and its mechanical environment[13]. The motor command acts to set the slack length of this equivalent spring and, in this way, influences motion of the limb. Feedback also gives rise to an unusual type of damping proportional to a low fractional power of velocity[14]. The individual motor servos interact with each other and with external loads via the trigonometric relations of the musculoskeletal matrix to produce resultant joint positions.

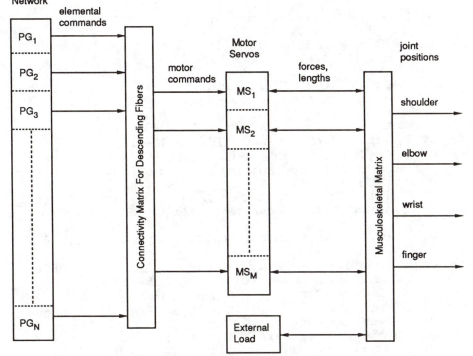

Figure 3: Schema for Motor Control Utilizing Pattern Generator Model of Cerebellum. An array of elemental
pattern generators (PG_n) operate in an adaptive, feedforward manner to produce motor commands. These outputs of the high-level stage are sent to the spinal
cord where they serve as inputs to a low-level array
of negative feedback systems called motor servos
(MS_m). The latter regulate the forces and lengths of
individual muscles to control joint angles.

While the schema for motor control presented here is
based on a considerable body of experimental data, and it
also seems plausible as a strategy for motor control, it
will be important to explore its capabilities for human
limb control with simulation studies. It may also be
fruitful to apply this schema to problems in robotics.
Since I am mainly an experimentalist, my authorship of this
paper is meant as an entré for collaborative work with
neural network modelers that may be interested in these
problems.

REFERENCES

1. M. Ito, The Cerebellum and Neural Control (Raven Press, N. Y., 1984).
2. J. C. Houk & A. R. Gibson, In: J. S. King, New Concepts in Cerebellar Neurobiology (Alan R. Liss, Inc., N. Y., 1987), p. 387.
3. J. C. Houk, In: M. Glickstein & C. Yeo, Cerebellum and Neuronal Plasticity (Plenum Press, N. Y., 1988), in press.
4. D. Marr, J. Physiol. (London) 202, 437 (1969).
5. J. S. Albus, Math. Biosci. 10, 25 (1971).
6. C. C. Boylls, A Theory of Cerebellar Function with Applications to Locomotion (COINS Tech. Rep., U. Mass. Amherst), 76-1.
7. J. C. Houk, In: J. E. Desmedt, Cerebral Motor Control in Man: Long Loop Mechanisms (Karger, Basel, 1978), p. 193.
8. C. C. Bell & R. J. Grimm, J. Neurophysiol., 32, 1044 (1969).
9 C.-F. Ekerot & M. Kano, Brain Res., 342, 357 (1985).
10. A. G. Barto, Human Neurobiol., 4, 229 (1985).
11. J. C. Houk, FASEB J., 2, 97-107 (1988).
12. J. C. Houk & W. Z. Rymer, In: V. B. Brooks, Handbook of Physiology, Vol. 1 of Sect. 1 (American Physiological Society, Bethesda, 1981), p.257.
13. J. C. Houk, Annu. Rev. Physiol., 41, 99 (1979).
14. C. C. A. M. Gielen & J. C. Houk, Biol. Cybern., 57, 217 (1987).

EXPERIMENTAL DEMONSTRATIONS OF OPTICAL NEURAL COMPUTERS

Ken Hsu, David Brady, and Demetri Psaltis
Department of Electrical Engineering
California Institute of Technology
Pasadena, CA 91125

ABSTRACT

We describe two expriments in optical neural computing. In the first a closed optical feedback loop is used to implement auto-associative image recall. In the second a perceptron-like learning algorithm is implemented with photorefractive holography.

INTRODUCTION

The hardware needs of many neural computing systems are well matched with the capabilities of optical systems[1,2,3]. The high interconnectivity required by neural computers can be simply implemented in optics because channels for optical signals may be superimposed in three dimensions with little or no cross coupling. Since these channels may be formed holographically, optical neural systems can be designed to create and maintain interconnections very simply. Thus the optical system designer can to a large extent avoid the analytical and topological problems of determining individual interconnections for a given neural system and constructing physical paths for these interconnections.

An archetypical design for a single layer of an optical neural computer is shown in Fig. 1. Nonlinear thresholding elements, neurons, are arranged on two dimensional planes which are interconnected via the third dimension by holographic elements. The key concerns in implementing this design involve the need for suitable nonlinearities for the neural planes and high capacity, easily modifiable holographic elements. While it is possible to implement the neural function using entirely optical nonlinearities, for example using etalon arrays[4], optoelectronic two dimensional spatial light modulators (2D SLMs) suitable for this purpose are more readily available. and their properties, i.e. speed and resolution, are well matched with the requirements of neural computation and the limitations imposed on the system by the holographic interconnections[5,6]. Just as the main advantage of optics in connectionist machines is the fact that an optical system is generally linear and thus allows the superposition of connections, the main disadvantage of optics is that good optical nonlinearities are hard to obtain. Thus most SLMs are optoelectronic with a non-linearity mediated by electronic effects. The need for optical nonlinearities arises again when we consider the formation of modifiable optical interconnections, which must be an all optical process. In selecting a holographic material for a neural computing application we would like to have the capability of real-time recording and slow erasure. Materials such as photographic film can provide this only with an impractical fixing process. Photorefractive crystals are nonlinear optical materials that promise to have a relatively fast recording response and long term memory[4,5,6,7,8].

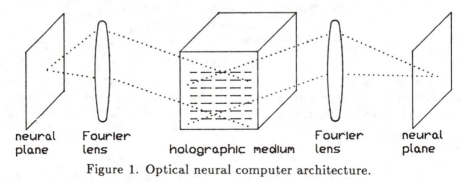

| neural | Fourier | | Fourier | neural |
| plane | lens | holographic medium | lens | plane |

Figure 1. Optical neural computer architecture.

In this paper we describe two experimental implementations of optical neural computers which demonstrate how currently available optical devices may be used in this application. The first experiment we describe involves an optical associative loop which uses feedback through a neural plane in the form of a pinhole array and a separate thresholding plane to implement associate regeneration of stored patterns from correlated inputs. This experiment demonstrates the input-output dynamics of an optical neural computer similar to that shown in Fig. 1, implemented using the Hughes Liquid Crystal Light Valve. The second experiment we describe is a single neuron optical perceptron implemented with a photorefractive crystal. This experiment demonstrates how the learning dynamics of long term memory may be controlled optically. By combining these two experiments we should eventually be able to construct high capacity adaptive optical neural computers.

OPTICAL ASSOCIATIVE LOOP

A schematic diagram of the optical associative memory loop is shown in Fig. 2. It is comprised of two cascaded Vander Lugt correlators[9]. The input section of the system from the threshold device P1 through the first hologram P2 to the pinhole array P3 forms the first correlator. The feedback section from P3 through the second hologram P4 back to the threshold device P1 forms the second correlator. An array of pinholes sits on the back focal plane of L2, which coincides with the front focal plane of L3. The purpose of the pinholes is to link the first and the second (reversed) correlator to form a closed optical feedback loop[10].

There are two phases in operating this optical loop, the learning phase and the recal phase. In the learning phase, the images to be stored are spatially multiplexed and entered simultaneously on the threshold device. The thresholded images are Fourier transformed by the lens L1. The Fourier spectrum and a plane wave reference beam interfere at the plane P2 and record a Fourier transform hologram. This hologram is moved to plane P4 as our stored memory. We then reconstruct the images from the memory to form a new input to make a second Fourier transform hologram that will stay at plane P2. This completes the learning phase. In the recalling phase an input is imaged on the threshold device. This image is correlated with the reference images in the hologram at P2. If the correlation between the input and one of the stored images is high a bright peak appears at one of the pinholes. This peak is sampled by the pinhole to reconstruct the stored image from the hologram at P4. The reconstructed beam is then imaged back to the threshold device to form a closed loop. If the overall optical gain in the loop exceeds the loss the loop signal will grow until the threshold device is saturated. In this case, we can cutoff the external input image and the optical loop will be latched at the stable memory.

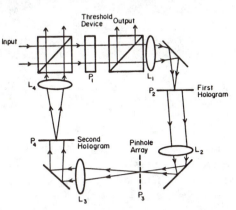

Figure. 2. All-optical associative loop. The threshold device is a LCLV, and the holograms are thermoplastic plates.

The key elements in this optical loop are the holograms, the pinhole array, and the threshold device. If we put a mirror[10] or a phase conjugate mirror[7,11] at the pinhole plane P3 to reflect the correlation signal back through the system then we only need one hologram to form a closed loop. The use of two holograms, however, improves system performance. We make the hologram at P2 with a high pass characteristic so that the input section of the loop has high spectral discrimination. On the other hand we want the images to be reconstructed with high fidelity to the original images. Thus the hologram at plane P4 must have broadband characteristics. We use a diffuser to achieve this when making this hologram. Fig. 3a shows the original images. Fig. 3b and Fig. 3c are the images reconstructed from first and second holograms, respectively. As desired, Fig. 3b is a high pass version of the stored image while Fig. 3c is broadband.

Each of the pinholes at the correlation plane P3 has a diameter of 60 μm. The separations between the pinholes correspond to the separations of the input images at plane P1. If one of the stored images appears at P1 there will be a bright spot at the corresponding pinhole on plane P3. If the input image shifts to the position of another image the correlation peak will also

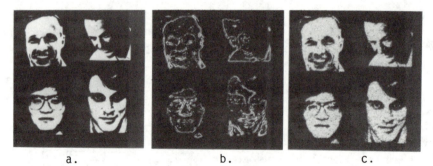

Figure 3. (a) The original images. (b)The reconstructed images from the high-pass hologram P2. (c) The reconstructed images from the band-pass hologram P4.

shift to another pinhole. But if the shift is not an exact image spacing the correlation peak can not pass the pinhole and we lose the feedback signal. Therefore this is a loop with "discrete" shift invariance. Without the pinholes the cross-correlation noise and the auto-correlation peak will be fed back to the loop together and the reconstructed images won't be recognizable. There is a compromise between the pinhole size and the loop performance. Small pinholes allow good memory discrimination and sharp reconstructed images, but can cut the signal to below the level that can be detected by the threshold device and reduce the tolerance of the system to shifts in the input. The function of the pinhole array in this system might also be met by a nonlinear spatial light modulator, in which case we can achieve full shift invariance[12].

The threshold device at plane P1 is a Hughes Liquid Crystal Light Valve. The device has a resolution of 16 lp/mm and uniform aperture of 1 inch diameter. This gives us about 160,000 neurons at P1. In order to compensate for the optical loss in the loop, which is on the order of 10^{-5}, we need the neurons to provide gain on the order of 10^5. In our system this is achieved by placing a Hamamatsu image intensifier at the write side of the LCLV. Since the microchannel plate of the image intensifier can give gains of 10^4, the combination of the LCLV and the image intensifier can give gains of 10^6 with sensitivity down to nW/cm^2. The optical gain in the loop can be adjusted by changing the gain of the image intensifier.

Since the activity of neurons and the dynamics of the memory loop is a continuously evolving phenomenon, we need to have a real time device to monitor and record this behavior. We do this by using a prism beam splitter to take part of the read out beam from the LCLV and image it onto a CCD camera. The output is displayed on a CRT monitor and also recorded on a video tape recorder. Unfortunately, in a paper we can only show static pictures taken from the screen. We put a window at the CCD plane so that each time we can pick up one of the stored images. Fig. 4a shows the read out image

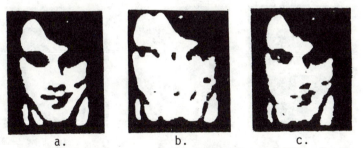

Figure 4. (a) The external input to the optical loop. (b) The feedback image superimposed with the input image. (c) The latched loop image.

from the LCLV which comes from the external input shifted away from its stored position. This shift moves its correlation peak so that it does not match the position of the pinhole. Thus there is no feedback signal going through the loop. If we cut off the input image the read out image will die out with a characteristic time on the order of 50 to 100 ms, corresponding to the response time of the LCLV. Now we shift the input image around trying to search for the correct position. Once the input image comes close enough to the correct position the correlation peak passes through the right pinhole, giving a strong feedback signal superimposed with the external input on the neurons. The total signal then goes through the feedback loop and is amplified continuously until the neurons are saturated. Depending on the optical gain of the neurons the time required for the loop to reach a stable state is between 100 ms and several seconds. Fig. 4b shows the superimposed images of the external input and the loop images. While the feedback signal is shifted somewhat with respect to the input, there is sufficient correlation to induce recall. If the neurons have enough gain then we can cut off the input and the loop stays in its stable state. Otherwise we have to increase the neuron gain until the loop can sustain itself. Fig. 4c shows the image in the loop with the input removed and the memory latched. If we enter another image into the system, again we have to shift the input within the window to search the memory until we are close enough to the correct position. Then the loop will evolve to another stable state and give a correct output.

The input images do not need to match exactly with the memory. Since the neurons can sense and amplify the feedback signal produced by a partial match between the input and a stored image, the stored memory can grow in the loop. Thus the loop has the capability to recall the complete memory from a partial input. Fig. 5a shows the image of a half face input into the system. Fig. 5b shows the overlap of the input with the complete face from the memory. Fig. 5c shows the stable state of the loop after we cut off the external input. In order to have this associative behavior the input must have enough correlation with the stored memory to yield a strong feedback signal. For instance, the loop does not respond to the the presentation of a picture of

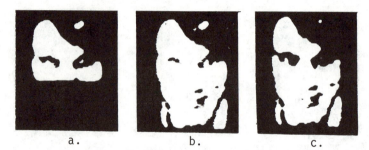

Figure 5. (a) Partial face used as the external input. (b) The superimposed images of the partial input with the complete face recalled by the loop. (c) The complete face latched in the loop.

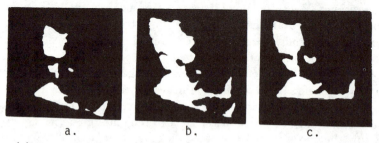

Figure 6. (a) Rotated image used as the external input. (b) The superimposed images of the input with the recalled image from the loop. (c) The image latched in the optical loop.

a person not stored in memory.

Another way to demonstrate the associative behavior of the loop is to use a rotated image as the input. Experiments show that for a small rotation the loop can recognize the image very quickly. As the input is rotated more, it takes longer for the loop to reach a stable state. If it is rotated too much, depending on the neuron gain, the input won't be recognizable. Fig. 6a shows the rotated input. Fig. 6b shows the overlap of loop image with input after we turn on the loop for several seconds. Fig. 6c shows the correct memory recalled from the loop after we cut the input. There is a trade-off between the degree of distortion at the input that the system can tolerate and its ability to discriminate against patterns it has not seen before. In this system the feedback gain (which can be adjusted through the image intensifier) controls this trade-off.

PHOTOREFRACTIVE PERCEPTRON

Holograms are recorded in photorefractive crystals via the electrooptic modulation of the index of refraction by space charge fields created by the migration of photogenerated charge[13,14]. Photorefractive crystals are attractive for optical neural applications because they may be used to store long term interactions between a very large number of neurons. While photorefractive recording does not require a development step, the fact that the response is not instantaneous allows the crystal to store long term traces of the learning process. Since the photorefractive effect arises from the reversible redistribution of a fixed pool of charge among a fixed set of optically addressable trapping sites, the photorefractive response of a crystal does not deteriorate with exposure. Finally, the fact that photorefractive holograms may extend over the entire volume of the crystal has previously been shown to imply that as many as 10^{10} interconnections may be stored in a single crystal with the independence of each interconnection guaranteed by an appropriate spatial arrangement of the interconnected neurons[6,5].

In this section we consider a rudimentary optical neural system which uses the dynamics of photorefractive crystals to implement perceptron-like learning. The architecture of this system is shown schematically in Fig. 7. The input to the system, $\bar{x}$, corresponds to a two dimensional pattern recorded from a video monitor onto a liquid crystal light valve. The light valve transfers this pattern on a laser beam. This beam is split into two paths which cross in a photorefractive crystal. The light propagating along each path is focused such that an image of the input pattern is formed on the crystal. The images along both paths are of the same size and are superposed on the crystal, which is assumed to be thinner than the depth of focus of the images. The intensity diffracted from one of the two paths onto the other by a hologram stored in the crystal is isolated by a polarizer and spatially integrated by a single output detector. The thresholded output of this detector corresponds to the output of a neuron in a perceptron.

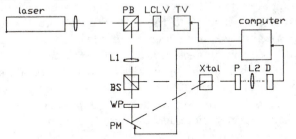

Figure 7. Photorefractive perceptron. PB is a polarizing beam splitter. L1 and L2 are imaging lenses. WP is a quarter waveplate. PM is a piezoelectric mirror. P is a polarizer. D is a detector. Solid lines show electronic control. Dashed lines show the optical path.

The ith component of the input to this system corresponds to the intensity in the ith pixel of the input pattern. The interconnection strength, w_i, between the ith input and the output neuron corresponds to the diffraction efficiency of the hologram taking one path into the other at the ith pixel of the image plane. While the dynamics of w_i can be quite complex in some geometries and crystals, it is possible to show from the band transport model for the photorefractive effect that under certain circumstances the time development of w_i may be modeled by

$$w_i(t) = w_{max} \left| \frac{e^{\frac{-t}{\tau}}}{\tau} \int_0^t m(s) e^{j\phi(s)} e^{\frac{s}{\tau}} ds \right|^2 \qquad (1)$$

where $m(s)$ and $\phi(s)$ are the modulation depth and phase, respectively, of the interference pattern formed in the crystal between the light in the two paths[15]. τ is a characteristic time constant for crystal. τ is inversely proportional to the intensity incident on the ith pixel of the crystal. Using Eqn. 1 it is possible to make $w_i(t)$ take any value between 0 and w_{max} by properly exposing the ith pixel of the crystal to an appropriate modulation depth and intensity. The modulation depth between two optical beams can be adjusted by a variety of simple mechanisms. In Fig. 7 we choose to control $m(t)$ using a mirror mounted on a piezoelectric crystal. By varying the frequency and the amplitude of oscillations in the piezoelectric crystal we can electronically set both $m(t)$ and $\phi(t)$ over a continuous range without changing the intensity in the optical beams or interrupting readout of the system. With this control over $m(t)$ it is possible via the dynamics described in Eqn. (1) to implement any learning algorithm for which w_i can be limited to the range $(0, w_{max})$.

The architecture of Fig. 7 classifies input patterns into two classes according to the thresholded output of the detector. The goal of a learning algorithm for this system is to correctly classify a set of training patterns. The perceptron learning algorithm involves simply testing each training vector and adding training vectors which yield too low an output to the weight vector and subtracting training vectors which yield too high an output from the weight vector until all training vectors are correctly classified[16]. This training algorithm is described by the equation $\Delta w_i = \alpha x_i$ where alpha is positive (negative) if the output for $\bar{x}$ is too low (high). An optical analog of this method is implemented by testing each training pattern and exposing the crystal with each incorrectly classified pattern. Training vectors that yield a high output when a low output is desired are exposed at zero modulation depth. Training vectors that yield a low output when high output is desired are exposed at a modulation depth of one.

The weight vector for the $k + 1^{th}$ iteration when erasure occurs in the kth iteration is given by

$$w_i(k+1) = e^{\frac{-2\Delta t}{\tau}} w_i(k) \approx \left(1 - \frac{2\Delta t}{\tau}\right) w_i(k) \qquad (2)$$

where we assume that the exposure time, Δt, is much less than τ. Note that since τ is inversely proportional to the intensity in the ith pixel, the change in

w_i is proportional to the ith input. The weight vector at the $k + 1^{th}$ iteration when recording occurs in the kth iteration is given by

$$w_i(k+1) = e^{\frac{-2\Delta t}{\tau}} w_i(k) + 2\sqrt{w_i(k)w_{max}} e^{\frac{-\Delta t}{\tau}}(1 - e^{\frac{-\Delta t}{\tau}}) + w_{max}(1 - e^{\frac{-\Delta t}{\tau}})^2 \quad (3)$$

To lowest order in $\frac{\Delta t}{\tau}$ and $\frac{w_i}{w_{max}}$, Eqn. (3) yields

$$w_i(k+1) = w_i(k) + 2\sqrt{w_i(k)w_{max}}(\frac{\Delta t}{\tau}) + w_{max}(\frac{\Delta t}{\tau})^2 \quad (4)$$

Once again the change in w_i is proportional to the ith input.

We have implemented the architecture of Fig. 7 using a SBN60:Ce crystal provided by the Rockwell International Science Center. We used the 488 nm line of an argon ion laser to record holograms in this crystal. Most of the patterns we considered were laid out on 10×10 grids of pixels, thus allowing 100 input channels. Ultimately, the number of channels which may be achieved using this architecture is limited by the number of pixels which may be imaged onto the crystal with a depth of focus sufficient to isolate each pixel along the length of the crystal.

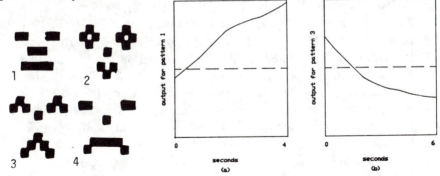

Figure 8. Training patterns.   Figure 9. Output in the second training cycle.

Using the variation on the perceptron learning algorithm described above with a fixed exposure times Δt_r and Δt_e for recording and erasing, we have been able to correctly classify various sets of input patterns. One particular set which we used is shown in Fig. 8. In one training sequence, we grouped patterns 1 and 2 together with a high output and patterns 3 and 4 together with a low output. After all four patterns had been presented four times, the system gave the correct output for all patterns. The weights stored in the crystal were corrected seven times, four times by recording and three by erasing. Fig. 9a shows the output of the detector as pattern 1 is recorded in the second learning cycle. The dashed line in this figure corresponds to the threshold level. Fig. 9b shows the output of the detector as pattern 3 is erased in the second learning cycle.

CONCLUSION

The experiments described in this paper demonstrate how neural network architectures can be implemented using currently available optical devices. By combining the recall dynamics of the first system with the learning capability of the second, we can construct sophisticated optical neural computers.

ACKNOWLEDGEMENTS

The authors thank Ratnakar Neurgaonkar and Rockwell International for supplying the SBN crystal used in our experiments and Hamamatsu Photonics K.K. for assistance with image intesifiers. We also thank Eung Gi Paek and Kelvin Wagner for their contributions to this research.

This research is supported by the Defense Advanced Research Projects Agency, the Army Research Office, and the Air Force Office of Scientific Research.

REFERENCES

1. Y. S. Abu-Mostafa and D. Psaltis, Scientific American, pp.88-95, March, 1987.
2. D. Psaltis and N. H. Farhat, Opt. Lett., **10**,(2), 98(1985).
3. A. D. Fisher, R. C. Fukuda, and J. N. Lee, Proc. SPIE **625**, 196(1986).
4. K. Wagner and D. Psaltis, Appl. opt., 26(23), pp.5061-5076(1987).
5. D. Psaltis, D. Brady, and K. Wagner, Applied optics, March 1988.
6. D. Psaltis, J. Yu, X. G. Gu, and H. Lee, Second Topical Meeting on Optical Computing, Incline Village, Nevada, March 16-18,1987.
7. A. Yariv, S.-K. Kwong, and K. Kyuma, SPIE proc. 613-01,(1986).
8. D. Z. Anderson, Proceedings of the International Conference on Neural Networks, San Diego, June 1987.
9. A. B. Vander Lugt, IEEE Trans. Inform. Theory, IT-10(2), pp.139-145(1964).
10. E. G. Paek and D. Psaltis, Opt. Eng., 26(5), pp.428-433(1987).
11. Y. Owechko, G. J. Dunning, E. Marom, and B. H. Soffer, Appl. Opt. **26**,(10),1900(1987).
12. D. Psaltis and J. Hong, Opt. Eng. **26**,10(1987).
13. N. V. Kuktarev, V. B. Markov, S. G. Odulov, M. S. Soskin, and V. L. Vinetskii, Ferroelectrics, **22**,949(1979).
14. J. Feinberg, D. Heiman, A. R. Tanguay, and R. W. Hellwarth, J. Appl. Phys. **51**,1297(1980).
15. T. J. Hall, R. Jaura, L. M. Connors, P. D. Foote, Prog. Quan. Electr. **10**,77(1985).
16. F. Rosenblatt, Principles of Neurodynamics: Perceptron and the Theory of Brain Mechanisms, Spartan Books, Washington,(1961).

Neural Net and Traditional Classifiers[1]

William Y. Huang and *Richard P. Lippmann*

MIT Lincoln Laboratory
Lexington, MA 02173, USA

Abstract. Previous work on nets with continuous-valued inputs led to generative procedures to construct convex decision regions with two-layer perceptrons (one hidden layer) and arbitrary decision regions with three-layer perceptrons (two hidden layers). Here we demonstrate that two-layer perceptron classifiers trained with back propagation can form both convex and disjoint decision regions. Such classifiers are robust, train rapidly, and provide good performance with simple decision regions. When complex decision regions are required, however, convergence time can be excessively long and performance is often no better than that of k-nearest neighbor classifiers. Three neural net classifiers are presented that provide more rapid training under such situations. Two use fixed weights in the first one or two layers and are similar to classifiers that estimate probability density functions using histograms. A third "feature map classifier" uses both unsupervised and supervised training. It provides good performance with little supervised training in situations such as speech recognition where much unlabeled training data is available. The architecture of this classifier can be used to implement a neural net k-nearest neighbor classifier.

1. INTRODUCTION

Neural net architectures can be used to construct many different types of classifiers [7]. In particular, multi-layer perceptron classifiers with continuous valued inputs trained with back propagation are robust, often train rapidly, and provide performance similar to that provided by Gaussian classifiers when decision regions are convex [12,7,5,8]. Generative procedures demonstrate that such classifiers can form convex decision regions with two-layer perceptrons (one hidden layer) and arbitrary decision regions with three-layer perceptrons (two hidden layers) [7,2,9]. More recent work has demonstrated that two-layer perceptrons can form non-convex and disjoint decision regions. Examples of hand crafted two-layer networks which generate such decision regions are presented in this paper along with Monte Carlo simulations where complex decision regions were generated using back propagation training. These and previous simulations [5,8] demonstrate that convergence time with back propagation can be excessive when complex decision regions are desired and performance is often no better than that obtained with k-nearest neighbor classifiers [4]. These results led us to explore other neural net classifiers that might provide faster convergence. Three classifiers called, "fixed weight," "hypercube," and "feature map" classifiers, were developed and evaluated. All classifiers were tested on illustrative problems with two continuous-valued inputs and two classes (A and B). A more restricted set of classifiers was tested with vowel formant data.

2. CAPABILITIES OF TWO LAYER PERCEPTRONS

Multi-layer perceptron classifiers with hard-limiting nonlinearities (node outputs of 0 or 1) and continuous-valued inputs can form complex decision regions. Simple constructive proofs demonstrate that a three-layer perceptron (two hidden layers) can

---

[1] This work was sponsored by the Defense Advanced Research Projects Agency and the Department of the Air Force. The views expressed are those of the authors and do not reflect the policy or position of the U. S. Government.

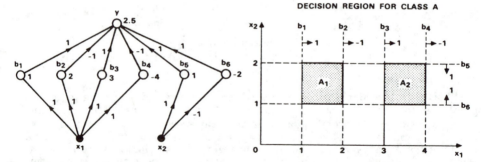

FIG. 1. *A two-layer perceptron that forms disjoint decision regions for class A (shaded areas). Connection weights and node offsets are shown in the left. Hyperplanes formed by all hidden nodes are drawn as dashed lines with node labels. Arrows on these lines point to the half plane where the hidden node output is "high".*

form arbitrary decision regions and a two-layer perceptron (one hidden layer) can form single convex decision regions [7,2,9]. Recently, however, it has been demonstrated that two-layer perceptrons can form decision regions that are not simply convex [14]. Fig. 1, for example, shows how disjoint decision regions can be generated using a two-layer perceptron. The two disjoint shaded areas in this Fig. represent the decision region for class A (output node has a "high" output, y = 1). The remaining area represents the decision region for class B (output node has a "low" output, y = 0). Nodes in this Fig. contain hard-limiting nonlinearities. Connection weights and node offsets are indicated in the left diagram. Ten other complex decision regions formed using two-layer perceptrons are presented in Fig. 2.

The above examples suggest that two-layer perceptrons can form decision regions with arbitrary shapes. We, however, know of no general proof of this capability. A 1965 book by Nilson discusses this issue and contains a proof that two-layer nets can divide a finite number of points into two arbitrary sets ([10] page 89). This proof involves separating M points using at most $M-1$ parallel hyperplanes formed by first-layer nodes where no hyperplane intersects two or more points. Proving that a given decision region can be formed in a two-layer net involves testing to determine whether the Boolean representations at the output of the first layer for all points within the decision region for class A are linearly separable from the Boolean representations for class B. One test for linear separability was presented in 1962 [13].

A problem with forming complex decision regions with two-layer perceptrons is that weights and offsets must be adjusted carefully because they interact extensively to form decision regions. Fig. 1 illustrates this sensitivity problem. Here it can be seen that weights to one hidden node form a hyperplane which influences decision regions in an entire half-plane. For example, small errors in first layer weights that results in a change in the slopes of hyperplanes b_5 and b_6 might only slightly extend the A_1 region but completely eliminate the A_2 region. This interdependence can be eliminated in three layer perceptrons.

It is possible to train two-layer perceptrons to form complex decision regions using back propagation and sigmoidal nonlinearities despite weight interactions. Fig. 3, for example, shows disjoint decision regions formed using back propagation for the problem of Fig. 1. In this and all other simulations, inputs were presented alternately from classes A and B and selected from a uniform distribution covering the desired decision region. In addition, the back propagation rate of descent term, η, was set equal to the momentum gain term, α and $\eta = \alpha = .01$. Small values for η and α were necessary to guarantee convergence for the difficult problems in Fig. 2. Other simulation details are

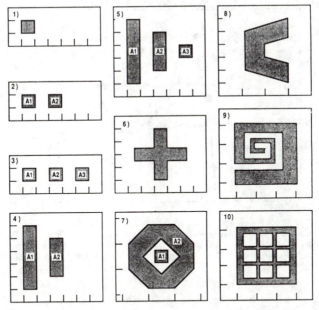

FIG. 2. *Ten complex decision regions formed by two-layer perceptrons. The numbers assigned to each case are the "case" numbers used in the rest of this paper.*

as in [5,8]. Also shown in Fig. 3 are hyperplanes formed by those first-layer nodes with the strongest connection weights to the output node. These hyperplanes and weights are similar to those in the networks created by hand except for sign inversions, the occurrence of multiple similar hyperplanes formed by two nodes, and the use of node offsets with values near zero.

3. COMPARATIVE RESULTS OF TWO-LAYERS VS. THREE-LAYERS

Previous results [5,8], as well as the weight interactions mentioned above, suggest that three-layer perceptrons may be able to form complex decision regions faster with back propagation than two-layer perceptrons. This was explored using Monte Carlo simulations for the first nine cases of Fig. 2. All networks have 32 nodes in the first hidden layer. The number of nodes in the second hidden layer was twice the number of convex regions needed to form the decision region (2, 4, 6, 4, 6, 6, 8, 6 and 6 for Cases 1 through 9 respectively). Ten runs were typically averaged together to obtain a smooth curve of percentage error vs. time (number of training trials) and enough trials were run (to a limit of 250,000) until the curve appeared to flatten out with little improvement over time. The error curve was then low-pass filtered to determine the convergence time. Convergence time was defined as the time when the curve crossed a value 5 percentage points above the final percentage error. This definition provides a framework for comparing the convergence time of the different classifiers. It, however, is not the time after which error rates do not improve. Fig. 4 summarizes results in terms of convergence time and final percentage error. In those cases with disjoint decision regions, back propagation sometimes failed to form separate regions after 250,000 trials. For example, the two disjoint regions required in Case 2 were never fully separated with

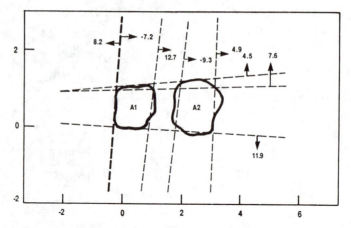

FIG. 3. *Decision regions formed using back propagation for Cases 2 of Fig. 2. Thick solid lines represent decision boundaries. Dashed lines and arrows have the same meaning as in Fig. 1. Only hyperplanes for hidden nodes with large weights to the output node are shown. Over 300,000 training trials were required to form separate regions.*

a two-layer perceptron but were separated with a three-layer perceptron. This is noted by the use of filled symbols in Fig. 4.

Fig. 4 shows that there is no significant performance difference between two and three layer perceptrons when forming complex decision regions using back propagation training. Both types of classifiers take an excessively long time ($> 100,000$ trials) to form complex decision regions. A minor difference is that in Cases 2 and 7 the two-layer network failed to separate disjoint regions after 250,000 trials whereas the three-layer network was able to do so. This, however, is not significant in terms of convergence time and error rate. Problems that are difficult for the two-layer networks are also difficult for the three-layer networks, and vice versa.

4. ALTERNATIVE CLASSIFIERS

Results presented above and previous results [5,8] demonstrate that multi-layer perceptron classifiers can take very long to converge for complex decision regions. Three alternative classifiers were studied to determine whether other types of neural net classifiers could provide faster convergence.

4.1. FIXED WEIGHT CLASSIFIERS

Fixed weight classifiers attempt to reduce training time by adapting only weights between upper layers of multi-layer perceptrons. Weights to the first layer are fixed before training and remain unchanged. These weights form fixed hyperplanes which can be used by upper layers to form decision regions. Performance will be good if the fixed hyperplanes are near the decision region boundaries that are required in a specific problem. Weights between upper layers are trained using back propagation as described above. Two methods were used to adjust weights to the first layer. Weights were adjusted to place hyperplanes randomly or in a grid in the region $(-1 < x_1, x_2 < 10)$. All decision regions in Fig. 2 fall within this region. Hyperplanes formed by first layer nodes for "fixed random" and "fixed grid" classifiers for Case 2 of Fig. 2 are shown as dashed lines in Fig. 5. Also shown in this Fig. are decision regions (shaded areas) formed

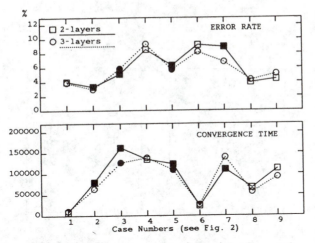

FIG. 4. *Percentage error (top) and convergence time (bottom) for Cases 1 through 9 of Fig. 2 for two-and three-layer perceptron classifiers trained using back propagation. Filled symbols indicate that separate disjoint regions were not formed after 250,000 trials.*

using back propagation to train only the upper network layers. These regions illustrate how fixed hyperplanes are combined to form decision regions. It can be seen that decision boundaries form along the available hyperplanes. A good solution is possible for the fixed grid classifier where desired decision region boundaries are near hyperplanes. The random grid classifier provides a poor solution because hyperplanes are not near desired decision boundaries. The performance of a fixed weight classifier depends both on the placement of hyperplanes and on the number of hyperplanes provided.

4.2. HYPERCUBE CLASSIFIER

Many traditional classifiers estimate probability density functions of input variables for different classes using histogram techniques [4]. Hypercube classifiers use this technique by fixing weights in the first two layers to break the input space into hypercubes (squares in the case of two inputs). Hypercube classifiers are similar to fixed weight classifiers, except weights to the first *two* layers are fixed, and only weights to output nodes are trained. Hypercube classifiers are also similar in structure to the CMAC model described by Albus [1]. The output of a second layer node is "high" only if the input is in the hypercube corresponding to that node. This is illustrated in Fig. 6 for a network with two inputs.

The top layer of a hypercube classifier can be trained using back propagation. A maximum likelihood approach, however, suggests a simpler training algorithm which consists of counting. The output of second layer node H_i is connected to the output node corresponding to that class with greatest frequency of occurrence of training inputs in hypercube H_i. That is, if a sample falls in hypercube H_i, then it is classified as class θ^* where

$$N_{i,\theta^*} > N_{i,\theta} \quad \text{for all} \quad \theta \neq \theta^*. \tag{1}$$

In this equation, $N_{i,\theta}$ is the number of training tokens in hypercube H_i which belong to class θ. This will be called maximum likelihood (ML) training. It can be implemented by connection second-layer node H_i only to that output node corresponding to class θ^* in Eq. (1). In all simulations hypercubes covered the area ($-1 < x_1, x_2 < 10$).

392

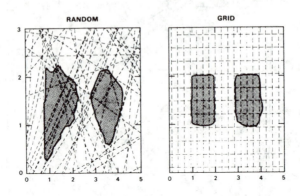

FIG. 5. *Decision regions formed with "fixed random" and "fixed grid" classifiers for Case 2 from Fig. 2 using back propagation training. Lines shown are hyperplanes formed by the first layer nodes. Shaded areas represent the decision region for class A.*

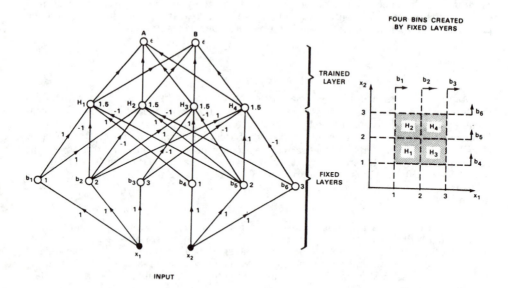

FIG. 6. *A hypercube classifier (left) is a three-layer perceptron with fixed weights to the first two layers, and trainable weights to output nodes. Weights are initialized such that outputs of nodes* H_1 *through* H_4 *(left) are "high" only when the input is in the corresponding hypercube (right).*

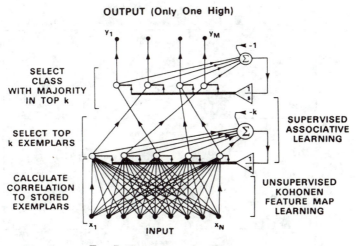

FIG. 7. *Feature map classifier.*

4.3.   FEATURE MAP CLASSIFIER

In many speech and image classification problems a large quantity of unlabeled training data can be obtained, but little labeled data is available. In such situations unsupervised training with unlabeled training data can substantially reduce the amount of supervised training required [3]. The feature map classifier shown in Fig. 7 uses combined supervised/unsupervised training, and is designed for such problems. It is similar to histogram classifiers used in discrete observation hidden Markov models [11] and the classifier used in [6]. The first layer of this classifier forms a feature map using a self organizing clustering algorithm described by Kohonen [6]. In all simulations reported in this paper 10,000 trials of unsupervised training were used. After unsupervised training, first-layer feature nodes sample the input space with node density proportional to the combined probability density of all classes. First layer feature map nodes perform a function similar to that of second layer hypercube nodes except each node has maximum output for input regions that are more general than hypercubes and only the output of the node with a maximum output is fed to the output nodes. Weights to output nodes are trained with supervision after the first layer has been trained. Back propagation, or maximum likelihood training can be used. Maximum likelihood training requires $N_{i,\theta}$ (Eq. 1) to be the number of times first layer node i has maximum output for inputs from class θ. In addition, during classification, the outputs of nodes with $N_{i,\theta} = 0$ for all θ (untrained nodes) are not considered when the first-layer node with the maximum output is selected. The network architecture of a feature map classifier can be used to implement a k-nearest neighbor classifier. In this case, the feedback connections in Fig. 7 (large circular summing nodes and triangular integrators) used to select those k nodes with the maximum outputs must be slightly modified. K is 1 for a feature map classifier and must be adjusted to the desired value of k for a k-nearest neighbor classifier.

5.   COMPARISON BETWEEN CLASSIFIERS

The results of Monte Carlo simulations using all classifiers for Case 2 are shown in Fig. 8. Error rates and convergence times were determined as in Section 3. All alter-

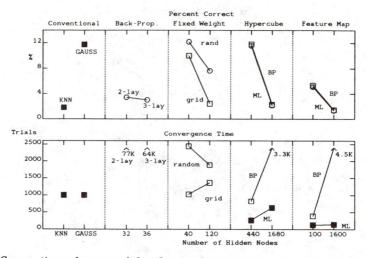

FIG. 8. *Comparative performance of classifiers for Case 2. Training time of the feature map classifiers does not include the 10,000 unsupervised training trials.*

native classifiers had shorter convergence times than multi-layer perceptron classifiers trained with back propagation. The feature map classifier provided best performance. With 1,600 nodes, its error rate was similar to that of the k-nearest neighbor classifiers but it required fewer than 100 supervised training tokens. The larger fixed weight and hypercube classifiers performed well but required more supervised training than the feature map classifiers. These classifiers will work well when the combined probability density function of all classes varies smoothly and the domain where this function is non-zero is known. In this case weights and offsets can be set such that hyperplanes and hypercubes cover the domain and provide good performance. The feature map classifier automatically covers the domain. Fixed weight "random" classifiers performed substantially worse than fixed weight "grid" classifiers. Back propagation training (BP) was generally much slower than maximum likelihood training (ML).

6. VOWEL CLASSIFICATION

Multi layer perceptron, feature map, and traditional classifiers were tested with vowel formant data from Peterson and Barney [11]. These data had been obtained by spectrographic analysis of vowels in /hVd/ context spoken by 67 men, women and children. First and second formant data of ten vowels was split into two sets, resulting in a total of 338 training tokens and 333 testing tokens. Fig. 9 shows the test data and the decision regions formed by a two-layer perceptron classifier trained with back propagation. The performance of classifiers is presented in Table I. All classifiers had similar error rates. The feature map classifier with only 100 nodes required less than 50 supervised training tokens (5 samples per vowel class) for convergence. The perceptron classifier trained with back propagation required more than 50,000 training tokens. The first stage of the feature map classifier and the multi-layer perceptron classifier were trained by randomly selecting entries from the 338 training tokens after labels had been removed and using tokens repetitively.

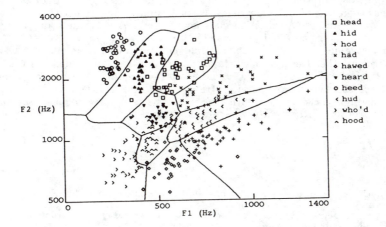

FIG. 9. *Decision regions formed by a two-layer network using BP after 200,000 training tokens from Peterson's steady state vowel data [Peterson, 1952]. Also shown are samples of the testing set. Legend show example of the pronunciation of the 10 vowels and the error within each vowel.*

ALGORITHM	TRAINING TOKENS	% ERROR
KNN	338	18.0
Gaussian	338	20.4
2-Layer Perceptron	50,000	19.8
Feature Map	< 50	22.8

TABLE I
Performance of classifiers on steady state vowel data.

7. Conclusions

Neural net architectures form a flexible framework that can be used to construct many different types of classifiers. These include Gaussian, k-nearest neighbor, and multi-layer perceptron classifiers as well as classifiers such as the feature map classifier which use unsupervised training. Here we first demonstrated that two-layer perceptrons (one hidden layer) can form non-convex and disjoint decision regions. Back propagation training, however, can be extremely slow when forming complex decision regions with multi-layer perceptrons. Alternative classifiers were thus developed and tested. All provided faster training and many provided improved performance. Two were similar to traditional classifiers. One (hypercube classifier) can be used to implement a histogram classifier, and another (feature map classifier) can be used to implement a k-nearest neighbor classifier. The feature map classifier provided best overall performance. It used combined supervised/unsupervised training and attained the same error rate as a k-nearest neighbor classifier, but with fewer supervised training tokens. Furthermore, it required fewer nodes then a k-nearest neighbor classifier.

REFERENCES

[1] J. S. Albus, *Brains, Behavior, and Robotics*. McGraw-Hill, Petersborough, N.H., 1981.

[2] D. J. Burr, "A neural network digit recognizer," in *Proceedings of the International Conference on Systems, Man, and Cybernetics*, IEEE, 1986.

[3] D. B. Cooper and J. H. Freeman, "On the asymptotic improvement in the outcome of supervised learning provided by additional nonsupervised learning," *IEEE Transactions on Computers*, vol. C-19, pp. 1055–63, November 1970.

[4] R. O. Duda and P. E. Hart, *Pattern Classification and Scene Analysis*. John-Wiley & Sons, New York, 1973.

[5] W. Y. Huang and R. P. Lippmann, "Comparisons between conventional and neural net classifiers," in *1st International Conference on Neural Network*, IEEE, June 1987.

[6] T. Kohonen, K. Makisara, and T. Saramaki, "Phonotopic maps — insightful representation of phonological features for speech recognition," in *Proceedings of the 7th International Conference on Pattern Recognition*, IEEE, August 1984.

[7] R. P. Lippmann, "An introduction to computing with neural nets," *IEEE ASSP Magazine*, vol. 4, pp. 4–22, April 1987.

[8] R. P. Lippmann and B. Gold, "Neural classifiers useful for speech recognition," in *1st International Conference on Neural Network*, IEEE, June 1987.

[9] I. D. Longstaff and J. F. Cross, "A pattern recognition approach to understanding the multi-layer perceptron," Mem. 3936, Royal Signals and Radar Establishment, July 1986.

[10] N. J. Nilsson, *Learning Machines*. McGraw Hill, N.Y., 1965.

[11] T. Parsons, *Voice and Speech Processing*. McGraw-Hill, New York, 1986.

[12] F. Rosenblatt, *Perceptrons and the Theory of Brain Mechanisms*. Spartan Books, 1962.

[13] R. C. Singleton, "A test for linear separability as applied to self-organizing machines," in *Self-Organization Systems, 1962*, (M. C. Yovits, G. T. Jacobi, and G. D. Goldstein, eds.), pp. 503–524, Spartan Books, Washington, 1962.

[14] A. Wieland and R. Leighton, "Geometric analysis of neural network capabilities," in *1st International Conference on Neural Networks*, IEEE, June 1987.

AN OPTIMIZATION NETWORK FOR MATRIX INVERSION

Ju-Seog Jang, Soo-Young Lee, and Sang-Yung Shin
Korea Advanced Institute of Science and Technology,
P.O. Box 150, Cheongryang, Seoul, Korea

ABSTRACT

Inverse matrix calculation can be considered as an optimization. We have demonstrated that this problem can be rapidly solved by highly interconnected simple neuron-like analog processors. A network for matrix inversion based on the concept of Hopfield's neural network was designed, and implemented with electronic hardware. With slight modifications, the network is readily applicable to solving a linear simultaneous equation efficiently. Notable features of this circuit are potential speed due to parallel processing, and robustness against variations of device parameters.

INTRODUCTION

Highly interconnected simple analog processors which mimic a biological neural network are known to excel at certain collective computational tasks. For example, Hopfield and Tank designed a network to solve the traveling salesman problem which is of the np-complete class,[1] and also designed an A/D converter of novel architecture[2] based on the Hopfield's neural network model.[3, 4] The network could provide good or optimum solutions during an elapsed time of only a few characteristic time constants of the circuit.

The essence of collective computation is the dissipative dynamics in which initial voltage configurations of neuron-like analog processors evolve simultaneously and rapidly to steady states that may be interpreted as optimal solutions. Hopfield has constructed the computational energy E (Liapunov function), and has shown that the energy function E of his network decreases in time when coupling coefficients are symmetric. At the steady state E becomes one of local minima.

In this paper we consider the matrix inversion as an optimization problem, and apply the concept of the Hopfield neural network model to this problem.

CONSTRUCTION OF THE ENERGY FUNCTIONS

Consider a matrix equation $\mathbf{AV}=\mathbf{I}$, where $\mathbf{A}$ is an input $n \times n$ matrix, $\mathbf{V}$ is the unknown inverse matrix, and $\mathbf{I}$ is the identity matrix. Following Hopfield we define n energy functions E_k, $k=1, 2, ..., n$,

$$E_1 = (1/2)[(\sum_{j=1}^{n} A_{1j}V_{j1}-1)^2 + (\sum_{j=1}^{n} A_{2j}V_{j1})^2 + \cdots + (\sum_{j=1}^{n} A_{nj}V_{j1})^2]$$

$$E_2 = (1/2)[(\sum_{j=1}^{n} A_{1j}V_{j2})^2 + (\sum_{j=1}^{n} A_{2j}V_{j2}-1)^2 + \cdots + (\sum_{j=1}^{n} A_{nj}V_{j2})^2]$$

. . . .

$$E_n = (1/2)[(\sum_{j=1}^{n} A_{1j}V_{jn})^2 + (\sum_{j=1}^{n} A_{2j}V_{jn})^2 + \cdots + (\sum_{j=1}^{n} A_{nj}V_{jn}-1)^2] \qquad (1)$$

where A_{ij} and V_{ij} are the elements of ith row and jth column of matrix $\mathbf{A}$ and $\mathbf{V}$, respectively. When $\mathbf{A}$ is a nonsingular matrix, the minimum value (=zero) of each energy function is unique and is located at a point in the corresponding hyperspace whose coordinates are $\{ V_{1k}, V_{2k}, \cdots, V_{nk} \}$, $k = 1, 2, ..., n$. At this minimum value of each energy function the values of $V_{11}, V_{12}, ..., V_{nn}$ become the elements of the inverse matrix $\mathbf{A}^{-1}$. When $\mathbf{A}$ is a singular matrix the minimum value (in general, not zero) of each energy function is not unique and is located on a contour line of the minimum value. Thus, if we construct a model network in which initial voltage configurations of simple analog processors, called neurons, converge simultaneously and rapidly to the minimum energy point, we can say the network have found the *optimum* solution of matrix inversion problem. The optimum solution means that when $\mathbf{A}$ is a nonsingular matrix the result is the inverse matrix that we want to know, and when $\mathbf{A}$ is a singular matrix the result is a solution that is optimal in a least-square sense of Eq. (1).

DESIGN OF THE NETWORK AND THE HOPFIELD MODEL

Designing the network for matrix inversion, we use the Hopfield model without inherent loss terms, that is,

$$\frac{du_{ik}}{dt} = -\frac{\partial}{\partial V_{ik}}E_k(V_{1k}, V_{2k}, \cdots, V_{nk})$$

$$V_{ik} = g_{ik}(u_{ik}), \qquad i, k = 1, 2, ..., n \qquad (2)$$

where u_{ik} is the input voltage of ith neuron in the kth network, V_{ik} is its output, and the function g_{ik} is the input-output relationship. But the neurons of this scheme operate in all the regions of g_{ik} differently from Hopfield's nonlinear 2-state neurons of associative memory models.[3, 4]

From Eq. (1) and Eq. (2), we can define coupling coefficients T_{ij} between ith and jth neurons and rewrite Eq. (2) as

$$\frac{du_{ik}}{dt} = -\sum_{j=1}^{n} T_{ij}V_{jk} + A_{ki}, \qquad T_{ij} = \sum_{l=1}^{n} A_{li}A_{lj} = T_{ji},$$

$$V_{ik} = g_{ik}(u_{ik}). \qquad (3)$$

It may be noted that T_{ij} is independent of k and only one set of hardware is needed for all k. The implemented network is shown in Fig. 1. The same set of hardware with bias levels, $\sum_{j=1}^{n} A_{ji}b_j$, can be used to solve a linear simultaneous equation represented by **Ax=b** for a given vector **b**.

INPUT

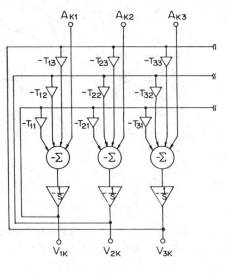

OUTPUT

Fig. 1. Implemented network for matrix inversion with externally
controllable coupling coefficients. Nonlinearity between
the input and the output of neurons is assumed to be
distributed in the adder and the integrator.

The application of the gradient Hopfield model to this problem gives the result
that is similar to the steepest descent method.[5] But the nonlinearity between the
input and the output of neurons is introduced. Its effect to the computational
capability will be considered next.

CHARACTERISTICS OF THE NETWORK

For a simple case of 3×3 input matrices the network is implemented with
electronic hardware and its dynamic behavior is simulated by integration of the
Eq. (3). For nonsingular input matrices, exact realization of T_{ij} connection and
bias A_{ki} is an important factor for calculation accuracy, but the initial condition
and other device parameters such as steepness, shape and uniformity of g_{ik} are
not. Even a complex g_{ik} function shown in Fig. 2 can not affect the computational capability. Convergence time of the output state is determined by the
characteristic time constant of the circuit. An example of experimental results is
shown in Fig. 3. For singular input matrices, the converged output voltage configuration of the network is dependent upon the initial state and the shape of g_{ik}.

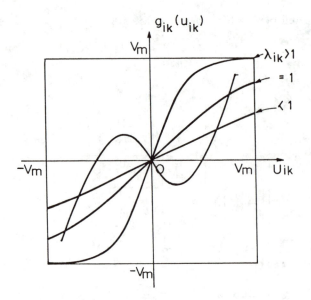

Fig. 2. g_{ik} functions used in computer simulations where λ_{ik} is the steepness of sigmoid function $tanh\,(\lambda_{ik}u_{ik})$.

input matrix $A = \begin{pmatrix} 1 & 2 & 1 \\ -1 & 1 & 1 \\ 1 & 0 & -1 \end{pmatrix}$ (cf) $A^{-1} = \begin{pmatrix} 0.5 & -1 & -0.5 \\ 0 & 1 & 1 \\ 0.5 & -1 & -1.5 \end{pmatrix}$

output matrix $V = \begin{pmatrix} 0.50 & -0.98 & -0.49 \\ 0.02 & 0.99 & 1.00 \\ 0.53 & -0.98 & -1.50 \end{pmatrix}$

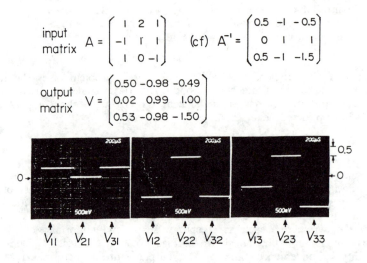

Fig. 3. An example of experimental results

COMPLEXITY ANALYSIS

By counting operations we compare the neural net approach with other well-known methods such as Triangular-decomposition and Gauss-Jordan elimination.[6]

(1) Triangular-decomposition or Gauss-Jordan elimination method takes $O(8n^3/3)$ multiplications/divisions and additions for large $n \times n$ matrix inversion, and $O(2n^3/3)$ multiplications/divisions and additions for solving the linear simultaneous equation $\mathbf{Ax=b}$.

(2) The neural net approach takes the number of operations required to calculate T_{ij} (nothing but matrix-matrix multiplication), that is, $O(n^3/2)$ multiplications and additions for both matrix inversion and solving the linear simultaneous equation. And the time required for output stablization is about a few times the characteristic time constant of the network. The calculation of coupling coefficients can be directly executed without multiple iterations by a specially designed optical matrix-matrix multiplier,[7] while the calculation of bias values in solving a linear simultaneous equation can be done by an optical vector-matrix multiplier.[8] Thus, this approach has a definite advantage in potential calculation speed due to global interconnection of simple parallel analog processors, though its calculation accuracy may be limited by the nature of analog computation. A large number of controllable T_{ij} interconnections may be easily realized with optoelectronic devices.[9]

CONCLUSIONS

We have designed and implemented a matrix inversion network based on the concept of the Hopfield's neural network model. This network is composed of highly interconnected simple neuron-like analog processors which process the information in parallel. The effect of sigmoid or complex nonlinearities on the computational capability is unimportant in this problem. Steep sigmoid functions reduce only the convergence time of the network. When a nonsingular matrix is given as an input, the network converges spontaneously and rapidly to the correct inverse matrix regardless of initial conditions. When a singular matrix is given as an input, the network gives a stable optimum solution that depends upon initial conditions of the network.

REFERENCES

1. J. J. Hopfield and D. W. Tank, Biol. Cybern. 52, 141 (1985).
2. D. W. Tank and J. J. Hopfield, IEEE Trans. Circ. Sys. CAS-33, 533 (1986).
3. J. J. Hopfield, Proc. Natl. Acad. Sci. U.S.A. 79, 2554 (1982).
4. J. J. Hopfield, Proc. Natl. Acad. Sci. U.S.A. 81, 3088 (1984).
5. G. A. Bekey and W. J. Karplus, Hybrid Computation (Wiley, 1968), P. 244.
6. M. J. Maron, Numerical Analysis: A Practical Approach (Macmillan, 1982), p. 138.
7. H. Nakano and K. Hotate, Appl. Opt. 26, 917 (1987).
8. J. W. Goodman, A. R. Dias, and I. M. Woody, Opt. Lett. 2, 1 (1978).
9. J. W. Goodman, F. J. Leonberg, S-Y. Kung, and R. A. Athale, IEEE Proc. 72, 850 (1984).

HOW THE CATFISH TRACKS ITS PREY: AN INTERACTIVE "PIPELINED" PROCESSING SYSTEM MAY DIRECT FORAGING VIA RETICULOSPINAL NEURONS.

Jagmeet S. Kanwal

Dept. of Cellular & Structural Biology, Univ. of Colorado, Sch. of Medicine, 4200 East, Ninth Ave., Denver, CO 80262.

ABSTRACT

Ictalurid catfish use a highly developed gustatory system to localize, track and acquire food from their aquatic environment. The neural organization of the gustatory system illustrates well the importance of the four fundamental ingredients (representation, architecture, search and knowledge) of an "intelligent" system. In addition, the "pipelined" design of architecture illustrates how a goal-directed system effectively utilizes interactive feedback from its environment. Anatomical analysis of neural networks involved in target-tracking indicated that reticular neurons within the medullary region of the brainstem, mediate connections between the gustatory (sensory) inputs and the motor outputs of the spinal cord. Electrophysiological analysis suggested that these neurons integrate selective spatio-temporal patterns of sensory input transduced through a rapidly adapting-type peripheral filter (responding tonically only to a continuously increasing stimulus concentration). The connectivity and response patterns of reticular cells and the nature of the peripheral taste response suggest a unique "gustation-seeking" function of reticulospinal cells, which may enable a catfish to continuously track a stimulus source once its directionality has been computed.

INTRODUCTION

Food search is an example of a broad class of behaviors generally classified as goal-directed behaviors. Goal-directed behavior is frequently exhibited by animals, humans and some machines. Although a preprogrammed, hard-wired machine may achieve a particular goal in a relatively short time, the general and heuristic nature of complex goal-directed tasks, however, is best exhibited by animals and best studied in some of the less advanced animal species, such as fishes, where anatomical, electro-physiological and behavioral analyses can be performed relatively accurately and easily.

Food search, which may lead to food acquisition and ingestion, is critical for the survival of an organism and, therefore, only highly successful systems are selected during the evolution of a species. The act of food search may be classified into two distinct phases, (i) orientation, and (ii) tracking (navigation and homing). In the channel catfish (the animal model utilized for this study), locomotion (swimming) is primarily controlled by the large forked caudal fin, which also mediates turning and directional swimming.

Both these forms of movement, which constitute the essential movements of target-tracking, involve control of the hypaxial/epiaxial muscles of the flank. The alternate contraction of these muscles causes caudal fin undulations. Each cycle of the caudal fin undulation provides either a symmetrical or an asymmetrical bilateral thrust. The former provides a net thrust forward, along the longitudinal axis of the fish causing it to move ahead, while the latter biases the direction of movement towards the right or left side of the fish.

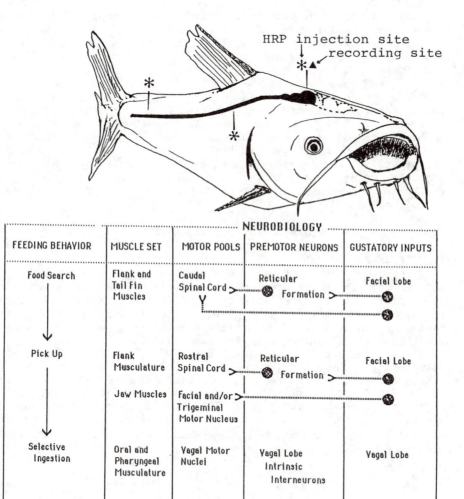

Fig. 1. Schematic representation of possible pathways for the gustatory modulation of foraging in the catfish.

Ictalurid catfishes possess a well developed gustatory system and use it to locate and acquire food from their aquatic environment[1,2,3]. Behavioral evidence also indicates that ictalurid catfishes can detect small intensity (stimulus concentration) differences across their barbels (interbarbel intensity differences), and may use this or other extraoral taste information to compute the directionality in space and track a gustatory stimulus source[1]. In other words, based upon the analysis of locomotion, it may be inferred that during food search, the gustatory sense of the catfish influences the duration and degree of asymmetrical or symmetrical undulations of the caudal fin, besides controlling reflex turns of the head and flank. Since directional swimming is ultimately dependent upon movement of the large caudal fin it may be postulated that, if the gustatory system is to coordinate food tracking, gustato-spinal connections exist upto the level of the caudal fin of the catfish (fig. 1).

The objectives of this study were (i) to reconsider the functional organization of the gustatory system within the costraints of the four fundamental ingredients (representation, architecture, search and knowledge) of a naturally or artificially "intelligent" agent, (ii) to test the existence of the postulated gustato-spinal connections, and (iii) to delineate as far as possible, using neuroanatomical and electrophysiological techniques, the neural mechanism/s involved in the control of goal-directed (foraging) behavior.

ORGANIZATIONAL CONSIDERATIONS

I. REPRESENTATION

Representation refers to the translation of a particular task into information structures and information processes and determines to a great extent the efficiency and efficacy with which a solution to the task can be generated[4]. The elaborate and highly sensitive taste system of an ictalurid catfish consists of an extensive array of chemo- and mechanosensory receptors distributed over most of the extraoral as well as oral regions of the epithelium[2,5]. Peripherally, branches of the facial nerve (which innervates all extraoral taste buds) respond to a wide range of stimulus (amino acids) concentrations[6,7,8] i.e. from 10^{-9}M to 10^{-3}M. The taste activity however, adapts rapidly (phasic response) to ongoing stimulation of the same concentration (Fig. 2) and responds tonically only to continuously increasing concentrations of stimuli, such as L-arginine and L-alanine.

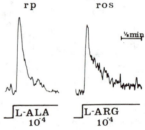

Fig. 1. Integrated, facial taste recordings to continuous application of amino acids to the palate and nasal barbel showing the phasic nature of the taste responses of the ramus palatinus (rp) and ramus ophthalmicus superficialis (ros), respectively.

Gustatory information from the extraoral and oral epithelium is "pipelined" into two separate subsystems, facial and glossopharyngeal-vagal, respectively. Each subsystem processes a subset of the incoming information (extraoral or oral) and coordinates a different component of food acquisition. Food search is accomplished by the extraoral subsystem, while selective ingestion is accomplished by the oral subsystem[2] (Fig. 3). The extraoral gustatory information terminates in the facial lobe where it is represented as a well-defined topographic map[9,10], while the oral information terminates in the adjacent vagal lobe where it is represented as a relatively diffuse map[11].

II. ARCHITECTURE

The information represented in an information structure eventually requires an operating frame (architecture) within which to select and carry out the various processes. In ictalurid catfish, partially processed information from the primary gustatory centers (facial and vagal lobes) in the medullary region of the brainstem converges along ascending and descending pathways (Fig. 4). One of the centers in the ascending pathways is the secondary gustatory nucleus in the isthmic region which is connected to the corresponding nucleus of the opposite side via a large commissure[12,13]. Facial and vagal gustatory information crosses over to the opposite side via this commissure thus making it possible for neurons to extract information about interbarbel or interflank intensity differences. Although neurons in this region are known to have large receptive fields[14], the exact function of this large commissural nucleus is not yet clearly established.

It is quite clear, however, that gustatory information is at first "pipelined" into separate regions where it is processed in parallel[15] before converging onto neurons in the ascending (isthmic) and descending (reticular) processors as well as other regions within the medulla. The "pipelined" architecture underscores the need for differential processing of subsets of sensory inputs which are consequently integrated to coordinate temporal transitions between the various components of goal-directed behavior.

III. SEARCH

An important task underlying all "intelligent" goal directed activity is that of search. In artificial systems this involves application of several general problem-solving methods such as means-end analysis, generate and test methods and heuristic search methods. No attempt, as yet, has been made to fit any of these models to the food-tracking behavior of the catfish. However, behavioral observations suggest that the catfish uses a combinatorial approach resulting in a different yet optimal foraging strategy each time[3].

What is interesting about biological models is that the intrinsic search strategy is expressed extrinsically by the behavior of the animal which, with a few precautions, can be observed quite easily. In addition, simple manipulations of either the animal[3] or its environment can provide interesting data about the search

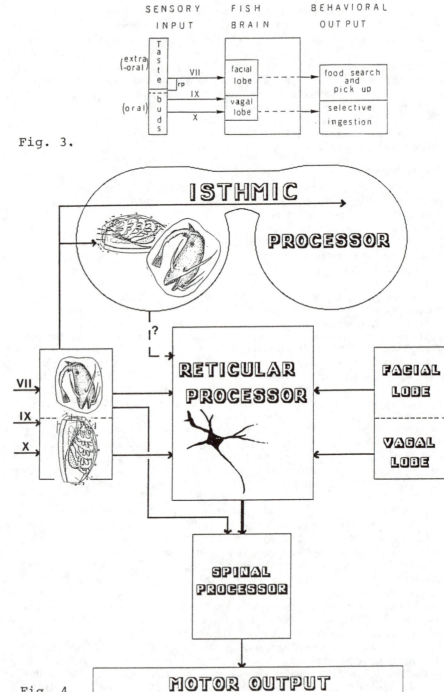

Fig. 3.

Fig. 4.

strategy/ies being used by the animal, which in turn can highlight some of the computational (neuronal) search strategies adopted by the brain e.g. the catfish seems to minimize the probability of failure by continuously interacting with the environment so as to be able to correct any computational or knowledge-based errors.

IV. KNOWLEDGE

If an "intelligent" goal-directed system resets to zero knowledge before each search trial, its success would depend entirely upon the information obtained over the time period of a search. Such a system would also require a labile architecture to process the varying sets of information generated during each search. For such a system, the solution space can become very large and given the constraints of time (generally an important criterion in biological systems) this can lead to continuous failure. For these reasons, knowledge becomes an important ingredient of an "intelligent" agent since it can keep the search under control.

For the gustatory system of the catfish too, randomly accessible knowledge, in combination with the immediately available information about the target, may play a critical role in the adoption of a successful search strategy. Although a significant portion of this knowledge is probably learned, it is not yet clear where and how this knowledge is stored in the catfish brain. The reduction in the solution space for a catfish which has gradually learned to find food in its environment may be attributed to the increase in the amount of knowledge, which to some extent may involve a restructuring of the neural networks during development.

EXPERIMENTAL METHODS

The methods employed for the present study are only briefly introduced here. Neuroanatomical tracing techniques exploit the phenomenon of axonal transport. Crystals of the enzyme, horseradish peroxidase (HRP) or some other substance, when injected at a small locus in the brain, are taken up by the damaged neurons and transported anterogradely and retrogradely from cell bodies and/or axons at the injection site. In the present study, small superficial injections of HRP (Sigma, Type VI) were made at various loci in the facial lobe (FL) in separate animals. After a survival period of 3 to 5 days, the animals were sacrificed and the brains sectioned and reacted for visualization of the neuronal tracer. In this manner, complex neural circuits can be gradually delineated.

Electrophysiological recordings from neurons in the central nervous system were obtained using heat-pulled glass micropipettes. These glass electrodes had a tip diameter of approximately 1 μm and an impedance of less than 1 megohm when filled with an electrolyte (3M KCl or 3M Nacl).

Chemical stimulation of the receptive fields was accomplished by injection of stimuli (amino acids, amino acid mixtures and liver or bait-extract solutions) into a continuous flow of well-water over the receptive epithelium. Tactile stimulation was performed by gentle strokes of a sable hair brush or a glass probe.

408

EXPERIMENTAL OBSERVATIONS

Injections of HRP into the spinal cord labelled two relevant populations of cells, (i) in the ipsilateral reticular formation at the level of the facial lobe (FL), and (ii) a few large scattered cells within the ipsilateral, rostral portion of the lateral lobule of the FL (Fig. 5). Injection of HRP at several sites within the FL resulted in the identification of a small region in the FL from where anterogradely filled fibers project to the reticular formation (Fig. 5). Superimposition of these injection sites onto the anatomical map of the extraoral surface of the catfish indicated that this small region, within the facial lobe, corresponds to the snout region of the extraoral surface.

FACIO-RETICULAR PROJECTIONS   FACIO- & RETICULO -SPINAL PROJECTIONS

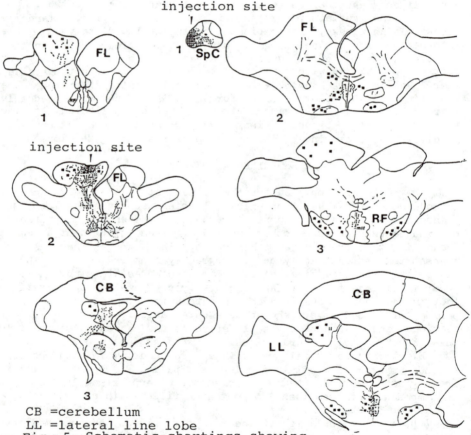

CB =cerebellum
LL =lateral line lobe
Fig. 5. Schematic chartings showing labelled cell bodies (squares) and fibers (dots) in transverse sections through the medulla.

409

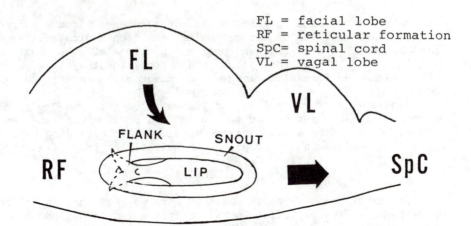

Fig. 6A.

FL = facial lobe
RF = reticular formation
SpC= spinal cord
VL = vagal lobe

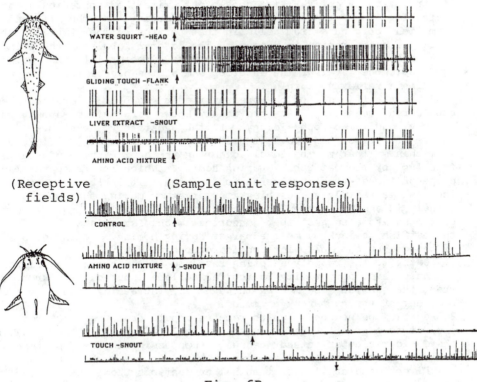

(Receptive fields)   (Sample unit responses)

Fig. 6B.

Multiunit electrophysiological recordings from various anteroposterior levels of the reticular formation indicated that the snout region (upper lip and proximal portion of the maxillary barbels) of the catfish project to a disproportionately large region of the reticular formation along with a mixed representation of the flank (Fig. 6A).

Single unit recordings indicated that some neurons have receptive fields restricted to a bilateral portion of the snout region, while others had large receptive fields extending over the whole flank or over an anteroposterior half of the body (Fig. 6B).

DISCUSSION

The experimental results obtained here suggest that facial lobe projections to the reticular formation form a functional connection. The reticular neurons project to the spinal cord and, most likely, influence the general cycle of swimming-related activity of motoneurons within the spinal cord [16].

The disproportionately large representation of the snout region within the medullary reticular formation, as determined electrophysiologically, is consistent with the anatomical data indicating that most of the fibers projecting to the reticular formation originate from cells in that portion of the facial lobe where the snout region is mapped. The lateral lobule of the spinal cord has a second pathway which projects directly into the spinal cord upto the level of the anterior end of the caudal fin and may coordinate reflexive turning.

The significance of the present results is best understood when considered together with previously known information about the anatomy and electrophysiology of the gustatory system. The information presented above is used to propose a model (Fig. 7) for a mechanism that may be involved during the homing phase of target tracking by the catfish. During homing, which refers to the last phase of target-tracking during food search, it may be assumed that the fish is rapidly approaching its target or moving through a steep signal intensity (stimulus concentration) gradient. The data presented above suggest that a neuronal mechanism exists which helps the catfish to lock on to the target during homing. This proposal is based upon the following considerations:

1. Owing to the rapidly adapting response of the peripheral filter, a tonic level of activity in the facial lobe input can occur only when the animal is moving through an increasing concentration gradient of the gustatory stimulus.

2. Facial lobe neurons, which receive inputs from the snout region, project to a group of cells in the reticular formation. Activity in the facio-reticular pathway causes a suppression in the spontaneous activity of the reticular neurons.

3. Direct and/or indirect spinal projections from the reticular neurons are involved in the modulation of activity of those spinal motoneurons which coordinate swimming. Thus, it may be hypothesized that during complete suppression of activity in a specific reticulo-spinal pathway, the fish swims straight ahead, but during excitation

of certain reticulospinal neurons the fish changes its direction as dictated by the pattern of activation.

Fig. 7. The snout region of the catfish has special significance because of its extensive representation in the reticular formation. In case the fish makes a random or computational error, while approaching its target, the snout is the first region to move out of the stimulus gradient.

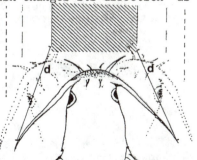

Thus, the spinal motoneurons, teleologically speaking, "seek" a gustatory stimulus in order to suppress activity of certain reticulospinal neurons, which in turn reduce variations in the pattern of activity of swimming-related spinal motoneurons. Accordingly, in a situation where the fish is rapidly approaching a target, ie. under the specific conditions of a continuously rising stimulus concentration at the snout region and an absence of a stimulus intensity difference across the barbels, there is a locking of the movement of the body (of the fish) towards the stationary or moving target (food or prey).

It should be pointed out, however, that the empirical data available so far, only offers clues to the target-tracking mechanism proposed here. Clearly, more research is needed to validate this proposal and to identify other mechanisms of target-tracking utilized by this biological system.

This research was supported in part by NIH Grant NS15258 to T.E. Finger.

REFERENCES

1. P. B. Johnsen and J. H. Teeter, J. Comp. Physiol. 140, 95 (1981).
2. J. Atema, Brain Behav. and Evol. 4, 273-294, (1971).
3. J. E. Bardach, et al., Science, 155, 1276-1278, (1967).
4. A. Newell, Mc-Graw Hill Encyclopedia of Electronics and Computers, (1984), p.71-74.
5. C. J. Herrick, Bull. US. Fish. Comm. 22, 237-272, (1904).
6. J. Caprio, Comp. Biochem. Physiol. 52A, 247-251, (1975).
7. C. J. Davenport and J. Caprio, J. Comp. Physiol. 147, 217 (1982).
8. J. S. Kanwal and J. Caprio, Brain Res. 406, 105-112, (1987).
9. T. E. Finger, J. Comp. Neurol. 165, 513-526 (1976).
10. T. Marui and J. Caprio, Brain Res. 231, 185-190 (1982).
11. J. S. Kanwal and J. Caprio, J. Neurobiol. in press, (1988).
12. C. J. Herrick, J. Comp. Neurol. 15, 375-456 (1905).
13. C. J. Herrick, J. Comp. Neurol. 16, 403-440 (1906).
14. C. F. Lamb and J. Caprio, ISOT, #P70, (1986).
15. T. E. Finger and Y. Morita, Science, 227, 776-778 (1985).
16. P. S. G. Stein, Handbook of the Spinal Cord, (Marcel Dekker Inc., N.Y., 1984), p. 647.

CAPACITY FOR PATTERNS AND SEQUENCES IN KANERVA'S SDM AS COMPARED TO OTHER ASSOCIATIVE MEMORY MODELS

James D. Keeler

*Chemistry Department, Stanford University, Stanford, CA 94305
and RIACS, NASA-AMES 230-5 Moffett Field, CA 94035.
e-mail: jdk@hydra.riacs.edu*

ABSTRACT

The information capacity of Kanerva's Sparse, Distributed Memory (SDM) and Hopfield-type neural networks is investigated. Under the approximations used here, it is shown that the total information stored in these systems is proportional to the number connections in the network. The proportionality constant is the same for the SDM and Hopfield-type models independent of the particular model, or the order of the model. The approximations are checked numerically. This same analysis can be used to show that the SDM can store sequences of spatiotemporal patterns, and the addition of time-delayed connections allows the retrieval of context dependent temporal patterns. A minor modification of the SDM can be used to store correlated patterns.

INTRODUCTION

Many different models of memory and thought have been proposed by scientists over the years. In (1943) McCulloch and Pitts proposed a simple model neuron with two states of activity (on and off) and a large number of inputs.[1] Hebb (1949) considered a network of such neurons and postulated mechanisms for changing synaptic strengths [2] to learn memories. The learning rule considered here uses the outer-product of patterns of +1s and -1s. Anderson (1977) discussed the effect of iterative feedback in such a system.[3] Hopfield (1982) showed that for symmetric connections,[4] the dynamics of such a network is governed by an energy function that is analogous to the energy function of a spin glass.[5] Numerous investigations have been carried out on similar models.[6-8]

Several limitations of these binary interaction, outer-product models have been pointed out. For example, the number of patterns that can be stored in the system (its capacity) is limited to a fraction of the length of the pattern vectors. Also, these models are not very successful at storing correlated patterns or temporal sequences.

Other models have been proposed to overcome these limitations. For example, one can allow higher-order interactions among the neurons.[9,10] In the following, I focus on a model developed by Kanerva (1984) called the Sparse, Distributed Memory (SDM) model.[11] The SDM can be viewed as a three layer network that uses an outer-product learning between the second and third layer. As discussed below, the SDM is more versatile than the above mentioned networks because the number of stored patterns can increased independent of the length of the pattern, and the SDM can be used to store spatiotemporal patterns with context retrieval, and store correlated patterns.

The capacity limitations of outer-product models can be alleviated by using higher-order interaction models or the SDM, but a price must be paid for this added capacity in terms of an increase in the number of connections. How much information is gained per connection? It is shown in the following that the total information stored in each system is proportional to the number of connections in the network, and that the proportionality constant is independent of the particular model or the order of the model. This result also holds if the connections are limited to one bit of precision (clipped weights). The analysis presented here requires certain simplifying assumptions. The approximate results are compared numerically to an exact calculation developed by Chou.[12]

SIMPLE OUTER-PRODUCT NEURAL NETWORK MODEL

As an example or a simple first-order neural network model, I consider in detail the model developed by Hopfield.[4] This model will be used to introduce the mathematics and the concepts that will be generalized for the analysis of the SDM. The "neurons" are simple two-state threshold devices: The state of the ith neuron, u_i, is either either +1 (on), or -1 (off). Consider a set of n such neurons with net input (local field), h_i, to the ith neuron given by

$$h_i = \sum_j^n T_{ij} \, u_j, \tag{1}$$

where T_{ij} represents the interaction strength between the ith neuron and the jth. The state of each neuron is updated asynchronously (at random) according to the rule

$$u_i \leftarrow g(h_i), \tag{2}$$

where the function g is a simple threshold function $g(x) = sign(x)$.

Suppose we are given M randomly chosen patterns (strings of length n of ±1s) which we wish to store in this system. Denote these M memory patterns as pattern vectors: $\mathbf{p}^\alpha = (p_1^\alpha, p_2^\alpha, \ldots, p_n^\alpha)$, $\alpha = 1,2,3, \ldots, M$. For example, $\mathbf{p}^1$ might look like $(+1,-1,+1,-1,-1,\ldots,+1)$. One method of storing these patterns is the outer-product (Hebbian) learning rule: Start with $T \equiv 0$, and accumulate the outer-products of the pattern vectors. The resulting connection matrix is given by

$$T_{ij} = \sum_{\alpha=1}^{M} p_i^\alpha p_j^\alpha, \quad T_{ii} = 0. \tag{3}$$

The system described above is a dynamical system with attracting fixed points. To obtain an approximate upper bound on the total information stored in this network, we sidestep the issue of the basins of attraction, and we check to see if each of the patterns stored by Eq. (3) is actually a fixed point of (2). Suppose we are given one of the patterns, $\mathbf{p}^\beta$, say, as the initial configuration of the neurons. I will show that $\mathbf{p}^\beta$ is expected to be a fixed point of Eq. (2). After inserting (3) for T into (1), the net input to the ith neuron becomes

$$h_i = \sum_{\alpha=1}^{M} p_i^\alpha [\sum_j^n p_j^\alpha p_j^\beta]. \tag{4}$$

The important term in the sum on α is the one for which $\alpha = \beta$. This term represents the "signal" between the input $\mathbf{p}^\beta$ and the desired output. The rest of the sum represents "noise" resulting from crosstalk with all of the other stored patterns. The expression for the net input becomes $h_i = signal_i + noise_i$ where

$$signal_i = p_i^\beta [\sum_j^n p_j^\beta \, p_j^\beta], \tag{5}$$

$$noise_i = \sum_{\alpha \neq \beta}^{M} p_i^\alpha [\sum_j^n p_j^\alpha \, p_j^\beta]. \tag{6}$$

Summing on all of the j_k in (6) yields $signal_i = (n-1)p_i^\beta$. Since n is positive, the sign of the signal term and p_i^β will be the same. Thus, if the noise term were exactly zero, the signal would give the same sign as p_i^β with a magnitude of $\approx n^d$, and $\mathbf{p}^\beta$ would be a fixed point of (2). Moreover, patterns close to $\mathbf{p}^\beta$ would give nearly the same signal, so that $\mathbf{p}^\beta$ should be an attracting fixed point.

For randomly chosen patterns, $<noise> = 0$, where $< >$ indicates statistical expectation, and its variance will be $\sigma^2 = (n-1)^d (M-1)$. The probability that there will be an error on recall of p_i^β is given by the probability that the noise is greater than the signal. For n large, the noise distribution is approximately gaussian, and the probability that there is an error in the ith bit is

$$P_e = \frac{1}{\sqrt{2\pi}\sigma} \int_{|signal|}^{\infty} e^{-x^2/2\sigma^2} dx. \tag{7}$$

INFORMATION CAPACITY

The number of patterns that can be stored in the network is known as its capacity.[13,14] However, for a fair comparison between all of the models discussed here, it is more relevant to compare the total number of bits (total information) stored in each model rather than the number of

414

patterns. This allows comparison of information storage in models with different lengths of the pattern vectors. If we view the memory model as a black box which receives input bit strings and outputs them with some small probability of error in each bit, then the definition of bit-capacity used here is exactly the definition of channel capacity used by Shannon.[15]

Define the *bit-capacity* as the number of bits that can be stored in a network with fixed probability of getting an error in a recalled bit, *i.e.* p_e = *constant* in (10). Explicitly, the bit-capacity is given by[16]

$$B = bit \ capacity = nM\eta, \tag{8}$$

where $\eta = (1 + p_e \log_2 p_e + (1-p_e)\log_2(1-p_e))$. Note that $\eta \approx 1$ for $p_e \approx 0$. Setting p_e to a constant is tantamount to keeping the signal-to-noise ratio (fidelity) constant, where the fidelity, R, is given by $R = |signal|/\sigma$. Explicitly, the relation between (constant) p_e and R, is just $R = \Phi^{-1}(1 - p_e)$ where

$$\Phi(R) = (1/2\pi)^{\frac{1}{2}} \int_{-\infty}^{R} e^{-t^2/2} dt. \tag{9}$$

Hence, the bit-capacity of these networks can be investigated by examining the fidelity of the models as a function of n, M, and R. From (8) and (9) the fidelity of the Hopfield model is $R^2 = n/(n(M-1))^{\frac{1}{2}}$ $(n \gg 1)$. Solving for M in terms of (fixed) R and η, the bit-capacity becomes $B = \eta[(n^2/R^2)+n]$.

The results above can be generalized to models with d^{th} order interactions.[17,18] The resulting expression for the bit-capacity for d^{th} order interaction models is just

$$B = \eta[\frac{n^{d+1}}{R^2}+n]. \tag{10}$$

Hence, we see that the number of bits stored in the system increases with the order d. However, to store these bits, one must pay a price by including more connections in the connection tensor. To demonstrate the relationship between the number of connections and the information stored, define the *information capacity*, γ, to be the total information stored in the network divided by the number of bits in the connection tensor (note that this is different than the definition used by Abu-Mostafa *et al.*).[19] Thus γ is just the bit-capacity divided by the number of bits in the tensor T and represents the efficiency with which information is stored in the network. Since T has n^{d+1} elements, the information capacity is found to be

$$\gamma = \frac{\eta}{R^2 b}, \tag{11}$$

where b is the number of bits of precision used per tensor element ($b \geq \log_2 M$ for no clipping of the weights). For large n, the information stored per neuronal connection is $\gamma = \eta/R^2 b$, independent of the order of the model (compare this result to that of Peretto, *et al.*).[20] To illustrate this point, suppose one decides that the maximum allowed probability of getting an error in a recalled bit is $p_e = 1/1000$, then this would fix the minimum value of R at 3.1. Thus, to store 10,000 bits with a probability of getting an error of a recalled bit of 0.001, equation (15) states that it would take $\approx 96,000b$ bits, independent of the order of the model, or $\approx 0.1n$ patterns can be stored with probability 1/1000 of getting an error in a recalled bit.

KANERVA'S SDM

Now we focus our attention on Kanerva's Sparse, Distributed Memory model (SDM).[11] The SDM can be viewed as a 3-layer network with the middle layer playing the role of hidden units. To get an autoassociative network, the output layer can be fed back into the input layer, effectively making this a two layer network. The first layer of the SDM is a layer of n, ±1 input units (the input address, a), the middle layer is a layer of m, hidden units, s, and the third layer consists of the n ±1 output units (the data, d). The connections between the input units and the hidden units are random weights of ±1 and are given by the $m \times n$ matrix A. The connections between the hidden units and the output units are given by the $n \times m$ connection matrix C, and these matrix elements are modified by an outer-product learning rule (C is analogous to the matrix T of the Hopfield model).

Given an input pattern **a**, the hidden unit activations are determined by

$$s = \theta_r(A\,a),\tag{12}$$

where θ_r is the Hamming-distance threshold function: The kth element is 1 if the input **a** is at most r Hamming units away from the kth row in A, and 0 if it is further than r units away, *i.e.*,

$$\theta_r(\mathbf{x})_i = \begin{cases} 1 & \text{if } \frac{1}{2}(n-x_i)\leq r \\ 0 & \text{if } \frac{1}{2}(n-x_i)> r \end{cases}\tag{13}$$

The hidden-units vector, or *select* vector, **s**, is mostly 0s with an average of δm 1s, where δ is some small number dependent on r; $\delta \ll 1$. Hence, **s** represents a large, sparsely coded vector of 0s and $\delta 1$s representing the input address. The net input, **h**, to the final layer can be simply expressed as the product of C with **s**:

$$h = C\,s.\tag{14}$$

Finally, the output data is given by $\mathbf{d} = \mathbf{g}(\mathbf{h})$, where $g_i(h_i) = sign(h_i)$.

To store the M patterns, $\mathbf{p}^1, \mathbf{p}^2, \cdots \mathbf{p}^M$, form the outer-product of these pattern vectors and their corresponding select vectors,

$$C = \sum_{\alpha=1}^{M} \mathbf{p}^\alpha \mathbf{s}^{\alpha T}.\tag{15}$$

where T denotes the transpose of the vector, and where each select vector is formed by the corresponding address, $\mathbf{s}^\alpha = \theta_r(A\,\mathbf{p}^\alpha)$. The storage algorithm (15) is an outer-product learning rule similar to (3).

Suppose that the M patterns $(\mathbf{p}^1, \mathbf{p}^2, \cdots \mathbf{p}^M)$ have been stored according to (15). Following the analysis presented for the Hopfield model, I show that if the system is presented with $\mathbf{p}^\beta$ as input, the output will be $\mathbf{p}^\beta$, (*i.e.* $\mathbf{p}^\beta$ is a fixed point). Setting $\mathbf{a} = \mathbf{p}^\beta$ in (16) and separating terms as before, the net input (18) becomes

$$h = \mathbf{d}^\beta(\mathbf{s}^\beta \cdot \mathbf{s}^\beta) + \sum_{\alpha \neq \beta}^{M} \mathbf{p}^\alpha(\mathbf{s}^\alpha \cdot \mathbf{s}^\beta).\tag{16}$$

where the first term represents the signal and the second is the noise. Recall that the select vectors have an average of δm 1s and the remainder 0s, so that the expected value of the signal is $\delta m\,\mathbf{s}^\beta$.

Assuming that the addresses and data are randomly chosen, the expected value of the noise is zero. To evaluate the fidelity, I make certain approximations. First, I assume that the select vectors are independent of each other. Second, I assume that the variance of the signal alone is zero or small compared to the variance of noise term alone. The first assumption will be valid for $m\delta^2 \ll 1$, and the second assumption will be valid for $M\delta \gg 1$. With these assumptions, we can easily calculate the variance of the noise term, because each of the select vectors are i.i.d. vectors of length m with mostly 0s and $\approx \delta m$ 1s. With these assumptions, the fidelity is given by

$$R^2 = \frac{m}{[(M-1)(1+\delta^2 m(1-1/m))]}.\tag{17}$$

In the limit of large m, with $\delta m \approx constant$, the number of stored bits scales as

$$B = \eta[\frac{mn}{R^2(1+\delta^2 m)} + n].\tag{18}$$

If we divide this by the number of elements in C, we find the information capacity, $\gamma = \eta/R^2 b$, just as before, so the information capacity is the same for the two models. (If we divide the bit capacity by the number of elements in C and A then we get $\gamma = \eta/R^2(b+1)$, which is about the same for large M.)

A few comments before we continue. First, it should be pointed out that the assumption made by Kanerva[11] and Keeler[17,18] that the variance of the signal term is much less than that of the noise is not valid over the entire range. If we took this into account, then the magnitude of the denominator would be increased by the variance of the signal term. Further, if we read at a distance l away from the write address, then it is easy to see that the signal changes to be $m\,\delta(l)$, where $\delta(l)$ the overlap of two spheres of radius r length l apart in the binomial space n

416

$(\delta \equiv \delta(0))$. The fidelity for reading at a distance l away from the write address is

$$R^2 = \frac{m^2\delta^2(l)}{m\,\delta(l)(1-\delta(l)) + (M-1)m\,\delta^2 + (M-1)\delta^4 m^2(1-1/m)},$$

(19)

Compare this to the formula derived by Chou,[12] for the exact signal-to-noise ratio:

$$R^2 = \frac{m^2\delta^2(l)}{m\,\delta(l)(1-\delta(l)) + (M-1)m\,\mu_{n,r} + (M-1)\sigma_{n,r}^2 m^2(1-1/m))},$$

(20)

where $\mu_{n,r}$ is the average overlap of the spheres of radius r binomially distributed with parameters $(n,1/2)$ and σ^2 is the square of this overlap. The difference in these two formulas lies in the denominator in the terms δ^2 verses $\mu_{n,r}$ and δ^4 vs. $\sigma_{n,r}^2$. The difference comes from the fact that Chou correctly calculates the overlap of the spheres without using the independence assumption.

How do these formula's differ? First of all, it is found numerically that δ^2 is identical with $\mu_{n,r}$. Hence, the only difference comes from δ^4 verses $\sigma_{n,r}^2$. For $m\delta^2 \ll 1$, the δ^4 term is negligible compared to the other terms in the denominator. In addition, δ^4 and σ^2 are approximately equal for large n and $r \approx n/2$. Hence, in the limit $n \to \infty$ the two formulas agree over most of the range if $M \approx 0.1m$, $m \ll 2^n$. However, for finite n, the two formulas can disagree when $m\delta^2 \approx 1$ (see Figure 1).

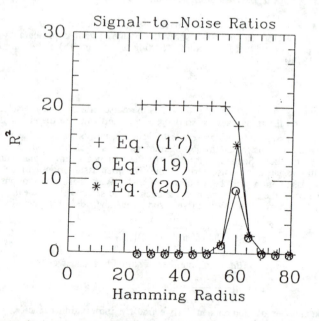

Figure 1: A comparison of the fidelity calculations of the SDM for typical n, M, and m values. Equation (17) was derived assuming no variance of the signal term, and is shown by the + line. Equation (19) uses the approximation that all of the select vectors are independent denoted by the o line. Equation (20) (*'s) is the exact derivation done by Chou[12]. The values used here were $n = 150$, $m = 2000$, $M = 100$.

Equation (20) suggests that there is a best read-write Hamming radius for the SDM. By setting $l = 0$ in (19) and by setting $\dfrac{dR^2}{d\delta} = 0$, we get an approximate expression for the best Hamming radius: $\delta_{best} \approx (2Mm)^{-1/3}$. This trend is qualitatively shown in Figure 2.

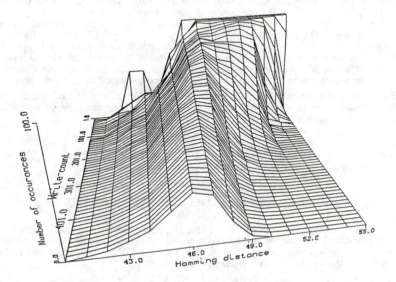

Figure 2: Numerical investigation of the capacity of the SDM. The vertical axis is the percent of recovered patterns with no errors. The x-axis (left to right) is the Hamming distance used for reading and writing. The y-axis (back to forward) is the number of patterns that were written into the memory. For this investigation, $n = 128$, $m = 1024$, and M ranges from 1 to 501. Note the similarity of a cross-section of this graph at constant M with Figure 1. This calculation was performed by David Cohn at RIACS, NASA-Ames.

Figure 1 indicates that the formula (17) that neglected the variance of the signal term is incorrect over much of the range. However, a variant of the SDM is to constrain the number of selected locations to be constant; circuitry for doing this is easily built.[21] The variance of the signal term would be zero in that case, and the approximate expression for the fidelity is given by Eq. (17). There are certain problems where it would be better to keep $\delta = constant$, as in the case of correlated patterns (see below).

The above analysis was done assuming that the elements (weights) in the outer-product matrix are not clipped *i.e.* that there are enough bits to store the largest value of any matrix element. It is interesting to consider what happens if we allow these values to be represented by only a few bits. If we consider the case case $b = 1$, *i.e.* the weights are clipped at one bit, it is easy to show[17] that $\gamma \approx 2\eta/\pi R^2$ for the dth order models and for the SDM, which yields $\gamma = 0.07$ for reasonable R, (this is substantially less than Willshaw's 0.69).

SEQUENCES

In an autoassociative memory, the system relaxes to one of the stored patterns and stay fixed in time until a new input is presented. However, there are many problems where the recalled patterns must change sequentially in time. For example, a song can be remembered as a string of notes played in the correct sequence; cyclic patterns of muscle contractions are essential for walking, riding a bicycle, or dribbling a basketball. As a first step we consider the very simplistic sequence production as put forth by Hopfield (1982) and Kanerva (1984).

Suppose that we wished to store a sequence of patterns in the SDM. Let the pattern vectors be given by $(\mathbf{p}^1, \mathbf{p}^2, \ldots, \mathbf{p}^M)$. This sequence of patterns could be stored by having each pattern point to the next pattern in the sequence. Thus, for the SDM, the patterns would be stored as input-output pairs $(\mathbf{a}^\alpha, \mathbf{d}^\alpha)$, where $\mathbf{a}^\alpha = \mathbf{p}^\alpha$ and $\mathbf{d}^\alpha = \mathbf{p}^{\alpha+1}$ for $\alpha = 1,2,3,\ldots,M-1$. Convergence to this sequence works as follows: If the SDM is presented with an address that is close to $\mathbf{p}^1$ the read data will be close to $\mathbf{p}^2$. Iterating the system with $\mathbf{p}^2$ as the new input address, the read data will be even closer to $\mathbf{p}^3$. As this iterative process continues, the read data will converge to the stored sequence, with the next pattern in the sequence being presented at each time step.

The convergence statistics are essentially the same for sequential patterns as that shown above for autoassociative patterns. Presented with $\mathbf{p}^\alpha$ as an input address, the signal for the stored sequence is found as before

$$<\text{signal}> = \delta m \, \mathbf{p}^{\alpha+1}. \tag{21}$$

Thus, given $\mathbf{p}^\alpha$, the read data is expected to be $\mathbf{p}^{\alpha+1}$. Assuming that the patterns in the sequence are randomly chosen, the mean value of the noise is zero, with variance

$$<\sigma^2> = (M-1)\delta^2 m \, (1+\delta^2(m-1)). \tag{22}$$

Hence, the length of a sequence that can be stored in the SDM increases linearly with m for large m.

Attempting to store sequences like this in the Hopfield model is not very successful due to the asynchronous updating use in the Hopfield model. A synchronously updated outer-product model (for example [6]) would work just as described for the SDM, but it would still be limited to storing fraction of the word size as the maximum sequence length.

Another method for storing sequences in Hopfield-like networks has been proposed independently by Kleinfeld[22] and Sompolinsky and Kanter.[23] These models relieve the problem created by asynchronous updating by using a time-delayed sequential term. This time-delay storage algorithm has different dynamics than the synchronous SDM model. In the time-delay algorithm, the system allows time for the units to relax to the first pattern before proceeding on to the next pattern whereas in the synchronous algorithms, the sequence is recalled imprecisely from imprecise input for the first few iterations and then correctly after that. In other words, convergence to the sequence takes place "on the fly" in the synchronous models — the system does not wait to zero in on the first pattern before proceeding on to recover the following patterns. This allows the synchronous algorithms to proceed k times as fast as the asynchronous time-delay algorithms with half as many (variable) matrix elements. This difference should be able to be detected in biological systems.

TIME DELAYS AND HYSTERESIS: FOLDS

The above scenario for storing sequences is inadequate to explain speech recognition or pattern generation. For example, the above algorithm cannot store sequences of the form $ABAC$, or overlapping sequences. In Kanerva's original work, he included the concept of time delays as a general way of storing sequences with hysteresis. The problem addressed by this is the following: Suppose we wish to store two sequences of patterns that overlap. For example, the two pattern sequences (a,b,c,d,e,f,...) and (x,y,z,d,w,v,...) overlap at the pattern $\mathbf{d}$. If the system only has knowledge of the present state, then when given the input $\mathbf{d}$, it cannot decide whether to output w or e. To store two such sequences, the system must have some knowledge of the immediate past. Kanerva incorporates this idea into the SDM by using "folds." A system with $F+1$ folds has a time history of F past states. These F states may be over the past F time steps or they may go even further back in time, skipping some time steps. The algorithm for reading from the SDM with folds becomes

$$\mathbf{d}(t+1) = g(C^0 \cdot \mathbf{s}(t) + C^1 \cdot \mathbf{s}(t-\tau_1) + \cdots + C^F \cdot \mathbf{s}(t-\tau_F)), \tag{23}$$

where $s(t-\tau_\beta) = \theta_r(A\,a(t-\tau_\beta))$. To store the Q pattern sequences $(\mathbf{p}_1^1, \mathbf{p}_1^2, \ldots, \mathbf{p}_1^{M_1})$, $(\mathbf{p}_2^1, \mathbf{p}_2^2, \ldots, \mathbf{p}_2^{M_2}), \ldots (\mathbf{p}_Q^1, \mathbf{p}_Q^2, \ldots, \mathbf{p}_Q^{M_Q})$, construct the matrix of the βth fold as follows:

$$\mathbf{C}^\beta = w_\beta \sum_{\alpha=1}^{Q} \sum_{\tau=1}^{M_\beta} \mathbf{p}_\alpha^{\tau+1} \times \mathbf{s}_\alpha^{\tau-\tau_\beta}, \tag{24}$$

where any vector with a superscript less than 1 is taken to be zero, $\mathbf{s}_\alpha^{\tau-\tau_x} = \theta_r(A\,\mathbf{p}_\alpha^{\tau-\tau_x})$, and w_β is a weighting factor that would normally decrease with increasing β.

Why do these folds work? Suppose that the system is presented with the pattern sequence $(\mathbf{p}_1^1, \mathbf{p}_1^2, \ldots, \mathbf{p}_1^{M_1})$, with each pattern presented sequentially as input until the τ_F time step. For simplicity, assume that $w_\beta = 1$ for all β. Each term in Eq. (39) will contribute a signal similar to the signal for the single-fold system. Thus, on the τth time step, the signal term coming from Eq. (39) is $\langle \text{signal}(t+1)\rangle = F\,\delta m\,\mathbf{p}_1^{\tau+1}$. The signal will have this value until the end of the pattern sequence is reached. The mean of the noise terms is zero, with variance $\langle \text{noise}^2\rangle = F(M-1)\delta^2 m(1+\delta^2(m-1))$. Hence, the signal-to-noise ratio is $\sqrt{F}$ times as strong as it is for the SDM without folds.

Suppose further that the second stored pattern sequence happens to match the first stored sequence at $t = \tau$. The signal term would then be

$$\text{signal}(t+1) = F\,\delta m\,\mathbf{p}_1^{\tau+1} + \delta m\,\mathbf{p}_2^{\eta+1}. \tag{25}$$

With no history of the past ($F = 1$) the signal is split between $\mathbf{p}_1^{\tau+1}$ and $\mathbf{p}_2^{\eta+1}$, and the output is ambiguous. However, for $F>1$, the signal for the first pattern sequence dominates and allows retrieval of the remainder of the correct sequence. This formulation allows context to aid in the retrieval of stored sequences, and can differentiate between overlapping sequences by using time delays.

The above formulation is still too simplistic in terms of being able to do real recognition problems such as speech recognition. First, the above algorithm can only recall sequences at a fixed time rate, whereas speech recognition occurs at widely varying rates. Second, the above algorithm does not allow for deletions in the incoming data. For example "seqnce" can be recognized as "sequence" even though some letters are missing. Third, as pointed out by Lashley[24] speech processing relies on hierarchical structures.

Although Kanerva's original algorithm is too simplistic, a straightforward modification allows retrieval at different rates with deletions. To achieve this, we can add on the time-delay terms with weights which are smeared out in time. Kanerva's (1984) formulation can thus be viewed as a discrete-time formulation of that put forth by Hopfield and Tank, (1987).[25] Explicitly we could write

$$\mathbf{h} = \sum_{\beta=1}^{F} \sum_{k=\beta-F}^{\beta} W_{\beta k}\, \mathbf{C}^\beta \mathbf{s}(t-\tau_{\beta-k}), \tag{26}$$

where the coefficients $W_{\beta k}$ are a discrete approximation to a smooth function which spreads the delayed signal out over time. As a further step, we could modify these weights dynamically to optimize the signal coming out. The time-delay patterns could also be placed in a hierarchical structure as in the matched filter avalanche structure put forth by Grossberg et al. (1986).[26]

CORRELATED PATTERNS

In the above associative memories, all of the patterns were taken to be randomly chosen, uniformly distributed binary vectors of length n. However, there are many applications where the set of input patterns is not uniformly distributed; the input patterns are correlated. In mathematical terms, the set κ of input patterns would not be uniformly distributed over the entire space of 2^n possible patterns. Let the probability distribution function for the Hamming distance between two randomly chosen vectors $\mathbf{p}^\alpha$ and $\mathbf{p}^\beta$ from the distribution κ be given by the function $\rho(d(\mathbf{p}^\alpha - \mathbf{p}^\beta))$, where $d(\mathbf{x}-\mathbf{y})$ is the Hamming distance between $\mathbf{x}$ and $\mathbf{y}$.

The SDM can be generalized from Kanerva's original formulation so that correlated input patterns can be associated with output patterns. For the moment, assume that the distribution set κ and the probability density function $\rho(x)$ are known a priori. Instead of constructing the rows of the matrix A from the entire space of 2^n patterns, construct the rows of A from the distribution κ. Adjust the Hamming distance r so that $\zeta = \delta m = constant$ number of locations are selected.

In other words, adjust r so that the value of δ is the same as given above, where δ is determined by

$$\delta = \frac{\int_0^r \rho(x)\,dx}{2^n}. \tag{27}$$

This implies that r would have to be adjusted dynamically. This could be done, for example, by a feedback loop. Circuitry for doing this is easily built,[21] and a similar structure appears in the Golgi cells in the Cerebellum.[27].

Using the same distribution for the rows of A as the distribution of the patterns in κ, and using (27) to specify the choice of r, all of the above analysis is applicable (assuming randomly chosen output patterns). If the outputs do not have equal 1s and -1s the mean of the noise is no 0. However, if the distribution of outputs is also known, the system can still be made to work by storing $1/p_+$ and $1/p_-$ for 1s and -1s respectively, where $p_\pm$ is the probability of getting a 1 or a -1 respectively. Using this storage algorithm, all of the above formulas hold, (as long as the distribution is smooth enough and not extremely dense). The SDM will be able to recover data stored with correlated inputs with a fidelity given by Equation (17).

What if the distribution function κ is not known *a priori*? In that case, we would need to have the matrix A learn the distribution $\rho(x)$. There are many ways to build A to mimic ρ. One such way is to start with a random A matrix and modify the entries of δ randomly chosen rows of A at each step according to the statistics of the most recent input patterns. Another method is to use competitive learning[78-30] to achieve the proper distribution of A_k.

The competitive learning algorithm is a method for adjusting the weights A_{ij} between the first and second layer to match this probability density function, $\rho(x)$. The ith row of the address matrix A can be viewd as a vector A_i. The competitive learning algorithm holds a competition between these vectors, and a few vectors that are the closest (within the Hamming sphere r) to the input pattern x are the winners. Each of these winners are then modified slightly in the direction of x. For large enough m, this algorithm almost always converges to a distribution of the A_i that is the same as $\rho(x)$.[XXX] The updating equation for the selected addresses is just

$$A_i^{new} = A_i^{old} - \lambda(A_i^{old} - x) \tag{28}$$

Note for $\lambda = 1$, this reduces to the so-called unary representation of Baum *et al.*[31] Which gives the maximum efficiency in terms of capacity.

DISCUSSION

The above analysis said nothing about the basins of attraction of these memory states. A measure of the performance of a content addressable memory should also say something about the average radius of convergence of the basin of attraction. The basins are in general quite complicated[32] and have been investigated numerically for the unclipped models and values of n and m ranging in the 100s.[21] The basins of attraction for the SDM and the $d=1$ model are very similar in their characteristics and their average radius of convergence. However, the above results give an upper bound on the capacity by looking at the fixed points of the system (if there is no fixed point, there is no basin).

In summary, the above arguments show that the total information stored in outer-product neural networks is a constant times the number of connections between the neurons. This constant is independent of the order of the model and is the same $(\eta/R^2 b)$ for the SDM as well as higher-order Hopfield-type networks. The advantage of going to an architecture like the SDM is that the number of patterns that can be stored in the network is independent of the size of the pattern, whereas the number of stored patterns is limited to a fraction of the word size for the Willshaw or Hopfield architecture. The point of the above analysis is that the efficiency of the SDM in terms of information stored per bit is the same as for Hopfield-type models.

It was also demonstrated how sequences of patterns can be stored in the SDM, and how time delays can be used to recover contextual information. A minor modification of the SDM could be used to recover time sequences at slightly different rates of presentation. Moreover, another minor modification allows the storage of correlated patterns in the SDM. With these modifications, the SDM presents a versatile and efficient tool for investigating properties of associative memory.

Acknowledgements: Discussions with John Hopfield and Pentti Kanerva are gratefully acknowledged. This work was supported by DARPA contract # 86-A227500-000.

REFERENCES

[1] McCulloch, W. S. & Pitts, W. (1943), *Bull. Math. Biophys.* **5**, 115-133.

[2] Hebb, D. O. (1949) *The Organization of Behavior.* John Wiley, New York.

[3] Anderson, J. A., Solverstein, J. W., Ritz, S. A. & Jones, R. S. (1977) *Psych. Rev.,* **84**, 412-451.

[4] Hopfield, J. J. (1982) *Proc. Natn'l. Acad. Sci. USA* **79** 2554-2558.

[5] Kirkpatrick, S. & Sherrington, D. (1978) *Phys Rev.* **17** 4384-4405.

[6] Little, W. A. & Shaw, G. L.(1978)*Math. Biosci.* **39**, 281-289.

[7] Nakano, K. (1972), Association - A model of associative memory, *IEEE Trans. Sys. Man Cyber.* **2**,

[8] Willshaw, D. J., Buneman, O. P. & Longuet-Higgins, H. C., (1969) *Nature*, **222** 960-962.

[9] Lee, Y. C.; Doolen, G.; Chen, H. H.; Sun, G. Z.; Maxwell, T.; Lee, H. Y.; & Giles, L. (1985) *Physica* , **22D**, 276-306.

[10] Baldi, P., and Venkatesh, S. S., (1987) Phys. Rev. Lett. **58**, 913-916.

[11] Kanerva, P. (1984) *Self-propagating Search: A Unified Theory of Memory,* Stanford University Ph.D. Thesis, and Bradford Books (MIT Press). In press (1987 est).

[12] Chou, P. A., *The capacity of Kanerva's Associative Memory* these proceedings.

[13] McEliece, R. J., Posner, E. C., Rodemich, E. R., & Venkatesh, S. S. (1986), *IEEE Trans. on Information Theory.*

[14] Amit, D. J., Gutfreund, H. & Sompolinsky, H. (1985) *Phys. Rev. Lett.* **55**, 1530-1533.

[15] Shannon, C. E., (1948), *Bell Syst. Tech. J.*, **27**, 379,623 (Reprinted in Shannon and Weaver 1949) .

[16] Kleinfeld, D. & Pendergraft, D. B., (1987) Biophys. J. **51**, 47-53.

[17] Keeler, J. D. (1986), *Comparison of Sparsely Distributed Memory and Hopfield-type Neural Network Models,* RIACS Technical Report 86-31, also submitted to *J. Cog. Sci.*

[18] Keeler, J. D. (1987) *Physics Letters* **124A**, 53-58.

[19] Abu-Mostafa, Y. & St. Jacques, (1985), IEEE Trans. on Info. Theor., **31**, 461.

[18] Keeler, J. D., *Basins of Attraction of Neural Network Models* AIP Conf. Proc. #151, Ed: John Denker, Neural Networks for Computing, Snowbird Utah, (1986).

[20] Peretto, P. & J.J. Niez, (1986) Biol. Cybern., **54**. 53-63.

[21] Keeler, J. D., Ph. D. Dissertation. *Collective phenomena of coupled lattice maps: Reaction-diffusion systems and neural networks.* Department of Physics, University of California, San Diego, (1987).

[22] Kleinfeld, D. (1986). *Proc. Nat. Acad. Sci.* **83** 9469-9473.

[23] Sompolinsky, H. & Kanter, I. (1986). *Physical Review Letters.*

[24] Lashley, K. S. (1951). *Cerebral Mechanisms in Behavior.* Edited by Jeffress, L. A. Wiley, New York, 112-136.

[25] Hopfield, J. J. & Tank, D. W. (1987). ICNN San Diego preprint.

[26] Grossberg, S. & Stone, G. (1986). *Psychological Review,* **93**, 46-74

[27] Marr, D. (1969). A *Journal of Phisiology,* **202**, 437-470.

[28] Grossberg, S. (1976). *Biological Cybernetics* **23**, 121-134.

[29] Kohonen, T. (1984) *Self-organization and associative memory.* Springer-Verlag, Berlin.

[30] Rumelhart, D. E. & Zipser, D. *J. Cognitive Sci.,* **9**, (1985), 75.

[31] Baum, E., Moody J., Wilczek F. (1987). Preprint for *Biological Cybernetics*

COMPUTING MOTION USING RESISTIVE NETWORKS

Christof Koch, Jin Luo, Carver Mead
California Institute of Technology, 216-76, Pasadena, Ca. 91125

James Hutchinson
Jet Propulsion Laboratory, California Institute of Technology
Pasadena, Ca. 91125

INTRODUCTION

To us, and to other biological organisms, vision seems effortless. We ope
our eyes and we "see" the world in all its color, brightness, and movemen
Yet, we have great difficulties when trying to endow our machines with simila
abilities. In this paper we shall describe recent developments in the theory o
early vision which lead from the formulation of the motion problem as an il
posed one to its solution by minimizing certain "cost" functions. These cos
or energy functions can be mapped onto simple analog and digital resistiv
networks. Thus, we shall see how the optical flow can be computed by injectin
currents into resistive networks and recording the resulting stationary voltag
distribution at each node. These networks can be implemented in cMOS VLS
circuits and represent plausible candidates for biological vision systems.

APERTURE PROBLEM AND SMOOTHNESS ASSUMPTION

In this study, we use intensity-based schemes for recovering motion. Let u
derive an equation relating the change in image brightness to the motion of th
image (see[1]). Let us assume that the brightness of the image is constant ove
time: $dI(x, y, t)/dt = 0$. On the basis of the chain rule of differentiation, thi
transforms into

$$\frac{\partial I}{\partial x}\frac{dx}{dt} + \frac{\partial I}{\partial y}\frac{dy}{dt} + \frac{\partial I}{\partial t} = I_x u + I_y v + I_t = \nabla I \cdot v + I_t = 0, \qquad (1$$

where we define the velocity v as $(u, v) = (dx/dt, dy/dt)$. Because we assum
that we can compute these spatial and temporal image gradients, we are now
left with a single linear equation in two unknowns, u and v, the two component
of the velocity vector (aperture problem). Any measuring system with a finit
aperture, whether biological or artificial, can only sense the velocity componen
perpendicular to the edge or along the spatial gradient $(-I_t/ \mid \nabla I \mid)$. Th
component of motion perpendicular to the gradient cannot, in principle, b
registered. The problem remains unchanged even if we measure these velocit
components at many points throughout the image.

How can this problem be made well-posed, that is, having a unique solu
tion depending continuously on the data? One form of "regularizing" ill-posed

problems is to restrict the class of admissible solutions by imposing appropriate constraints[2]. Applying this method to motion, we shall argue that in general objects are smooth—except at isolated discontinuities—undergoing smooth movements. Thus, in general, neighboring points in the world will have similar velocities and the projected velocity field should reflect this fact. We therefore impose on the velocity field the constraint that it should be the smoothest as well as satisfying the data. As measure of smoothness we choose, the square of the velocity field gradient. The final velocity field (u, v) is the one that minimizes

$$E(u, v) = \iint (I_x u + I_y v + I_t)^2 +$$

$$\lambda \iint \left[\left(\frac{\partial u}{\partial x} \right)^2 + \left(\frac{\partial u}{\partial y} \right)^2 + \left(\frac{\partial v}{\partial x} \right)^2 + \left(\frac{\partial v}{\partial y} \right)^2 \right] dx\, dy \qquad (2)$$

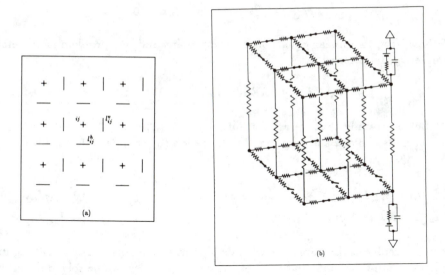

(a)

(b)

Fig. 1. (a) The location of the horizontal (l_{ij}^h) and vertical (l_{ij}^v) line processes relative to the motion field nngrid. (b) The hybrid resistive network, computing the optical flow in the presence of discontinuities. The conductances T_{c-ij} connecting both grids depend on the brightness gradient, as do the conductances g_{ij}^u and g_{ij}^v connecting each node with the battery. For clarity, only two such elements are shown. The battery E_{ij} depends on both the temporal and the spatial gradient and is zero if no brightness change occurs. The x (resp. y) component of the velocity is given by the voltage in the top (resp. bottom) network. Binary switches, which make or break the resistive connections between nodes, implement motion discontinuities. These switches could be under the control of distributed digital processors. Analog cMOS implementations are also feasible[?]

The first term implements the constraint that the final solution should follow as closely as possible the measured data whereas the second term imposes the smoothness constraint on the solution. The degree to which one or the other terms are minimized is governed by the parameter λ. If the data is very accurate, it should be "expensive" to violate the first term and λ will be small. If, conversely, the data is unreliable (low signal-to-noise), much more emphasis will be placed on the smoothness term. Horn and Schunck[1] first formulated this variational approach to the motion problem.

The energy $E(u, v)$ is quadratic in the unknown u and v. It then follows from standard calculus of variation that the associated Euler-Lagrange equations will be linear in u and v:

$$I_x^2 u + I_x I_y v - \lambda \nabla^2 u + I_x I_t = 0$$
$$I_x I_y u + I_y^2 v - \lambda \nabla^2 v + I_y I_t = 0. \tag{3}$$

We now have two linear equations at every point and our problem is therefore completely determined.

ANALOG RESISTIVE NETWORKS

Let us assume that we are formulating eqs. (2) and (3) on a discrete 2-D grid, such as the one shown in fig. 1a. Equation (3) then transforms into

$$I_{xij}^2 u_{ij} + I_{xij} I_{yij} v_{ij} - \lambda \left(u_{i+1j} + u_{ij+1} - 4u_{ij} + u_{i-1j} + u_{ij-1} \right) + I_{xij} I_{tij} = 0$$
$$I_{xij} I_{yij} u_{ij} + I_{yij}^2 v_{ij} - \lambda \left(v_{i+1j} + v_{ij+1} - 4v_{ij} + v_{i-1j} + v_{ij-1} \right) + I_{yij} I_{tij} = 0 \tag{4}$$

where we replaced the Laplacian with its 5 point approximation on a rectangular grid. We shall now show that this set of linear equations can be solved naturally using a particular simple resistive network. Let us apply Kirchhoff's current law to the nodne i, j in the top layer of the resistive network shown in fig. 1b. We then have the following update equation:

$$C \frac{du_{ij}}{dt} = T \left(u_{i+1j} + u_{ij+1} - 4u_{ij} + u_{i-1j} + u_{ij-1} \right)$$
$$+ g_{ij}^u \left(E_{ij} - u_{ij} \right) + T_{c-ij} (v_{ij} - u_{ij}). \tag{5}$$

where v_{ij} is the voltage at node i, j in the bottom network. Once $du_{ij}/dt = 0$ and $dv_{ij}/dt = 0$, this equation is seen to be identical with eq. (4), if we identify

$$T \longrightarrow \lambda$$
$$T_{c-ij} \longrightarrow -I_{xij}I_{yij}$$
$$g_{ij}^u \longrightarrow I_{xij}(I_{xij} + I_{yij})$$
$$g_{ij}^v \longrightarrow I_{yij}(I_{xij} + I_{yij})$$
$$E_{ij} \longrightarrow \frac{-I_t}{I_{xij} + I_{yij}}.$$

(6)

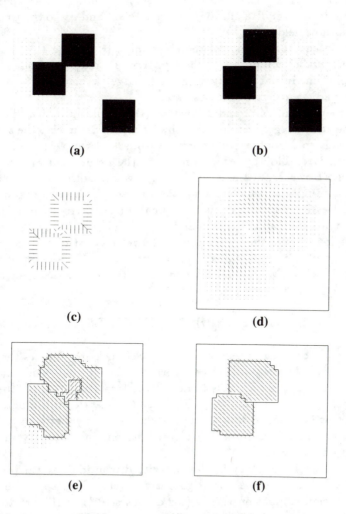

(a) (b)

(c) (d)

(e) (f)

Fig. 2. Motion sequence using synthetic data. (a) and (b) Two images of three high contrast squares on a homogeneous background. (c) The initial velocity data. The inside of both squares contain no data. (d) The final state

of the network after 240 iterations, corresponding to the smooth optical flo
field. (e) Optical flow in the presence of motion discontinuities (indicated b
solid lines). (f) Discontinuities are strongly encouraged to form at the locatio
of intensity edges[4]. Both (e) and (f) show the state of the hybrid network aft
six analog-digital cycles.

Once we set the batteries and the conductances to the values indicated i
eq. (6), the network will settle—following Kirchhoff's laws—into the state ϵ
least power dissipation. The associated stationary voltages correspond to th
sought solution: u_{ij} is equivalent to the x component and v_{ij} to the y componer
of the optical flow field.

We simulated the behavior of these networks by solving the above circu
equations on parallel computers of the Hypercube family. As boundary cond
tions we copied the initial velocity data at the edge of the image into the node
lying directly adjacent but outside the image.

The sequences in figs. 2 and 3 illustrate the resulting optical flow for syr
thetic and natural images. As discussed by Horn and Schunck[1], the smoothne
constraint leads to a qualitatively correct estimate of the velocity field. Thu
one undifferentiated blob appears to move to the lower right and one blob t
the upper left. However, at the occluding edge where both squares overlap, th
smoothness assumption results in a spatial average of the two opposing veloc
ities, and the estimated velocity is very small or zero. In parts of the imag
where the brightness gradient is zero and thus no initial velocity data exists (fc
instance, the interiors of the two squares), the velocity estimates are simply th
spatial average of the neighboring velocity estimates. These empty areas wi
eventually fill in from the boundary, similar to the flow of heat for a uniforr
flat plate with "hot" boundaries.

MOTION DISCONTINUITIES

The smoothness assumption of Horn and Schunck[1] regularizes the apertur
problem and leads to the qualitatively correct velocity field inside moving ok
jects. However, this approach fails to detect the locations at which the velocit
changes abruptly or discontinuously. Thus, it smoothes over the figure-groun
discontinuity or completely fails to detect the boundary between two object
with differing velocities because the algorithm combines velocity informatio
across motion boundaries.

A quite successful strategy for dealing with discontinuities was proposed b
Geman and Geman[5]. We shall not rigorously develop their approach, which i
based on Bayesian estimation theory (for details see[5,6]). Suffice it to say tha
a priori knowledge, for instance, that the velocity field should in general b
smooth, can be formulated in terms of a Markov Random Field model of th
image. Given such an image model, and given noisy data, we then estimat
the "best" flow field by some likelihood criterion. The one we will use her

is the maximum a posteriori estimate, although other criteria are possible and have certain advantages[6]. This can be shown to be equivalent to minimizing an expression such as eq. (2).

In order to reconstruct images consisting of piecewise constant segments, Geman and Geman[5] further introduced the powerful idea of a line process l. For our purposes, we will assume that a line process can be in either one of two states: "on" ($l = 1$) or "off" ($l = 0$). They are located on a regular lattice set between the original pixel lattice (see fig. 1a), such that each pixel i, j has a horizontal l_{ij}^h and a vertical l_{ij}^v line process associated with it. If the appropriate line process is turned on, the smoothness term between the two adjacent pixels will be set to zero. In order to prevent line processes from forming everywhere and, furthermore, in order to incorporate additional knowledge regarding discontinuities into the line processes, we must include an additional term $V_c(l)$ into the new energy function:

$$E(u, v, l^h, l^v) = \sum_{i,j} (I_x u_{ij} + I_y v_{ij} + I_t)^2 +$$

$$\lambda \sum_{i,j} (1 - l_{ij}^h) \left[(u_{i+1j} - u_{ij})^2 + (v_{i+1j} - v_{ij})^2 \right] + \qquad (7)$$

$$\lambda \sum_{i,j} (1 - l_{ij}^v) \left[(u_{ij+1} - u_{ij})^2 + (v_{ij+1} - v_{ij})^2 \right] + V_c(l).$$

V_c contains a number of different terms, penalizing or encouraging specific configurations of line processes:

$$V_c(l) = C_c \sum_{i,j} l_{ij}^h + C_p \sum_{i,j} l_{ij}^h (l_{ij+1}^h + l_{ij+2}^h) + C_I V_I(1), \qquad (8)$$

plus the corresponding expression for the vertical line process l_{ij}^v (obtained by interchanging i with j and l_{ij}^v with l_{ij}^h). The first term penalizes each introduction of a line process, since the cost C_c has to be "payed" every time a line process is turned on. The second term prevents the formation of parallel lines: if either l_{ij+1}^h or l_{ij+2}^h is turned on, this term will tend to prevent l_{ij}^h from turning on. The third term, $C_I V_I$, embodies the fact that in general, motion discontinuities occur along extended contours and rarely intersect (for more details see[7]).

We obtain the optical flow by minimizing the cost function in eq. (7) with respect to both the velocity v and the line processes l^h and l^v. To find an optimal solution to this non-quadratic minimization problem, we follow Koch et al.[7] and use a purely deterministic algorithm, based on solving Kirchhoff's equations for a mixed analog/digital network (see also [8]). Our algorithm exploits the fact that for a fixed distribution of line processes, the energy function (7) is quadratic. Thus, we first initialize the analog resistive network (see fig. 2b) according to eq. (6) and with no line processes on. The network then converges to the smoothest solution. Subsequently, we update the line processes by deciding at each site of the line process lattice whether the overall energy can be lowered by setting or breaking the line process; that is, l_{ij}^h will be turned on if $E(u, v, l_{ij}^h = l, l^v) < E(u, v, l_{ij}^h = 0, l^v)$; otherwise, $l_{ij}^h = 0$. Line processes are switched on by breaking the appropriate resistive connection between the two neighboring nodes. After the completion of one such analog-digital cycle, we reiterate and compute—for the newly updated distribution of line processes—the smoothest state of the analog network. Although there is no guarantee that the system will converge to the global minimum, since we are using a gradient descent rule, it seems to find next-to-optimal solutions in about 10 to 15 analog-digital cycles.

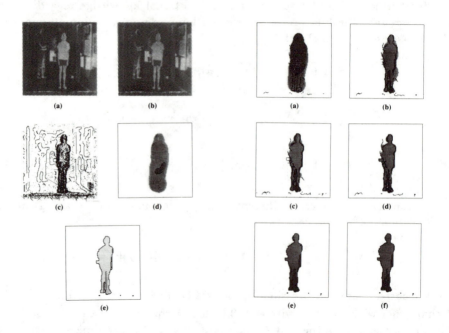

Figure 3. Optical flow of a moving person. (a) and (b) Two 128 by 128 pixel images captured by a video camera. The person in the foreground is moving toward the right while the person in the background is stationary. The noise in the lower part of the image is a camera artifact. (c) Zero-crossings superimposed on the initial velocity data. (d) The smooth optical flow after 1000 iterations. Note that the noise in the lower part of both images is completely smoothed away. (e) The final piecewise smooth optical flow. The velocity field is subsampled to improve visibility. The evolution of the hybrid network is shown after the 1. (a), 3. (b), 5. (c), 7. (d), 10. (e), and 13. (f) analog-digital cycle in the right part of the figure.

The synthetic motion sequence in fig. 2 demonstrates the effect of the line processes. The optical flow outside the discontinuities approximately delineating the boundaries of the moving squares is zero, as it should be (fig. 2e). However, where the two squares overlap the velocity gradient is high and multiple intersecting discontinuities exist. To restrict further the location of discontinuities, we adopt a technique used by Gamble and Poggio[4] to locate depth discontinuities by requiring that depth discontinuities coincide with the location of intensity edges. Our rationale behind this additional constraint is that with very few exceptions, the physical processes and the geometry of the 3-dimensional scene giving rise to the motion discontinuity will also give rise to an intensity edge. As edges we use the zero-crossings of a Laplacian of a Gaussian convolved with the original image[9]. We now add a new term $V_{Z-C_{ij}}$ to our energy function E, such that $V_{Z-C_{ij}}$ is zero if l_{ij} is off or if l_{ij} is on and a zero-crossing exists between locations i and j. If $l_{ij} = 1$ in the absence of a zero-crossing, $V_{Z-C_{ij}}$ is set to 1000. This strategy effectively prevents motion discontinuities from forming at locations where no zero-crossings exist, unless the data strongly suggest it. Conversely, however, zero-crossings by themselves will not induce the formation of discontinuities in the absence of motion gradients (figs. 2f and 3).

ANALOG VLSI NETWORKS

Even with the approximations and optimizations described above, the computations involved in this and similar early vision tasks require minutes to hours on computers. It is fortunate then that modern integrated circuit technology gives us a medium in which extremely complex, analog real-time implementations of these computational metaphors can be realized[3].

We can achieve a very compact implementation of a resistive network using an ordinary cMOS process, provided the transistors are run in the sub-threshold range where their characterstics are ideal for implementing low-current analog functions. The effect of a resistor is achieved by a circuit configuration, such as the one shown in fig. 4, rather than by using the resistance of a special layer in the process. The value of the resulting resistance can be controlled over three orders of magnitude by setting the bias voltages on the upper and lower current source transistors. The current-voltage curve saturates above about 100 mV; a feature that can be used to advantage in many applications. When the voltage gradients are small, we can treat the circuit just as if it were a linear resistor. Resistances with an effective negative resistance value can easily be realized.

In two dimensions, the ideal configuration for a network implementation is shown in fig. 4. Each point on the hexagonal grid is coupled to six equivalent neighbors. Each node includes the resistor apparatus, and a set of sample-and-hold circuits for setting the confidence and signal the input and output voltages. Both the sample-and-hold circuits and the output buffer are addressed by a scanning mechanism, so the stored variables can be refreshed or updated, and the map of node voltages read out in real time.

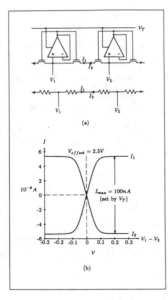

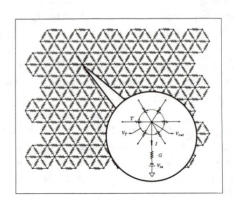

Figure 4. Circuit design for a resistive network for interpolating and smoothing noisy and sparsely sampled depth measurements. (a) Circuit—consisting of transistors—implementing a variable nonlinear resistance. (b) If the voltage gradient is below 100 mV its approximates a linear resistance. The voltage V controls the maximum current and thus the slope of the resistance, which can vary between 1 $M\Omega$ and 1 $G\Omega$ [3]. This cMOS circuit contains 20 by 20 grid points on a hexagonal lattice. The individual resistive elements with a variable slope controlled by V_T correspond to the term governing the smoothness, λ. At those locations where a depth measurement d_{ij} is present, the battery is set to this value ($V_{in} = d_{ij}$) and the value of the conductance G is set to some fixed value. If no depth data is present at that node, G is set to zero. The voltage at each node corresponds to the discrete values of the smoothed surface fitted through the noisy and sparse measurements[7].

A 48 by 48 silicon retina has been constructed that uses the hexagonal network of fig. 4 as a model for the horizontal cell layer in the vertebrate retina[10]. In this application, the input potentials were the outputs of logarithmic photoreceptors—implemented via phototransistors—and the potential difference across the conductance T formed an excellent approximation to the Laplacian operator.

DISCUSSION

We have demonstrated in this study that the introduction of binary motion discontinuities into the algorithm of Horn and Schunck[1] leads to a dramatically improved performance of their method, in particular for the optical flow in the presence of a number of moving non-rigid objects. Moreover, we have shown that the appropriate computations map onto simple resistive networks. We are now implementing these resistive networks into VLSI circuits, using subtheshold cMOS technology. This approach is of general interest, because a great number of problems in early vision can be formulated in terms of similar non-convex energy functions that need to be minimized, such as binocular stereo, edge detection, surface interpolation, structure from motion, etc.[2,6,8].

These networks share several features with biological neural networks. Specifically, they do not require a system-wide clock, they rely on many connections between simple computational nodes, they converge rapidly—within several time constants—and they are quite robust to hardware errors. Another interesting feature is that our networks only consume very moderate amounts of power; the entire retina chip requires about 100 μW [10]

Acknowledgments: An early version of this model was developed and implemented in collaboration with A. L. Yuille[8]. M. Avalos and A. Hsu wrote the code for the Imaging Technology system and E. Staats for the NCUBE. C.K. is supported by an ONR Research Young Investigator Award and by the Sloan and the Powell Foundations. C.M. is supported by ONR and by the System Development Foundation. A portion of this research was carried out at the Jet Propulsion Laboratory and was sponsored by NSF grant No. EET-8714710, and by NASA.

REFERENCES

1. Horn, B. K. P. and Schunck, B. G. Artif. Intell. 17, 185–203 (1981).
2. Poggio, T., Torre, V. and Koch, C. Nature 317, 314–319 (1985).
3. Mead, C. Analog VLSI and Neural Systems. Addison-Wesley: Reading, MA (1988).
4. Gamble, E. and Poggio, T. Artif. Intell. Lab. Memo. No. 970, MIT, Cambridge MA (1987).
5. Geman, S. and Geman, D. IEEE Trans. PAMI 6, 721–741 (1984).
6. Marroquin, J., Mitter, S. and Poggio, T. J. Am. Stat. Assoc. 82, 76–89 (1987).
7. Koch, C., Marroquin, J. and Yuille, A. Proc. Natl. Acad. Sci. USA 83, 4263–4267 (1986).
8. Yuille, A. L. Artif. Intell. Lab. Memo. No. 987, MIT, Cambridge, MA (1987).
9. Marr, D. and Hildreth, E. C. Proc. R. Soc. Lond. B 207, 187–217 (1980).
10. Sivilotti, M. A., Mahowald, M. A. and Mead, C. A. In: 1987 Stanford VLSI Conference, ed. P. Losleben, pp. 295–312 (1987).

Performance Measures for Associative Memories that Learn and Forget

Anthony Kuh
Department of Electrical Engineering
University of Hawaii at Manoa
Honolulu HI, 96822

ABSTRACT

Recently, many modifications to the McCulloch/Pitts model have been propose where both learning and forgetting occur. Given that the network never saturates (ceas to function effectively due to an overload of information), the learning updates can co tinue indefinitely. For these networks, we need to introduce performance measures in add tion to the information capacity to evaluate the different networks. We mathematical define quantities such as the plasticity of a network, the efficacy of an information vect and the probability of network saturation. From these quantities we analytically compa different networks.

1. Introduction

Work has recently been undertaken to quantitatively measure the computation aspects of network models that exhibit some of the attributes of neural networks. T McCulloch/Pitts model discussed in [1] was one of the earliest neural network models to analyzed. Some computational properties of what we call a Hopfield Associative Memc Network (HAMN) similar to the McCulloch/Pitts model was discussed by Hopfield in [The HAMN can be measured quantitatively by defining and evaluating the informati capacity as [2-6] have shown, but this network fails to exhibit more complex computatio capabilities that neural network have due to its simplified structure. The HAMN belor to a class of networks which we call static. In static networks the learning and recall pi cedures are separate. The network first learns a set of data and after learning is comple recall occurs. In dynamic networks, as opposed to static networks, updated learning a associative recall are intermingled and continual. In many applications such as in adapti communications systems, image processing, and speech recognition dynamic networks a needed to adaptively learn the changing information data. This paper formally develc and analyzes some dynamic models for neural networks. Some existing models [7-10] a analyzed, new models are developed, and measures are formulated for evaluating the p formance of different dynamic networks.

In [2-6], the asymptotic information capacity of the HAMN is defined and evaluat In [4-5], this capacity is found by first assuming that the information vectors (IVs) to stored have components that are chosen randomly and independently of all other co ponents in all IVs. The information capacity then gives the maximum number of IVs t can be stored in the HAMN such that IVs can be recovered with high probability duri retrieval. At or below capacity, the network with high probability, successfully recov the desired IVs. Above capacity, the network quickly degrades and eventually fails recover any of the desired IVs. This phenomena is sometimes referred to as the "forgetti catastrophe" [10]. In this paper we will refer to this phenomena as network saturation.

There are two ways to avoid this phenomena. The first method involves learning limited number of IVs such that this number is below capacity. After this learning tal place, no more learning is allowed. Once learning has stopped, the network does i change (defined as static) and therefore lacks many of the interesting computatio capabilities that adaptive learning and neural network models have. The second method is to incorporate some type of forgetting mechanism in the learning structure so that the information stored in the network can never exceed capacity. This type of network would be able to adapt to the changing statistics of the IVs and the network would only be able to recall the most recently learned IVs. This paper focuses on analyzing dynamic networks that adaptively learn new information and do not exhibit network saturation phenomena by selectively forgetting old data. The emphasis is on developing simple models and much of the analysis is performed on a dynamic network that uses a modified Hebbian learning rule.

Section 2 introduces and qualitatively discusses a number of network models that are classified as dynamic networks. This section also defines some pertinent measures for evaluating dynamic network models. These measures include the plasticity of a network, the probability of network saturation, and the efficacy of stored IVs. A network with no plasticity cannot learn and a network with high plasticity has interconnection weights that exhibit large changes. The efficacy of a stored IV as a function of time is another important parameter as it is used in determining the rate at which a network forgets information.

In section 3, we mathematically analyze a simple dynamic network referred to as the Attenuated Linear Updated Learning (ALUL) network that uses linear updating and a modified Hebbian rule. Quantities introduced in section 3 are analytically determined for the ALUL network. By adjusting the attenuation parameter of the ALUL network, the forgetting factor is adjusted. It is shown that the optimal capacity for a large ALUL network in steady state defined by (2.13,3.1) is a factor of e less than the capacity of a HAMN. This is the tradeoff that must be paid for having dynamic capabilities. We also conjecture that no other network can perform better than this network when a worst case criterion is used. Finally, section 4 discusses further directions for this work along with possible applications in adaptive signal processing.

2. Dynamic Associative Memory Networks

The network models discussed in this paper are based on the concept of associative memory. Associative memories are composed of a collection of interconnected elements that have data storage capabilities. Like other memory structures, there are two operations that occur in associative memories. In the learning operation (referred to as a write operation for conventional memories), information is stored in the network structure. In the recall operation (referred to as a read operation for conventional memories), information is retrieved from the memory structure. Associative memories recall information on the basis of data content rather than by a specific address. The models that we consider will have learning and recall operations that are updated in discrete time with the activation state $X(j)$ consisting of N cells that take on the values $\{-1,1\}$.

2.1. Dynamic Network Measures

General associative memory networks are described by two sets of equations. If we let $X(j)$ represent the activation state at time j and $W(k)$ represent the weight matrix or interconnection state at time k then the activation or recall equation is described by

$$X(j+1) = f(X(j),W(k)), \qquad j \geq 0,\ k \geq 0,\ X(0) = \hat{X} \tag{2.1}$$

where $\hat{X}$ is the data probe vector used for recall. The learning algorithm or interconnection equation is described by

$$W(k+1) = g(V(i), 0 \leq i \leq k, W(0)) \tag{2.2}$$

where $\{V(i)\}$ are the information vectors (IVs) to be stored and $W(0)$ is the initial state of the interconnection matrix. Usually the learning algorithm time scale is much longer than

434

the recall equation time scale so that W in (2.1) can be considered time invariant. Ofte (2.1) is viewed as the equation governing short term memory and (2.2) is the equatic governing long term memory. From the Hebbian hypothesis we note that the data prob vectors should have an effect on the interconnection matrix W. If a number of data prob vectors recall an IV $V(i)$, the strength of recall of the IV $V(i)$ should be increased b appropriate modification of W. If another IV is never recalled, it should gradually be fo gotten by again adjusting terms of W. Following the analysis in [4,5] we assume that a components of IVs introduced are independent and identically distributed Bernoulli randor variables with the probability of a 1 or -1 being chosen equal to $\frac{1}{2}$.

Our analysis focuses on learning algorithms. Before describing some dynamic learnir algorithms we present some definitions. A network is defined as dynamic if given som period of time the rate of change of W is never nonzero. In addition we will primarily di cuss networks where learning is gradual and updated at discrete times as shown in (2.2 By gradual, we want networks where each update usually consists of one IV being learne and/or forgotten. IVs that have been introduced recently should have a high probability recovery. The probability of recall for one IV should also be a monotonic decreasing func tion of time, given that the IV is not repeated. The networks that we consider should als have a relatively low probability of network saturation.

Quantitatively, we let $e(k,l,i)$ be the event that an IV introduced at time l can b recovered at time k with a data probe vector which is of Hamming distance i from th desired IV. The efficacy of network recovery is then given as $p(k,l,i) = Pr(e(k,l,i))$. I the analysis performed we say a a vector V can recover $V(l)$, if $V(l) = \Delta(V)$ where $\Delta($ is a synchronous activation update of all cells in the network. The capacity for dynam networks is then given by

$$C(k,i,\epsilon) = \max m \ni Pr(r(e(k,l,i),0 \le l < k) = m) > 1 - \epsilon \qquad 0 \le i < \frac{N}{2} \qquad (2.$$

where $r(X)$ gives the cardinality of the number of events that occur in the set X. Closel related to the capacity of a network is network saturation. Saturation occurs when th network is overloaded with IVs such that few or none of the IVs can be successfull recovered. When a network at time 0 starts to learn IVs, at some time $l < j$ we have tha $C(l,i,\epsilon) \ge C(j,i,\epsilon)$. For $k \ge l$ the network saturation probability is defined by $S(k,m$ where S describes the probability that the network cannot recover m IVs.

Another important measure in analyzing the performance of dynamic networks is th plasticity of the interconnections of the weight matrix W. Following definitions that a similar to [10], define

$$h(k) = \frac{\sum_{i \ne j} \sum_{j=1}^{N} \text{VAR}\{W_{i,j}(k) - W_{i,j}(k-1)\}}{N(N-1)} \qquad (2.$$

as the incremental synaptic intensity and

$$H(k) = \frac{\sum_{i \ne j} \sum_{j=1}^{N} \text{VAR}\{W_{i,j}(k)\}}{N(N-1)} \qquad (2.$$

as the cumulative synaptic intensity. From these definitions we can define the plasticity the network as

$$P(k) = \frac{h(k)}{H(k)} \qquad (2.$$

When network plasticity is zero, the network does not change and no learning takes plac When plasticity is high, the network interconnections exhibit large changes.

When analyzing dynamic networks we are often interested if the network reaches a steady state. We say a dynamic network reaches steady state if

$$\lim_{k \to \infty} H(k) = H \qquad (2.7)$$

where H is a finite nonzero constant. If the IVs have stationary statistics and given that the learning operations are time invariant, then if a network reaches steady state, we have that

$$\lim_{k \to \infty} P(k) = P \qquad (2.8)$$

where P is a finite constant. It is also easily verified from (2.6) that if the plasticity converges to a nonzero constant in a dynamic network, then given the above conditions on the IVs and the learning operations the network will eventually reach steady state.

Let us also define the synaptic state at time k for activation state V as

$$s(k, V) = W(k) V \qquad (2.9)$$

From the synaptic state, we can define the SNR of V, which we show in section 3 is closely related to the efficacy of an IV and the capacity of the network.

$$\text{SNR}(k, V, i) = \frac{(\text{E}(s_i(k, V)))^2}{\text{VAR}(s_i(k, V))} \qquad (2.10)$$

Another quantity that is important in measuring dynamic networks is the complexity of implementation. Quantities dealing with network complexity are discussed in [12] and this paper focuses on networks that are memoryless. A network is memoryless if (2.2) can be expressed in the following form:

$$W(k+1) = g^*(W(k), V(k)) \qquad (2.11)$$

Networks that are not memoryless have the disadvantage that all IVs need to be saved during all learning updates. The complexity of implementation is greatly increased in terms of space complexity and very likely increased in terms of time complexity.

2.2. Examples of Dynamic Associative Memory Networks

The previous subsection discussed some quantities to measure dynamic networks. This subsection discusses some examples of dynamic associative memory networks and qualitatively discusses advantages and disadvantages of different networks. All the networks considered have the memoryless property.

The first network that we discuss is described by the following difference equation

$$W(k+1) = a(k) W(k) + b(k) L(V(k)) \qquad k \geq 1 \qquad (2.12)$$

with $W(0)$ being the initial value of weights before any learning has taken place. Networks with these learning rules will be labeled as Linear Updated Learning (LUL) networks and in addition if $0 < a(k) < 1$ for $k \geq 0$ the network is labeled as an Attenuated Linear Updated Learning (ALUL) network. We will primarily deal with ALUL where $0 < a(k) < 1$ and $b(k)$ do not depend on the position in W. This model is a specialized version of Grossberg's Passive Decay LTM equation discussed in [11]. If the learning algorithm is of the correlation type then

$$L(V(k)) = V(k) V(k)^T - I \qquad k \geq 1 \qquad (2.13)$$

This learning scheme has similarities to the marginalist learning schemes introduced in [10]. One of the key parameters in the ALUL network is the value of the attenuation coefficient a. From simulations and intuition we know that if the attenuation coefficient is to high, the network will saturate and if the attenuation parameter is to low, the network will forget all but the most recently introduced IVs. Fig. 1 uses Monte Carlo methods to show a plot of the number of IVs recoverable in a 64 cell network when $a=1$, (the HAMN) as function of the learning time scale. From this figure we clearly see that network saturation is exhibited and for the time $k \geq 25$ no IV are recoverable with high probability. Section further analyzes the ALUL network and derives the value of different measures introduced in section 2.1.

Another learning scheme called bounded learning (BL) can be described by

$$L(V(k)) = \begin{cases} V(k)V(k)^T - I & F(W(k) \geq \overline{A} \\ 0 & F(W(k)) < \overline{A} \end{cases} \tag{2.14}$$

By setting the attenuation parameter $a=1$ and letting

$$F(W(k)) = \max_{i,j} W_{i,j}(k) \tag{2.15}$$

this is identical to the learning with bounds scheme discussed in [10]. Unfortunately there is a serious drawbacks to this model. If $\overline{A}$ is too large the network will saturate with high probability. If $\overline{A}$ is set such that the probability of network saturation is low then the network has the characteristic of not learning for almost all values of $k > k(\overline{A}) = \min l \ni F(W(l)) \geq \overline{A}$. Therefore we have that the efficacy of network recovery, $p(k,l,0) \approx 0$ for all $k \geq l \geq k(\overline{A})$.

In order for the (BL) scheme to be classified as dynamic learning, the attenuation parameter a must have values between 0 and 1. This learning scheme is just a more complex version of the learning scheme derived from (2.10,2.11). Let us qualitatively analyze the learning scheme when a and b are constant. There are two cases to consider. When $\overline{A} > H$, then the network is not affected by the bounds and the network behaves as the ALUL network. When $\overline{A} < H$, then the network accepts IVs until the bound is reached. When the bound is reached, the network waits until the values of the interconnection matrix have attenuated to the prescribed levels where learning can continue. If $\overline{A}$ is judiciously chosen, BL with $a \leq 1$ provides a means for a network to avoid saturation. By holding an IV until $H(k) < \overline{A}$, it is not too difficult to show that this learning scheme equivalent to an ALUL network with $b(k)$ time varying.

A third learning scheme called refresh learning (RL) can be described by (2.12) with $b(k)=1$, $W(0)=0$, and

$$a(k) = 1 - \delta(k \bmod(l)) \tag{2.16}$$

This learning scheme learns a set of IV and periodically refreshes the weighting matrix so that all interconnections are 0. RL can be classified as dynamic learning, but learning not gradual during the periodic refresh cycle. Another problem with this learning scheme that the efficacy of the IVs depend on where during the period they were learned. IV learned late in a period are quickly forgotten where as IVs learned early in a period have longer time in which they are recoverable.

In all the learning schemes introduced, the network has both learning and forgetting capabilities. A network introduced in [7,8] separates the learning and forgetting tasks by using the standard HAMN algorithm to learn IV and a random selective forgetting algorithm to unlearn excess information. The algorithm which we call random selective forgetting (RSF) can be described formally as follows.

$$W(k+1) = Y(k) + L(V(k)) \qquad k \geq 1 \tag{2.17}$$

where

$$Y(k) = W(k) - \mu(k) \sum_{i=1}^{n(F(W(k)))} (V(k,i)V(k,i)^T - n(F(W(k)))I) \tag{2.18}$$

Each of the vectors $V(k,i)$ are obtained by choosing a random vector V in the same manner IVs are chosen and letting V be the initial state of the HAMN with interconnection matrix $W(k)$. The recall operation described by (2.1) is repeated until the activation has settled into a local minimum state. $V(k,i)$ is then assigned this state. $\mu(k)$ is the rate at which the randomly selected local minimum energy states are forgotten, $W(k)$ is given by (2.15), and $n(X)$ is a nonnegative integer valued function that is a monotonically increasing function of X.

The analysis of the RSF algorithm is difficult, because the energy manifold that describes the energy of each activation state and the updates allowable for (2.1) must be well understood. There is a simple transformation between the weighting matrix and the energy of an activation state given below,

$$E(X(k)) = -\tfrac{1}{2}\sum_i\sum_j W_{i,j}X_i(j)X_j(k) \quad k\geq 0 \tag{2.19}$$

but aggregately analyzing all local minimum energy activation states is complex. Through computer simulations and simplified assumptions [7,8] have come up with a qualitative explanation of the RSF algorithm based on an eigenvalue approach.

3. Analysis of the ALUL Network

Section 2 focused on defining properties and analytical measures for dynamic AMN along with presenting some examples of some learning algorithms for dynamic AMN. This section will focus on the analysis of one of the simpler algorithms, the ALUL network. From (2.12) we have that the time invariant ALUL network can be described by the following interconnection state equation.

$$W(k+1) = aW(k) + bL(V(k)) \quad k\geq 1 \tag{3.1}$$

where a and b are nonnegative real numbers. Many of the measures introduced in section 2 can easily be determined for the ALUL network.

To calculate the incremental synaptic intensity $h(k)$ and the cumulative synaptic intensity $H(k)$ let the initial condition of the interconnection state $W_{i,j}(0)$ be independent of all other interconnections states and independent of all IVs. If $E W_{i,j}(0) = 0$ and $\mathbf{VAR}\, W_{i,j}(0) = \gamma$ then

$$h(k) = (1-a)^2\left\{b^2\frac{1-a^{2(k-1)}}{1-a^2} + a^{2(k-1)}\gamma\right\}+ b^2 \tag{3.2}$$

and

$$H(k) = b^2\frac{1-a^{2k}}{1-a^2} + a^{2k}\gamma \tag{3.3}$$

In steady state when $a<1$ we have that

$$P = 2(1-a) \tag{3.4}$$

From this simple relationship between the attenuation parameter a and the plasticity measure P, we can directly relate plasticity to other measures such as the capacity of the network.

We define the steady state capacity as $C(i,\epsilon)= \lim_{k\to\infty} C(k,i,\epsilon)$ for networks where steady state exists. To analytically determine the capacity first assume that $S(k,V(j)) = S(k-j)$ is a jointly Gaussian random vector. Further assume that $S_i(l)$ for $1\leq i\leq N$, $1\leq l\leq m$ are all independent and identically distributed. Then for N sufficiently large, $f(a) = \overline{a^{2(k-j-1)}(1-a^2)}$, and

$$\mathbf{SNR}(k, V(j)) = \mathbf{SNR}(k-j) = \frac{(N-1)f(a)}{1-f(a)}$$

$$= c(a)\log N \gg 1 \quad j < k \qquad (3.\text{?})$$

we have that

$$p(k,j,0) = \left(1 - \frac{N^{\frac{-c(a)}{2}}}{\sqrt{2\pi c(a)\log N}}\right)^N$$

$$\approx 1 - \frac{N^{1-\frac{c(a)}{2}}}{\sqrt{2\pi c(a)\log N}} \quad j < k \qquad (3.\text{?})$$

Given a we first find the largest $m = k-j > 0$ where $\lim_{N\to\infty} p(k,j,0) \approx 1$. Note that $\lim_{N\to\infty} p(k,j,0) = 1$ when $c(a) \geq 2$. By letting $c(a) = 2$ the maximum m is given when

$$\frac{f(a)}{1-f(a)} = \frac{2\log N}{N} \qquad (3.\text{?})$$

Solving for m we get that

$$m = \frac{1}{2}\frac{\log\left[\frac{2\log N}{(N+2\log N)(1-a^2)}\right]}{\log a} + 1 \qquad (3.\text{?})$$

It is also possible to find the value of a that maximizes m. If we let $\epsilon = 1 - a^2$, then

$$m \approx \frac{\log\left[\frac{2\log N}{(N+2\log N)\epsilon}\right]}{\epsilon} \qquad (3.\text{?})$$

m is at a maximum value when $\epsilon \approx \frac{2e\log N}{N}$ or when $m \approx \frac{N}{2e\log N}$. This corresponds to $a \approx \frac{2m-1}{2m}$. Note that this is a factor of e less than the maximum number of IVs allowable in a static HAMN [4,5], such that one of the IVs is recoverable. By following the analysis in [5], the independence assumption and the Gaussian assumptions used earlier can be removed. The arguments involve using results from exchangeability theory and normal approximation theory.

A similar and somewhat more cumbersome analysis can be performed to show that at steady state the maximum capacity achievable is when $a \approx \frac{2m-1}{2m}$ and given by

$$\lim_{N\to\infty} C(k,0,\epsilon) = \frac{N}{4e\log N} \qquad (3.1\text{?})$$

This again is a factor of e less than the maximum number of IVs allowable in a static HAMN [4,5], such that all IVs are recoverable. Fig. 2 shows a Monte Carlo simulation of the number of IVs recoverable in a 64 cell network versus the learning time scale for a varying between .5 and .99. We can see that the network reaches approximate steady state when $k \geq 35$. The maximum capacity achievable is when $a \approx .9$ and the capacity is around 5. This is slightly more than the theoretical value predicted by the analysis just shown when we compare to Fig. 1. For smaller simulations conducted with larger networks the simulated capacity was closer to the predicted value. From the simulations and the analysis we observe that when a is too small IVs are forgotten at too high a rate and when

a is too high network saturation occurs.

Using the same arguments, it is possible to analyze the capacity of the network and efficacy of IVs when k is small. Assuming zero initial conditions and $a \approx \dfrac{2m-1}{2m}$ we can summarize the learning behavior of the ALUL network. The learning behavior can be divided into three phases. In the first phase for $k \leq \dfrac{N}{4e\log N}$ all IVs are remembered and the characteristics of the network are similar to the HAMN below saturation. In the second phase some IVs are forgotten as the rate of forgetting becomes nonzero. During this phase the maximum capacity is reached as shown in fig. 2. At this capacity the network cannot dynamically recall all IVs so the network starts to forget more information then it receives. This continues until steady state is reached where the learning and forgetting rates are equal. If initial conditions are nonzero the network starts in phase 1 or the beginning of phase 2 if $H(k)$ is below the value corresponding to the maximum capacity and at the end of phase 2 for larger $H(k)$.

The calculation of the network saturation probabilities $S(k,m)$ is trivial for large networks when the capacity curves have been found. When $m \leq C(k,0,\epsilon)$ then $S(k,m) \approx 0$ otherwise $S(k,m) \approx 1$.

Before leaving this section let us briefly examine ALUL networks where $a(k)$ and $b(k)$ are time varying. An example of a time varying network is the marginalist learning scheme introduced in [10]. The network is defined by fixing the value of the $\mathbf{SNR}(k,k-1,i) = D(N)$ for all k. This value is fixed by setting $a = 1$ and varying b. Since the $\mathbf{VAR}S_i(k,V(k-1))$ is a monotonic increasing function of k, $b(k)$ must also be a monotonic increasing function of k. It is not too difficult to show that when k is large, the marginalist learning scheme is equivalent to the steady state ALUL defined by (3.1). The argument is based on noting that the steady state $\mathbf{SNR}$ depends not on the update time, but on the difference between the update time and when the IV was stored as is the case with the marginalist learning scheme. The optimal value of $D(N)$ giving the highest capacity is when $D(N) = 4e\log N$ and

$$b(k+1) = \frac{2m}{2m-1}b(k) \tag{3.11}$$

where $m = \dfrac{N}{4e\log N}$.

If performance is defined by a worst case criterion with the criterion being

$$J(l,N) = \min(C(k,0,\epsilon), k \geq l) \tag{3.12}$$

then we conjecture that for l large, no ALUL as defined in (2.12,2.13) can have larger $J(l,N)$ than the optimal ALUL defined by (3.1). If we consider average capacity, we note that the RL network has an average capacity of $\dfrac{N}{8\log N}$ which is larger than the optimal ALUL network defined in (3.1). However, for most envisioned applications a worst case criterion is a more accurate measure of performance than a criterion based on average capacity.

4. Summary

This paper has introduced a number of simple dynamic neural network models and defined several measures to evaluate the performance of these models. All parameters for the steady state ALUL network described by (3.1) were evaluated and the attenuation parameter a giving the largest capacity was found. This capacity was found to be a factor of e less than the static HAMN capacity. Furthermore we conjectured that if we consider a worst case performance criteria that no ALUL network could perform better than the optimal ALUL network defined by (3.1). Finally, a number of other dynamic mode including BL, RL, and marginalist learning were stated to be equivalent to ALUL networ under certain conditions.

The network models that were considered in this paper all have binary vector value activation states and may be to simplistic to be considered in many signal processing appl cation. By generalizing the analysis to more complicated models with analog vector value activation states and continuous time updating it may be possible to use these generalize models in speech and image processing. A specific example would be a controller for moving robot. The generalized network models would learn the input data by adaptive changing the interconnections of the network. Old data would be forgotten and data th was repeatedly being recalled would be reinforced. These network models could also b used when the input data statistics are nonstationary.

References

[1] W. S. McCulloch and W. Pitts, *"A Logical Calculus of the Ideas Iminent in Nervo Activity"*, Bulletin of Mathematical Biophysics, 5, 115-133, 1943.

[2] J. J. Hopfield, *"Neural Networks and Physical Systems with Emergent Collective Con putational Abilities"*, Proc. Natl. Acad. Sci. USA 79, 2554-2558, 1982.

[3] Y. S. Abu-Mostafa and J. M. St. Jacques, *"The Information Capacity of the Hopfie Model"*, IEEE Trans. Inform. Theory, vol. IT-31, 461-464, 1985.

[4] R. J. McEliece, E. C. Posner, E. R. Rodemich and S. S. Venkatesh, *"The Capacity the Hopfield Associative Memory"*, IEEE Trans. Inform. Theory, vol. IT-33, 461-48 1987.

[5] A. Kuh and B. W. Dickinson, *"Information Capacity of Associative Memories"*, to b published IEEE Trans. Inform. Theory.

[6] D. J. Amit, H. Gutfreund, and H. Sompolinsky, *"Spin-Glass Models of Neural Ne works"*, Phys. Rev. A, vol. 32, 1007-1018, 1985.

[7] J. J. Hopfield, D. I. Feinstein, and R. G. Palmer, *" 'Unlearning' has a Stabilizin effect in Collective Memories"*, Nature, vol. 304, 158-159, 1983.

[8] R. J. Sasiela, *"Forgetting as a way to Improve Neural-Net Behavior"* , AIP Confe ence Proceedings 151, 386-392, 1986.

[9] J. D. Keeler, *"Basins of Attraction of Neural Network Models"*, AIP Conferenc Proceedings 151, 259-265, 1986.

[10] J. P. Nadal, G. Toulouse, J. P. Changeux, and S. Dehaene, *"Networks of Formo Neurons and Memory Palimpsests"*, Europhysics Let., Vol. 1, 535-542, 1986.

[11] S. Grossberg, *"Nonlinear Neural Networks: Principles, Mechanisms, and Architec tures"*, Neural Networks in press.

[12] S. S. Venkatesh and D. Psaltis, *"Information Storage and Retrieval in Two Associa tive Nets"*, California Institute of Technology Pasadena, Dept. of Elect. Eng., pre print, 1986.

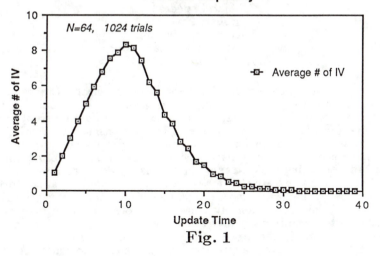

Fig. 1

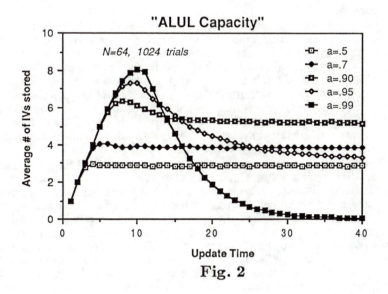

Fig. 2

How Neural Nets Work

Alan Lapedes
Robert Farber
Theoretical Division
Los Alamos National Laboratory
Los Alamos, NM 87545

Abstract:

There is presently great interest in the abilities of neural networks to mimic "qualitative reasoning" by manipulating neural incodings of symbols. Less work has been performed on using neural networks to process floating point numbers and it is sometimes stated that neural networks are somehow inherently inaccurate and therefore best suited for "fuzzy" qualitative reasoning. Nevertheless, the potential speed of massively parallel operations make neural net "number crunching" an interesting topic to explore. In this paper we discuss some of our work in which we demonstrate that for certain applications neural networks can achieve significantly higher numerical accuracy than more conventional techniques. In particular, prediction of future values of a chaotic time series can be performed with exceptionally high accuracy. We analyze how a neural net is able to do this , and in the process show that a large class of functions from $R^n \to R^m$ may be accurately approximated by a backpropagation neural net with just two "hidden" layers. The network uses this functional approximation to perform either interpolation (signal processing applications) or extrapolation (symbol processing applications). Neural nets therefore use quite familiar methods to perform their tasks. The geometrical viewpoint advocated here seems to be a useful approach to analyzing neural network operation and relates neural networks to well studied topics in functional approximation.

1. Introduction

Although a great deal of interest has been displayed in neural network's capabilities to perform a kind of qualitative reasoning, relatively little work has been done on the ability of neural networks to process floating point numbers in a massively parallel fashion. Clearly, this is an important ability. In this paper we discuss some of our work in this area and show the relation between numerical, and symbolic processing. We will concentrate on the the subject of accurate prediction in a time series. Accurate prediction has applications in many areas of signal processing. It is also a useful, and fascinating ability, when dealing with natural, physical systems. Given some data from the past history of a system, can one accurately predict what it will do in the future?

Many conventional signal processing tests, such as correlation function analysis, cannot distinguish deterministic chaotic behavior from from stochastic noise. Particularly difficult systems to predict are those that are nonlinear and chaotic. Chaos has a technical definition based on nonlinear, dynamical systems theory, but intuitivly means that the system is deterministic but "random," in a rather similar manner to deterministic, pseudo random number generators used on conventional computers. Examples of chaotic systems in nature include turbulence in fluids (D. Ruelle, 1971; H. Swinney, 1978), chemical reactions (K. Tomita, 1979), lasers (H. Haken, 1975), plasma physics (D. Russel, 1980) to name but a few. Typically, chaotic systems also display the full range of non-linear behavior (fixed points, limit cycles etc.) when parameters are varied, and therefore provide a good testbed in which to investigate techniques of nonlinear signal processing. Clearly, if one can uncover the underlying, deterministic algorithm from a chaotic time series, then one may be able to predict the future time series quite accurately.

In this paper we review and extend our work (Lapedes and Farber,1987) on predicting the behavior of a particular dynamical system, the Glass-Mackey equation. We feel that the method will be fairly general, and use the Glass-Mackey equation solely for illustrative purposes. The Glass-Mackey equation has a strange attractor with fractal dimension controlled by a constant parameter appearing in the differential equation. We present results on a neural network's ability to predict this system at two values of this parameter, one value corresponding to the onset of chaos, and the other value deeply in the chaotic regime. We also present the results of more conventional predictive methods and show that a neural net is able to achieve significantly better numerical accuracy. This particular system was chosen because of D. Farmer's and J. Sidorowich's (D. Farmer, J. Sidorowich, 1987) use of it in developing a new, non-neural net method for predicting chaos. The accuracy of this non-neural net method, and the neural net method, are roughly equivalent, with various advantages or disadvantages accruing to one method or the other depending on one's point of view. We are happy to acknowledge many valuable discussions with Farmer and Sidorowich that has led to further improvements in each method.

We also show that a neural net never needs more than two hidden layers to solve most problems . This statement arises from a more general argument that a neural net can approximate functions from $R^n \rightarrow R^m$ with only two hidden layers, and that the accuracy of the approximation is controlled by the number of neurons in each layer. The argument assumes that the global minimum to the backpropagation minimization problem may be found, or that a local minima very close in value to the global minimum may be found. This seems to be the case in the examples we considered, and in many examples considered by other researchers, but is never guaranteed. The conclusion of an upper bound of two hidden layers is related to a similar conclusion of R. Lipman (R. Lipman, 1987) who has previously analyzed the number of hidden layers needed to form arbitrary decision regions for symbolic processing problems. Related issues are discussed by J. Denker (J. Denker et.al. 1987) It is easy to extend the argument to draw similar conclusions about an upper bound of two hidden layers for symbol processing and to place signal processing, and symbol processing in a common theoretical framework.

2. Backpropagation

Backpropagation is a learning algorithm for neural networks that seeks to find weights, T_{ij}, such that given an input pattern from a training set of pairs of Input/Output patterns, the network will produce the Output of the training set given the Input. Having learned this mapping between I and O for the training set, one then applies a new, previously unseen Input, and takes the Output as the "conclusion" drawn by the neural net based on having learned fundamental relationships between Input and Output from the training set. A popular configuration for backpropagation is a totally feedforward net (Figure 1) where Input feeds up through "hidden layers" to an Output layer.

444

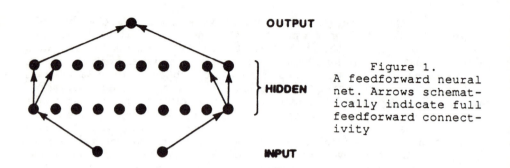

OUTPUT

HIDDEN

INPUT

Figure 1.
A feedforward neural
net. Arrows schemat-
ically indicate full
feedforward connect-
ivity

Each neuron forms a weighted sum of the inputs from previous layers to which it is connected, adds a threshold value, and produces a nonlinear function of this sum as its output value. This output value serves as input to the future layers to which the neuron is connected, and the process is repeated. Ultimately a value is produced for the outputs of the neurons in the Output layer. Thus, each neuron performs:

$$X_i^{out} = g\left(\sum_j T_{ij} X_j^{in} + \theta_i\right) \tag{1}$$

where T_{ij} are continuous valued, positive or negative weights, θ_i is a constant, and $g(x)$ is a nonlinear function that is often chosen to be of a sigmoidal form. For example, one may choose

$$g(x) = \frac{1}{2}\left(1 + tanh x\right) \tag{2}$$

where tanh is the hyperbolic tangent, although the exact formula of the sigmoid is irrelevant to the results.
If $t_i^{(p)}$ are the target output values for the pth Input pattern then ones trains the network by minimizing

$$E = \sum_p \sum_i \left(t_i^{(p)} - 0_i^{(p)}\right)^2 \tag{3}$$

where $t_i^{(p)}$ is the target output values (taken from the training set) and $0_i^{(p)}$ is the output of the network when the pth Input pattern of the training set is presented on the Input layer. i indexes the number of neurons in the Output layer.

An iterative procedure is used to minimize E. For example, the commonly used steepest descents procedure is implemented by changing T_{ij} and θ_i by ΔT_{ij} and $\Delta\theta_i$ where

$$\Delta T_{ij} = -\frac{\partial E}{\partial T_{ij}} \cdot \epsilon \qquad (4a)$$

$$\Delta \theta_i = -\frac{\partial E}{\partial \theta_i} \cdot \epsilon \qquad (4b)$$

This implies that $\Delta E < 0$ and hence E will decrease to a local minimum. Use of the chain rule and definition of some intermediate quantities allows the following expressions for ΔT_{ij} to be obtained (Rumelhart, 1987):

$$\Delta T_{ij} = \sum_p \epsilon \delta_i^{(p)} 0_j^{(p)} \qquad (5a)$$

$$\Delta \theta_i = \epsilon \sum_{\wedge} \delta_i^{(p)} \qquad (5b)$$

where

$$\delta_i^{(p)} = \left(t_i^{(p)} - 0_i^{(p)} \right) 0_i^{(p)} (1 - 0_i^{(p)}) \qquad (6)$$

if i is labeling a neuron in the Output layer; and

$$\delta_i^{(p)} = 0_i^{(p)} (1 - 0_i^{(p)}) \sum_j T_{ij} \delta_j^{(p)} \qquad (7)$$

if i labels a neuron in the hidden layers. Therefore one computes $\delta_i^{(p)}$ for the Output layer first, then uses Eqn. (7) to computer $\delta_i^{(p)}$ for the hidden layers, and finally uses Eqn. (5) to make an adjustment to the weights. We remark that the steepest descents procedure in common use is extremely slow in simulation, and that a better minimization procedure, such as the classic conjugate gradient procedure (W. Press, 1986), can offer quite significant speedups. Many applications use bit representations (0,1) for symbols, and attempt to have a neural net learn fundamental relationships between the symbols. This procedure has been successfully used in converting text to speech (T. Sejnowski, 1986) and in determining whether a given fragment of DNA codes for a protein or not (A. Lapedes, R. Farber, 1987).

There is no fundamental reason, however, to use integer's as values for Input and Output. If the Inputs and Outputs are instead a collection of floating point numbers, then the network, after training, yields a specific continuous function in n variables (for n inputs) involving g(x) (i.e. hyperbolic tanh's) that provides a type of nonlinear, least mean square interpolant formula for the discrete set of data points in the training set. Use of this formula $0 = f(I_1, I_2, \ldots I_n)$ when given a new input not in the training set, is then either interpolation or extrapolation.

Since the Output values, when assumed to be floating point numbers may have a dynamic range great than [0,1], one may modify the g(x) on the Output layer to be a linear function, instead of sigmoidal, so as to encompass the larger dynamic range. Dynamic range of the Input values is not so critical, however we have found that numerical problems may be avoided by scaling the Inputs (and also the Outputs) to [0,1], training the network, and then rescaling the T_{ij}, θ_i to encompass the original dynamic range. The point is that scale changes in I and O may, for feedforward networks, always be absorbed in the T_{ij}, θ_i and vice versa. We use this procedure (backpropagation, conjugate gradient, linear outputs and scaling) in the following section to predict points in a chaotic time series.

3. Prediction

Let us consider situations in Nature where a system is described by nonlinear differential equations. This is faily generic. We choose a particular nonlinear equation that has an infinite dimensional phase space, so that it is similar to other infinite dimensional systems such as partial differential equations. A differential equation with an infinite dimensional phase space (i.e. an infinite number of values are necessary to describe the initial condition) is a delay, differential equation. We choose to consider the time series generated by the Glass-Mackey equation:

$$\dot{x} = \frac{ax(t-\tau)}{1 + x^{10}(t-\tau)} - bx(t) \tag{8}$$

This is a nonlinear differential, delay equation with an initial condition specified by an initial function defined over a strip of width τ (hence the infinite dimensional phase space i.e. initial functions, not initial constants are required). Choosing this function to be a constant function, and a = .2, b = .1, and $\tau = 17$ yields a time series, x(t), (obtained by integrating Eqn. (8)), that is chaotic with a fractal attractor of dimension 2.1. Increasing τ to 30 yields more complicated evolution and a fractal dimension of 3.5. The time series for 500 time steps for $\tau=30$ (time in units of τ) is plotted in Figure 2. The nonlinear evolution of the system collapses the infinite dimensional phase space down to a low (approximately 2 or 3 dimensional) fractal, attracting set. Similar chaotic systems are not uncommon in Nature.

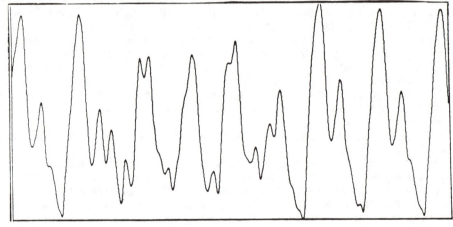

Figure 2. Example time series at tau = 30.

The goal is to take a set of values of x() at discrete times in some time window containing times less than t, and use the values to accurately predict $x(t + P)$, where P is some prediction time step into the future. One may fix P, collect statistics on accuracy for many prediction times t (by sliding the window along the time series), and then increase P and again collect statistics on accuracy. This one may observe how an average index of accuracy changes as P is increased. In terms of Figure 2 we will select various prediction time steps, P, that correspond to attempting to predict within a "bump," to predicting a couple of "bumps" ahead. The fundamental nature of chaos dictates that prediction accuracy will decrease as P is increased. This is due to inescapable inaccuracies of finite precision in specifying the x(t) at discrete times in the past that are used for predicting the future. Thus, all predictive methods will degrade as P is increased – the question is "How rapidly does the error increase with P?" We will demonstrate that the neural net method can be orders of magnitude more accurate than conventional methods at large prediction time steps, P.

Our goal is to use backpropagation, and a neural net, to construct a function

$$0(t + P) = f\left(I_1(t), I_2(t - \Delta) \ldots I_m(t - m\Delta)\right) \tag{9}$$

where $0(t + P)$ is the output of a single neuron in the Output layer, and $I_1 \to I_m$ are input neurons that take on values $x(t), x(t - \Delta) \ldots x(t - m\Delta)$, where Δ is a time delay. $0(t + P)$ takes on the value $x(t + P)$. We chose the network configuration of Figure 1.

We construct a training set by selecting a set of input values:

$$I_1 = x(t_p)$$

$$I_2 = x(t_p - \Delta) \tag{10}$$

$$I_m = x(t_p - m\Delta)$$

with associated output values $0 = x(t_p + P)$, for a collection of discrete times that are labelled by t_p. Typically we used 500 I/O pairs in the training set so that p ranged from $1 \to 500$. Thus we have a collection of 500 sets of $\{I_1^{(p)}, I_2^{(p)}, \ldots, I_m^{(p)}; 0^{(p)}\}$ to use in training the neural net. This procedure of using delayed sampled values of x(t) can be implemented by using tapped delay lines, just as is normally done in linear signal processing applications, (B. Widrow, 1985). Our prediction procedure is a straightforward nonlinear extension of the linear Widrow Hoff algorithm. After training is completed, prediction is performed on a new set of times, t_p, not in the training set i.e. for p = 500.

We have not yet specified what m or Δ should be, nor given any indication why a formula like Eqn. (9) should work at all. An important theorem of Takens (Takens, 1981) states that for flows evolving to compact attracting manifolds of dimension d_A, that a functional relation like Eqn. (9) does exist, and that m lies in the range $d_A < m + 1 < 2d_A + 1$. We therefore choose m = 4, for $\tau = 30$. Takens provides no information on Δ and we chose $\Delta = 6$ for both cases. We found that a few different choices of m and Δ can affect accuracy by a factor of 2 - a somewhat significant but not overwhelming sensitivity, in view of the fact that neural nets tend to be orders of magnitude more accurate than other methods. Takens theorem gives no information on the form of f() in Eqn. (9). It therefore is necessary to show that neural nets provide a robust approximating procedure for continuous f(), which we do in the following section. It is interesting to note that attempts to predict future values of a time series using past values of x(t) from a tapped delay line is a common procedure in signal processing, and yet there is little, if any, reference to results of nonlinear dynamical systems theory showing why any such attempt is reasonable.

After training the neural net as described above, we used it to predict 500 new values of x(t) in the future and computed the average accuracy for these points. The accuracy is defined to be the average root mean square error, divided by a constant scale factor, which we took to be the standard deviation of the data. It is necessary to remove the scale dependence of the data and dividing by the standard deviation of the data provides a scale to use. Thus the resulting "index of accuracy" is insensitive to the dynamic range of x(t).

As just described, if one wanted to use a neural net to continuously predict x(t) values at, say, 6 time steps past the last observed value (i.e. wanted to construct a net predicting x(t + 6)) then one would train one network, at P = 6, to do this. If one wanted to always predict 12 time steps past the last observed x(t) then a separate, P = 12, net would have to be trained. We, in fact, trained separate networks for P ranging between 6 and 100 in steps of 6. The index of accuracy for these networks (as obtained by computing the index of accuracy in the prediction phase) is plotted as curve D in Figure 3. There is however an alternate way to predict. If one wished to predict, say, x(t + 12) using a P = 6 net, then one can iterate the P = 6 net. That is, one uses the P = 6 net to predict the x(t +6) values, and then feeds x(t +6) back into the input line to predict x(t + 12) using the **predicted** x(t + 6) value instead of the **observed** x(t + 6) value. In fact, one can't use the observed x(t +6) value, because it hasn't been observed yet – the rule of the game is to use only data occurring at time t and before, to predict x(t +12). This procedure corresponds to iterating the map given by Eqn. (9) to perform prediction at multiples of P. Of course, the delays, Δ, must be chosen commensurate with P.

This iterative method of prediction has potential dangers. Because (in our example of iterating the P = 6 map) the predicted x(t + 6) is always made with some error, then this error is compounded in iteration, because predicted, and not observed values, are used on the input lines. However, one may predict more accurately for smaller P, so it may be the case that choosing a very accurate small P prediction, and iterating, can ultimately achieve higher accuracy at the larger P's of interest. This turns out to be true, and the iterated net method is plotted as curve E in Figure 3. It is the best procedure to use. Curves A,B,C are alternative methods (iterated polynomial,Widrow-Hoff, and non-iterated polynomial respectively. More information on these conventional methods is in (Lapedes and Farber, 1987)).

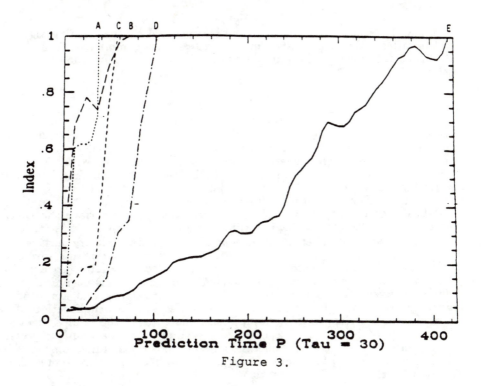

Figure 3.

4. Why It Works

Consider writing out explicitly Eqn. (9) for a two hidden layer network where the output is assumed to be a linear neuron. We consider Input connects to Hidden Layer 1, Hidden Layer 1 to Hidden Layer 2, and Hidden Layer 2 to Output. Therefore:

$$0_\ell = \sum_{k \in H_2} T_{\ell k} g \left(\sum_{i \in H_1} T_{ki} g \left(\sum_{j \in I} T_{ij} I_j + \theta_j \right) + \theta_k \right) + \theta_\ell \qquad (11)$$

Recall that the output neurons a linear computing element so that only two g()s occur in formula (11), due to the two nonlinear hidden layers. For ease in later analysis, let us rewrite this formula as

$$0_\ell = \sum_{k \in H_2} T_{\ell k} g \left(SUM_k + \theta_k \right) + \theta_\ell \qquad (12a)$$

where

$$SUM_k = \sum_{i \in H_1} T_{ki} g \left(\sum_{j \in I} T_{ij} I_j + \theta_j \right). \qquad (12b)$$

450

The T's and θ's are specific numbers specified by the training algorithm, so that after training is finished one has a relatively complicated formula (12a, 12b) that expresses the Output value as a specific, known, function of the Input values:

$$0_\ell = f\left(I_1, I_2, \ldots I_m\right).$$

A functional relation of this form, when there is only one output, may be viewed as surface in m + 1 dimensional space, in exactly the same manner one interprets the formula z = f(x,y) as a two dimensional surface in three dimensional space. The general structure of f() as determined by Eqn. (12a, 12b) is in fact quite simple. From Eqn. (12b) we see that one first forms a sum of g() functions (where g() is s sigmoidal function) and then from Eqn. (12a) one forms yet another sum involving g() functions. It may at first be thought that this special, simple form of f() restricts the type of surface that may be represented by $0_\ell = f(I_j)$. This initial thought is wrong – the special form of Eqn. (12) is actually a general representation for quite arbitrary surfaces.

To prove that Eqn. (12) is a reasonable representation for surfaces we first point out that surfaces may be approximated by adding up a series of "bumps" that are appropriately placed. An example of this occurs in familiar Fourier analysis, where wave trains of suitable frequency and amplitude are added together to approximate curves (or surfaces). Each half period of each wave of fixed wavelength is a "bump," and one adds all the bumps together to form the approximant. Let us now see how Eqn. (12) may be interpreted as adding together bumps of specified heights and positions. First consider SUM$_k$ which is a sum of g( ) functions. In Figure (4) we plot an example of such a g() function for the case of two inputs.

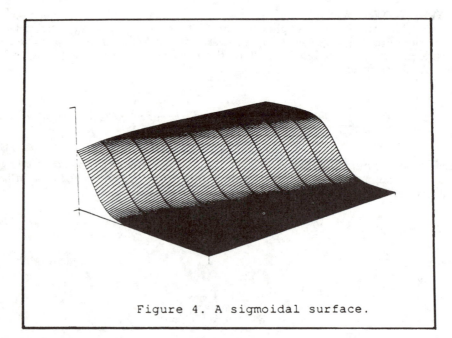

Figure 4. A sigmoidal surface.

The orientation of this sigmoidal surface is determined by T_{ij}, the position by θ_j, and height by T_{ki}. Now consider another g() function that occurs in SUM_k. The θ_j of the second g() function is chosen to displace it from the first, the T_{ij} is chosen so that it has the same orientation as the first, and T_{ki} is chosen to have opposite sign to the first. These two g( ) functions occur in SUM_k, and so to determine their contribution to SUM_k we sum them together and plot the result in Figure (5). The result is a ridged surface.

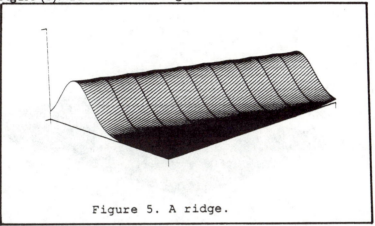

Figure 5. A ridge.

Since our goal is to obtain localized bumps we select another pair of g() functions in SUM_k, add them together to get a ridged surface perpendicular to the first ridged surface, and then add the two perpendicular ridged surfaces together to see the contribution to SUM_k. The result is plotted in Figure (6).

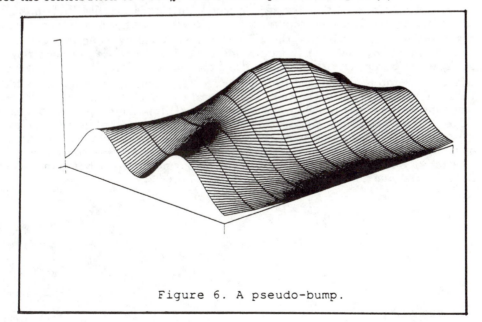

Figure 6. A pseudo-bump.

We see that this almost worked, in so much as one obtains a local maxima by this procedure. However there are also saddle-like configurations at the corners which corrupt the bump we were trying to obtain. Note that one way to fix this is to take $g(\text{SUM}_k + \theta_k)$ which will, if θ_k is chosen appropriately, depress the local minima and saddles to zero while simultaneously sending the central maximum towards 1. The result is plotted in Figure (7) and is the sought after bump.

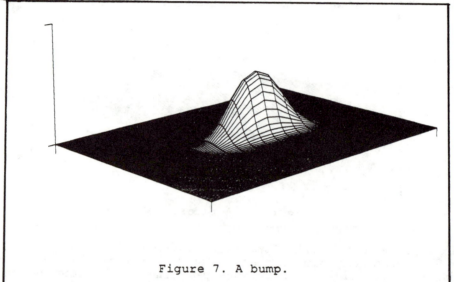

Figure 7. A bump.

Furthermore, note that the necessary g() function is supplied by Eqn. (12). Therefore Eqn. (12) is a procedure to obtain localized bumps of arbitrary height and position. For two inputs, the kth bump is obtained by using four g() functions from SUM_k (two g() functions for each ridged surface and two ridged surfaces per bump) and then taking g() of the result in Eqn. (12a). The height of the kth bump is determined by $T_{\ell k}$ in Eqn. (12a) and the k bumps are added together by that equation as well. The general network architecture which corresponds to the above procedure of adding two g() functions together to form a ridge, two perpendicular ridges together to form a pseudo-bump, and the final g() to form the final bump is represented in Figure (8). To obtain any number of bumps one adds more neurons to the hidden layers by repeatedly using the connectivity of Figure (8) as a template (i.e. four neurons per bump in Hidden Layer 1, and one neuron per bump in Hidden Layer 2).

Figure 8. Connectivity needed to obtain one bump. Add four more neurons to Hidden layer 1, and one more neuron to Hidden Layer 2, for each additional bump.

One never needs more than two layers, or any other type of connectivity than that already schematically specified by Figure (8). The accuracy of the approximation depends on the number of bumps, which in turn is specified, by the number of neurons per layer. This result is easily generalized to higher dimensions (more than two Inputs) where one needs 2m hiddens in the first hidden layer, and one hidden neuron in the second layer for each bump.

The argument given above also extends to the situation where one is processing symbolic information with a neural net. In this situation, the Input information is coded into bits (say 0s and 1s) and similarly for the Output. Or, the Inputs may still be real valued numbers, in which case the binary output is attempting to group the real valued Inputs into separate classes. To make the Output values tend toward 0 and 1 one takes a third and final g() on the output layer, i.e. each output neuron is represented by $g(O_\ell)$ where O_ℓ is given in Eqn. (11). Recall that up until now we have used linear neurons on the output layer. In typical backpropagation examples, one never actually achieves a hard 0 or 1 on the output layers but achieves instead some value between 0.0 and 1.0. Then typically any value over 0.5 is called 1, and values under 0.5 are called 0. This "postprocessing" step is not really outside the framework of the network formalism, because it may be performed by merely increasing the slope of the sigmoidal function on the Output layer. Therefore the only effect of the third and final g() function used on the Output layer in symbolic information processing is to pass a hyperplane through the surface we have just been discussing. This plane cuts the surface, forming "decision regions," in which high values are called 1 and low values are called 0. Thus we see that the heart of the problem is to be able to form surfaces in a general manner, which is then cut by a hyperplane into general decision regions. We are therefore able to conclude that the network architecture consisting of just two hidden layers is sufficient for learning any symbol processing training set. For Boolean symbol mappings one need not use the second hidden layer to remove the saddles on the bump (c.f. Fig. 6). The saddles are lower than the central maximum so one may choose a threshold on the output layer to cut the bump at a point over the saddles to yield the correct decision region. Whether this representation is a reasonable one for subsequently achieving good prediction on a prediction set, as opposed to "memorizing" a training set, is an issue that we address below.

We also note that use of Sigma II; units (Rummelhart, 1986) or high order correlation nets (Y.-C. Lee, 1987) is an attempt to construct a surface by a general polynomial expansion, which is then cut by a hyperplane into decision regions, as in the above. Therefore the essential element of all these neural net learning algorithms are identical (i.e. surface construction), only the particular method of parameterizing the surface varies from one algorithm to another. This geometrical viewpoint, which provides a unifying framework for many neural net algorithms, may provide a useful framework in which to attempt construction of new algorithms.

Adding together bumps to approximate surfaces is a reasonable procedure to use when dealing with real valued inputs. It ties in to general approximation theory (c.f. Fourier series, or better yet, B splines), and can be quite successful as we have seen. Clearly some economy is gained by giving the neural net bumps to start with, instead of having the neural net form its own bumps from sigmoids. One way to do this would be to use multidimensional Gaussian functions with adjustable parameters.

The situation is somewhat different when processing symbolic (binary valued) data. When input symbols are encoded into N bit bit-strings then one has well defined input values in an N dimensional input space. As shown above, one can learn the training set of input patterns by appropriately forming and placing bump surfaces over this space. This is an effective method for **memorizing** the training set, but a very poor method for obtaining correct predictions on new input data. The point is that, in contrast to real valued inputs that come from, say, a chaotic time series, the input points in symbolic processing problems are widely separated and the bumps do not add together to form smooth surfaces. Furthermore, each input bit string is a corner of an 2^N vertex hypercube, and there is no sense in which one corner of a hypercube is surrounded by the other corners. Thus the commonly used input representation for symbolic processing problems requires that the neural net **extrapolate** the surface to make a new prediction for a new input pattern (i.e. new corner of the hypercube) and not interpolate, as is commonly the case for real valued inputs. Extrapolation is a farmore dangerous procedure than interpolation, and in view of the separated bumps of the training set one might expect on the basis of this argument that neural nets would fail dismally at symbol processing. This is not the case.

The solution to this apparent conundrum, of course, is that although it is sufficient for a neural net to learn a symbol processing training set by forming bumps it is not necessary for it to operate in this manner. The simplest example of this occurs in the XOR problem. One can implement the input/output mapping for this problem by duplicating the hidden layer architecture of Figure (8) appropiately for two bumps (i.e. 8 hiddens in layer 1, 2 hiddens in layer 2). As discussed above, for Boolean mappings, one can even eliminate the second hidden layer. However the architecture of Figure (9) will also suffice.

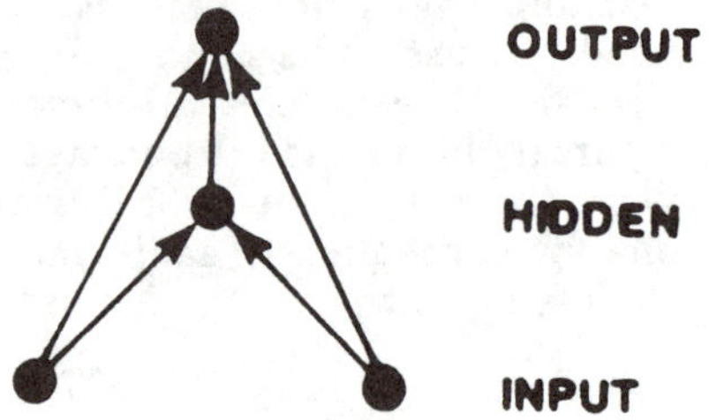

Figure 9. Connectivity for XOR

OUTPUT

HIDDEN

INPUT

Plotting the output of this network, Figure(9), as a function of the two inputs yields a ridge orientated to run between (0,1) and (1,0) Figure(10). Thus a neural net may learn a symbolic training set without using bumps, and a high dimensional version of this process takes place in more complex symbol processing tasks. Ridge/ravine representations of the training data are considerably more efficient than bumps (less hidden neurons and weights) and the extended nature of the surface allows reasonable predictions i.e. extrapolations.

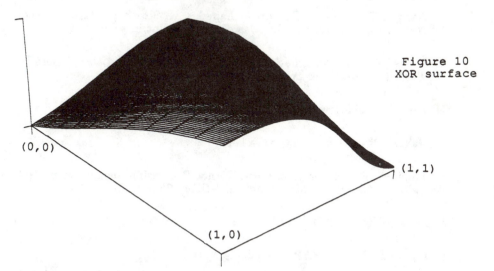

Figure 10
XOR surface

(0,0)

(1,1)

(1,0)

5. Conclusions

Neural nets, in contrast to popular misconception, are capable of quite accurate number crunching, with an accuracy for the prediction problem we considered that exceeds conventional methods by orders of magnitude. Neural nets work by constructing surfaces in a high dimensional space, and their operation when performing signal processing tasks on real valued inputs, is closely related to standard methods of functional approximation. One does not need more than two hidden layers for processing real valued input data, and the accuracy of the approximation is controlled by the number of neurons per layer, and not the number of layers. We emphasize that although two layers of hidden neurons are sufficient they may not be efficient. Multilayer architectures may provide very efficient networks (in the sense of number of neurons and number of weights) that can perform accurately and with minimal cost.

Effective prediction for symbolic input data is achieved by a slightly different method than that used for real value inputs. Instead of forming localized bumps (which would accurately represent the training data but would not predict well on new inputs) the network can use ridge/ravine like surfaces (and generalizations thereof) to efficiently represent the scattered input data. While neural nets generally perform prediction by interpolation for real valued data, they must perform extrapolation for symbolic data if the usual bit representations are used. An outstanding problem is why do tanh representations seem to extrapolate well in symbol processing problems? How do other functional bases do? How does the representation for symbolic inputs affect the ability to extrapolate? This geometrical viewpoint provides a unifying framework for examing many neural net algorithms, for suggesting questions about neural net operatio and for relating current neural net approaches to conventional methods.

Acknowledgments

We thank Y. C. Lee, J. D. Farmer, and J. Sidorovich for a number valuable discussions.

References

C. Barnes, C. Burks, R. Farber, A. Lapedes, K. Sirotkin, "Pattern Recogn by Neural Nets in Genetic Databases", manuscript in preparation

J. Denker et. al.,"Automatic Learning, Rule Extraction,and Generalizati ATT, Bell Laboratories preprint, 1987

D. Farmer, J.Sidorowich, Phys.Rev. Lett., 59(8), p. 845,1987

H. Haken, Phys. Lett. A53, p77 (1975)

A. Lapedes, R. Farber "Nonlinear Signal Processing Using Neural Netw Prediction and System Modelling", LA-UR87-2662,1987

Y.C. Lee, Physica 22D,(1986)

R. Lippman, IEEE ASAP magazine,p.4, 1987

D. Ruelle, F. Takens, Comm. Math. Phys. 20, p167 (1971)

D. Rummelhart, J. McClelland in "Parallel Distributed Processing" Vo M.I.T. Press Cambridge, MA (1986)

D. Russel et al., Phys. Rev. Lett. 45, p1175 (1980)

T. Sejnowski et al., "Net Talk: A Parallel Network that Learns to Read Al Johns Hopkins Univ. preprint (1986)

H. Swinney et al., Physics Today 31 (8), p41 (1978)

F. Takens, "Detecting Strange Attractor in Turbulence," Lecture Notes in N ematics, D. Rand, L. Young (editors), Springer Berlin, p366 (1981)

K. Tomita et al., J. Stat. Phys. 21, p65 (1979)

DISTRIBUTED NEURAL INFORMATION PROCESSING
IN THE VESTIBULO-OCULAR SYSTEM

Clifford Lau
Office of Naval Research Detachment
Pasadena, CA 91106

Vicente Honrubia*
UCLA Division of Head and Neck Surgery
Los Angeles, CA 90024

ABSTRACT

A new distributed neural information-processing
model is proposed to explain the response characteristics
of the vestibulo-ocular system and to reflect more
accurately the latest anatomical and neurophysiological
data on the vestibular afferent fibers and vestibular nuclei.
In this model, head motion is sensed topographically by hair
cells in the semicircular canals. Hair cell signals are then
processed by multiple synapses in the primary afferent
neurons which exhibit a continuum of varying dynamics. The
model is an application of the concept of "multilayered"
neural networks to the description of findings in the
bullfrog vestibular nerve, and allows us to formulate
mathematically the behavior of an assembly of neurons
whose physiological characteristics vary according to their
anatomical properties.

INTRODUCTION

Traditionally the physiological properties of
individual vestibular afferent neurons have been modeled as
a linear time-invariant system based on Steinhausen's
description of cupular motion.[1] The vestibular nerve input
to different parts of the central nervous system is usually
represented by vestibular primary afferents that have

---

*Work supported by grants NS09823 and NS08335 from the National
Institutes of Health (NINCDS) and grants from the Pauley Foundation and the
Hope for Hearing Research Foundation.

response properties defined by population averages from individual neurons.[2]

A new model of vestibular nerve organization is proposed to account for the observed variabilities in the primary vestibular afferent's anatomical and physiological characteristics. The model is an application of the concept of "multilayered" neural networks,[3,4] and it attempts to describe the behavior of the entire assembly of vestibular neurons based on new physiological and anatomical findings in the frog vestibular nerve. It was found that primary vestibular afferents show systematic differences in sensitivity and dynamics and that there is a correspondence between the individual neuron's physiological properties and the location of innervation in the area of the crista and also the sizes of the neuron's fibers and somas. This new view of topological organization of the receptor and vestibular nerve afferents is not included in previous models of vestibular nerve function. Detailed findings from this laboratory on the anatomical and physiological properties of the vestibular afferents in the bullfrog have been published.[5,6]

REVIEW OF THE ANATOMY AND PHYSIOLOGY OF THE VESTIBULAR NERVE

The most pertinent anatomical and physiological data on the bullfrog vestibular afferents are summarized here. In the vestibular nerve from the anterior canal four major branches (bundles) innervate different parts of the crista (Figure 1). From serial histological sections it has been shown that fibers in the central bundle innervate hair cells at the center of the crista, and the lateral bundles project to the periphery of the crista. In each nerve there is an average of 1170 ± 171 (n = 5) fibers, of which the thick fibers (diameter > 7.0 microns, large dots) constitute 8% and the thin fibers (< 4.0 microns, small dots) 76%. The remaining fibers (16%) fall into the range between 4.0 and 7.0 microns. We found that the thick fibers innervate only the center of the crista, and the thinner ones predominantly innervate the periphery.

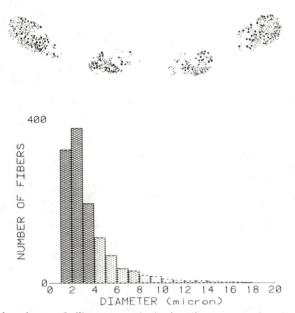

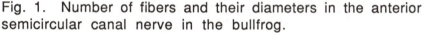

Fig. 1. Number of fibers and their diameters in the anterior semicircular canal nerve in the bullfrog.

There appears to be a physiological and anatomical correlation between fiber size and degree of regularity of spontaneous activity. By recording from individual neurons and subsequently labeling them with horseradish peroxidase intracellularly placed in the axon, it is possible to visualize and measure individual ganglion cells and axons and to determine the origin of the fiber in the crista as well as the projections in different parts of the vestibular nuclei. Figure 2 shows an example of three neurons of different sizes and degrees of regularity of spontaneous activity. In general, fibers with large diameters tend to be more irregular with large coefficients of variation (CV) of the interspike intervals, whereas thin fibers tend to be more regular. There is also a relationship for each neuron between CV and the magnitude of the response to physiological rotatory stimuli, that is, the response gain. (Gain is defined as the ratio of the response in spikes per second to the stimulus in degrees per second.) Figure 3 shows a plot of gain as a function of CV as well as of fiber diameter. For the more regular fibers (CV < 0.5), the gain tends to increase as the diameter of the fiber increases.

460

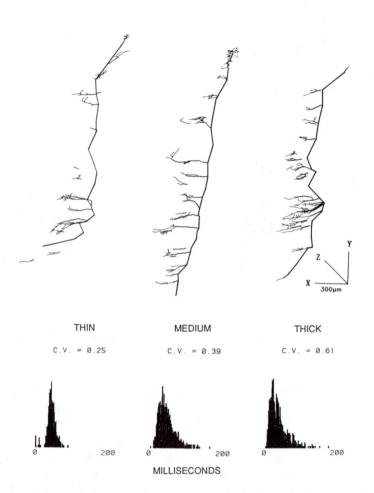

Fig. 2. Examples of thin, medium and thick fibers and their spontaneous activity. CV - coefficient of variation.

For the more irregular fibers (CV > 0.5), the gain tends to remain the same with increasing fiber diameter (4.9 ± 1.9 spikes/second/degrees/second).

Figure 4 shows the location of projection of the afferent fibers at the vestibular nuclei from the anterior, posterior, and horizontal canals and saccule. There is an overall organization in the pattern of innervation from the afferents of each vestibular organ to the vestibular nuclei, with fibers from different receptors overlapping in various

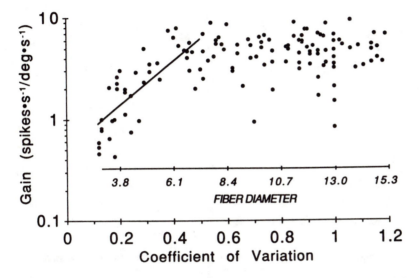

Fig. 3. Gain versus fiber diameters and CV. Stimulus was a sinusoidal rotation of 0.05 Hz at 22 degrees/second peak velocity.

parts of the vestibular nuclei. Fibers from the anterior semicircular canal tend to travel ventrally, from the horizontal canal dorsally, and from the posterior canal the most dorsally.

For each canal nerve the thick fibers (indicated by large dots) tend to group together to travel lateral to the thin fibers (indicated by diffused shading); thus, the topographical segregation between thick and thin fibers at the periphery is preserved at the vestibular nuclei.

In following the trajectories of individual neurons in the central nervous system, however, we found that each fiber innervates all parts of the vestibular nuclei, caudally to rostrally as well as transversely, and because of the spread of the large number of branches, as many as 200 from each neuron, there is a great deal of overlap among the projections.

DISTRIBUTED NEURAL INFORMATION-PROCESSING MODEL

Figure 5 represents a conceptual organization, based on the above anatomical and physiological data, of Scarpa's

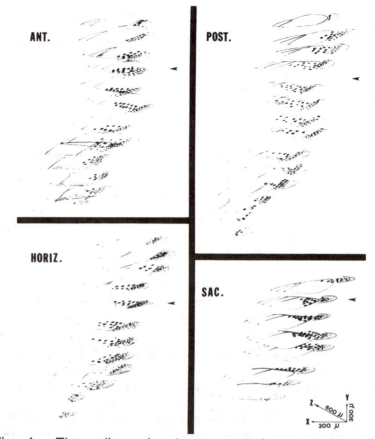

Fig. 4. Three-dimensional reconstruction of the primary afferent fibers' location in the vestibular nuclei.

ganglion cells of the vestibular nerve and their innervation of the hair cells and of the vestibular nuclei. The diagram depicts large Scarpa's ganglion cells with thick fibers innervating restricted areas of hair cells near the center of the crista (top) and smaller Scarpa's ganglion cells with thin fibers on the periphery of the crista innervating multiple hair cells with a great deal of overlap among fibers. At the vestibular nuclei, both thick and thin fibers innervate large areas with a certain gradient of overlapping among fibers of different diameters.

The new distributed neural information-processing model for the vestibular system is based on this anatomical organization, as shown in Figure 6. The response

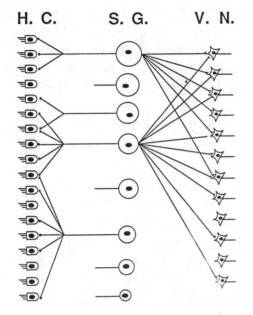

Fig. 5. Anatomical organization of the vestibular nerve. H.C. - hair cells. S.G. - Scarpa's ganglion cells. V.N. - vestibular nuclei.

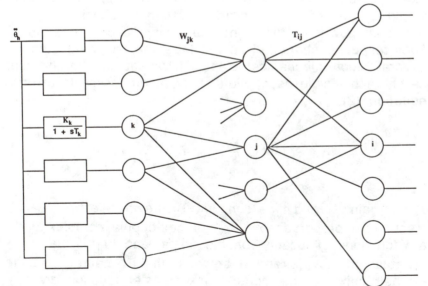

Fig. 6. Distributed neural information-processing model of the vestibular nerve.

characteristic of the primary afferent fiber is represented by the transfer function $SG_j(s)$. This transfer function serves as a description of the gain and phase response of individual neurons to angular rotation. The simplest model would be a first-order system with d.c. gain K_j (spikes/second over head acceleration) and a time constant T_j (seconds) for the jth fiber as shown in equation (1):

$$SG_j(s) = \frac{K_j}{1 + sT_j}. \tag{1}$$

For the bullfrog, K_j can range from about 3 to 25 spikes/second/degree/second^2, and T_j from about 10 to 0.5 second. The large and high-gain neurons are more phasic than the small neurons and tend to have shorter time constants. As described above, K_j and T_j for the jth neuron are functions of location and fiber diameter. Bode plots (gain and phase versus frequency) of experimental data seem to indicate, however, that a better transfer function would consist of a higher-order system that includes fractional power. This is not surprising since the afferent fiber response characteristic must be the weighted sum of several electromechanical steps of transduction in the hair cells. A plausible description of these processes is given in equation (2):

$$SG_j(s) = \sum_k W_{jk} \frac{K_k}{1 + sT_k}, \tag{2}$$

where gain K_k and time constant T_k are the electro-mechanical properties of the hair cell-cupula complex and are functions of location on the crista, and W_{jk} is the synaptic efficacy (strength) between the jth neuron and the kth hair cell. In this context, the transfer function given in equation (1) provides a measure of the "weighted average" response of the multiple synapses given in equation (2).

We also postulate that the responses of the vestibular nuclei neurons reflect the weighted sums of the responses of the primary vestibular afferents, as follows:

$$VN_i = f \left(\sum_j T_{ij}\, SG_j \right), \tag{3}$$

where f(.) is a sigmoid function describing the change in firing rates of individual neurons due to physiological stimulation. It is assumed to saturate between 100 to 300 spikes/second, depending on the neuron. T_{ij} is the synaptic efficacy (strength) between the i*th* vestibular neuron and the j*th* afferent fiber.

CONCLUSIONS

Based on anatomical and physiological data from the bullfrog we presented a description of the organization of the primary afferent vestibular fibers. The responses of the afferent fibers represent the result of summated excitatory processes. The information on head movement in the assemblage of neurons is codified as a continuum of varying physiological responses that reflect a sensoritopic organization of inputs from the receptor to the central nervous system. We postulated a new view of the organization in the peripheral vestibular organs and in the vestibular nuclei. This view does not require unnecessary simplification of the varying properties of the individual neurons. The model is capable of extracting the weighted average response from assemblies of large groups of neurons while the unitary contribution of individual neurons is preserved. The model offers the opportunity to incorporate further developments in the evaluation of the different roles of primary afferents in vestibular function. Large neurons with high sensitivity and high velocity of propagation are more effective in activating reflexes that require quick responses such as vestibulo-spinal and vestibulo-ocular reflexes. Small neurons with high thresholds for the generation of action potentials and lower sensitivity are more tuned to the maintenance of posture and muscle tonus. We believe the physiological differences reflect the different physiological roles.

In this emerging scheme of vestibular nerve organization it appears that information about head movement, topographically filtered in the crista, is distributed through multiple synapses in the vestibular centers. Consequently, there is also reason to believe that different neurons in the vestibular nuclei preserve the variability in response characteristics and the topological discrimination observed in the vestibular nerve. Whether this idea of the organization and function of the vestibular system is valid remains to be proven experimentally.

REFERENCES

1. W. Steinhausen, Arch. Ges. Physiol. 217, 747 (1927).
2. J. M. Goldberg and C. Fernandez, in: Handbook of Physiology, Sect. 1, Vol. III, Part 2 (I. Darian-Smith, ed., Amer. Physiol. Soc., Bethesda, MD, 1984), p. 977.
3. D. E. Rumelhart, G. E. Hinton and J. L. McClelland, in: Parallel Distributed Processing: Explorations in the Microstructure of Cognition, Vol. 1: Foundations (D. E. Rumelhart, J. L. McClelland and the PDP Research Group, eds., MIT Press, Cambridge, MA, 1986), p. 45.
4. J. Hopfield, Proc. Natl. Acad. Sci. 79, 2554 (1982).
5. V. Honrubia, S. Sitko, J. Kimm, W. Betts and I. Schwartz, Intern. J. Neurosci. 15, 197 (1981).
6. V. Honrubia, S. Sitko, R. Lee, A. Kuruvilla and I. Schwartz, Laryngoscope 94, 464 (1984).

SPONTANEOUS AND INFORMATION-TRIGGERED SEGMENTS OF SERIES OF HUMAN BRAIN ELECTRIC FIELD MAPS

D. Lehmann, D. Brandeis*, A. Horst, H. Ozaki* and I. Pal*
Neurology Department, University Hospital, 8091 Zürich, Switzerland

ABSTRACT

The brain works in a state-dependent manner: processing strategies and access to stored information depends on the momentary functional state which is continuously re-adjusted. The state is manifest as spatial configuration of the brain electric field. Spontaneous and information-triggered brain electric activity is a series of momentary field maps. Adaptive segmentation of spontaneous series into spatially stable epochs (states) exhibited 210 msec mean segments, discontinuous changes. Different maps imply different active neural populations, hence expectedly different effects on information processing: Reaction time differed between map classes at stimulus arrival. Segments might be units of brain information processing (content/mode/step), possibly operationalizing consciousness time. Related units (e.g. triggered by stimuli during figure perception and voluntary attention) might specify brain sub-mechanisms of information treatment.

BRAIN FUNCTIONAL STATES AND THEIR CHANGES

The momentary functional state of the brain is reflected by the configuration of the brain's electro-magnetic field. The state manifests the strategy, mode, step and content of brain information processing, and the state constrains the choice of strategies and modes and the access to memory material available for processing of incoming information (1). The constraints include the available range of changes of state in PAVLOV's classical "orienting reaction" as response to new or important informations. Different states might be viewed as different functional connectivities between the neural elements.

The orienting reaction (see 1,2) is the result of the first ("pre-attentive") stage of information processing. This stage operates automatically (no involvement of consciousness) and in a parallel mode, and quickly determines whether (a) the information is important or unknown and hence requires increased attention and alertness, i.e. an orienting reaction which means a re-adjustment of functional state in order to deal adequately with the information invoking consciousness for further processing, or whether (b) the information is known or unimportant and hence requires no re-adjustment of state, i.e. that it can be treated further with well-

* Present addresses: D.B. at Psychiat. Dept., V.A. Med. Center, San Francisco CA 94121; H.O. at Lab. Physiol. for the Developmentally Handicapped, Ibaraki Univ., Mito, Japan 310; I.P. at BioLogic Systems Corp., Mundelein IL 60060.

established ("automatic") strategies. Conscious strategies are slow but flexible (offer wide choice), automatic strategies are fast but rigid.

Examples for functional states on a gross scale are wakefulness, drowsiness and sleep in adults, or developmental stages as infancy, childhood and adolescence, or drug states induced by alcohol or other psychoactive agents. The different states are associated with distinctly different ways of information processing. For example, in normal adults, reality-close, abstracting strategies based on causal relationships predominate during wakefulness, whereas in drowsiness and sleep (dreams), reality-remote, visualizing, associative concatenations of contents are used. Other well-known examples are drug states.

HUMAN BRAIN ELECTRIC FIELD DATA AND STATES

While alive, the brain produces an ever-changing electromagnetic field, which very sensitively reflects global and local states as effected by spontaneous activity, incoming information, metabolism, drugs, and diseases. The electric component of the brain's electro-magnetic field as non-invasively measured from the intact human scalp shows voltages between 0.1 and 250 microVolts, temporal frequencies between 0.1 and 30, 100 or 3000 Hz depending on the examined function, and spatial frequencies up to 0.2 cycles/cm.

Brain electric field data are traditionally viewed as time series of potential differences between two scalp locations (the electroencephalogram or EEG). Time series analysis has offered an effective way to class different gross brain functional states, typically using EEG power spectral values. Differences between power spectra during different gross states typically are greater than between different locations. States of lesser functional complexity such as childhood vs adult states, sleep vs wakefulness, and many drug-states vs non-drug states tend to increased power in slower frequencies (e.g. 1,4).

Time series analyses of epochs of intermediate durations between 30 and 10 seconds have demonstrated (e.g. 1,5,6) that there are significant and reliable relations between spectral power or coherency values of EEG and characteristics of human mentation (reality-close thoughts vs free associations, visual vs non-visual thoughts, positive vs negative emotions).

Viewing brain electric field data as series of momentary field maps (7,8) opens the possibility to investigate the temporal microstructure of brain functional states in the sub-second range. The rationale is that the momentary configuration of activated neural elements represents a given brain functional state, and that the spatial pattern of activation is reflected by the momentary brain electric field which is recordable on the scalp as a momentary field map. Different configurations of activation (different field maps) are expected to be associated with different modes, strategies, steps and contents of information processing.

SEGMENTATION OF BRAIN ELECTRIC MAP SERIES INTO STABLE SEGMENTS

When viewing brain electric activity as series of maps of momentary potential distributions, changes of functional state are recognizable as changes of the "electric landscapes" of these maps. Typically, several successive maps show similar landscapes, then quickly change to a new configuration which again tends to persist for a number of successive maps, suggestive of stable states concatenated by non-linear transitions (9,10). Stable map landscapes might be hypothesized to indicate the basic building blocks of information processing in the brain, the "atoms of thoughts". Thus, the task at hand is the recognition of the landscape configurations; this leads to the adaptive segmentation of time series of momentary maps into segments of stable landscapes during varying durations.

We have proposed and used a method which describes the configuration of a momentary map by the locations of its maximal and minimal potential values, thus invoking a dipole model. The goal here is the phenomenological recognition of different momentary functional states using a very limited number of major map features as classifiers, and we suggest conservative interpretion of the data as to real brain locations of the generating processes which always involve millions of neural elements.

We have studied (11) map series recorded from 16 scalp locations over posterior skull areas from normal subjects during relaxation with closed eyes. For adaptive segmentation, the maps at the times of maximal map relief were selected for optimal signal/noise conditions. The locations of the maximal and minimal (extrema) potentials were extracted in each map as descriptors of the landscape; taking into account the basically periodic nature of spontaneous brain electric activity (Fig. 1), extrema locations were treated disregarding polarity information. If over time an extreme left its pre-set spatial window (say, one electrode distance), the segment was terminated. The map series showed stable map configurations for varying durations (Fig. 2), and discontinuous, step-wise changes. Over 6 subjects, resting alpha-type EEG showed 210 msec mean segment duration; segments longer than 323 msec covered 50% of total time; the most prominent segment class (1.5% of all classes) covered 20% of total time (prominence varied strongly over classes; not all possible classes occurred). Spectral power and phase of averages of adaptive and pre-determined segments demonstrated the adequacy of the strategy and the homogeneity of adaptive segment classes by their reduced within-class variance. Segmentation using global map dissimilarity (sum of Euklidian difference vs average reference at all measured points) emulates the results of the extracted-characteristics-strategy.

FUNCTIONAL SIGNIFICANCE OF MOMENTARY MICRO STATES

Since different maps of momentary EEG fields imply activity of different neural populations, different segment classes must manifest different brain functional states with expectedly different

470

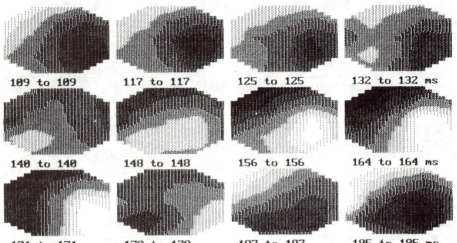

189 to 189   117 to 117   125 to 125   132 to 132 ms

140 to 140   148 to 148   156 to 156   164 to 164 ms

171 to 171   179 to 179   187 to 187   195 to 195 ms
RECORD=1   FILE=A:VP3EC2A   NORMAL SUBJECT, EYES CLOSED

Fig. 1. Series of momentary potential distribution maps of the brain
field recorded from the scalp of a normal human during relaxation
with closed eyes. Recording with 21 electrodes (one 5-electrode row
added to the 16-electrode array in Fig. 2) using 128 samples/sec/
channel. Head seen from above, left ear left; white positive, dark
negative, 8 levels from +32 to -32 microVolts. Note the periodic
reversal of field polarity within the about 100 msec (one cycle of
the 8-12Hz so-called "EEG alpha" activity) while the field configuration remains largely constant. - This recording and display was
done with a BRAIN ATLAS system (BioLogic Systems, Mundelein, IL).

effects on ongoing information processing. This was supported by
measurements of selective reaction time to acoustic stimuli which
were randomly presented to eight subjects during different classes
of EEG segments (323 responses for each subject). We found
significant reaction time differences over segment classes (ANOVA p
smaller than .02), but similar characteristics over subjects. This
indicates that the momentary sub-second state as manifest in the
potential distribution map significantly influences the behavioral
consequence of information reaching the brain.
    Presentation of information is followed by a sequence of
potential distribution maps ("event-related potentials" or ERP's,
averaged over say, 100 presentations of the same stimulus, see 12).
The different spatial configurations of these maps (12) are thought
to reflect the sequential stages of information processing
associated with "components" of event-related brain activity (see
e.g. 13) which are traditionally defined as times of maximal
voltages after information input (maximal response strength).

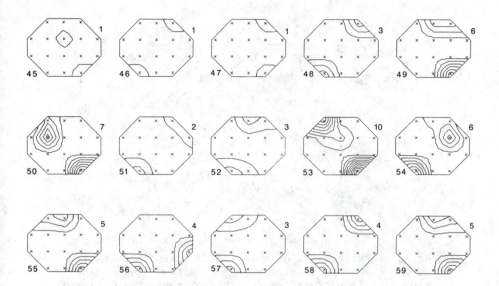

Fig. 2. Sequence of spatially stable segments during a spontaneous
series of momentary EEG maps of 3.1 sec duration in a normal
volunteer. Each map shows the occurrence of the extreme potential
values during one adaptively determined segment: the momentary maps
were searched for the locations of the two extreme potentials; these
locations were accumulated, and linearly interpolated between
electrodes to construct the present maps. (The number of iso-
frequency-of-occurrence lines therefore is related to the number of
searched maps). - Head seen from above, left ear left, electrode
locations indicated by crosses, most forward electrode at vertex.
Data FIR filtered to 8-12Hz (alpha EEG). The figure to the left
below each map is a running segment number. The figure to the right
above each map multiplied by 50 indicates the segment duration in
msec.

Application of the adaptive segmentation procedure described above
for identification of functional components of event-related brain
electric map sequences requires the inclusion of polarity
information (14); such adaptive segmentation permits to separate
different brain functional states without resorting to the strength
concept of processing stages.

An example (12) might illustrate the type of results obtained
with this analysis: Given segments of brain activity which were
triggered by visual information showed different map configurations
when subjects paid attention vs when they paid no attention to the
stimulus, and when they viewed figures vs meaningless shapes as

472

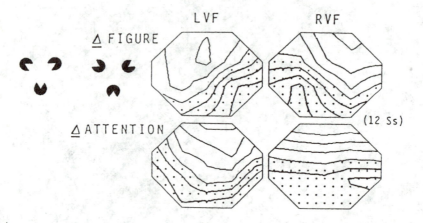

Fig. 3. Four difference maps, computed as differences between maps
obtained during (upper row) perception of a visual "illusionary"
triangle figure (left picture) minus a visual non-figure (right)
shown to the left and right visual hemi-fields (LVF, RVF), and
obtained during (lower row) attending minus during ignoring the
presented display. The analysed segment covered the time from 168 to
200 msec after stimulus presentations. - Mean of 12 subjects. Head
seen from above, left ear left, 16 electrodes as in Fig. 2,
isopotential contour lines at 0.1 microVolt steps, dotted negative
referred to mean of all values. The "illusionary" figure stimulus
was studied by Kanisza (16); see also (12). - Note that the mirror
symmetric configuration of the difference maps for LVF and RVF is
found for the "figure" effect only, not for the "attention" effect,
but that the anterior-posterior difference is similar for both cases.

stimuli. Fig. 3 illustrates such differences in map configuration.
The "attention"-induced and "figure"-induced changes in map
configuration showed certain similarities e.g. in the illustrated
segment 168-200 msec after information arrival, supporting the
hypothesis that brain mechanisms for figure perception draw on brain
resources which in other circumstances are utilized in voluntary
attention.
     The spatially homogeneous temporal segments might be basic
building blocks of brain information processing, possibly
operationalizing consciousness time (15), and offering a common
concept for analysis of brain spontaneous activity and event related
brain potentials. The functional significance of the segments might
be types/ modes/ steps of brain information processing or
performance. Identification of related building blocks during
different brain functions accordingly could specify brain sub-
mechanisms of information treatment.

Acknowledgement: Financial support by the Swiss National Science
Foundation (including Fellowships to H.O. and I.P.) and by the EMDO,
the Hartmann Muller and the SANDOZ Foundation is gratefully
acknowledged.

REFERENCES

1. M. Koukkou and D. Lehmann, Brit. J. Psychiat. 142, 221-231
    (1983).
2. A. Ohman, In: H.D. Kimmel, E.H. von Olst and J.F. Orlebeke
    (Eds.), Drug-Discrimination and State Dependent Learning
    (Academic Press, New York, 1979), pp. 283-318.
3. A. Katada, H. Ozaki, H. Suzuki and K. Suhara, Electroenceph.
    Clin. Neurophysiol. 52, 192-201 (1981).
4. M. Koukkou and D. Lehmann, Biol. Psychiat. 11, 663-677 (1976).
5. J. Berkhout, D.O. Walter and W.R. Adey, Electroenceph. clin.
    Neurophysiol. 27, 457-469 (1969).
6. P. Grass, D. Lehmann, B. Meier, C.A. Meier and I. Pal, Sleep
    Res. 16, 231 (1987).
7. D. Lehmann, Electroenceph. Clin. Neurophysiol. 31, 439-449
    (1971).
8. D. Lehmann, In: H.H. Petsche and M.A.B. Brazier (eds.),
    Synchronization of EEG Activity in Epilepsies (Springer, Wien,
    1972), pp. 307-326.
9. H. Haken, Advanced Synergetics (Springer, Heidelberg, 1983).
10. J.J. Wright, R.R. Kydd and G.L. Lees, Biol. Cybern., 1985, 53,
    11-17.
11. D. Lehmann, H. Ozaki and I. Pal, Electroenceph. Clin.
    Neurophysiol. 67, 271-288 (1987).
12. D. Brandeis and D. Lehmann, Neuropsychologia 24, 151-168 (1986).
13. A.S. Gevins, N.H. Morgan, S.L. Bressler, B.A. Cutillo, R.M.
    White, J. Illes, D.S. Greer, J.C.Doyle and M. Zeitlin, Science
    235, 580-585 (1987).
14. D. Lehmann and W. Skrandies, Progr. Neurobiol. 23, 227-250
    (1984).
15. B. Libet, Human Neurobiol. 1, 235-242 (1982).
16. G. Kanisza, Organization of Vision (Praeger, New York, 1979).

474

OPTIMIZATION WITH ARTIFICIAL NEURAL NETWORK SYSTEMS:
A MAPPING PRINCIPLE
AND
A COMPARISON TO GRADIENT BASED METHODS [†]

Harrison MonFook Leong
Research Institute for Advanced Computer Science
NASA Ames Research Center 230-5
Moffett Field, CA, 94035

ABSTRACT
General formulae for mapping optimization problems into systems of ordinary differential equations associated with artificial neural networks are presented. A comparison is made to optimization using gradient-search methods. The performance measure is the settling time from an initial state to a target state. A simple analytical example illustrates a situation where dynamical systems representing artificial neural network methods would settle faster than those representing gradient-search. Settling time was investigated for a more complicated optimization problem using computer simulations. The problem was a simplified version of a problem in medical imaging: determining loci of cerebral activity from electromagnetic measurements at the scalp. The simulations showed that gradient based systems typically settled 50 to 100 times faster than systems based on current neural network optimization methods.

INTRODUCTION

Solving optimization problems with systems of equations based on neurobiological principles has recently received a great deal of attention. Much of this interest began when an artificial neural network was devised to find near-optimal solutions to an np-complete problem [13]. Since then, a number of problems have been mapped into the same artificial neural network and variations of it [10, 13, 14, 17, 18, 19, 21, 23, 24]. In this paper, a unifying principle underlying these mappings is derived for systems of first to n^{th}-order ordinary differential equations. This mapping principle bears similarity to the mathematical tools used to generate optimization methods based on the gradient. In view of this, it seemed important to compare the optimization efficiency of dynamical systems constructed by the neural network mapping principle with dynamical systems constructed from the gradient.

THE PRINCIPLE

This paper concerns itself with networks of computational units having a state variable v, a function f that describes how a unit is driven by inputs, a linear ordinary differential operator with constant coefficients $D(v)$ that describes the dynamical response of each unit, and a function g that describes how the output of a computational unit is determined from its state v. In particular, the paper explores how outputs of the computational units evolve with time in terms of a scalar function E, a single state variable for the whole network. Fig. 1 summarizes the relationships between variables, functions, and operators associated with each computational unit. Eq. (1) summarizes the equations of motion for a network composed of such units:

$$\vec{D}^{(M)}(v) = \vec{f}(g_1(v_1), \ldots, g_N(v_N)) \tag{1}$$

where the ith element of $\vec{D}^{(M)}$ is $D^{(M)}(v_i)$, superscript (M) denotes that operator D is M^{th} order, the ith element of $\vec{f}$ is $f_i(g_1(v_1), \ldots, g_N(v_N))$, and the network is comprised of N computational units. The network of Hopfield [12] has $M=1$, functions $\vec{f}$ are weighted linear sums, and functions $\vec{g}$ (where the ith element of $\vec{g}$ is $g_i(v_i)$) are all the same sigmoid function. We will examine two ways of defining functions $\vec{f}$ given a function F. Along with these definitions will be

---

† Work supported by NASA Cooperative Agreement No. NCC 2-408

defined corresponding functions E that will be used to describe the dynamics of Eq. (1).

The first method corresponds to optimization methods introduced by artificial neural network research. It will be referred to as method $\nabla_{\vec{g}}$ ("dell g"):

$$\vec{f} \equiv \nabla_{\vec{g}} F \tag{2a}$$

with associated E function

$$E_{\vec{g}} = F(\vec{g}) - \int^t \sum_i^N \left[D^{(M)}(v_i(s)) - \frac{dv_i(s)}{dt} \right] \frac{dg_i(s)}{dt} \, ds. \tag{2b}$$

Here, $\nabla_{\vec{x}} H$ denotes the gradient of H, where partials are taken with respect to variables of $\vec{x}$, and $E_{\vec{g}}$ denotes the E function associated with gradient operator $\nabla_{\vec{g}}$. With appropriate operator D and functions $\vec{f}$ and $\vec{g}$, $E_{\vec{g}}$ is simply the "energy function" of Hopfield [12]. Note that Eq. (2a) makes explicit that we will only be concerned with $\vec{f}$ that can be derived from scalar potential functions. For example, this restriction excludes artificial neural networks that have connections between excitatory and inhibitory units such as that of Freeman [8]. The second method corresponds to optimization methods based on the gradient. It will be referred to as method $\nabla_{\vec{v}}$ ("dell v"):

$$\vec{f} \equiv \nabla_{\vec{v}} F \tag{3a}$$

with associated E function

$$E_{\vec{v}} = F(\vec{g}) - \int^t \sum_i^N \left[D^{(M)}(v_i(s)) - \frac{dv_i(s)}{dt} \right] \frac{dv_i(s)}{dt} \, ds \tag{3b}$$

where notation is analogous to that for Eqs. (2).

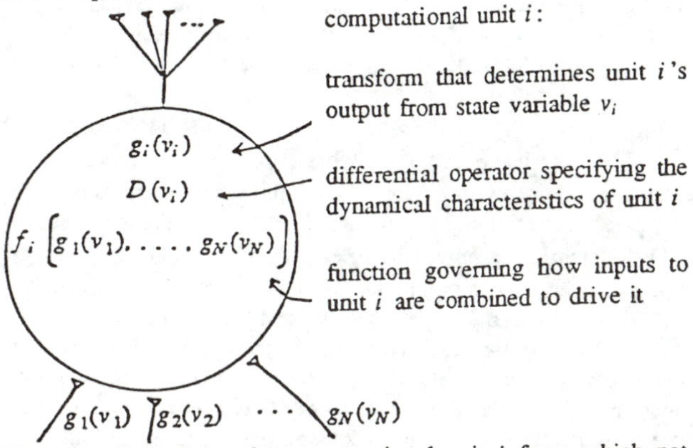

The critical result that allows us to map optimization problems into networks described by Eq. (1) is that conditions on the constituents of the equation can be chosen so that along any solution trajectory, the E function corresponding to the system will be a monotonic function of time. For method $\nabla_{\vec{g}}$, here are the conditions: all functions $\vec{g}$ are 1) differentiable and 2) monotonic in the same sense. Only the first condition is needed to make a similar assertion for

computational unit i:

transform that determines unit i's output from state variable v_i

differential operator specifying the dynamical characteristics of unit i

function governing how inputs to unit i are combined to drive it

Figure 1: Schematic of a computational unit i from which networks considered in this paper are constructed. Triangles suggest connections between computational units.

method $\nabla_{\vec{v}}$. When these conditions are met and when solutions of Eq. (1) exist, the dynamical systems can be used for optimization. The appendix contains proofs for the monotonicity of function E along solution trajectories and references necessary existence theorems. In conclusion, mapping optimization problems onto dynamical systems summarized by Eq. (1) can be reduced to a matter of differentiation if a scalar function representation of the problem can be found and the integrals of Eqs. (2b) and (3b) are ignorable. This last assumption is certainly upheld for the case where operator D has no derivatives less than M^{th} order. In simulations below, it will be observed to hold for the case $M=1$ with a nonzero 0th order derivative in D. (Also see Lapedes and Farber [19].)

PERSPECTIVES OF RECENT WORK

The formulations above can be used to classify the neural network optimization techniques used in several recent studies. In these studies, the functions $\vec{g}$ were all identical. For the most part, following Hopfield's formulation, researchers [10, 13, 14, 17, 23, 24] have used method $\nabla_{\vec{g}}$ to derive forms of Eq. (1) that exhibit the ability to find extrema of $E_{\vec{g}}$ with $E_{\vec{g}}$ quadratic in functions $\vec{g}$ and all functions $\vec{g}$ describable by sigmoid functions such as $tanh(x)$. However, several researchers have written about artificial neural networks associated with non-quadratic E functions. Method $\nabla_{\vec{g}}$ has been used to derive systems capable of finding extrema of non-quadratic $E_{\vec{g}}$ [19]. Method $\nabla_{\vec{v}}$ has been used to derive systems capable of optimizing $E_{\vec{v}}$ where $E_{\vec{v}}$ were not necessarily quadratic in variables $\vec{v}$ [21]. A sort of hybrid of the two methods was used by Jeffery and Rosner [18] to find extrema of functions that were not quadratic. The important distinction is that their functions $\vec{f}$ were derived from a given function F using Eq. (3a) where, in addition, a sign definite diagonal matrix was introduced; the left side of Eq. (3a) was left multiplied by this matrix. A perspective on the relationship between all three methods to construct dynamical systems for optimization is summarized by Eq. (4) which describes the relationship between methods $\nabla_{\vec{g}}$ and $\nabla_{\vec{v}}$:

$$\nabla_{\vec{g}}F = diag \left[\frac{\partial g(v_i)}{\partial v_i} \right]^{-1} \nabla_{\vec{v}}F \tag{4}$$

where $diag[\, x_i \,]$ is a diagonal matrix with x_i as the diagonal element of row i. (A similar equation has been derived for quadratic F [5].) The relationship between the method of Jeffery and Rosner and $\nabla_{\vec{v}}$ is simply Eq. (4) with the time dependent diagonal matrix replaced by a constant diagonal matrix of free parameters. It is noted that Jeffery and Rosner presented timing results that compared simulated annealing, conjugate-gradient, and artificial neural network methods for optimization. Their results are not comparable to the results reported below since they used computation time as a performance measure, not settling times of analog systems. The perspective provided by Eq. (4) will be useful for anticipating the relative performance of methods $\nabla_{\vec{g}}$ and $\nabla_{\vec{v}}$ in the analytical example below and will aid in understanding the results of computer simulations.

COMPARISON OF METHODS $\nabla_{\vec{g}}$ AND $\nabla_{\vec{v}}$

When $M=1$ and operator D has no 0th order derivatives, method $\nabla_{\vec{v}}$ is the basis of gradient-search methods of optimization. Given the long history of of such methods, it is important to know what possible benefits could be achieved by the relatively new optimization scheme, method $\nabla_{\vec{g}}$. In the following, the optimization efficiency of methods $\nabla_{\vec{g}}$ and $\nabla_{\vec{v}}$ is compared by comparing settling times, the time required for dynamical systems described by Eq. (1) to traverse a continuous path to local optima. To qualify this performance measure, this study anticipates application to the creation of analog devices that would instantiate Eq. (1); hence, we are not interested in estimating the number of discrete steps that would be required to find local optima, an appropriate performance measure if the point was to develop new numerical methods. An analytical example will serve to illustrate the possibility of improvements in settling time by using method $\nabla_{\vec{g}}$ instead of method $\nabla_{\vec{v}}$. Computer simulations will be reported for more complicated problems following this example.

For the analytical example, we will examine the case where all functions $\vec{g}$ are identical and

$$g(v) = tanhG(v - Th) \tag{5}$$

where $G > 0$ is the gain and Th is the threshold. Transforms similar to this are widely used in artificial neural network research. Suppose we wish to use such computational units to search a multi-dimensional binary solution space. We note that

$$\frac{dg}{dv} = G \, sech^2 G(v - Th) \tag{6}$$

is near 0 at valid solution states (corners of a hypercube for the case of binary solution spaces). We see from Eq. (4) that near a valid solution state, a network based on method $\nabla_{\vec{g}}$ will allow computational units to recede from incorrect states and approach correct states comparatively faster. Does this imply faster settling time for method $\nabla_{\vec{g}}$?

To obtain an analytical comparison of settling times, consider the case where $M=1$ and operator D has no 0th order derivatives and

$$F = \frac{1}{2}\sum_{i,j}S_{ij}(tanhGv_i)(tanhGv_j) \tag{7}$$

where matrix S is symmetric. Method $\nabla_{\vec{g}}$ gives network equations

$$\frac{d\vec{v}}{dt} = S\,tanhG\vec{v} \tag{8}$$

and method $\nabla_{\vec{v}}$ gives network equations

$$\frac{d\vec{v}}{dt} = diag\,[G\,sech^2Gv_i]\,S\,tanhG\vec{v} \tag{9}$$

where $tanhG\vec{v}$ denotes a vector with i^{th} component $tanhGv_i$. For method $\nabla_{\vec{g}}$, there is one stable point, i.e. where $\frac{d\vec{v}}{dt} = 0$, at $\vec{v} = 0$. For method $\nabla_{\vec{v}}$ the stable points are $\vec{v} = 0$ and $\vec{v} \epsilon V$ where V is the set of vectors with component values that are either $+\infty$ or $-\infty$. Further trivialization allows for comparing estimates of settling times: Suppose S is diagonal. For this case, if $v_i = 0$ is on the trajectory of any computational unit i for one method, $v_i = 0$ is on the trajectory of that unit for the other method; hence, a comparison of settling times can be obtained by comparing time estimates for a computational unit to evolve from near 0 to near an extremum or, equivalently, the converse. Specifically, let the interval be $[\delta_0, 1-\delta]$ where $0 < \delta_0 < 1-\delta$ and $0 < \delta < 1$. For method $\nabla_{\vec{v}}$, integrating velocity over time gives the estimate

$$T_{\nabla_{\vec{v}}} = \frac{1}{G}\left[\frac{1}{2}\left[\frac{1}{\delta(2-\delta)} - \frac{1}{1-\delta_0^2}\right] + \ln\left[\frac{1-\delta}{\sqrt{\delta(2-\delta)}}\frac{\sqrt{1-\delta_0^2}}{\delta_0}\right]\right] \tag{10}$$

and for method $\nabla_{\vec{g}}$ the estimate is

$$T_{\nabla_{\vec{g}}} = \frac{1}{G}\ln\left[\frac{1-\delta}{\sqrt{\delta(2-\delta)}}\frac{\sqrt{1-\delta_0^2}}{\delta_0}\right] \tag{11}$$

From these estimates, method $\nabla_{\vec{v}}$ will always take longer to satisfy the criterion for convergence: Note that only with the largest value for δ_0, $\delta_0 = 1-\delta$, is the first term of Eq. (10) zero; for any smaller δ_0, this term is positive. Unfortunately, this simple analysis cannot be generalized to non-diagonal S. With diagonal S, all computational units operate independently. Hence, the derivation of $\frac{d\vec{v}}{dt}$ is irrelevant with respect to convergence rates; convergence rate depends only on the diagonal element of S having the smallest magnitude. In this sense, the problem is one dimensional. But for non-diagonal S, the problem would be, in general, multi-dimensional and, hence, the direction of $\frac{d\vec{v}}{dt}$ becomes relevant. To compare settling times for non-diagonal S, computer simulations were done. These are described below.

COMPUTER SIMULATIONS

Methods

The problem chosen for study was a much simplified version of a problem in medical imaging: Given electromagnetic field measurements taken from the human scalp, identify the location and magnitude of cerebral activity giving rise to the fields. This problem has received much attention in the last 20 years [3,6,7]. The problem, sufficient for our purposes here, was reduced to the following problem: given a few samples of the electric potential field at the surface of a spherical conductor within which reside several static electric dipoles, identify the dipole locations and moments. For this situation, there is a closed form solution for electric potential fields at the spherical surface:

$$\Phi(\vec{x}_{sample}) = \sum_{\substack{all\ dipoles \\ i}} \vec{p}_i \bullet \left[\frac{2\hat{d}_i}{d_i} + \frac{\hat{x}_{sample} + \hat{d}_i}{1 + \hat{x}_{sample} \bullet \hat{d}_i} \frac{1}{x_{sample}} \right] \frac{1}{d_i} \qquad (12)$$

where Φ is the electric potential at the spherical conductor surface, $\vec{x}_{sample}$ is the location of the sample point ($\vec{x}$ denotes a vector, $\hat{x}$ the corresponding unit vector, and x the corresponding vector magnitude), $\vec{p}_i$ is the dipole moment of dipole i, and $\vec{d}_i$ is the vector from dipole i to $\vec{x}_{sample}$ (This equation can be derived from one derived by Brody, Terry, and Ideker [4]). Fig. 2 facilitates picturing these relationships.

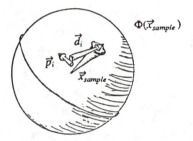

$\Phi(\vec{x}_{sample})$

Figure 2: Vectors of Eq. (12).

With this analytical solution, the problem was formulated as a least squares minimization problem where the variables were dipole moments. In short, the following process was used: A dipole model was chosen. This model was used with Eq. (12) to calculate potentials at points on a sphere which covered about 60% of the surface. A cluster of internal locations that encompassed the locations of the model was specified. The two optimization techniques were then required to determine dipole moment values at cluster locations such that the collection of dipoles at cluster locations accurately reflected the dipole distribution specified by the model.

This was to be done given only the potential values at the sample points and an initial guess of dipole moments at cluster locations. The optimization systems were to accomplish the task by minimizing the sum of squared differences between potentials calculated using the dipole model and potentials calculated using a guess of dipole moments at cluster locations where the sum is taken over all sample points. Further simplifications of the problem included

1) choosing the dipole model locations to correspond exactly to various locations of the cluster,
2) requiring dipole model moments to be 1, 0, or -1, and
3) representing dipole moments at cluster locations with two bit binary numbers.

To describe the dynamical systems used, it suffices to specify operator D and functions $\vec{g}$ of Eq. (1) and function F used in Eqs. (2a) and (3a). Operator D was

$$D = \frac{d}{dt} + 1. \qquad (13)$$

Eq. (5) with a multiplicative factor of ½ was used for all functions $\vec{g}$. Hence, regarding simplification 3) above, each cluster location was associated with two computational units. Considering simplification 2) above, dipole moment magnitude 1 would be represented by both computational units being in the high state, for -1, both in the low state, and for 0, one in the high state and one in the low state. Regarding function F,

$$F = \sum_{\substack{all\ sample \\ points\ s}} \left[\Phi_{measured}(\vec{x}_s) - \Phi_{cluster}(\vec{x}_s) \right]^2 - c \sum_{\substack{all\ computational \\ units\ j}} g(v_j)^2 \qquad (14)$$

where $\Phi_{measured}$ is calculated from the dipole model and Eq. (12) (The subscript *measured* is used because the role of the dipole model is to simulate electric potentials that would be measured in a real world situation. In real world situations, we do not know the source distribution underlying $\Phi_{measured}$), c is an experimentally determined constant (.002 was used), and $\Phi_{cluster}$ is Eq. (12) where the sum of Eq. (12) is taken over all cluster locations and the kth coordinate of the ith cluster location dipole moment is

$$p_{ik} = \sum_{all\ bits\ b} g(v_{ikb}). \qquad (15)$$

Index j of Eq. (14) corresponds to one combination of indices *ikb*.

Sample points, 100 of them, were scattered semi-uniformly over the spherical surface emphasized by horizontal shading in Fig. 3. Cluster locations, 11, and model dipoles, 5, were scattered within the subset of the sphere emphasized by vertical shading. For the dipole model used, 10 dipole moment components were non-zero; hence, optimization techniques needed to hold 56 dipole moment components at zero and set 10 components to correct non-zero values in order to correctly identify the dipole model underlying $\Phi_{measured}$.

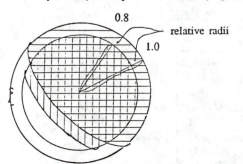

0.8 relative radii

1.0

Figure 3: Illustration of the distribution of sample points on the surface of the spherical conductor (horizontal shading) and the distribution of model dipole locations and cluster locations within the conductor (vertical shading).

The dynamical systems corresponding to methods $\nabla_{\vec{g}}$ and $\nabla_{\vec{v}}$ were integrated using the forward Euler method (e.g. Press, Flannery, Teukolsky, and Vetterling [22]). Numerical methods were observed to be convergent experimentally: settling time and path length were observed to asymtotically approach stable values as step size of the numerical integrator was decreased over two orders of magnitude.

Settling times, path lengths, and relative directions of travel were calculated for the two optimization methods using several different initial bit patterns at the cluster locations. In other words, the search was started at different corners of the hypercube comprising the space of acceptable solutions. One corner of the hypercube was chosen to be the target solution. (Note that a zero dipole moment has a degenerate two bit representation in the dynamical systems explored; the target corner was arbitrarily chosen to be one of the degenerate solutions.) Note from Eq. (5) that for the network to reach a hypercube corner, all elements of $\vec{v}$ would have to be singular. For this reason, settling time and other measures were studied as a function of the proximity of the computational units to their extremum states.

Computations were done on a Sequent Balance.

Results

Graph 1 shows results for exploring settling time as a function of *extremum depth*, the minimum of the deviations of variables $\vec{v}$ from the threshold of functions $\vec{g}$. Extremum depth is reported in multiples of the width of functions $\vec{g}$. The term *transition*, used in the caption of Graph 1 and below, refers to the movement of a computational unit from one extremum state to the other. The calculations were done for two initial states, one where the output of 1 computational unit was set to zero and one where outputs of 13 computational units were set to zero; hence, 1 and 13, respectively, half transitions were required to reach the target hypercube corner. It can be observed that settling time increases faster for method $\nabla_{\vec{v}}$ than that for method $\nabla_{\vec{g}}$ just as we would expect from considering Eqs. (4) and (5). However, it can be observed that method $\nabla_{\vec{v}}$ is still an order of magnitude faster even when extremum depth is 3 widths of functions $\vec{g}$. For the purpose of unambiguously identifying what hypercube corner the dynamical system settles

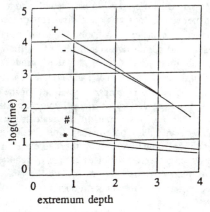

extremum depth

Graph 1: settling time as a function of extremum depth. #: method $\nabla_{\vec{v}}$, 1 half transition required. *: method $\nabla_{\vec{v}}$, 13 half transitions required. +: method $\nabla_{\vec{g}}$, 1 half transition required. -: $\nabla_{\vec{g}}$, 13 half transitions required.

480

to, this extremum depth is more than adequate.

Table 1 displays results for various initial conditions. Angles are reported in degrees. These measures refer to the angle between directions of travel in $\vec{v}$-space as specified by the two optimization methods. The average angle reported is taken over all trajectory points visited by the numerical integrator. Initial angle is the angle at the beginning of the path. *Parasite cost percentage* is a measure that compares parasite cost, the integral in Eqs. (2b) and (3b), to the range of function F over the path:

$$parasite\ cost\ \% = 100\times\frac{parasite\ cost}{|F_{final}-F_{initial}|} \tag{16}$$

transitions required	time	relative time	path length	initial angle	Mean angle (std dev)	extremum depth	parasite cost %
1	0.16	100	6.1	68	76 (3.8)	2.3	0.22
	0.0016		1.9		76 (3.5)	2.3	0.039
2	0.14	78	4.7	75	72 (4.3)	2.5	0.055
	0.0018		1.9		73 (4.1)	2.5	0.016
3	0.15	71	4.7	74	71 (3.7)	2.3	0.051
	0.0021		2.1		72 (3.0)	2.5	0.0093
7	0.19	59	4.6	63	69 (4.1)	2.4	0.058
	0.0032		2.4		71 (7.0)	2.7	0.0033
10	0.17	49	3.8	60	63 (2.8)	2.5	0.030
	0.0035		2.5		64 (4.7)	2.8	0.00060
13	0.80	110	9.2	39	77 (11)	2.3	0.076
	0.0074		3.2		71 (8.9)	2.7	0.0028

Table 1: Settling time and other measurements for various required transitions. For each transition case, the upper row is for $\nabla_{\vec{v}}$ and the lower row is for $\nabla_{\vec{g}}$. *Std dev* denotes standard deviation. See text for definition of measurement terms and units.

Noting the differences in path length and angles reported, it is clear that the path taken to the target hypercube corner was quite different for the two methods. Method $\nabla_{\vec{v}}$ settles from 1 to 2 orders of magnitude faster than method $\nabla_{\vec{g}}$ and usually takes a path less than half as long. These relationships did not change significantly for different values for c of Eq. (14) and coefficients of Eq. (13) (both unity in Eq. (13)). Values used favored method $\nabla_{\vec{g}}$. Parasite cost is consistently less significant for method $\nabla_{\vec{v}}$ and is quite small for both methods.

To further compare the ability of the optimization methods to solve the brain imaging problem, a large variety of initial hypercube corners were tested. Table 2 displays results that suggest the ability of each method to locate the target corner or to converge to a solution that was consistent with the dipole model. Initial corners were chosen by randomly selecting a number of computational units and setting them to extremum states opposite to that required by the target solution. Five cases were run for each case of required transitions. It can be observed that the system based on method $\nabla_{\vec{v}}$ is better at finding the target corner and is much better at finding a solution that is consistent with the dipole model.

DISCUSSION

The simulation results seem to contradict settling time predictions of the second analytical example. It is intuitively clear that there is no contradiction when considering the analytical example as a one dimensional search and the simulations as multi-dimensional searches. Consider Fig. 4 which illustrates one dimensional search starting at point I. Since both optimization methods must decrease function E monotonically, both must head along the same path to the minimum point A. Now consider Fig. 5 which illustrates a two dimensional search starting at point I: Here, the two methods needn't follow the same paths. The two dashed paths suggest that method $\nabla_{\vec{g}}$ can still be

transitions required	$\nabla_{\vec{g}}$			$\nabla_{\vec{v}}$		
	different dipole solution	different corner	target corner	different dipole solution	different corner	target corner
3	1	0	4	0	0	5
4	1	1	3	0	1	4
5	0	1	4	0	1	4
6	2	1	2	0	1	4
7	4	0	1	0	1	4
13	5	0	0	1	3	1
20	5	0	0	0	5	0
26	5	0	0	2	3	0
33	5	0	0	3	2	0
40	5	0	0	3	2	0
46	5	0	0	2	3	0
53	5	0	0	4	1	0

Table 2: Solutions found starting from various initial conditions, five cases for each transition case. *Different dipole solution* indicates that the system assigned non-zero dipole moments at cluster locations that did not correspond to locations of the dipole model sources. *Different corner* indicates the solution was consistent with the dipole model but was not the target hypercube corner. *Target corner* indicates that the solution was the target solution.

monotonically decreasing E while traversing a more circuitous route to minimum B or traversing a path to minimum A. The longer path lengths reported in Table 1 for method $\nabla_{\vec{g}}$ suggest the occurrence of the former. The data of Table 2 verifies the occurrence of the latter: Note that for many cases where the system based on method $\nabla_{\vec{v}}$ settled to the target corner, the system based on method $\nabla_{\vec{g}}$ settled to some other minimum.

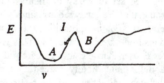

Figure 4: One dimensional search for minima.

Would we observe similar differences in optimization efficiency for other optimization problems that also have binary solution spaces? A view that supports the plausibility of the affirmative is the following: Consider Eq. (4) and Eq. (5). We have already made the observation that method $\nabla_{\vec{v}}$ would slow convergence into extrema of functions $\vec{g}$. We have observed this experimentally via Graph 1. These observations suggest that computational units of $\nabla_{\vec{v}}$ systems tend to stay closer to the transition regions of functions $\vec{g}$ compared to computational units of $\nabla_{\vec{g}}$ systems. It seems plausible that this property may allow $\nabla_{\vec{v}}$ systems to avoid advancing too deeply toward ineffective solutions and, hence, allow the systems to approach effective solutions more efficiently. This behavior might also be the explanation for the comparative success of method $\nabla_{\vec{v}}$ revealed in Table 2.

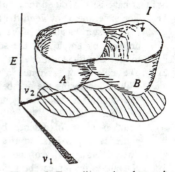

Figure 5: Two dimensional search for minima.

Regarding the construction of electronic circuitry to instantiate Eq. (1), systems based on method $\nabla_{\vec{v}}$ would require the introduction of a component implementing multiplication by the derivative of functions $\vec{g}$. This additional complexity may hinder the use of method $\nabla_{\vec{v}}$ for the

(a)   (b)

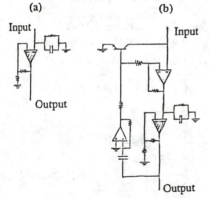

Figure 6: Schematized circuits for a computational unit. Notation is consistent with Horowitz and Hill [15]. Shading of amplifiers is to earmark components referred to in the text. a) Computational unit for method $\nabla_{\vec{g}}$. b) Computational unit for method $\nabla_{\vec{v}}$.

construction of analog circuits for optimization. To illustrate the extent of this additional complexity, Fig. 6a shows a schematized circuit for a computational unit of method $\nabla_{\vec{g}}$ and Fig. 6b shows a schematized circuit for a computational unit of method $\nabla_{\vec{v}}$. The simulations reported above suggest that there may be problems for which improvements in settling time may offset complications that might come with added circuit complexity.

On the problem of imaging cerebral activity, the results above suggest the possibility of constructing analog devices to do the job. Consider the problem of analyzing electric potentials from the scalp of one person: It is noted that the measured electric potentials, $\Phi_{measured}$, appear as linear coefficients in F of Eq. (14); hence, they would appear as constant terms in $\vec{f}$ of Eq. (1). Thus, $\Phi_{measured}$ would be implemented as amplifier biases in the circuits of Figs. 6. This is a significant benefit. To understand this, note that function f_i of Fig. 1 corresponding to the optimization of function F of Eq. (14) would involve a weighted linear sum of inputs $g_1(v_1),...,g_N(v_N)$. The weights would be the nonlinear coefficients of Eq. (14) and correspond to the strengths of the connections shown in Fig. 1. These connection strengths need only be calculated once for the person and can then be set in hardware using, for example, a resistor network. Electric potential measurements could then be analyzed by simply using the measurements to bias the input to shaded amplifiers of Figs. 6. For initialization, the system can be initialized with all dipole moments at zero (the 10 transition case in Table 1). This is a reasonable first guess if it is assumed that cluster locations are far denser than the loci of cerebral activity to be observed. For subsequent measurements, the solution for immediately preceding measurements would be a reasonable initial state if it is assumed that cerebral activity of interest waxes and wanes continuously.

Might non-invasive real time imaging of cerebral activity be possible using such optimization devices? Results of this study are far from adequate for answering this question. Many complexities that have been avoided may nullify the practicality of the idea. Among these problems are:

1)   The experiment avoided the possibility of dipole sources actually occurring at locations other than cluster locations. The minimization of function F of Eq. (14) may circumvent this problem by employing the superposition of dipole moments at neighboring cluster locations to give a sufficient model in the mean.

2)   The experiment assumed a very restricted range of dipole strengths. This might be dealt with by increasing the number of bits used to represent dipole moments.

3)   The conductor model, a homogeneously conducting sphere, may not be sufficient to model the human head [16]. Non-sphericity and major inhomogeneities in conductivity can be dealt with, to a certain extent, by replacing Eq. (12) with a generalized equation based on a numerical approximation of a boundary integral equation [20]

4)   The cerebral activity of interest may not be observable at the scalp.

5)   Not all forms of cerebral activity give rise to dipolar sources. (For example, this is well known in olfactory cortex [8].)

6)   Activity of interest may be overwhelmed by irrelevant activity. Many methods have been devised to contend with this problem (For example, Gevins and Morgan [9].)

Clearly, much theoretical work is left to be done.

CONCLUDING REMARKS

In this study, the mapping principle underlying the application of artificial neural networks to the optimization of multi-dimensional scalar functions has been stated explicitly. Hopfield [12] has shown that for some scalar functions, i.e. functions F quadratic in functions $\vec{g}$, this mapping can lead to dynamical systems that can be easily implemented in hardware, notably, hardware that requires electronic components common to semiconductor technology. Here, mapping principles that have been known for a considerably longer period of time, those underlying gradient based optimization, have been shown capable of leading to dynamical systems that can also be implemented using semiconductor hardware. A problem in medical imaging which requires the search of a multi-dimensional surface full of local extrema has suggested the superiority of the latter mapping principle with respect to settling time of the corresponding dynamical system. This advantage may be quite significant when searching for global extrema using techniques such as iterated descent [2] or iterated genetic hill climbing [1] where many searches for local extrema are required. This advantage is further emphasized by the brain imaging problem: volumes of measurements can be analyzed without reconfiguring the interconnections between computational units; hence, the cost of developing problem specific hardware for finding local extrema may be justifiable. Finally, simulations have contributed plausibility to a possible scheme for non-invasively imaging cerebral activity.

APPENDIX

To show that for a dynamical system based on method $\nabla_{\vec{g}}$, $E_{\vec{g}}$ is a monotonic function of time given that all functions $\vec{g}$ are differentiable and monotonic in the same sense, we need to show that the derivative of $E_{\vec{g}}$ with respect to time is semi-definite:

$$\frac{dE_{\vec{g}}}{dt} = \sum_i^N \frac{\partial F_{\vec{g}}}{\partial g_i} \frac{dg_i}{dt} - \sum_i^N \left[D^{(M)}(v_i) - \frac{dv_i}{dt} \right] \frac{dg_i}{dt}. \tag{A1a}$$

Substituting Eq. (2a),

$$\frac{dE_{\vec{g}}}{dt} = \sum_i^N \left[f_i - D^{(M)}(v_i) + \frac{dv_i}{dt} \right] \frac{dg_i}{dt}. \tag{A1b}$$

Using Eq. (1),

$$\frac{dE_{\vec{g}}}{dt} = \sum_i^N \left[\frac{dv_i}{dt} \right]^2 \frac{\partial g_i}{\partial v_i} \gtrless 0 \tag{A1c}$$

as needed. The appropriate inequality depends on the sense in which functions $\vec{g}$ are monotonic. In a similar manner, the result can be obtained for method $\nabla_{\vec{v}}$. With the condition that functions $\vec{g}$ are differentiable, we can show that the derivative of $E_{\vec{v}}$ is semi-definite:

$$\frac{dE_{\vec{v}}}{dt} = \sum_i^N \frac{\partial F_{\vec{v}}}{\partial v_i} \frac{dv_i}{dt} - \sum_i^N \left[D^{(M)}(v_i) - \frac{dv_i}{dt} \right] \frac{dv_i}{dt}. \tag{A2a}$$

Using Eqs. (3a) and (1),

$$\frac{dE_{\vec{v}}}{dt} = \sum_i^N \left[\frac{dv_i}{dt} \right]^2 \gtrless 0 \tag{A2b}$$

as needed.

In order to use the results derived above to conclude that Eq. (1) can be used for optimization of functions $E_{\vec{v}}$ and $E_{\vec{g}}$ in the vicinity of some point $\vec{v}_0$, we need to show that there exists a neighborhood of $\vec{v}_0$ in which there exist solution trajectories to Eq. (1). The necessary existence theorems and transformations of Eq. (1) needed in order to apply the theorems can be found in many texts on ordinary differential equations; e.g. Guckenheimer and Holmes [11]. Here, it is mainly important to state that the theorems require that functions $\vec{f} \epsilon C^{(1)}$, functions $\vec{g}$ are differentiable, and initial conditions are specified for all derivatives of lower order than M.

ACKNOWLEDGEMENTS

I would like to thank Dr. Michael Raugh and Dr. Pentti Kanerva for constructive criticism and support. I would like to thank Bill Baird and Dr. James Keeler for reviewing this work. I would like to thank Dr. Derek Fender, Dr. John Hopfield, and Dr. Stanley Klein for giving me opportunities that fostered this conglomeration of ideas.

REFERENCES

[1] Ackley D.H., "Stochastic iterated genetic hill climbing", PhD. dissertation, Carnegie Mellon U., 1987.

[2] Baum E., Neural Networks for Computing, ed. Denker J.S. (AIP Confrnc. Proc. 151, ed. Lerner R.G.), p53-58, 1986.

[3] Brody D.A., IEEE Trans. vBME-32, n2, p106-110, 1968.

[4] Brody D.A., Terry F.H., Ideker R.E., IEEE Trans. vBME-20, p141-143, 1973.

[5] Cohen M.A., Grossberg S., IEEE Trans. vSMC-13, p815-826, 1983.

[6] Cuffin B.N., IEEE Trans. vBME-33, n9, p854-861, 1986.

[7] Darcey T.M., Ary J.P., Fender D.H., Prog. Brain Res., v54, p128-134, 1980.

[8] Freeman W.J., "Mass Action in the Nervous System", Academic Press, Inc., 1975.

[9] Gevins A.S., Morgan N.H., IEEE Trans., vBME-33, n12, p1054-1068, 1986.

[10] Goles E., Vichniac G.Y., Neural Networks for Computing, ed. Denker J.S. (AIP Confrnc. Proc. 151, ed. Lerner R.G.), p165-181, 1986.

[11] Guckenheimer J., Holmes P., "Nonlinear Oscillations, Dynamical Systems, and Bifurcations of Vector Fields", Springer Verlag, 1983.

[12] Hopfield J.J., Proc. Natl. Acad. Sci., v81, p3088-3092, 1984.

[13] Hopfield J.J., Tank D.W., Bio. Cybrn., v52, p141-152, 1985.

[14] Hopfield J.J., Tank D.W., Science, v233, n4764, p625-633, 1986.

[15] Horowitz P., Hill W., "The art of electronics", Cambridge U. Press, 1983.

[16] Hosek R.S., Sances A., Jodat R.W., Larson S.J., IEEE Trans., vBME-25, n5, p405-413, 1978.

[17] Hutchinson J.M., Koch C., Neural Networks for Computing, ed. Denker J.S. (AIP Confrnc. Proc. 151, ed. Lerner R.G.), p235-240, 1986.

[18] Jeffery W., Rosner R., Astrophys. J., v310, p473-481, 1986.

[19] Lapedes A., Farber R., Neural Networks for Computing, ed. Denker J.S. (AIP Confrnc. Proc. 151, ed. Lerner R.G.), p283-298, 1986.

[20] Leong H.M.F., "Frequency dependence of electromagnetic fields: models appropriate for the brain", PhD. dissertation, California Institute of Technology, 1986.

[21] Platt J.C., Hopfield J.J., Neural Networks for Computing, ed. Denker J.S. (AIP Confrnc. Proc. 151, ed. Lerner R.G.), p364-369, 1986.

[22] Press W.H., Flannery B.P., Teukolsky S.A., Vetterling W.T., "Numerical Recipes", Cambridge U. Press, 1986.

[23] Takeda M., Goodman J.W., Applied Optics, v25, n18, p3033-3046, 1986.

[24] Tank D.W., Hopfield J.J., "Neural computation by concentrating information in time", preprint, 1987.

TOWARDS AN ORGANIZING PRINCIPLE FOR
A LAYERED PERCEPTUAL NETWORK

Ralph Linsker
IBM Thomas J. Watson Research Center, Yorktown Heights, NY 10598

Abstract

An information-theoretic optimization principle is proposed for the development of each processing stage of a multilayered perceptual network. This principle of "maximum information preservation" states that the signal transformation that is to be realized at each stage is one that maximizes the information that the output signal values (from that stage) convey about the input signals values (to that stage), subject to certain constraints and in the presence of processing noise. The quantity being maximized is a Shannon information rate. I provide motivation for this principle and -- for some simple model cases -- derive some of its consequences, discuss an algorithmic implementation, and show how the principle may lead to biologically relevant neural architectural features such as topographic maps, map distortions, orientation selectivity, and extraction of spatial and temporal signal correlations. A possible connection between this information-theoretic principle and a principle of minimum entropy production in nonequilibrium thermodynamics is suggested.

Introduction

This paper describes some properties of a proposed information-theoretic organizing principle for the development of a layered perceptual network. The purpose of this paper is to provide an intuitive and qualitative understanding of how the principle leads to specific feature-analyzing properties and signal transformations in some simple model cases. More detailed analysis is required in order to apply the principle to cases involving more realistic patterns of signaling activity as well as specific constraints on network connectivity.

This section gives a brief summary of the results that motivated the formulation of the organizing principle, which I call the principle of "maximum information preservation." In later sections the principle is stated and its consequences studied.

In previous work[1] I analyzed the development of a layered network of model cells with feedforward connections whose strengths change in accordance with a Hebb-type synaptic modification rule. I found that this development process can produce cells that are selectively responsive to certain input features, and that these feature-analyzing properties become progressively more sophisticated as one proceeds to deeper cell layers. These properties include the analysis of contrast and of edge orientation, and are qualitatively similar to properties observed in the first several layers of the mammalian visual pathway.[2]

Why does this happen? Does a Hebb-type algorithm (which adjusts synaptic strengths depending upon correlations among signaling activities[3]) cause a developing perceptual network to optimize some property that is deeply connected with the mature network's functioning as an information processing system?

Further analysis[4,5] has shown that a suitable Hebb-type rule causes a linear-response cell in a layered feedforward network (without lateral connections) to develop so that the statistical variance of its output activity (in response to an ensemble of inputs from the previous layer) is maximized, subject to certain constraints. The mature cell thus performs an operation similar to principal component analysis (PCA), an approach used in statistics to expose regularities (e.g., clustering) present in high-dimensional input data. (Oja[6] had earlier demonstrated a particular form of Hebb-type rule that produces a model cell that implements PCA exactly.)

Furthermore, given a linear device that transforms inputs into an output, and given any particular output value, one can use optimal estimation theory to make a "best estimate" of the input values that gave rise to that output. Of all such devices, I have found that an appropriate Hebb-type rule generates that device for which this "best estimate" comes closest to matching the input values.[4,5] Under certain conditions, such a cell has the property that its output preserves the maximum amount of information about its input values.[5]

Maximum Information Preservation

The above results have suggested a possible organizing principle for the development of each layer of a multilayered perceptual network.[5] The principle can be applied even if the cells of the network respond to their inputs in a nonlinear fashion, and even if lateral as well as feedforward connections are present. (Feedback from later to earlier layers, however, is absent from this formulation.) This principle of "maximum information preservation" states that for a layer of cells L that is connected to and provides input to another layer M, the connections should develop so that the transformation of signals from L to M (in the presence of processing noise) has the property that the set of output values *M* conveys the <u>maximum amount of information</u> about the input values *L*, subject to various constraints on, e.g., the range of lateral connections and the processing power of each cell. The statistical properties of the ensemble of inputs *L* are assumed stationary, and the particular L-to-M transformation that achieves this maximization depends on those statistical properties. The quantity being maximized is a Shannon information rate.[7]

An equivalent statement of this principle is: The L-to-M transformation is chosen so as to <u>minimize</u> the amount of information that would be conveyed by the input values *L* to someone who already knows the output values *M*.

We shall regard the set of input signal values *L* (at a given time) as an input "message"; the message is processed to give an output message *M*. Each message is in general a set of real-valued signal activities. Because noise is introduced during the processing, a given input message may generate any of a range of different output messages when processed by the same set of connections.

The Shannon information rate (i.e., the average information transmitted from L to M per message) is[7]

$$R = \Sigma_L \, \Sigma_M \, P(L,M) \, \log \, [P(L,M)/P(L)P(M)]. \tag{1}$$

For a discrete message space, $P(L)$ [resp. $P(M)$] is the probability of the input (resp. output) message being L (resp. M), and $P(L,M)$ is the joint probability of the input being L and the output being M. [For a continuous message space, probabilities are replaced by probability densities, and sums (over states) by integrals.] This rate can be written as

$$R = I_L - I_{L|M} \tag{2}$$

where

$$I_L \equiv - \Sigma_L P(L) \log P(L) \tag{3}$$

is the average information conveyed by message L and

$$I_{L|M} \equiv - \Sigma_M P(M) \Sigma_L P(L|M) \log P(L|M) \tag{4}$$

is the average information conveyed by message L to someone who already knows M. Since I_L is fixed by the properties of the input ensemble, maximizing R means minimizing $I_{L|M}$, as stated above.

The information rate R can also be written as

$$R = I_M - I_{M|L} \tag{5}$$

where I_M and $I_{M|L}$ are defined by interchanging L and M in Eqns. 3 and 4. This form is heuristically useful, since it suggests that one can attempt to make R large by (if possible) simultaneously making I_M large and $I_{M|L}$ small. The term I_M is largest when each message M occurs with equal probability. The term $I_{M|L}$ is smallest when each L is transformed into a unique M, and more generally is made small by "sharpening" the $P(M|L)$ distribution, so that for each L, $P(M|L)$ is near zero except for a small set of messages M.

How can one gain insight into biologically relevant properties of the $L \rightarrow M$ transformation that may follow from the principle of maximum information preservation (which we also call the "infomax" principle)? In a network, this $L \rightarrow M$ transformation may be a function of the values of one or a few variables (such as a connection strength) for each of the allowed connections between and within layers, and for each cell. The search space is quite large, particularly from the standpoint of gaining an intuitive or qualitative understanding of network behavior. We shall therefore consider a simple model in which the dimensionalities of the L and M signal spaces are greatly reduced, yet one for which the infomax analysis exhibits features that may also be important under more general conditions relevant to biological and synthetic network development.

The next four sections are organized as follows. (i) A model is introduced in which the L and M messages, and the L-to-M transformation, have simple forms. The infomax principle is found to be satisfied when some simple geometric conditions (on the transformation) are met. (ii) I relate this model to the analysis of signal processing and noise in an interconnection network. The formation of topographic maps is discussed. (iii) The model is applied to simplified versions of biologically relevant problems, such as the emergence of orientation selectivity. (iv) I show that the main properties of the infomax principle for this model can be realized by certain local algorithms that have been proposed to generate topographic maps using lateral interactions.

A Simple Geometric Model

In this model, each input message L is described by a point in a low-dimensional vector space, and the output message M is one of a number of discrete states. For definiteness, we will take the L space to be two-dimensional (the extension to higher dimensionality is straightforward). The $L \to M$ transformation consists of two steps. (i) A noise process alters L to a message L' lying within a neighborhood of radius ν centered on L. (ii) The altered message L' is mapped deterministically onto one of the output messages M.

A given $L' \to M$ mapping corresponds to a partitioning of the L space into regions labeled by the output states M. (We do not exclude a priori the possibility that multiple disjoint regions may be labeled by the same M.) Let A denote the total area of the L state space. For each M, let $A(M)$ denote the area of L space that is labeled by M. Let $s(M)$ denote the total border length that the region(s) labeled M share with regions of unlike M-label. A point L lying within distance ν of a border can be mapped onto either M-value (because of the noise process $L \to L'$). Call this a "borderline" L. A point L that is more than a distance ν from every border can only be mapped onto the M-value of the region containing it.

Suppose ν is sufficiently small that (for the partitionings of interest) the area occupied by borderline L states is small compared to the total area of the L space. Consider first the case in which $P(L)$ is uniform over L. Then the information rate R (using Eqn. 5) is given approximately (through terms of order ν) by

$$R = - \Sigma_M [A(M)/A] \log[A(M)/A] - (\gamma\nu/A) \Sigma_M s(M). \tag{6}$$

To see this, note that $P(M) = A(M)/A$ and that $P(M|L) \log P(M|L)$ is zero except for borderline L (since $0 \log 0 = 1 \log 1 = 0$). Here γ is a positive number whose value depends upon the details of the noise process, which determines $P(M|L)$ for borderline L as a function of distance from the border.

For small ν (low noise) the first term (I_M) on the RHS of Eqn. 6 dominates. It is maximized when the $A(M)$ [and hence the $P(M)$] values are equal for all M. The second term (with its minus sign), which equals ($-I_{M|L}$), is maximized when the sum of the border lengths of all M regions is minimized. This corresponds to "sharpening" the $P(M|L)$ distribution in our earlier, more general, discussion. This suggests that the infomax solution is obtained by partitioning the L space into M-regions (one for each M value) that are of substantially equal area, with each M-region tending to have near-minimum border length.

Although this simple analysis applies to the low-noise case, it is plausible that even when ν is comparable to the spatial scale of the M regions, infomax will favor making the M regions have approximately the same extent in all directions (rather than be elongated), in order to "sharpen" $P(M|L)$ and reduce the probability of the noise process mapping L onto many different M states.

What if $P(L)$ is nonuniform? Then the same result (equal areas, minimum border) is obtained except that both the area and border-length elements must now be weighted by the local value of $P(L)$. Therefore the infomax principle tends to produce maps in which greater representation in the output space is given to regions of the input signal space that are activated more frequently.

To see how lateral interactions within the M layer can affect these results, let us suppose that the $L \to M$ mapping has three, not two, process steps: $L \to L'$

$\rightarrow M' \rightarrow M$, where the first two steps are as above, and the third step changes the output M' into any of a number of states M (which by definition comprise the "M-neighborhood" of M'). We consider the case in which this M-neighborhood relation is symmetric.

This type of "lateral interaction" between M states causes the infomax principle to favor solutions for which M regions sharing a border in L space are M-neighbors in the sense defined. For a simple example in which each state M' has n M-neighbors (including itself), and each M-neighbor has an equal chance of being the final state (given M'), infomax tends to favor each M-neighborhood having similar extent in all directions (in L space).

Relation Between the Geometric Model and Network Properties

The previous section dealt with certain classes of transformations from one message space to another, and made no specific reference to the implementation of these transformations by an interconnected network of processor cells. Here we show how some of the features discussed in the previous section are related to network properties.

For simplicity suppose that we have a two-dimensional layer of uniformly distributed cells, and that the signal activity of each cell at any given time is either 1 (active) or 0 (quiet). We need to specify the ensemble of input patterns. Let us first consider a simple case in which each pattern consists of a disk of activity of fixed radius, but arbitrary center position, against a quiet background. In this case the pattern is fully defined by specifying the coordinates of the disk center. In a two-dimensional L state space (previous section), each pattern would be represented by a point having those coordinates.

Now suppose that each input pattern consists not of a sharply defined disk of activity, but of a "fuzzy" disk whose boundary (and center position) are not sharply defined. [Such a pattern could be generated by choosing (from a specified distribution) a position $\mathbf{x}_c$ as the nominal disk center, then setting the activity of the cell at position $\mathbf{x}$ to 1 with a probability that decreases with distance $|\mathbf{x} - \mathbf{x}_c|$.] Any such pattern can be described by giving the coordinates of the "center of activity" along with many other values describing (for example) various moments of the activity pattern relative to the center.

For the noise process $L \rightarrow L'$ we suppose that the activity of an L cell can be "misread" (by the cells of the M layer) with some probability. This set of distorted activity values is the "message" L'. We then suppose that the set of output activities M is a deterministic function of L'.

We have constructed a situation in which (for an appropriate choice of noise level) two of the dimensions of the L state space -- namely, those defined by the disk center coordinates -- have large variance compared to the variance induced by the noise process, while the other dimensions have variance comparable to that induced by noise. In other words, the center position of a pattern is changed only a small amount by the noise process (compared to the typical difference between the center positions of two patterns), whereas the values of the other attributes of an input pattern differ as much from their noise-altered values as two typical input patterns differ from each other. (Those attributes are "lost in the noise.")

Since the distance between L states in our geometric model (previous section) corresponds to the likelihood of one L state being changed into the other by the noise process, we can heuristically regard the L state space (for the present example) as a "slab" that is elongated in two dimensions and very thin in all other dimensions. (In general this space could have a much more complicated topology, and the noise process which we here treat as defining a simple metric structure on the L state space need not do so. These complications are beyond the scope of the present discussion.)

This example, while simple, illustrates a feature that is key to understanding the operation of the infomax principle: The character of the ensemble statistics and of the noise process jointly determine which attributes of the input pattern are statistically most significant; that is, have largest variance relative to the variance induced by noise. We shall see that the infomax principle selects a number of these most significant attributes to be encoded by the $L \rightarrow M$ transformation.

We turn now to a description of the output state space M. We shall assume that this space is also of low dimensionality. For example, each M pattern may also be a disk of activity having a center defined within some tolerance. A discrete set of discriminable center-coordinate values can then be used as the M-region "labels" in our geometric model.

Restricting the form of the output activity in this particular way restricts us to considering positional encodings $L \rightarrow M$, rather than encodings that make use of the shape of the output pattern, its detailed activity values, etc. However, this restriction on the form of the output does not determine which features of the input patterns are to be encoded, nor whether or not a topographic (neighbor-preserving) mapping is to be used. These properties will be seen to emerge from the operation of the infomax principle.

In the previous section we saw that the infomax principle will tend to lead to a partitioning of the L space into M regions having equal areas [if $P(L)$ is uniform in the coordinates of the L disk center] and minimum border length. For the present case this means that the M regions will tend to "tile" the two long dimensions of the L state space "slab," and that a single M value will represent all points in L space that differ only in their low-variance coordinates. If $P(L)$ is nonuniform, then the area of the M region at L will tend to be inversely proportional to $P(L)$. Furthermore, if there are local lateral connections between M cells, then (depending upon the particular form of such interaction) M states corresponding to nearby localized regions of layer-M activity can be M-neighbors in the sense of the previous section. In this case the mapping from the two high-variance coordinates of L space to M space will tend to be topographic.

Examples: Orientation Selectivity and Temporal Feature Maps

The simple example in the previous section illustrates how infomax can lead to topographic maps, and to map distortions [which provide greater M-space representation for regions of L having large $P(L)$]. Let us now consider a case in which information about input features is positionally encoded in the output layer as a result of the infomax principle.

Consider a model case in which an ensemble of patterns is presented to the input layer L. Each pattern consists of a rectangular bar of activity (of fixed length and width) against a quiet background. The bar's center position and orientation are chosen for each pattern from uniform distributions over some spatial interval for the position, and over all orientation angles (i.e., from $0°$ to $180°$). The bar need not be sharply defined, but can be "fuzzy" in the sense described above. We assume, however, that all properties that distinguish different patterns of the ensemble -- except for center position and orientation -- are "lost in the noise" in the sense we discussed.

To simplify the representation of the solution, we further assume that only one coordinate is needed to describe the center position of the bar for the given ensemble. For example, the ensemble could consist of bar patterns all of which have the same y coordinate of center position, but differ in their x coordinate and in orientation θ.

We can then represent each input state by a point in a rectangle (the L state space defined in a previous section) whose abscissa is the center-position coordinate x and whose ordinate is the angle θ. The horizontal sides of this rectangle are identified with each other, since orientations of $0°$ and $180°$ are identical. (The interior of the rectangle can thus be thought of as the surface of a horizontal cylinder.)

The number N_x of different x positions that are discriminable is given by the range of x values in the input ensemble divided by the tolerance with which x can be measured (given the noise process $L \rightarrow L'$); similarly for N_θ. The relative lengths Δx and $\Delta \theta$ of the sides of the L state space rectangle are given by $\Delta x / \Delta \theta = N_x / N_\theta$. We discuss below the case in which $N_x >> N_\theta$; if N_θ were $>> N_x$ the roles of x and θ in the resulting mappings would be reversed.

There is one complicating feature that should be noted, although in the interest of clarity we will not include it in the present analysis. Two horizontal bar patterns that are displaced by a horizontal distance that is small compared with the bar length, are more likely to be rendered indiscriminable by the noise process than are two vertical bar patterns that are displaced by the same horizontal distance (which may be large compared with the bar's width). The Hamming distance, or number of binary activity values that need to be altered to change one such pattern into the other, is greater in the latter case than in the former. Therefore, the distance in L state space between the two

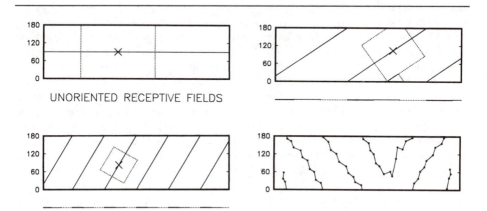

Figure 1.   Orientation Selectivity in a Simple Model: As the input domain size (see text) is reduced [from (a) upper left, to (b) upper right, to (c) lower left figure], infomax favors the emergence of an orientation-selective $L \rightarrow M$ mapping. (d) Lower right figure shows a solution obtained by applying Kohonen's relaxation algorithm with 50 M-points (shown as dots) to this mapping problem.

states should be greater in the latter case. This leads to a "warped" rather than simple rectangular state space. We ignore this effect here, but it must be taken into account in a fuller treatment of the emergence of orientation selectivity.

Consider now an $L \rightarrow M$ transformation that consists of the three-step process (discussed above) (i) noise-induced $L \rightarrow L'$; (ii) deterministic $L' \rightarrow M'$; (iii) lateral-interaction-induced $M' \rightarrow M$. Step (ii) maps the two-dimensional L state space of points (x, θ) onto a one-dimensional M state space. For the present discussion, we consider $L' \rightarrow M'$ maps satisfying the following Ansatz: Points corresponding to the M states are spaced uniformly, and in topographic order, along a helical line in L state space (which we recall is represented by the surface of a horizontal cylinder). The pitch of the helix (or the slope $d\theta/dx$) remains to be determined by the infomax principle. Each M-neighborhood of M states (previous section) then corresponds to an interval on such a helix. A state L' is mapped onto a state in a particular M-neighborhood if L' is closer (in L space) to the corresponding interval of the helix than to any other portion of the helix. We call this set of L states (for an M-neighborhood centered on M) the "input domain" of M. It has rectangular shape and lies on the cylindrical surface of the L space.

We have seen (previous sections) that infomax tends to produce maps having (i) equal M-region areas, (ii) topographic organization, and (iii) an input domain (for each M-neighborhood) that has similar extent in all directions (in L space). Our choice of Ansatz enforces (i) and (ii) explicitly. Criterion (iii) is satisfied by choosing $d\theta/dx$ such that the input domain is square (for a given M-neighborhood size).

Figure 1a (having $d\theta/dx = 0$) shows a map in which the output M encodes only information about bar center position x, and is independent of bar orientation θ. The size of the M-neighborhood is relatively large in this case. The input domain of the state M denoted by the 'x' is shown enclosed by dotted lines. (The particular θ value at which we chose to draw the M line in Fig. 1a is irrelevant.) For this M-neighborhood size, the length of the border of the input domain is as small as it can be.

As the M-neighborhood size is reduced, the dotted lines move closer together. A vertically oblong input domain (which would result if we kept $d\theta/dx = 0$) would not satisfy the infomax criterion. The helix for which the input domain is square (for this smaller choice of M-neighborhood size) is shown in Fig. 1b. The M states for this solution encode information about bar orientation as well as center position. If each M state corresponds to a localized output activity pattern centered at some position in a one-dimensional array of M cells, then this solution corresponds to orientation-selective cells organized in "orientation columns" (really "orientation intervals" in this one-dimensional model). A "labeling" of the linear array of cells according to whether their orientation preferences lie between 0 and 60, 60 and 120, or 120 and 180 degrees is indicated by the bold, light, and dotted line segments beneath the rectangle in Fig. 1b (and 1c).

As the M-neighborhood size is decreased still further, the mapping shown in Fig. 1c becomes favored over that of either Fig. 1a or 1b. The "orientation columns" shown in the lower portion of Fig. 1c are narrower than in Fig. 1b.

A more detailed analysis of the information rate function for various mappings confirms the main features we have here obtained by a simple geometric argument.

The same type of analysis can be applied to different types of input pattern ensembles. To give just one other example, consider a network that receives an ensemble of simple patterns of acoustic input. Each such pattern consists of a tone of some frequency that is sensed by two "ears" with some interaural time delay. Suppose that the initial network layers organize the information from each ear (separately) into tonotopic maps, and that (by means of connections having a range of different time delays) the signals received by both ears over some time interval appear as patterns of cell activity at some intermediate layer L. We can then apply the infomax principle to the signal transformation from layer L to the next layer M. The L state space can (as before) be represented as a rectangle, whose axes are now frequency and interaural delay (rather than spatial position and bar orientation). Apart from certain differences (the density of L states may be nonuniform, and states at the top and bottom of the rectangle are no longer identical), the infomax analysis can be carried out as it was for the simplified case of orientation selectivity.

Local Algorithms

The information rate (Eqn. 1), which the infomax principle states is to be maximized subject to constraints (and possibly as part of an optimization function containing other cost terms not discussed here), has a very complicated mathematical form. How might this optimization process, or an approximation to it, be implemented by a network of cells and connections each of which has limited computational power? The geometric form in which we have cast the infomax principle for some very simple model cases, suggests how this might be accomplished.

An algorithm due to Kohonen [8] demonstrates how topographic maps can emerge as a result of lateral interactions within the output layer. I applied this algorithm to a one-dimensional M layer and a two-dimensional L layer, using a Euclidean metric and imposing periodic boundary conditions on the short dimension of the L layer. A resulting map is shown in Fig. 1d. This map is very similar to those of Figs. 1b and 1c, except for one reversal of direction. The reversal is not surprising, since the algorithm involves only local moves (of the M-points) while the infomax principle calls for a globally optimal solution.

More generally, Kohonen's algorithm tends empirically[8] to produce maps having the property that if one constructs the Voronoi diagram corresponding to the positions of the M-points (that is, assigns each point L to an M region based on which M-point L is closest to), one obtains a set of M regions that tend to have areas inversely proportional to $P(L)$, and neighborhoods (corresponding to our input domains) that tend to have similar extent in all directions rather than being elongated.

The Kohonen algorithm makes no reference to noise, to information content, or even to an optimization principle. Nevertheless, it appears to implement, at least in a qualitative way, the geometric conditions that infomax imposes in some simple cases. This suggests that local algorithms along similar lines may be capable of implementing the infomax principle in more general situations.

Our geometric formulation of the infomax principle also suggests a connection with an algorithm proposed by von der Malsburg and Willshaw[9] to generate topographic maps. In their "tea trade" model, neighborhood relationships are postulated within the source and the target spaces, and the algorithm's operation leads to the establishment of a neighborhood-preserving mapping from source to target space. Such neighborhood relationships arise naturally in our analysis when the infomax principle is applied to our three-step $L \rightarrow L' \rightarrow M' \rightarrow M$ transformation. The noise process induces a

494

neighborhood relation on the L space, and lateral connections in the M cell layer can induce a neighborhood relation on the M space.

More recently, Durbin and Willshaw[10] have devised an approach to solving certain geometric optimization problems (such as the traveling salesman problem) by a gradient descent method bearing some similarity to Kohonen's algorithm.

There is a complementary relationship between the infomax principle and a local algorithm that may be found to implement it. On the one hand, the principle may explain what the algorithm is "for" -- that is, how the algorithm may contribute to the generation of a useful perceptual system. This in turn can shed light on the system-level role of lateral connections and synaptic modification mechanisms in biological networks. On the other hand, the existence of such a local algorithm is important for demonstrating that a network of relatively simple processors -- biological or synthetic -- can in fact find global near-maxima of the Shannon information rate.

A Possible Connection Between Infomax and a Thermodynamic Principle

The principle of "maximum preservation of information" can be viewed equivalently as a principle of "minimum dissipation of information." When the principle is satisfied, the loss of information from layer to layer is minimized, and the flow of information is in this sense as "nearly reversible" as the constraints allow. There is a resemblance between this principle and the principle of "minimum entropy production" [11] in nonequilibrium thermodynamics. It has been suggested by Prigogine and others that the latter principle is important for understanding self-organization in complex systems. There is also a resemblance, at the algorithmic level, between a Hebb-type modification rule and the autocatalytic processes[12] considered in certain models of evolution and natural selection. This raises the possibility that the connection I have drawn between synaptic modification rules and an information-theoretic optimization principle may be an example of a more general relationship that is important for the emergence of complex and apparently "goal-oriented" structures and behaviors from relatively simple local interactions, in both neural and non-neural systems.

References

[1]   R. Linsker, *Proc. Natl. Acad. Sci. USA* **83** , 7508, 8390, 8779 (1986).
[2]   D. H. Hubel and T. N. Wiesel, *Proc. Roy. Soc. London* **B198** , 1 (1977).
[3]   D. O. Hebb, *The Organization of Behavior* (Wiley, N. Y., 1949).
[4]   R. Linsker, in: R. Cotterill (ed.), *Computer Simulation in Brain Science* (Copenhagen, 20-22 August 1986; Cambridge Univ. Press, in press), p. 416.
[5]   R. Linsker, *Computer* (March 1988, in press).
[6]   E. Oja, *J. Math. Biol.* **15** , 267 (1982).
[7]   C. E. Shannon, *Bell Syst. Tech. J.* **27** , 623 (1948).
[8]   T. Kohonen, *Self-Organization and Associative Memory* (Springer-Verlag, N. Y., 1984).
[9]   C. von der Malsburg and D. J. Willshaw, *Proc. Natl. Acad. Sci. USA* **74** , 5176 (1977).
[10]   R. Durbin and D. J. Willshaw, *Nature* **326** , 689 (1987).
[11]   P. Glansdorff and I. Prigogine, *Thermodynamic Theory of Structure, Stability, and Fluctuations* (Wiley-Interscience, N. Y., 1971).
[12]   M. Eigen and P. Schuster, *Die Naturwissenschaften* **64** , 541 (1977).

REFLEXIVE ASSOCIATIVE MEMORIES

Hendricus G. Loos

Laguna Research Laboratory, Fallbrook, CA 92028-9765

ABSTRACT

In the synchronous discrete model, the average memory capacity of bidirectional associative memories (BAMs) is compared with that of Hopfield memories, by means of a calculation of the percentage of good recall for 100 random BAMs of dimension 64x64, for different numbers of stored vectors. The memory capacity is found to be much smaller than the Kosko upper bound, which is the lesser of the two dimensions of the BAM. On the average, a 64x64 BAM has about 68 % of the capacity of the corresponding Hopfield memory with the same number of neurons. Ortho-normal coding of the BAM increases the effective storage capacity by only 25 %. The memory capacity limitations are due to spurious stable states, which arise in BAMs in much the same way as in Hopfield memories. Occurrence of spurious stable states can be avoided by replacing the thresholding in the backlayer of the BAM by another nonlinear process, here called "Dominant Label Selection" (DLS). The simplest DLS is the winner-take-all net, which gives a fault-sensitive memory. Fault tolerance can be improved by the use of an orthogonal or unitary transformation. An optical application of the latter is a Fourier transform, which is implemented simply by a lens.

INTRODUCTION

A reflexive associative memory, also called bidirectional associative memory, is a two-layer neural net with bidirectional connections between the layers. This architecture is implied by Dana Anderson's optical resonator[1], and by similar configurations[2,3]. Bart Kosko[4] coined the name "Bidirectional Associative Memory" (BAM), and investigated several basic properties[4-6]. We are here concerned with the memory capacity of the BAM, with the relation between BAMs and Hopfield memories[7], and with certain variations on the BAM.

BAM STRUCTURE

We will use the discrete model in which the state of a layer of neurons is described by a bipolar vector. The Dirac notation[8] will be used, in which |> and <| denote respectively column and row vectors. <a| and |a> are each other transposes, <a|b> is a scalar product, and |a><b| is an outer product. As depicted in Fig. 1, the BAM has two layers of neurons, a front layer of N neurons with state vector |f>, and a back layer of P neurons with state vector |b>. The bidirectional connections between the layers allow signal flow in two directions. The front stroke gives |b>= s(B|f>), where B is the connection matrix, and s( ) is a threshold function, operating at

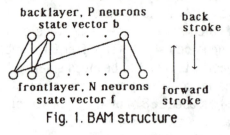

backlayer, P neurons
state vector b

back stroke

frontlayer, N neurons
state vector f

forward stroke

Fig. 1. BAM structure

zero. The back stroke results in an upgraded front state <f'|=s(<b|B), which also may be written as |f'>=s(B^T|b>), where the superscript T denotes transposition. We consider the synchronous model, where all neurons of a layer are updated simultaneously, but the front and back layers are updated at different times. The BAM action is shown in Fig. 2. The forward stroke entails taking scalar products between a front state vector |f> and the rows of B, and entering the thresholded results as elements of the back state vector |b>. In the back stroke we take

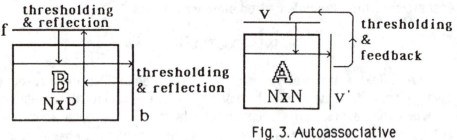

f

thresholding & reflection

B
NxP

thresholding & reflection

b

v

thresholding & feedback

A
NxN

v'

Fig. 2. BAM action

Fig. 3. Autoassociative memory action

scalar products of |b> with column vectors of B, and enter the thresholded results as elements of an upgraded state vector |f'>. In contrast, the action of an autoassociative memory is shown in Figure 3. The BAM may also be described as an autoassociative memory[5] by concatenating the front and back vectors into a single state vector $|v\rangle = |f,b\rangle$, and by taking the $(N+P)\times(N+P)$ connection matrix as shown in Fig. 4. This autoassociative memory has the same number of neurons as our BAM, viz. $N+P$. The BAM operation where initially only the front state is specified may be obtained with the corresponding autoassociative memory by initially specifying $|b\rangle$ as zero, and by arranging the thresholding operation such that $s(0)$ does not alter the state vector component. For a Hopfield memory[7] the connection matrix is

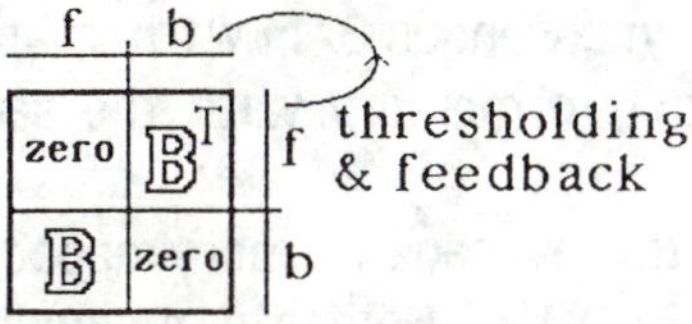

Fig. 4. BAM as autoasso-
ciative memory

$$H = (\sum_{m=1}^{M} |m\rangle\langle m|) - MI \; , \tag{1}$$

where $|m\rangle$, $m=1$ to M, are stored vectors, and I is the identity matrix. Writing the $N+P$ dimensional vectors $|m\rangle$ as concatenations $|d_m,c_m\rangle$, (1) takes the form

$$H = (\sum_{m=1}^{M} (|d_m\rangle\langle d_m| + |c_m\rangle\langle c_m| + |d_m\rangle\langle c_m| + |c_m\rangle\langle d_m|)) - MI \; , \tag{2}$$

with proper block placing of submatrices understood. Writing

$$K = \sum_{m=1}^{M} |c_m\rangle\langle d_m| \; , \tag{3}$$

$$H_d = (\sum_{m=1}^{M} |d_m\rangle\langle d_m|) - MI \; , \qquad H_c = (\sum_{m=1}^{M} |c_m\rangle\langle c_m|) - MI, \tag{4}$$

where the I are identities in appropriate subspaces, the Hopfield matrix H may be partitioned as shown in Fig. 5. K is just the BAM matrix given by Kosko[5], and previously used by Kohonen[9] for linear heteroassociative memories. Comparison of Figs. 4 and 5 shows that in the synchronous discrete model the BAM with connection matrix (3) is equivalent to a Hopfield memory in which the diagonal blocks H_d and H_c have been deleted. Since the Hopfield memory is robust, this "pruning" may not affect much the associative recall of stored vectors, if M is small; however, on the average, pruning will not improve the memory capacity. It follows that, on the average, a discrete synchronous BAM with matrix (3) can at best have the capacity of a Hopfield memory with the same number of neurons.

We have performed computations of the average memory capacity for 64x64 BAMs and for corresponding 128x128 Hopfield memories. Monte Carlo calculations were done for 100 memories, each of which stores M random bipolar vectors. The straight recall of all these vectors was checked, allowing for 24 iterations. For the BAMs, the iterations were started with a forward stroke in which one of the stored vectors $|d_m\rangle$ was used as input. The percentage of good recall and its standard deviation were calculated. The results plotted in Fig. 6 show that the square BAM has about 68% of the capacity of the corresponding Hopfield memory. Although the total number of neurons is the same, the BAM only needs 1/4 of the number of connections of the Hopfield memory. The storage capacity found is much smaller than the Kosko [6] upper bound, which is min (N,P).

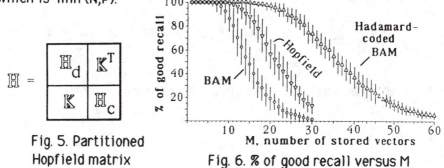

$$\mathbb{H} = \begin{array}{|c|c|} \hline \mathbb{H}_d & \mathbb{K}^T \\ \hline \mathbb{K} & \mathbb{H}_c \\ \hline \end{array}$$

Fig. 5. Partitioned Hopfield matrix

Fig. 6. % of good recall versus M

CODED BAM

So far, we have considered both front and back states to be used for data. There is another use of the BAM in which only front states are used as data, and the back states are seen as providing a code, label, or pointer for the front state . Such use was anticipated in our expression (3) for the BAM matrix which stores data vectors $|d_m\rangle$ and their labels or codes $|c_m\rangle$. For a square BAM, such an arrangement cuts the information contained in a single stored data vector in half. However, the freedom of choosing the labels $|c_m\rangle$ may perhaps be put to good use. Part of the problem of spurious stable states, which plagues BAMs as well as Hopfield memories as they are loaded up, is due to the lack of orthogonality of the stored vectors. In the coded BAM we have the opportunity to remove part of this problem by choosing the labels as orthonormal. Such labels have been used previously by Kohonen[9] in linear heteroassociative memories. The question whether memory capacity can be improved in this manner was explored by taking 64x64 BAMs in which the labels are chosen as Hadamard vectors. The latter are bipolar vectors with Euclidean norm $\sqrt{P}$, which form an orthonormal set. These vectors are rows of a PxP Hadamard matrix; for a discussion see Harwit and Sloane[10]. The storage capacity of such Hadamard-coded BAMs was calculated as function of the number M of stored vectors for 100 cases for each value of M, in the manner discussed before. The percentage of good recall and its standard deviation are shown in Fig. 6. It is seen that the Hadamard coding gives about a factor 2.5 in M, compared to the ordinary 64x64 BAM. However, the coded BAM has only half the stored data vector dimension. Accounting for this factor 2 reduction of data vector dimension, the effective storage capacity advantage obtained by Hadamard coding comes to only 25 %.

HALF BAM WITH HADAMARD CODING

For the coded BAM there is the option of deleting the threshold operation in the front layer. The resulting architecture may be called "half BAM". In the half BAM, thresholding is only done on the labels, and consequently, the data may be taken as analog vectors. Although such an arrangement diminishes the robustness of the memory somewhat, there are applications of interest. We have calculated the percentage of good recall for 100 cases, and found that giving up the data thresholding cuts the storage capacity of the Hadamard-coded BAM by about 60 %.

SELECTIVE REFLEXIVE MEMORY

The memory capacity limitations shown in Fig. 6 are due to the occurence of spurious states when the memories are loaded up.

Consider a discrete BAM with stored data vectors $|m\rangle$, m=1 to M, orthonormal labels $|c_m\rangle$, and the connection matrix

$$K = \sum_{m=1}^{M} |c_m\rangle\langle m| \ . \tag{5}$$

For an input data vector $|v\rangle$ which is closest to the stored data vector $|1\rangle$, one has in the forward stroke

$$|b\rangle = s(c|c_1\rangle + \sum_{m=2}^{M} a_m|c_m\rangle) \ , \tag{6}$$

where

$$c = \langle 1|v\rangle, \quad \text{and} \quad a_m = \langle m|v\rangle \ . \tag{7}$$

Although for $m \neq 1$ $a_m < c$, for some vector component the sum $\sum_{m=2}^{M} a_m|c_m\rangle$ may accumulate to such a large value as to affect the thresholded result $|b\rangle$. The problem would be avoided if the thresholding operation $s(\ )$ in the back layer of the BAM were to be replaced by another nonlinear operation which selects, from the linear combination

$$c|c_1\rangle + \sum_{m=2}^{M} a_m|c_m\rangle \tag{8}$$

the dominant label $|c_1\rangle$. The hypothetical device which performs this operation is here called the "Dominant Label Selector" (DLS)[11], and we call the resulting memory architecture "Selective Reflexive Memory" (SRM). With the back state selected as the dominant label $|c_1\rangle$, the back stroke gives $\langle f'| = s(\langle c_1|K) = s(P\langle 1|) = \langle 1|$, by the orthogonality of the labels $|c_m\rangle$. It follows[11] that the SRM gives perfect associative recall of the nearest stored data vector, for any number of vectors stored. Of course, the linear independence of the P-dimensional label vectors $|c_m\rangle$, $m=1$ to M, requires $P \geq M$.

The DLS must select, from a linear combination of orthonormal labels, the dominant label. A trivial case is obtained by choosing the labels $|c_m\rangle$ as basis vectors $|u_m\rangle$, which have all components zero except for the mth component, which is unity. With this choice of labels, the DLS may be taken as a winner-take-all net W, as shown in Fig. 7.

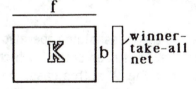

Fig.7. Simplest reflexive memory with DLS

This case appears to be included in Adaptive Resonance Theory (ART)[12] as a special simplified case. A relationship between the ordinary BAM and ART was pointed out by Kosko[5]. As in ART, there is considerable fault sensitivity in this memory, because the stored data vectors appear in the connection matrix as rows.

A memory with better fault tolerance may be obtained by using orthogonal labels other than basis vectors. The DLS can then be taken as an orthogonal transformation G followed by a winner-take-all net, as shown in Fig. 8. G is to be chosen such that it transforms the labels $|c_m\rangle$ into vectors proportional to the basis vectors $|u_m\rangle$. This can always be done by taking

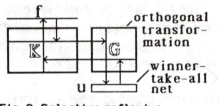

Fig. 8. Selective reflexive memory

$$G = \sum_{p=1}^{P} |u_p\rangle\langle c_p| \, , \qquad (9)$$

where the $|c_p\rangle$, p=1 to P, form a complete orthonormal set which contains the labels $|c_m\rangle$, m=1 to M. The neurons in the DLS serve as grandmother cells. Once a single winning cell has been activated, i.e., the state of the layer is a single basis vector, say $|u_1\rangle$, this vector must be passed back, after application of the transformation G^{-1}, such as to produce the label $|c_1\rangle$ at the back of the BAM. Since G is orthogonal, we have $G^{-1} = G^T$, so that the required inverse transformation may be accomplished simply by sending the basis vector back through the transformer; this gives

$$\langle u_1|G = \sum_{p=1}^{P} \langle u_1|u_p\rangle\langle c_p| = \langle c_1| \qquad , \qquad (10)$$

as required.

HALF SRM

The SRM may be modified by deleting the thresholding operation in the front layer. The front neurons then have a linear output, which is reflected back through the SRM, as shown in Fig. 9. In this case, the

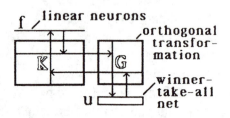

Fig. 9. Half SRM with linear
neurons in front layer

stored data vectors and the input data vectors may be taken as analog vectors, but we require all the stored vectors to have the same norm. The action of the SRM proceeds in the same way as described above, except that we now require the ortho- normal labels to have unit norm. It follows that, just like the full SRM, the half SRM gives perfect associative recall to the nearest stored vector, for any number of stored vectors up to the dimension P of the labels. The latter condition is due to the fact that a P-dimensional vector space can at most contain P orthonormal vectors.

In the SRM the output transform G is introduced in order to improve the fault tolerance of the connection matrix K. This is accomplished at the cost of some fault sensitivity of G, the extent of which needs to be investigated. In this regard it is noted that in certain optical implemen- tations of reflexive memories, such as Dana Anderson's resonator[1] and similar configurations[2,3], the transformation G is a Fourier transform, which is implemented simply as a lens. Such an implementation is quite insentive to the common semiconductor damage mechanisms.

EQUIVALENT AUTOASSOCIATIVE MEMORIES

Concatenation of the front and back state vectors allows descrip- tion of the SRMs in terms of autoassociative memories. For the SRM which uses basis vectors as labels the corresponding autoassociative memory is shown in Fig. 10. This connection matrix structure was also proposed by Guest et. al.[13]. The winner-take-all net W needs to be

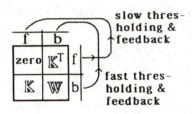

Fig. 10. Equivalent auto-associative memory

given time to settle on a basis vector state before the state |b⟩ can influence the front state |f⟩. This may perhaps be achieved by arranging the W network to have a thresholding and feedback which are fast compared with that of the K network. An alternate method may be to equip the W network with an output gate which is opened only after the W net has settled. These arrangements present a complication and cause a delay, which in some applications may be inappropriate, and in others may be acceptable in a trade between speed and memory density.

For the SRM with output transformer and orthonormal labels other than basis vectors, a corresponding autoassociative memory may be composed as shown in Fig.11. An output gate in the w layer is chosen as the device which prevents the backstroke through the BAM to take place before the winner-take-al net has settled. The same effect may perhaps be achieved by choosing different response times for the neuron layers f and w. These matters require investigation. Unless the output transform G is already required for other reasons, as in some optical resonators, the DLS with output transform is clumsy. It would far better to combine the transformer G and the net W into a single network. To find such a DLS should be considered a challenge.

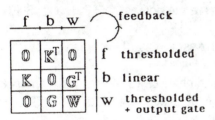

Fig. 11. Autoassociative memory equivalent to SRM with transform

output gate
w winner-take-all output
b back layer, linear
f front layer

$\mathbb{K}$ = BAM connections
$\mathbb{G}$ = orthogonal transformation
$\mathbb{W}$ = winner-take-all net

Fig. 12. Structure of SRM

The work was partly supported by the Defense Advanced Research Projects Agency, ARPA order #5916, through Contract DAAHO1-86-C -0968 with the U.S. Army Missile Command.

REFERENCES

1. D. Z. Anderson, "Coherent optical eigenstate memory", Opt. Lett. 11, 56 (1986).

2. B. H. Soffer, G. J. Dunning, Y. Owechko, and E. Marom, "Associative holographic memory with feedback using phase-conjugate mirrors", Opt. Lett. 11, 118 (1986).

3. A. Yarriv and S. K. Wong, "Associative memories based on message-bearing optical modes in phase-conjugate resonators", Opt. Lett. 11, 186 (1986).

4. B. Kosko, "Adaptive Cognitive Processing", NSF Workshop for Neural Networks and Neuromorphic Systems, Boston, Mass., Oct. &-8, 1986.

5. B. Kosko, "Bidirectional Associative Memories", IEEE Trans. SMC, in press, 1987.

6. B. Kosko, "Adaptive Bidirectional Associative Memories", Appl. Opt., in press, 1987.

7. J. J. Hopfield, "Neural networks and physical systems with emergent collective computational abilities", Proc. Natl. Acad. Sci. USA 79, 2554 (1982).

8. P. A. M. Dirac, THE PRINCIPLES OF QUANTUM MECHANICS, Oxford, 1958.

9. T. Kohonen, "Correlation Matrix Memories", Helsinski University of Technology Report TKK-F-A130, 1970.

10. M. Harwit and N. J. A. Sloane, HADAMARD TRANSFORM OPTICS, Academic Press, New York, 1979.

11. H. G. Loos, "Adaptive Stochastic Content-Addressable Memory", Final Report, ARPA Order 5916, Contract DAAHO1-86-C-0968, March 1987.

12. G. A. Carpenter and S. Grossberg, "A Massively Parallel Architecture for a Self-Organizing Neural Pattern Recognition Machine", Computer Vision, Graphics, and Image Processing, 37, 54 (1987).

13. R. D. TeKolste and C. C. Guest , "Optical Cohen-Grossberg System with All-Optical Feedback", IEEE First Annual International Conference on Neural Networks, San Diego, June 21-24, 1987.

CONNECTING TO THE PAST

Bruce A. MacDonald, Assistant Professor
Knowledge Sciences Laboratory, Computer Science Department
The University of Calgary, 2500 University Drive NW
Calgary, Alberta T2N 1N4

ABSTRACT

Recently there has been renewed interest in neural-like processing systems, evidenced for example in the two volumes *Parallel Distributed Processing* edited by Rumelhart and McClelland, and discussed as parallel distributed systems, connectionist models, neural nets, value passing systems and multiple context systems. Dissatisfaction with symbolic manipulation paradigms for artificial intelligence seems partly responsible for this attention, encouraged by the promise of massively parallel systems implemented in hardware. This paper relates simple neural-like systems based on multiple context to some other well-known formalisms—namely production systems, k-length sequence prediction, finite-state machines and Turing machines—and presents earlier sequence prediction results in a new light.

1 INTRODUCTION

The revival of neural net research has been very strong, exemplified recently by Rumelhart and McClelland[1], new journals and a number of meetings[a]. The nets are also described as parallel distributed systems[1], connectionist models[2], value passing systems[3] and multiple context learning systems[4,5,6,7,8,9]. The symbolic manipulation paradigm for artificial intelligence does not seem to have been as successful as some hoped[1], and there seems at last to be real promise of massively parallel systems implemented in hardware. However, in the flurry of new work it is important to consolidate new ideas and place them solidly alongside established ones. This paper relates simple neural-like systems to some other well-known notions—namely production systems, k-length sequence prediction, finite-state machines and Turing machines—and presents earlier results on the abilities of such networks in a new light.

The general form of a connectionist system[10] is simplified to a three layer net with binary fixed weights in the hidden layer, thereby avoiding many of the difficulties—and challenges—of the recent work on neural nets. The hidden unit weights are regularly patterned using a template. Sophisticated, expensive learning algorithms are avoided, and a simple method is used for determining output unit weights. In this way we gain some of the advantages of multi-layered nets, while retaining some of the simplicity of two layer net training methods. Certainly nothing is lost in computational power—as I will explain—and the limitations of two layer nets are not carried over to the simplified three layer one. Biological systems may similarly avoid the need for learning algorithms such as the "simulated annealing" method commonly used in connectionist models[11]. For one thing, biological systems do not have the same clearly distinguished training phase.

Briefly, the simplified net[b] is a production system implemented as three layers of neuron-like units; an output layer, an input layer, and a hidden layer for the productions themselves. Each hidden production unit potentially connects a predetermined set of inputs to any output. A k-length sequence predictor is formed once k levels of delay unit are introduced into the input layer. k-length predictors are unable to distinguish simple sequences such as $ba\ldots a$ and $aa\ldots a$ since after k or more characters the system has forgotten whether an a or b appeared first. If the k-length predictor is augmented with "auxiliary" actions, it is able to learn this and other regular languages, since the auxiliary actions can be equivalent to states, and can be inputs to

---

[a]Among them the 1st International Conference on Neural Nets, San Diego,CA, June 21-24, 1987, and this conference.

[b]Roughly equivalent to a single context system in Andreae's multiple context system[4,5,6,7,8,9]. See also MacDonald[12].

Figure 1: The general form of a connectionist system[10].

(a) Form of a unit

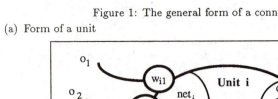

(a) Operations within a unit

the production units enabling predictions to depend on previous states[7]. By combining several augmented sequence predictors a Turing machine tape can be simulated along with a finite-state controller[9], giving the net the computational power of a Universal Turing machine. Relatively simple neural-like systems do not lack computational ability. Previous implementations[7,9] of this ability are production system equivalents to the simplified nets.

1.1 Organization of the paper

The next section briefly reviews the general form of connectionist systems. Section 2 simplifies this, then section 3 explains that the result is equivalent to a production system dealing only with inputs and outputs of the net. Section 4 extends the simplified version, enabling it to learn to predict sequences. Section 5 explains how the computational power of the sequence predictor can be increased to that of a Turing machine if some input units receive auxiliary actions; in fact the system can *learn* to be a Turing machine. Section 6 discusses the possibility of a number of nets combining their outputs, forming an overall net with "association areas".

1.2 General form of a connectionist system

Figure 1 shows the general form of a connectionist system unit, neuron or cell[10]. In the figure unit i has inputs, which are the outputs o_j of possibly all units in the network, and an output of its own, o_i. The *net input excitation, net_i*, is the weighted sum of inputs, where w_{ij} is the weight connecting the output from unit j as an input to unit i. The *activation*, a_i of the unit is some function F_i of the net input excitation. Typically F_i is *semilinear*, that is non-decreasing and differentiable[13], and is the same function for all, or at least large groups of units. The output is a function f_i of the activation; typically some kind of threshold function. I will assume that the quantities vary over discrete time steps, so for example the activation at time $t+1$ is $a_i(t+1)$ and is given by $F_i((net_i(t))$.

In general there is no restriction on the connections that may be made between units. Units not connected directly to inputs or outputs are *hidden* units. In more complex nets than those described in this paper, there may be more than one *type* of connection. Figure 2 shows a common connection topology, where there are three layers of units—input, hidden and output—with no cycles of connection.

The net is trained by presenting it with input combinations, each along with the desired output combination. Once trained the system should produce the desired outputs given just

Figure 2: The basic structure of a three layer connectionist system.

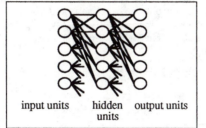

inputs. During training the weights are adjusted in some fashion that reduces the discrepancy between desired and actual output. The general method is[10]:

$$\Delta w_{ij} = g(a_i, t_i)\, h(o_j, w_{ij}), \tag{1}$$

where t_i is the desired, "training" activation. Equation 1 is a general form of Hebb's classic rule for adjusting the weight between two units with high activations[10]. The weight adjustment is the product of two functions, one that depends on the desired and actual activations—often just the difference—and another that depends on the input to that weight and the weight itself. As a simple example suppose g is the difference and h as just the output o_j. Then the weight change is the product of the output error and the input excitation to that weight:

$$\Delta w_{ij} = \eta o_j (t_i - a_i)$$

where the constant η determines the learning rate. This is the Widrow-Hoff or Delta rule which may be used in nets without hidden units.[10]

The important contribution of recent work on connectionist systems is how to implement equation 1 in hidden units; for which there are no training signals t_i directly available. The Boltzmann learning method iteratively varies both weights and hidden unit training activations using the controlled, gradually decreasing randomizing method "simulated annealing"[14]. Back-propagation[13] is also iterative, performing gradient descent by propagating training signal errors back through the net to hidden units. I will avoid the need to determine training signals for hidden units, by fixing the weights of hidden units in section 2 below.

2   SIMPLIFIED SYSTEM

Assume these simplifications are made to the general connectionist system of section 1.2:

1. The system has three layers, with the topology shown in Figure 2 (ie no cycles)

2. *All* hidden layer unit weights are fixed, say at unity or zero

3. Each unit is a linear threshold unit[10], which means the activation function for all units is the identity function, giving just net_i, a weighted sum of the inputs, and the output function is a simple binary threshold of the form:

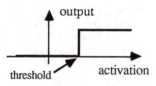

so that the output is binary; on or off. Hidden units will have thresholds requiring all inputs to be active for the output to be active (like an AND gate) while output units will have thresholds requiring only 1 or two active highly weighted inputs for an output to be generated (like an OR gate). This is in keeping with the production system view of the net, explained in section 3.

4. Learning—which now occurs only at the output unit weights—gives weight adjustments according to:

$$w_{ij} = 1 \quad if \ a_i = o_j = 1$$
$$w_{ij} = 0 \quad otherwise$$

so that weights are turned on if their input and the unit output are on, and off otherwise. That is, $w_{ij} = a_i \wedge o_j$. A simple example is given in Figure 3 in section 3 below.

This simple form of net can be made probabilistic by replacing 4 with 4' below:

4'. Adjust weights so that w_{ij} estimates the conditional probability of the unit i output being on when output j is on. That is,

$$w_{ij} = \text{estimate of } P(o_i|o_j).$$

Then, assuming independence of the inputs to a unit, an output unit is turned on when the conditional probability of occurrence of that output exceeds the threshold of the output function.

Once these simplifications are made, there is no need for learning in the hidden units. Also no iterative learning is required; weights are either assigned binary values, or estimate conditional probabilities. This paper presents some of the characteristics of the simplified net. Section 6 discusses the motivation for simplifying neural nets in this way.

3   PRODUCTION SYSTEMS

The simplified net is a kind of simple production system. A production system comprises a global database, a set of production rules and a control system[15]. The database for the net is the system it interacts with, providing inputs as reactions to outputs from the net. The hidden units of the network are the production rules, which have the form

IF *precondition* THEN *action*

The precondition is satisfied when the input excitation exceeds the threshold of a hidden unit. The actions are represented by the output units which the hidden production units activate. The control system of a production system chooses the rule whose action to perform, from the set of rules whose preconditions have been met. In a neural net the control system is distributed throughout the net in the output units. For example, the output units might form a winner-take-all net. In production systems more complex control involves forward and backward chaining to choose actions that seek goals. This is discussed elsewhere[4,12,16]. Figure 3 illustrates a simple production implemented as a neural net. As the figure shows, the inputs to hidden units are just the elements of the precondition. When the appropriate input combination is present the associated hidden (production) unit is fired. Once weights have been learned connecting hidden units to output units, firing a production results in output. The simplified neural net is directly equivalent to a production system whose elements are inputs and outputs[c].

Some production systems have symbolic elements, such as variables, which can be given values by production actions. The neural net cannot directly implement this, since it can have outputs only from a predetermined set. However, we will see later that extensions to the framework enable this and other abilities.

---

[c]This might be referred to as a "sensory-motor" production system, since when implemented in a real system such as a robot, it deals only with sensed inputs and executable motor actions, which may include the auxiliary actions of section 4.3.

Figure 3: A production implemented in a simplified neural net.

(a) A production rule

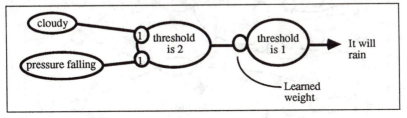

(b) The rule implemented as a hidden unit. The threshold of the hidden unit is 2 so it is an AND gate. The threshold of the output unit is 1 so it is an OR gate. The learned weight will be 0 or 1 if the net is not probabilistic, otherwise it will be an estimate of P(it will rain|clouds AND pressure falling)

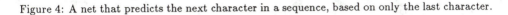

Figure 4: A net that predicts the next character in a sequence, based on only the last character.

(a) The net. Production units (hidden units) have been combined with input units. For example this net could predict the sequence $abcabcabc\ldots$. Productions have the form: IF last character is ...THEN next character will be .... The learning rule is $w_{ij} = 1$ if (input$_j$ AND output$_i$). Output is $a_i = \underset{j}{OR} \; w_{ij}o_j$

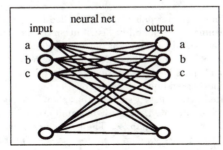

(b) Learning procedure.

1. Clamp inputs and outputs to desired values

2. System calculates weight values

3. Repeat 4 and 4 for all required input/output combinations

4 SEQUENCE PREDICTION

A production system or neural net can predict sequences. Given examples of a repeating sequence, productions are learned which predict future events on the basis of recent ones. Figure 4 shows a trivially simple sequence predictor. It predicts the next character of a sequence based on the previous one. The figure also gives the details of the learning procedure for the simplified net. The net need be trained only once on each input combination, then it will "predict" as an output every character seen after the current one. The probabilistic form of the net would estimate conditional probabilities for the next character, conditional on the current one. Many

Figure 5: Using delayed inputs, a neural net can implement a k-length sequence predictor. (a) A net with the last three characters as input.

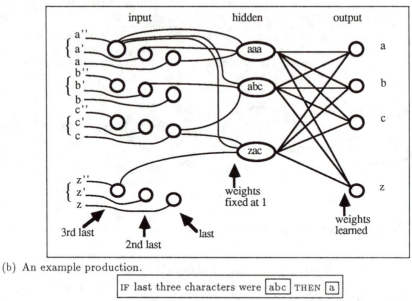

(b) An example production.

IF last three characters were $\boxed{abc}$ THEN $\boxed{a}$

presentations of each possible character pair would be needed to properly estimate the probabilities. The net would be learning the probability distribution of character pairs. A predictor like the one in Figure 4 can be extended to a general k-length[17] predictor so long as inputs delayed by $1, 2, \ldots, k$ steps are available. Then, as illustrated in Figure 5 for 3-length prediction, hidden production units represent all possible combinations of k symbols. Again output weights are trained to respond to previously seen input combinations, here of three characters. These delays can be provided by dedicated neural nets[d], such as that shown in Figure 6. Note that the net is assumed to be synchronously updated, so that the input from feedback around units is not changed until one step after the output changes. There are various ways of implementing delay in neurons, and Andreae[4] investigates some of them for the same purpose—delaying inputs—in a more detailed simulation of a similar net.

4.1 Other work on sequence prediction in neural nets

Feldman and Ballard[2] find connectionist systems initially not suited to representing changes with time. One form of change is sequence, and they suggest two methods for representing sequence in nets. The first is by units connected to each other in sequence so that sequential tasks are represented by firing these units in succession. The second method is to buffer the inputs in time so that inputs from the recent past are available as well as current inputs; that is, delayed inputs are available as suggested above. An important difference is the necessary length of the buffer; Feldman and Ballard suggest the buffer be long enough to hold a phrase of natural language, but I expect to use buffers no longer than about 7, after Andreae[4]. Symbolic inputs can represent more complex information effectively giving the length seven buffers more information than the most recent seven simple inputs, as discussed in section 5.

The method of back-propagation[13] enables recurrent networks to learn sequential tasks in a

---

[d]Feldman and Ballard[2] give some dedicated neural net connections for a variety of functions

Figure 6: Inputs can be delayed by dedicated neural subnets. A two stage delay is shown.
(a) Delay network.

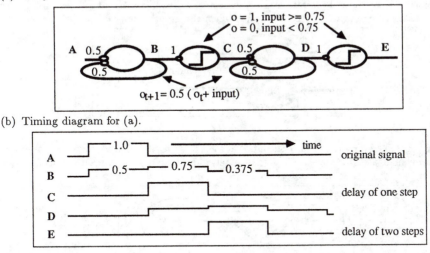

(b) Timing diagram for (a).

manner similar to the first suggestion in the last paragraph, where sequences of connected units represent sequenced events. In one example a net learns to complete a sequence of characters; when given the first two characters of a six character sequence the next four are output. Errors must be propagated around cycles in a recurrent net a number of times.

Seriality may also be achieved by a sequence of states of distributed activation[18]. An example is a net playing both sides of a tic-tac-toe game[18]. The sequential nature of the net's behavior is derived from the sequential nature of the responses to the net's actions; tic-tac-toe moves. A net can model sequence internally by modeling a sequential part of its environment. For example, a tic-tac-toe playing net can have a model of its opponent.

k-length sequence predictors are unable to learn sequences which do not repeat more frequently that every k characters. Their k-length context includes only information about the last k events. However, there are two ways in which information from before the kth last input can be retained in the net. The first method latches some inputs, while the second involves auxiliary actions.

4.2 Latch units

Inputs can be latched and held indefinitely using the combination shown in Figure 7. Not all inputs would normally be latched. Andreae[4] discusses this technique of "threading" latched events among non-latched events, giving the net both information arbitrarily far back in its input-output history and information from the immediate past. Briefly, the sequence $ba \ldots a$ can be distinguished from $aa \ldots a$ if the first character is latched. However, this is an *ad hoc* solution to this problem[e].

4.3 Auxiliary actions

When an output is fed back into the net as an input signal, this enables the system to choose the next output at least partly based on the previous one, as indicated in Figure 8. If a particular fed back output is also one without external manifestation, or whose external manifestation is independent of the task being performed, then that output is an *auxiliary* action. It has

---

[e]The interested reader should refer to Andreae[4] where more extensive analysis is given.

Figure 7: Threading. A latch circuit remembers an event until another comes along. This is a two input latch, e.g. for two letters a and b, but any number of units may be similarly connected. It is formed from a mutual inhibition layer, or winner-take-all connection, along with positive feedback to keep the selected output activated when the input disappears.

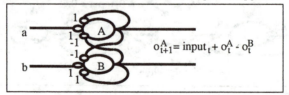

$$o^A_{t+1} = \text{input}_t + o^A_t - o^B_t$$

Figure 8: Auxiliary actions—the S outputs—are fed back to the inputs of a net, enabling the net to remember a state. Here both part of a net and an example of a production are shown. There are two types of action, characters and S actions.

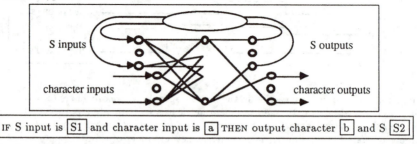

IF S input is $\boxed{S1}$ and character input is $\boxed{a}$ THEN output character $\boxed{b}$ and S $\boxed{S2}$

no direct effect on the task the system is performing since it evokes no relevant inputs, and so can be used by the net as a *symbolic* action. If an auxiliary action is latched at the input then the symbolic information can be remembered indefinitely, being lost only when another auxiliary action of that kind is input and takes over the latch. Thus auxiliary actions can act like remembered states; the system performs an action to "remind" itself to be in a particular state. The figure illustrates this for a system that predicts characters and state changes given the previous character and state. An obvious candidate for auxiliary actions is speech. So the blank oval in the figure would represent the net's environment, through which its own speech actions are heard. Although it is externally manifested, speech has no direct effect on our physical interactions with the world. Its symbolic ability not only provides the power of auxiliary actions, but also includes other speakers in the interaction.

5   SIMULATING ABSTRACT AUTOMATA

The example in Figure 8 gives the essence of simulating a finite state automaton with a production system or its neural net equivalent. It illustrates the transition function of an automaton; the new state and output are a function of the previous state and input. Thus a neural net can simulate a finite state automaton, so long as it has additional, auxiliary actions.

A Turing machine is a finite state automaton controller plus an unbounded memory. A neural net could simulate a Turing machine in two ways, and both ways have been demonstrated with production system implementations—equivalent to neural nets—called "multiple context learning systems"[f], briefly explained in section 6. The first Turing machine simulation[7] has the system simulate only the finite state controller, but is able to use an unbounded external memory

[f] See John Andreae's and his colleagues' work[4,5,6,7,8,9,12,16]

Figure 9: Multiple context learning system implementation as multiple neural nets. Each 3 layer net has the simplified form presented above, with a number of elaborations such as extra connections for goal-seeking by forward and backward chaining.

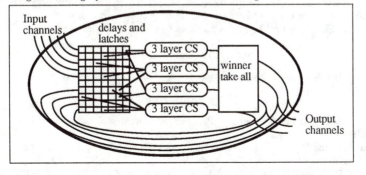

from the real world, much like the paper of Turing's original work[19]. The second simulation[9,12] embeds the memory in the multiple context learning system, along with a counter for accessing this simulated memory. Both learn all the productions—equivalent to learning output unit weights—required for the simulations. The second is able to add internal memory as required, up to a limit dependent on the size of the network (which can easily be large enough to allow 70 years of computation!). The second could also employ external memory as the first did. Briefly, the second simulation comprised multiple sequence predictors which predicted auxiliary actions for remembering the state of the controller, and the current memory position. The memory element is updated by relearning the production representing that element; the precondition is the address and the production action the stored item.

6   MULTIPLE SYSTEMS FORM ASSOCIATION AREAS

A multiple context learning system is production system version of a multiple neural net, although a simple version has been implemented as a simulated net[4,20]. It effectively comprises several nets—or "association" areas—which may have outputs and inputs in common, as indicated in Figure 9. Hidden unit weights are specified by templates; one for each net. A template gives the inputs to have a zero weight for the hidden units of a net and the inputs to have a weight of unity. Delayed and latched inputs are also available. The actual outputs are selected from the combined predictions of the nets in a winner-take-all fashion.

I see the design for real neural nets, say as controllers for real robots, requiring a large degree of predetermined connectivity. A robot controller could not be one three layer net with every input connected to every hidden unit in turn connected to every output. There will need to be some connectivity constraints so the net reflects the functional specialization in the control requirements[g]. The multiple context learning system has all the hidden layer connections predetermined, but allows output connections to be learned. This avoids the "credit assignment" problem and therefore also the need for learning algorithms such as Boltzmann learning and back-propagation. However, as the multiple context learning system has auxiliary actions, and delayed and latched inputs, it does not lack computational power. Future work in this area should investigate, for example, the ability of different kinds of nets to learn auxiliary actions. This may be difficult as symbolic actions may not be provided in training inputs and outputs.

---

[g]For example a controller for a robot body would have to deal with vision, manipulation, motion, *etc.*

7 CONCLUSION

This paper has presented a simplified three layer connectionist model, with fixed weights for hidden units, delays and latches for inputs, sequence prediction ability, auxiliary "state" actions, and the ability to use internal and external memory. The result is able to learn to simulate a Turing machine. Simple neural-like systems do not lack computational power.

ACKNOWLEDGEMENTS

This work is supported by the Natural Sciences and Engineering Council of Canada.

REFERENCES

1. Rumelhart,D.E. and McClelland,J.L. Parallel distributed processing. Volumes 1 and 2. MIT Press. (1986)
2. Feldman,J.A. and Ballard,D.H. Connectionist models and their properties. *Cognitive Science 6,* pp.205-254. (1982)
3. Fahlman,S.E. Three Flavors of Parallelism. Proc.4th Nat.Conf. CSCSI/SCSEIO, Saskatoon. (1982)
4. Andreae,J.H. Thinking with the teachable machine. Academic Press. (1977)
5. Andreae,J.H. Man-Machine Studies Progress Reports UC-DSE/1-28. Dept Electrical and Electronic Engineering, Univ. Canterbury, Christchurch, New Zealand. editor. (1972-87) (Also available from NTIS, 5285 Port Royal Rd, Springfield, VA 22161)
6. Andreae,J.H. and Andreae,P.M. Machine learning with a multiple context. Proc.9th Int.Conf.on Cybernetics and Society. Denver. October. pp.734-9. (1979)
7. Andreae,J.H. and Cleary,J.G. A new mechanism for a brain. *Int.J.Man-Machine Studies 8*(1): pp.89-119. (1976)
8. Andreae,P.M. and Andreae,J.H. A teachable machine in the real world. *Int.J.Man-Machine Studies 10:* pp.301-12. (1978)
9. MacDonald,B.A. and Andreae,J.H. The competence of a multiple context learning system. *Int.J.Gen.Systems 7:* pp.123-37. (1981)
10. Rumelhart,D.E., Hinton,G.E. and McClelland,J.L. A general framework for parallel distributed processing. chapter 2 in Rumelhart and McClelland[1], pp.45-76. (1986)
11. Hinton,G.E. and Sejnowski,T.L. Learning and relearning in Boltzmann machines. chapter 7 in Rumelhart and McClelland[1], pp.282-317. (1986)
12. MacDonald,B.A. Designing teachable robots. PhD thesis, University of Canterbury, Christchurch, New Zealand. (1984)
13. Rumelhart,D.E., Hinton,G.E. and Williams,R.J. Learning Internal Representations by Error Propagation. chapter 8 in Rumelhart and McClelland[1], pp.318-362. (1986)
14. Ackley,D.H., Hinton,G.E. and Sejnowski,T.J. A Learning Algorithm for Boltzmann Machines. *Cognitive Science 9,* pp.147-169. (1985)
15. Nilsson,N.J. Principles of Artificial Intelligence. Tioga. (1980)
16. Andreae,J.H. and MacDonald,B.A. Expert control for a robot body. Research Report 87/286/34 Dept. of Computer Science, University of Calgary, Alberta, Canada, T2N-1N4. (1987)
17. Witten,I.H. Approximate, non-deterministic modelling of behaviour sequences. *Int. J. General Systems, vol. 5* pp.1-12. (1979)
18. Rumelhart,D.E.,Smolensky,P.,McClelland,J.L. and Hinton,G.E. Schemata and Sequential thought Processes in PDP Models. chapter 14, vol 2 in Rumelhart and McClelland[1]. pp.7-57. (1986)
19. Turing,A.M. On computable numbers, with an application to the entscheidungsproblem. *Proc. London Math. Soc. vol 42*(3). pp. 230-65. (1936)
20. Dowd,R.B. A digital simulation of mew-brain. Report no. UC-DSE/10^5. pp.25-46. (1977)

MICROELECTRONIC IMPLEMENTATIONS OF CONNECTIONIST NEURAL NETWORKS

Stuart Mackie, Hans P. Graf, Daniel B. Schwartz, and John S. Denker

AT&T Bell Labs, Holmdel, NJ 07733

Abstract

In this paper we discuss why special purpose chips are needed for useful implementations of connectionist neural networks in such applications as pattern recognition and classification. Three chip designs are described: a hybrid digital/analog programmable connection matrix, an analog connection matrix with adjustable connection strengths, and a digital pipelined best-match chip. The common feature of the designs is the distribution of arithmetic processing power amongst the data storage to minimize data movement.

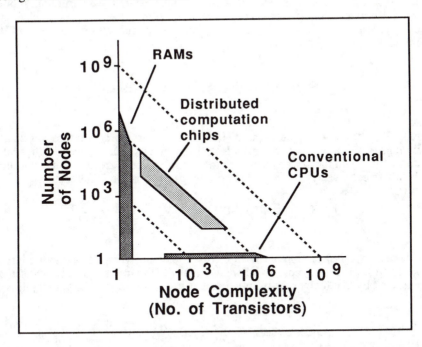

Figure 1. A schematic graph of addressable node complexity and size for conventional computer chips. Memories can contain millions of very simple nodes each with a very few transistors but with no processing power. CPU chips are essentially one very complex node. Neural network chips are in the distributed computation region where chips contain many simple fixed instruction processors local to data storage. (After Reece and Treleaven[1])

Introduction

It is clear that conventional computers lag far behind organic computers when it comes to dealing with very large data rates in problems such as computer vision and speech recognition. Why is this? The reason is that the brain performs a huge number of operations in parallel whereas in a conventional computer there is a very fast processor that can perform a variety of instructions very quickly, but operates on only two pieces of data at a time.

The rest of the many megabytes of RAM is idle during any instruction cycle. The duty cycle of the processor is close to 100%, but that of the stored data is very close to zero. If we wish to make better use of the data, we have to distribute processing power amongst the stored data, in a similar fashion to the brain. Figure 1 illustrates where distributed computation chips lie in comparison to conventional computer chips as regard number and complexity of addressable nodes per chip.

In order for a distributed strategy to work, each processing element must be small in order to accommodate many on a chip, and communication must be local and hard-wired. Whereas the processing element in a conventional computer may be able to execute many hundred different operations, in our scheme the processor is hard-wired to perform just one. This operation should be tailored to some particular application. In neural network and pattern recognition algorithms, the dot products of an input vector with a series of stored vectors (referred to as features or memories) is often required. The general calculation is:

$$\text{Sum of Products} \qquad V \cdot F(i) = \sum_j v_j f_{ij}$$

where V is the input vector and $F(i)$ is one of the stored feature vectors. Two variations of this are of particular interest. In feature extraction, we wish to find all the features for which the dot product with the input vector is greater than some threshold T, in which case we say that such features are *present* in the input vector.

$$\text{Feature Extraction} \qquad V \cdot F(i) = \sum_j v_j f_{ij}$$

In pattern classification we wish to find the stored vector that has the largest dot product with the input vector, and we say that the the input is a *member of the class* represented by that feature, or simply that that stored vector is *closest* to input vector.

$$\text{Classification} \qquad \max(V \cdot F(i) = \sum_j v_j f_{ij}$$

The chips described here are each designed to perform one or more of the above functions with an input vector and a number of feature vectors in parallel. The overall strategy may be summed up as follows: we recognize that in typical pattern recognition applications, the feature vectors need to be changed infrequently compared to the input vectors, and the calculation that is performed is fixed and low-precision, we therefore distribute simple fixed-instruction processors throughout the data storage area, thus minimizing the data movement and optimizing the use of silicon. Our ideal is to have every transistor on the chip doing something useful during every instruction cycle.

Analog Sum-of-Products

Using an idea slightly reminiscent of synapses and neurons from the brain, in two of the chips we store elements of features as connections from input wires on which the elements of the input vectors appear as voltages to summing wires where a sum-of-products is performed. The voltage resulting from the current summing is applied to the input of an amplifier whose output is then read to determine the result of the calculation. A schematic arrangement is shown in Figure 2 with the vertical inputs connected to the horizontal summing wires through resistors chosen such that the conductance is proportional to the magnitude of the feature element. When both positive and negative values are required, inverted input lines are also necessary. Resistor matrices have been fabricated using amorphous silicon connections and metal linewidths. These were programmed during fabrication by electron beam lithography to store names using the distributed feedback method described by Hopfield[2,3]. This work is described more fully elsewhere.[4,5] Hard-wired resistor matrices are very compact, but also very inflexible. In many applications it is desirable to be able to reprogram the matrix without having to fabricate a new chip. For this reason, a series of programmable chips has been designed.

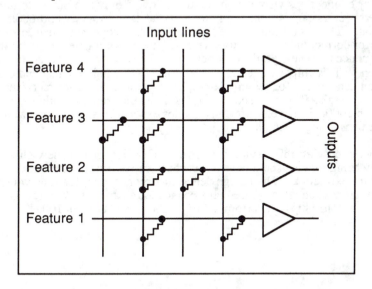

Figure 2. A schematic arrangement for calculating parallel sum-of-products with a resistor matrix. Features are stored as connections along summing wires and the input elements are applied as voltages on the input wires. The voltage generated by the current summing is thresholded by the amplifier whose output is read out at the end of the calculation. Feedback connections may be

made to give mutual inhibition and allow only one feature amplifier to turn on, or allow the matrix to be used as a distributed feedback memory.

Programmable Connection Matrix

Figure 3 is a schematic diagram of a programmable connection using the contents of two RAM cells to control current sinking or sourcing into the summing wire. The switches are pass transistors and the 'resistors' are transistors with gates connected to their drains. Current is sourced or sunk if the appropriate RAM cell contains a '1' and the input V_i is high thus closing both switches in the path. Feature elements can therefore take on values (a,0,-b) where the values of a and b are determined by the conductivities of the n- and p-transistors obtained during processing. A matrix with 2916 such connections allowing full interconnection of the inputs and outputs of 54 amplifiers was designed and fabricated in 2.5μm CMOS (Figure 4). Each connection is about 100×100μm, the chip is 7×7mm and contains about 75,000 transistors. When loaded with 49 49-bit features (7×7 kernel), and presented with a 49-bit input vector, the chip performs 49 dot products in parallel in under 1μs. This is equivalent to 2.4 billion bit operations/sec. The flexibility of the design allows the chip to be operated in several modes. The chip was programmed as a distributed feedback memory (associative memory), but this did not work well because the current sinking capability of the n-type transistors was 6 times that of the p-types. An associative memory was implemented by using a 'grandmother cell' representation, where the memories were stored along the input lines of amplifiers, as for feature extraction, but mutually inhibitory connections were also made that allowed only one output to turn on. With 10 stored vectors each 40 bits long, the best match was found in 50-600ns, depending on the data. The circuit can also be programmed to recognize sequences of vectors and to do error correction when vectors were omitted or wrong vectors were inserted into the sequences. The details of operation of the chip are described more fully elsewhere[6]. This chip has been interfaced to a UNIX minicomputer and is in everyday use as an accelerator for feature extraction in optical character recognition of hand-written numerals. The chip speeds up this time consuming calculation by a factor of more than 1000. The use of the chip enables experiments to be done which would be too time consuming to simulate.

Experience with this device has led to the design of four new chips, which are currently being tested. These have no feedback capability and are intended exclusively for feature extraction. The designs each incorporate new features which are being tested separately, but all are based on a connection matrix which stores 46 vectors each 96 bits long. The chip will perform a full parallel calculation in 100ns.

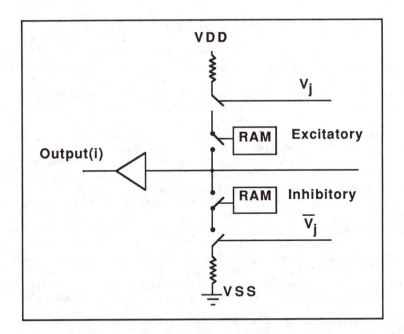

Figure 3. Schematic diagram of a programmable connection. A current sourcing or sinking connection is made if a RAM cell contains a '1' and the input V_i is high. The currents are summed on the input wire of the amplifier.

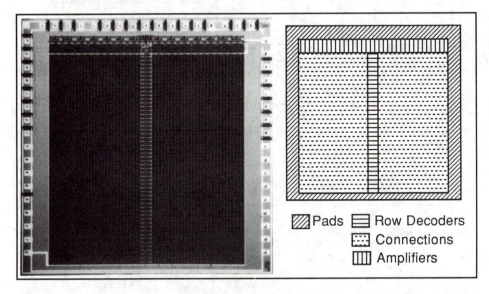

Figure 4. Programmable connection matrix chip. The chip contains 75,000 transistors in 7x7mm, and was fabricated using 2.5µm design rules.

Adaptive Connection Matrix

Many problems require analog depth in the connection strengths, and this is especially important if the chip is to be used for learning, where small adjustments are required during training. Typical approaches which use transistors sized in powers of two to give conductance variability take up an area equivalent to the same number of minimum sized transistors as the dynamic range, which is expensive in area and enables only a few connections to be put on a chip. We have designed a fully analog connection based on a DRAM structure that can be fabricated using conventional CMOS technology. A schematic of a connection and a connection matrix is shown in Figure 5. The connection strength is represented by the difference in voltages stored on two MOS capacitors. The capacitors are 33μm on edge and lose about 1% of their charge in five minutes at room temperature. The leakage rate can be reduced by three orders of magnitude by cooling the the capacitors to –50°C and by five orders of magnitude by cooling to –100°C. The output is a current proportional to the product of the input voltage and the connection strength. The output currents are summed on a wire and are sent off chip to external amplifiers. The connection strengths can be adjusted using transferring charge between the capacitors through a chain of transistors. The connections strengths may be of either polarity and it is expected that the connections will have about 7 bits of analog depth. A chip has been designed in 1.25μm CMOS containing 1104 connections in an array with 46 inputs and 24 outputs.

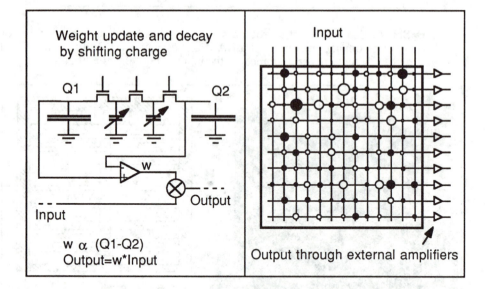

Figure 5. Analog connection. The connection strength is represented by the difference in voltages stored on two capacitors. The output is a current proprtional to the product of the input voltage and the connection strength.

Each connection is 70×240μm. The design has been sent to foundry, and testing is expected to start in April 1988. The chip has been designed to perform a network calculation in <30ns, i.e., the chip will perform at a rate of 33 billion multiplies/sec. It can be used simply as a fast analog convolver for feature extraction, or as a learning engine in a gradient descent algorithm using external logic for connection strength adjustment . Because the inputs and outputs are true analog, larger networks may be formed by tiling chips, and layered networks may be made by cascading through amplifiers acting as hidden units.

Digital Classifier Chip

The third design is a digital implementation of a classifier whose architecture is not a connectionist matrix. It is nearing completion of the design stage, and will be fabricated using 1.25μm CMOS. It calculates the largest five $V\cdot F(i)$ using an all-digital pipeline of identical processors, each attached to one stored word. Each processor is also internally pipelined to the extent that no stage contains more than two gate delays. This is important, since the throughput of the processor is limited by the speed of the slowest stage. Each processor calculates the Hamming distance (number of difference bits) between an input word and its stored word, and then compares that distance with each of the smallest 5 values previously found for that input word. An updated list of 5 best matches is then passed to the next processor in the pipeline. At the end of the pipeline the best 5 matches overall are output.

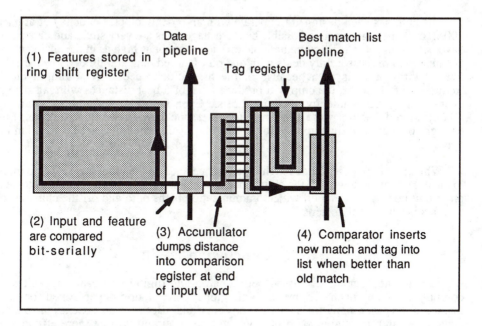

Fig. 6   Schematic of one of the 50 processors in the digital classifier chip. The Hamming distance of the input vector to the feature vector is calculated, and if better than one of the five best matches found so far, is inserted into the match list together with the tag and passed onto the next processor. At the end of the pipeline the best five matches overall are output.

The data paths on chip are one bit wide and all calculations are bit serial. This means that the processing elements and the data paths are compact and maximizes the number of stored words per chip. The layout of a single processor is shown in Fig. 6. The features are stored as 128-bit words in 8 16-bit ring shift registers and associated with each feature is a 14-bit tag or name string that is stored in a static register. The input vector passes through the chip and is compared bit-by-bit to each stored vector, whose shift registers are cycled in turn. The total number of bits difference is summed in an accumulator. After a vector has passed through a processor, the total Hamming distance is loaded into the comparison register together with the tag. At this time, the match list for the input vector arrives at the comparator. It is an ordered list of the 5 lowest Hamming distances found in the pipeline so far, together with associated tag strings. The distance just calculated is compared bit-serially with each of the values in the list in turn. If the current distance is smaller than one of the ones in the list, the output streams of the comparator are switched, having the effect of inserting the current match and tag into the list and deleting the previous fifth best match. After the last processor in the pipeline, the list stream contains the best five distances overall, together with the tags of the stored vectors that generated them. The data stream and the list stream are loaded into 16-bit wide registers ready for output. The design enables chips to be connected together to extend the pipeline if more than 50 stored vectors are required. The throughput is constant, irrespective of the number of chips connected together; only the latency increases as the number of chips increases.

The chip has been designed to operate with an on-chip clock frequency of at least 100MHz. This high speed is possible because stage sizes are very small and data paths have been kept short. The computational efficiency is not as high as in the analog chips because each processor only deals with one bit of stored data at a time. However, the overall throughput is high because of the high clock speed. Assuming a clock frequency of 100MHz, the chip will produce a list of 5 best distances with tag strings every 1.3μs, with a latency of about 2.5μs. Even if a thousand chips containing 50,000 stored vectors were pipelined together, the latency would be 2.5ms, low enough for most real time applications. The chip is expected to perform 5 billion bit operation/sec.

While it is important to have high clock frequencies on the chip, it is also important to have them much lower off the chip, since frequencies above 50MHz are hard to deal on circuit boards. The 16-bit wide communication paths onto and off the chip ensure that this is not a problem here.

Conclusion

The two approaches discussed here, analog and digital, represent opposites in computational approach. In one, a single global computation is performed for each match, in the other many local calculations are done. Both the approaches have their advantages and it remains to be seen which type of circuit will be more efficient in applications, and how closely an electronic implementation of a neural network should resemble the highly interconnected nature of a biological network.

These designs represent some of the first distributed computation chips. They are characterized by having simple processors distributed amongst data storage. The operation performed by the processor is tailored to the application. It is interesting to note some of the reasons why these designs can now be made: minimum linewidths on circuits are now small enough that enough processors can be put on one chip to make these designs of a useful size, sophisticated design tools are now available that enable a single person to design and simulate a complete circuit in a matter of months, and fabrication costs are low enough that highly speculative circuits can be made without requiring future volume production to offset prototype costs.

We expect a flurry of similar designs in the coming years, with circuits becoming more and more optimized for particular applications. However, it should be noted that the impressive speed gain achieved by putting an algorithm into custom silicon can only be done once. Further gains in speed will be closely tied to mainstream technological advances in such areas as transistor size reduction and wafer-scale integration. It remains to be seen what influence these kinds of custom circuits will have in useful technology since at present their functions cannot even be simulated in reasonable time. What can be achieved with these circuits is very limited when compared with a three dimensional, highly complex biological system, but is a vast improvement over conventional computer architectures.

The authors gratefully acknowledge the contributions made by L.D. Jackel, and R.E. Howard

References

1 M. Reece and P.C. Treleaven, "Parallel Architectures for Neural Computers", Neural Computers, R. Eckmiller and C. v.d. Malsburg, eds (Springer-Verlag, Heidelberg, 1988)

2 J.J. Hopfield, Proc. Nat. Acad. Sci. **79**, 2554 (1982).

3 J.S. Denker, Physica **22D**, 216 (1986).

4 R.E. Howard, D.B. Schwartz, J.S. Denker, R.W. Epworth, H.P. Graf, W.E. Hubbard, L.D. Jackel, B.L. Straughn, and D.M. Tennant, IEEE Trans. Electron Devices **ED-34**, 1553, (1987)

5 H.P. Graf and P. deVegvar, "A CMOS Implementation of a Neural Network Model", in "Advanced Research in VLSI", Proceedings of the 1987 Stanford Conference, P. Losleben (ed.), (MIT Press 1987).

6 H.P. Graf and P. deVegvar, "A CMOS Associative Memory Chip Based on Neural Networks", Tech. Digest, 1987 IEEE International Solid-State Circuits Conference.

BASINS OF ATTRACTION FOR
ELECTRONIC NEURAL NETWORKS

C. M. Marcus
R. M. Westervelt
Division of Applied Sciences and Department of Physics
Harvard University, Cambridge, MA 02138

ABSTRACT

We have studied the basins of attraction for fixed point and oscillatory attractors in an electronic analog neural network. Basin measurement circuitry periodically opens the network feedback loop, loads raster-scanned initial conditions and examines the resulting attractor. Plotting the basins for fixed points (memories), we show that overloading an associative memory network leads to irregular basin shapes. The network also includes analog time delay circuitry, and we have shown that delay in symmetric networks can introduce basins for oscillatory attractors. Conditions leading to oscillation are related to the presence of frustration; reducing frustration by diluting the connections can stabilize a delay network.

(1) - INTRODUCTION

The dynamical system formed from an interconnected network of nonlinear neuron-like elements can perform useful parallel computation[1-5]. Recent progress in controlling the dynamics has focussed on algorithms for encoding the location of fixed points[1,4] and on the stability of the flow to fixed points[3,5-8]. An equally important aspect of the dynamics is the structure of the basins of attraction, which describe the location of all points in initial condition space which flow to a particular attractor[10,22].

In a useful associative memory, an initial state should lead reliably to the "closest" memory. This requirement suggests that a well-behaved basin of attraction should evenly surround its attractor and have a smooth and regular shape. One dimensional basin maps plotting "pull in" probability against Hamming distance from an attractor do not reveal the shape of the basin in the high dimensional space of initial states[9,19]. Recently, a numerical study of a Hopfield network with discrete time and two-state neurons showed rough and irregular basin shapes in a two dimensional Hamming space, suggesting that the high dimensional basin has a complicated structure[10]. It is not known how the basin shapes change with the size of the network and the connection rule.

We have investigated the basins of attraction in a network with continuous state dynamics by building an electronic neural network with eight variable gain sigmoid neurons and a three level $(+,0,-)$ interconnection matrix. We have also built circuitry that can map out the basins of attraction in two dimensional slices of initial state space (Fig.1). The network and the basin measurements are described in section 2.

In section 3, we show that the network operates well as an associative memory and can retrieve up to four memories (eight fixed points) without developing spurious attractors, but that for storage of three or more memories, the basin shapes become irregular.

In section 4, we consider the effects of time delay. Real network components cannot switch infinitely fast or propagate signals instantaneously, so that delay is an intrinsic part of any hardware implementation of a neural network. We have included a controllable CCD (charge coupled device) analog time delay in each neuron to investigate how time delay affects the dynamics of a neural network. We find that networks with symmetric interconnection matrices, which are guaranteed to converge to fixed points for no delay, show collective sustained oscillations when time delay is present. By discovering which configurations are maximally unstable to oscillation, and looking at how these configurations appear in networks, we are able to show that by diluting the interconnection matrix, one can reduce or eliminate the oscillations in neural networks with time delay.

(2) – NETWORK AND BASIN MEASUREMENT

A block diagram of the network and basin measurement circuit is shown in fig.1.

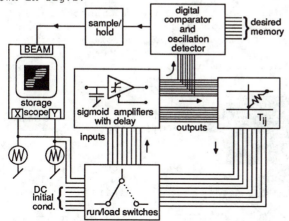

Fig.1 Block diagram of the network and basin measurement system.

The main feedback loop consists of non-linear amplifiers ("neurons", see fig.2) with capacitive inputs and a resistor matrix allowing interconnection strengths of $-1/R$, 0, $+1/R$ ($R = 100$ kΩ). In all basin measurements, the input capacitance was 10 nF, giving a time constant of 1 ms. A charge coupled device (CCD) analog time delay[11] was built into each neuron, providing an adjustable delay per neuron over a range 0.4 – 8 ms.

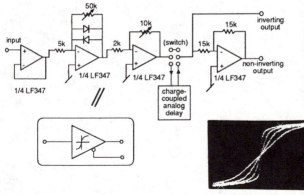

Fig.2 Electronic neuron. Non-linear gain provided by feedback diodes. Inset: Nonlinear behavior at several different values of gain.

Analog switches allow the feedback path to be periodically disconnected and each neuron input charged to an initial voltage. The network is then reconnected and settles to the attractor associated with that set of initial conditions. Two of the initial voltages are raster scanned (on a time scale that is long compared to the load/run switching time) with function generators that are also connected to the X and Y axes of a storage scope. The beam of the scope is activated when the network settles into a desired attractor, producing an image of the basin for that attractor in a two-dimensional slice of initial condition space. The "attractor of interest" can be one of the 2^8 fixed points or an oscillatory attractor.

A simple example of this technique is the case of three neurons with symmetric non-inverting connection shown in fig.3.

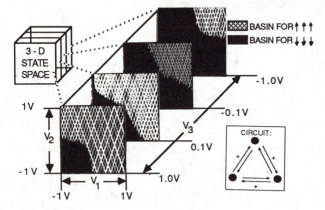

Fig.3 Basin of attraction for three neurons with symmetric non-inverting coupling. Slices are in the plane of initial voltages on neurons 1 and 2. The two fixed points are all neurons saturated positive or all negative. The data are photographs of the scope screen.

(3) BASINS FOR FIXED POINTS - ASSOCIATIVE MEMORY

Two dimensional slices of the eight dimensional initial condition space (for the full network) reveal important qualitative features about the high dimensional basins. Fig. 4 shows a typical slice for a network programmed with three memories according to a clipped Hebb rule[1,12]:

$$T_{ij} = 1/R \; Sgn(\Sigma_{\alpha=1,m} \; \xi_i^{\alpha} \; \xi_j^{\alpha}); \quad T_{ii} = 0 \qquad (1)$$

where ξ is an N-component memory vector of 1's and -1's, and m is the number of memories. The memories were chosen to be orthogonal $(\xi^{\alpha} \cdot \xi^{\beta} = N \; \delta_{\alpha\beta})$.

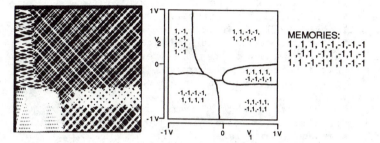

Fig. 4 A slice of initial condition space shows the basins of attraction for five of the six fixed points for three memories in eight-neuron Hopfield net. Learning rule was clipped Hebb (Eq.1). Neuron gain = 15.

Because the Hebb rule (eq.1) makes ξ^{α} and $-\xi^{\alpha}$ stable attractors, a three-memory network will have six fixed point attractors. In fig.4, the basins for five of these attractors are visible, each produced with a different rastering pattern to make it distinctive. Several characteristic features should be noted:

    -- All initial conditions lead to one of the memories (or inverses), no spurious attractors were seen for three or four memories. This is interesting in light of the well documented emergence of spurious attractors at m/N ~15% in larger networks with discrete time[2,18].

    -- The basins have smooth and continuous edges.

    -- The shapes of the basins as seen in this slice are irregular. Ideally, a slice with attractors at each of the corners should have rectangular basins, one basin in each quadrant of the slice and the location of the lines dividing quadrants determined by the initial conditions on the other neurons (the "unseen" dimensions). With three or more memories the actual basins do not resemble this ideal form.

(4) TIME DELAY, FRUSTRATION AND SUSTAINED OSCILLATION

    Arguments defining conditions which guarantee convergence to fixed points[3,5,6] (based, for example, on the construction of a Liapunov function) generally assume instantaneous communication between elements of the network. In any hardware implementation, these assumptions break down due to the finite switching speed of amplifiers and the charging time of long interconnect lines.[13] It is the ratio of delay/RC which is important for stability, so keeping this ratio small limits how fast a neural network chip can be designed to run. Time delay is also relevant to biological neural nets where propagation and response times are comparable.[14,15]

Our particular interest in this section is how time delay can lead to sustained oscillation in networks which are known to be stable when there is no delay. We therefore restrict our attention to networks with symmetric interconnection matrices ($T_{ij} = T_{ji}$).

An obvious ingredient in producing oscillations in a delay network is feedback, or stated another way, a graph representing the connections in a network must contain loops.

The simplest oscillatory structure made of delay elements is the ring oscillator (fig.5a). Though not a symmetric configuration, the ring oscillator illustrates an important point: the ring will oscillate only when there is negative feedback at dc – that is, when the product of interconnection around the loop is negative. Positive feedback at dc (loop product of connections > 0) will lead to saturation.

Observing various symmetric configurations (e.g. fig.5b) in the delayed-neuron network, we find that a negative product of connections around a loop is also a necessary condition for sustained oscillation in symmetric circuits. An important difference between the ring (fig.5a) and the symmetric loop (fig.5b) is that the period of oscillation for the ring is the total accumulated delay around the ring – the larger the ring the longer the period. In contrast, for those symmetric configurations which have oscillatory attractors, the period of oscillation is roughly twice the delay, regardless of the size of the configuration or the value of delay. This indicates that for symmetric configurations the important feedback path is local, not around the loop.

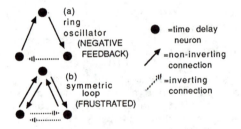

(a) ring oscillator (NEGATIVE FEEDBACK)

(b) symmetric loop (FRUSTRATED)

● =time delay neuron

↗ =non-inverting connection

↗ =inverting connection

Fig.5 (a) A ring oscillator: needs negative feedback at dc to oscillate. (b) Symmetrically connected triangle. This configuration is "frustrated" (defined in text), and has both oscillatory and fixed point attractors when neurons have delay.

Configurations with loop connection product < 0 are important in the theory of spin glasses[16], where such configurations are called "frustrated." Frustration in magnetic (spin) systems, gives a measure of "serious" bond disorder (disorder that cannot be removed by a change of variables) which can lead to a spin glass state.[16,17] Recent results based on the similarity between spin glasses and symmetric neural networks has shown that storage capacity limitations can be understood in terms of this bond disorder.[18,19] Restating our observation above: We only find stable oscillatory modes in symmetric networks with delay when there is frustration. A similar result for a sign-symmetric network (T_{ij}, T_{ji} both ≥ 0 or ≤ 0) with no delay is described by Hirsch.[6]

We can set up the basin measurement system (fig.1) to plot the basin of attraction for the oscillatory mode. Fig.6 shows a slice of the oscillatory basin for a frustrated triangle of delay neurons.

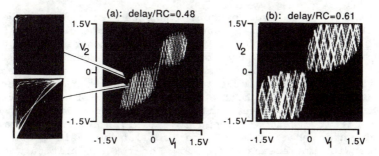

Fig.6 Basin for oscillatory attractor (cross-hatched region) in frustrated triangle of delay-neurons. Connections were all symmetric and inverting; other frustrated configurations (e.g. two non-inverting, one inverting, all symmetric) were similar.(6a): delay = 0.48RC, inset shows trajectory to fixed point and oscillatory mode for two close-lying initial conditions. (6b): delay = 0.61RC, basin size increases.

A fully connected feedback associative network with more that one memory will contain frustration. As more memories are added, the amount of frustration will increases until memory retrieval disappears. But before this point of memory saturation is reached, delay could cause an oscillatory basin to open. In order to design out this possibility, one must understand how frustration, delay and global stability are related. A first step in determining the stability of a delay network is to consider which small configurations are most prone to oscillation, and then see how these "dangerous" configurations show up in the network. As described above, we only need to consider frustrated configurations.

A frustrated configuration of neurons can be sparsely connected, as in a loop, or densely connected, with all neurons connected to all others, forming what is called in graph theory a "clique." Representing a network with inverting and non-inverting connections as a signed graph (edges carry + and -), we define a *frustrated clique* as a fully connected set of vertices (r vertices; $r(r-1)/2$ edges) with all sets of three vertices in the clique forming frustrated triangles. Some examples of frustrated loops and cliques are shown in fig.7. Notice that neurons connected with all inverting symmetric connections, a configuration that is useful as a "winner-take-all" circuit, is a frustrated clique.

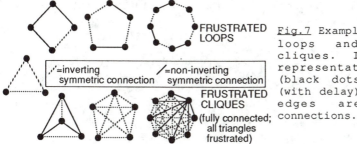

Fig.7 Examples of frustrated loops and frustrated cliques. In the graph representation vertices (black dots) are neurons (with delay) and undirected edges are symmetric connections.

530

We find that delayed neurons connected in a frustrated loop
longer than three neurons do not show sustained oscillation for any
value of delay (tested up to delay = 8RC). In contrast, when delayed
neurons are connected in any frustrated clique configuration, we do
find basins of attraction for sustained oscillation as well as fixed
point attractors, and that the larger the frustrated clique, the more
easily it oscillates in the following ways: (1) For a given value of
delay/RC, the size of the oscillatory basin increases with r, the
size of the frustrated clique (fig.8). (2) The critical value of
delay at which the volume of the oscillatory basin goes to zero
decreases with increasing r (fig.9); For r=8 the critical delay is
already less than 1/30 RC.

  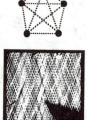

Fig.8   Size of basin
for oscillatory mode
increases with size of
frustrated clique.   The
delay is 0.46RC per
neuron in each picture.
Slices are in the space
of initial voltages on
neurons 1 and 2, other
initial voltages near
zero.

Fig.9 The critical value of delay
where the oscillatory mode vanishes.
Measured by reducing delay until
system leaves oscillatory attractor.
Delay plotted in units of the
characteristic time $R_{in}C$, where R_{in}
$=(\Sigma_j 1/R_{ij})^{-1}=10^5\Omega/(r-1)$ and C=10nF,
indicating that the critical delay
decreases faster than 1/(r-1).

Having identified frustrated cliques as the maximally unstable
configuration of time delay neurons, we now ask how many cliques of a
given size do we expect to find in a large network.

A set of r vertices (neurons) can be fully connected by r(r-1)/2
edges of two types (+ or -) to form $2^{r(r-1)/2}$ different cliques. Of
these, $2^{(r-1)}$ will be frustrated cliques. Fig.10 shows all $2^{(4-1)}=8$
cases for r=4.

Fig.10 All graphs of size r=4 that are frustrated cliques
(fully connected, every triangle frustrated.) Solid lines =
positive edges, dashed lines = negative edges.

For a randomly connected network, this result combined with results from random graph theory[20] gives an expected number of frustrated cliques of size r in a network of size N, $E_N(r)$:

$$E_N(r) = \binom{N}{r} c(r,p) \qquad (2)$$

$$c(r,p) = 2^{-(r-1)(r-2)/2} \, p^{r(r-1)/2} \qquad (3)$$

where $\binom{N}{r}$ is the binomial coefficient and $c(r,p)$ is defined as the concentration of frustrated cliques. p is the connectance of the network, defined as the probability that any two neurons are connected. Eq.3 is the special case where + and – edges (non-inverting, inverting connections) are equally probable. We have also generalized this result to the case $p(+) \neq p(-)$.

Fig.11 shows the dramatic reduction in the concentration of all frustrated configurations in a diluted random network. For the general case $(p(+) \neq p(-))$ we find that the negative connections affect the concentrations of frustrated cliques more strongly than the positive connections, as expected (Frustration requires negatives, not positives, see fig.10).

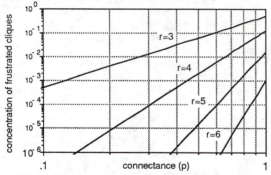

Fig.11 Concentration of frustrated cliques of size r=3,4,5,6 in an unbiased random network, from eq.3. Concentrations decrease rapidly as the network is diluted, especially for large cliques (note: log scale).

When the interconnections in a network are specified by a learning rule rather than at random, the expected numbers of any configuration will differ from the above results. We have compared the number of frustrated triangles in large three-valued (+1,0,-1) Hebb interconnection matrices (N=100,300,600) to the expected number in a random matrix of the same size and connectance. The Hebb matrix was constructed according to the rule:

$$T_{ij} = Z_k \left(\Sigma_{\alpha=1,m} \, \xi_i^\alpha \, \xi_j^\alpha \right) \; ; \; T_{ii} = 0 \qquad (4a)$$
$$Z_k(x) = +1 \text{ for } x > k; \; 0 \text{ for } -k \leq x \leq k; \; -1 \text{ for } x < -k; \qquad (4b)$$

m is the number of memories, Z_k is a threshold function with cutoff k, and ξ^α is a random string of 1's and -1's. The matrix constructed by eq.4 is roughly unbiased (equal number of positive and negative connections) and has a connectance p(k). Fig.12 shows the ratio of frustrated triangles in a diluted Hebb matrix to the expected number in a random graph with the same connectance for different numbers of memories stored in the Hebb matrix. At all values of connectance, the Hebb matrix has fewer frustrated triangles than the random matrix by a ratio that is decreased by diluting the matrix or storing fewer memories. The curves do not seem to depend on the size of the matrix, N. This result suggests that diluting a Hebb matrix breaks up frustration even more efficiently than diluting a random matrix.

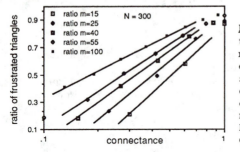

Fig.12 The number of frustrated triangles in a $(+,0,-)$ Hebb rule matrix (300x300) divided by the expected number in a random signed graph with equal connectance. The different sets of points are for different numbers of random memories in the Hebb matrix. The lines are guides to the eye.

The sensitive dependence of frustration on connectance suggests that oscillatory modes in a large neural network with delay can be eliminated by diluting the interconnection matrix. As an example, consider a unbiased random network with delay = RC/10. From fig.9, only frustrated cliques of size r=5 or larger have oscillatory basins for this value of delay; frustration in smaller configurations in the network cannot lead to sustained oscillation in the network. Diluting the connectance to 60% will reduce the concentration of frustrated cliques with r=5 by a factor of over 100 and r=6 by a factor of 2000. The reduction would be even greater for a clipped Hebb matrix.

Results from spin glass theory[21] suggest that diluting a clipped Hebb matrix can actually improve the storage capacity for moderated dilution, with a maximum in the capacity at a connectance of 61%. To the extent this treatment applies to an analog continuous-time network, we should expect that by diluting connections, oscillatory modes can be killed before memory capacity is compromised.

We have confirmed the stabilizing effect of dilution in our network: For a fully connected eight neuron network programmed with three orthogonal memories according to eq.1, adding a delay of 0.4RC opens large basins for sustained oscillation. By randomly diluting the interconnections to $p \sim 0.85$, we were able to close the oscillatory basins and recover a useful associative memory.

SUMMARY

We have investigated the structure of fixed point and oscillatory basins of attraction in an electronic network of eight non-linear amplifiers with controllable time delay and a three value $(+,0,-)$ interconnection matrix.

For fixed point attractors, we find that the network performs well as an associative memory - no spurious attractors were seen for up to four stored memories - but for three or more memories, the shapes of the basins of attraction became irregular.

A network which is stable with no delay can have basins for oscillatory attractors when time delay is present. For symmetric networks with time delay, we only observe sustained oscillation when there is frustration. Frustrated cliques (fully connected configurations with all triangles frustrated), and not loops, are most prone to oscillation, and the larger the frustrated clique, the more easily it oscillates. The number of these "dangerous" configurations in a large network can be greatly reduced by diluting the connections. We have demonstrated that a network with a large basin for an oscillatory attractor can be stabilized by dilution.

ACKNOWLEDGEMENTS

We thank K.L.Babcock, S.W.Teitsworth, S.Strogatz and P.Horowitz for useful discussions. One of us (C.M.M) acknowledges support as an AT&T Bell Laboratories Scholar. This work was supported by JSEP contract no. N00014-84-K-0465.

REFERENCES

1) J.S.Denker, Physica 22D, 216 (1986).
2) J.J. Hopfield, Proc.Nat.Acad.Sci. 79, 2554 (1982).
3) J.J. Hopfield, Proc.Nat.Acad.Sci. 81, 3008 (1984).
4) J.S. Denker, Ed. Neural Networks for Computing, AIP Conf. Proc. 151 (1986).
5) M.A. Cohen, S. Grossberg, IEEE Trans. SMC-13, 815 (1983).
6) M.W.Hirsch, Convergence in Neural Nets, IEEE Conf.on Neural Networks, 1987.
7) K.L. Babcock, R.M. Westervelt, Physica 23D,464 (1986).
8) K.L. Babcock, R.M. Westervelt, Physica 28D,305 (1987).
9) See, for example: D.B.Schwartz, et al, Appl.Phys.Lett.,50 (16), 1110 (1987); or M.A.Silviotti,et al, in Ref.4, pg.408.
10) J.D. Keeler in Ref.4, pg.259.
11) CCD analog delay: EG&G Reticon RD5106A.
12) D.O.Hebb, The Organization of Behavior, (J.Wiley, N.Y., 1949).
13) Delay in VLSI discussed in: A. Muhkerjee, Introduction to nMOS and CMOS VLSI System Design, (Prentice Hall, N.J.,1985).
14) U. an der Heiden, J.Math.Biology, 8, 345 (1979).
15) M.C. Mackey, U. an der Heiden, J.Math.Biology,19, 221 (1984).
16) Theory of spin glasses reviewed in: K. Binder, A.P. Young, Rev. Mod. Phys.,58 (4),801, (1986).
17) E. Fradkin,B.A. Huberman,S.H. Shenker, Phys.Rev.B18 (9),4789 (1978).
18) D.J. Amit, H. Gutfreund, H. Sompolinski, Ann.Phys. 173, 30, (1987) and references therein.
19) J.L. van Hemmen, I. Morgenstern, Editors, Heidelberg Colloquium on Glassy Dynamics, Lecture Notes in Physics 275, (Springer-Verlag, Heidelberg, 1987).
20) P.Erdos, A.Renyi, Pub.Math.Inst.Hung.Acad.Sci., 5,17, (1960).
21) I.Morgenstern in Ref.19, pg.399;H.Sompolinski in Ref.19, pg.485.
22) J. Guckenheimer, P.Holmes, Nonlinear Oscillations,Dynamical Systems and Bifurcations of Vector Fields (Springer,N.Y.1983).

The Performance of Convex Set Projection Based Neural Networks

Robert J. Marks II, Les E. Atlas, Seho Oh and James A. Ritcey

Interactive Systems Design Lab, FT-10
University of Washington, Seattle, Wa 98195.

ABSTRACT

We consider a class of neural networks whose performance can be analyzed and geometrically visualized in a signal space environment. Alternating projection neural networks (APNN's) perform by alternately projecting between two or more constraint sets. Criteria for desired and unique convergence are easily established. The network can be configured in either a homogeneous or layered form. The number of patterns that can be stored in the network is on the order of the number of input and hidden neurons. If the output neurons can take on only one of two states, then the trained layered APNN can be easily configured to converge in one iteration. More generally, convergence is at an exponential rate. Convergence can be improved by the use of sigmoid type nonlinearities, network relaxation and/or increasing the number of neurons in the hidden layer. The manner in which the network responds to data for which it was not specifically trained (i.e. how it *generalizes*) can be directly evaluated analytically.

1. INTRODUCTION

In this paper, we depart from the performance analysis techniques normally applied to neural networks. Instead, a signal space approach is used to gain new insights via ease of analysis and geometrical interpretation. Building on a foundation laid elsewhere[1-3], we demonstrate that alternating projecting neural network's (APNN's) formulated from such a viewpoint can be configured in layered form or homogeneously.

Significantly, APNN's have advantages over other neural network architectures. For example,
(a) APNN's perform by alternatingly projecting between two or more constraint sets. Criteria can be established for proper iterative convergence for both synchronous and asynchronous operation. This is in contrast to the more conventional technique of formulation of an energy metric for the neural networks, establishing a lower energy bound and showing that the energy reduces each iteration[4-7]. Such procedures generally do not address the accuracy of the final solution. In order to assure that such networks arrive at the desired globally minimum energy, computationally lengthly procedures such as simulated annealing are used[8-10]. For synchronous networks, steady state oscillation can occur between two states of the same energy[11]

(b) Homogeneous neural networks such as Hopfield's content addressable memory[4,12-14] do not scale well, i.e. the capacity

of Hopfield's neural networks less than doubles when the number
of neurons is doubled [15-16]. Also, the capacity of previously
proposed layered neural networks[17,18] is not well understood.
The capacity of the *layered* APNN's, on the other hand, is
roughly equal to the number of input and hidden neurons[19].

(c) The speed of backward error propagation learning [17-18] can be
painfully slow. Layered APNN's, on the other hand, can be
trained on only one pass through the training data[2]. If the
network memory does not saturate, new data can easily be
learned without repeating previous data. Neither is the
effectiveness of recall of previous data diminished. Unlike
layered back propagation neural networks, the APNN recalls by
iteration. Under certain important applications, however, the
APNN will recall in one iteration.

(d) The manner in which layered APNN's *generalizes* to data for
which it was not trained can be analyzed straightforwardly.

The outline of this paper is as follows. After establishing the
dynamics of the APNN in the next section, sufficient criteria for
proper convergence are given. The convergence dynamics of the APNN
are explored. Wise use of nonlinearities, e.g. the sigmoidal type
nonlinearities[2], improve the network's performance. Establishing a
hidden layer of neurons whose states are a nonlinear function of
the input neurons' states is shown to increase the network's
capacity and the network's convergence rate as well. The manner in
which the networks respond to data outside of the training set is
also addressed.

2. THE ALTERNATING PROJECTION NEURAL NETWORK

In this section, we established the notation for the APNN.
Nonlinear modificiations to the network made to impose certain
performance attributes are considered later.

Consider a set of N continuous level linearly independent
library vectors (or patterns) of length $L > N$: { $\vec{f}_n$ | $0 \leq n \leq N$ }. We form
the library matrix $\underline{F} = [\vec{f}_1 | \vec{f}_2 | ... | \vec{f}_N]$ and the neural network
interconnect matrix[a] $\underline{T} = \underline{F} (\underline{F}^T \underline{F})^{-1} \underline{F}^T$ where the superscript T
denotes transposition. We divide the L neurons into two sets: one
in which the states are known and the remainder in which the states
are unknown. This partition may change from application to
application. Let $s_k(M)$ be the state of the kth node at time M. If
the kth node falls into the known category, its state is *clamped* to
the known value (i.e. $s_k(M) = f_k$ where $\vec{f}$ is some library vector).
The states of the remaining *floating* neurons are equal to the sum
of the inputs into the node. That is, $s_k(M) = i_k$, where

$$i_k = \sum_{p=1}^{L} t_{pk} s_p \qquad (1)$$

---

[a] The interconnect matrix is better trained iteratively[2]. To include
a new library vector $\vec{f}$, the interconnects are updated as
$\underline{T} + (\vec{\epsilon} \vec{\epsilon}^T) / (\vec{\epsilon}^T \vec{\epsilon})$ where $\vec{\epsilon} = (\underline{I} - \underline{T}) \vec{f}$.

If all neurons change state simultaneously (i.e. $s_p = s_p(M-1)$), then the net is said to operate synchronously. If only one neuron changes state at a time, the network is operating asynchronously.

Let P be the number of clamped neurons. We have proven[1] that the neural states converge strongly to the extrapolated library vector if the first P rows of $\underline{F}$ (denoted $\underline{F}_p$) form a matrix of full column rank. That is, no column of $\underline{F}_p$ can be expressed as a linear combination of those remaining. By strong convergence[b], we mean $\lim_{M \to \infty} \| \vec{s}(M) - \vec{\ell} \| = 0$ where $\| \vec{x} \|^2 = \vec{x}^T \vec{x}$.

Lastly, note that subsumed in the criterion that $\underline{F}_p$ be full rank is the condition that the number of library vectors not exceed the number of known neural states ($P \geq N$). Techniques to bypass this restriction by using hidden neurons are discussed in section 5.

Partition Notation: Without loss of generality, we will assume that neurons 1 through P are clamped and the remaining neurons are floating. We adopt the vector partitioning notation

$$\vec{i} = \begin{bmatrix} \vec{i}_P \\ \vec{i}_Q \end{bmatrix}$$

where $\vec{i}_P$ is the P-tuple of the first P elements of $\vec{i}$ and $\vec{i}_Q$ is a vector of the remaining $Q = L-P$. We can thus write, for example, $\underline{F}_p = [\ \vec{f}_1^P\ |\vec{f}_2^P\ |...|\vec{f}_N^P\ ]$. Using this partition notation, we can define the neural clamping operator by:

$$\eta \vec{i} = \begin{bmatrix} \vec{\ell}^P \\ \vec{i}_Q \end{bmatrix}$$

Thus, the first P elements of $\vec{i}$ are clamped to $\vec{\ell}^P$. The remaining Q nodes "float".

Partition notation for the interconnect matrix will also prove useful. Define

$$\underline{T} = \begin{bmatrix} \underline{T}_2 & \underline{T}_1 \\ \hline \underline{T}_3 & \underline{T}_4 \end{bmatrix}$$

where $\underline{T}_2$ is a P by P and $\underline{T}_4$ a Q by Q matrix.

3. STEADY STATE CONVERGENCE PROOFS

For purposes of later reference, we address convergence of the network for synchronous operation. Asynchronous operation is addressed in reference 2. For proper convergence, both cases require that $\underline{F}_p$ be full rank. For synchronous operation, the network iteration in (1) followed by clamping can be written as:

$$\vec{s}(M+1) = \eta\ \underline{T}\ \vec{s}(M) \qquad (2)$$

As is illustrated in[1-3], this operation can easily be visualized in an L dimensional signal space.

---

[b] The referenced convergence proofs prove strong convergence in an infinite dimensional Hilbert space. In a discrete finite dimensional space, both strong and weak convergence imply uniform convergence[19,20], i.e. $\vec{s}(M) \to \vec{\ell}$ as $M \to \infty$.

For a given partition with P clamped neurons, (2) can be written in partitioned form as

$$
\left[\begin{array}{c} \vec{\ell}^{\,P} \\ \hline \vec{s}^{\,Q} \, (M+1) \end{array} \right] = \eta \left[\begin{array}{c|c} \underline{T}_2 & \underline{T}_1 \\ \hline \underline{T}_3 & \underline{T}_4 \end{array} \right] \left[\begin{array}{c} \vec{\ell}^{\,P} \\ \hline \vec{s}^{\,Q} \, (M) \end{array} \right] \tag{3}
$$

The states of the P clamped neurons are not affected by their input sum. Thus, there is no contribution to the iteration by $\underline{T}_1$ and $\underline{T}_2$. We can equivalently write (3) as

$$
\vec{s}^{\,Q} \, (M+1) = \underline{T}_3 \, \vec{\ell}^{\,P} + \underline{T}_4 \, \vec{s}^{\,Q} \, (M) \tag{4}
$$

We show in that if $\underline{F}_P$ is full rank, then the spectral radius (magnitude of the maximum eigenvalue) of $\underline{T}_4$ is strictly less than one[19]. It follows that the steady state solution of (4) is:

$$
\vec{\ell}^{\,Q} = (\underline{I} - \underline{T}_4)^{-1} \, \underline{T}_3 \, \vec{\ell}^{\,P} \tag{5}
$$

where, since $\underline{F}_P$ is full rank, we have made use of our claim that

$$
\vec{s}^{\,Q} \, (\infty) = \vec{\ell}^{\,Q} \tag{6}
$$

4. CONVERGENCE DYNAMICS

In this section, we explore different convergence dynamics of the APNN when $\underline{F}_P$ is full column rank. If the library matrix displays certain orthogonality characteristics, or if there is a single output (floating) neuron, convergence can be achieved in a single iteration. More generally, convergence is at an exponential rate. Two techniques are presented to improve convergence. The first is standard relaxation. Use of nonlinear convex constraint at each neuron is discussed elsewhere[2,19].

One Step Convergence: There are at least two important cases where the APNN converges other than uniformly in one iteration. Both require that the output be bipolar (±1). Convergence is in one step in the sense that

$$
\vec{\ell}^{\,Q} = \underline{\text{sign}} \, \vec{s}^{\,Q} \, (1) \tag{7}
$$

where the vector operation sign takes the sign of each element of the vector on which it operates.

CASE 1: If there is a single output neuron, then, from (4),(5) and (6), $s^Q(1) = (1 - t_{LL}) \, \ell^Q$. Since the eigenvalue of the (scalar) matrix, $\underline{T}_4 = t_{LL}$ lies between zero and one[19], we conclude that $1 - t_{LL} > 0$. Thus, if ℓ^Q is restricted to ±1, (7) follows immediately. A technique to extend this result to an arbitrary number of output neurons in a layered network is discussed in section 7.

CASE 2: For certain library matrices, the APNN can also display one step convergence. We showed that if the columns of $\underline{F}$ are orthogonal and the columns of $\underline{F}_P$ are also orthogonal, then one synchronous iteration results in floating states proportional to the steady state values[19]. Specifically, for the floating neurons,

$$\vec{s}^Q (1) = \frac{\| \vec{f}^P \|^2}{\| \vec{f} \|^2} \vec{f}^Q \tag{8}$$

An important special case of (8) is when the elements of $\underline{F}$ are all ± 1 and orthogonal. If each element were chosen by a 50-50 coin flip, for example, we would expect (in the statistical sense) that this would be the case.

Exponential Convergence: More generally, the convergence rate of the APNN is exponential and is a function of the eigenstructure of $\underline{T}_4$. Let $\{\vec{p}_r \mid 1 \le r \le Q\}$ denote the eigenvectors of $\underline{T}_4$ and $\{\lambda_r\}$ the corresponding eigenvalues. Define $\underline{P} = [\ \vec{p}_1 | \vec{p}_2 | \dots | \vec{p}_Q]$ and the diagonal matrix $\underline{\Lambda}_4$ such that $\text{diag}\ \underline{\Lambda}_4 = [\lambda_1\ \lambda_2 \dots \lambda_Q]^T$. Then we can write $\underline{T}_4 = \underline{P}\ \underline{\Lambda}_4\ \underline{P}^T$. Define $\vec{x}(M) = \underline{P}^T \vec{s}(M)$. Since $\underline{P}\ \underline{P}^T = \underline{I}$, it follows from the difference equation in (4) that $\vec{x}(M+1) = \underline{P}^T \underline{T}_4 \underline{P}\ \underline{P}^T \vec{s}(M) + \underline{P}^T \underline{T}_3 \vec{f}^P$ $= \underline{\Lambda}_4\ \vec{x}(M) + \vec{g}$ where $\vec{g} = \underline{P}^T \underline{T}_3 \vec{f}^P$. The solution to this difference equation is

$$x_k (M) = \sum_{r=0}^{M} \lambda_k^r\ g_k = [\ 1 - \lambda_k^{M+1}\ ]\ (\ 1 - \lambda_k\ )^{-1}\ g_k \tag{9}$$

Since the spectral radius of $\underline{T}_4$ is less than one[19], $\lambda_k^M \to 0$ as $M \to \infty$. Our steady state result is thus $x_k(\infty) = (1 - \lambda_k)^{-1}\ g_k$. Equation (9) can therefore be written as $x_k(M) = [\ 1 - \lambda_k^{M+1}]\ x_k(\infty)$. The equivalent of a "time constant" in this exponential convergence is $1/\ell n(1/|\lambda_k|)$. The speed of convergence is thus dictated by the spectral radius of $\underline{T}_4$. As we have shown[19] later, adding neurons in a hidden layer in an APNN can significiantly reduce this spectral radius and thus improve the convergence rate.

Relaxation: Both the projection and clamping operations can be relaxed to alter the network's convergence without affecting its steady state[20-21]. For the interconnects, we choose an appropriate value of the relaxation parameter θ in the interval $(0,2)$ and redefine the interconnect matrix as $\underline{T}^\theta = \theta \underline{T} + (1 - \theta)\underline{I}$ or equivalently,

$$t_{nm}^\theta = \begin{cases} \theta\ (t_{nn} - 1) + 1 & ;\ n = m \\ \theta\ t_{nm} & ;\ n \ne m \end{cases}$$

To see the effect of such relaxation on convergence, we need simply examine the resulting eigenvalues. If $\underline{T}_4$ has eigenvalues $\{\lambda_r\}$, then $\underline{T}_4^\theta$ has eigenvalues $\lambda_r^\theta = 1 + \theta(\lambda_r - 1)$. A wise choice of θ reduces the spectral radius of $\underline{T}^\theta$ with respect to that of $\underline{T}_4$, and thus decreases the time constant of the network's convergence.

Any of the operators projecting onto convex sets can be relaxed without affecting steady state convergence[19-20]. These include the η operator[2] and the sigmoid-type neural operator that projects onto a box. Choice of stationary relaxation parameters without numerical and/or empirical study of each specific case, however, generally remains more of an art than a science.

5. LAYERED APNN'S

The networks thus far considered are homogeneous in the sense that any neuron can be clamped or floating. If the partition is such that the same set of neurons always provides the network stimulus and the remainder respond, then the networks can be simplified. Clamped neurons, for example, ignore the states of the other neurons. The corresponding interconnects can then be deleted from the neural network architecture. When the neurons are so partitioned, we will refer the APNN as *layered*.

In this section, we explore various aspects of the layered APNN and in particular, the use of a so called hidden layer of neurons to increase the storage capacity of the network. An alternate architecture for a homogeneous APNN that require only Q neurons has been reported by Marks[2].

Hidden Layers: In its generic form, the APNN cannot perform a simple exclusive or (XOR). Indeed, failure to perform this same operation was a nail in the coffin of the perceptron[22]. Rumelhart et. al.[17-18] revived the perceptron by adding additional layers of neurons. Although doing so allowed nonlinear discrimination, the iterative training of such networks can be painfully slow. With the addition of a hidden layer, the APNN likewise generalizes. In contrast, the APNN can be trained by looking at each data vector only once[1].

Although neural networks will not likely be used for performing XOR's, their use in explaining the role of hidden neurons is quite instructive. The library matrix for the XOR is

$$\underline{F} = \begin{bmatrix} 0 & 0 & 1 & 1 \\ 0 & 1 & 0 & 1 \\ 0 & 1 & 1 & 0 \end{bmatrix}$$

The first two rows of $\underline{F}$ do not form a matrix of full column rank. Our approach is to augment $\underline{F}_p$ with two more rows such that the resulting matrix is full rank. Most any *nonlinear* combination of the first two rows will in general increase the matrix rank. Such a procedure, for example, is used in Φ-classifiers[23]. Possible nonlinear operations include multiplication, a logical "AND" and running a weighted sum of the clamped neural states through a memoryless nonlinearity such as a sigmoid. This latter alteration is particularly well suited to neural architectures.

To illustrate with the exclusive or (XOR), a new hidden neural state is set equal to the exponentiation of the sum of the first two rows. A second hidden neurons will be assigned a value equal to the cosine of the sum of the first two neural states multiplied by $\pi/2$. (The choice of nonlinearities here is arbitrary.) The augmented library matrix is

$$\underline{F}_+ = \begin{bmatrix} 0 & 0 & 1 & 1 \\ 0 & 1 & 0 & 1 \\ 1 & e & e & e^2 \\ 1 & 0 & 0 & -1 \\ 0 & 1 & 1 & 0 \end{bmatrix}$$

In either the training or look-up mode, the states of the hidden neurons are clamped indirectly as a result of clamping the input neurons.

The playback architecture for this network is shown in Fig.1. The interconnect values for the dashed lines are unity. The remaining interconnects are from the projection matrix formed from $\underline{F}_+$.

Geometrical Interpretation : In lower dimensions, the effects of hidden neurons can be nicely illustrated geometrically. Consider the library matrix

$$\underline{F} = \left[\begin{array}{cc} 1/2 & 1 \\ 1 & 1/2 \end{array} \right]$$

Clearly $\underline{F}_p = [1/2 \; 1]$. Let the neurons in the hidden layer be determined by the nonlineariy x^2 where x denotes the elements in the first row of $\underline{F}$. Then

$$\underline{F}_+ = \left[\begin{array}{c|c} \vec{\ell}_1^+ & \vec{\ell}_2^+ \end{array} \right] = \left[\begin{array}{cc} 1/2 & 1 \\ 1/4 & 1 \\ 1 & 1/2 \end{array} \right]$$

The corresponding geometry is shown in Fig.2 for x the input neuron, y the output and h the hidden neuron. The augmented library vectors are shown and a portion of the generated subspace is shown lightly shaded. The surface of $h = x^2$ resembles a cylindrical lens in three dimensions. Note that the linear variety corresponding to $x = 1/2$ intersects the cylindrical lens and subspace only at $\vec{\ell}_1^+$. Similarly, the $x = 1$ plane intersects the lens and subspace at $\vec{\ell}_2^+$. Thus, in both cases, clamping the input corresponding to the first element of one of the two library vectors uniquely determines the library vector.

Convergence Improvement: Use of additional neurons in the hidden layer will improve the convergence rate of the APNN[19]. Specifically, the spectral radius of the $\underline{T}_4$ matrix is decreased as additional neurons are added. The dominant time constant controlling convergence is thus decreased.

Capacity: Under the assumption that nonlinearities are chosen such that the augmented $\underline{F}_p$ matrix is of full rank, the number of vectors which can be stored in the layered APNN is equal to the sum of the number of neurons in the input and hidden layers. Note, then, that interconnects between the input and output neurons are not needed if there are a sufficiently large number of neurons in the hidden layer.

6. GENERALIZATION

We are assured that the APNN will converge to the desired result if a portion of a training vector is used to stimulate the network. What, however, will be the response if an initialization is used that is not in the training set or, in other words, how does the network *generalize* from the training set ?

To illustrate generalization, we return to the XOR problem. Let $s_5(M)$ denote the state of the output neuron at the Mth (synchronous)

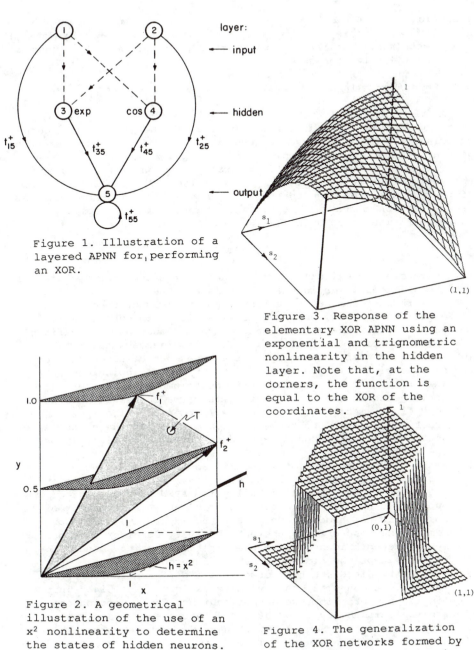

Figure 1. Illustration of a layered APNN for performing an XOR.

Figure 3. Response of the elementary XOR APNN using an exponential and trignometric nonlinearity in the hidden layer. Note that, at the corners, the function is equal to the XOR of the coordinates.

Figure 2. A geometrical illustration of the use of an x^2 nonlinearity to determine the states of hidden neurons.

Figure 4. The generalization of the XOR networks formed by thresholding the function in Fig.3 at 3/4. Different hidden layer nonlinearities result in different generalizations.

iteration. If s_1 and s_2 denote the input clamped value, then $s_5(m+1)=t_{15}s_1 + t_{25}s_2 + t_{35}s_3 + t_{45}s_4 + t_{55}s_5(m)$ where $s_3=\exp(s_1+s_2)$ and $s_4=\cos[\pi(s_1 + s_2)/2]$ To reach steady state, we let m tend to infinity and solve for $s_5(\infty)$:

$$s_5(\infty) = \frac{1}{1 - t_{55}} [t_{15}s_1 + t_{25}s_2 + t_{35}\exp(s_1+s_2) + t_{45}\cos\frac{\pi}{2}(s_1+s_2)] \tag{10}$$

A plot of $s_5(\infty)$ versus (s_1,s_2) is shown in Figure 3. The plot goes through 1 and zero according to the XOR of the corner coordinates. Thresholding Figure 3 at 3/4 results in the generalization perspective plot shown in Figure 4.

To analyze the network's generalization when there are more than one output neuron, we use (5) of which (10) is a special case. If conditions are such that there is one step convergence, then generalization plots of the type in Figure 4 can be computed from one network iteration using (7).

7. NOTES

(a) There clearly exists a great amount of freedom in the choice of the nonlinearities in the hidden layer. Their effect on the network performance is currently not well understood. One can envision, however, choosing nonlinearities to enhance some network attribute such as interconnect reduction, classification region shaping (generalization) or convergence acceleration.

(b) There is a possibility that for a given set of hidden neuron nonlinearities, augmentation of the F_p matrix coincidentally will result in a matrix of deficent column rank, proper convergence is then not assured. It may also result in a poorly conditioned matrix, convergence will then be quite slow. A practical solution to these problems is to pad the hidden layer with additional neurons. As we have noted, this will improve the convergence rate.

(c) We have shown in section 4 that if an APNN has a single bipolar output neuron, the network converges in one step in the sense of (7). Visualize a layered APNN with a single output neuron. If there are a sufficiently large number of neurons in the hidden layer, then the input layer does not need to be connected to the output layer. Consider a second neural network identical to the first in the input and hidden layers except the hidden to output interconnects are different. Since the two networks are different only in the output interconnects, the two networks can be combined into a singlee network with two output neurons. The interconnects from the hidden layer to the output neurons are identical to those used in the single output neurons architectures. The new network will also converge in one step. This process can clearly be extended to an arbitrary number of output neurons.

REFERENCES

1. R.J. Marks II, "A Class of Continuous Level Associative Memory Neural Nets," Appl. Opt., vol.26, no.10, p.2005, 1987.

2. K.F. Cheung *et. al.*, "Neural Net Associative Memories Based on Convex Set Projections," Proc. IEEE 1st International Conf. on Neural Networks, San Diego, 1987.

3. R.J. Marks II *et. al.*, "A Class of Continuous Level Neural Nets," Proc. 14th Congress of International Commission for Optics Conf., Quebec, Canada, 1987.

4. J.J. Hopfield, "Neural Networks and Physical Systems with Emergent Collective Computational Abilities," Proceedings Nat. Acad. of Sciences, USA, vol.79, p.2554, 1982.

5. J.J. Hopfield *et. al.*, "Neural Computation of Decisions in Optimization Problem," Biol. Cyber., vol.52, p.141, 1985.

6. D.W. Tank *et. al.*, "Simple Neurel Optimization Networks: an A/D Converter, Signal Decision Circuit and a Linear Programming Circuit," IEEE Trans. Cir. Sys., vol. CAS-33, p.533, 1986.

7. M. Takeda *et. al*, "Neural Networks for Computation: Number Representation and Programming Complexity," Appl. Opt., vol. 25, no. 18, p.3033, 1986.

8. S. Geman *et. al.*, "Stochastic Relaxation, Gibb's Distributions, and the Bayesian Restoration of Images," IEEE Trans. Pattern Recog. & Machine Intelligence., vol. PAMI-6, p.721, 1984.

9. S. Kirkpatrick *et. al.* ,"Optimization by Simulated Annealing," Science, vol. 220, no. 4598, p.671, 1983.

10. D.H. Ackley *et. al.*, "A Learning Algorithm for Boltzmann Machines," Cognitive Science, vol. 9, p.147, 1985.

11. K.F. Cheung *et. al.*, "Synchronous vs. Asynchronous Behaviour of Hopfield's CAM Neural Net," to appear in Applied Optics.

12. R.P. Lippmann, "An Introduction to Computing With Neural nets," IEEE ASSP Magazine, p.7, Apr 1987.

13. N. Farhat *et. al.*, "Optical Implementation of the Hopfield Model," Appl. Opt., vol. 24, pp.1469, 1985.

14. L.E. Atlas, "Auditory Coding in Higher Centers of the CNS," IEEE Eng. in Medicine and Biology Magazine, p.29, Jun 1987.

15. Y.S. Abu-Mostafa *et. al.*, "Information Capacity of the Hopfield Model, " IEEE Trans. Inf. Theory, vol. IT-31, p.461, 1985.

16. R.J. McEliece *et. al.*,"The Capacity of the Hopfield Associative Memory, " IEEE Trans. Inf. Theory (submitted), 1986.

17. D.E. Rumelhart *et. al.*, **Parallel Distributed Processing**, vol. I & II, Bradford Books, Cambridge, MA, 1986.

18. D.E. Rumelhart *et. al.*, "Learning Representations by Back-Propagation Errors," Nature. vol. 323, no. 6088, p.533, 1986.

19. R.J. Marks II *et. al.*,"Alternating Projection Neural Networks," ISDL report #11587, Nov. 1987 (Submitted for publication).

20. D.C. Youla *et. al*, "Image Restoration by the Method of Convex Projections: Part I-Theory," IEEE Trans. Med. Imaging, vol. MI-1, p.81, 1982.

21. M.I. Sezan and H. Stark. "Image Restoration by the Method of Convex Projections: Part II-Applications and Numerical Results," IEEE Trans. Med. Imaging, vol. MI-1, p.95, 1985.

22. M. Minsky *et. al.*, **Perceptrons**, MIT Press, Cambridge, MA, 1969.

23. J. Sklansky *et. al.*, **Pattern Classifiers and Trainable Machines**, Springer-Verlag, New York, 1981.

MURPHY: A Robot that Learns by Doing

Bartlett W. Mel

Center for Complex Systems Research
University of Illinois
508 South Sixth Street
Champaign, IL 61820

January 2, 1988

Abstract

MURPHY consists of a camera looking at a robot arm, with a connectionist network architecture situated in between. By moving its arm through a small, representative sample of the 1 billion possible joint configurations, MURPHY learns the relationships, backwards and forwards, between the positions of its joints and the state of its visual field. MURPHY can use its internal model in the forward direction to "envision" sequences of actions for planning purposes, such as in grabbing a visually presented object, or in the reverse direction to "imitate", with its arm, autonomous activity in its visual field. Furthermore, by taking explicit advantage of continuity in the mappings between visual space and joint space, MURPHY is able to learn non-linear mappings with only a single layer of modifiable weights.

Background

Current Focus Of Learning Research

Most connectionist learning algorithms may be grouped into three general catagories, commonly referred to as *supervised, unsupervised*, and *reinforcement* learning. Supervised learning requires the explicit participation of an intelligent teacher, usually to provide the learning system with task-relevant input-output pairs (for two recent examples, see [1,2]). Unsupervised learning, exemplified by "clustering" algorithms, are generally concerned with detecting structure in a stream of input patterns [3,4,5,6,7]. In its final state, an unsupervised learning system will typically represent the discovered structure as a set of categories representing regions of the input space, or, more generally, as a mapping from the input space into a space of lower dimension that is somehow better suited to the task at hand. In reinforcement learning, a "critic" rewards or penalizes the learning system, until the system ultimately produces the correct output in response to a given input pattern [8].

It has seemed an inevitable tradeoff that systems needing to rapidly learn specific, behaviorally useful input-output mappings must necessarily do so under the auspices of an intelligent teacher with a ready supply of task-relevant training examples. This state of affairs has seemed somewhat paradoxical, since the processes of perceptual and cognitive development in human infants, for example, do not depend on the moment by moment intervention of a teacher of any sort.

Learning by Doing

The current work has been focused on a fourth type of learning algorithm, i.e. *learning-by-doing*, an approach that has been very little studied from either a connectionist perspective or in the context of more traditional work in machine learning. In its basic form, the learning agent

- begins with a repertoire of actions and some form of perceptual input,

- exercises its repertoire of actions, learning to predict i) the detailed sensory consequences of its actions, and, in the other direction, ii) its actions that are associated with incoming sensory patterns, and

- runs its internal model (in one or both directions) in a variety of behaviorally-relevant tasks, e.g. to "envision" sequences of actions for planning purposes, or to internally "imitate", via its internal action representation, an autonomously generated pattern of perceptual activity.

In comparison to standard *supervised* learning algorithms, the crucial property of learning-by-doing is that *no intelligent teacher is needed to provide input-output pairs for learning*. Laws of physics simply translate actions into their resulting percepts, both of which are represented internally. The learning agent need only notice and record this relationship for later use. In contrast to traditional unsupervised learning approaches, learning-by-doing allows the acquisition of specific, task-relevant mappings, such as the relationship between a simultaneously represented visual and joint state. Learning-by-doing differs as well from reinforcement paradigms in that it can operate in the absence of a critic, i.e. in situations where reward or penalty for a particular training instance may be inappropriate.

Learning by doing may therefore by described as an *unsupervised associative* algorithm, capable of acquiring rich, task-relevant associations, but without an intelligent teacher or critic.

Abridged History of the Approach

The general concept of leaning by doing may be attributed at least to Piaget from the 1940's (see [9] for review). Piaget, the founder of the "constructivist" school of cognitive development, argued that knowledge is not given to a child as a passive observer, but is rather discovered and constructed by the child, through active manipulation of the environment. A handful of workers in artificial intelligence have addressed the issue of learning-by-doing, though only in highly schematized, simulated domains, where actions and sensory states are represented as logical predicates [10,11,12,13].

Barto & Sutton [14] discuss learning-by-doing in the context of system identification and motor control. They demonstrated how a simple simulated automaton with two actions and three sensory states can build a model of its environment through exploration, and subsequently use it to choose among behavioral alternatives. In a similar vein, Rumelhart [15] has suggested this same approach could be used to learn the behavior of a robot arm or a set of speech articulators. Furthermore, the forward-going "mental model", once learned, could be used internally to train an *inverse* model using back-propagation.

In previous work, this author [16] described a connectionist model (VIPS) that learned to perform 3-D visual transformations on simulated wire-frame objects. Since in complex sensory-motor environments it is not possible, in general, to learn a direct relationship between an outgoing command state and an incoming sensory state, VIPS was designed to predict *changes* in the state of its visual field as a function of its outgoing motor command. VIPS could then use its generic knowledge of motor-driven visual transformations to "mentally rotate" objects through a series of steps.

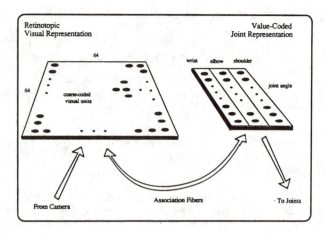

Figure 1: MURPHY's Connectionist Architecture. 4096 coarsely-tuned visual units are organized in a square, retinotopic grid. These units are bi-directionally interconnected with a population of 273 joint units. The joint population is subdivided into 3 subpopulations, each one a value-coded representation of joint angle for one of the three joints. During training, activity in the joint unit population determines the physical arm configuration.

Inside MURPY

The current work has sought to further explore the process of learning-by-doing in a complex sensory-motor domain, extending previous work in three ways. First, the learning of mappings between sensory and command (e.g. motor) representations should be allowed to proceed in both directions *simultaneously* during exploratory behavior, where each mapping may ultimately subserve a very different behavioral goal. Secondly, MURPHY has been implemented with a real camera and robot arm in order to insure representational realism to the greatest degree possible. Third, while the specifics of MURPHY's internal structures are not intended as a model of a specific neural system, a serious attempt has been made to adhere to architectural components and operations that have either been directly suggested by nervous system structures, or are at least compatible with what is currently known. Detailed biological justification on this point awaits further work.

MURPHY's Body

MURPHY consists of a 512 x 512 JVC color video camera pointed at a Rhino XR-3 robotic arm. Only the shoulder, elbow, and wrist joints are used, such that the arm can move only in the image plane of the camera. (A fourth, waist joint is not used). White spots are stuck to the arm in convenient places; when the image is thresholded, only the white spots appear in the image. This arrangement allows continuous control over the complexity of the visual image of the arm, which in turn affects time spent both in computing visual features and processing weights during learning. A Datacube image processing system is used for the thresholding operation and to "blur" the image in real time with a gaussian mask. The degree of blur is variable and can be used to control the degree of coarse-coding (i.e. receptive field overlap) in the camera-topic array of visual units. The arm is software controllable, with a stepper motor for each joint. Arm dynamics are not considered in this work.

MURPHY's Mind

MURPHY is currently based on two interconnected populations of neuron-like units. The first is organized as a rectangular, visuotopically-mapped 64 x 64 grid of coarsely-tuned visual units that each responds when a visual feature (such as a white spot on the arm) falls into its receptive field (fig. 1). Coarse coding insures that a single visual feature will activate a small population of units whose receptive fields overlap the center of stimulation. The upper trace in figure 2 shows the unprocessed camera view, and the center trace depicts the resulting pattern of activation over the grid of visual units.

The second population of 273 units consists of three subpopulations, representing the angles of each of the three joints. The angle of each joint is value-coded in a line of units dedicated to that joint (fig. 1). Each unit in the population is "centered" at a some joint angle, and is maximally activated when the joint is to be sent to that angle. Neighboring joint units within a joint subpopulation have overlapping "projective fields" and progressively increasing joint-angle centers.

It may be noticed that both populations of units are coarsely tuned, that is, the units have overlapping receptive fields whose centers vary in an orderly fashion from unit to neighboring unit. This style of representation is extremely common in biological sensory systems [17,18,19], and has been attributed a number of representational advantages (e.g. fewer units needed to encode range of stimuli, increased immunity to noise and unit malfunction, and finer stimulus discriminations). A number of *additional* advantages of this type of encoding scheme are discussed in section , in relation to ease of learning, speed of learning, and efficacy of generalization.

MURPHY's Education

By moving its arm through a small, representative sample (approximately 4000) of the 1 billion possible joint configurations, MURPHY learns the relationships, backwards and forwards, between the positions of its joints and the state of its visual field. During training, the physical environment to which MURPHY's visual and joint representations are wired *enforces* a particular mapping between the states of these two representations. The mapping comprises both the kinematics of the arm as well as the optical parameters and global geometry of the camera/imaging system. It is incrementally learned as each unit in population B comes to "recognize", through a process of weight modification, the states of population A in which it has been strongly activated. After sufficient training experience therefore, the state of population A is sufficient to generate a "mental image" on population B, that is, to *predictively* activate the units in B via the weighted interconnections developed during training.

In its current configuration, MURPHY steps through its entire joint space in around 1 hour, developing a total of approximately 500,000 weights between the two populations.

The Learning Rule

Tradeoffs in Learning and Representation

It is well known in the folklore of connectionist network design that a tradeoff exists between the choice of representation (i.e. the "semantics") at the single unit level and the consequent ease or difficulty of learning within that representational scheme.

At one extreme, the single-unit representation might be completely *decoded*, calling for a separate unit for each possible input pattern. While this scheme requires a combinatorially explosive number of units, and the system must "see" every possible input pattern during training, the actual weight modification rule is rendered very simple. At another extreme, the single unit representation might be chosen in a highly *encoded* fashion with complex interactions among input units. In this case, the activation of an output unit may be a highly non-linear or discontinuous function of the input pattern, and must be learned and represented in multiple layers of weights.

Research in connectionism has often focused on Boolean functions [20,21,1,22,23], typified by the encoder problem [20], the shifter problem [21] and n-bit parity [22]. Since Boolean functions are in general discontinuous, such that two input patterns that are close in the sense of Hamming distance do not in general result in similar outputs, much effort has been directed toward the development of sophisticated, multilayer weight-modification rules (e.g. back-propagation) capable of learning arbitrary discontinuous functions. The complexity of such learning procedures has raised troubling questions of scaling behavior and biological plausibility.

The assumption of *continuity* in the mappings to be learned, however, can act to significationly simplify the learning problem while still allowing for full generalization to novel input patterns. Thus, by relying on the continuity assumption, MURPHY's is able to learn continuous non-linear functions using a weight modification procedure that is simple, locally computable, and confined to a single layer of modifiable weights.

How MURPHY learns

For sake of concrete illustration, MURPHY's representation and learning scheme will be described in terms of the mapping learned from joint units to visual units during training. The output activity of a given visual unit may be described as a function over the 3-dimensional joint space, whose shape is determined by the kinematics of the arm, the location of visual features (i.e. white spots) on the arm, the global properties of the camera/imaging system, and the location of the visual unit's receptive field. In order for the function to be learned, a visual unit must learn to "recognize" the regions of joint space in which it has been visually activated during training. In effect, each visual unit learns to recognize the global arm configurations that happen to put a white spot in its receptive field.

It may be recalled that MURPHY's joint unit population is value-coded by dimension, such that each unit is centered on a range of joint angles (overlapping with neighboring units) for one of the 3 joints. In this representation, a global arm configuration can be represented as the *conjunctive* activity of the k (where $k = 3$) most active joint units. MURPHY's visual units can therefore learn to recognize the regions of joint space in which they are strongly activated by simply "memorizing" the relevant global joint configurations as conjunctive clusters of input connections from the value-coded joint unit population.

To realize this conjunctive learning scheme, MURPHY's uses sigma-pi units (see [24]), as described below. At training step S, the set of k most active joint units are first identified. Some subset of visual units is also strongly activated in state S, each one signalling the presence of a visual feature (such as a white spot) in its receptive fields. At the input to each active visual unit, connections from the k most highly active joint units are formed as a multiplicative k-tuple of synaptic weights. The weights w_i on these connections are initially chosen to be of unit strength. The output c_j of a given synaptic conjunction is computed by multiplying the k joint unit activation values x_i together with their weights:

$$c_j = \prod_i w_i x_i.$$

The output y of the entire unit is computed as a weighted sum of the outputs of each conjunction and then applying a sigmoidal nonlinearity:

$$y = \sigma(\sum_j W_j c_j).$$

Sigma-pi units of this kind may be thought of as a generalization of a logical disjunction of conjunctions (OR of AND's). The multiplicative nature of the conjunctive clusters insures that every input to the conjunct is active in order for the conjunct to have an effect on the unit as a whole. If only a single input to a conjunct is *inactive*, the effect of the conjunction is nullified.

Specific-Instance Learning in Continuous Domains

MURPHY's learning scheme is directly reminiscent of *specific-instance* learning as discussed by Hampson & Volper [23] in their excellent review of Boolean learning and representational schemes. Specific-instance learning requires that each unit simply "memorize" all relevant input states, i.e. those states in which the unit is intended to fire. Unfortunately, simple specific-instance learning allows for no generalization to novel inputs, implying that each desired system responses will have to have been explicitly seen during training. Such a state of affairs is clearly impractical in natural learning contexts. Hampson & Volper [23] have further shown that random Boolean functions will require an exponential number of weights in this scheme.

For *continous* functions on the other hand, two kinds of generalization *are* possible within this type of specific-instance learning scheme. We consider each in turn, once again from the perspective of MURPHY's visual units learning to recognize the regions in joint space in which they are activated.

Generalization by Coarse-Coding

When a visual unit is activated in a given joint configuration, and acquires an appropriate conjunct of weights from the set of highly active units in the joint population, *by continuity* the unit may assume that it should be at least partially activated in nearby joint configurations as well. Since MURPHY's joint units are coarse-coded in joint angle, this will happen automatically: as the joints are moved a small distance away from the specific training configuration, the output of the conjunct encoding that training configuration will decay smoothly from its maximum. Thus, a visual unit can "fire" predictively in joint configurations that it has never specifically seen during training, by interpolating among conjuncts that encode *nearby* joint configurations.

This scheme suggests that training must be sufficiently dense in joint space to have seen configurations "nearby" to all points in the space by some criterion. In practice, the training step size is related to the degree of coarse-coding in the joint population, which is chosen in turn such that a joint pertubation equal to the radius of a joint unit's projective field (i.e. the range of joint angles over which the unit is active) should on average push a feature in the visual field a distance of about one *visual* receptive field radius. As a rule of thumb, the visual receptive field radius is chosen small enough so as to contain only a single feature on average.

Generalization by Extrapolation

The second type of generalization is based on a heuristic principle, again illustrated in terms of learning in the visual population. If a visual unit has during training been very often activated over a large, easy-to-specify region of joint space, such as a hyper-rectangular region, then it may be assumed that the unit is activated over the *entire* region of joint space, i.e. even at points not yet seen. At the synaptic level, "large regions" can be represented as conjuncts with fewer terms. In its simplest form, this kind of generalization amounts to simply throwing out one or more joints as irrelevant to the activation of a given visual unit. What synaptic mechanism can achieve this effect? Competition among joint unit afferents can be used to drive irrelevant variables from the sigma-pi conjuncts. Thus, if a visual unit is activated repeatedly during training, and the elbow and shoulder angle units are constantly active while the most active wrist unit varies from step to step, then the weighted connections from the repeatedly activated elbow and shoulder units

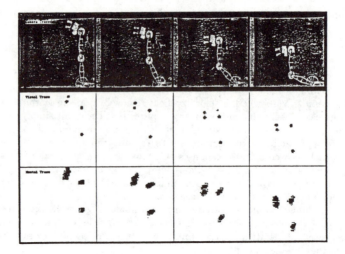

Figure 2: Three Visual Traces. The top trace shows the unprocessed camera view of MURPHY's arm. White spots have been stuck to the arm at various places, such that a thresholded image contains *only* the white spots. This allows continuous control over the visual complexity of the image. The center trace represents the resulting pattern of activation over the 64 x 64 grid of coarsely-tuned visual units. The bottom trace depicts an internally-produced "mental" image of the arm in the same configuration as driven by weighted connections from the joint population. Note that the "mental" trace is a sloppy, but highly recognizable approximation to the camera-driven trace.

will become progressively and mutually reinforced at the expense of the *set* of wrist unit connections, each of which was only activated a single time.

This form of generalization is similar in function to a number of "covering" algorithms designed to discover optimal hyper-rectangular decompositions (with possible overlap) of a set of points in a multi-dimensional space (e.g. [25,26]). The competitive feature has not yet been implemented explicitly at the synaptic level, rather, the full set of conjuncts acquired during training are currently collapsed *en masse* into a more compact set of conjuncts, according to the above heuristics. In a typical run, MURPHY is able to eliminate between 30% and 70% of its conjuncts in this way.

What MURPHY Does

Grabbing A Visually Presented Target

Once MURPHY has learned to image its arm in an arbitrary joint configuration, it can use heuristic search to guide its arm "mentally" to a visually specified goal. Figure 3(a) depicts a hand trajectory from an initial position to the location of a visually presented target. Each step in the trajectory represents the position of the hand (large blob) at an intermediate joint configuration. MURPHY can visually evaluate the remaining distance to the goal at each position and use best-first search to reduce the distance. Once a complete trajectory has been found, MURPHY can move its arm to the goal in a single physical step, dispensing with all backtracking dead-ends, and other wasted operations (fig. 3(b)). It would also be possible to use the *inverse* model, i.e. the map from a desired *visual* into an internal *joint* image, to send the arm directly to its final position. Unfortunately, MURPHY has no means in its current early state of development to generate a full-blown

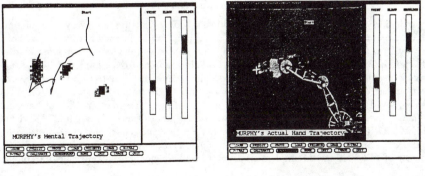

(a) (b)

Figure 3: Grabbing an Object. (a) Irregular trajectory represents sequence of "mental" steps taken by MURPHY in attempting to "grab" a visually-presented target (shown in (b) as white cross). Mental image depicts MURPHY's arm in its final goal configuration, i.e. with hand on top of object. Coarse-coded joint activity is shown at right. (b) Having mentally searched and found the target through a series of steps, MURPHY moves its arm *physically* in a single step to the target, discarding the intermediate states of the trajectory that are not relevant in this simple problem.

visual image of its arm in one of the final goal positions, of which there are many possible.

Sending the tip of a robot arm to a given point in space is a classic task in robotics. The traditional approach involves first writing explicit kinematic equations for the arm based on the specific geometric details of the given arm. These equations take joint angles as inputs and produce manipulator coordinates as outputs. In general, however, it is most often useful to specify the coordinates of the manipulator tip (i.e. its desired final position), and compute the joint angles necessary to achieve this goal. This involves the *solution* of the kinematic equations to generate an inverse kinematic model. Deriving such expressions has been called "the most difficult problem we will encounter" in vision-based robotics [27]. For this reason, it is highly desirable for a mobile agent to *learn* a model of its sensory-motor environment from scratch, in a way that depends little or not at all on the specific parameters of the motor apparatus, the sensory apparatus, or their mutual interactions. It is interesting to note that in this reaching task, MURPHY appears from the outside to be driven by an inverse kinematic model of its arm, since its first official act after training is to reach directly for a visually-presented object.

While it is clear that best-first search is a weak method whose utility is limited in complex problem solving domains, it may be speculated that given the ability to rapidly image arm configurations, combined with a set of simple visual heuristics and various mechanism for escaping local minima (e.g. send the arm home), a number of more interesting visual planning problems may be within MURPHY's grasp, such as grabbing an object in the presence of obstacles. Indeed, for problems that are either difficult to invert, or for which the goal state is not fully known *a priori*, the technique of iterating a forward-going model has a long history (e.g. Newton's Method).

Imitating Autonomous Arm Activity

A particularly interesting feature of "learning-by-doing" is that for every pair of unit populations present in the learning system, a mapping can be learned between them both backwards and forwards. Each such mapping may enable a unique and interesting kind of behavior. In MURPHY's case, we have seen that the mapping from a joint state to a visual image is useful for planning arm trajectories. The reverse mapping from a visual state to a joint image has an altogether different use, i.e. that of "imitation". Thus, if MURPHY's arm is moved passively, the model can be used to "follow" the motion with an internal command (i.e. joint) trace. Or, if a substitute arm is positioned in MURPHY's visual field, MURPHY can "assume the position", i.e. imitate the model arm configuration by mapping the afferent visual state into a joint image, and using the joint image to move the arm. As of this writing, the implementation of this behavior is still somewhat unreliable.

Discussion and Future Work

In short, this work has been concerned with learning-by-doing in the domain of vision-based robotics. A number of features differentiate MURPHY from most other learning systems, and from other approaches to vision-based robotics:

- No intelligent teacher is needed to provide input-ouput pairs. MURPHY learns by exercising its repertoire of actions and learning the relationship between these actions and the sensory images that result.

- Mappings between populations of units, regardless of modality, can be learned in both directions simultaneously during exploratory behavior. Each mapping learned can support a distinct class of behaviorally useful functions.

- MURPHY uses its internal models to first solve problems "mentally". Plans can therefore be developed and refined before they are actually executed.

- By taking explicit advantage of *continuity* in the mappings between visual and joint spaces, and by using a variant of specific-instance learning in such a way as to allow generalization to novel inputs, MURHPY can learn "difficult" non-linear mappings with only a single layer of modifiable weights.

Two future steps in this endeavor are as follows:

- Provide MURPHY with direction-selective visual and joint units both, so that it may learn to predict relationships between *rates of change* in the visual and joint domains. In this way, MURPHY can learn how to perturb its joints in order to send its hand in a particular direction, greatly reducing the current need to *search* for hand trajectories.

- Challenge MURPHY to grab actual objects, possibly in the presence of obstacles, where path of approach is crucial. The ability to readily envision intermediate arm configurations will become critical for such tasks.

Acknowledgements

Particular thanks are due to Stephen Omohundro for his unrelenting scientific and moral support, and for suggesting vision and robotic kinematics as ideal domains for experimentation.

References

[1] T.J. Sejnowski & C.R. Rosenberg, Complex Systems, *1*, 145, (1969).

[2] G.J. Tesauro & T.J. Sejnowski. A parallel network that learns to play backgammon. Submitted for publication.

[3] S. Grossberg, Biol. Cybern., *23*, 187, (1976).

[4] T. Kohonen, *Self organization and associative memory.*, (Springer-Verlag, Berlin 1984).

[5] D.E. Rumelhart & D. Zipser. In *Parallel distributed processing: explorations in the microstructure of cognition, vol. 1*, D.E. Rumelhart, J.L. McClelland, Eds., (Bradford: Cambridge, MA, 1986), p. 151.

[6] R. Linsker, Proc. Natl. Acad. Sci., *83*, 8779, (1986).

[7] G.E. Hinton & J.L. McClelland. Learning representations by recirculation. Oral presentation, IEEE conference on Neural Information Processing Systems, Denver, 1987.

[8] A.G. Barto, R.S. Sutton, & C.W. Anderson, IEEE Trans. on Sys., Man, Cybern., *smc-13*, 834, (1983).

[9] H. Ginsburg & S. Opper, *Piaget's theory of intellectual development.*, (Prentice Hall, New Jersey, 1969).

[10] J.D. Becker. In *Computer models for thought and language.*, R. Schank & K.M. Colby, Eds., (Freeman, San Francisco, 1973).

[11] R.L. Rivest & R.E. Schapire. In Proc. of the 4th int. workshop on machine learning, 364-375, (1987).

[12] J.G. Carbonell & Y. Gil. In Proc. of the 4th int. workshop on machine learning, 256-266, (1987).

[13] K. Chen, Tech Report, Dept. of Computer Science, University of Illinois, 1987.

[14] A.G. Barto & R.S. Sutton, AFWAL-TR-81-1070, Avionics Laboratory, Air Force Wright Aeronautical Laboratories, Wright-Patterson AFB, Ohio 45433, 1981.

[15] D.E. Rumelhart, "On learning to do what you want". Talk given at CMU Connectionist Summer School, 1986a.

[16] B.W. Mel In Proc. of 8th Ann. Conf. of the Cognitive Science Soc., 562-571, (1986).

[17] D.H. Ballard, G.E. Hinton, & T.J Sejnowski, Nature, *306*, 21, (1983).

[18] R.P. Erikson, American Scientist, May-June 1984, p. 233.

[19] G.E. Hinton, J.L. McClelland, & D.E. Rumelhart. In *Parallel distributed processing: explorations in the microstructure of cognition, vol. 1*, D.E. Rumelhart, J.L. McClelland, Eds., (Bradford, Cambridge, 1986), p. 77.

[20] D.H. Ackley, G.E. Hinton, & T.J. Sejnowski, Cognitive Science, *9*, 147, (1985).

[21] G.E. Hinton & T.J. Sejnowski. In *Parallel distributed processing: explorations in the microstructure of cognition, vol. 1*, D.E. Rumelhart, J.L. McClelland, Eds., (Bradford, Cambridge, 1986), p. 282.

[22] G.J. Tesauro, Complex Systems, *1*, 367, (1987).

[23] S.E. Hampson & D.J. Volper, Biological Cybernetics, *56*, 121, (1987).

[24] D.E. Rumelhart, G.E. Hinton, & J.L. McClelland. In *Parallel distributed processing: explorations in the microstructure of cognition, vol. 1*, D.E. Rumelhart, J.L. McClelland, Eds., (Bradford, Cambridge, 1986), p. 3.

[25] R.S. Michalski, J.G. Carbonell, & T.M. Mitchell, Eds., *Machine learning: an artificial intelligence approach*, Vols. I and II, (Morgan Kaufman, Los Altos, 1986).

[26] S. Omohundro, Complex Systems, *1*, 273, (1987).

[27] Paul, R. *Robot manipulators: mathematics, programming, and control.*, (MIT Press, Cambridge, 1981).

STABILITY RESULTS FOR NEURAL NETWORKS

A. N. Michel[1] , J. A. Farrell[1] , and W. Porod[2]

Department of Electrical and Computer Engineering
University of Notre Dame
Notre Dame, IN 46556

ABSTRACT

In the present paper we survey and utilize results from the qualitative theory of large scale interconnected dynamical systems in order to develop a qualitative theory for the Hopfield model of neural networks. In our approach we view such networks as an interconnection of many single neurons. Our results are phrased in terms of the qualitative properties of the individual neurons and in terms of the properties of the interconnecting structure of the neural networks. Aspects of neural networks which we address include asymptotic stability, exponential stability, and instability of an equilibrium; estimates of trajectory bounds; estimates of the domain of attraction of an asymptotically stable equilibrium; and stability of neural networks under structural perturbations.

INTRODUCTION

In recent years, neural networks have attracted considerable attention as candidates for novel computational systems[1-3]. These types of large-scale dynamical systems, in analogy to biological structures, take advantage of distributed information processing and their inherent potential for parallel computation[4,5]. Clearly, the design of such neural-network-based computational systems entails a detailed understanding of the dynamics of large-scale dynamical systems. In particular, the stability and instability properties of the various equilibrium points in such networks are of interest, as well as the extent of associated domains of attraction (basins of attraction) and trajectory bounds.

In the present paper, we apply and survey results from the qualitative theory of large scale interconnected dynamical systems[6-9] in order to develop a qualitative theory for neural networks. We will concentrate here on the popular Hopfield model[3], however, this type of analysis may also be applied to other models. In particular, we will address the following problems: (i) determine the stability properties of a given equilibrium point; (ii) given that a specific equilibrium point of a neural network is asymptotically stable, establish an estimate for its domain of attraction; (iii) given a set of initial conditions and external inputs, establish estimates for corresponding trajectory bounds; (iv) give conditions for the instability of a given equilibrium point; (v) investigate stability properties under structural perturbations. The present paper contains local results. A more detailed treatment of local stability results can be found in Ref. 10, whereas global results are contained in Ref. 11.

In arriving at the results of the present paper, we make use of the method of analysis advanced in Ref. 6. Specifically, we view high dimensional neural network as an

[1]The work of A. N. Michel and J. A. Farrell was supported by NSF under grant ECS84-19918.
[2]The work of W. Porod was supported by ONR under grant N00014-86-K-0506.

interconnection of individual subsystems (neurons). This *interconnected systems view-point* makes our results distinct from others derived in the literature[1,12]. Our results are phrased in terms of the qualitative properties of the free subsystems (individual neurons, disconnected from the network) and in terms of the properties of the interconnecting structure of the neural network. As such, these results may constitute useful design tools. This approach makes possible the systematic analysis of high dimensional complex systems and it frequently enables one to circumvent difficulties encountered in the analysis of such systems by conventional methods.

The structure of this paper is as follows. We start out by defining the Hopfield model and we then introduce the interconnected systems viewpoint. We then present representative stability results, including estimates of trajectory bounds and of domains of attraction, results for instability, and conditions for stability under structural perturbations. Finally, we present concluding remarks.

THE HOPFIELD MODEL FOR NEURAL NETWORKS

In the present paper we consider neural networks of the Hopfield type[3]. Such systems can be represented by equations of the form

$$\dot{u}_i = -b_i u_i + \sum_{j=1}^{N} A_{ij} \, G_j(u_j) + U_i(t), \quad for \ \ i = 1, \ldots, N, \tag{1}$$

where $A_{ij} = \frac{T_{ij}}{C_i}, U_i(t) = \frac{I_i(t)}{C_i}$ and $b_i = \frac{1}{\tau_i C_i}$. As usual, $C_i > 0, T_{ij} = \frac{1}{R_{ij}}, R_{ij} \epsilon R = (-\infty, \infty), \frac{1}{\tau_i} = \frac{1}{R_i} + \sum_{j=1}^{N} |T_{ij}|, \ R_i > 0, I_i : R^+ = [0, \infty) \rightarrow R, I_i$ is continuous, $\dot{u}_i = \frac{du_i}{dt}, G_i : R \rightarrow (-1, 1), G_i$ is continuously differentiable and strictly monotonically increasing (i.e., $G_i(u_i') > G_i(u_i'')$ if and only if $u_i' > u_i''$), $u_i G_i(u_i) > 0$ for all $u_i \neq 0$, and $G_i(0) = 0$. In (1), C_i denotes capacitance, R_{ij} denotes resistance (possibly including a sign inversion due to an inverter), $G_i(\cdot)$ denotes an amplifier nonlinearity, and $I_i(\cdot)$ denotes an external input.

In the literature it is frequently assumed that $T_{ij} = T_{ji}$ for all $i, j = 1, \ldots, N$ and that $T_{ii} = 0$ for all $i = 1, \ldots, N$. We will make these assumptions only when explicitly stated.

We are interested in the qualitative behavior of solutions of (1) near equilibrium points (rest positions where $\dot{u}_i \equiv 0, \ for \ i = 1, \ldots, N$). By setting the external inputs $U_i(t), \ i = 1, \ldots, N,$ equal to zero, we define $u^* = [u_1^*, \ldots, u_N^*]^T \epsilon R^N$ to be an *equilibrium* for (1) provided that $-b_i u_i^* + \sum_{j=1}^{N} A_{ij} \, G_j(u_j^*) = 0, \ for \ i = 1, \ldots, N$. The locations of such equilibria in R^N are determined by the interconnection pattern of the neural network (i.e., by the parameters $A_{ij}, i, j = 1, \ldots, N$) as well as by the parameters b_i and the nature of the nonlinearities $G_i(\cdot), i = 1, \ldots, N$.

Throughout, we will assume that a given equilibrium u^* being analyzed is an *isolated* equilibrium for (1), i.e., there exists an $r > 0$ such that in the neighborhood $B(u^*, r) = \{(u - u^*) \epsilon R^N : |u - u^*| < r\}$ no equilibrium for (1), other than $u = u^*$, exists.

When analyzing the stability properties of a given equilibrium point, we will be able to assume, without loss of generality, that this equilibrium is located at the origin $u = 0$ of R^N. If this is not the case, a trivial transformation can be employed which shifts the equilibrium point to the origin and which leaves the structure of (1) the same.

INTERCONNECTED SYSTEMS VIEWPOINT

We will find it convenient to view system (1) as an interconnection of N *free subsystems* (or *isolated subsystems*) described by equations of the form

$$\dot{p}_i = -b_i p_i + A_{ii}\, G_i(p_i) + U_i(t). \tag{2}$$

Under this viewpoint, the *interconnecting structure* of the system (1) is given by

$$G_i(x_1,\ldots,x_n) \triangleq \sum_{\substack{j=1 \\ i \neq j}}^{N} A_{ij} G_j(x_j), \quad i = 1,\ldots,N. \tag{3}$$

Following the method of analysis advanced in[6], we will establish stability results which are phrased in terms of the qualitative properties of the free subsystems (2) and in terms of the properties of the interconnecting structure given in (3). This method of analysis makes it often possible to circumvent difficulties that arise in the analysis of complex high-dimensional systems. Furthermore, results obtained in this manner frequently yield insight into the dynamic behavior of systems in terms of system components and interconnections.

GENERAL STABILITY CONDITIONS

We demonstrate below an example of a result for exponential stability of an equilibrium point. The principal Lyapunov stability results for such systems are presented, e.g., in Chapter 5 of Ref. 7.

We will utilize the following hypotheses in our first result.

(A-1) For system (1), the external inputs are all zero, i.e.,

$$U_i(t) \equiv 0, \quad i = 1,\ldots,N.$$

(A-2) For system (1), the interconnections satisfy the estimate

$$x_i A_{ij}\, G_j(x_j) \leq x_i\, a_{ij} x_j$$

for all $|x_i| < r_i$, $|x_j| < r_j$, $i,j = 1,\ldots,N$, where the a_{ij} are real constants.

(A-3) There exists an N-vector $\alpha > 0$ (i.e., $\alpha^T = (\alpha_1,\ldots,\alpha_N)$ and $\alpha_i > 0$, *for all* $i = 1,\ldots,N$) such that the *test matrix* $S = [s_{ij}]$

$$s_{ij} = \begin{cases} \alpha_i(-b_i + a_{ii}), & i = j \\ (\alpha_i\, a_{ij} + \alpha_j\, a_{ji})/2, & i \neq j \end{cases}$$

is negative definite, where the b_i are defined in (1) and the a_{ij} are given in (A-2).

We are now in a position to state and prove the following result.

Theorem 1 *The equilibrium $x = 0$ of the neural network (1) is* **exponentially stable** *if hypotheses (A-1), (A-2) and (A-3) are satisfied.*

Proof. For (1) we choose the Lyanpunov function

$$v(x) = \sum_{i=1}^{N} \frac{1}{2} \alpha_i x_i^2 \tag{4}$$

where the α_i are given in (A-3). This function is clearly positive definite. The time derivative of v along the solutions of (1) is given by

$$Dv_{(1)}(x) = \sum_{i=1}^{N} \frac{1}{2} \alpha_i (2x_i)[-b_i x_i + \sum_{j=1}^{N} A_{ij} \, G_j(x_j)]$$

where (A-1) has been invoked. In view of (A-2) we have

$$
\begin{aligned}
Dv_{(1)}(x) &\leq \sum_{i=1}^{N} \alpha_i (-b_i x_i^2 + x_i \sum_{j=1}^{N} a_{ij} x_j) \\
&= x^T R x \quad \text{for all } |x|_2 < r
\end{aligned}
$$

where $r = \min_i(r_i)$, $|x|_2 = \left(\sum_{i=1}^{N} x_i^2 \right)^{1/2}$, and the matrix $R = [r_{ij}]$ is given by

$$
r_{ij} = \begin{cases} \alpha_i(-b_i + a_{ii}), & i = j \\ \alpha_i \, a_{ij}, & i \neq j. \end{cases}
$$

But it follows that

$$x^T R x = x^T \left(\frac{R + R^T}{2} \right) x = x^T S x \leq \lambda_M(S) \, |x|_2^2 \tag{5}$$

where S is the matrix given in (A-3) and $\lambda_M(S)$ denotes the largest eigenvalue of the real symmetric matrix S. Since S is by assumption negative definite, we have $\lambda_M(S) < 0$. It follows from (4) and (5) that in some neighborhood of the origin $x = 0$, we have $c_1 |x|_2^2 \leq v(x) \leq c_2 |x|_2^2$ and $Dv_{(1)}(x) \leq -c_3 |x|_2^2$, where $c_1 = \frac{1}{2} \min_i \alpha_i > 0$, $c_2 = \frac{1}{2} \max_i \alpha_i > 0$, and $c_3 = -\lambda_M(S) > 0$. Hence, the equilibrium $x = 0$ of the neural network (1) is exponentially stable (c.f. Theorem 9.10 in Ref. 7).

Consistent with the philosophy of viewing the neural network (1) as an interconnection of N free subsystems (2), we think of the Lyapunov function (4) as consisting of a weighted sum of Lyapunov functions for each free subsystem (2) (with $U_i(t) \equiv 0$). The weighting vector $\alpha > 0$ provides flexibility to emphasize the relative importance of the qualitative properties of the various individual subsystems. Hypothesis $(A - 2)$ provides a measure of interaction between the various subsystems (3). Furthermore, it is emphasized that Theorem 1 does not require that the parameters A_{ij} in (1) form a symmetric matrix.

WEAK COUPLING CONDITIONS

The test matrix S given in hypothesis $(A - 3)$ has off-diagonal terms which may be positive or nonpositive. For the special case where the off-diagonal terms of the test matrix $S = [s_{ij}]$ are non-negative, equivalent stability results may be obtained which are much easier to apply than Theorem 1. Such results are called *weak-coupling conditions* in the literature[6,9]. The conditions $s_{ij} \geq 0$ for all $i \neq j$ may reflect properties of the system (1) or they may be the consequence of a majorization process.

In the proof of the subsequent result, we will make use of some of the properties of M- matrices (see, for example, Chapter 2 in Ref. 6). In addition we will use the following assumptions.

(A-4) For system (1), the nonlinearity $G_i(x_i)$ satisfies the sector condition

$$0 < \sigma_{i1} \leq \frac{G_i(x_i)}{x_i} \leq \sigma_{i2}, \quad for \; all \; |x_i| < r_i, \quad i = 1, \ldots, N.$$

(A-5) The successive principal minors of the $N \times N$ *test matrix* $D = [d_{ij}]$

$$d_{ij} = \begin{cases} \frac{b_i}{\sigma_{i2}} - A_{ii}, & i = j \\ -|A_{ij}|, & i \neq j \end{cases}$$

are all positive where, the b_i and A_{ij} are defined in (1) *and σ_{i2} is defined in* $(A-4)$.

Theorem 2 *The equilibrium $x = 0$ of the neural network (1) is* **asymptotically stable** *if hypotheses (A-1), (A-4) and (A-5) are true.*

Proof. The proof proceeds[10] along lines similar to the one for Theorem 1, this time with the following Lyapunov function

$$v(x) = \sum_{i=1}^{N} \alpha_i |x_i|. \tag{6}$$

The above Lyapunov function again reflects the interconnected nature of the whole system. Note that this Lyapunov function may be viewed as a generalized Hamming distance of the state vector from the origin.

ESTIMATES OF TRAJECTORY BOUNDS

In general, one is not only interested in questions concerning the stability of an equilibrium of the system (1), but also in performance. One way of assessing the qualitative properties of the neural system (1) is by investigating solution bounds near an equilibrium of interest. We present here such a result by assuming that the hypotheses of Theorem 2 are satisfied.

In the following, we will not require that the external inputs $U_i(t)$, $i = 1, \ldots, N$ be zero. However, we will need to make the additional assumptions enumerated below.

(A-6) Assume that there exist $\lambda_i > 0$, *for* $i = 1, \ldots, N$, and an $\epsilon > 0$ such that

$$\left(\frac{b_i}{\sigma_{i2}} - A_{ii}\right) - \sum_{\substack{j=1 \\ i \neq j}}^{N} \left(\frac{\lambda_j}{\lambda_i}\right) |A_{ji}| \geq \epsilon > 0, \quad i = 1, \ldots, N$$

where b_i and A_{ij} are defined in (1) and σ_{i2} is defined in (A-4).

(A-7) Assume that for system (1),

$$\sum_{i=1}^{N} \lambda_i |U_i(t)| \leq k \quad \text{for all} \quad t \geq 0$$

for some constant $k > 0$ where the λ_i, $i = 1, \ldots, N$ are defined in (A-6).

In the proof of our next theorem, we will make use of a comparison result. We consider a scalar comparison equation of the form $\dot{y} = G(y)$ where $y \epsilon R, G : B(r) \to R$ for some $r > 0$, and G is continuous on $B(r) = \{x \epsilon R : |x| < r\}$. We can then prove the following auxiliary theorem: Let $p(t)$ denote the maximal solution of the comparison equation with $p(t_0) = y_0 \epsilon B(r)$, $t \geq t_0 > 0$. If $r(t)$, $t \geq t_0 \geq 0$ is a continuous function such that $r(t_0) \leq y_0$, and if $r(t)$ satisfies the differential inequality $Dr(t) = \lim_{k \to 0^+} \frac{1}{k} \sup[r(t + k) - r(t)] \leq G(r(t))$ almost everywhere, then $r(t) \leq p(t)$ for $t \geq t_0 \geq 0$, for as long as both $r(t)$ and $p(t)$ exist. For the proof of this result, as well as other comparison theorems, see e.g., Refs. 6 and 7.

For the next theorem, we adopt the following notation. We let $\delta = \min_i \sigma_{i1}$ where σ_{i1} is defined in $(A - 4)$, we let $c = \epsilon \delta$, where ϵ is given in (A-6), and we let $\phi(t, t_0, x_0) = [\phi_1(t, t_0, x_0), \ldots, \phi_N(t, t_0, x_0)]^T$ denote the solution of (1) with $\phi(t_0, t_0, x_0) = x_0 = (x_{10}, \ldots, x_{N0})^T$ for some $t_0 \geq 0$.

We are now in a position to prove the following result, which provides bounds for the solution of (1).

Theorem 3 *Assume that hypotheses (A-6) and (A-7) are satisfied. Then*

$$\|\phi(t, t_0, x_0)\| \triangleq \sum_{i=1}^{N} \lambda_i |\phi_i(t, t_0, x_0)| \leq \left(\alpha - \frac{k}{c}\right)e^{-c(t-t_0)} + \frac{k}{c}, \quad t \geq t_0 \geq 0$$

provided that $\alpha > k/c$ *and* $\|x_0\| = \sum_{i=1}^{N} \lambda_i |x_{i0}| \leq \alpha$, *where the* λ_i, $i = 1, \ldots, N$ *are given in (A-6) and* k *is given in (A-7).*

Proof. For (1) we choose the Lyapunov function

$$v(x) = \sum_{i=1}^{N} \lambda_i |x_i|. \tag{7}$$

Along the solutions of (1), we obtain

$$Dv_{(1)}(x) \leq \lambda^T Dw + \sum_{i=1}^{N} \lambda_i |U_i(t)| \tag{8}$$

where $w^T = \left[\frac{G_1(x_1)}{x_1} |x_1|, \ldots, \frac{G_N(x_N)}{x_N} |x_N| \right]$, $\lambda = (\lambda_1, \ldots, \lambda_N)^T$, and $D = [d_{ij}]$ is the test matrix given in (A-5). Note that when (A-6) is satisfied, as in the present theorem, then (A-5) is automatically satisfied. Note also that $w \geq 0$ (i.e., $w_i \geq 0$, $i = 1, \ldots, N$) and $w = 0$ if and only if $x = 0$.

Using manipulations involving (A-6), (A-7) and (8), it is easy to show that $Dv_{(1)}(x) \leq -cv(x) + k$. This inequality yields now the comparison equation $\dot{y} = -cy + k$, whose unique solution is given by

$$p(t, t_0, p_0) = \left(p_0 - \frac{k}{c} \right) e^{-c(t-t_0)} + \frac{k}{c}, \quad for \ all \ t \geq t_0.$$

If we let $r = v$, then we obtain from the comparison result

$$p(t) \geq r(t) = v(\phi(t, t_0, x_0)) = \sum_{i=1}^{N} \lambda_i |\phi_i(t, t_0, x_0)| = ||\phi(t, t_0, x_0)||,$$

i.e., the desired estimate is true, provided that $|r(t_0)| = \sum_{i=1}^{N} \lambda_i |x_{i0}| = ||x_0|| \leq \alpha$ and $\alpha > k/c$.

ESTIMATES OF DOMAINS OF ATTRACTION

Neural networks of the type considered herein have many equilibrium points. If a given equilibrium is asymptotically stable, or exponentially stable, then the extent of this stability is of interest. As usual, we assume that $x = 0$ is the equilibrium of interest. If $\phi(t, t_0, x_0)$ denotes a solution of the network (1) with $\phi(t_0, t_0, x_0) = x_0$, then we would like to know for which points x_0 it is true that $\phi(t, t_0, x_0)$ tends to the origin as $t \to \infty$. The set of all such points x_0 makes up the *domain of attraction* (the *basin of attraction*) of the equilibrium $x = 0$. In general, one cannot determine such a domain in its entirety. However, several techniques have been devised to estimate subsets of a domain of attraction. We apply one such method to neural networks, making use of Theorem 1. This technique is applicable to our other results as well, by making appropriate modifications.

We assume that the hypotheses (A-1), (A-2) and (A-3) are satisfied and for the free subsystem (2) we choose the Lyapunov function

$$v_i(p_i) = \frac{1}{2} p_i^2. \tag{9}$$

Then $Dv_{i_{(2)}}(p_i) \leq (-b_i + a_{ii})p_i^2$, $|p_i| < r_i$ for some $r_i > 0$. If (A-3) is satisfied, we must have $(-b_i + a_{ii}) < 0$ and $Dv_{i_{(2)}}(p_i)$ is negative definite over $B(r_i)$.

Let $C_{v_{0i}} = \{p_i \epsilon R : v_i(p_i) = \frac{1}{2} p_i^2 < \frac{1}{2} r_i^2 \triangleq v_{0i}\}$. Then $C_{v_{0i}}$ is contained in the domain of attraction of the equilibrium $p_i = 0$ for the free subsystem (2).

To obtain an estimate for the domain of attraction of $x = 0$ for the whole neural network (1), we use the Lyapunov function

$$v(x) = \sum_{i=1}^{N} \frac{1}{2}\alpha_i x_i^2 = \sum_{i=1}^{N} \alpha_i v_i(x_i). \tag{10}$$

It is now an easy matter to show that the set

$$C_\lambda = \{x\epsilon R^N : v(x) = \sum_{i=1}^{N} \alpha_i v_i(x_i) < \lambda\}$$

will be a subset of the domain of attraction of $x = 0$ for the neural network (1), where

$$\lambda = \min_{1\leq i\leq N}(\alpha_i v_{0i}) = \min_{1\leq i\leq N}\left(\frac{1}{2}\alpha_i r_i^2\right).$$

In order to obtain the best estimate of the domain of attraction of $x = 0$ by the present method, we must choose the α_i in an optimal fashion. The reader is referred to the literature[9,13,14] where several methods to accomplish this are discussed.

INSTABILITY RESULTS

Some of the equilibrium points in a neural network may be unstable. We present here a sample instability theorem which may be viewed as a counterpart to Theorem 2. Instability results, formulated as counterparts to other stability results of the type considered herein may be obtained by making appropriate modifications.

(A-8) For system (1), the interconnections satisfy the estimates

$$x_i A_{ii} G_i(x_i) \leq \delta_i A_{ii} x_i^2,$$
$$|x_i A_{ij} G_j(x_j)| \leq |x_i||A_{ij}|\sigma_{j2}|x_i|, \ i \neq j$$

where $\delta_i = \sigma_{i1}$ when $A_{ii} < 0$ and $\delta_i = \sigma_{i2}$ when $A_{ii} > 0$ for all $|x_i| < r_i$, and for all $|x_j| < r_j, i,j = 1,\ldots,N$.

(A-9) The successive principal minors of the $N \times N$ test matrix $D = [d_{ij}]$ given by

$$d_{ij} = \begin{cases} \sigma_i, & i = j \\ -|A_{ij}|, & i \neq j \end{cases}$$

are positive, where $\sigma_i = \frac{b_i}{\sigma_{i2}} - A_{ii}$ when $i\epsilon F_s$ (i.e., stable subsystems) and $\sigma_i = -\frac{b_i}{\sigma_{i1}} + A_{ii}$ when $i\epsilon F_u$ (i.e., unstable subsystems) with $F = F_s \cup F_u$ and $F = \{1,\ldots,N\}$ and $F_u \neq \phi$.

We are now in a position to prove the following result.

Theorem 4 *The equilibrium $x = 0$ of the neural network (1) is* **unstable** *if hypotheses (A-1), (A-8) and (A-9) are satisfied. If in addition, $F_s = \phi$ (ϕ denotes the empty set), then the equilibrium $x = 0$ is* **completely unstable**.

Proof. We choose the Lyapunov function

$$v(x) = \sum_{i \epsilon F_u} \alpha_i(-|x_i|) + \sum_{i \epsilon F_s} \alpha_i |x_i| \tag{11}$$

where $\alpha_i > 0$, $i = 1, \ldots, N$. Along the solutions of (1) we have (following the proof of Theorem 2), $Dv_{(1)}(x) \le -\alpha^T Dw$ for all $x \epsilon B(r)$, $r = \min_i r_i$ where $\alpha^T = (\alpha_1, \ldots, \alpha_N)$, D is defined in (A-9), and $w^T = \left[\frac{G_1(x_1)}{x_1} |x_1|, \ldots, \frac{G_N(x_N)}{x_N} |x_N| \right]$. We conclude that $Dv_{(1)}(x)$ is negative definite over $B(r)$. Since every neighborhood of the origin $x = 0$ contains at least one point x' where $v(x') < 0$, it follows that the equilibrium $x = 0$ for (1) is unstable. Moreover, when $F_s = \phi$, then the function $v(x)$ is negative definite and the equilibrium $x = 0$ of (1) is in fact completely unstable (c.f. Chapter 5 in Ref. 7).

STABILITY UNDER STRUCTURAL PERTURBATIONS

In specific applications involving adaptive schemes for learning algorithms in neural networks, the interconnection patterns (and external inputs) are changed to yield an evolution of different sets of desired asymptotically stable equilibrium points with appropriate domains of attraction. The present diagonal dominance conditions (see, e.g., hypothesis (A-6)) can be used as constraints to guarantee that the desired equilibria always have the desired stability properties.

To be more specific, we assume that a given neural network has been designed with a set of interconnections whose strengths can be varied from zero to some specified values. We express this by writing in place of (1),

$$\dot{x}_i = -b_i x_i + \sum_{j=1}^{N} \theta_{ij} A_{ij} G_j(x_j) + U_i(t), \quad for \ i = 1, \ldots, N, \tag{12}$$

where $0 \le \theta_{ij} \le 1$. We also assume that in the given neural network things have been arranged in such a manner that for some given desired value $\Delta > 0$, it is true that $\Delta = \min_i \left(\frac{b_i}{\sigma_{i2}} - \theta_{ii} A_{ii} \right)$. From what has been said previously, it should now be clear that if $U_i(t) \equiv 0$, $i = 1, \ldots, N$ and if the diagonal dominance conditions

$$\Delta - \sum_{\substack{j=1 \\ i \ne j}}^{N} \left(\frac{\lambda_j}{\lambda_i} \right) |\theta_{ij} A_{ij}| > 0, \quad for \ i = 1, \ldots, N \tag{13}$$

are satisfied for some $\lambda_i > 0$, $i = 1, \ldots, N$, then the equilibrium $x = 0$ for (12) will be asymptotically stable. It is important to recognize that condition (13) constitutes a single stability condition for the neural network under structural perturbations. Thus, the strengths of interconnections of the neural network may be rearranged in any manner to achieve some desired set of equilibrium points. If (13) is satisfied, then these equilibria will be asymptotically stable. (Stability under structural perturbations is nicely surveyed in Ref. 15.)

CONCLUDING REMARKS

In the present paper we surveyed and applied results from the qualitative theory of large scale interconnected dynamical systems in order to develop a qualitative theory for neural networks of the Hopfield type. Our results are local and use as much information as possible in the analysis of a given equilibrium. In doing so, we established criteria for the exponential stability, asymptotic stability, and instability of an equilibrium in such networks. We also devised methods for estimating the domain of attraction of an asymptotically stable equilibrium and for estimating trajectory bounds for such networks. Furthermore, we showed that our stability results are applicable to systems under structural perturbations (e.g., as experienced in neural networks in adaptive learning schemes).

In arriving at the above results, we viewed neural networks as an interconnection of many single neurons, and we phrased our results in terms of the qualitative properties of the free single neurons and in terms of the network interconnecting structure. This viewpoint is particularly well suited for the study of hierarchical structures which naturally lend themselves to implementations[16] in VLSI. Furthermore, this type of approach makes it possible to circumvent difficulties which usually arise in the analysis and synthesis of complex high dimensional systems.

REFERENCES

[1] For a review, see, *Neural Networks for Computing*, J. S. Denker, Editor, American Institute of Physics Conference Proceedings **151**, Snowbird, Utah, 1986.

[2] J. J. Hopfield and D. W. Tank, *Science* **233**, 625 (1986).

[3] J. J. Hopfield, *Proc. Natl. Acad. Sci. U.S.A.* **79**, 2554 (1982), and *ibid.* **81**, 3088 (1984).

[4] G. E. Hinton and J. A. Anderson, Editors, *Parallel Models of Associative Memory*, Erlbaum, 1981.

[5] T. Kohonen, *Self-Organization and Associative Memory*, Springer-Verlag, 1984.

[6] A. N. Michel and R. K. Miller, *Qualitative Analysis of Large Scale Dynamical Systems*, Academic Press, 1977.

[7] R. K. Miller and A. N. Michel, *Ordinary Differential Equations*, Academic Press, 1982.

[8] I. W. Sandberg, *Bell System Tech. J.* **48**, 35 (1969).

[9] A. N. Michel, *IEEE Trans. on Automatic Control* **28**, 639 (1983).

[10] A. N. Michel, J. A. Farrell, and W. Porod, submitted for publication.

[11] J.-H. Li, A. N. Michel, and W. Porod, *IEEE Trans. Circ. and Syst.*, in press.

[12] G. A. Carpenter, M. A. Cohen, and S. Grossberg, *Science* **235**, 1226 (1987).

[13] M. A. Pai, *Power System Stability*, Amsterdam, North Holland, 1981.

[14] A. N. Michel, N. R. Sarabudla, and R. K. Miller, *Circuits, Systems and Signal Processing* **1**, 171 (1982).

[15] Lj. T. Grujic, A. A. Martynyuk and M. Ribbens-Pavella, *Stability of Large-Scale Systems Under Structural and Singular Perturbations*, Nauka Dumka, Kiev, 1984.

[16] D. K. Ferry and W. Porod, *Superlattices and Microstructures* **2**, 41 (1986).

564

PROGRAMMABLE SYNAPTIC CHIP FOR
ELECTRONIC NEURAL NETWORKS

A. Moopenn, H. Langenbacher, A.P. Thakoor, and S.K. Khanna
Jet Propulsion Laboratory
California Institute of Technology
Pasadena, CA 91009

ABSTRACT

A binary synaptic matrix chip has been developed for electronic neural networks. The matrix chip contains a programmable 32X32 array of "long channel" NMOSFET binary connection elements implemented in a 3-um bulk CMOS process. Since the neurons are kept off-chip, the synaptic chip serves as a "cascadable" building block for a multi-chip synaptic network as large as 512X512 in size. As an alternative to the programmable NMOSFET (long channel) connection elements, tailored thin film resistors are deposited, in series with FET switches, on some CMOS test chips, to obtain the weak synaptic connections. Although deposition and patterning of the resistors require additional processing steps, they promise substantial savings in silcon area. The performance of a synaptic chip in a 32-neuron breadboard system in an associative memory test application is discussed.

INTRODUCTION

The highly parallel and distributive architecture of neural networks offers potential advantages in fault-tolerant and high speed associative information processing. For the past few years, there has been a growing interest in developing electronic hardware to investigate the computational capabilities and application potential of neural networks as well as their dynamics and collective properties[1-5]. In an electronic hardware implementation of neural networks[6,7], the neurons (analog processing units) are represented by threshold amplifiers and the synapses linking the neurons by a resistive connection network. The synaptic strengths between neurons (the electrical resistance of the connections) represent the stored information or the computing function of the neural network.

Because of the massive interconectivity of the neurons and the large number of the interconnects required with the increasing number of neurons, implementation of a synaptic network using current LSI/VLSI technology can become very difficult. A synaptic network based on a multi-chip architecture would lessen this difficulty. We have designed, fabricated, and successfully tested CMOS-based programmable synaptic chips which could serve as basic "cascadable" building blocks for a multi-chip electronic neural network. The synaptic chips feature complete programmability of 1024, (32X32) binary synapses. Since the neurons are kept off-chip, the synaptic chips can be connected in parallel, to obtain multiple grey levels of the connection strengths, as well as "cascaded" to form larger synaptic arrays for an expansion to a 512-neuron system in a feedback or feed-forward architecture. As a research tool, such a system would offer a significant speed improvement over conventional software-based neural network simulations since convergence times for the parallel hardware system would be significantly smaller.

In this paper, we describe the basic design and operation of synaptic CMOS chips incorporating MOSFET's as binary connection elements. The design and fabrication of synaptic test chips with tailored thin film resistors as ballast resistors for controlling power dissipation are also described. Finally, we describe a synaptic chip-based 32-neuron breadboard system in a feedback configuration and discuss its performance in an associative memory test application.

BINARY SYNAPTIC CMOS CHIP WITH MOSFET CONNECTION ELEMENTS

There are two important design requirements for a binary connection element in a high density synaptic chip. The first requirement is that the connection in the ON state should be "weak" to ensure low overall power dissipation. The required degree of "weakness" of the ON connection largely depends on the synapse density of the chip. If, for example, a synapse density larger than 1000 per chip is desired, a dynamic resistance of the ON connection should be greater than ~100 K-ohms. The second requirement is that to obtain grey scale synapses with up to four bits of precision from binary connections, the consistency of the ON state connection resistance must be better than +/-5 percent, to ensure proper threshold operation of the neurons. Both of the requirements are generally difficult to satisfy simultaneously in conventional VLSI CMOS technology. For example, doped-polysilicon resistors could be used to provide the weak connections, but they are difficult to fabricate with a resistance uniformity of better than 5 percent.

We have used NMOSFET's as connection elements in a multi-chip synaptic network. By designing the NMOSFET's with long channel, both the required high uniformity and high ON state resistance have been obtained. A block diagram of a binary synaptic test chip incorporating NMOSFET's as programmable connection elements is shown in Fig. 1. A photomicrograph of the chip is shown in Fig. 2. The synaptic chip was fabricated through MOSIS (MOS Implementation Service) in a 3-micron, bulk CMOS, two-level metal, P-well technology. The chip contains 1024 synaptic cells arranged in a 32X32 matrix configuration. Each cell consists of a long channel NMOSFET connected in series with another NMOSFET serving as a simple ON/OFF switch. The state of the FET switch is controlled by the output of a latch which can be externally addressed via the ROW/COL address decoders. The 32 analog input lines (from the neuron outputs) and 32 analog output lines (to the neuron inputs) allow a number of such chips to be connected together to form larger connection matrices with up to 4-bit planes.

The long channel NMOSFET can function as either a purely resistive or a constant current source connection element, depending

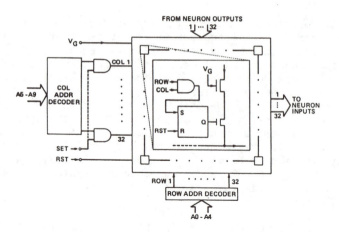

Figure 1. Block diagram of a 32X32 binary synnaptic chip with long channel NMOSFETs as connection elements.

Figure 2. Photomicrographs of a 32X32 binary connection CMOS chip. The blowup on the right shows several synaptic cells; the "S"-shape structures are the long-channel NMOSFETs.

on whether analog or binary output neurons are used. As a resistive connection, the NMOSFET's must operate in the linear region of the transistor's drain I-V characteristics. In the linear region, the channel resistance is approximately given by[8]

$$R_{ON} = (1/K) \ (L/W) \ (V_G - V_{TH})^{-1}.$$

Here, K is a proportionality constant which depends on process parameters, L and W are the channel length and width respectively, V_G is the gate voltage, and V_{TH} is the threshold voltage. The transistor acts as a linear resistor provided the voltage across the channel is much less than the difference of the gate and threshold voltages, and thus dictates the operating voltage range of the connection. The NMOSFET's presently used in our synaptic chip design have a channel length of 244 microns and width of 12 microns. At a gate voltage of 5 volts, a channel resistance of about 200 K-ohms was obtained over an operating voltage range of 1.5 volts. The consistency of the transistor I-V characteristics has been verified to be within +/-3 percent in a single chip and +/-5 percent for chips from different fabrication runs. In the latter case, the transistor characteristics in the linear region can be further matched to within +/-3% by the fine adjustment of their common gate bias.

With two-state neurons, current source connections may be used by operating the transistor in the saturation mode. Provided the voltage across the channel is greater than $(V_G - V_{TH})$, the transistor behaves almost as a constant current source with the saturation current given approximately by[8]

$$I_{ON} = K \ (W/L) \ (V_G - V_{TH})^2 \ .$$

With the appropriate selection of L, W, and V_G, it is possible to obtain ON-state currents which vary by two orders of magnitude in values. Figure 3 shows a set of measured I-V curves for a NMOSFET with the channel dimensions, L= 244 microns and W=12 microns and applied gate voltages from 2 to 4.5 volts. To ensure constant current source operation, the neuron's ON-state output should be greater than 3.5 volts. A consistency of the ON-state currents to within +/-5 percent has similarly been observed in a set of chip samples. With current source connections therefore, quantized grey scale synapses with up to 16 grey levels (4 bits) can be realized using a network of binary weighted current sources.

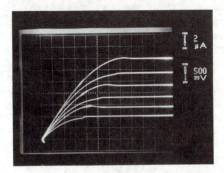

Figure 3. I-V characteristics of an NMOSFET connection element. Channel dimension: L=244 um, W=12um

For proper operation of the NMOSFET connections, the analog output lines (to neuron inputs) should always be held close to ground potential. Moreover, the voltages at the analog input lines must be at or above ground potential. Since the current normally flows from the analog input to the output, the NMOSFET's may be used as either all excitatory or inhibitory type connections. However, the complementary connection function can be realized using long channel PMOSFET's in series with PMOSFET switches. For a PMOSFET connection, the voltage of an analog input line would be at or below ground. Furthermore, due to the difference in the mobilites of electrons and holes in the channel, a PMOSFET used as a resistive connection has a channel resistance about twice as large as an NMOSFET with the same channel dimension. This fact results in a subtantial reduction in the size of PMOSFET needed.

THIN FILM RESISTOR CONNECTIONS

The use of MOSFET's as connection elements in a CMOS synaptic matrix chip has the major advantage that the complete device can be readily fabricated in a conventional CMOS production run. However, the main disadvantages are the large area (required for the long channel) for the MOSFET's connections and their non-symmetrical inhibitory/excitatory functional characteristics. The large overall gate area not only substantially limits the number of synapses that can be fabricated on a single chip, but the transistors are more susceptible to processing defects which can lead to excessive gate leakage and thus reduce chip yield considerably. An alternate approach is simply to use resistors in place of MOSFET's. We have investigated one such approach where thin film resistors are deposited on top of the passivation layer of CMOS-processed chips as an additional special processing step to the normal CMOS fabrication run. With an appropriate choice of resistive materials, a dense array of resistive connections with highly uniform resistance of up to 10 M-ohms appears feasible.

Several candidate materials, including a cermet based on platinum/aluminum oxide, and amorphous semiconductor/metal alloys such as a-Ge:Cu and a-Ge:Al, have been examined for their applicability as thin film resistor connections. These materials are of particular interest since their resistivity can easily be tailored in the desired semiconducting range of 1-10 ohm-cm by controlling the metal content[9]. The a-Ge/metal films are deposited by thermal evaporation of presynthesized alloys of the desired composition in high vacuum, whereas platinum/aluminum oxide films are deposited by co-sputtering from platinum and aluminum oxide targets in a high purity argon and oxygen gas mixture. Room temperature resistivities in the 0.1 to 100 ohm-cm range have been obtained by varying the metal content in these materials. Other factors which would also determine their suitability include their device processing and material compatibilities and their stability with time, temperature, and extended application of normal operating electric current. The temperature coefficient of resistance (TCR) of these materials at room temperature has been measured to be in the 2000 to 6000 ppm range. Because of their relatively high TCR's, the need for weak connections to reduce the effect of localized heating is especially important here. The a-Ge/metal alloy films are observed to be relatively stable with exposure to air for temperatures below 130° C.

The platinum/aluminum oxide film stabilize with time after annealing in air for several hours at 130° C.

Sample test arrays of thin film resistors based on the described materials have been fabricated to test their consistency. The resistors, with a nominal resistance of 1 M-ohm, were deposited on a glass substrate in a 40X40 array over a 0.4cm by 0.4cm area. Variation in the measured resistance in these test arrays has been found to be from +/- 2-5 percent for all three materials. Smaller test arrays of a-Ge:Cu thin film resistors on CMOS test chips have also been fabricated. A photo-micrograph of a CMOS synaptic test chip containing a 4X4 array of a-Ge:Cu thin film resistors is shown in Fig. 4. Windows in the passivation layer of silicon nitride (SiN) were opened in the final processing step of a normal CMOS fabrication run to provide access to the aluminum metal for electrical contacts. A layer of resistive material was deposited and patterned by lift-off. A layer of buffer metal of platinum or nickel was then deposited by RF sputtering and also patterned by lift-off. The buffer metal pads serve as a conducting bridges for connecting the aluminum electrodes to the thin film resistors. In addition to providing a reliable ohmic contact to the aluminum and resistor, it also provides conformal step coverage over the silicon nitride window edge. The resistor elements on the test chip are 100 micron long, 10 micron wide with a thickness of about 1500 angstroms and a nominal resistance of 250 K-ohms. Resistance variations from 10-20 percent have been observed in several such test arrays. The unusually large variation is largely due to the surface roughness of the chip passivation layer. As one possible solution, a thin spin-

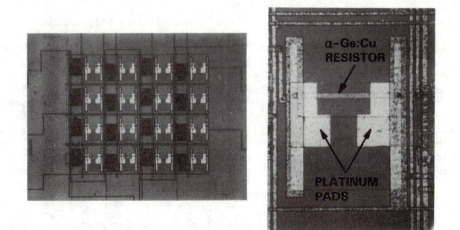

Figure 4. Photomicrographs of a CMOS synaptic test chip with a 4X4 array of a-Ge:Cu thin film resistors. The nominal resistance was 250 K-ohms.

on coating of an insulating material such as polyimide to smooth out the surface of the passivation layer prior to depositing the resistors is under investigation.

SYNAPTIC CHIP-BASED 32-NEURON BREADBOARD SYSTEM

A 32-neuron breadboard system utilizing an array of discrete neuron electronics has been fabricated to evaluate the operation of 32X32 binary synaptic CMOS chips with NMOSFET connection elements. Each neuron consists of an operational amplifier configured as a current to voltage converter (with virtual ground input) followed by a fixed-gain voltage difference amplifier. The overall time constant of the neurons is approximately 10 microseconds. The neuron array is interfaced directly to the synaptic chip in a full feedback configuration. The system also contains prompt electronics consisting of a programmable array of RC discharging circuits with a relaxation time of approximately 5 microseconds. The prompt hardware allows the neuron states to be initialized by precharging the selected capacitors in the RC circuits. A microcomputer interfaced to the breadboard system is used for programming the synaptic matrix chip, controlling the prompt electronics, and reading the neuron outputs.

The stability of the breadboard system is tested in an associative memory feedback configuration[6]. A dozen random dilute-coded binary vectors are stored using the following simplified outer-product storage scheme:

$$T_{ij} = \begin{cases} -1 & \text{if } \sum_{s} V_i^s V_j^s = 0 \\ 0 & \text{otherwise.} \end{cases}$$

In this scheme, the feedback matrix consists of only inhibitory (-1) or open (0) connections. The neurons are set to be normally ON and are driven OFF when inhibited by another neuron via the feedback matrix. The system exhibits excellent stability and associative recall performance. Convergence to a nearest stored memory in Hamming distance is always observed for any given input cue. Figure 5 shows some typical neuron output traces for a given test prompt and a set of stored memories. The top traces show the response of two neurons that are initially set ON; the bottom traces for two other neurons initially set OFF. Convergence times of 10-50 microseconds have been observed, depending on the prompt conditions, but are primarily governed by the speed of the neurons.

CONCLUSIONS

Synaptic CMOS chips containing 1024 programmable binary synapses in a 32X32 array have been designed, fabricated, and tested. These synaptic chips are designed to serve as basic building blocks for large multi-chip synaptic networks. The use of long channel MOSFET's as either resistive or current source connection elements meets the "weak" connection and consistency

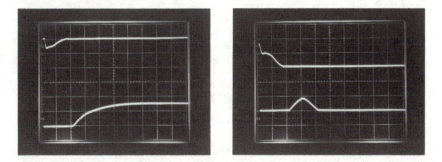

Figure 5. Typical neuron response curves for a test prompt input. (Horiz scale: 10 microseconds per div)

requirements. Alternately, CMOS-based synaptic test chips with specially deposited thin film high-valued resistors, in series with FET switches, offer an attractive approach to high density programmable synaptic chips. A 32-neuron breadboard system incorporating a 32X32 NMOSFET synaptic chip and a feedback configuration exhibits excellent stability and associative recall performance as an associative memory. Using discrete neuron array, convergence times of 10-50 microseconds have been demonstrated. With optimization of the input/output wiring layout and the use of high speed neuron electronics, convergence times can certainly be reduced to less than a microsecond.

ACKNOWLEDGEMENTS

This work was performed by the Jet Propulsion Laboratory, California Institute of Technology, and was sponsored by the Joint Tactical Fusion Program Office, through an agreement with the National Aeronautics and Space Administration. The authors would like to thank John Lambe for his invaluable suggestions, T. Duong for his assistance in the breadboard hardware development, J. Lamb and S. Thakoor for their help in the thin film resistor deposition, and R. Nixon and S. Chang for their assistance in the chip layout design.

REFERENCES

1. J. Lambe, A. Moopenn, and A.P. Thakoor, Proc. AIAA/ACM/NASA/-IEEE Computers in Aerospace V, 160 (1985)
2. A.P. Thakoor, J.L. Lamb, A. Moopenn, and S.K. Khanna, MRS Proc. 95, 627 (1987)
3. W. Hubbard, D. Schwartz, J. Denker, H.P. Graf, R. Howard, L. Jackel, B. Straughn, and D. Tennant, AIP Conf. Proc. 151, 227 (1986)
4. M.A. Sivilotti, M.R. Emerling, and C. Mead, AIP Conf. Proc. 151, 408 (1986)
5. J.P. Sage, K. Thompson, and R.S. Withers, AIP Conf. Proc. 151,

381 (1986)

6. J.J. Hopfield, Proc. Nat. Acad. Sci., 81, 3088 (1984)
7. J.J. Hopfield, Proc. Nat. Acad. Sci., 79, 2554 (1982)
8. S.M. Sze, "Semiconductor Devices-Physics and Technology," (Wiley, New York, 1985) p.205
9. J.L. Lamb, A.P. Thakoor, A. Moopenn, and S.K. Khanna, J. Vac. Sci. Tech., A 5 (4) , 1407 (1987)

BIT - SERIAL NEURAL NETWORKS

Alan F. Murray, Anthony V. W. Smith and Zoe F. Butler.
Department of Electrical Engineering, University of Edinburgh,
The King's Buildings, Mayfield Road, Edinburgh,
Scotland, EH9 3JL.

ABSTRACT

A bit - serial VLSI neural network is described from an initial architecture for a synapse array through to silicon layout and board design. The issues surrounding bit - serial computation, and analog/digital arithmetic are discussed and the parallel development of a hybrid analog/digital neural network is outlined. Learning and recall capabilities are reported for the bit - serial network along with a projected specification for a 64 - neuron, bit - serial board operating at 20 MHz. This technique is extended to a 256 (256^2 synapses) network with an update time of 3ms, using a "paging" technique to time - multiplex calculations through the synapse array.

1. INTRODUCTION

The functions a synthetic neural network may aspire to mimic are the ability to consider many solutions simultaneously, an ability to work with corrupted data and a natural fault tolerance. This arises from the parallelism and distributed knowledge representation which gives rise to gentle degradation as faults appear. These functions are attractive to implementation in VLSI and WSI. For example, the natural fault - tolerance could be useful in silicon wafers with imperfect yield, where the network degradation is approximately proportional to the non-functioning silicon area.

To cast neural networks in engineering language, a neuron is a state machine that is either "on" or "off", which in general assumes intermediate states as it switches smoothly between these extrema. The synapses *weighting* the signals from a transmitting neuron such that it is more or less excitatory or inhibitory to the receiving neuron. The set of synaptic weights determines the stable states and represents the learned information in a system.

The neural state, V_i, is related to the total neural activity stimulated by inputs to the neuron through an *activation function*, F. Neural activity is the level of excitation of the neuron and the activation is the way it reacts in a response to a change in activation. The neural output state at time t, V_i^t, is related to x_i^t by

$$V_i^t = F(x_i^t) \tag{1}$$

The activation function is a "squashing" function ensuring that (say) V_i is 1 when x_i is large and -1 when x_i is small. The neural update function is therefore straightforward:

$$x_i^{t+1} = x_i^t \; ..... + \delta \sum_{j=0}^{j=n-1} T_{ij} \, V_j^t \tag{2}$$

where δ represents the rate of change of neural activity, T_{ij} is the synaptic weight and n is the number of terms giving an n - neuron array [1].

Although the *neural* function is simple enough, in a totally interconnected n - neuron network there are n^2 synapses requiring n^2 multiplications and summations and a large number of interconnects. The challenge in VLSI is therefore to design a simple, compact synapse that can be repeated to build a VLSI neural network with manageable interconnect. In a network with fixed functionality, this is relatively straightforward. If the network is to be able to learn, however, the synaptic weights must be programmable, and therefore more complicated.

2. DESIGNING A NEURAL NETWORK IN VLSI

There are fundamentally two approaches to implementing any function in silicon - digital and analog. Each technique has its advantages and disadvantages, and these are listed below, along with the merits and demerits of bit - serial architectures in digital (synchronous) systems.

Digital vs. analog: The primary advantage of digital design for a synapse array is that digital memory is well understood, and can be incorporated easily. Learning networks are therefore possible without recourse to unusual techniques or technologies. Other strengths of a digital approach are that design techniques are advanced, automated and well understood and noise immunity and computational speed can be high. Unattractive features are that digital circuits of this complexity need to be synchronous and all states and activities are quantised, while real neural networks are asynchronous and unquantised. Furthermore, digital multipliers occupy a large silicon area, giving a low synapse count on a single chip.

The advantages of analog circuitry are that asynchronous behaviour and smooth neural activation are automatic. Circuit elements can be small, but noise immunity is relatively low and arbitrarily high precision is not possible. Most importantly, no reliable analog, non - volatile memory technology is as yet readily available. For this reason, learning networks lend themselves more naturally to digital design and implementation.

Several groups are developing neural chips and boards, and the following listing does not pretend to be exhaustive. It is included, rather, to indicate the spread of activity in this field. Analog techniques have been used to build resistor / operational amplifier networks [2, 3] similar to those proposed by Hopfield and Tank [4]. A large group at Caltech is developing networks implementing early vision and auditory processing functions using the intrinsic nonlinearities of MOS transistors in the subthreshold regime [5, 6]. The problem of implementing analog networks with electrically programmable synapses has been addressed using CCD/MNOS technology [7]. Finally, Garth [8] is developing a digital neural accelerator board ("Netsim") that is effectively a fast SIMD processor with supporting memory and communications chips.

Bit - serial vs. bit - parallel: Bit - serial arithmetic and communication is efficient for computational processes, allowing good communication within and between VLSI chips and tightly pipelined arithmetic structures. It is ideal for neural networks as it minimises the interconnect requirement by eliminating multi - wire busses. Although a bit - parallel design would be free from computational latency (delay between input and output), pipelining makes optimal use of the high bit - rates possible in serial systems, and makes for efficient circuit usage.

2.1 An asynchronous pulse stream VLSI neural network:

In addition to the digital system that forms the substance of this paper, we are developing a hybrid analog/digital network family. This work is outlined here, and has been reported in greater detail elsewhere [9, 10, 11]. The generic (logical and layout) architecture of a single network of n totally *interconnected* neurons is shown schematically in figure 1. Neurons are represented by circles, which signal their states, V_i upward into a matrix of synaptic operators. The state signals are connected to a n - bit horizontal bus running through the synaptic array, with a connection to each synaptic operator in every column. All columns have n operators (denoted by squares) and each operator adds its synaptic contribution, $T_{ij}V_j$, to the running total of activity for the neuron i at the foot of the column. The synaptic function is therefore to *multiply* the signalling neuron state, V_j, by the synaptic weight, T_{ij}, and to *add* this product to the running total. This architecture is common to both the bit - serial and pulse - stream networks.

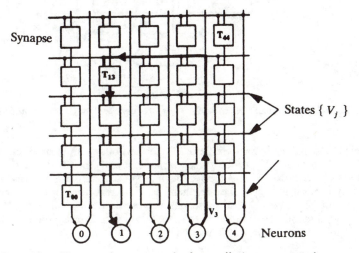

Figure 1. *Generic architecture for a network of n totally interconnected neurons.*

This type of architecture has many attractions for implementation in 2 - dimensional silicon as the summation $\sum_{j=0}^{j=n-1} T_{ij}V_j$ is distributed in space. The interconnect requirement (n inputs to each neuron) is therefore distributed through a column, reducing the need for long - range wiring. The architecture is modular, regular and can be easily expanded.

In the hybrid analog/digital system, the circuitry uses a "pulse stream" signalling method similar to that in a natural neural system. Neurons indicate their state by the presence or absence of pulses on their outputs, and synaptic weighting is achieved by time - chopping the presynaptic pulse stream prior to adding it to the postsynaptic activity summation. It is therefore asynchronous and imposes no fundamental limitations on the activation or neural state. Figure 2 shows the pulse stream mechanism in more detail. The synaptic weight is stored in digital memory local to the operator. Each synaptic operator has an excitatory and inhibitory pulse stream input and output. The resultant product of a synaptic operation, $T_{ij}V_j$, is added to the running total propagating down either the excitatory or inhibitory channel. One binary bit (the MSBit) of the stored T_{ij} determines whether the contribution is excitatory or inhibitory.

The incoming excitatory and inhibitory pulse stream inputs to a neuron are integrated to give a neural activation potential that varies smoothly from 0 to 5 V. This potential controls a feedback loop with an odd number of logic inversions and

576

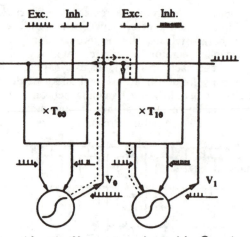

Figure 2. *Pulse stream arithmetic. Neurons are denoted by ○ and synaptic operators by □.*

thus forms a switched "ring - oscillator". If the inhibitory input dominates, the feedback loop is broken. If excitatory spikes subsequently dominate at the input, the neural activity rises to 5V and the feedback loop oscillates with a period determined by a delay around the loop. The resultant periodic waveform is then converted to a series of voltage spikes, whose pulse rate represents the neural state, V_i. Interestingly, a not dissimilar technique is reported elsewhere in this volume, although the synapse function is executed differently [12].

3. A 5 - STATE BIT - SERIAL NEURAL NETWORK

The overall architecture of the 5 - state bit - serial neural network is identical to that of the pulse stream network. It is an array of n^2 interconnected synchronous synaptic operators, and whereas the pulse stream method allowed V_j to assume all values between "off" and "on", the 5 - state network V_j is constrained to 0, ± 0.5 or ± 1. The resultant activation function is shown in Figure 3. Full digital multiplication is costly in silicon area, but multiplication of T_{ij} by $V_j = 0.5$ merely requires the synaptic weight to be right - shifted by 1 bit. Similarly, multiplication by 0.25 involves a further right - shift of T_{ij}, and multiplication by 0.0 is trivially easy. $V_j < 0$ is not problematic, as a switchable adder/subtractor is not much more complex than an adder. Five neural states are therefore feasible with circuitry that is only slightly more complex than a simple serial adder. The neural state expands from a 1 bit to a 3 bit (5 - state) representation, where the bits represent "add/subtract?", "shift?" and "multiply by 0?".

Figure 4 shows part of the synaptic array. Each synaptic operator includes an 8 bit shift register memory block holding the synaptic weight, T_{ij}. A 3 bit bus for the 5 neural states runs horizontally above each synaptic row. Single phase dynamic CMOS has been used with a clock frequency in excess of 20 MHz [13]. Details of a synaptic operator are shown in figure 5. The synaptic weight T_{ij} cycles around the shift register and the neural state V_j is present on the state bus. During the first clock cycle, the synaptic weight is multiplied by the neural state and during the second, the most significant bit (MSBit) of the resultant $T_{ij}V_j$ is sign - extended for

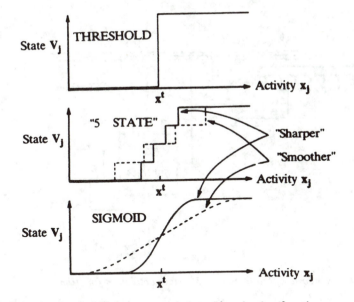

Figure 3. *"Hard - threshold", 5 - state and sigmoid activation functions.*

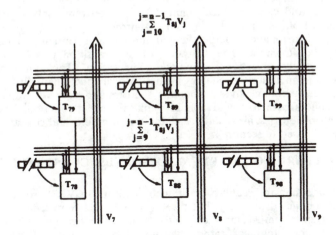

Figure 4. *Section of the synaptic array of the 5 - state activation function neural network.*

8 bits to allow for word growth in the running summation. A least significant bit (LSBit) signal running down the synaptic columns indicates the arrival of the LSBit of the x_i running total. If the neural state is ± 0.5 the synaptic weight is right shifted by 1 bit and then added to or subtracted from the running total. A multiplication of ± 1 adds or subtracts the weight from the total and multiplication by 0

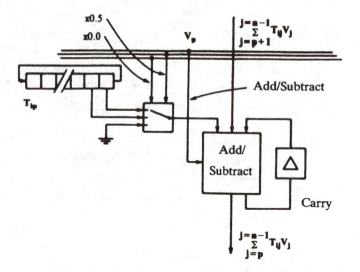

Figure 5. *The synaptic operator with a 5 - state activation function.*

does not alter the running summation.

The final summation at the foot of the column is thresholded externally according to the 5 - state activation function in figure 3. As the neuron activity x_j, increases through a threshold value x_t, ideal sigmoidal activation represents a smooth switch of neural state from -1 to 1. The 5 - state "staircase" function gives a superficially much better approximation to the sigmoid form than a (much simpler to implement) threshold function. The sharpness of the transition can be controlled to "tune" the neural dynamics for learning and computation. The control parameter is referred to as temperature by analogy with statistical functions with this sigmoidal form. High "temperature" gives a smoother staircase and sigmoid, while a temperature of 0 reduces both to the "Hopfield" - like threshold function. The effects of temperature on both learning and recall for the threshold and 5 - state activation options are discussed in section 4.

4. LEARNING AND RECALL WITH VLSI CONSTRAINTS

Before implementing the reduced - arithmetic network in VLSI, simulation experiments were conducted to verify that the 5 - state model represented a worthwhile enhancement over simple threshold activation. The "benchmark" problem was chosen for its ubiquitousness, rather than for its intrinsic value. The implications for learning and recall of the 5 - state model, the threshold (2 - state) model and smooth sigmoidal activation (∞ - state) were compared at varying temperatures with a restricted dynamic range for the weights T_{ij}. In each simulation a totally interconnected 64 node network attempted to learn 32 random patterns using the delta rule learning algorithm (see for example [14]). Each pattern was then corrupted with 25% noise and recall attempted to probe the content addressable memory properties under the three different activation options.

During learning, individual weights can become large (positive or negative). When weights are "driven" beyond the maximum value in a hardware implementation, which is determined by the size of the synaptic weight blocks, some limiting mechanism must be introduced. For example, with eight bit weight registers, the limitation is $-128 \leq T_{ij} \leq 127$. With integer weights, this can be seen to be a problem of *dynamic range*, where it is the relationship between the smallest possible weight (± 1) and the largest ($+127/-128$) that is the issue.

Results: Fig. 6 shows examples of the results obtained, studying *learning* using 5 - state activation at different temperatures, and recall using both 5 - state and threshold activation. At temperature $T=0$, the 5 - state and threshold models are degenerate, and the results identical. Increasing smoothness of activation (temperature) during learning improves the *quality* of learning regardless of the activation function used in recall, as more patterns are recognised successfully. Using 5 - state activation in recall is more effective than simple threshold activation. The effect of dynamic range restrictions can be assessed from the horizontal axis, where T_{ij}^{max} is shown. The results from these and many other experiments may be summarised as follows:-

5 - State activation vs. threshold:

1) Learning with 5 - state activation was protracted over the threshold activation, as *binary* patterns were being learnt, and the inclusion of intermediate values added extra degrees of freedom.

2) Weight sets learnt using the 5 - state activation function were "better" than those learnt via threshold activation, as the recall properties of both 5 - state and threshold networks using such a weight set were more robust against noise.

3) Full sigmoidal activation was better than 5 - state, but the enhancement was less significant than that incurred by moving from threshold → 5 - state. This suggests that the law of diminishing returns applies to addition of levels to the neural state V_j. This issue has been studied mathematically [15], with results that agree qualitatively with ours.

Weight Saturation:

Three methods were tried to deal with weight saturation. Firstly, inclusion of a decay, or "forgetting" term was included in the learning cycle [1]. It is our view that this technique *can* produce the desired weight limiting property, but in the time available for experiments, we were unable to "tune" the rate of decay sufficiently well to confirm it. Renormalisation of the weights (division to bring large weights back into the dynamic range) was very unsuccessful, suggesting that information distributed throughout the numerically small weights was being destroyed. Finally, the weights were allowed to "clip" (ie any weight outside the dynamic range was set to the maximum allowed value). This method proved very successful, as the learning algorithm adjusted the weights over which it still had control to compensate for the saturation effect. It is interesting to note that other experiments have indicated that Hopfield nets can "forget" in a different way, under different learning control, giving preference to recently acquired memories [16]. The results from the saturation experiments were:-

1) For the 32 pattern/64 node problem, integer weights with a dynamic range greater than ± 30 were necessary to give enough storage capability.

2) For weights with maximum values $T_{ij}^{max} = 50 \rightarrow 70$, "clipping" occurs, but network performance is not seriously degraded over that with an unrestricted weight set.

580

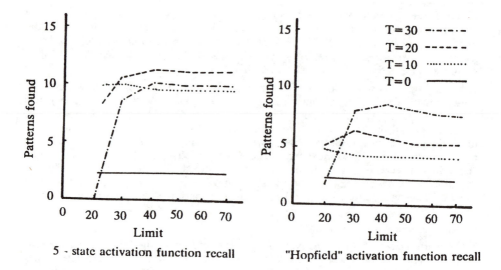

Figure 6. *Recall of patterns learned with the 5 - state activation function and subsequently restored using the 5-state and the hard - threshold activation functions. T is the "temperature", or smoothness of the activation function, and "limit" the value of T_{ij}^{max}.*

These results showed that the 5 - state model was worthy of implementation as a VLSI neural board, and suggested that 8 - bit weights were sufficient.

5. PROJECTED SPECIFICATION OF A HARDWARE NEURAL BOARD

The specification of a 64 neuron board is given here, using a 5 - state bit - serial 64 x 64 synapse array with a derated clock speed of 20 MHz. The synaptic weights are 8 bit words and the word length of the running summation x_i is 16 bits to allow for growth. A 64 synapse column has a computational latency of 80 clock cycles or bits, giving an update time of 4µs for the network. The time to load the weights into the array is limited to 60µs by the supporting RAM, with an access time of 120ns. These load and update times mean that the network is executing 1 x 10^9 operations/second, where one operation is ± $T_{ij}V_j$. This is much faster than a natural neural network, and much faster than is necessary in a hardware accelerator. We have therefore developed a "paging" architecture, that effectively "trades - off" some of this excessive speed against increased network size.

A "moving - patch" neural board: An array of the 5 - state synapses is currently being fabricated as a VLSI integrated circuit. The shift registers and the adder/subtractor for each synapse occupy a disappointingly large silicon area, allowing only a 3 x 9 synaptic array. To achieve a suitable size neural network from this array, several chips need to be included on a board with memory and control circuitry. The "moving patch" concept is shown in figure 7, where a small array of synapses is passed over a much larger n x n synaptic array.

Each time the array is "moved" to represent another set of synapses, new weights must be loaded into it. For example, the first set of weights will be T_{11} ... T_{ij} ... T_{21} ... T_{2j} to T_{jj}, the second set $T_{j+1,1}$ to T_{ss} etc.. The final weight to be loaded will be

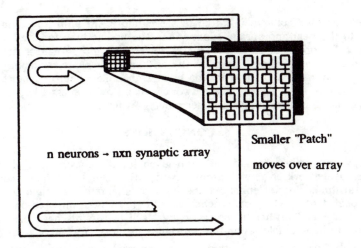

Figure 7. *The "moving patch" concept, passing a small synaptic "patch" over a larger nxn synapse array.*

T_{nn}. Static, off - the - shelf RAM is used to store the weights and the whole operation is pipelined for maximum efficiency. Figure 8 shows the board level design for the network.

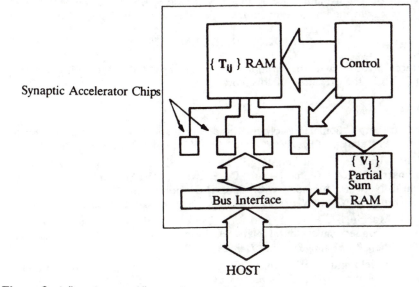

Figure 8. *A "moving patch" neural network board.*

The small "patch" that moves around the array to give n neurons comprises 4 VLSI synaptic accelerator chips to give a 6 x 18 synaptic array. The number of neurons to be simulated is 256 and the weights for these are stored in 0.5 Mb of RAM with a load time of 8ms. For each "patch" movement, the partial running summation ς

582

calculated for each column, is stored in a separate RAM until it is required to be added into the next appropriate summation. The update time for the board is 3ms giving 2×10^7 operations/second. This is slower than the 64 neuron specification, but the network is 16 times larger, as the arithmetic elements are being used more efficiently. To achieve a network of greater than 256 neurons, more RAM is required to store the weights. The network is then slower unless a larger number of accelerator chips is used to give a larger moving "patch".

6. CONCLUSIONS

A strategy and design method has been given for the construction of bit - serial VLSI neural network chips and circuit boards. Bit - serial arithmetic, coupled to a reduced arithmetic style, enhances the level of integration possible beyond more conventional digital, bit - parallel schemes. The restrictions imposed on both synaptic weight size and arithmetic precision by VLSI constraints have been examined and shown to be tolerable, using the associative memory problem as a test.

While we believe our digital approach to represent a good compromise between arithmetic accuracy and circuit complexity, we acknowledge that the level of integration is disappointingly low. It is our belief that, while digital approaches may be interesting and useful in the medium term, essentially as hardware accelerators for neural simulations, analog techniques represent the best ultimate option in 2 - dimensional silicon. To this end, we are currently pursuing techniques for analog pseudo - static memory, using standard CMOS technology. In any event, the full development of a nonvolatile analog memory technology, such as the MNOS technique [7], is key to the long - term future of VLSI neural nets that can learn.

7. ACKNOWLEDGEMENTS

The authors acknowledge the support of the Science and Engineering Research Council (UK) in the execution of this work.

References

1. S. Grossberg, "Some Physiological and Biochemical Consequences of Psychological Postulates," *Proc. Natl. Acad. Sci. USA*, vol. 60, pp. 758 - 765, 1968.

2. H. P. Graf, L. D. Jackel, R. E. Howard, B. Straughn, J. S. Denker, W. Hubbard, D. M. Tennant, and D. Schwartz, "VLSI Implementation of a Neural Network Memory with Several Hundreds of Neurons," *Proc. AIP Conference on Neural Networks for Computing, Snowbird*, pp. 182 - 187, 1986.

3. W. S. Mackie, H. P. Graf, and J. S. Denker, "Microelectronic Implementation of Connectionist Neural Network Models," *IEEE Conference on Neural Information Processing Systems, Denver*, 1987.

4. J. J. Hopfield and D. W. Tank, "Neural" Computation of Decisions in Optimisation Problems," *Biol. Cybern.*, vol. 52, pp. 141 - 152, 1985.

5. M. A. Sivilotti, M. A. Mahowald, and C. A. Mead, *Real - Time Visual Computations Using Analog CMOS Processing Arrays*, 1987. To be published

6. C. A. Mead, "Networks for Real - Time Sensory Processing," *IEEE Conference on Neural Information Processing Systems, Denver*, 1987.

7. J. P. Sage, K. Thompson, and R. S. Withers, "An Artificial Neural Network Integrated Circuit Based on MNOS/CCD Principles," *Proc. AIP Conference on Neural Networks for Computing, Snowbird*, pp. 381 - 385, 1986.

8. S. C. J. Garth, "A Chipset for High Speed Simulation of Neural Network Systems," *IEEE Conference on Neural Networks, San Diego*, 1987.

9. A. F. Murray and A. V. W. Smith, "A Novel Computational and Signalling Method for VLSI Neural Networks," *European Solid State Circuits Conference* , 1987.

10. A. F. Murray and A. J. W. Smith, "Asynchronous Arithmetic for VLSI Neural Systems," *Electronics Letters*, vol. 23, no. 12, p. 642, June, 1987.

11. A. F. Murray and A. V. W. Smith, "Asynchronous VLSI Neural Networks using Pulse Stream Arithmetic," *IEEE Journal of Solid-State Circuits and Systems*, 1988. To be published

12. M. E. Gaspar, "Pulsed Neural Networks : Hardware, Software and the Hopfield A/D Converter Example," *IEEE Conference on Neural Information Processing Systems, Denver*, 1987.

13. M. S. McGregor, P. B. Denyer, and A. F. Murray, "A Single - Phase Clocking Scheme for CMOS VLSI," *Advanced Research in VLSI : Proceedings of the 1987 Stanford Conference*, 1987.

14. D. E. Rumelhart, G. E. Hinton, and R. J. Williams, "Learning Internal Representations by Error Propagation," *Parallel Distributed Processing : Explorations in the Microstructure of Cognition*, vol. 1, pp. 318 - 362, 1986.

15. M. Fleisher and E. Levin, "The Hopfiled Model with Multilevel Neurons Models," *IEEE Conference on Neural Information Processing Systems, Denver*, 1987.

16. G. Parisi, "A Memory that Forgets," *J. Phys. A : Math. Gen.*, vol. 19, pp. L617 - L620, 1986.

PHASOR NEURAL NETWORKS

André J. Noest, N.I.B.R., NL-1105 AZ Amsterdam, The Netherlands.

ABSTRACT

A novel network type is introduced which uses unit-length 2-vectors
for local variables. As an example of its applications, associative
memory nets are defined and their performance analyzed. Real systems
corresponding to such 'phasor' models can be e.g. (neuro)biological
networks of limit-cycle oscillators or optical resonators that have
a hologram in their feedback path.

INTRODUCTION

Most neural network models use either binary local variables or
scalars combined with sigmoidal nonlinearities. Rather awkward coding
schemes have to be invoked if one wants to maintain linear relations
between the local signals being processed in e.g. associative memory
networks, since the nonlinearities necessary for any nontrivial
computation act directly on the range of values assumed by the local
variables. In addition, there is the problem of representing signals
that take values from a space with a different topology, e.g. that
of the circle, sphere, torus, etc. Practical examples of such a
signal are the orientations of edges or the directions of local optic
flow in images, or the phase of a set of (sound or EM) waves as they
arrive on an array of detectors. Apart from the fact that 'circular'
signals occur in technical as well as biological systems, there are
indications that some parts of the brain (e.g. olfactory bulb, cf.
Dr.B.Baird's contribution to these proceedings) can use limit-cycle
oscillators formed by local feedback circuits as functional building
blocks, even for signals without circular symmetry. With respect to
technical implementations, I had speculated before the conference
whether it could be useful to code information in the phase of the
beams of optical neurocomputers, avoiding slow optical switching
elements and using only (saturating) optical amplification and a

hologram encoding the (complex) 'synaptic' weight factors. At the conference, I learnt that Prof. Dana Anderson had independently developed an optical device (cf. these proceedings) that basically works this way, at least in the slow-evolution limit of the dynamic hologram. Hopefully, some of the theory that I present here can be applied to his experiment. In turn, such implementations call for interesting extensions of the present models.

BASIC ELEMENTS OF GENERAL PHASOR NETWORKS

Here I study the perhaps simplest non-scalar network by using unit-length 2-vectors (phasors) as continuous local variables. The signals processed by the network are represented in the relative phaseangles. Thus, the nonlinearities (unit-length 'clipping') act orthogonally to the range of the variables coding the information. The behavior of the network is invariant under any rigid rotation of the complete set of phasors, representing an arbitrary choice of a global reference phase. Statistical physicists will recognize the phasor model as a generalization of O_2-spin models to include vector-valued couplings.

All 2-vectors are treated algebraically as complex numbers, writing $|x|$ for the length, $/x/$ for the phase-angle, and $\bar{x}$ for the complex conjugate of a 2-vector x.

A phasor network then consists of $N \gg 1$ phasors s_i , with $|s_i|=1$, interacting via couplings c_{ij}, with $c_{ii}= 0$. The c_{ij} are allowed to be complex-valued quantities. For optical implementations this is clearly a natural choice, but it may seem less so for biological systems. However, if the coupling between two limitcycle oscillators with frequency f is mediated via a path having propagationdelay d, then that coupling in fact acquires a phaseshift of $f.d.2\pi$ radians. Thus, complex couplings can represent such systems more faithfully than the usual models which neglect propagationdelays altogether. Only 2-point couplings are treated here, but multi-point couplings c_{ijk}, etc., can be treated similarly.

The dynamics of each phasor depends only on its local field

$$h_i= \frac{1}{z} \sum_j c_{ij}s_j + n_i \quad , \qquad \text{where z is the number of inputs}$$

$c_{ij} \neq 0$ per cell and n_i is a local noise term (complex and Gaussian). Various dynamics are possible, and yield largely similar results: Continuous-time, parallel evolution: ("type A")

$$\frac{d}{dt}(/s_i/) = |h_i|.\sin(/h_i/ - /s_i/)$$

Discrete-time updating: $s_i(t+\delta t) = h_i/|h_i|$, either serially in random i-sequence ("type B"), or in parallel for all i ("type C"). The natural time scale for type-B dynamics is obtained by scaling the discrete time-interval δt as $\sim 1/N$; type-C dynamics has $\delta t=1$.

LYAPUNOV FUNCTION   (alias "ENERGY", or "HAMILTONIAN")

If one limits the attention temporarily to purely deterministic ($n_i=0$) models, then the question suggests itself whether a class of couplings exists for which one can easily find a Lyapunov function i.e. a function of the network variables that is monotonic under the dynamics. A well-known example[1] is the 'energy' of the binary and scalar Hopfield models with symmetric interactions. It turns out that a very similar function exists for phasor networks with type-A or B dynamics and a Hermitian matrix of couplings.

$$-H = \sum_i \bar{s}_i h_i = (1/z)\sum_{i,j} \bar{s}_i c_{ij} s_j$$

Hermiticity ($c_{ij} = \bar{c}_{ji}$) makes H real-valued and non-increasing in time. This can be shown as follows, e.g. for the serial dynamics (type B). Suppose, without loss of generality, that phasor i=1 is updated.

$$\text{Then} \quad -z H = z \bar{s}_1 h_1 + \sum_{i>1} \bar{s}_i c_{i1} s_1 + \sum_{i,j>1} \bar{s}_i c_{ij} s_j$$

$$= z \bar{s}_1 h_1 + s_1.\sum_{i>1} c_{i1} \bar{s}_i + \text{constant.}$$

With Hermitian couplings, H becomes real-valued, and one also has

$$\sum_{i>1} c_{i1} \bar{s}_i = \sum_{i>1} \bar{c}_{1i} \bar{s}_i = z \bar{h}_1 .$$

Thus, $- H - \text{constant} = \bar{s}_1 h_1 + s_1 \bar{h}_1 = 2 \text{Re}(s_1 \bar{h}_1)$.
Clearly, H is minimized with respect to s_1 by $s_1(t+1) = h_1/|h_1|$.
Type-A dynamics has the same Lyapunovian, but type C is more complex. The existence of Hermitian interactions and the corresponding energy function simplifies greatly the understanding and design of phasor networks, although non-Hermitian networks can still have a Lyapunov-

function, and even networks for which such a function is not readily
found can be useful, as will be illustrated later.

AN APPLICATION : ASSOCIATIVE MEMORY.

A large class of collective computations, such as optimisations
and content-addressable memory, can be realised with networks having
an energy function. The basic idea is to define the relevant penalty
function over the solution-space in the form of the generic 'energy'
of the net, and simply let the network relax to minima of this energy.
As a simple example, consider an associative memory built within the
framework of Hermitian phasor networks.

In order to store a set of patterns in the network, i.e. to make
a set of special states (at least approximatively) into attractive
fixed points of the dynamics, one needs to choose an appropriate
set of couplings. One particularly simple way of doing this is via
the phasor-analog of "Hebb's rule" (note the Hermiticity)

$$c_{ij} = \sum_{k}^{p} s_i^{(k)} \bar{s}_j^{(k)}, \text{ where } s_i^{(k)} \text{ is phasor i in learned pattern k.}$$

The rule is understood to apply only to the input-sets $\eth$i of each i.
Such couplings should be realisable as holograms in optical networks,
but they may seem unrealistic in the context of biological networks
of oscillators since the phase-shift (e.g. corresponding to a delay)
of a connection may not be changeable at will. However, the required
coupling can still be implemented naturally if e.g. a few paths with
different fixed delays exist between pairs of cells. The synaps in
each path then simply becomes the projection of the complex coupling
on the direction given by the phase of its path, i.e. it is just a
classical Hebb-synapse that computes the correlation of its pre- and
post-synaptic (imposed) signals, which now are phase-shifted versions
of the phasors $s_i^{(k)}$. The required complex c_{ij} are then realised as the
vector sum over at least two signals arriving via distinct paths with
corresponding phase-shift and real-valued synaps. Two paths suffice
if they have orthogonal phase-shifts, but random phases will do as
well if there are a reasonable number of paths.

We need to have a concise way of expressing how 'near' any state
of the net is to one or more of the stored patterns. A natural way

of doing this is via a set of p order parameters called "overlaps"

$$M_k = \frac{1}{N} \left| \sum_i^N s_i \cdot \bar{s}_i^{(k)} \right| \quad ; \quad 0 \leq M_k \leq 1 \quad ; \quad 1 \leq k \leq p .$$

Note the constraint on the p overlaps $\sum_k^p M_k^2 \leq 1$ if all the patterns are orthogonal, or merely random in the limit $N \to \infty$. This will be assumed from now on. Also, one sees at once that the whole behaviour of the network does not depend on any rigid rotation of all phasors over some angle since H, M_k, c_{ij} and the dynamics are invariant under multiplication of all s_i by a fixed phasor : $s_i' = S \cdot s_i$ with $|S|=1$.

Let us find the performance at low loading: $N, p, z \to \infty$, with $p/z \to 0$ and zero local noise. Also assume an initial overlap $m > 0$ with only one pattern, say with $k=1$. Then the local field is

$$h_i = \frac{1}{z} \sum_{j \in \partial i} s_j \cdot \sum_k^p s_i^{(k)} \cdot \bar{s}_j^{(k)} = h_i^{(1)} + h_i^{\star} , \text{ where}$$

$$h_i^{(1)} = \frac{1}{z} s_i^{(1)} \cdot \sum_{j \in \partial i} \bar{s}_j^{(1)} \cdot s_j = m_1 \cdot s_i^{(1)} \cdot S + O(1/\sqrt{z}) , \text{ with } S \neq f(i); |S|=1,$$

and

$$h_i^{\star} = \frac{1}{z} \sum_{k=2}^p s_i^{(k)} \cdot \sum_{j \in \partial i} \bar{s}_j^{(k)} \cdot s_j = O(\sqrt{(p-1)/z}) .$$

Thus, perfect recall ($M_1 = 1$) occurs in one 'pass' at loadings $p/z \to 0$.

EXACTLY SOLVABLE CASE: SPARSE and ASYMMETRIC couplings

Although it would be interesting to develop the full thermodynamics of Hermitian phasor networks with p and z of order N (analogous to the analysis of the finite-T Hopfield model by the teams of Amit[2] and van Hemmen[3]), I will analyse here instead a model with sparse, asymmetric connectivity, which has the great advantages of being exactly solvable with relative ease, and of being arguably more realistic biologically and more easily scalable technologically. In neurobiological networks a cell has up to $z \cong 10^4$ asymmetric connections, whereas $N \cong 10^{11}$. This probably has the same reason as applies to most VLSI chips, namely to alleviate wiring problems. For my present purposes, the theoretical advantage of getting some exact results is of primary interest[4].

Suppose each cell has z incoming connections from randomly selected other cells. The state of each cell at time t depends on at most z^t cells at time $t=0$. Thus, if $z^t \ll N^{1/2}$ and N large, then the respective trees of 'ancestors' of any pair cells have no cells in common[4]. In particular, if $z \sim (\log N)^x$, for any finite x, then there are no common ancestors for any finite time t in the limit $N \to \infty$. For fundamental information-theoretic reasons, one can hope to be able to store p patterns with p at most of order z for any sort of 2-point couplings. Important questions to be settled are: What are the accuracy and speed of the recall process, and how large are the basins of the attractors representing recalled patterns?

Take again initial conditions (t=0) with, say, $m(t) = M_1 > M_{>1} = 0$. Allowing again local random Gaussian (complex) noise n_i, the local fields become, in now familiar notation, $h_i = h_i^{(1)} + h_i^* + n_i$. As in the previous section, the $h_i^{(1)}$ term consists of the 'signal' $m(t).s_i$ (modulo the rigid rotation S) and a random term of variance at most $1/z$. For $p \sim z$, the h_i^* term becomes important. Being sums of $z(p-1)$ phasors oriented randomly relative to the signal, the h_i^* are independent Gaussian zero-mean 2-vectors with variance $(p-1)/z$, as p,z and $N \to \infty$. Finally, let the local noises n_i have variance r^2. Then the distribution of the $s_i(t+1)$ phasors can be found in terms of the signal m(t) and the total variance $a=(p/z)+r^2$ of the random $h_i^*+n_i$. After somewhat tedious algebraic manipulations (to be reported in detail elsewhere) one obtains the dynamic behaviour of m(t) :

$$m(t+1) = F(m(t),a) \qquad \text{for discrete parallel (type-C) dynamics,}$$
and
$$\frac{d}{dt} m(t) = F(m(t),a) - m(t) \qquad \text{for type-A or type-B dynamics ,}$$
where the function $F(m,a) =$

$$\frac{m}{4\sqrt{a\pi}} \int_{-\pi}^{+\pi} dx.(1+\cos 2x).\exp[-(m.\sin x)^2/a].(1+\mathrm{erf}[(m.\cos x)/\sqrt{a}])$$

The attractive fixed points $M^*(a) = F(M^*,a)$ represent the retrieval accuracy when the loading-plus-noise factor equals a. See figure 1. For $a \ll 1$ one obtains the expansion $1-M^*(a) = a/4 + 3a^2/32 + O(a^3)$. The recall solutions vanish continuously as $M^* \sim (a_c - a)^{1/2}$ at $a_c = \pi/4$.

One also obtains (at any t) the distribution of the phase scatter of the phasors around the ideal values occurring in the stored pattern.

590

$$P(/u_i/) = (1/2\pi).\exp(-m^2/a).(1+\sqrt{\pi}.L.\exp(L^2).(1+erf(L))\ ,$$

where $\quad L = (m/\sqrt{a}).\cos(/u_i/)\ $, and $\quad u_i = s_i\ \bar{s}_i^{(k)}$(modulo S).

Useful approximations for the high, respectively low M regimes are:

$$M \gg \sqrt{a}\ :\ P(/u_i/) = (M/\sqrt{a\pi}).\exp[-(M./u_i/)^2/a]\quad;\quad |/u_i/| \ll \pi/2$$

$$M \ll \sqrt{a}\ :\ P(/u_i/) = (1/2\pi).(1+L.\sqrt{\pi})$$

Figure 1.

RETRIEVAL-ERROR and BASIN OF ATTRACTION versus LOADING + NOISE.

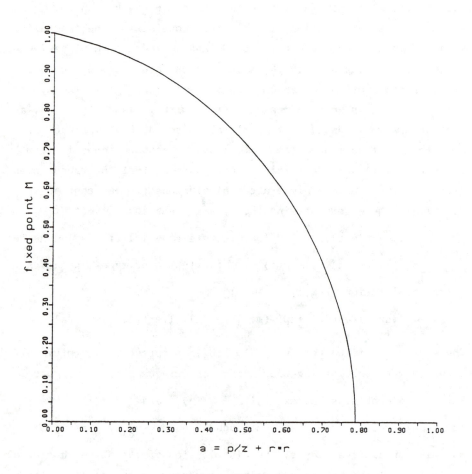

DISCUSSION

It has been shown that the usual binary or scalar neural networks can be generalized to phasor networks, and that the general structure of the theoretical analysis for their use as associative memories can be extended accordingly. This suggests that many of the other useful applications of neural nets (back-prop, etc.) can also be generalized to a phasor setting. This may be of interest both from the point of view of solving problems naturally posed in such a setting, as well as from that of enabling a wider range of physical implementations, such as networks of limit-cycle oscillators, phase-encoded optics, or maybe even Josephson-junctions.

The performance of phasor networks turns out to be roughly similar to that of the scalar systems; the maximum capacity $p/z = \pi/4$ for phasor nets is slightly larger than its value $2/\pi$ for binary nets, but there is a seemingly faster growth of the recall error $1-M$ at small a (linear for phasors, against $\exp(-1/(2a))$ for binary nets). However, the latter measures cannot be compared directly since they stem from quite different order parameters. If one reduces recalled phasor patterns to binary information, performance is again similar. Finally, the present methods and results suggest several roads to further generalizations, some of which may be relevant with respect to natural or technical implementations. The first class of these involves local variables ranging over the k-sphere with $k>1$. The other generalizations involve breaking the $O(n)$ (here $n=2$) symmetry of the system, either by forcing the variables to discrete positions on the circle (k-sphere), and/or by taking the interactions between two variables to be a more general function of the angular distance between them. Such models are now under development.

REFERENCES

1. J.J.Hopfield, Proc.Nat.Acad.Sci.USA 79, 2554 (1982) and idem, Proc.Nat.Acad.Sci.USA 81, 3088 (1984).
2. D.J.Amit, H.Gutfreund and H.Sompolinski, Ann.Phys. 173, 30 (1987).
3. D.Grensing, R.Kuhn and J.L. van Hemmen, J.Phys.A 20, 2935 (1987).
4. B.Derrida, E.Gardner and A.Zippelius, Europhys.Lett. 4, 167 (1987)

A Trellis-Structured Neural Network*

Thomas Petsche[†] and Bradley W. Dickinson
Princeton University, Department of Electrical Engineering
Princeton, NJ 08544

Abstract

We have developed a neural network which consists of cooperatively inter-connected Grossberg on-center off-surround subnets and which can be used to optimize a function related to the log likelihood function for decoding convolu-tional codes or more general FIR signal deconvolution problems. Connections in the network are confined to neighboring subnets, and it is representative of the types of networks which lend themselves to VLSI implementation. Analytical and experimental results for convergence and stability of the network have been found. The structure of the network can be used for distributed representation of data items while allowing for fault tolerance and replacement of faulty units.

1 Introduction

In order to study the behavior of locally interconnected networks, we have focused on a class of "trellis-structured" networks which are similar in structure to multilayer networks [5] but use symmetric connections and allow every neuron to be an output. We are studying such locally interconnected neural networks because they have the potential to be of great practical interest. Globally interconnected networks, e.g., Hopfield networks [3], are difficult to implement in VLSI because they require many long wires. Locally connected networks, however, can be designed to use fewer and shorter wires.

In this paper, we will describe a subclass of trellis-structured networks which op-timize a function that, near the global minimum, has the form of the log likelihood function for decoding convolutional codes or more general finite impulse response sig-nals. Convolutional codes, defined in section 2, provide an alternative representation scheme which can avoid the need for global connections. Our network, described in section 3, can perform maximum likelihood sequence estimation of convolutional coded sequences in the presence of noise. The performance of the system is optimal for low error rates.

The specific application for this network was inspired by a signal decomposition network described by Hopfield and Tank [6]. However, in our network, there is an emphasis on local interconnections and a more complex neural model, the Grossberg on-center off-surround network [2], is used. A modified form of the Gorssberg model is defined in section 4. Section 5 presents the main theoretical results of this paper. Although the deconvolution network is simply a set of cooperatively interconnected

*Supported by the Office of Naval Research through grant N00014-83-K-0577 and by the National Science Foundation through grant ECS84-05460.
[†]Permanent address: Siemens Corporate Research and Support, Inc., 105 College Road East, Princeton, NJ 08540.

on-center off-surround subnetworks, and absolute stability for the individual subnetworks has been proven [1], the cooperative interconnections between these subnets make a similar proof difficult and unlikely. We have been able, however, to prove equiasymptotic stability in the Lyapunov sense for this network given that the gain of the nonlinearity in each neuron is large. Section 6 will describe simulations of the network that were done to confirm the stability results.

2 Convolutional Codes and MLSE

In an error correcting code, an input sequence is transformed from a b-dimensional input space to an M-dimensional output space, where $M \geq b$ for error correction and/or detection. In general, for the b-bit input vector $\mathbf{U} = (u_1, \ldots, u_b)$ and the M-bit output vector $\mathbf{V} = (v_1, \ldots, v_M)$, we can write $\mathbf{V} = F(u_1, \ldots, u_b)$. A convolutional code, however, is designed so that relatively short subsequences of the input vector are used to determine subsequences of the output vector. For example, for a rate $1/3$ convolutional code (where $M \approx 3b$), with input subsequences of length 3, we can write the output, $\mathbf{V} = (\mathbf{v}_1, \ldots, \mathbf{v}_b)$ for $\mathbf{v}_i = (v_{i,1}, v_{i,2}, v_{i,3})$, of the encoder as a convolution of the input vector $\mathbf{U} = (u_1, \ldots, u_b, 0, 0)$ and three generator sequences

$$\mathbf{g}_0 = (1\ 1\ 1) \qquad \mathbf{g}_1 = (1\ 1\ 0) \qquad \mathbf{g}_2 = (0\ 1\ 1).$$

This convolution can be written, using modulo-2 addition, as

$$\mathbf{v}_i = \sum_{k=\max(1, i-2)}^{i} u_k \mathbf{g}_{i-k} \tag{1}$$

In this example, each 3-bit output subsequence, $\mathbf{v}_i$, of $\mathbf{V}$ depends only on three bits of the input vector, i.e., $\mathbf{v}_i = f(u_{i-2}, u_{i-1}, u_i)$. In general, for a rate $1/n$ code, the *constraint length*, K, is the number of bits of the input vector that uniquely determine each n-bit output subsequence. In the absence of noise, any subsequences in the input vector separated by more than K bits (i.e., that do not overlap) will produce subsequences in the output vector that are independent of each other.

If we view a convolutional code as a special case of block coding, this rate $1/3$, $K = 3$ code converts a b-bit input word into a codeword of length $3(b + 2)$ where the 2 is added by introducing two zeros at the end of every input to "zero-out" the code. Equivalently, the coder can be viewed as embedding 2^b memories into a $2^{3(b+2)}$-dimensional space. The minimum distance between valid memories or codewords in this space is the *free distance* of the code, which in this example is 7. This implies that the code is able to correct a minimum of three errors in the received signal.

For a convolutional code with constraint length K, the encoder can be viewed as a finite state machine whose state at time i is determined by the $K - 1$ input bits, $u_{i-k}, \ldots, u_{i-1}$. The encoder can also be represented as a trellis graph such as the one shown in figure 1 for a $K = 3$, rate $1/3$ code. In this example, since the constraint length is three, the two bits u_{i-2} and u_{i-1} determine which of four possible states the encoder is in at time i. In the trellis graph, there is a set of four nodes arranged in a vertical column, which we call a stage, for each time step i. Each node is labeled with the associated values of u_{i-2} and u_{i-1}. In general, for a rate $1/n$ code, each stage of the trellis graph contains 2^{K-1} nodes, representing an equal number of possible states. A trellis graph which contains S stages therefore fully describes the operation of the encoder for time steps 1 through S. The graph is read from left to right and the upper edge leaving the right side of a node in stage i is followed if u_i is a zero; the lower edge

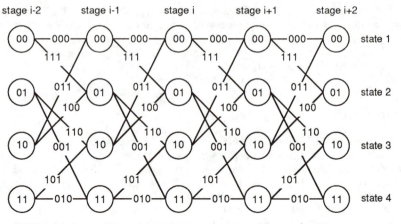

stage i-2   stage i-1   stage i   stage i+1   stage i+2

Figure 1: Part of the trellis-code representation for a rate 1/3, $K = 3$ convolutional code.

if u_i is a one. The label on the edge determined by u_i is $\mathbf{v}_i$, the output of the encoder given by equation 1 for the subsequence u_{i-2}, u_{i-1}, u_i.

Decoding a noisy sequence that is the output of a convolutional coder plus noise is typically done using a maximum likelihood sequence estimation (MLSE) decoder which is designed to accept as input a possibly noisy convolutional coded sequence, $\mathbf{R}$, and produce as output the maximum likelihood estimate, $\hat{\mathbf{V}}$, of the original sequence, $\mathbf{V}$. If the set of possible $n(b+2)$-bit encoder output vectors is $\{\mathbf{X}_m : m = 1, ..., 2^{n(b+2)}\}$ and $\mathbf{x}_{m,i}$ is the ith n-bit subsequence of $\mathbf{X}_m$ and $\mathbf{r}_i$ is the ith n-bit subsequence of $\mathbf{R}$ then

$$\hat{\mathbf{V}} = \arg\max_{\mathbf{X}_m} \prod_{i=1}^{b} P(\mathbf{r}_i \mid \mathbf{x}_{m,i}) \tag{2}$$

That is, the decoder chooses the $\mathbf{X}_m$ that maximizes the conditional probability, given $\mathbf{X}_m$, of the received sequence.

A binary symmetric channel (BSC) is an often used transmission channel model in which the decoder produces output sequences formed from an alphabet containing two symbols and it is assumed that the probability of either of the symbols being affected by noise so that the other symbol is received is the same for both symbols. In the case of a BSC, the log of the conditional probability, $P(\mathbf{r}_i \mid \mathbf{x}_{m,i})$, is a linear function of the Hamming distance between $\mathbf{r}_i$ and $\mathbf{x}_{m,i}$ so that maximizing the right side of equation 2 is equivalent to choosing the $\mathbf{X}_m$ that has the most bits in common with $\mathbf{R}$. Therefore, equation 2 can be rewritten as

$$\hat{\mathbf{V}} = \arg\max_{\mathbf{X}_m} \sum_{i=1}^{b} \sum_{l=1}^{n} I_{r_{i,l}}(x_{m,i,l}) \tag{3}$$

where $\mathbf{x}_{m,i,l}$ is the lth bit of the ith subsequence of $\mathbf{X}_m$ and $I_a(b)$ is the indicator function: $I_a(b) = 1$ if and only if a equals b.

For the general case, maximum likelihood sequence estimation is very expensive since the number of possible input sequences is exponential in b. The Viterbi algorithm [7], fortunately, is able to take advantage of the structure of convolutional codes and their trellis graph representations to reduce the complexity of the decoder so that it is only exponential in K (in general $K \ll b$). An optimum version of the Viterbi algorithm examines all b stages in the trellis graph, but a more practical and very nearly optimum version typically examines approximately $5K$ stages, beginning at stage i, before making a decision about u_i.

3 A Network for MLSE Decoding

The structure of the network that we have defined strongly reflects the structure of a trellis graph. The network usually consists of $5K$ subnetworks, each containing 2^{K-1} neurons. Each subnetwork corresponds to a stage in the trellis graph and each neuron to a state. Each stage is implemented as an "on-center off-surround" competitive network [2], described in more detail in the next section, which produces as output a contrast enhanced version of the input. This contrast enhancement creates a "winner take all" situation in which, under normal circumstances, only one neuron in each stage —the neuron receiving the input with greatest magnitude — will be on. The activation pattern of the network after it reaches equilibrium indicates the decoded sequence as a sequence of "on" neurons in the network. If the j-th neuron in subnet i, $\mathcal{N}_{i,j}$ is on, then the node representing state j in stage i lies on the network's estimate of the most likely path.

For a rate $1/n$ code, there is a symmetric cooperative connection between neurons $\mathcal{N}_{i,j}$ and $\mathcal{N}_{i+1,k}$ if there is an edge between the corresponding nodes in the trellis graph. If $(x_{i,j,k,1}, \ldots, x_{i,j,k,n})$ are the encoder output bits for the transition between these two nodes and $(r_{i,1}, \ldots, r_{i,n})$ are the received bits, then the connection weight for the symmetric cooperative connection between $\mathcal{N}_{i,j}$ and $\mathcal{N}_{i+1,k}$ is

$$m_{i,j,k} = \frac{1}{n} \sum_{l=1}^{n} I_{r_{i,l}}(x_{i,j,k,l}) \tag{4}$$

If there is no edge between the nodes, then $m_{i,j,k} = 0$.

Intuitively, it is easiest to understand the action of the entire network by examining one stage. Consider the nodes in stage i of the trellis graph and assume that the conditional probabilities of the nodes in stages $i-1$ and $i+1$ are known. (All probabilities are conditional on the received sequence.) Then the conditional probability of each node in stage i is simply the sum of the probabilities of each node in stages $i-1$ and $i+1$ weighted by the conditional transition probabilities. If we look at stage i in the network, and let the outputs of the neighboring stages $i-1$ and $i+1$ be fixed with the output of each neuron corresponding to the "likelihood" of the corresponding state at that stage, then the final outputs of the neurons $\mathcal{N}_{i,j}$ will correspond to the "likelihood" of each of the corresponding states. At equilibrium, the neuron corresponding to the most likely state will have the largest output.

4 The Neural Model

The "on-center off-surround" network[2] is used to model each stage in our network. This model allows the output of each neuron to take on a range of values, in this case between zero and one, and is designed to support contrast enhancement and competition between neurons. The model also guarantees that the final output of each neuron is a function of the relative intensity of its input as a fraction of the total input provided to the network.

Using the "on-center off-surround" model for each stage and the interconnection weights, $m_{i,j,k}$, defined in equation 4, the differential equation that governs the instantaneous activity of the neurons in our deconvolution network with S stages and N states in each stage can be written as

$$
\begin{aligned}
\dot{u}_{i,j} = {} & -Au_{i,j} + (B - u_{i,j})\left(f(u_{i,j}) + \sum_{k=1}^{N}[m_{i-1,k,j}f(u_{i-1,k}) + m_{i,j,k}f(u_{i+1,k})]\right) \\
& - (C + u_{i,j})\sum_{k \neq j}^{N}\left(f(u_{i,k}) + \sum_{l=1}^{N}[m_{i-1,k,l}f(u_{i-1,k}) + m_{i,l,k}f(u_{i+1,k})]\right)
\end{aligned}
\tag{5}
$$

where $f(x) = (1 + e^{-\lambda x})^{-1}$, λ is the gain of the nonlinearity, and A, B, and C are constants

For the analysis to be presented in section 5, we note that equation 5 can be rewritten more compactly in a notation that is similar to the equation for additive analog neurons given in [4]:

$$
\dot{u}_{i,j} = -Au_{i,j} - \sum_{k=1}^{S}\sum_{l=1}^{N}(u_{i,j}S_{i,j,k,l}f(u_{k,l}) - T_{i,j,k,l}f(u_{k,l}))
\tag{6}
$$

where, for $1 \leq l \leq N$,

$$
\begin{aligned}
S_{i,j,i,l} &= 1 & T_{i,j,i,j} &= B \\
S_{i,j,i-1,l} &= \sum_{q} m_{i-1,l,q} & T_{i,j,i,l} &= -C \quad \forall\, l \neq j \\
S_{i,j,i+1,l} &= \sum_{q} m_{i,q,l} & T_{i,j,i-1,l} &= Bm_{i-1,l,j} - C\sum_{q \neq j} m_{i-1,l,q} \\
S_{i,j,k,l} &= 0 \quad \forall\, k \notin \{i-1,i,i+1\} & T_{i,j,i+1,l} &= Bm_{i,j,l} - C\sum_{q \neq j} m_{i,q,l}
\end{aligned}
\tag{7}
$$

To eliminate the need for global interconnections within a stage, we can add two summing elements to calculate

$$
X_i = \sum_{j=1}^{N} f(x_{i,j}) \quad \text{and} \quad J_i = \sum_{j=1}^{N}\sum_{k=1}^{N}[m_{i-1,k,j}f(u_{i-1,k}) + m_{i,j,k}f(u_{i+1,k})]
\tag{8}
$$

Using these two sums allows us to rewrite equation 5 as

$$
\dot{u}_{i,j} = -Au_{i,j} + (B + C)(f(u_{i,j}) + I_{i,j}) - u_{i,j}(X_i + J_i)
\tag{9}
$$

This form provides a more compact design for the network that is particularly suited to implementation as a digital filter or for use in simulations since it greatly reduces the calculations required.

5   Stability of the Network

The end of section 3 described the desired operation of a single stage, given that the outputs of the neighboring stages are fixed. It is possible to show that in this situation a single stage is stable. To do this, fix $f(u_{k,l})$ for $k \in \{i-1, i+1\}$ so that equation 6 can be written in the form originally proposed by Grossberg [2]:

$$
\dot{u}_{i,j} = -Au_{i,j} + (B - u_{i,j})(I_{i,j} + f(u_{i,j})) - (u_{i,j} + C)\left(\sum_{k=1}^{N} I_{i,k} + \sum_{k=1}^{N} f(u_{i,k})\right)
\tag{10}
$$

where $I_{i,j} = \sum_{k=1}^{N} [m_{i-1,k,j}f(u_{i-1,k}) + m_{i,j,k}f(u_{i+1,k})]$.

Equation 10 is a special case of the more general nonlinear system

$$\dot{x}_i = a_i(x_i)\left(b_i(x_i) - \sum_{k=1}^{n} c_{i,k}d_k(x_k)\right) \tag{11}$$

where: (1) $a_i(x_i)$ is continuous and $a_i(x_i) > 0$ for $x_i \geq 0$; (2) $b_i(x_i)$ is continuous for $x_i \geq 0$; (3) $c_{i,k} = c_{k,i}$; and (4) $d_i(x_i) \geq 0$ for all $x_i \in (-\infty, \infty)$. Cohen and Grossberg [1] showed that such a system has a global Lyapunov function:

$$V(\mathbf{x}) = -\sum_{i=1}^{n} \int_0^{x_i} b_i(\xi_i)d_i'(\xi_i)d(\xi_i) + \frac{1}{2}\sum_{j=1}^{n}\sum_{k=1}^{n} c_{j,k}d_j(x_j)d_k(x_k) \tag{12}$$

and that, therefore, such a system is equiasymptotically stable for all constants and functions satisfying the four constraints above. In our case, this means that a single stage has the desired behavior when the neighboring stages are fixed. If we take the output of each neuron to correspond to the likelihood of the corresponding state then, if the two neighboring stages are fixed, stage i will converge to an equilibrium point where the neuron receiving the largest input will be on and the others will be off, just as it should according to section 2.

It does not seem possible to use the Cohen-Grossberg stability proof for the entire system in equation 5. In fact, Cohen and Grossberg note that networks which allow cooperative interactions define systems for which no stability proof exists [1].

Since an exact stability proof seems unlikely, we have instead shown that in the limit as the gain, λ, of the nonlinearity gets large the system is asymptotically stable. Using the notation in [4], define $V_i = f(u_i)$ and a normalized nonlinearity $\bar{f}(\cdot)$ such that $\bar{f}^{-1}(V_i) = \lambda u_i$. Then we can define an energy function for the deconvolution network to be

$$E = -\frac{1}{2}\sum_{i,j,k,l} T_{i,j,k,l}V_{i,j}V_{k,l} - \sum_{i,j}\frac{1}{\lambda}\left(-A - \sum_{k,l} S_{i,j,k,l}V_{k,l}\right)\int_{\frac{1}{2}}^{V_{k,l}} \bar{f}^{-1}(\zeta)\,d\zeta \tag{13}$$

The time derivative of E is

$$\dot{E} = -\sum_{i,j}\frac{dV_{i,j}}{dt}\left(-Au_{i,j} - u_{i,j}\sum_{k,l} S_{i,j,k,l}V_{k,l} + \sum_{k,l} T_{i,j,k,l}V_{k,l}\right.$$
$$\left. - \frac{1}{\lambda}\sum_{k,l} S_{i,j,k,l}\int_{\frac{1}{2}}^{V_{k,l}} \bar{f}^{-1}(\zeta)d\zeta\right) \tag{14}$$

It is difficult to prove that $\dot{E}$ is nonpositive because of the last term in the parentheses. However, for large gain, this term can be shown to have a negligible effect on the derivative.

It can be shown that for $f(u) = (1 + e^{-\lambda u})^{-1}$, $\int_{\frac{1}{2}}^{V_i} \bar{f}^{-1}(\zeta)\,d\zeta$ is bounded above by $\log(2)$. In this deconvolution network, there are no connections between neurons unless they are in the same or neighboring stages, i.e., $S_{i,j,k,l} = 0$ for $|i - k| > 1$ and l is restricted so that $0 \leq l \leq S$, so there are no more than $3S$ non-zero terms in the problematical summation. Therefore, we can write that

$$\lim_{\lambda\to\infty} -\frac{1}{\lambda}\sum_{k,l} S_{i,j,k,l}\int_{\frac{1}{2}}^{V_{k,l}} \bar{f}^{-1}(\zeta)\,d\zeta = 0$$

Then, in the limit as $\lambda \to \infty$, the terms in parentheses in equation 14 converge to $\dot{u}_i$ in equation 6, so that $\lim_{\lambda \to \infty} \dot{E} = \sum_{i,j} \frac{dV_{i,j}}{dt} \dot{u}_i$. Using the chain rule, we can rewrite this as

$$\lim_{\lambda \to \infty} \dot{E} = -\sum_{i,j} \left(\frac{dV_{i,j}}{dt} \right)^2 \left(\frac{d}{dV_{i,j}} f^{-1}(V_{i,j}) \right)$$

It can also be shown that that, if $f(\cdot)$ is a monotonically increasing function then $\frac{d}{dV_i} f^{-1}(V_i) > 0$ for all V_i. This implies that for all $\mathbf{u} = (u_{i,1}, \ldots, u_{N,S})$, $\lim_{\lambda \to \infty} \dot{E} \leq 0$, and, therefore, for large gains, E as defined in equation 13 is a Lyapunov function for the system described by equation 5 and the network is equiasymtotically stable.

If we apply a similar asymptotic argument to the energy function, equation 13 reduces to

$$E = -\frac{1}{2} \sum_{i,j,k,l} T_{i,j,k,l} V_{i,j} V_{k,l} \tag{15}$$

which is the Lyapunov function for a network of discontinuous on-off neurons with interconnection matrix $\mathbf{T}$. For the binary neuron case, it is fairly straight forward to show that the energy function has minima at the desired decoder outputs if we assume that only one neuron in each stage may be on and that B and C are appropriately chosen to favor this. However, since there are $O(S^2 N)$ terms in the disturbance summation in equation 15, convergence in this case is not as fast as for the derivative of the energy function in equation 13, which has only $O(S)$ terms in the summation.

6   Simulation Results

The simulations presented in this section are for the rate $1/3$, $K = 3$ convolutional code illustrated in figure 1. Since this code has a constraint length of 3, there are 4 possible states in each stage and an MLSE decoder would normally examine a minimum of $5K$ subsequences before making a decision, we will use a total of 16 stages. In these simulations, the first and last stage are fixed since we assume that we have prior knowledge or a decision about the first stage and zero knowledge about the last stage. The transmitted codeword is assumed to be all zeros.

The simulation program reads the received sequence from standard input and uses it to define the interconnection matrix W according to equation 4. A relaxation subroutine is then called to simulate the performance of the network according to an Euler discretization of equation 5. Unit time is then defined as one RC time constant of the unforced system. All variables were defined to be single precision (32 bit) floating point numbers.

Figure 2a shows the evolution of the network over two unit time intervals with the sampling time $T = 0.02$ when the received codeword contains no noise. To interpret the figure, recall that there are 16 stages of 4 neurons each. The output of each stage is a vertical set of 4 curves. The upper-left set is the output of the first stage; the upper-most curve is the output of the first neuron in the stage. For the first stage, the first neuron has a fixed output of 1 and the other neurons have a fixed output of 0. The outputs of the neurons in the last stages are fixed at an intermediate value to represent zero *a priori* knowledge about these states. Notice that the network reaches an equilibrium point in which only the top neurons in each state (representing the "00" node in figure 1) are on and all others are off. This case illustrates that the network can correctly decode an unerrored input and that it does so rapidly, i.e., in about one time constant. In this case, with no errors in the input, the network performs the

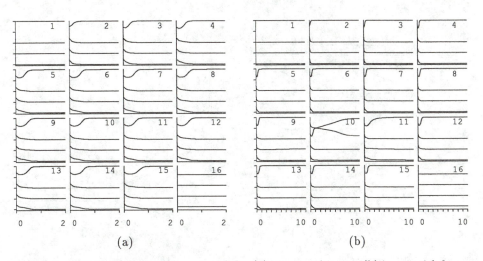

Figure 2: Evolution of the trellis network for (a) unerrored input, (b) input with burst errors: $\mathbf{R}$ is 000 000 000 000 000 000 000 000 111 000 000 000 000 000 000. $\lambda = 10.$, $A = 1.0$, $B = 1.0$, $C = 0.75$, $T = 0.02$. The initial conditions are $x_{1,1} = 1.$, $x_{1,j} = 0.0$, $x_{16,j} = 0.2$, all other $x_{i,j} = 0.0$.

same function as Hopfield and Tank's network and does so quite well. Although we have not been able to prove it analytically, all our simulations support the conjecture that if $x_{i,j}(0) = \frac{1}{2}$ for all i and j then the network will always converge to the global minimum.

One of the more difficult decoding problems for this network is the correction of a burst of errors in a transition subsequence. Figure 2b shows the evolution of the network when three errors occur in the transition between stages 9 and 10. Note that 10 unit time intervals are shown since complete convergence takes much longer than in the first example. However, the network has correctly decoded many of the stages far from the burst error in a much shorter time.

If the received codeword contains scattered errors, the convolutional decoder should be able to correct more than 3 errors. Such a case is shown in figure 3a in which the received codeword contains 7 errors. The system takes longest to converge around two transitions, 5–6 and 11–12. The first is in the midst of consecutive subsequences which each have one bit errors and the second transition contains two errors.

To illustrate that the energy function shown in equation 13 is a good candidate for a Lyapunov function for this network, it is plotted in figure 3b for the three cases described above. The nonlinearity used in these simulations has a gain of ten, and, as predicted by the large gain limit, the energy decreases monotonically.

To more thoroughly explore the behavior of the network, the simulation program was modified to test many possible error patterns. For one and two errors, the program exhaustively tested each possible error pattern. For three or more errors, the errors were generated randomly. For four or more errors, only those errored sequences for which the MLS estimate was the sequence of all zeros were tested. The results of this simulation are summarized in the column labeled "two-nearest" in figure 4. The performance of the network is optimum if no more than 3 errors are present in the received sequence, however for four or more errors, the network fails to correctly decode some sequences that the MLSE decoder can correctly decode.

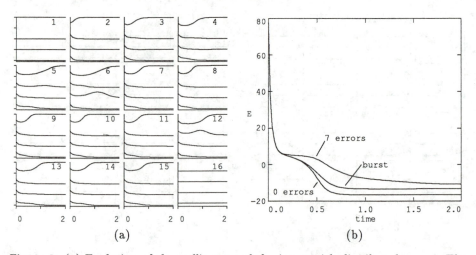

(a)   (b)

Figure 3: (a) Evolution of the trellis network for input with distributed errors. The input, **R**, is 000 010 010 010 100 001 000 000 000 000 110 000 000 000 000. The constants and initial conditions are the same as in figure 2. (b) The energy function defined in equation 13 evaulated for the three simulations discussed.

errored bits	number of test vectors	number of errors	
		two-nearest	four-nearest
0	1	0	0
1	39	0	0
2	500	0	0
3	500	0	0
4	500	7	0
5	500	33	20
6	500	72	68
7	500	132	103
Total	2500	244	191

Figure 4: Simulation results for a deconvolution network for a $K = 3$, rate 1/3 code. The network parameters were: $\lambda = 15$, $A = 6$, $B = 1$, $C = 0.45$, and $T = 0.025$.

For locally interconnected networks, the major concern is the flow of information through the network. In the simulations presented until now, the neurons in each stage are connected only to neurons in neighboring stages. A modified form of the network was also simulated in which the neurons in each stage are connected to the neurons in the four nearest neighboring stages. To implement this network, the subroutine to initialize the connection weights was modified to assign a non-zero value to $w_{i,j,i+2,k}$. This is straight-forward since, for a code with a constraint length of three, there is a single path connecting two nodes a distance two apart.

The results of this simulation are shown in the column labeled "four-nearest" in figure 4. It is easy to see that the network with the extra connections performs better than the previous network. Most of the errors made by the nearest neighbor network occur for inputs in which the received subsequences r_i and r_{i+1} or r_{i+2} contain a total of four or more errors. It appears that the network with the additional connections is, in effect, able to communicate around subsequences containing errors that block communications for the two-nearest neighbor network.

7 Summary and Conclusions

We have presented a locally interconnected network which minimizes a function that is analogous to the log likelihood function near the global minimum. The results of simulations demonstrate that the network can successfully decode input sequences containing no noise at least as well as the globally connected Hopfield-Tank [6] decomposition network. Simulations also strongly support the conjecture that in the noiseless case, the network can be guaranteed to converge to the global minimum. In addition, for low error rates, the network can also decode noisy received sequences.

We have been able to apply the Cohen-Grossberg proof of the stability of "on-center off-surround" networks to show that each stage will maximize the desired local "likelihood" for noisy received sequences. We have also shown that, in the large gain limit, the network as a whole is stable and that the equilibrium points correspond to the MLSE decoder output. Simulations have verified this proof of stability even for relatively small gains. Unfortunately, a proof of strict Lyapunov stability is very difficult, and may not be possible, because of the cooperative connections in the network.

This network demonstrates that it is possible to perform interesting functions even if only localized connections are allowed, although there may be some loss of performance. If we view the network as an associative memory, a trellis structured network that contains NS neurons can correctly recall 2^S memories. Simulations of trellis networks strongly suggest that it is possible to guarantee a non-zero minimum radius of attraction for all memories. We are currently investigating the use of trellis structured layers in multilayer networks to explicitly provide the networks with the ability to tolerate errors and replace faulty neurons.

References

[1] M. Cohen and S. Grossberg, "Absolute stability of global pattern formation and parallel memory storage by competitive neural networks," *IEEE Trans. Sys., Man, and Cyber.*, vol. 13, pp. 815–826, Sep.–Oct. 1983.

[2] S. Grossberg, "How does a brain build a cognitive code," in *Studies of Mind and Brain*, pp. 1–52, D. Reidel Pub. Co., 1982.

[3] J. Hopfield, "Neural networks and physical systems with emergent collective computational abilities," *Proceedings of the National Academy of Sciences USA*, vol. 79, pp. 2554–2558, 1982.

[4] J. Hopfield, "Neurons with graded response have collective computational properties like those of two-state neurons," *Proceedings of the National Academy of Science, USA*, vol. 81, pp. 3088–3092, May 1984.

[5] J. McClelland and D. Rumelhart, *Parallel Distributed Processing, Vol. 1*. The MIT Press, 1986.

[6] D. Tank and J. Hopfield, "Simple 'neural' optimization networks: an A/D converter, signal decision circuit and a linear programming circuit," *IEEE Trans. on Circuits and Systems*, vol. 33, pp. 533–541, May 1986.

[7] A. Viterbi and J. Omura, *Principles of Digital Communications and Coding*. McGraw-Hill, 1979.

GENERALIZATION OF BACKPROPAGATION
TO
RECURRENT AND HIGHER ORDER NEURAL NETWORKS

Fernando J. Pineda

Applied Physics Laboratory, Johns Hopkins University
Johns Hopkins Rd., Laurel MD 20707

Abstract

A general method for deriving backpropagation algorithms for networks with recurrent and higher order networks is introduced. The propagation of activation in these networks is determined by dissipative differential equations. The error signal is backpropagated by integrating an associated differential equation. The method is introduced by applying it to the recurrent generalization of the feedforward backpropagation network. The method is extended to the case of higher order networks and to a constrained dynamical system for training a content addressable memory. The essential feature of the adaptive algorithms is that adaptive equation has a simple outer product form.

Preliminary experiments suggest that learning can occur very rapidly in networks with recurrent connections. The continuous formalism makes the new approach more suitable for implementation in VLSI.

Introduction

One interesting class of neural networks, typified by the Hopfield neural networks [1,2] or the networks studied by Amari[3,4] are dynamical systems with three salient properties. First, they posses very many degrees of freedom, second their dynamics are nonlinear and third, their dynamics are dissipative. Systems with these properties can have complicated attractor structures and can exhibit computational abilities.

The identification of attractors with computational objects, e.g. memories ar d rules, is one of the foundations of the neural network paradigm. In this paradign, programming becomes an excercise in manipulating attractors. A learning algorithm is a rule or dynamical equation which changes the locations of fixed points to encode information. One way of doing this is to minimize, by gradient descent, some function of the system parameters. This general approach is reviewed by Amari[4] and forms the basis of many learning algorithms. The formalism described here is a specific case of this general approach.

The purpose of this paper is to introduce a formalism for obtaining adaptive dynamical systems which are based on backpropagation[5,6,7]. These dynamical systems are expressed as systems of coupled first order differential equations. The formalism will be illustrated by deriving adaptive equations for a recurrent network with first order neurons, a recurrent network with higher order neurons and finally a recurrent first order associative memory.

Example 1: Recurrent backpropagation with first order units

Consider a dynamical system whose state vector **x** evolves according to the following set of coupled differential equations

$$dx_i/dt = -x_i + g_i(\Sigma_j w_{ij}x_j) + I_i \qquad (1)$$

where i=1,...,N. The functions g_i are assumed to be differentiable and may have different forms for various populations of neurons. In this paper we shall make no other requirements on g_i. In the neural network literature it is common to take these functions to be sigmoid shaped functions. A commonly used form is the logistic function,

$$g(\xi) = (1 + e^{-\xi})^{-1}. \qquad (2)$$

This form is biologically motivated since it attempts to account for the refractory phase of real neurons. However, it is important to stress that there is nothing in the mathematical content of this paper which requires this form -- any differentiable function will suffice in the formalism presented in this paper. For example, a choice which may be of use in signal processing is $\sin(\xi)$.

A necessary condition for the learning algorithms discussed here to exist is that the system posesses stable isolated attractors, i.e. fixed points. The attractor structure of (1) is the same as the more commonly used equation

$$du_i/dt = -u_i + \Sigma_j w_{ij}g(u_j) + K_i. \qquad (3)$$

Because (1) and (3) are related by a simple linear transformation. Therefore results concerning the stability of (3) are applicable to (1). Amari[3] studied the dynamics of equation (3) in networks with random conections. He found that collective variables corresponding to the mean activation and its second moment must exhibit either stable or bistable behaviour. More recently, Hopfield[2] has shown how to construct content addressable memories from symmetrically connected networks with this same dynamical equation. The symmetric connections in the network gaurantee global stability. The solution of equation (1) is also globally asymptotically stable if **w** can be transformed into a lower triangular matrix by row and column exchange operations. This is because in such a case the network is a simply a feedforward network and the output can be expressed as an explicit function of the input. No Liapunov function exists for arbitrary weights as can be demonstrated by constructing a set of weights which leads to oscillation. In practice, it is found that oscillations are not a problem and that the system converges to fixed points unless special weights are chosen. Therefore it shall be assumed, for the purposes of deriving the backpropagation equations, that the system ultimately settles down to a fixed point.

Consider a system of N neurons, or units, whose dynamics is determined by equation (1). Of all the units in the network we will arbitrarily define some subset of them (A) as *input* units and some other subset of them (Ω) as *output* units. Units which are neither members of A nor Ω are denoted *hidden* units. A unit may be simultaneously an input unit and an output unit. The external environment influences the system through the source term, **I**. If a unit is an input unit, the corresponding component of **I** is nonzero. To make this more precise it is useful to introduce a notational convention. Suppose that Φ represent some subset of units in the network then the function $\Theta_{i\Phi}$ is defined by

$$\Theta_{i\Phi} = \left\{ \begin{array}{ll} 1 & \text{if i-th unit is a member of } \Phi \\ 0 & \text{otherwise} \end{array} \right. \qquad (4)$$

In terms of this function, the components of the **I** vector are given by

$$I_i = \xi_i\Theta_{iA} , \qquad (5)$$

where ξ_i is determined by the external environment.

Our goal will be to find a local algorithm which adjusts the weight matrix $\mathbf{w}$ so that a given initial state $\mathbf{x}^0 = \mathbf{x}(t_0)$, and a given input $\mathbf{I}$ result in a fixed point, $\mathbf{x}^\infty = \mathbf{x}(t_\infty)$, whose components have a desired set of values T_i along the output units . This will be accomplished by minimizing a function E which measures the distance between the desired fixed point and the actual fixed point i.e.,

$$E = \frac{1}{2} \sum_{i=1}^{N} J_i^2 \qquad (6)$$

where

$$J_i = (T_i - x^\infty_i) \, \Theta_{i\Omega} \; . \qquad (7)$$

E depends on the weight matrix $\mathbf{w}$ through the fixed point $\mathbf{x}^\infty(\mathbf{w})$. A learning algorithm drives the fixed points towards the manifolds which satisfy $x_i^\infty = T_i$ on the output units. One way of accomplishing this with dynamics is to let the system evolve in the weight space along trajectories which are antiparallel to the gradient of E. In other words,

$$dw_{ij}/dt = - \eta \frac{\partial E}{\partial w_{ij}} \qquad (8)$$

where η is a numerical constant which defines the (slow) time scale on which $\mathbf{w}$ changes. η *must be small so that* $\mathbf{x}$ *is always essentially at steady state* , i.e. $\mathbf{x}(t) \cong \mathbf{x}^\infty$. It is important to stress that the choice of gradient descent for the learning dynamics is by no means unique, nor is it necessarily the best choice. Other learning dynamics which employ second order time derivatives (e.g. the momentum method[5]) or which employ second order space derivatives (e.g. second order backpropagation[8]) may be more useful in particular applications. However, equation (8) does have the virtue of being the simplest dynamics which minimizes E.

On performing the differentiations in equation (8), one immediately obtains

$$dw_{rs}/dt = \eta \sum_k J_k \frac{\partial x^\infty_k}{\partial w_{rs}} \; . \qquad (9)$$

The derivative of x^∞_k with respect to w_{rs} is obtained by first noting that the fixed points of equation (1) satisfy the nonlinear algebraic equation

$$x^\infty_i = g_i(\sum_j w_{ij} x^\infty_j) + I_i \; , \qquad (10)$$

differentiating both sides of this equation with respect to w_{rs} and finally solving for $\partial x^\infty_k / \partial w_{rs}$. The result is

$$\frac{\partial x^\infty_k}{\partial w_{rs}} = (L^{-1})_{kr} \, g_r'(u_r) x^\infty_s \qquad (11)$$

where g_r' is the derivative of g_r and where the matrix $\mathbf{L}$ is given by

$$L_{ij} = \delta_{ij} - g_i'(u_i) w_{ij} \; . \qquad (12)$$

δ_{ij} is the Kroneker δ function ($\delta_{ij} = 1$ if i=j, otherwise $\delta_{ij} = 0$). On substituting (11) into (9) one obtains the remarkably simple form

$$dw_{rs}/dt = \eta \, y_r x^\infty_{\ s} \tag{13}$$

where

$$y_r = g_r{'}(u_r) \, \Sigma J_k(L^{-1})_{kr} \quad . \tag{14}$$
$$k=$$

Equations (13) and (14) specify a formal learning rule. Unfortunately, equation (14) requires a matrix inversion to calculate the error signals y_k. Direct matrix inversions are necessarily nonlocal calculations and therefore this learning algorithm is not suitable for implementation as a neural network. Fortunately, a local method for calculating y_r can be obtained by the introduction of an associated dynamical system. To obtain this dynamical system first rewrite equation (14) as

$$\Sigma L_{rk}\{y_r / g_r{'}(u_r)\} = J_k \quad . \tag{15}$$
$$r$$

Then multiply both sides by $f_k{'}(u_k)$, substitute the explicit form for L and finally sum over r. The result is

$$0 = -y_k + g_k{'}(u_k)\{\Sigma w_{rk}y_r + J_k\} \quad . \tag{16}$$
$$r$$

One now makes the observation that the solutions of this linear equation are the fixed points of the dynamical system given by

$$dy_k/dt = -y_k + g_k{'}(u_k)\{\Sigma w_{rk}y_r + J_k\} \quad . \tag{17}$$
$$r$$

This last step is not unique, equation (16) could be transformed in various ways leading to related differential equations, cf. Pineda[9]. It is not difficult to show that the first order finite difference approximation (with a time step $\Delta t = 1$) of equations (1), (13) and (17) has the same form as the conventional backpropagation algorithm.

Equations (1), (13) and (17) completely specify the dynamics for an adaptive neural network, provided that (1) and (17) converge to stable fixed points and provided that both quantities on the right hand side of equation (13) are the steady state solutions of (1) and (17).

It was pointed out by Almeida[10] that the local stability of (1) is a sufficient condition for the local stability of (17). To prove this it suffices to linearize equation (1) about a stable fixed point. The resulting linearized equation depends on the same matrix L whose transpose appears in the derivation of equation (17), cf. equation (15). But L and L^T have the same eigenvalues, hence it follows that the fixed points of (17) must also be locally stable if the fixed points of (1) are locally stable.

Learning multiple associations

It is important to stress that up to this point the entire discussionhas assumed that I and T are constant in time, thus no mechanism has been obtained for learning multiple input/output associations. Two methods for training the network to learn multiple associations are now discussed. These methods lead to qualitatively different learning behaviour.

Suppose that each input/output pair is labeled by a pattern label α, i.e. $\{I^\alpha, T^\alpha\}$. Then the energy function which is minimized in the above discussion must also depend on this label since it is an implicit function of the I^α, T^α pairs. In order to learn multiple input/output associations it is necessary to minimize all the $E[\alpha]$ simultaniously. In otherwords the function to minimize is

$$E_{total} = \Sigma E[\alpha] \tag{18}$$
$$\alpha$$

where the sum is over all input/output associations. From (18) it follows that the gradient for E_{total} is simply the sum of the gradients for each association, hence the corresponding gradient descent equation has the form,

$$dw_{ij}/dt \;=\; \eta \sum_{\alpha} y^{\infty}_i[\alpha]\, x^{\infty}_j[\alpha] \;. \qquad (19)$$

In numerical simulations, each time step of (19) requires relaxing (1) and (17) for each pattern and accumulating the gradient over all the patterns. This form of the algorithm is deterministic and is guaranteed to converge because, by construction, E_{total} is a Liapunov function for equation (19). However, the system may get stuck in a local minimum. This method is similar to the master/slave approach of Lapedes and Farber[11]. Their adaptive equation, which plays the same role as equation (19), also has a gradient form, although it is not strictly descent along the gradient. For a randomly or fully connected network it can be shown that the number of operations required per weight update in the master/slave formalism is proportional to N^3 where N is the number of units. This is because there are $O(N^2)$ update equations and each equation requires $O(N)$ operations (assuming some precomputation). On the other hand, in the backpropagation formalism each update equation requires only $O(1)$ operations because of their trivial outer product form. Also $O(N^2)$ operations are required to precompute x^{∞} and y^{∞}. The result is that each weight update requires only $O(N^2)$ operations. It is not possible to conclude from this argument that one or the other approach will be more efficient in a particular application because there are other factors to consider such as the number of patterns and the number of time steps required for x and y to converge. A detailed comparison of the two methods is in preparation.

A second approach to learning multiple patterns is to use (13) and to change the patterns *randomly* on each time step. The system therefore receives a sequence of random impulses each of which attempts to minimize $E[\alpha]$ for a single pattern. One can then define $L(w)$ to be the mean $E[\alpha]$ (averaged over the distribution of patterns).

$$L(\mathbf{w}) = <E\,[\mathbf{w}, \mathbf{I}^{\alpha}, \mathbf{T}^{\alpha}\,]> \qquad (20)$$

Amari[4] has pointed out that if the sequence of random patterns is stationary and if $L(w)$ has a unique minimum then the theory of stochastic approximation guarantees that the solution of (13) $w(t)$ will converge to the minimum point $\mathbf{w}_{min}$ of $L(w)$ to within a small fluctuating term which vanishes as η tends to zero. Evidently η is analogous to the temperature parameter in simulated annealing. This second approach generally converges more slowly than the first, but it will ultimately converge (in a statistical sense) to the *global* minimum.

In principle the fixed points, to which the solutions of (1) and (17) eventually converge, depend on the initial states. Indeed, Amari's[3] results imply that equation (1) is bistable for certain choices of weights. Therefore the presentation of multiple patterns might seem problematical since in both approaches the final state of the previous pattern becomes the initial state of the new pattern. The safest approach is to reinitialize the network to the same initial state each time a new pattern is presented, e.g. $x_i(t_0) = 0.5$ for all i. In practice the system learns robustly even if the initial conditions are chosen randomly.

Example 2: Recurrent higher order networks

It is straightforward to apply the technique of the previous section to a dynamical system with higher order units. Higher order systems have been studied by Sejnowski [12] and Lee et al.[13]. Higher order networks may have definite advantages over networks with first order units alone A detailed discussion of the backpropagation formalism applied to higher order networks is beyond the scope of this paper. Instead, the adaptive equations for a network with purely n-th order units will be presented as an example of the formalism. To this end consider a dynamical system of the form

$$dx_i/dt = -x_i + g_i(u_i) + I_i \tag{21}$$

where

$$u_i = \sum_j \cdots \sum_k w^{(n)}_{ij\cdots k} f_j(x_j) \cdots f_k(x_k) \ . \tag{22}$$

and where there are n+1 indices and the summations are over all indices except i. The superscript on the weight tensor indicates the order of the correlation. Note that an additional nonlinear function f has been added to illustrate a further generalization. Both f and g must be differentiable and may be chosen to be sigmoids. It is not difficult, although somewhat tedious, to repeat the steps of the previous example to derive the adaptive equations for this system. The objective function in this case is the same as was used in the first example, i.e. equation (6). The n-th order gradient descent equation has the form

$$dw^{(n)}_{ij\cdots k}/dt = \eta y^{(n)\infty}_i f(x^\infty_j) \cdots f(x^\infty_k) \ . \tag{23}$$

Equation (23) illustrates the major feature of backpropagation which distinguishes it from other gradient descent algorithms or similar algorithms which make use of a gradient. Namely, that the gradient of the objective function has a very trivial outer product form. $y^{(n)\infty}$ is the steady state solution of

$$dy^{(n)}_k/dt = -y^{(n)}_k + g_k'(u_k)\{f_k'(x_k)\sum_r V^{(n)}_{rk} y^{(n)}_r + J_k\} \ . \tag{24}$$

The matrix $V^{(n)}$ plays the role of w in the previous example, however $V^{(n)}$ now depends on the state of the network according to

$$V^{(n)}_{ij} = \sum_k \cdots \sum_l S^{(n)}_{ijk\cdots l} \{ f(x_k) \cdots f(x_l)\} \tag{25}$$

where is $S^{(n)}$ a tensor which is symmetric with respect to the exchange of the second index and all the indices to the right, i.e.

$$S^{(n)}_{ijk\cdots l} = w^{(n)}_{ijk\cdots l} + w^{(n)}_{ikj\cdots l} + \cdots + w^{(n)}_{ijl\cdots k} \ . \tag{26}$$

Finally, it should be noted that: 1) If the polynomial u_i is not homogenous, the adaptive equations are more complicated and involve cross terms between the various orders and that: 2) The local stability of the n-th order backpropagation equations now depends on the eigenvalues of the matrix

$$L_{ij} = \delta_{ij} - g_i'(u_i) f_i'(x_i) V^{(n)}_{ij} \ . \tag{27}$$

As before, if the forward propagation converges so will the backward propagation.

Example 3: Adaptive content addressable memory

In this section the adaptive equations for a content addressable memory (CAM) are derived as a final illustration of the generality of the formalism. Perhaps the best known (and best studied) examples of dynamical systems which exhibit CAM behaviour are the systems discussed by Hopfield[1,2]. Hopfield used a nonadaptive method for programming the symmetric weight matrix. More recently Lapedes and Farber[11] have demonstrated how to contruct a master dynamical system which can be used to train the weights of a slave system which has the Hopfield form. This slave system then performs the CAM operation. The resulting weights are not symmetric.

The learning proceedure presented in this section is most closely related to the method of Lapedes and Farber in that a master network is used to adjust the weights of a slave network. In contrast to the afforementioned formalism, which requires a very large associated weight matrix for the master network, both the master and slave networks of the following approach make use of the same weight matrix. The CAM under consideration is based on equation (1). However, the interpretation of the dynamics will be somewhat different from the first section. The main difference is that the dynamics in the learning phase is constrained. The constrained dynamical system is denoted the master network. The unconstrained system is denoted the slave network. The units in the network are divided into only two sets: the set of visible units (V) and the set of internal or hidden units (H). There will be no distinction made between input and output units. Thus, I will generally be zero unless an input bias is needed in some application.

The dynamical system will be used as an autoassociative memory, thus the memory recall is performed by starting the network at a particular initial state which represents partial information about a stored memory. More precisely, suppose that there exists a subset K of the visible units whose states are known to have values T_i. Then the initial state of the network is

$$x_i(t_o) = T_i \, \Theta_{iK} + b_i \, (1 - \Theta_{iK}), \qquad (28)$$

where the b_i are arbitrary. The CAM relaxes to the previously stored memory whose basin of attraction contains this partial state.

Memories are stored by a master network whose topology is exactly the same as the slave network, but whose dynamics is somewhat modified. The state vector z of the master network evolves according to the equation

$$dz_i/dt = -z_i + g_i(\sum_{k=1}^{N} w_{ik} Z_k) + I_i \qquad (29)$$

where Z is defined by

$$Z_i = T_i \, \Theta_{iV} + z_i \, \Theta_{iH} . \qquad (30)$$

The components of Z along the visible units are just the target value specified by T. This equation is useful as a master equation because if the weights can be chosen so that the z_i of the visible units relax to the target values T_i, then a fixed point of (29) is also a fixed point of (1). It can be concluded therefore, that by training the weights of the master network one is also training the weights of the slave network. Note that the form of Z implies that equation (29) can be rewritten as

$$dz_i/dt = -z_i + g_i(\sum_{k \varepsilon H} w_{ik} z_k - \theta_i) + I_i \qquad (31)$$

where

$$\theta_i = - \sum_{k \varepsilon V} w_{ik} T_k . \qquad (32)$$

From equations (31) and (32) it is clear that the dynamics of the master system is driven by the thresholds which depend on the targets.

To derive the adaptive equations consider the objective function

$$E_{master} = \frac{1}{2} \sum_{i=1}^{N} J_i^2 \; . \tag{33}$$

where

$$J_i = Z^{\infty}_i - z^{\infty}_i \; . \tag{34}$$

It is straightforward to apply the steps discussed in previous sections to E_{master}. This results in adaptive equations for the weights. The mathematical details will be omitted since they are essentially the same as before, the gradient descent equation is

$$dw_{ij}/dt = \eta y^{\infty}_i Z^{\infty}_j \tag{35}$$

where y^{∞} is the steady state solution of

$$dy_k/dt = - y_k + g'_k(v_k)\{\Theta_{iH}\sum_r w_{rk}y_r + J_k\} \tag{36}$$

where

$$v_i = \sum_{i \, \varepsilon H} w_{ik} Z^{\infty}_k \; . \tag{37}$$

Equations (31), and (35)-37) define the dynamics of the master network. To train the slave network to be an autoassociative memory it is necessary to use the stored memories as the initial states of the master network, i.e.

$$z_i(t_o) = T_i \, \Theta_{iV} + b_i \, \Theta_{iH} \tag{39}$$

where b_i is an arbitrary value as before. The previous discussions concerning the stability of the three equations (1), (13) and (17) apply to equations (31) (35) and (36) as well. It is also possible to derive the adaptive equations for a higher order associative network, but this will not be done here.

Only preliminary computer simulations have been performed with this algorithm to verify their validity, but more extensive experiments are in progress. The first simulation was with a fully connected network with 10 visible units and 5 hidden units. The training set consisted of four random binary vectors with the magnitudes of the vectors adjusted so that $0.1 \leq T_i \leq 0.9$. The equations were approximated by first order finite difference equations with $\Delta t = 1$ and $\eta = 1$. The training was performed with the deterministic method for learning multiple associations. Figure 1. shows E_{total} as a function of the number of updates for both the master and slave networks. E_{total} for the slave exhibits discontinous behaviour because the trajectory through the weight space causes $x(t_o)$ to cut across the basins of attraction for the fixed points of equation (1).

The number of updates required for the network to learn the patterns is relatively modest and can be reduced further by increasing η. This suggests that learning can occur very rapidly in this type of network.

Discussion

The algorithms presented here by no means exhaust the class of possible adaptive algorithms which can be obtained with this formalism. Nor is the choice of gradient descent a crucial feature in this formalism. The key idea is that it is possible to express the gradient of an objective function as the outer product of vectors which can be calculated by dynamical systems. This outer product form is also responsible for the fact that the gradient can be calculated with only $O(N^2)$ operations in a fully connected or randomly connected network. In fact the number of operations per

610

weight update is proportional to the number of connections in the network. The methods used here will generalize to calculate higher order derivatives of the objective function as well.

The fact that the algorithms are expressed as differential equations suggests that they may be implemented in analog electronic or optical hardware.

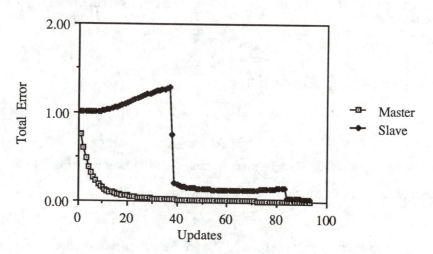

figure 1. E_{total} as a function of the the number of updates.

References

(1) J. J. Hopfield, *Neural Networks as Physical Systems with Emergent Collective Computational Abilities*, Proc. Nat. Acad. Sci. USA, Bio.79, 2554-2558, (1982)

(2) J. J. Hopfield, *Neurons with graded response have collective computational properties like those of two-state neurons*, Proc. Nat. Acad. Sci. USA, Bio. 81, 3088-3092, (1984)

(3) Shun-Ichi Amari, IEEE Trans. on Systems Man and Cybernetics, 2, 643-657, (1972)

(4) Shun-Ichi Amari, in Systems Neuroscience, ed. Jacqueline Metzler, Academic press, (1977)

(5) D. E. Rumelhart, G. E. Hinton and R.J. Williams, in Parallel Distributed Processing, edited by D. E. Rumelhart and J. L. McClelland, M.I.T. press, (1986)

(6) David B. Parker, *Learning-Logic*, Invention Report, S81-64, File 1, Office of Technology Licensing, Stanford University, October, 1982

(7) Y. LeChun, Proceedings of Cognitiva, 85, p. 599, (1985)

(8) David B. Parker, *Second Order Backpropagation: Implementing an Optimal O(n) Approximation to Newton's Method as an Artificial Neural Network*, submitted to Computer, (1987)

(9) Fernando J. Pineda, *Generalization of backpropagation to recurrent neural networks*, Phys. Rev. Lett., 18, 2229-2232, (1987)

(10) Luis B. Almeida, in the Proceedings of the IEEE First Annual International Conference on Neural Networks, San Diego, California, June 1987, edited by

M. Caudil and C. Butler (to be published This is a discrete version of the algorithm presented as the first example

(11) Alan Lapedes and Robert Farber, *A self-optimizing, nonsymmetrical neural net for content addressable memory and pattern recognition*, Physica, D22, 247-259, (1986), see also, *Programming a Massively Parallel, Computation Universal System: Static Behaviour*, in Neural Networks for Computing Snowbird, UT 1986, AIP Conference Proceedings, 151, (1986), edited by John S. Denker

(12) Terrence J. Sejnowski, *Higher-order Boltzmann Machines*, Draft preprint obtained from author

(13) Y.C. Lee, Gary Doolen, H.H. Chen, G.Z. Sun, Tom Maxwell, H.Y. Lee and C. Lee Giles, *Machine Learning using a higher order correlation network*, Physica D22, 276-306, (1986)

Constrained Differential Optimization

John C. Platt

Alan H. Barr

California Institute of Technology, Pasadena, CA 91125

Abstract

Many optimization models of neural networks need constraints to restrict the space of outputs to a subspace which satisfies external criteria. Optimizations using energy methods yield "forces" which act upon the state of the neural network. The penalty method, in which quadratic energy constraints are added to an existing optimization energy, has become popular recently, but is not guaranteed to satisfy the constraint conditions when there are other forces on the neural model or when there are multiple constraints. In this paper, we present the *basic differential multiplier method* (BDMM), which satisfies constraints exactly; we create forces which gradually apply the constraints over time, using "neurons" that estimate Lagrange multipliers.

The basic differential multiplier method is a differential version of the method of multipliers from Numerical Analysis. We prove that the differential equations locally converge to a constrained minimum.

Examples of applications of the differential method of multipliers include enforcing permutation codewords in the analog decoding problem and enforcing valid tours in the traveling salesman problem.

1. Introduction

Optimization is ubiquitous in the field of neural networks. Many learning algorithms, such as back-propagation,[18] optimize by minimizing the difference between expected solutions and observed solutions. Other neural algorithms use differential equations which minimize an energy to solve a specified computational problem, such as associative memory,[9] differential solution of the traveling salesman problem,[5,10] analog decoding,[15] and linear programming.[19] Furthermore, Lyapunov methods show that various models of neural behavior find minima of particular functions.[4,9]

Solutions to a constrained optimization problem are restricted to a subset of the solutions of the corresponding unconstrained optimization problem. For example, a mutual inhibition circuit[6] requires one neuron to be "on" and the rest to be "off". Another example is the traveling salesman problem,[13] where a salesman tries to minimize his travel distance, subject to the constraint that he must visit every city exactly once. A third example is the curve fitting problem, where elastic splines are as smooth as possible, while still going through data points.[3] Finally, when digital decisions are being made on analog data, the answer is constrained to be bits, either 0 or 1.[14]

A constrained optimization problem can be stated as

$$\text{minimize } f(\underline{x}),$$
$$\text{subject to } g(\underline{x}) = 0, \tag{1}$$

where $\underline{x}$ is the state of the neural network, a position vector in a high-dimensional space; $f(\underline{x})$ is a scalar energy, which can be imagined as the height of a landscape as a function of position $\underline{x}$; $g(\underline{x}) = 0$ is a scalar equation describing a subspace of the state space. During constrained optimization, the state should be attracted to the subspace $g(\underline{x}) = 0$, then slide along the subspace until it reaches the locally smallest value of $f(\underline{x})$ on $g(\underline{x}) = 0$.

In section 2 of the paper, we describe classical methods of constrained optimization, such as the penalty method and Lagrange multipliers.

Section 3 introduces the basic differential multiplier method (BDMM) for constrained optimization, which calculates a good local minimum. If the constrained optimization problem is convex, then the local minimum is the global minimum; in general, finding the global minimum of non-convex problems is fairly difficult.

In section 4, we show a Lyapunov function for the BDMM by drawing on an analogy from physics.

In section 5, augmented Lagrangians, an idea from optimization theory, enhances the convergence properties of the BDMM.

In section 6, we apply the differential algorithm to two neural problems, and discuss the insensitivity of BDMM to choice of parameters. Parameter sensitivity is a persistent problem in neural networks.

2. Classical Methods of Constrained Optimization

This section discusses two methods of constrained optimization, the penalty method and Lagrange multipliers. The penalty method has been previously used in differential optimization. The basic differential multiplier method developed in this paper applies Lagrange multipliers to differential optimization.

2.1. The Penalty Method

The penalty method is analogous to adding a rubber band which attracts the neural state to the subspace $g(\underline{x}) = 0$. The penalty method adds a quadratic energy term which penalizes violations of constraints. [8] Thus, the constrained minimization problem (1) is converted to the following unconstrained minimization problem:

$$\text{minimize } \mathcal{E}_{\text{penalty}}(\underline{x}) = f(\underline{x}) + c(g(\underline{x}))^2. \tag{2}$$

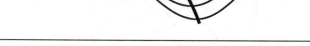

$g(\underline{x}) = 0$

$g(\underline{x})^2$

Figure 1. The penalty method makes a trough in state space

The penalty method can be extended to fulfill multiple constraints by using more than one rubber band. Namely, the constrained optimization problem

$$\begin{aligned} &\text{minimize } f(\underline{x}), \\ &\text{subject to } g_\alpha(\underline{x}) = 0; \quad \alpha = 1, 2, \ldots, n; \end{aligned} \tag{3}$$

is converted into unconstrained optimization problem

$$\text{minimize } \mathcal{E}_{\text{penalty}}(\underline{x}) = f(\underline{x}) + \sum_{\alpha=1}^{n} c_\alpha(g_\alpha(\underline{x}))^2. \tag{4}$$

The penalty method has several convenient features. First, it is easy to use. Second, it is globally convergent to the correct answer as $c_\alpha \to \infty$.[8] Third, it allows compromises between constraints. For example, in the case of a spline curve fitting input data, there can be a compromise between fitting the data and making a smooth spline.

However, the penalty method has a number of disadvantages. First, for finite constraint strengths c_α, it doesn't fulfill the constraints exactly. Using multiple rubber band constraints is like building a machine out of rubber bands: the machine would not hold together perfectly. Second, as more constraints are added, the constraint strengths get harder to set, especially when the size of the network (the dimensionality of $\underline{x}$) gets large.

In addition, there is a dilemma to the setting of the constraint strengths. If the strengths are small, then the system finds a deep local minimum, but does not fulfill all the constraints. If the strengths are large, then the system quickly fulfills the constraints, but gets stuck in a poor local minimum.

2.2. Lagrange Multipliers

Lagrange multiplier methods also convert constrained optimization problems into unconstrained extremization problems. Namely, a solution to the equation (1) is also a critical point of the energy

$$\mathcal{E}_{\text{Lagrange}}(\underline{x}) = f(\underline{x}) + \lambda g(\underline{x}).\tag{5}$$

λ is called the Lagrange multiplier for the constraint $g(\underline{x}) = 0$.[8]

A direct consequence of equation (5) is that the gradient of f is collinear to the gradient of g at the constrained extrema (see Figure 2). The constant of proportionality between ∇f and ∇g is $-\lambda$:

$$\nabla \mathcal{E}_{\text{Lagrange}} = 0 = \nabla f + \lambda \nabla g.\tag{6}$$

We use the collinearity of ∇f and ∇g in the design of the BDMM.

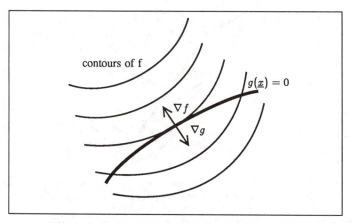

contours of f

$g(\underline{x}) = 0$

∇f

∇g

Figure 2. At the constrained minimum, $\nabla f = -\lambda \nabla g$

A simple example shows that Lagrange multipliers provide the extra degrees of freedom necessary to solve constrained optimization problems. Consider the problem of finding a point (x, y) on the line $x + y = 1$ that is closest to the origin. Using Lagrange multipliers,

$$\mathcal{E}_{\text{Lagrange}} = x^2 + y^2 + \lambda(x + y - 1)\tag{7}$$

Now, take the derivative with respect to all variables, x, y, and λ.

$$\frac{\partial \mathcal{E}_{\text{Lagrange}}}{\partial x} = 2x + \lambda = 0$$

$$\frac{\partial \mathcal{E}_{\text{Lagrange}}}{\partial y} = 2y + \lambda = 0\tag{8}$$

$$\frac{\partial \mathcal{E}_{\text{Lagrange}}}{\partial \lambda} = x + y - 1 = 0$$

With the extra variable λ, there are now three equations in three unknowns. In addition, the last equation is precisely the constraint equation.

3. The Basic Differential Multiplier Method for Constrained Optimization

This section presents a new "neural" algorithm for constrained optimization, consisting of differential equations which estimate Lagrange multipliers. The neural algorithm is a variation of the method of multipliers, first presented by Hestenes[9] and Powell[16].

3.1. Gradient Descent does not work with Lagrange Multipliers

The simplest differential optimization algorithm is *gradient descent*, where the state variables of the network slide downhill, opposite the gradient. Applying gradient descent to the energy in equation (5) yields

$$\dot{x}_i = -\frac{\partial \mathcal{E}_{\text{Lagrange}}}{\partial x_i} = -\frac{\partial f}{\partial x_i} - \lambda \frac{\partial g}{\partial x_i},$$
$$\dot{\lambda} = -\frac{\partial \mathcal{E}_{\text{Lagrange}}}{\partial \lambda} = -g(\underline{x}). \tag{9}$$

Note that there is a auxiliary differential equation for λ, which is an additional "neuron" necessary to apply the constraint $g(\underline{x}) = 0$. Also, recall that when the system is at a constrained extremum, $\nabla f = -\lambda \nabla g$, hence, $\dot{x}_i = 0$.

Energies involving Lagrange multipliers, however, have critical points which tend to be saddle points. Consider the energy in equation (5). If $\underline{x}$ is frozen, the energy can be decreased by sending λ to $+\infty$ or $-\infty$.

Gradient descent does not work with Lagrange multipliers, because a critical point of the energy in equation (5) need not be an attractor for (9). A stationary point must be a local minimum in order for gradient descent to converge.

3.2. The New Algorithm: the Basic Differential Multiplier Method

We present an alternative to differential gradient descent that estimates the Lagrange multipliers, so that the constrained minima are attractors of the differential equations, instead of "repulsors." The differential equations that solve (1) is

$$\dot{x}_i = -\frac{\partial f}{\partial x_i} - \lambda \frac{\partial g}{\partial x_i},$$
$$\dot{\lambda} = +g(\underline{x}). \tag{10}$$

Equation (10) is similar to equation (9). As in equation (9), constrained extrema of the energy (5) are stationary points of equation (10). Notice, however, the sign inversion in the equation for λ, as compared to equation (9). The equation (10) is performing gradient *ascent* on λ. The sign flip makes the BDMM stable, as shown in section 4.

Equation (10) corresponds to a neural network with anti-symmetric connections between the λ neuron and all of the $\underline{x}$ neurons.

3.3. Extensions to the Algorithm

One extension to equation (10) is an algorithm for constrained minimization with multiple constraints. Adding an extra neuron for every equality constraint and summing all of the constraint forces creates the energy

$$\mathcal{E}_{\text{multiple}} = f(\underline{x}) + \sum_{\alpha} \lambda_\alpha g_\alpha(\underline{x}), \tag{11}$$

which yields differential equations

$$\dot{x}_i = -\frac{\partial f}{\partial x_i} - \sum_{\alpha} \lambda_\alpha \frac{\partial g\alpha}{\partial x_i},$$
$$\dot{\lambda}_\alpha = +g_\alpha(\underline{x}). \tag{12}$$

Another extension is constrained minimization with inequality constraints. As in traditional optimization theory,[8] one uses extra slack variables to convert inequality constraints into equality constraints. Namely, a constraint of the form $h(\underline{x}) \geq 0$ can be expressed as

$$g(\underline{x}) = h(\underline{x}) - z^2. \tag{13}$$

Since z^2 must always be positive, then $h(\underline{x})$ is constrained to be positive. The slack variable z is treated like a component of $\underline{x}$ in equation (10). An inequality constraint requires two extra neurons, one for the slack variable x and one for the Lagrange multiplier λ.

Alternatively, the inequality constraint can be represented as an equality constraint. For example, if $h(\underline{x}) \geq 0$, then the optimization can be constrained with $g(\underline{x}) = h(\underline{x})$, when $h(\underline{x}) \geq 0$; and $g(\underline{x}) = 0$ otherwise.

4. Why the algorithm works

The system of differential equations (10) (the BDMM) gradually fulfills the constraints. Notice that the function $g(\underline{x})$ can be replaced by $kg(\underline{x})$, without changing the location of the constrained minimum. As k is increased, the state begins to undergo damped oscillation about the constraint subspace $g(\underline{x}) = 0$. As k is increased further, the frequency of the oscillations increase, and the time to convergence increases.

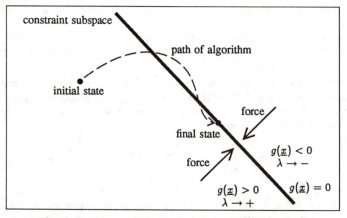

Figure 3. The state is attracted to the constraint subspace

The damped oscillations of equation (10) can be explained by combining both of the differential equations into one second-order differential equation.

$$\ddot{x}_i + \sum_j \left(\frac{\partial^2 f}{\partial x_i \partial x_j} + \lambda \frac{\partial^2 g}{\partial x_i \partial x_j} \right) \dot{x}_j + g \frac{\partial g}{\partial x_i} = 0. \tag{14}$$

Equation (14) is the equation for a damped mass system, with an inertia term $\ddot{x}_i$, a damping matrix

$$A_{ij} = \frac{\partial^2 f}{\partial x_i \partial x_j} + \lambda \frac{\partial^2 g}{\partial x_i \partial x_j}, \tag{15}$$

and an internal force, $g \partial g / \partial x_i$, which is the derivative of the internal energy

$$U = \frac{1}{2}(g(\underline{x}))^2. \tag{16}$$

If the system is damped and the state remains bounded, the state falls into a constrained minima.

As in physics, we can construct a total energy of the system, which is the sum of the kinetic and potential energies.

$$E = T + U = \sum_i \frac{1}{2}(\dot{x}_i)^2 + \frac{1}{2}(g(\underline{x}))^2. \tag{17}$$

If the total energy is decreasing with time and the state remains bounded, then the system will dissipate any extra energy, and will settle down into the state where

$$g(\underline{x}) = 0,$$
$$\dot{x}_i = \frac{\partial f}{\partial x_i} + \lambda \frac{\partial g}{\partial x_i} = 0. \tag{18}$$

which is a constrained extremum of the original problem in equation (1).

The time derivative of the total energy in equation (17) is

$$\dot{E} = \sum_i \ddot{x}_i \dot{x}_i + g(\underline{x}) \frac{\partial g}{\partial x_i} \dot{x}_i$$
$$= -\sum_{i,j} \dot{x}_i A_{ij} \dot{x}_j. \tag{19}$$

If damping matrix A_{ij} is positive definite, the system converges to fulfill the constraints.

BDMM always converges for a special case of constrained optimization: quadratic programming. A quadratic programming problem has a quadratic function $f(\underline{x})$ and a piecewise linear continuous function $g(\underline{x})$ such that

$$\frac{\partial^2 f}{\partial x_i \partial x_j} \text{ is positive definite}; \qquad \frac{\partial^2 g}{\partial x_i \partial x_j} = 0. \tag{20}$$

Under these circumstances, the damping matrix A_{ij} is positive definite for all $\underline{x}$ and λ, so that the system converges to the constraints.

4.1. Multiple constraints

For the case of multiple constraints, the total energy for equation (12) is

$$E = T + U = \sum_i \frac{1}{2}(\dot{x}_i)^2 + \sum_\alpha \frac{1}{2}g_\alpha(\underline{x})^2. \tag{21}$$

and the time derivative is

$$\dot{E} = \sum_i \ddot{x}_i \dot{x}_i + \sum_\alpha g_\alpha(\underline{x}) \frac{\partial g_\alpha}{\partial x_i} \dot{x}_i$$
$$= -\sum_{i,j} \dot{x}_i \left(\frac{\partial^2 f}{\partial x_i \partial x_j} + \sum_\alpha \lambda_\alpha \frac{\partial^2 g_\alpha}{\partial x_i \partial x_j} \right) \dot{x}_j. \tag{22}$$

Again, BDMM solves a quadratic programming problem, if a solution exists. However, it is possible to pose a problem that has contradictory constraints. For example,

$$g_1(x) = x = 0, \qquad g_2(x) = x - 1 = 0 \tag{23}$$

In the case of conflicting constraints, the BDMM compromises, trying to make each constraint g_α as small as possible. However, the Lagrange multipliers λ_α goes to $\pm\infty$ as the constraints oppose each other. It is possible, however, to arbitrarily limit the λ_α at some large absolute value.

LaSalle's invariance theorem[12] is used to prove that the BDMM eventually fulfills the constraints. Let G be an open subset of R^n. Let F be a subset of G^*, the closure of G, where the system of differential equations (12) is at an equilibrium.

$$F = \{\underline{x}, \underline{\lambda} \mid \dot{x}_i = 0, \dot{\lambda}_\alpha = 0, \underline{x}, \underline{\lambda} \in G^*\} \tag{24}$$

If the damping matrix

$$\frac{\partial^2 f}{\partial x_i \partial x_j} + \sum_\alpha \lambda_\alpha \frac{\partial^2 g_\alpha}{\partial x_i \partial x_j} \tag{25}$$

is positive definite in G, if $x_i(t)$ and $\lambda_\alpha(t)$ are bounded, and remain in G for all time, and if F is non-empty, then F is the largest invariant set in G^*, hence, by LaSalle's invariance theorem, the system $x_i(t), \lambda_\alpha(t)$ approaches F as $t \to \infty$.

5. The Modified Differential Method of Multipliers

This section presents the *modified differential multiplier method* (MDMM), which is a modification of the BDMM with more robust convergence properties. For a given constrained optimization problem, it is frequently necessary to alter the BDMM to have a region of positive damping surrounding the constrained minima. The non-differential method of multipliers from Numerical Analysis also has this difficulty. [2] Numerical Analysis combines the multiplier method with the penalty method to yield a modified multiplier method that is locally convergent around constrained minima. [2]

The BDMM is completely compatible with the penalty method. If one adds a penalty force to equation (10) corresponding to an quadratic energy

$$E_{\text{penalty}} = \frac{c}{2}(g(\underline{x}))^2. \tag{26}$$

then the set of differential equations for MDMM is

$$\dot{x}_i = -\frac{\partial f}{\partial x_i} - \lambda \frac{\partial g}{\partial x_i} - cg \frac{\partial g}{\partial x_i}, \tag{27}$$

$$\dot{\lambda} = g(\underline{x}).$$

The extra force from the penalty does *not* change the position of the stationary points of the differential equations, because the penalty force is 0 when $g(\underline{x}) = 0$. The damping matrix is modified by the penalty force to be

$$A_{ij} = \frac{\partial^2 f}{\partial x_i \partial x_j} + \lambda \frac{\partial^2 g}{\partial x_i \partial x_j} + c \frac{\partial g}{\partial x_i} \frac{\partial g}{\partial x_j} + cg \frac{\partial^2 g}{\partial x_i \partial x_j}. \tag{28}$$

There is a theorem [1] that states that there exists a $c^* > 0$ such that if $c > c^*$, the damping matrix in equation (28) is positive definite at constrained minima. Using continuity, the damping matrix is positive definite in a region R surrounding each constrained minimum. If the system starts in the region R and remains bounded and in R, then the convergence theorem at the end of section 4 is applicable, and MDMM will converge to a constrained minimum.

The minimum necessary penalty strength c for the MDMM is usually much less than the strength needed by the penalty method alone.[2]

6. Examples

This section contains two examples which illustrate the use of the BDMM and the MDMM. First, the BDMM is used to find a good solution to the planar traveling salesman problem. Second, the MDMM is used to enforcing mutual inhibition and digital results in the task of analog decoding.

6.1. Planar Traveling Salesman

The traveling salesman problem (TSP) is, given a set of cities lying in the plane, find the shortest closed path that goes through every city exactly once. Finding the shortest path is NP-complete.

Finding a nearly optimal path, however, is much easier than finding a globally optimal path. There exist many heuristic algorithms for approximately solving the traveling salesman problem.[5,10,11,13] The solution presented in this section is moderately effective and illustrates the independence of BDMM to changes in parameters.

Following Durbin and Willshaw,[5] we use an elastic snake to solve the TSP. A snake is a discretized curve which lies on the plane. The elements of the snake are points on the plane, (x_i, y_i). A snake is a locally connected neural network, whose neural outputs are positions on the plane.

The snake minimizes its length

$$\sum_i (x_{i+1} - x_i)^2 - (y_{i+1} - y_i)^2, \tag{29}$$

subject to the constraint that the snake must lie on the cities:

$$k(x^* - x_c) = 0, \qquad k(y^* - y_c) = 0, \tag{30}$$

where (x^*, y^*) are city coordinates, (x_c, y_c) is the closest snake point to the city, and k is the constraint strength.

The minimization in equation (29) is quadratic and the constraints in equation (30) are piecewise linear, corresponding to a C^0 continuous potential energy in equation (21). Thus, the damping is positive definite, and the system converges to a state where the constraints are fulfilled.

In practice, the snake starts out as a circle. Groups of cities grab onto the snake, deforming it. As the snake gets close to groups of cities, it grabs onto a specific ordering of cities that locally minimize its length (see Figure 4).

The system of differential equations that solve equations (29) and (30) are piecewise linear. The differential equations for x_i and y_i are solved with implicit Euler's method, using tridiagonal LU decomposition to solve the linear system.[17] The points of the snake are sorted into bins that divide the plane, so that the computation of finding the nearest point is simplified.

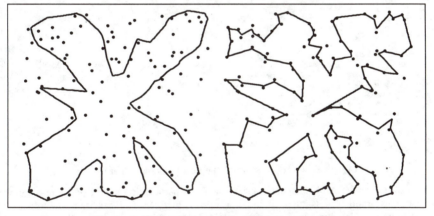

Figure 4. The snake eventually attaches to the cities

The constrained minimization in equations (29) and (30) is a reasonable method for approximately solving the TSP. For 120 cities distributed in the unti square, and 600 snake points, a numerical step size of 100 time units, and a constraint strength of 5×10^{-3}, the tour lengths are $6\% \pm 2\%$ longer than that yielded by simulated annealing[11]. Empirically, for 30 to 240 cities, the time needed to compute the final city ordering scales as $N^{1.6}$, as compared to the Kernighan-Lin method[13], which scales roughly as $N^{2.2}$.

The constraint strength is usable for both a 30 city problem and a 240 city problem. Although changing the constraint strength affects the performance, the snake attaches to the cities for any non-zero constraint strength. Parameter adjustment does not seem to be an issue as the number of cities increases, unlike the penalty method.

6.2. Analog Decoding

Analog decoding uses analog signals from a noisy channel to reconstruct codewords. Analog decoding has been performed neurally,[15] with a code space of permutation matrices, out of the possible space of binary matrices.

To perform the decoding of permutation matrices, the nearest permutation matrix to the signal matrix must be found. In other words, find the nearest matrix to the signal matrix, subject to the constraint that the matrix has on/off binary elements, and has exactly one "on" per row and one "on" per column. If the signal matrix is I_{ij} and the result is V_{ij}, then minimize

$$-\sum_{i,j} V_{ij} I_{ij} \tag{31}$$

subject to constraints

$$V_{ij}(1 - V_{ij}) = 0; \qquad \sum_i V_{ij} - 1 = 0; \qquad \sum_j V_{ij} - 1 = 0. \tag{32}$$

In this example, the first constraint in equation (32) forces crisp digital decisions. The second and third constraints are mutual inhibition along the rows and columns of the matrix.

The optimization in equation (31) is not quadratic, it is linear. In addition, the first constraint in equation (32) is non-linear. Using the BDMM results in undamped oscillations. In order to converge onto a constrained minimum, the MDMM must be used. For both a 5×5 and a 20×20 system, a $c = 0.2$ is adequate for damping the oscillations. The choice of c seems to be reasonably insensitive to the size of the system, and a wide range of c, from 0.02 to 2.0, damps the oscillations.

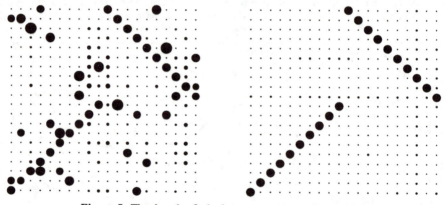

Figure 5. The decoder finds the nearest permutation matrix

In a test of the MDMM, a signal matrix which is a permutation matrix plus some noise, with a signal-to-noise ratio of 4 is supplied to the network. In figure 5, the system has turned on the correct neurons but also many incorrect neurons. The constraints start to be applied, and eventually the system reaches a permutation matrix. The differential equations do not need to be reset. If a new signal matrix is applied to the network, the neural state will move towards the new solution.

7. Conclusions

In the field of neural networks, there are differential optimization algorithms which find local solutions to non-convex problems. The basic differential multiplier method is a modification of a standard constrained optimization algorithm, which improves the capability of neural networks to perform constrained optimization.

The BDMM and the MDMM offer many advantages over the penalty method. First, the differential equations (10) are much less stiff than those of the penalty method. Very large quadratic terms are not needed by the MDMM in order to strongly enforce the constraints. The energy terrain for the penalty method looks like steep canyons, with gentle floors; finding minima of these types of energy surfaces is numerically difficult. In addition, the steepness of the penalty terms is usually sensitive to the dimensionality of the space. The differential multiplier methods are promising techniques for alleviating stiffness.

The differential multiplier methods separate the speed of fulfilling the constraints from the accuracy of fulfilling the constraints. In the penalty method, as the strengths of a constraint goes to ∞, the constraint is fulfilled, but the energy has many undesirable local minima. The differential multiplier methods allow one to choose how quickly to fulfill the constraints.

The BDMM fulfills constraints exactly and is compatible with the penalty method. Addition of penalty terms in the MDMM does not change the stationary points of the algorithm, and sometimes helps to damp oscillations and improve convergence.

Since the BDMM and the MDMM are in the form of first-order differential equations, they can be directly implemented in hardware. Performing constrained optimization at the raw speed of analog VLSI seems like a promising technique for solving difficult perception problems.[14]

There exist Lyapunov functions for the BDMM and the MDMM. The BDMM converges globally for quadratic programming. The MDMM is provably convergent in a local region around the constrained minima. Other optimization algorithms, such as Newton's method,[17] have similar local convergence properties. The global convergence properties of the BDMM and the MDMM are currently under investigation.

In summary, the differential method of multipliers is a useful way of enforcing constraints on neural networks for enforcing syntax of solutions, encouraging desirable properties of solutions, and making crisp decisions.

Acknowledgments

This paper was supported by an AT&T Bell Laboratories Fellowship (JCP).

References

1. K. J. Arrow, L. Hurwicz, H. Uzawa, *Studies in Linear and Nonlinear Programming*, (Stanford University Press, Stanford, CA, 1958).

2. D. P. Bertsekas, *Automatica*, **12**, 133-145, (1976).

3. C. de Boor, *A Practical Guide to Splines*, (Springer-Verlag, NY, 1978).

4. M. A. Cohen, S. Grossberg, *IEEE Trans. Systems, Man, and Cybernetics*, , 815-826, (1983).

5. R. Durbin, D. Willshaw, *Nature*, **326**, 689-691, (1987).

6. J. C. Eccles, *The Physiology of Nerve Cells*, (Johns Hopkins Press, Baltimore, 1957).

7. M. R. Hestenes, *J. Opt. Theory Appl.*, **4**, 303-320, (1969).

8. M. R. Hestenes, *Optimization Theory*, (Wiley & Sons, NY, 1975).

9. J. J. Hopfield, *PNAS*, **81**, 3088, (1984).

10. J. J. Hopfield, D. W. Tank, *Biological Cybernetics*, **52**, 141, (1985).

11. S. Kirkpatrick, C. D. Gelatt, C. M. Vecchi, *Science*, **220**, 671-680, (1983).

12. J. LaSalle, *The Stability of Dynamical Systems*, (SIAM, Philadelphia, 1976).

13. S. Lin, B. W. Kernighan, *Oper. Res.*, **21**, 498-516 (1973).

14. C. A. Mead, *Analog VLSI and Neural Systems*, (Addison-Wesley, Reading, MA, TBA).

15. J. C. Platt, J. J. Hopfield, in *AIP Conf. Proc. 151: Neural Networks for Computing* (J. Denker ed.) 364-369, (American Institute of Physics, NY, 1986).

16. M. J. Powell, in *Optimization*, (R. Fletcher, ed.), 283-298, (Academic Press, NY, 1969).

17. W. H. Press, B. P. Flannery, S. A. Teukolsky, W. T. Vetterling, *Numerical Recipes*, (Cambridge University Press, Cambridge, 1986).

18. D. Rumelhart, G. Hinton, R. Williams, in *Parallel Distributed Processing*, (D. Rumelhart, ed.), **1**, 318-362, (MIT Press, Cambridge, MA, 1986).

19. D. W. Tank, J. J. Hopfield, *IEEE Trans. Cir. & Sys.*, **CAS-33**, no. 5, 533-541 (1986).

LEARNING A COLOR ALGORITHM FROM EXAMPLES

Anya C. Hurlbert and Tomaso A. Poggio
Artificial Intelligence Laboratory and Department of Brain and Cognitive Sciences,
Massachusetts Institute of Technology, Cambridge, Massachusetts 02139, USA

ABSTRACT

A lightness algorithm that separates surface reflectance from illumination in a
Mondrian world is synthesized automatically from a set of examples, pairs of input
(image irradiance) and desired output (surface reflectance). The algorithm, which resembles a new lightness algorithm recently proposed by Land, is approximately equivalent to filtering the image through a center-surround receptive field in individual chromatic channels. The synthesizing technique, optimal linear estimation, requires only
one assumption, that the operator that transforms input into output is linear. This
assumption is true for a certain class of early vision algorithms that may therefore be
synthesized in a similar way from examples. Other methods of synthesizing algorithms
from examples, or "learning", such as backpropagation, do not yield a significantly different or better lightness algorithm in the Mondrian world. The linear estimation and
backpropagation techniques both produce simultaneous brightness contrast effects.

The problems that a visual system must solve in decoding two-dimensional images
into three-dimensional scenes (inverse optics problems) are difficult: the information
supplied by an image is not sufficient by itself to specify a unique scene. To reduce
the number of possible interpretations of images, visual systems, whether artificial
or biological, must make use of natural constraints, assumptions about the physical
properties of surfaces and lights. Computational vision scientists have derived effective
solutions for some inverse optics problems (such as computing depth from binocular
disparity) by determining the appropriate natural constraints and embedding them in
algorithms. How might a visual system discover and exploit natural constraints on its
own? We address a simpler question: Given only a set of examples of input images and
desired output solutions, can a visual system synthesize, or "learn", the algorithm that
converts input to output? We find that an algorithm for computing color in a restricted
world can be constructed from examples using standard techniques of optimal linear
estimation.

The computation of color is a prime example of the difficult problems of inverse
optics. We do not merely discriminate between different wavelengths of light; we assign

roughly constant colors to objects even though the light signals they send to our eyes change as the illumination varies across space and chromatic spectrum. The computational goal underlying color constancy seems to be to extract the invariant surface spectral reflectance properties from the image irradiance, in which reflectance and illumination are mixed[1].

Lightness algorithms [2-8], pioneered by Land, assume that the color of an object can be specified by its lightness, or relative surface reflectance, in each of three independent chromatic channels, and that lightness is computed in the same way in each channel. Computing color is thereby reduced to extracting surface reflectance from the image irradiance in a single chromatic channel.

The image irradiance, s', is proportional to the product of the illumination intensity e' and the surface reflectance r' in that channel:

$$s'(x,y) = r'(x,y)e'(x,y). \tag{1}$$

This form of the image intensity equation is true for a Lambertian reflectance model, in which the irradiance s' has no specular components, and for appropriately chosen color channels [9]. Taking the logarithm of both sides converts it to a sum:

$$s(x,y) = r(x,y) + e(x,y), \tag{2}$$

where $s = log(s')$, $r = log(r')$ and $e = log(e')$.

Given $s(x,y)$ alone, the problem of solving Eq. 2 for $r(x,y)$ is underconstrained. Lightness algorithms constrain the problem by restricting their domain to a world of Mondrians, two-dimensional surfaces covered with patches of random colors[2] and by exploiting two constraints in that world: (i) $r'(x,y)$ is uniform within patches but has sharp discontinuities at edges between patches and (ii) $e'(x,y)$ varies smoothly across the Mondrian. Under these constraints, lightness algorithms can recover a good approximation to $r(x,y)$ and so can recover lightness triplets that label roughly constant colors [10].

We ask whether it is possible to synthesize from examples an algorithm that extracts reflectance from image irradiance, and whether the synthesized algorithm will resemble existing lightness algorithms derived from an explicit analysis of the constraints. We make one assumption, that the operator that transforms irradiance into reflectance is linear. Under that assumption, motivated by considerations discussed later, we use optimal linear estimation techniques to synthesize an operator from examples. The examples are pairs of images: an input image of a Mondrian under illumination that varies smoothly across space and its desired output image that displays the reflectance of the Mondrian without the illumination. The technique finds the linear estimator that best maps input into desired output, in the least squares sense.

For computational convenience we use one-dimensional "training vectors" that represent vertical scan lines across the Mondrian images (Fig. 1). We generate many

624

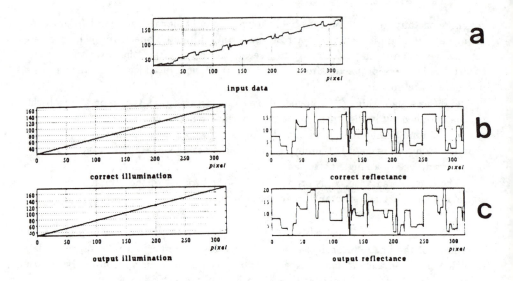

input data

correct illumination

correct reflectance

output illumination

output reflectance

a

b

c

Fig. 1. (a) The input data, a one-dimensional vector 320 pixels long. Its random Mondrian reflectance pattern is superimposed on a linear illumination gradient with a random slope and offset. (b) shows the corresponding output solution, on the left the illumination and on the right reflectance. We used 1500 such pairs of input-output examples (each different from the others) to train the operator shown in Fig. 2. (c) shows the result obtained by the estimated operator when it acts on the input data (a), not part of the training set. On the left is the illumination and on the right the reflectance, to be compared with (b). This result is fairly typical: in some cases the prediction is even better, in others it is worse.

different input vectors s by adding together different random r and e vectors, according to Eq. 2. Each vector r represents a pattern of step changes across space, corresponding to one column of a reflectance image. The step changes occur at random pixels and are of random amplitude between set minimum and maximum values. Each vector e represents a smooth gradient across space with a random offset and slope, corresponding to one column of an illumination image. We then arrange the training vectors s and r as the columns of two matrices S and R, respectively. Our goal is then to compute the optimal solution L of

$$LS = R \qquad (3)$$

where L is a linear operator represented as a matrix.

It is well known that the solution of this equation that is optimal in the least squares sense is

$$L = RS^+ \qquad (4)$$

where S^+ is the Moore-Penrose pseudoinverse [11]. We compute the pseudoinverse by overconstraining the problem – using many more training vectors than there are number of pixels in each vector – and using the straightforward formula that applies in the overconstrained case [12]: $S^+ = S^T(SS^T)^{-1}$.

The operator L computed in this way recovers a good approximation to the correct output vector r when given a new s, not part of the training set, as input (Fig. 1c). A second operator, estimated in the same way, recovers the illumination e. Acting on a random two-dimensional Mondrian L also yields a satisfactory approximation to the correct output image.

Our estimation scheme successfully synthesizes an algorithm that performs the lightness computation in a Mondrian world. *What* is the algorithm and what is its relationship to other lightness algorithms? To answer these questions we examine the structure of the matrix L. We assume that, although the operator is *not* a convolution operator, it should approximate one far from the boundaries of the image. That is, in its central part, the operator should be space-invariant, performing the same action on each point in the image. Each row in the central part of L should therefore be the same as the row above but displaced by one element to the right. Inspection of the matrix confirms this expectation. To find the form of L in its center, we thus average the rows there, first shifting them appropriately. The result, shown in Fig. 2, is a space-invariant filter with a narrow positive peak and a broad, shallow, negative surround.

Interestingly, the filter our scheme synthesizes is very similar to Land's most recent retinex operator [5], which divides the image irradiance at each pixel by a weighted average of the irradiance at all pixels in a large surround and takes the logarithm of that result to yield lightness [13]. The lightness triplets computed by the retinex operator agree well with human perception in a Mondrian world. The retinex operator and our matrix L both differ from Land's earlier retinex algorithms, which require a non-linear thresholding step to eliminate smooth gradients of illumination.

The shape of the filter in Fig. 2, particularly of its large surround, is also suggestive of the "nonclassical" receptive fields that have been found in V4, a cortical area implicated in mechanisms underlying color constancy [14-17].

The form of the space-invariant filter is similar to that derived in our earlier formal analysis of the lightness problem [8]. It is qualitatively the same as that which results from the direct application of regularization methods exploiting the spatial constraints on reflectance and illumination described above [9,18,19]. The Fourier transform of the filter of Fig. 2 is approximately a bandpass filter that cuts out low frequencies due

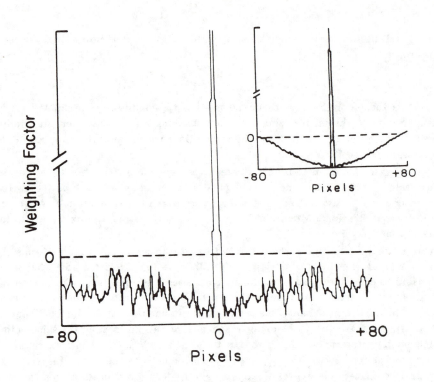

Fig. 2. The space-invariant part of the estimated operator, obtained by shifting and averaging the rows of a 160-pixel-wide central square of the matrix L, trained on a set of 1500 examples with linear illumination gradients (see Fig. 1). When logarithmic illumination gradients are used, a qualitatively similar receptive field is obtained. In a separate experiment we use a training set of one-dimensional Mondrians with either linear illumination gradients or slowly varying sinusoidal illumination components with random wavelength, phase and amplitude. The resulting filter is shown in the inset. The surrounds of both filters extend beyond the range we can estimate reliably, the range we show here.

to slow gradients of illumination and preserves intermediate frequencies due to step changes in reflectance. In contrast, the operator that recovers the illumination, e, takes the form of a low-pass filter. We stress that the entire operator L is not a space-invariant filter.

In this context, it is clear that the shape of the estimated operator should vary with the type of illumination gradient in the training set. We synthesize a second operator using a new set of examples that contain equal numbers of vectors with random, sinusoidally varying illumination components and vectors with random, linear illumination gradients. Whereas the first operator, synthesized from examples with strictly linear illumination gradients, has a broad negative surround that remains virtually constant throughout its extent, the new operator's surround (Fig. 2, inset) has a smaller extent and decays smoothly towards zero from its peak negative value in its center.

We also apply the operator in Fig. 2 to new input vectors in which the density and amplitude of the step changes of reflectance differ greatly from those on which the operator is trained. The operator performs well, for example, on an input vector representing one column of an image of a small patch of one reflectance against a uniform background of a different reflectance, the entire image under a linear illumination gradient. This result is consistent with psychophysical experiments that show that color constancy of a patch holds when its Mondrian background is replaced by an equivalent grey background [20].

The operator also produces simultaneous brightness contrast, as expected from the shape and sign of its surround. The output reflectance it computes for a patch of fixed input reflectance decreases linearly with increasing average irradiance of the input test vector in which the patch appears. Similarly, to us, a dark patch appears darker when against a light background than against a dark one.

This result takes one step towards explaining such illusions as the Koffka Ring [21]. A uniform gray annulus against a bipartite background (Fig. 3a) appears to split into two halves of different lightnesses when the midline between the light and dark halves of the background is drawn across the annulus (Fig. 3b). The estimated operator acting on the Koffka Ring of Fig. 3b reproduces our perception by assigning a lower output reflectance to the left half of the annulus (which appears darker to us) than to the right half [22]. Yet the operator gives this brightness contrast effect whether or not the midline is drawn across the annulus (Fig. 3c). Because the operator can perform only a linear transformation between the input and output images, it is not surprising that the addition of the midline in the input evokes so little change in the output. These results demonstrate that the linear operator alone cannot compute lightness in all worlds and suggest that an additional operator might be necessary to mark and guide it within bounded regions.

Our estimation procedure is motivated by our previous observation [9,23,18] that standard regularization algorithms [19] in early vision define linear mappings between input and output and therefore can be estimated associatively under certain conditions. The technique of optimal linear estimation that we use is closely related to optimal Bayesian estimation [9]. If we were to assume from the start that the optimal linear operator is space-invariant, we could considerably simplify (and streamline) the computation by using standard correlation techniques [9,24].

How does our estimation technique compare with other methods of "learning" a lightness algorithm? We can compute the regularized pseudoinverse using gradient descent on a "neural" network [25] with linear units. Since the pseudoinverse is the unique best linear approximation in the L_2 norm, a gradient descent method that

628

minimizes the square error between the actual output and desired output of a fully connected linear network is guaranteed to converge, albeit slowly. Thus gradient descent in weight space converges to the same result as our first technique, the global minimum.

a

b

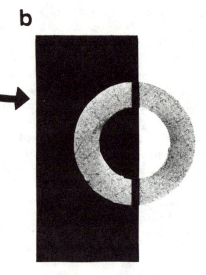

c

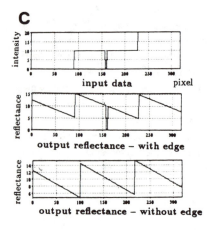

Fig. 3. (a) Koffka Ring. (b) Koffka Ring with midline drawn across annulus. (c) Horizontal scan lines across Koffka Ring. Top: Scan line starting at arrow in (b). Middle: Scan line at corresponding location in the output of linear operator acting on (b). Bottom: Scan line at same location in the output of operator acting on (a).

We also compare the linear estimation technique with a "backpropagation" network: gradient descent on a 2-layer network with sigmoid units [25] (32 inputs, 32 "hidden units", and 32 linear outputs), using training vectors 32 pixels long. The network requires an order of magnitude more time to converge to a stable configuration than does the linear estimator for the same set of 32-pixel examples. The network's performance is slightly, yet consistently, better, measured as the root-mean-square error in output, averaged over sets of at least 2000 new input vectors. Interestingly, the backpropagation network and the linear estimator err in the same way on the same input vectors. It is possible that the backpropagation network may show considerable inprovement over the linear estimator in a world more complex than the Mondrian one. We are presently examining its performance on images with real-world features such as shading, shadows, and highlights[26].

We do not think that our results mean that color constancy may be learned during a critical period by biological organisms. It seems more reasonable to consider them simply as a demonstration on a toy world that in the course of evolution a visual system may recover and exploit natural constraints hidden in the physics of the world. The significance of our results lies in the facts that a simple statistical technique may be used to synthesize a lightness algorithm from examples; that the technique does as well as other techniques such as backpropagation; and that a similar technique may be used for other problems in early vision. Furthermore, the synthesized operator resembles both Land's psychophysically-tested retinex operator and a neuronal nonclassical receptive field. The operator's properties suggest that simultaneous color (or brightness) contrast might be the result of the visual system's attempt to discount illumination gradients [27].

REFERENCES AND NOTES

1. Since we do not have perfect color constancy, our visual system must not extract reflectance exactly. The limits on color constancy might reveal limits on the underlying computation.

2. E.H. Land, *Am. Sci.* **52**, 247 (1964).

3. E.H. Land and J.J. McCann, *J. Opt. Soc. Am.* **61**, 1 (1971).

4. E.H. Land, in *Central and Peripheral Mechanisms of Colour Vision*, T. Ottoson and S. Zeki, Eds., (Macmillan, New York, 1985), pp. 5-17.

5. E.H. Land, *Proc. Nat. Acad. Sci. USA* **83**, 3078 (1986).

6. B.K.P. Horn, *Computer Graphics and Image Processing* **3**, 277 (1974).

7. A. Blake, in *Central and Peripheral Mechanisms of Colour Vision*, T. Ottoson and S. Zeki, Eds., (Macmillan, New York, 1985), pp. 45-59.

8. A. Hurlbert, *J. Opt. Soc. Am. A* **3**, 1684 (1986).

9. A. Hurlbert and T. Poggio, *Artificial Intelligence Laboratory Memo 909*, (M.I.T., Cambridge, MA, 1987).

10. $r'(x,y)$ can be recovered at best only to within a constant, since Eq. 1 is invariant under the transformation of r' into ar' and e' into $a^{-1}e'$, where a is a constant.

11. A. Albert, *Regression and the Moore-Penrose Pseudoinverse*, (Academic Press, New York, 1972).

12. The pseudoinverse, and therefore L, may also be computed by recursive techniques that improve its form as more data become available[11].

13. Our synthesized filter is not exactly identical with Land's: the filter of Fig. 2 subtracts from the value at each point the average value of the logarithm of irradiance at all pixels, rather than the logarithm of the average values. The estimated operator is therefore linear in the logarithms, whereas Land's is not. The numerical difference between the outputs of the two filters is small in most cases (Land, personal communication), and both agree well with psychophysical results.

14. R. Desimone, S.J. Schein, J. Moran and L.G. Ungerleider, *Vision Res.* **25**, 441 (1985).

15. H.M. Wild, S.R. Butler, D. Carden and J.J. Kulikowski, *Nature (London)* **313**, 133 (1985).

16. S.M. Zeki, *Neuroscience* **9**, 741 (1983).

17. S.M. Zeki, *Neuroscience* **9**, 767 (1983).

18. T. Poggio, et. al., in *Proceedings Image Understanding Workshop*, L. Baumann, Ed., (Science Applications International Corporation, McLean, VA, 1985), pp. 25-39.

19. T. Poggio, V. Torre and C. Koch, *Nature (London)* **317**, 314 (1985).

20. A. Valberg and B. Lange-Malecki, *Investigative Ophthalmology and Visual Science Supplement* **28**, 92 (1987).

21. K. Koffka, *Principles of Gestalt Psychology*, (Harcourt, Brace and Co., New York, 1935).

22. Note that the operator achieves this effect by subtracting a non-existent illumination gradient from the input signal.

23. T. Poggio and A. Hurlbert, *Artificial Intelligence Laboratory Working Paper 264*, (M.I.T., Cambridge, MA, 1984).

24. Estimation of the operator on two-dimensional examples is possible, but computationally very expensive if done in the same way. The present computer simulations require several hours when run on standard serial computers. The two-dimensional case

will need much more time (our one-dimensional estimation scheme runs orders of magnitude faster on a CM-1 Connection Machine System with 16K-processors).

25. D. E. Rumelhart, G.E. Hinton and R.J. Williams, *Nature (London)* **323**, 533 (1986).

26. A. Hurlbert, *The Computation of Color*, Ph.D. Thesis, M.I.T., Cambridge, MA, in preparation.

27. We are grateful to E. Land, E. Hildreth, J. Little, F. Wilczek and D. Hillis for reading the draft and for useful discussions. A. Rottenberg developed the routines for matrix operations that we used on the Connection Machine. T. Breuel wrote the backpropagation simulator.

STATIC AND DYNAMIC ERROR PROPAGATION NETWORKS WITH APPLICATION TO SPEECH CODING

A J Robinson, F Fallside
Cambridge University Engineering Department
Trumpington Street, Cambridge, England

Abstract

Error propagation nets have been shown to be able to learn a variety of tasks in which a static input pattern is mapped onto a static output pattern. This paper presents a generalisation of these nets to deal with time varying, or dynamic patterns, and three possible architectures are explored. As an example, dynamic nets are applied to the problem of speech coding, in which a time sequence of speech data are coded by one net and decoded by another. The use of dynamic nets gives a better signal to noise ratio than that achieved using static nets.

1. INTRODUCTION

This paper is based upon the use of the error propagation algorithm of Rumelhart, Hinton and Williams[1] to train a connectionist net. The net is defined as a set of units, each with an activation, and weights between units which determine the activations. The algorithm uses a gradient descent technique to calculate the direction by which each weight should be changed in order to minimise the summed squared difference between the desired output and the actual output. Using this algorithm it is believed that a net can be trained to make an arbitrary non-linear mapping of the input units onto the output units if given enough intermediate units. This 'static' net can be used as part of a larger system with more complex behaviour.

The static net has no memory for past inputs, but many problems require the context of the input in order to compute the answer. An extension to the static net is developed, the 'dynamic' net, which feeds back a section of the output to the input, so creating some internal storage for context, and allowing a far greater class of problems to be learned. Previously this method of training time dependence into nets has suffered from a computational requirement which increases linearly with the time span of the desired context. The three architectures for dynamic nets presented here overcome this difficulty.

To illustrate the power of these networks a general coder is developed and applied to the problem of speech coding. The non-linear solution found by training a dynamic net coder is compared with an established linear solution, and found to have an increased performance as measured by the signal to noise ratio.

2. STATIC ERROR PROPAGATION NETS

A static net is defined by a set of units and links between the units. Denoting o_i as the value of the ith unit, and $w_{i,j}$ as the weight of the link between o_i and o_j, we may divide up the units into input units, hidden units and output units. If we assign o_0 to a constant to form a bias, the input units run from o_1 up to $o_{n_{inp}}$, followed by the hidden units to $o_{n_{hid}}$ and then the output units to $o_{n_{out}}$. The values of the input units are defined by the problem and the values of the remaining units are defined by:

$$net_i \;=\; \sum_{j=0}^{i-1} w_{i,j} o_j \qquad\qquad (2.1)$$

$$o_i \;=\; f(net_i) \qquad\qquad (2.2)$$

where $f(x)$ is any continuous monotonic non-linear function and is known as the activation function. The function used the application is:

$$f(x) \;=\; \frac{2}{1+e^{-2x}} - 1 \qquad\qquad (2.3)$$

These equations define a net which has the maximum number of interconnections. This arrangement is commonly restricted to a layered structure in which units are only connected to the immediately preceding layer. The architecture of these nets is specified by the number of input, output and hidden units. Diagrammatically the static net is transformation of an input u, onto the output y, as in figure 1.

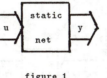

figure 1

The net is trained by using a gradient descent algorithm which minimises an energy term, E, defined as the summed squared error between the actual outputs, o_i, and the target outputs, t_i. The algorithm also defines an error signal, δ_i, for each unit:

$$E \;=\; \frac{1}{2} \sum_{i=n_{hid}+1}^{n_{out}} (t_i - o_i)^2 \qquad\qquad (2.4)$$

$$\delta_i \;=\; f'(net_i)(t_i - o_i) \qquad n_{hid} < i \le n_{out} \qquad (2.5)$$

$$\;=\; f'(net_i) \sum_{j=i+1}^{n_{out}} \delta_i w_{j,i} \qquad n_{inp} < i \le n_{hid} \qquad (2.6)$$

where $f'(x)$ is the derivative of $f(x)$. The error signal and the activations of the units define the change in each weight, $\Delta w_{i,j}$.

$$\Delta w_{i,j} \;=\; \eta \delta_i o_j \qquad\qquad (2.7)$$

where η is a constant of proportionality which determines the learning rate. The above equations define the error signal, δ_i, for the input units as well as for the hidden units. Thus any number of static nets can be connected together, the values of δ_i being passed from input units of one net to output units of the preceding net. It is this ability of error propagation nets to be 'glued' together in this way that enables the construction of dynamic nets.

3. DYNAMIC ERROR PROPAGATION NETS

The essential quality of the dynamic net is is that its behaviour is determined both by the external input to the net, and also by its own internal state. This state is represented by the activation of a group of units. These units form part of the output of a static net and also part of the input to another copy of the same static net in the next time period. Thus the state units link multiple copies of static nets over time to form a dynamic net.

3.1. DEVELOPMENT FROM LINEAR CONTROL THEORY

The analogy of a dynamic net in linear systems[2] may be stated as:

$$x_{p+1} \;=\; Ax_p + Bu_p \tag{3.1.1}$$

$$y_p \;=\; Cx_p \tag{3.1.2}$$

where u_p is the input vector, x_p the state vector, and y_p the output vector at the integer time p. A, B and C are matrices.

The structure of the linear systems solution may be implemented as a non-linear dynamic net by substituting the matrices A, B and C by static nets, represented by the non-linear functions $A[.]$, $B[.]$ and $C[.]$. The summation operation of Ax_p and Bu_p could be achieved using a net with one node for each element in x and u and with unity weights from the two inputs to the identity activation function $f(x) = x$. Alternatively this net can be incorporated into the $A[.]$ net giving the architecture of figure 2.

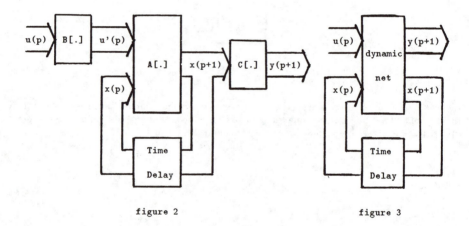

figure 2   figure 3

The three networks may be combined into one, as in figure 3. Simplicity of architecture is not just an aesthetic consideration. If three nets are used then each one must have enough computational power for its part of the task, combining the nets means that only the combined power must be sufficient and it allows common computations can be shared.

The error signal for the output y_{p+1}, can be calculated by comparison with the desired output. However, the error signal for the state units, x_p, is only given by the net at time $p+1$, which is not known at time p. Thus it is impossible to use a single backward pass to train this net. It is this difficulty which introduces the variation in the architectures of dynamic nets.

3.2. THE FINITE INPUT DURATION (FID) DYNAMIC NET

If the output of a dynamic net, y_p, is dependent on a finite number of previous inputs, u_{p-P} to u_p, or if this assumption is a good approximation, then it is possible to formulate the learning algorithm by expansion of the dynamic net for a finite time, as in figure 4. This formulation is simlar to a restricted version of the recurrent net of Rumelhart, Hinton and Williams.[1]

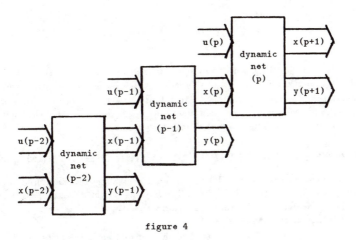

figure 4

Consider only the component of the error signal in past instantiations of the nets which is the result of the error signal at time t. The error signal for y_p is calculated from the target output and the error signal for x_p is zero. This combined error signal is propagated back though the dynamic net at p to yield the error signals for u_p and x_p. Similarly these error signals can then be propagated back through the net at $t-p$, and so on for all relevant inputs. The summed error signal is then used to change the weights as for a static net.

Formalising the FID dynamic net for a general time q, $q \leq p$:

n_s is the number of state units

$o_{q,i}$ is the output value of unit i at time q

$t_{q,i}$ is the target value of unit i at time q

$\delta_{q,i}$ is the error value of unit i at time q

$w_{i,j}$ is the weight between o_i and o_j

$\Delta w_{q,i,j}$ is the weight change for this iteration at time q

$\Delta w_{i,j}$ is the total weight change for this iteration

These values are calculated in the same way as in a static net,

$$\text{net}_{q,i} \; = \; \sum_{j=0}^{i-1} w_{i,j} o_{q,j} \tag{3.2.1}$$

$$o_{q,i} \; = \; f(\text{net}_{q,i}) \tag{3.2.2}$$

$$\delta_{q,i} \; = \; f'(\text{net}_{q,i})(t_{q,i} - o_{q,i}) \qquad n_{\text{hid}} + n_s < i \leq n_{\text{out}} \tag{3.2.3}$$

$$\; = \; \delta_{q+1,i-n_{\text{hid}}+n_{\text{inp}}-n_s} \qquad n_{\text{hid}} < i \leq n_{\text{hid}} + n_s \tag{3.2.4}$$

$$\; = \; f'(\text{net}_{q,i}) \sum_{j=i+1}^{n_{\text{out}}} \delta_{q,j} w_{j,i} \qquad n_{\text{inp}} < i \leq n_{\text{hid}} \tag{3.2.5}$$

$$\Delta w_{q,i,j} \; = \; \eta \delta_{q,i} o_{q,j} \tag{3.2.6}$$

and the total weight change is given by the summation of the partial weight changes for all previous times.

$$\Delta w_{i,j} \;=\; \sum_{q=p-P}^{p} \Delta w_{q,i,j} \tag{3.2.7}$$

$$=\; \sum_{q=p-P}^{p} \eta \delta_{q,i} o_{q,j} \tag{3.2.8}$$

Thus, it is possible to train a dynamic net to incorporate the information from any time period of finite length, and so learn any function which has a finite impulse response.[*]

In some situations the approximation to a finite length may not be valid, or the storage and computational requirements of such a net may not be feasible. In such situations another approach is possible, the infinite input duration dynamic net.

3.3. THE INFINITE INPUT DURATION (IID) DYNAMIC NET

Although the forward pass of the FID net of the previous section is a non-linear process, the backward pass computes the effect of small variations on the forward pass, and is a linear process. Thus the recursive learning procedure described in the previous section may be compressed into a single operation.

Given the target values for the output of the net at time p, equations (3.2.3) and (3.2.4) define values of $\delta_{p,i}$ at the outputs. If we denote this set of $\delta_{p,i}$ by D_p then equation (3.2.5) states that any $\delta_{p,i}$ in the net at time p is simply a linear transformation of D_p. Writing the transformation matrix as S:

$$\delta_{p,i} \;=\; S_{p,i} D_p \tag{3.3.1}$$

In particular the set of $\delta_{p,i}$ which is to be fed back into the network at time $p-1$ is also a linear transformation of D_p

$$D_{p-1} \;=\; T_p D_p \tag{3.3.2}$$

or for an arbitrary time q:

$$D_q \;=\; \left(\prod_{r=q+1}^{p} T_r \right) D_p \tag{3.3.3}$$

so substituting equations (3.3.1) and (3.3.3) into equation (3.2.8):

$$\Delta w_{i,j} \;=\; \eta \sum_{q=-\infty}^{p} S_{q,i} \left(\prod_{r=q+1}^{p} T_r \right) D_p o_{q,j} \tag{3.3.4}$$

$$=\; \eta M_{p,i,j} D_p \tag{3.3.5}$$

where:

$$M_{p,i,j} \;=\; \sum_{q=-\infty}^{p} S_{q,i} \left(\prod_{r=q+1}^{p} T_r \right) o_{q,j} \tag{3.3.6}$$

---

[*] This is a restriction on the class of functions which can be learned, the output will always be affected in some way by all previous inputs giving an infinite impulse response performance.

and note that $M_{p,i,j}$ can be written in terms of $M_{p-1,i,j}$:

$$M_{p,i,j} = S_{p,i} \left(\prod_{r=p+1}^{t} T_r \right) o_{p,j} + \left(\sum_{q=-\infty}^{p-1} S_{q,i} \left(\prod_{r=q+1}^{p-1} T_r \right) o_{q,j} \right) T_p \qquad (3.3.7)$$

$$= S_{p,i} o_{p,j} + M_{p-1,i,j} T_p \qquad (3.3.8)$$

Hence we can calculate the weight changes for an infinite recursion using only the finite matrix M.

3.3. THE STATE COMPRESSION DYNAMIC NET

The previous architectures for dynamic nets rely on the propagation of the error signal back in time to define the format of the information in the state units. An alternative approach is to use another error propagation net to define the format of the state units. The overall architecture is given in figure 5.

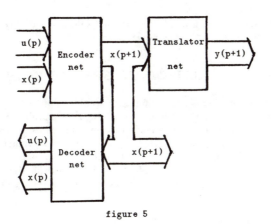

figure 5

The encoder net is trained to code the current input and current state onto the next state, while the decoder net is trained to do the reverse operation. The translator net codes the next state onto the desired output. This encoding/decoding attempts to represent the current input and the current state in the next state, and by the recursion, it will try to represent all previous inputs. Feeding errors back from the translator directs this coding of past inputs to those which are useful in forming the output.

3.4. COMPARISON OF DYNAMIC NET ARCHITECTURES

In comparing the three architectures for dynamic nets, it is important to consider the computational and memory requirements, and how these requirements scale with increasing context.

To train an FID net the net must store the past activations of the all the units within the time span of the necessary context. Using this minimal storage, the computational load scales proportionally to the time span considered, as for every new input/output pair the net must propagate an error signal back though all the past nets. However, if more sets of past activations are stored in a buffer, then it is possible to wait until this buffer is full before computing the weight changes. As the buffer size increases the computational load in calculating the weight changes tends to that of a single backward pass through the units, and so becomes independent of the amount of context.

The largest matrix required to compute the IID net is M, which requires a factor of the number of outputs of the net more storage than the weight matrix. This must be updated on each iteration, a computational requirement larger than that of the FID net for small problems[3]. However, if this architecture were implemented on a parallel machine it would be possible to store the matrix M in a distributed form over the processors, and locally calculate the weight changes. Thus, whilst the FID net requires the error signal to be propagated back in time in a strictly sequential manner, the IID net may be implemented in parallel, with possible advantages on parallel machines.

The state compression net has memory and computational requirements independent of the amount of context. This is achieved at the expense of storing recent information in the state units whether it is required to compute the output or not. This results in an increased computational and memory load over the more efficient FID net when implemented with a buffer for past outputs. However, the exclusion of external storage during training gives this architecture more biological plausibility, constrained of course by the plausibility of the error propagation algorithm itself.

With these considerations in mind, the FID net was chosen to investigate a 'real world' problem, that of the coding of the speech waveform.

4. APPLICATION TO SPEECH CODING

The problem of speech coding is one of finding a suitable model to remove redundancy and hence reduce the data rate of the speech. The Boltzmann machine learning algorithm has already been extended to deal to the dynamic case and applied to speech recognition[4]. However, previous use of error propagation nets for speech processing has mainly been restricted to explicit presentation of the context[5,6] or explicit feeding back the output units to the input[7,8], with some work done in using units with feedback links to themselves[9]. In a similar area, static error propagation nets have been used to perform image coding as well as conventional techniques[10].

4.1. THE ARCHITECTURE OF A GENERAL CODER

The coding principle used in this section is not restricted to coding speech data. The general problem is one of encoding the present input using past input context to form the transmitted signal, and decoding this signal using the context of the coded signals to regenerate the original input. Previous sections have shown that dynamic nets are able to represent context, so two dynamic nets in series form the architecture of the coder, as in figure 6.

This architecture may be specified by the number of input, state, hidden and transmission units. There are as many output units as input units and, in this application, both the transmitter and receiver have the same number of state and hidden units.

The input is combined with the internal state of the transmitter to form the coded signal, and then decoded by the receiver using its internal state. Training of the net involves the comparison of the input and output to form the error signal, which is then propagated back through past instantiations of the receiver and transmitter in the same way as a for a FID dynamic net.

It is useful to introduce noise into the coded signal during the training to reduce the information capacity of the transmission line. This forces the dynamic nets to incorporate time information, without this constraint both nets can learn a simple transformation without any time dependence. The noise can be used to simulate quantisation of the coded signal so

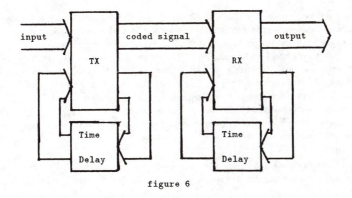

figure 6

quantifying the transmission rate. Unfortunately, a straight implementation of quantisation violates the requirement of the activation function to be continuous, which is necessary to train the net. Instead quantisation to n levels may be simulated by adding a random value distributed uniformly in the range $+1/n$ to $-1/n$ to each of the channels in the coded signal.

4.2. TRAINING OF THE SPEECH CODER

The chosen problem was to present a single sample of digitised speech to the input, code to a single value quantised to fifteen levels, and then to reconstruct the original speech at the output. Fifteen levels was chosen as the point where there is a marked loss in the intelligibility of the speech, so implementation of these coding schemes gives an audible improvement. Two version of the coder net were implemented, both nets had eight hidden units, with no state units for the static time independent case and four state units for the dynamic time dependent case.

The data for this problem was 40 seconds of speech from a single male speaker, digitised to 12 bits at 10kHz and recorded in a laboratory environment. The speech was divided into two halves, the first was used for training and the second for testing.

The static and the dynamic versions of the architecture were trained on about 20 passes through the training data. After training the weights were frozen and the inclusion of random noise was replaced by true quantisation of the coded representation. A further pass was then made through the test data to yield the performance measurements.

The adaptive training algorithm of Chan[11] was used to dynamically alter the learning rates during training. Previously these machines were trained with fixed learning rates and weight update after every sample[3], and the use of the adaptive training algorithm has been found to result in a substantially deeper energy minima. Weights were updated after every 1000 samples, that is about 200 times in one pass of the training data.

4.3. COMPARISON OF PERFORMANCE

The performance of a coding schemes can be measured by defining the noise energy as half the summed squared difference between the actual output and the desired output. This energy is the quantity minimised by the error propagation algorithm. The lower the noise energy in relation to the energy of the signal, the higher the performance.

Three non-connectionist coding schemes were implemented for comparison with the static and dynamic net coders. In the first the signal is linearly quantised within the dynamic range of the original signal. In the second the quantiser is restricted to operate over a reduced dynamic range, with values outside that range thresholded to the maximum and minimum outputs of the quantiser. The thresholds of the quantiser were chosen to optimise the signal to noise ratio. The third scheme used the technique of Differential Pulse Code Modulation (DPCM)[12] which involves a linear filter to predict the speech waveform, and the transmitted signal is the difference between the real signal and the predicted signal. Another linear filter reconstructs the original signal from the difference signal at the receiver. The filter order of the DPCM coder was chosen to be the same as the number of state units in the dynamic net coder, thus both coders can store the same amount of context enabling a comparison with this established technique.

The resulting noise energy when the signal energy was normalised to unity, and the corresponding signal to noise ratio are given in table 1 for the five coding techniques.

coding method	normalised noise energy	signal to noise ratio in dB
linear, original thresholds	0.071	11.5
linear, optimum thresholds	0.041	13.9
static net	0.049	13.1
DPCM, optimum thresholds	0.037	14.3
dynamic net	0.028	15.5

table 1

The static net may be compared with the two forms of the linear quantiser. Firstly note that a considerable improvement in the signal to noise ratio may be achieved by reducing the thresholds of the quantiser from the extremes of the input. This improvement is achieved because the distribution of samples in the input is concentrated around the mean value, with very few values near the extremes. Thus many samples are represented with greater accuracy at the expense of a few which are thresholded. The static net has a poorer performance than the linear quantiser with optimum thresholds. The form of the linear quantiser solution is within the class of problems which the static net can represent. It's failure to do so can be attributed to finding a local minima, a plateau in weight space, or corruption of the true steepest descent direction by noise introduced by updating the weights more than once per pass through the training data.

The dynamic net may be compared with the DPCM coding. The output from both these coders is no longer constrained to discrete signal levels and the resulting noise energy is lower than all the previous examples. The dynamic net has a significantly lower noise energy than any other coding scheme, although, from the static net example, this is unlikely to be an optimal solution. The dynamic net achieves a lower noise energy than the DPCM coder by virtue of the non-linear processing at each unit, and the flexibility of data storage in the state units.

As expected from the measured noise energies, there is an improvement in signal quality and intelligibility from the linear quantised speech through to the DCPM and dynamic net quantised speech.

5. CONCLUSION

This report has developed three architectures for dynamic nets. Each architecture can be formulated in a way where the computational requirement is independent of the degree of context necessary to learn the solution. The FID architecture appears most suitable for implementation on a serial processor, the IID architecture has possible advantages for implementation on parallel processors, and the state compression net has a higher degree of biological plausibility.

Two FID dynamic nets have been coupled together to form a coder, and this has been applied to speech coding. Although the dynamic net coder is unlikely to have learned the optimum coding strategy, it does demonstrate that dynamic nets can be used to achieve an improved performance in a real world task over an established conventional technique.

One of the authors, A J Robinson, is supported by a maintenance grant from the U.K. Science and Engineering Research Council, and gratefully acknowledges this support.

References

[1] D. E. Rumelhart, G. E. Hinton, and R. J. Williams. Learning internal representations by error propagation. In D. E. Rumelhart and J. L. McClelland, editors, *Parallel Distributed Processing: Explorations in the Microstructure of Cognition. Vol. 1: Foundations.*, Bradford Books/MIT Press, Cambridge, MA, 1986.

[2] O. L. R. Jacobs. *Introduction to Control Theory.* Clarendon Press, Oxford, 1974.

[3] A. J. Robinson and F. Fallside. *The Utility Driven Dynamic Error Propagation Network.* Technical Report CUED/F-INFENG/TR.1, Cambridge University Engineering Department, 1987.

[4] R. W. Prager, T. D. Harrison, and F. Fallside. Boltzmann machines for speech recognition. *Computer Speech and Language*, 1:3–27, 1986.

[5] J. L. Elman and D. Zipser. *Learning the Hidden Structure of Speech.* ICS Report 8701, University of California, San Diego, 1987.

[6] A. J. Robinson. *Speech Recognition with Associative Networks.* M.Phil Computer Speech and Language Processing thesis , Cambridge University Engineering Department, 1986.

[7] M. I. Jordan. *Serial Order: A Parallel Distributed Processing Approach.* ICS Report 8604, Institute for Cognitive Science, University of California, San Diego, May 1986.

[8] D. J. C. MacKay. *A Method of Increasing the Contextual Input to Adaptive Pattern Recognition Systems.* Technical Report RIPRREP/1000/14/87, Research Initiative in Pattern Recognition, RSRE, Malvern, 1987.

[9] R. L. Watrous, L. Shastri, and A. H. Waibel. Learned phonetic discrimination using connectionist networks. In J. Laver and M. A. Jack, editors, *Proceedings of the European Conference on Speech Technology*, CEP Consultants Ltd, Edinburgh, September 1987.

[10] G. W. Cottrell, P. Munro, and D Zipser. *Image Compression by Back Propagation: An Example of Existential Programming.* ICS Report 8702, Institute for Cognitive Science, University of California, San Diego, Febuary 1986.

[11] L. W. Chan and F. Fallside. *An Adaptive Learning Algorithm for Back Propagation Networks.* Technical Report CUED/F-INFENG/TR.2, Cambridge University Engineering Department, 1987, submitted to *Computer Speech and Language.*

[12] L. R. Rabiner and R. W. Schefer. *Digital Processing of Speech Signals.* Prentice Hall, Englewood Cliffs, New Jersey, 1978.

642

LEARNING BY STATE RECURRENCE DETECTION

Bruce E. Rosen, James M. Goodwin[†], and Jacques J. Vidal
University of California, Los Angeles, Ca. 90024

ABSTRACT

This research investigates a new technique for unsupervised learning of nonlinear control problems. The approach is applied both to Michie and Chambers BOXES algorithm and to Barto, Sutton and Anderson's extension, the ASE/ACE system, and has significantly improved the convergence rate of stochastically based learning automata.

Recurrence learning is a new nonlinear reward-penalty algorithm. It exploits information found during learning trials to reinforce decisions resulting in the recurrence of nonfailing states. Recurrence learning applies positive reinforcement during the exploration of the search space, whereas in the BOXES or ASE algorithms, only negative weight reinforcement is applied, and then only on failure. Simulation results show that the added information from recurrence learning increases the learning rate.

Our empirical results show that recurrence learning is faster than both basic failure driven learning and failure prediction methods. Although recurrence learning has only been tested in failure driven experiments, there are goal directed learning applications where detection of recurring oscillations may provide useful information that reduces the learning time by applying negative, instead of positive reinforcement.

Detection of cycles provides a heuristic to improve the balance between evidence gathering and goal directed search.

INTRODUCTION

This research investigates a new technique for unsupervised learning of nonlinear control problems with delayed feedback. Our approach is compared to both Michie and Chambers BOXES algorithm[1], to the extension by Barto, et al., the ASE (Adaptive Search Element) and to their ASE/ACE (Adaptive Critic Element) system[2], and shows an improved learning time for stochastically based learning automata in failure driven tasks.

We consider adaptively controlling the behavior of a system which passes through a sequence of states due to its internal dynamics (which are not assumed to be known a priori) and due to choices of actions made in visited states. Such an adaptive controller is often referred to as a learning automaton. The decisions can be deterministic or can be made according to a stochastic rule. A learning automaton has to discover which action is best in each circumstance by producing actions and observing the resulting information.

This paper was motivated by the previous work of Barto, et al. to investigate neuronlike adaptive elements that affect and learn from their environment. We were inspired by their current work and the recent attention to neural networks and connectionist systems, and have chosen to use the cart-pole control problem[2], to enable a comparison of our results with theirs.

[†]Permanent address: California State University, Stanislaus; Turlock, California.

THE CART-POLE PROBLEM

In their work on the cart-pole problem, Barto, Sutton and Anderson considered a learning system composed of an automaton interacting with an environment. The problem requires the automaton to balance a pole acting as an inverted pendulum hinged on a moveable cart. The cart travels left or right along a bounded one dimensional track; the pole may swing to the left or right about a pivot attached to the cart. The automaton must learn to keep the pole balanced on the cart, and to keep the cart within the bounds of the track. The parameters of the cart/pole system are the cart position and velocity, and the pole angle and angular velocity. The only actions available to the automaton are the applications of a fixed impulsive force to the cart in either right or left direction; one of these actions *must* be taken.

This balancing is an extremely difficult problem if there is no a priori knowledge of the system dynamics, if these dynamics change with time, or if there is no preexisting controller that can be imitated (e.g. Widrow and Smith's[3] ADALINE controller). We assumed no a priori knowledge of the dynamics nor any preexisting controller and anticipate that the system will be able to deal with any changing dynamics.

Numerical simulations of the cart-pole solution via recurrence learning show substantial improvement over the results of Barto et al., and of Michie and Chambers, as is shown in figure 1. The algorithms used, and the results shown in figure 1, will be discussed in detail below.

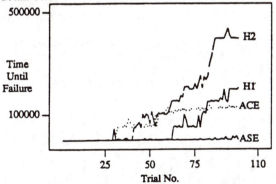

Figure 1: Performance of the ASE, ASE/ACE, Constant Recurrence (H1) and Short Recurrence (H2) Algorithms.

THE GENERAL PROBLEM: ASSIGNMENT OF CREDIT

The cart-pole problem is one of a class of problems known as "credit assignment"[4], and in particular *temporal* credit assignment. The recurrence learning algorithm is an approach to the general temporal credit assignment problem. It is characterized by seeking to improve learning by making decisions about early actions. The goal is to find actions responsible for improved or degraded performance at a much later time.

An example is the bucket brigade algorithm[5]. This is designed to assign credit to rules in the system according to their overall usefulness in attaining their goals. This is done by adjusting the strength value (weight) of each rule. The problem is of modifying these strengths is to permit rules activated early in the sequence to result in successful actions later.

Samuels considered the credit assignment problem for his checkers playing program[6]. He noted that it is easy enough to credit the rules that combine to produce a triple jump at some point in the game; it is much harder to decide which rules active earlier were responsible for changes that made the later jump possible.

State recurrence learning assigns a strength to an individual rule or action and modifies that action's strength (while the system accumulates experience) on the basis of the action's overall usefulness in the situations in which it has been invoked. In this it follows the bucket brigade paradigm of Holland.

PREVIOUS WORK

The problems of learning to control dynamical systems have been studied in the past by Widrow and Smith[3], Michie and Chambers[1], Barto, Sutton, and Anderson[2], and Connell[7]. Although different approaches have been taken and have achieved varying degrees of success, each investigator used the cart/pole problem as the basis for empirically measuring how well their algorithms work.

Michie and Chambers[1] built BOXES, a program that learned to balance a pole on a cart. The BOXES algorithm choose an action that had the highest average time until failure. After 600 trials (a trial is a run ending in eventual failure or by some time limit expiration), the program was able to balance the pole for 72,000 time steps. Figure 2a describes the BOXES learning algorithm. States are penalized (after a system failure) according to recency. Active states immediately preceding a system failure are punished most.

Barto, Sutton and Anderson[2] used two neuronlike adaptive elements to solve the control problem. Their ASE/ACE algorithm chose the action with the highest probability of keeping the pole balanced in the region, and was able to balance the pole for over 60,000 time steps before completion of the 100th trial.

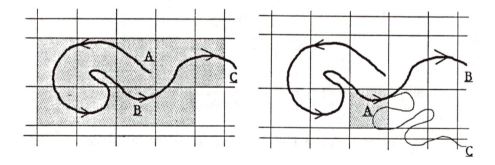

Figure 2a and 2b: The BOXES and ASE/ACE (Associative Search Elelement - Adpative Critic Element) algorithms

Figure 2a shows the BOXES (and ASE) learning algorithm paradigm When the automaton enters a failure state (C), all states that it has traversed (shaded rectangles) are punished, although state B is punished more than state A. (Failure states are those at the edges of the diagram.) Figure 2b describes the ASE/ACE learning algorithm. If a system failure occurs before a state's expected failure time, the state is penalized. If a system failure occurs after its expected failure time, the state is rewarded. State A is penalized because a failure occurred at B sooner than expected. State A's expected failure time is the time for the automaton to traverse from state A to failure point C. When leaving state A, the weights are updated if the new state's expected failure time differs from that of state A.

Anderson[8] used a connectionist system to learn to balance the pole. Unlike the previous experiments, the system did provide well-chosen states a priori. On the average, 10,000 trials were necessary to learn to balance the pole for 7000 time steps.

Connell and Utgoff[7] developed an approach that did not depend on partitioning the state space into discrete regions. They used Shepard's function[9,10] to interpolate the degree of desirability of a cart-pole state. The system learned the control task after 16 trials. However, their system used a knowledge representation that had a priori information about the system.

OTHER RELATED WORK

Klopf[11] proposed a more neurological class of differential learning mechanisms that correlates earlier changes of inputs with later changes of outputs. The adaptation formula used multiplies the change in outputs by the weighted sum of the absolute value of the t previous inputs weights (Δw_j), the τ previous differences in inputs (Δx_j), and the τ previous time coefficients (c_j).

Sutton's temporal differences (TD)[12] approach is one of a class of adaptive prediction methods. Elements of this class use the sum of previously predicted output values multiplied by the gradient and an exponentially decaying coefficient to modify the weights. Barto and Sutton [13] used temporal differences as the underlying learning procedure for classical conditioning.

THE RECURRENCE LEARNING METHOD

DEFINITIONS

A *state* is the set of values (or ranges) of parameters sufficient to specify the instantaneous condition of the system.

The *input decoder* groups the environmental states into equivalence classes: elements of one class have identical system responses. Every environmental input is mapped into one of n input states. (All further references to "states" assumes that the input values fall into the discrete ranges determined by the decoder, unless otherwise specified.)

States returned to after visiting one or more alternate states *recur*.

An *action* causes the modification of system parameters, which may change the system state. However, no change of state need occur, since the altered parameter values may be decoded within the same ranges.

A *weight*, w(t), is associated with each action for each state, with the probability of an allowed action dependent on the current value of its weight.

A *rule* determines which of the allowable actions is taken. The rule is not deterministic. It chooses an action stochastically, based on the weights.

Weight *changes*, $\Delta w(t)$, are made to reduce the likelihood of choosing an action which will cause an eventual failure. These changes are made based on the idea that the previous action of an element, when presented with input x(t), had some influence in causing a similar pattern to occur again. Thus, weight changes are made to increase the likelihood that an element produces the same action f(t) when patterns similar to x(t) occur in the future.

For example, consider the classic problem of balancing a pole on a moving cart. The *state* is specified by the positions and velocities of both the cart and the pole. The allowable *actions* are fixed velocity increments to the right or to the left, and the *rule* determines which action to take, based on the current weights.

THE ALGORITHM

The recurrence learning algorithm presented here is a nonlinear reward-penalty method[14]. Empirical results show that it is successful for stationary environments. In contrast to other methods, it also may be applicable to nonstationary environments'. Our efforts have been to develop algorithms that reward decision choices that lead the controller/environment to quasi-stable cycles that avoid failure (such as limit cycles, converging oscillations and absorbing points).

Our technique exploits recurrence information obtained during learning trials. The system is rewarded upon return to a previous state, however weight changes are only permitted when a state transition occurs. If the system returns to a state, it has avoided failure. A recurring state is rewarded. A sequence of recurring states can be viewed as evidence for a (possibly unstable) cycle. The algorithm forms temporal "cause and effect" associations.

To optimize performance, dynamic search techniques must balance between choosing a search path with known solution costs, and exploring new areas of the search space to find better or cheaper solutions. This is known as the two armed bandit problem[15], i.e. given a two handed slot machine with one arm's *observed* reward probabilities higher than the other, one should not exclude playing with the arm with the lesser payoff. Like the ASE/ACE system, recurrence learning learns while searching, in contrast to the BOXES and ASE algorithms which learn only upon failure.

RANGE DECODING

In our work, as in Barto and others, the real valued input parameters are analyzed as members of ranges. This reduces computing resource demands. Only a limited number of ranges are allowed for each parameter. It is possible for these ranges to overlap, although this aspect of range decoding is not discussed in this paper, and the ranges were considered nonoverlapping. When the parameter value falls into one of the ranges that range is *active*. The specification of a state consists of one of the active ranges for each of the parameters. If the ranges do not overlap, then the set of parameter values specify one unique state; otherwise the set of parameter values may specify several states. Thus, the parameter values at any time determine one or several active states S_i from the set of n possible states.

The value of each environmental parameter falls into one of a number of ranges, which may be different for different parameters. A state is specified by the active range for each parameter.

The set of input parameter values are decoded into one (or more) of n ranges S_i, $0 <= i <= n$. For this problem, boolean values are used to describe the activity level of a state S_i. The activity value of a state is 1 if the state is active, or 0 if it is inactive.

ACTION DECISIONS

Our model is the same as that of the BOXES and ASE/ACE systems, where only one input (and state) is active at any given time. All states were nonoverlapping and mutually exclusive, although there was no reason to preclude them from overlapping other than for consistency with the two previous models. In the ASE/ACE system and in ours as well, the output decision rule for the controller is based on the weighted sum of its inputs plus some stochastic noise. The action (output) decision of the controller is either +1 or -1, as given by:

$$y_i(t) = f\left(\sum_{i=1}^{n} w_i(t)\, x_i(t) + \text{noise}(t) \right) \tag{1}$$

where

$$f(z) = \begin{bmatrix} +1 \text{ if } z \geq 0 \\ -1 \text{ if } z < 0 \end{bmatrix} \tag{2}$$

and noise is a real randomly (Gaussian) distributed value with some mean μ and variance σ^2. An output, $f(z)$, for the car/pole controller is interpreted as a push to the left if $f(z) = -1$ or to the right if $f(z) = +1$.

RECURRENCE LEARNING

The goal of the recurrence learning algorithm is to avoid failure by moving toward states that are part of cycles if such states exist, or quasi-stable oscillations if they don't. This concept can be compared to juggling. As long as all the balls are in the air, the juggler is judged a success and rewarded. No consideration is given to whether the balls are thrown high or low, left or right; the controller, like the juggler, tries for the most stable cycles. Optimum performance is not demanded from recurrence learning.

Two heuristics have been devised. The fundamental basis of each of them is to reward a state for being repeatedly visited (or repeatedly activated). The first heuristic is to reward a state when it is revisited, as part of a cycle in which no failure had occurred. The second heuristic augments the first by giving more reward to states which participate in shorter cycles. These heuristics are discussed below in detail.

HEURISTIC H1: *If a state has been visited more than once during one trial, reward it by reinforcing its weight.*

RATIONALE

This heuristic assumes that states that are visited more than once have been part of a cycle in which no failure had occurred. The action taken in the previous visit is assumed to have had some influence on the recurrence. By reinforcing a weight upon state revisitation, it is assumed to increase the likelihood that the cycle will occur again. No assumptions are made as to whether other states were responsible for the cycle.

RESTRICTION

An action may not immediately cause the environment to change to a different state. There may be some delay before a transition, since small changes of parameters may be decoded into the same input ranges, and hence the same state. This inertia is incorporated into the heuristics. When the same state appears twice in succession, its weight is not reinforced, since that would assume that the *action*, rather than inertia, directly caused the state's immediate recurrence.

THE RECURRENCE EQUATIONS

The recurrence learning equations stem in part from the weight alteration formula used in the ASE system. The weight of a state is a sum of its previous weight, and the product of the learning rate (α), the reward (r), and the state's eligibility (e).

$$w_i(t+1) = w_i(t) + \alpha r(t)e_i(t) \qquad\qquad r(t) \in \{-1,0\} \qquad\qquad (3)$$

The eligibility index $e_i(t)$ is an exponentially decaying trace function.

$$e_i(t+1) = \beta e_i(t) + (1-\beta)y_i(t)x_i(t) \qquad\qquad (4)$$

where $0<=\beta<=1$, $x_i \in \{0,1\}$, and $y_i \in \{-1,1\}$. The output value y_i is the last output decision, and β determines the decay rate.
The reward function is:

$$r(t) = \left\{ \begin{matrix} -1 & \text{when the system fails at time t} \\ 0 & \text{otherwise} \end{matrix} \right\} \qquad\qquad (5)$$

REINFORCEMENT OF CYCLES

Equations (1) through (5) describe the basic ASE system. Our algorithm extends the weight updating procedure as follows:

$$w_i(t+1) = w_i(t) + \alpha r(t)e_i(t) + \alpha_2 r_2(t)e_{2,i}(t) \qquad\qquad (6)$$

The term $\alpha r(t)e_i(t)$ is the same as in (3), providing failure reinforcement. The term $\alpha_2 r_2(t)e_{2,i}(t)$ provides reinforcement for success. When state i is eligible (by virtue of $x_i > 0$), there is a weight change by the amount: α_2 multiplied by the reward value, $r_2(t)$, and the current eligibility $e_{2,i}(t)$. For simplicity, the reward value, $r_2(t)$, may be taken to be some positive constant, although it need not be; any environmental feedback, yielding a reinforcement value as a function of time could be used instead. The second eligibility function $e_{2,i}(t)$ yields one of three constant values for **H1**: $-\beta_2$, 0, or β_2 according to formula (7) below:

$$e_{2,i}(t) = \left\{ \begin{matrix} 0 & \text{if } t-t_{i,last} = 1 \text{ or } t_{i,last} = 0 \\ \beta_2 x_i(t)y(t_{i,last}) & \text{otherwise} \end{matrix} \right\} \qquad\qquad (7)$$

where $t_{i,last}$ is the last time that state was active. If a state has not previously been active (i.e. $x_i(t) = 0$ for all t) then $t_{i,last}=0$. As the formula shows, $e_{2,i}(t) = 0$ if the state has not been previously visited or if no state transition occurred in the last time step; otherwise, $e_{2,i}(t) = \beta_2 x_i(t)y(t_{i,last})$.
The direction (increase or decrease) of the weight change due to the final term in (6) is that of the last action taken, $y(t_{i,last})$.

Heuristic **H1** is called *constant recurrence learning* because the eligibility function is designed to reinforce *any* cycle.

HEURISTIC H2: *Reward a short cycle more than a longer one.*

Heuristic **H2** is called *short recurrence learning* because the eligibility function is designed to reinforce *shorter* cycle more than longer cycles.

REINFORCEMENT OF SHORTER CYCLES

The basis of the second heuristic is the conjecture that short cycles converge more easily to absorbing points than long ones, and that long cycles diverge more easily than shorter ones, although any cycle can "grow" or diverge to a larger cycle. The following extension to the our basic heuristic is proposed.

The formula for the recurrence eligibility function is:

$$
e_{2,i}(t) = \begin{cases} 0 & \text{if } t\text{-}t_{i,last} = 1 \text{ or } t_{i,last} = 0 \\ \dfrac{\beta_2}{(\beta_2 + t\text{-}t_{i,last})} x_i(t) \, y(t_{i,last}) \text{ otherwise} \end{cases} \qquad (8)
$$

The current eligibility function $e_{2,i}(t)$ is similar to the previous failure eligibility function in (7); however, $e_{2,i}(t)$ reinforces shorter cycles more, because the eligibility decays with time. The value returned from $e_{2,i}(t)$ is inversely proportional to the period of the cycle from $t_{i,last}$ to t. **H2** reinforces converging oscillations; the term $\alpha_2 r_2(t) e_{2,i}(t)$ in (6) ensures weight reinforcement for returning to an already visited state.

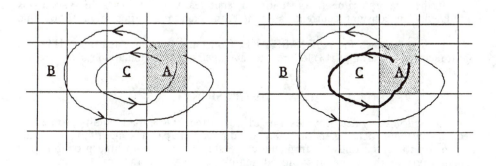

Figure 3a and 3b: The Constant Recurrence algorithm and Short Recurrence algorithms

Figure 3A shows the Constant Recurrence algorithm (H1). A state is rewarded when it is reactivated by a transition from another state. In the example below, state A is reward by a constant regardless of weather the cycle traversed states B or C. Figure 3b describes the Short Recurrence algorithm (H2). A state is rewarded according to the difference between the current time and its last activation time. Small differences are rewarded more than large differences In the example below, state A is rewarded more when the cycle (through state C) traverses the states shown by the dark heavy line rather than when the cycle (through state B) traverses the lighter line, since state A recurs sooner when traversing the darker line.

SIMULATION RESULTS

We simulated four algorithms: ASE, ASE/ACE and the two recurrence algorithms. Each experiment consisted of ten runs of the cart-pole balancing task, each consisting of 100 trials. Each trial lasted for 500,000 time steps or until the cart-pole system failed (i.e. the pole fell or the cart went beyond the track boundaries). In an effort to conserve cpu time, simulations were also terminated when the system achieved two consecutive trials each lasting for over 500,000 time steps; all remaining trials were assumed to also last 500,000 time steps. This assumption was reasonable: the resulting weight space causes the controller to become deterministic regardless of the influence of stochastic noise. Because of the long time require to run simulations, no attempts were made to optimize parameters of the algorithm.

As in Barto[2], each trial began with the cart centered, and the pole upright. No assumptions were made as to the state space configuration, the desirability of the initial states, or the continuity of the state space.

The first experiment consisted of failure and recurrence reward learning. The ASE failure learning runs averaged 1578 time steps until failure after 100 trials[*]. Next, the predictive ASE/ACE system was run as a comparative metric, and it was found that this method caused the controller to average 131,297 time steps until failure; the results are comparable to that described by Barto, Sutton and Anderson.

In the next experiment, short recurrence learning system was added to the ASE system. Again, ten 100 trial learning session were executed. On the average, the short recurrence learning algorithm ran for over 400,736 time steps after 100th trial, bettering the ASE/ACE system by 205%.

In the final experiment, constant recurrence learning with the ASE system was simulated. The constant recurrence learning eliminated failure after only 207,562 time steps.

Figure 1 shows the ASE, ASE/ACE, Constant recurrence learning (**H1**) and Short recurrence learning (**H2**) failure rates averaged over 10 simulation runs.

DISCUSSION

Detection of cycles provides a heuristic for the "two armed bandit" problem to decide between evidence gathering, and goal directed search. The algorithm allows the automaton to search outward from the cycle states (states with high probability of revisitation) to the more unexplored search space. The rate of exploration is proportional to the recurrence learning parameter α_2; as α_2 is decreased, the influence of the cycles governing the decision process also decreases and the algorithm explores more of the search space that is not part of any cycle or oscillation path.

---

[*] However, there was a relatively large degree of variance in the final trials. The last 10 trails (averaged over each of the 10 simulations) ranged from 607 to 15,459 time steps until failure

THE FUTURE

Our future experiments will study the effects of rewarding predictions of cycle lengths in a manner similar to the prediction of failure used by the ASE/ACE system. The effort will be to minimize the differences of predicted time of cycles in order to predict their period. Results of this experiment will be shown in future reports. We hope to show that this recurrence prediction system is generally superior to either the ASE/ACE predictive system or the short recurrence system operating alone.

CONCLUSION

This paper presented an extension to the failure driven learning algorithm based on reinforcing decisions that cause an automaton to enter environmental states more than once. The controller learns to synthesize the best values by reinforcing areas of the search space that produce recurring state visitation. Cycle states, which under normal failure driven learning algorithms do not learn, achieve weight alteration from success. Simulations show that recurrence reward algorithms show improved overall learning of the cart-pole task with a substantial decrease in learning time.

REFERENCES

1.   D. Michie and R. Chambers, *Machine Intelligence*, E. Dale and D. Michie, Ed.: (Oliver and Boyd, Edinburgh, 1968), p. 137.
2.   A. Barto, R. Sutton, and C. Anderson, *Coins Tech. Rept.*, No. 82-20, 1982.
3.   B. Widrow and F. Smith, in *Computer and Information Sciences*, J. Tou and R. Wilcox Eds., (Clever Hume Press, 1964).
4.   M. Minsky, in *Proc. IRE*, **49**, 8, (1961).
5.   J. Holland, in *Proc. Int. Conf., Genetic Algs. and their Appl.*, 1985, p. 1.
6.   A. Samuel, *IBM Journ. Res.and Dev.* **3**, 211, (1959)
7.   M. Connell and P. Utgoff, in *Proc. AAAI-87* (Seattle, 1987), p. 456.
8.   C. Anderson, *Coins Tech. Rept.*, No. 86-50: Amherst, MA. 1986.
9.   R. Barnhill, in *Mathematical Software III,* (Academic Press, 1977).
10.   L. Schumaker, in *Approximation Theory II.* (Academic Press, 1976).
11.   A. H. Klopf, in *IEEE Int. Conf. Neural Networks,,* June 1987.
12.   R. Sutton, *GTE Tech. Rept.TR87-509.1*, GTE Labs. Inc., Jan. 1987
13.   R. Sutton and A. G. Barto, *Tech. Rept. TR87-5902.2* March 1987
14.   A. Barto and P. Anandan, *IEEE Trans. SMC* **15**, 360 (1985).
15.   M. Sato, K. Abe, and H. Takeda, *IEEE Trans.SMC* **14** , 528 (1984).

Scaling Properties of Coarse-Coded Symbol Memories

Ronald Rosenfeld
David S. Touretzky

Computer Science Department
Carnegie Mellon University
Pittsburgh, Pennsylvania 15213

Abstract: Coarse-coded symbol memories have appeared in several neural network symbol processing models. In order to determine how these models would scale, one must first have some understanding of the mathematics of coarse-coded representations. We define the general structure of coarse-coded symbol memories and derive mathematical relationships among their essential parameters: *memory size, symbol-set size* and *capacity*. The computed capacity of one of the schemes agrees well with actual measurements of the coarse-coded working memory of DCPS, Touretzky and Hinton's distributed connectionist production system.

1 Introduction

A *distributed representation* is a memory scheme in which each entity (concept, symbol) is represented by a pattern of activity over many units [3]. If each unit participates in the representation of many entities, it is said to be *coarsely tuned*, and the memory itself is called *a coarse-coded memory*.

Coarse-coded memories have been used for storing symbols in several neural network symbol processing models, such as Touretzky and Hinton's distributed connectionist production system DCPS [8,9], Touretzky's distributed implementation of linked list structures on a Boltzmann machine, BoltzCONS [10], and St. John and McClelland's PDP model of case role defaults [6]. In all of these models, memory capacity was measured empirically and parameters were adjusted by trial and error to obtain the desired behavior. We are now able to give a mathematical foundation to these experiments by analyzing the relationships among the fundamental memory parameters.

There are several paradigms for coarse-coded memories. In a *feature-based representation*, each unit stands for some semantic feature. Binary units can code features with binary values, whereas more complicated units or groups of units are required to code more complicated features, such as multi-valued properties or numerical values from a continuous scale. The units that form the representation of a concept define an intersection of features that constitutes that concept. Similarity between concepts composed of binary features can be measured by the Hamming distance between their representations. In a neural network implementation, relationships between concepts are implemented via connections among the units forming their representations. Certain types of generalization phenomena thereby emerge automatically.

A different paradigm is used when representing points in a multidimensional continuous space [2,3]. Each unit encodes values in some subset of the space. Typically the subsets are hypercubes or hyperspheres, but they may be more coarsely tuned along some dimensions than others [1]. The point to be represented is in the subspace formed by the intersection of all active units. As more units are turned on, the accuracy of the representation improves. The density and degree of overlap of the units' receptive fields determines the system's resolution [7].

Yet another paradigm for coarse-coded memories, and the one we will deal with exclusively, does not involve features. Each concept, or symbol, is represented by an arbitrary subset of the units, called its *pattern*. Unlike in feature-based representations, the units in the pattern bear no relationship to the meaning of the symbol represented. A symbol is stored in memory by turning on all the units in its pattern. A symbol is deemed present if all the units in its pattern are active.[1] The *receptive field* of each unit is defined as the set of all symbols in whose pattern it participates. We call such memories *coarse-coded symbol memories* (CCSMs). We use the term "symbol" instead of "concept" to emphasize that the internal structure of the entity to be represented is not involved in its representation. In CCSMs, a short Hamming distance between two symbols does not imply semantic similarity, and is in general an undesirable phenomenon.

The efficiency with which CCSMs handle sparse memories is the major reason they have been used in many connectionist systems, and hence the major reason for studying them here. The unit-sharing strategy that gives rise to efficient encoding in CCSMs is also the source of their major weakness. Symbols share units with other symbols. As more symbols are stored, more and more of the units are turned on. At some point, some symbol may be deemed present in memory because all of its units are turned on, even though it was not explicitly stored: a "ghost" is born. Ghosts are an unwanted phenomenon arising out of the overlap among the representations of the various symbols. The emergence of ghosts marks the limits of the system's *capacity*: the number of symbols it can store simultaneously and reliably.

2 Definitions and Fundamental Parameters

A coarse coded symbol memory in its most general form consists of:

- A set of N binary state **units**.

- An alphabet of α **symbols** to be represented. Symbols in this context are atomic entities: they have no constituent structure.

- A **memory scheme**, which is a function that maps each symbol to a subset of the units – its **pattern**. The **receptive field** of a unit is defined as the set of all symbols to whose pattern it belongs (see Figure 1). The exact nature of the

---

[1]This criterion can be generalized by introducing a *visibility threshold*: a fraction of the pattern that should be on in order for a symbol to be considered present. Our analysis deals only with a visibility criterion of 100%, but can be generalized to accommodate noise.

	S_1	S_2	S_3	S_4	S_5	S_6	S_7	$S8$
U_1	•			•	•		•	
U_2		•	•		•	•		
U_3		•		•	•			•
U_4	•					•	•	
U_5			•					•
U_6	•	•		•	•		•	

Figure 1: A memory scheme ($N = 6$, $\alpha = 8$) defined in terms of units U_i and symbols S_j. The columns are the symbols' *patterns*. The rows are the units' *receptive fields*.

memory scheme mapping determines the properties of the memory, and is the central target of our investigation.

As symbols are stored, the memory fills up and ghosts eventually appear. It is not possible to detect a ghost simply by inspecting the contents of memory, since there is no general way of distinguishing a symbol that was stored from one that emerged out of overlaps with other symbols. (It is sometimes possible, however, to conclude that there are no ghosts.) Furthermore, a symbol that emerged as a ghost at one time may not be a ghost at a later time if it was subsequently stored into memory. Thus the definition of a ghost depends not only on the state of the memory but also on its history.

Some memory schemes guarantee that no ghost will emerge as long as the number of symbols stored does not exceed some specified limit. In other schemes, the emergence of ghosts is an ever-present possibility, but its probability can be kept arbitrarily low by adjusting other parameters. We analyze systems of both types. First, two more bits of notation need to be introduced:

P_{ghost}: **Probability of a ghost.** The probability that at least one ghost will appear after some number of symbols have been stored.

k: Capacity. The maximum number of symbols that can be stored simultaneously before the probability of a ghost exceeds a specified threshold. If the threshold is 0, we say that the capacity is **guaranteed.**

A localist representation, where every symbol is represented by a single unit and every unit is dedicated to the representation of a single symbol, can now be viewed as a special case of coarse-coded memory, where $k = N = \alpha$ and $P_{\text{ghost}} = 0$. Localist representations are well suited for memories that are not sparse. In these cases, coarse-coded memories are at a disadvantage. In designing coarse-coded symbol memories we are interested in cases where $k \ll N \ll \alpha$. The permissible probability for a ghost in these systems should be low enough so that its impact can be ignored.

3 Analysis of Four Memory Schemes

3.1 Bounded Overlap (guaranteed capacity)

If we want to construct the memory scheme with the largest possible α (given N and k) while guaranteeing $P_{\text{ghost}} = 0$, the problem can be stated formally as:

> Given a set of size N, find the largest collection of subsets of it such that no union of k such subsets subsumes any other subset in the collection.

This is a well known problem in Coding Theory, in slight disguise. Unfortunately, no complete analytical solution is known. We therefore simplify our task and consider only systems in which all symbols are represented by the same number of units (i.e. all patterns are of the same size). In mathematical terms, we restrict ourselves to constant weight codes. The problem then becomes:

> Given a set of size N, find the largest collection of subsets of size *exactly* L such that no union of k such subsets subsumes any other subset in the collection.

There are no known complete analytical solutions for the size of the largest collection of patterns even when the patterns are of a fixed size. Nor is any efficient procedure for constructing such a collection known. We therefore simplify the problem further. We now restrict our consideration to patterns whose pairwise overlap is bounded by a given number. For a given pattern size L and desired capacity k, we require that no two patterns overlap in more than m units, where:

$$m = \left\lfloor \frac{L-1}{k} \right\rfloor \tag{1}$$

Memory schemes that obey this constraint are guaranteed a capacity of at least k symbols, since any k symbols taken together can overlap at most $L-1$ units in the pattern of any other symbol – one unit short of making it a ghost. Based on this constraint, our mathematical problem now becomes:

> Given a set of size N, find the largest collection of subsets of size exactly L such that the intersection of any two such subsets is of size $\leq m$ (where m is given by equation 1.)

Coding theory has yet to produce a complete solution to this problem, but several methods of deriving upper bounds have been proposed (see for example [4]). The simple formula we use here is a variant of the Johnson Bound. Let α_{bo} denote the maximum number of symbols attainable in memory schemes that use bounded overlap. Then

$$\alpha_{bo}(N, L, m) \leq \frac{\binom{N}{m+1}}{\binom{L}{m+1}} \tag{2}$$

The Johnson bound is known to be an *exact* solution asymptotically (that is, when $N, L, m \to \infty$ and their ratios remain finite).

Since we are free to choose the pattern size, we optimize our memory scheme by maximizing the above expression over all possible values of L. For the parameter subspace we are interested in here ($N < 1000$, $k < 50$) we use numerical approximation to obtain:

$$\alpha_{bo}(N,k) = \max_{L \in [1,N]} \frac{\binom{N}{m+1}}{\binom{L}{m+1}} < \max_{L \in [1,N]} \left(\frac{N}{L-m}\right)^{m+1} < e^{0.367 \frac{N}{k}} \qquad (3)$$

(Recall that m is a function of L and k.) Thus the upper bound we derived depicts a simple exponential relationship between α and N/k. Next, we try to construct memory schemes of this type. A Common Lisp program using a modified depth-first search constructed memory schemes for various parameter values, whose α's came within 80% to 90% of the upper bound. These results are far from conclusive, however, since only a small portion of the parameter space was tested.

In evaluating the viability of this approach, its apparent optimality should be contrasted with two major weaknesses. First, this type of memory scheme is hard to construct computationally. It took our program several minutes of CPU time on a Symbolics 3600 to produce reasonable solutions for cases like $N = 200, k = 5, m = 1$, with an exponential increase in computing time for larger values of m. Second, if CC-SMs are used as models of memory in naturally evolving systems (such as the brain), this approach places too great a burden on developmental mechanisms.

The importance of the bounded overlap approach lies mainly in its role as an upper bound for all possible memory schemes, subject to the simplifications made earlier. All schemes with guaranteed capacities can be measured relative to equation 3.

3.2   Random Fixed Size Patterns (a stochastic approach)

Randomly produced memory schemes are easy to implement and are attractive because of their naturalness. However, if the patterns of two symbols coincide, the guaranteed capacity will be zero (storing one of these symbols will render the other a ghost). We therefore abandon the goal of guaranteeing a certain capacity, and instead establish a tolerance level for ghosts, P_{ghost}. For large enough memories, where stochastic behavior is more robust, we may expect reasonable capacity even with very small P_{ghost}.

In the first stochastic approach we analyze, patterns are randomly selected subsets of a fixed size L. Unlike in the previous approach, choosing k does not bound α. We may define as many symbols as we wish, although at the cost of increased probability of a ghost (or, alternatively, decreased capacity). The probability of a ghost appearing after k symbols have been stored is given by Equation 4:

$$P_{\text{ghost}}(N, L, k, \alpha) = 1 - \sum_{c=L}^{\min(N,kL)} T_{N,L}(k,c) \cdot \left[1 - \frac{\binom{C}{L}}{\binom{N}{L}}\right]^{\alpha-k} \qquad (4)$$

$T_{N,L}(k,c)$ is the probability that exactly c units will be active after k symbols have been stored. It is defined recursively by Equation 5:

$$T_{N,L}(0,0) = 1$$
$$T_{N,L}(k,c) = 0 \quad \text{for either } k=0 \text{ and } c \neq 0, \text{ or } k>0 \text{ and } c < L \qquad (5)$$
$$T_{N,L}(k,c) = \sum_{a=0}^{L} T(k-1,c-a) \cdot \binom{N-(c-a)}{a} \cdot \binom{c-a}{L-a} / \binom{N}{L}$$

We have constructed various coarse-coded memories with random fixed-size receptive fields and measured their capacities. The experimental results show good agreement with the above equation.

The optimal pattern size for fixed values of N, k, and α can be determined by binary search on Equation 4, since $P_{\text{ghost}}(L)$ has exactly one maximum in the interval $[1, N]$. However, this may be expensive for large N. A computational shortcut can be achieved by estimating the optimal L and searching in a small interval around it. A good initial estimate is derived by replacing the summation in Equation 4 with a single term involving $E[c]$: the expected value of the number of active units after k symbols have been stored. The latter can be expressed as:

$$E[c] = N \cdot \left[1 - (1 - L/N)^k \right]$$

The estimated L is the one that maximizes the following expression:

$$\binom{E[c]}{L} \bigg/ \binom{N}{L}$$

An alternative formula, developed by Joseph Tebelskis, produces very good approximations to Eq. 4 and is much more efficient to compute. After storing k symbols in memory, the probability P_x that a single arbitrary symbol x has become a ghost is given by:

$$P_x(N,L,k,\alpha) = \sum_{j=0}^{L} (-1)^j \binom{L}{j} \binom{N-j}{L}^k / \binom{N}{L}^k \qquad (6)$$

If we now assume that each symbol's P_x is independent of that of any other symbol, we obtain:

$$P_{\text{ghost}} \approx 1 - (1 - P_x)^{\alpha-k} \qquad (7)$$

This assumption of independence is not strictly true, but the relative error was less than 0.1% for the parameter ranges we considered, when P_{ghost} was no greater than 0.01.

We have constructed the two-dimensional table $T_{N,L}(k,c)$ for a wide range of (N, L) values ($70 \leq N \leq 1000$, $7 \leq L \leq 43$), and produced graphs of the relationships between N, k, α, and P_{ghost} for optimum pattern sizes, as determined by Equation 4. The results show an approximately exponential relationship between α and N/k [5]. Thus, for a fixed number of symbols, the capacity is proportional to the number of units. Let α_{rfp} denote the maximum number of symbols attainable in memory schemes that use random fixed-size patterns. Some typical relationships, derived from the data, are:

$$\alpha_{rfp}(P_{\text{ghost}} = 0.01) \approx 0.0086 \cdot e^{0.468\frac{N}{k}}$$
$$\alpha_{rfp}(P_{\text{ghost}} = 0.001) \approx 0.0008 \cdot e^{0.473\frac{N}{k}} \tag{8}$$

3.3 Random Receptors (a stochastic approach)

A second stochastic approach is to have each unit assigned to each symbol with an independent fixed probability s. This method lends itself to easy mathematical analysis, resulting in a closed-form analytical solution.

After storing k symbols, the probability that a given unit is active is $1 - (1 - s)^k$ (independent of any other unit). For a given symbol to be a ghost, every unit must either be active or else not belong to that symbol's pattern. That will happen with a probability $\left[1 - s \cdot (1 - s)^k\right]^N$, and thus the probability of a ghost is:

$$P_{\text{ghost}}(\alpha, N, k, s) = 1 - \left[1 - \left[1 - s \cdot (1 - s)^k\right]^N\right]^{\alpha - k} \tag{9}$$

Assuming $P_{\text{ghost}} \ll 1$ and $k \ll \alpha$ (both hold in our case), the expression can be simplified to:

$$P_{\text{ghost}}(\alpha, N, k, s) = \alpha \cdot \left[1 - s \cdot (1 - s)^k\right]^N$$

from which α can be extracted:

$$\alpha_{rr}(N, k, s, P_{\text{ghost}}) = \frac{P_{\text{ghost}}}{\left[1 - s \cdot (1 - s)^k\right]^N} \tag{10}$$

We can now optimize by finding the value of s that maximizes α, given any desired upper bound on the expected value of P_{ghost}. This is done straightforwardly by solving $\partial\alpha/\partial s = 0$. Note that $s \cdot N$ corresponds to L in the previous approach. The solution is $s = 1/(k + 1)$, which yields, after some algebraic manipulation:

$$\alpha_{rr} = P_{\text{ghost}} \cdot e^{N \log[(k+1)^{k+1}/((k+1)^{k+1} - k^k)]} \tag{11}$$

A comparison of the results using the two stochastic approaches reveals an interesting similarity. For large k, with $P_{\text{ghost}} = 0.01$ the term $0.468/k$ of Equation 8 can be seen as a numerical approximation to the log term in Equation 11, and the multiplicative factor of 0.0086 in Equation 8 approximates P_{ghost} in Equation 11. This is hardly surprising, since the Law of Large Numbers implies that in the limit ($N, k \rightarrow \infty$, with s fixed) the two methods are equivalent.

Finally, it should be noted that the stochastic approaches we analyzed generate a family of memory schemes, with non-identical ghost-probabilities. P_{ghost} in our formulas is therefore better understood as *an expected value,* averaged over the entire family.

3.4 Partitioned Binary Coding (a reference point)

The last memory scheme we analyze is not strictly distributed. Rather, it is somewhere in between a distributed and a localist representation, and is presented for comparison with the previous results. For a given number of units N and desired capacity k, the units are partitioned into k equal-size "slots," each consisting of N/k units (for simplicity we assume that k divides N). Each slot is capable of storing exactly one symbol.

The most efficient representation for all possible symbols that may be stored into a slot is to assign them binary codes, using the N/k units of each slot as bits. This would allow $2^{N/k}$ symbols to be represented. Using binary coding, however, will not give us the required capacity of 1 symbol, since binary patterns subsume one another. For example, storing the code '10110' into one of the slots will cause the codes '10010', '10100' and '00010' (as well as several other codes) to become ghosts.

A possible solution is to use only half of the bits in each slot for a binary code, and set the other half to the binary complement of that code (we assume that N/k is even). This way, the codes are guaranteed not to subsume one another. Let α_{pbc} denote the number of symbols representable using a partitioned binary coding scheme. Then,

$$\alpha_{pbc} = 2^{N/2k} = e^{0.347\frac{N}{k}} \tag{12}$$

Once again, α is exponential in N/k. The form of the result closely resembles the estimated upper bound on the Bounded Overlap method given in Equation 3. There is also a strong resemblance to Equations 8 and 11, except that the fractional multiplier in front of the exponential, corresponding to P_{ghost}, is missing. P_{ghost} is 0 for the Partitioned Binary Coding method, but this is enforced by dividing the memory into disjoint sets of units rather than adjusting the patterns to reduce overlap among symbols.

As mentioned previously, this memory scheme is not really distributed in the sense used in this paper, since there is no one pattern associated with a symbol. Instead, a symbol is represented by any one of a set of k patterns, each N/k bits long, corresponding to its appearance in one of the k slots. To check whether a symbol is present, all k slots must be examined. To store a new symbol in memory, one must scan the k slots until an empty one is found. Equation 12 should therefore be used only as a point of reference.

4 Measurement of DCPS

The three distributed schemes we have studied all use unstructured patterns, the only constraint being that patterns are at least roughly the same size. Imposing more complex structure on any of these schemes may is likely to reduce the capacity somewhat. In

Memory Scheme	Result
Bounded Overlap	$\alpha_{bo}(N,k) \quad < \quad e^{0.367\frac{N}{k}}$
Random Fixed-size Patterns	$\alpha_{rfp}(P_{\text{ghost}} = \quad 0.01) \approx 0.0086 \cdot e^{0.468\frac{N}{k}}$
	$\alpha_{rfp}(P_{\text{ghost}} = 0.001) \approx 0.0008 \cdot e^{0.473\frac{N}{k}}$
Random Receptors	$\alpha_{rr} = P_{\text{ghost}} \cdot e^{N \cdot \log(k+1)^{k+1}/((k+1)^{k+1}-k^k)}$
Partitioned Binary Coding	$\alpha_{pbc} = \quad e^{0.347\frac{N}{k}}$

Table 1 Summary of results for various memory schemes.

order to quantify this effect, we measured the memory capacity of DCPS (BoltzCONS uses the same memory scheme) and compared the results with the theoretical models analyzed above.

DCPS' memory scheme is a modified version of the Random Receptors method [5]. The symbol space is the set of all triples over a 25 letter alphabet. Units have fixed-size receptive fields organized as $6 \times 6 \times 6$ subspaces. Patterns are manipulated to minimize the variance in pattern size across symbols. The parameters for DCPS are: $N = 2000$, $\alpha = 25^3 = 15625$, and the mean pattern size is $(6/25)^3 \times 2000 = 27.65$ with a standard deviation of 1.5. When $P_{\text{ghost}} = 0.01$ the measured capacity was $k = 48$ symbols. By substituting for N in Equation 11 we find that the highest k value for which $\alpha_{rr} \geq 15625$ is 51. There does not appear to be a significant cost for maintaining structure in the receptive fields.

5   Summary and Discussion

Table 1 summarizes the results obtained for the four methods analyzed. Some differences must be emphasized:

- α_{bo} and α_{pbc} deal with guaranteed capacity, whereas α_{rfp} and α_{rr} are meaningful only for $P_{\text{ghost}} > 0$.

- α_{bo} is only an upper bound.

- α_{rfp} is based on numerical estimates.

- α_{pbc} is based on a scheme which is not strictly coarse-coded.

The similar functional form of all the results, although not surprising, is aesthetically pleasing. Some of the functional dependencies among the various parameters can be derived informally using qualitative arguments. Only a rigorous analysis, however, can provide the definite answers that are needed for a better understanding of these systems and their scaling properties.

Acknowledgments

We thank Geoffrey Hinton, Noga Alon and Victor Wei for helpful comments, and Joseph Tebelskis for sharing with us his formula for approximating P_{ghost} in the case of fixed pattern sizes.

This work was supported by National Science Foundation grants IST-8516330 and EET-8716324, and by the Office of Naval Research under contract number N00014-86-K-0678. The first author was supported by a National Science Foundation graduate fellowship.

References

[1] Ballard, D H. (1986) Cortical connections and parallel processing: structure and function. *Behavioral and Brain Sciences* 9(1).

[2] Feldman, J. A., and Ballard, D. H. (1982) Connectionist models and their properties. *Cognitive Science* 6, pp. 205-254.

[3] Hinton, G. E., McClelland, J. L., and Rumelhart, D. E. (1986) Distributed representations. In D. E. Rumelhart and J. L. McClelland (eds.), *Parallel Distributed Processing: Explorations in the Microstructure of Cognition*, volume 1. Cambridge, MA: MIT Press.

[4] Macwilliams, F.J., and Sloane, N.J.A. (1978). *The Theory of Error-Correcting Codes*, North-Holland.

[5] Rosenfeld, R. and Touretzky, D. S. (1987) Four capacity models for coarse-coded symbol memories. Technical report CMU-CS-87-182, Carnegie Mellon University Computer Science Department, Pittsburgh, PA.

[6] St. John, M. F. and McClelland, J. L. (1986) Reconstructive memory for sentences: a PDP approach. *Proceedings of the Ohio University Inference Conference.*

[7] Sullins, J. (1985) Value cell encoding strategies. Technical report TR-165, Computer Science Department, University of Rochester, Rochester, NY.

[8] Touretzky, D. S., and Hinton, G. E. (1985) Symbols among the neurons: details of a connectionist inference architecture. *Proceedings of IJCAI-85*, Los Angeles, CA, pp. 238-243.

[9] Touretzky, D. S., and Hinton, G. E. (1986) A distributed connectionist production system. Technical report CMU-CS-86-172, Computer Science Department, Carnegie Mellon University, Pittsburgh, PA.

[10] Touretzky, D. S. (1986) BoltzCONS: reconciling connectionism with the recursive nature of stacks and trees. *Proceedings of the Eighth Annual Conference of the Cognitive Science Society*, Amherst, MA, pp. 522-530.

662

AN ADAPTIVE AND HETERODYNE FILTERING PROCEDURE FOR THE IMAGING OF MOVING OBJECTS

F. H. Schuling, H. A. K. Mastebroek and W. H. Zaagman
Biophysics Department, Laboratory for General Physics
Westersingel 34, 9718 CM Groningen, The Netherlands

ABSTRACT

Recent experimental work on the stimulus velocity dependent time resolving power of the neural units, situated in the highest order optic ganglion of the blowfly, revealed the at first sight amazing phenomenon that at this high level of the fly visual system, the time constants of these units which are involved in the processing of neural activity evoked by moving objects, are -roughly spoken- inverse proportional to the velocity of those objects over an extremely wide range. In this paper we will discuss the implementation of a two dimensional heterodyne adaptive filter construction into a computer simulation model. The features of this simulation model include the ability to account for the experimentally observed stimulus-tuned adaptive temporal behaviour of time constants in the fly visual system. The simulation results obtained, clearly show that the application of such an adaptive processing procedure delivers an improved imaging technique of moving patterns in the high velocity range.

A FEW REMARKS ON THE FLY VISUAL SYSTEM

The visual system of the diptera, including the blowfly *Calliphora erythrocephala (Mg.)* is very regularly organized and allows therefore very precise optical stimulation techniques. Also, long term electrophysiological recordings can be made relatively easy in this visual system. For these reasons the blowfly (which is well-known as a very rapid and 'clever' pilot) turns out to be an extremely suitable animal for a systematic study of basic principles that may underlie the detection and further processing of movement information at the neural level.

In the fly visual system the input retinal mosaic structure is precisely mapped onto the higher order optic ganglia (lamina, medulla, lobula). This means that each neural column in each ganglion in this visual system corresponds to a certain optical axis in the visual field of the compound eye. In the lobula complex a set of wide-field movement sensitive neurons is found, each of which integrates the input signals over the whole visual field of the entire eye. One of these wide field neurons, that has been classified as H1 by Hausen[1] has been extensively studied both anatomically[2, 3, 4] as well as electrophysiologically[5, 6, 7]. The obtained results generally agree very well with those found in behavioral optomotor experiments on movement detection[8] and can be understood in terms of Reichardts correlation model[9, 10].

The H1 neuron is sensitive to horizontal movement and directionally selective: very high rates of action potentials (*spikes*) up to 300 per second can be recorded from this element in the case of visual stimuli which move horizontally inward, i.e. from back to front in the visual field (*preferred direction*), whereas movement horizontally outward, i.e. from front to back (*null direction*) suppresses its activity.

EXPERIMENTAL RESULTS AS A MODELLING BASE

When the H1 neuron is stimulated in its preferred direction with a step wise pattern displacement, it will respond with an increase of neural activity. By repeating this stimulus step over and over one can obtain the averaged response: after a 20 ms latency period the response manifests itself as a sharp increase in average firing rate followed by a much slower decay to the spontaneous activity level. Two examples of such averaged responses are shown in the Post Stimulus Time Histograms (PSTH's) of figure 1. Time to peak and peak height are related and depend on modulation depth, stimulus step size and spatial extent of the stimulus. The tail of the responses can be described adequately by an exponential decay toward a constant spontaneous firing rate:

$$R(t)=c+a \cdot e^{(-t/\tau)} \tag{1}$$

For each setting of the stimulus parameters, the response parameters, defined by equation (1), can be estimated by a least-squares fit to the tail of the PSTH. The smooth lines in figure 1 are the results of two such fits.

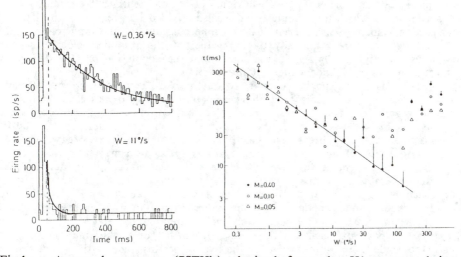

Fig.1   Averaged responses (PSTH's) obtained from the H1 neuron, being adapted to smooth stimulus motion with velocities 0.36°/s (top) and 11°/s (bottom) respectively. The smooth lines represent least-squares fits to the PSTH's of the form $R(t)=c+a \cdot e^{(-t/\tau)}$. Values of τ for the two PSTH's are 331 and 24 ms respectively (de Ruyter van Steveninck et al.[7]).

Fig.2   Fitted values of τ as a function of adaptation velocity for three modulation depths M. The straight line is a least-squares fit to represent the data for M=0.40 in the region w=0.3-100°/s. It has the form $\tau=\alpha \cdot w^{-\beta}$ with α=150 ms and β=0.7 (de Ruyter van Steveninck et al.[7]).

Figure 2 shows fitted values of the response time constant τ as a function of the angular velocity of a moving stimulus (a square wave grating in most experiments) which was presented to the animal during a period long enough to let its visual system adapt to this moving pattern and before the step wise pattern displacement (which reveals τ) was given. The straight line, described by

$$\tau = \alpha \cdot W^{-\beta} \tag{2}$$

(with W in $°/s$ and τ in ms) represents a least-squares fit to the data over the velocity range from 0.36 to 125 $°/s$. For this range, τ varies from 320 to roughly 10 ms, with $\alpha=150\pm10$ ms and $\beta=0.7\pm0.05$. Defining the adaptation range of τ as that interval of velocities for which τ decreases with increasing velocity, we may conclude from figure 2 that within the adaptation range, τ is not very sensitive to the modulation depth.

The outcome of similar experiments with a constant modulation depth of the pattern (M=0.40) and a constant pattern velocity but with four different values of the contrast frequency f_c (i.e. the number of spatial periods per second that traverse an individual visual axis as determined by the spatial wavelength λ_s of the pattern and the pattern velocity v according to $f_c=v/\lambda_s$) reveal also an almost complete independency of the behaviour of τ on contrast frequency. Other experiments in which the stimulus field was subdivided into regions with different adaptation velocities, made clear that the time constants of the input channels of the H1 neuron were set locally by the values of the stimulus velocity in each stimulus sub-region. Finally, it was found that the adaptation of τ is driven by the stimulus velocity, independent of its direction.

These findings can be summarized qualitatively as follows: in steady state, the response time constants τ of the neural units at the highest level in the fly visual system are found to be tuned locally within a large velocity range exclusively by the magnitude of the velocity of the moving pattern and not by its direction, despite the directional selectivity of the neuron itself. We will not go into the question of how this amazing adaptive mechanism may be hard-wired in the fly visual system. Instead we will make advantage of the results derived thus far and attempt to fit the experimental observations into an image processing approach. A large number of theories and several distinct classes of algorithms to encode velocity and direction of movement in visual systems have been suggested by, for example, Marr and Ullman[11] and van Santen and Sperling[12].

We hypothesize that the adaptive mechanism for the setting of the time constants leads to an optimization for the overall performance of the visual system by realizing a velocity independent representation of the moving object. In other words: within the range of velocities for which the time constants are found to be tuned by the velocity, the representation of that stimulus at a certain level within the visual circuitry, should remain independent of any variation in stimulus velocity.

OBJECT MOTION DEGRADATION: MODELLING

Given the physical description of motion and a linear space invariant model, the motion degradation process can be represented by the following convolution integral:

$$g(x,y) = \int\limits_{-\infty}^{\infty} \int\limits_{-\infty}^{\infty} (h(x-u,y-v) \cdot f(u,v))\, du\, dv \tag{3}$$

where f(u,v) is the object intensity at position (u,v) in the object coordinate frame, h(x-u,y-v) is the Point Spread Function (PSF) of the imaging system, which is the response at (x,y) to a unit pulse at (u,v) and g(x,y) is the image intensity at the spatial position (x,y) as blurred by the imaging system. Any possible additive white noise degradation of the already motion blurred image is neglected in the present considerations.

For a review of principles and techniques in the field of digital image degradation and restoration, the reader is referred to Harris[13], Sawchuk[14], Sondhi[15], Nahi[16], Aboutalib et al.[17, 18], Hildebrand[19], Rajala de Figueiredo[20]. It has been demonstrated first by Aboutalib et al.[17] that for situations in which the motion blur occurs in a straight line along one spatial coordinate, say along the horizontal axis, it is correct to look at the blurred image as a collection of degraded line scans through the entire image. The dependence on the vertical coordinate may then be dropped and eq. (3) reduces to:

$$g(x)= \int_{-\infty}^{\infty} h(x-u) \cdot f(u)du \qquad (4)$$

Given the mathematical description of the relative movement, the corresponding PSF can be derived exactly and equation (4) becomes:

$$g(x)= \int_R h(x-u) \cdot f(u)du \qquad (5)$$

where R is the extent of the motion blur. Typically, a discrete version of (5), applicable for digital image processing purposes, is described by:

$$g(k)=\sum_1^L h(k-l) \cdot f(l) \qquad ; k=1,...,N \qquad (6)$$

where k and l take on integer values and L is related to the motion blur extent.

According to Aboutalib et al.[18] a scalar difference equation model (M,a,b,c) can then be derived to model the motion degradation process:

$$x(k+1) = M \cdot x(k)+a \cdot f(k)$$

$$g(k) = b \cdot x(k)+c \cdot f(k) \qquad ; k=1,...,N \qquad (7)$$

$$h(i) = c_0\Delta(i)+c_1\Delta(i-1)+ ...... +c_m\Delta(i-m)$$

where x(k) is the m-dimensional state vector at position k along a scan line, f(k) is the input intensity at position k, g(k) is the output intensity, m is the blur extent, N is the number of elements in a line, c is a scalar, M, a and b are constant matrices of order (mxm), (mx1) and (1xm) respectively, containing the discrete values c_j of the blurring PSF h(j) for j=0,...,m and $\Delta(.)$ is the Kronecker delta function.

INFLUENCE OF BOTH TIME CONSTANT AND VELOCITY ON THE AMOUNT OF MOTION BLUR IN AN ARTIFICIAL RECEPTOR ARRAY

To start with, we incorporate in our simulation model a PSF, derived from equation (1), to model the performance of all neural columnar arranged filters in the lobula complex, with the restriction that the time constants τ remain fixed throughout the whole range of stimulus velocities. Realization of this PSF can easily be achieved via the just mentioned state space model.

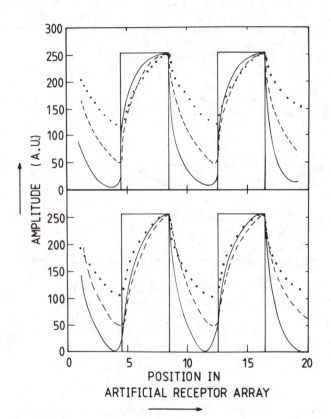

Fig.3   **upper part.** Demonstration of the effect that an increase in magnitude of the time constants of an one-dimensional array of filters will result in increase in motion blur (while the pattern velocity remains constant). Original pattern shown in solid lines is a square-wave grating with a spatial wavelength equal to 8 artificial receptor distances. The three other wave forms drawn, show that for a gradual increase increase in magnitude of the time constants, the representation of the original square-wave will consequently degrade. **lower part.** A gradual increase in velocity of the moving square-wave (while the filter time constants are kept fixed) results also in a clear increase of degradation.

First we demonstrate the effect that an increase in time constant (while the pattern velocity remains the same) will result in an increase in blur. Therefore we introduce an one dimensional array of filters all being equipped with the same time constant in their impulse response. The original pattern shown in square and solid lines in the upper part of figure 3 consists of a square wave grating with a spatial period overlapping 8 artificial receptive filters. The 3 other patterns drawn there show that for the same constant velocity of the moving grating, an increase in the magnitude of the time constants of the filters results in an increased blur in the representation of that grating. On the other hand, an increase in velocity (while the time constants of the artificial receptive units remain the same) also results in a clear increase in motion blur, as demonstrated in the lower part of figure 3.

Inspection of the two wave forms drawn by means of the dashed lines in both upper and lower half of the figure, yields the conclusion, that (apart from rounding errors introduced by the rather small number of artificial filters available), equal amounts of smear will be produced when the product of time constant and pattern velocity is equal. For the upper dashed wave form the velocity was four times smaller but the time constant four times larger than for its equivalent in the lower part of the figure.

ADAPTIVE SCHEME

In designing a proper image processing procedure our next step is to incorporate the experimentally observed flexibility property of the time constants in the imaging elements of our device. In figure 4^a a scheme is shown, which filters the information with fixed time constants, not influenced by the pattern velocity. In figure 4^b a network is shown where the time constants also remain fixed no matter what pattern movement is presented, but now at the next level of information processing, a spatially differential network is incorporated in order to enhance blurred contrasts.

In the filtering network in figure 4^c, first a measurement of the magnitude of the velocity of the moving objects is done by thus far hypothetically introduced movement processing algorithms, modelled here as a set of receptive elements sampling the environment in such a manner that proper estimation of local pattern velocities can be done. Then the time constants of the artificial receptive elements will be tuned according to the estimated velocities and finally the same differential network as in scheme 4^b, is used.

The actual tuning mechanism used for our simulations is outlined in figure 5: once given the range of velocities for which the model is supposed to be operational, and given a lower limit for the time constant τ_{min} (τ_{min} can be the smallest value which physically can be realized), the time constant will be tuned to a new value according to the experimentally observed reciprocal relationship, and will, for all velocities within the adaptive range, be larger than the fixed minimum value. As demonstrated in the previous section the corresponding blur in the representation of the moving stimulus will thus always be larger than for the situation in which the filtering is done with fixed and smallest time constants τ_{min}. More important however is the fact that due to this tuning mechanism the blur will be constant since the product of velocity and time constant is kept constant. So, once the information has been processed by such a system, a velocity independent representation of the image will be the result, which can serve as the input for the spatially differentiating network as outlined in figure 4^c.

The most elementary form for this differential filtering procedure is the one

668

in which the gradient of two filters K-1 and K+1 which are the nearest neighbors of filter K, is taken and then added with a constant weighing factor to the central output K as drawn in figure 4^b and 4^c, where the sign of the gradient depends on the direction of the estimated movement. Essential for our model is that we claim that this weighing factor should be constant throughout the whole set of filters and for the whole high velocity range in which the heterodyne imaging has to be performed. Important to notice is the existence of a so-called settling time, i.e. the minimal time needed for our movement processing device to be able to accurately measure the object velocity. [Note: this time can be set equal to zero in the case that the relative stimulus velocity is known a priori, as demonstrated in figure 3]. Since, without doubt, within this settling period estimated velocity values will come out erroneously and thus no optimal performance of our imaging device can be expected, in all further examples, results after this initial settling procedure will be shown.

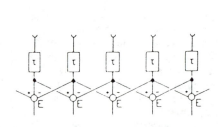

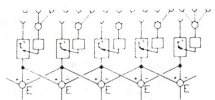

Fig. 4   Pattern movement in this figure is to the right.
A: Network consisting of a set of filters with a fixed, pattern velocity independent, time constant in their impulse response.
B: Identical network as in figure 4A now followed by a spatially differentiating circuitry which adds the weighed gradients of two neighboring filter outputs K-1 and K+1 to the central filter output K.
C: The time constants of the filtering network are tuned by a hypothetical movement estimating mechanism, visualized here as a number of receptive elements, of which the combined output tunes the filters. A detailed description of this mechanism is shown in figure 5. This tuned network is followed by an identical spatially differentiating circuit as described in figure 4B.

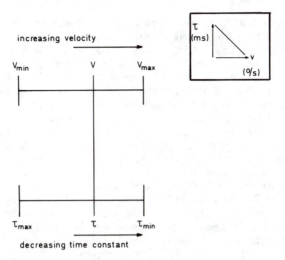

Fig. 5   Detailed description of the mechanism used to tune the time constants. The time constant τ of a specific neural channel is set by the pattern velocity according to the relationship shown in the insert, which is derived from eq. (2) with $\alpha=1$ and $\beta=1$.

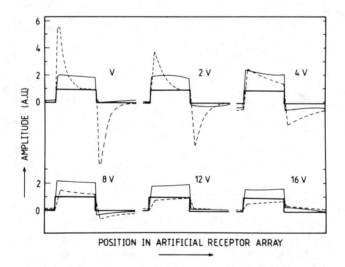

Fig.6   **Thick lines:** square-wave stimulus pattern with a spatial wavelength overlapping 32 artificial receptive elements. **Thick lines:** responses for 6 different pattern velocities in a system consisting of paralleling neural filters equipped with time constants, tuned by this velocity, and followed by a spatially differentiating network as described.
Dashed lines: responses to the 6 different pattern velocities in a filtering system with fixed time constants, followed by the same spatial differentiating circuitry as before. Note the sharp over- and under shoots for this case.

Results obtained with an imaging procedure as drawn in figure 4[b] and 4[c] are shown in figure 6. The pattern consists of a square wave, overlapping 32 picture elements. The pattern moves (to the left) with 6 different velocities v, 2v, 4v, 8v, 12v, 16v. At each velocity only one wavelength is shown. Thick lines: square wave pattern. Dashed lines: the outputs of an imaging device as depicted in figure 4[b]: constant time constants and a constant weighing factor in the spatial processing stage. Note the large differences between the several outputs. Thin continuous lines: the outputs of an imaging device as drawn in figure 4[c]: tuned time constants according to the reciprocal relationship between pattern velocity and time constant and a constant weighing factor in the spatial processing stage. For further simulation details the reader is referred to Zaagman et al.[21]. Now the outputs are almost completely the same and in good agreement with the original stimulus throughout the whole velocity range.

Figure 7 shows the effect of the gradient weighing factor on the overall filter performance, estimated as the improvement of the deblurred images as compared with the blurred image, measured in dB. This quantitative measure has been determined for the case of a moving square wave pattern with motion blur

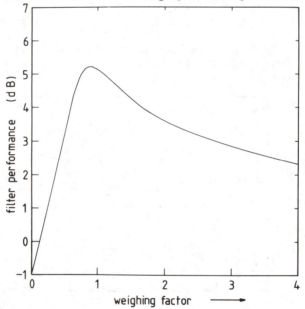

Fig. 7   Effect of the weighing factor on the overall filter performance. Curve measured for the case of a moving square-wave grating. Filter performance is estimated as the improvement in signal to noise ratio:

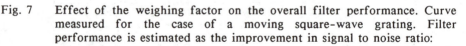

$$I = 10 \cdot {}^{10}\!\log \left(\frac{\Sigma_i \Sigma_j ((v(i,j) - u(i,j))^2}{\Sigma_i \Sigma_j ((\hat{u}(i,j) - u(i,j))^2} \right)$$

where u(i,j) is the original intensity at position (i,j) in the image, v(i,j) is the intensity at the same position (i,j) in the motion blurred image and û(i,j) is the intensity at (i,j) in the image, generated with the adaptive tuning procedure.

extents comparable to those used for the simulations to be discussed in section IV. From this curve it is apparent that for this situation there is an optimum value for this weighing factor. Keeping the weight close to this optimum value will result in a constant output of our adaptive scheme, thus enabling an optimal deblurring of the smeared image of the moving object.

On the other hand, starting from the point of view that the time constants should remain fixed throughout the filtering process, we should had have to tune the gradient weights to the velocity in order to produce a constant output as demonstrated in figure 6 where the dashed lines show strongly differing outputs of a fixed time constant system with spatial processing with constant weight (figure 4^b). In other words, tuning of the time constants as proposed in this section results in: 1) the realization of the blur-constancy criterion as formulated previously, and 2) -as a consequence- the possibility to deblur the obtained image optimally with one and the same weighing factor of the gradient in the final spatial processing layer over the whole heterodyne velocity range.

COMPUTER SIMULATION RESULTS AND CONCLUSIONS

The image quality improvement algorithm developed in the present contribution has been implemented on a general purpose DG Eclipse S/140 mini-computer for our two dimensional simulations. Figure 8^a shows an undisturbed image, consisting of 256 lines of each 256 pixels, with 8 bit intensity resolution. Figure 8^b shows what happens with the original image if the PSF is modelled according to the exponential decay (2). In this case the time constants of all spatial information processing channels have been kept fixed. Again, information content in the higher spatial frequencies has been reduced largely. The implementation of the heterodyne filtering procedure was now done as follows: first the adaptation range was defined by setting the range of velocities. This means that our adaptive heterodyne algorithm is supposed to operate adequately only within the thus defined velocity range and that -in that range- the time constants are tuned according to relationship (2) and will always come out larger than the minimum value τ_{min}. For demonstration purposes we set $\alpha=1$ and $\beta=1$ in eq. (2), thus introducing the phenomenon that for any velocity, the two dimensional set of spatial filters with time constants tuned by that velocity, will always produce a constant output, independent of this velocity which introduces the motion blur. Figure 8^c shows this representation. It is important to note here that this constant output has far more worse quality than any set of filters with smallest and fixed time constants τ_{min} would produce for velocities within the operational range. The advantage of a velocity independent output at this level in our simulation model, is that in the next stage a differential scheme can be implemented as discussed in detail in the preceding paragraph. Constancy of the weighing factor which is used in this differential processing scheme is guaranteed by the velocity independency of the obtained image representation.

Figure 8^d shows the result of the differential operation with an optimized gradient weighing factor. This weighing factor has been optimized based on an almost identical performance curve as described previously in figure 7. A clear and good restoration is apparent from this figure, though close inspection reveals fine structure (especially for areas with high intensities) which is unrelated with the original intensity distribution. These artifacts are caused by the phenomenon that for these high intensity areas possible tuning errors will show up much more pronounced than for low intensities.

672

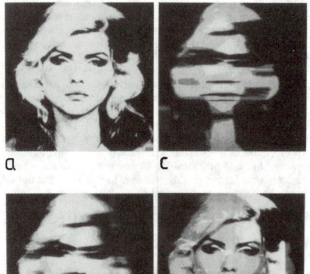

Fig. 8a   Original 256x256x8 bit picture.

Fig. 8b   Motion degraded image with a PSF derived from $R(t)=c+a \cdot e^{(-t/\tau)}$, where τ is kept fixed to 12 pixels and the motion blur extent is 32 pixels.

Fig. 8c   Worst case, i.e. the result of motion degradation of the original image with a PSF as in figure 8b, but with tuning of the time constants based on the velocity.

Fig. 8d   Restored version of the degraded image using the heterodyne adaptive processing scheme.

In conclusion: a heterodyne adaptive image processing technique, inspired by the fly visual system, has been presented as an imaging device for moving objects. A scalar difference equation model has been used to represent the motion blur degradation process. Based on the experimental results described and on this state space model, we developed an adaptive filtering scheme, which produces at a certain level within the system a constant output, permitting further differential operations in order to produce an optimally deblurred representation of the moving object.

ACKNOWLEDGEMENTS

The authors wish to thank mr. Eric Bosman for his expert programming assistance, mr. Franco Tommasi for many inspiring discussions and advises during the implementation of the simulation model and dr. Rob de Ruyter van Steveninck for experimental help. This research was partly supported by the *Netherlands Organization for the Advancement of Pure Research (Z.W.O.)* through the foundation *Stichting voor Biofysica*.

REFERENCES

1. K. Hausen, Z. Naturforschung **31c**, 629-633 (1976).
2. N. J. Strausfeld, Atlas of an insect brain (Springer Verlag, Berlin, Heidelberg, New York, 1976).
3. K. Hausen, Biol. Cybern. **45**, 143-156 (1982).
4. R. Hengstenberg, J. Comp. Physiol. **149**, 179-193 (1982).
5. W. H. Zaagman, H. A. K. Mastebroek, J. W. Kuiper, Biol. Cybern. **31**, 163-168 (1978).
6. H. A. K. Mastebroek, W. H. Zaagman, B. P. M. Lenting, Vision Res. **20**, 467-474 (1980)
7. R. R. de Ruyter van Steveninck, W. H. Zaagman, H. A. K. Mastebroek, Biol. Cybern., **54**, 223-236 (1986).
8. W. Reichardt, T. Poggio, Q. Rev. Biophys. **9**, 311-377 (1976).
9. W. Reichardt, *in* Reichardt, W. (Ed.) Processing of optical Data by Organisms and Machines (Academic Press, New York, 1969), pp. 465-493.
10. T. Poggio, W. Reichardt, Q. Rev. Bioph. **9**, 377-439 (1976).
11. D. Marr, S. Ullman, Proc. R. Soc. Lond. **211**, 151-180 (1981).
12. J. P. van Santen, G. Sperling, J. Opt. Soc. Am. A **1**, 451-473 (1984).
13. J. L. Harris SR., J. Opt. Soc. Am. **56**, 569-574 (1966).
14. A. A. Sawchuk, Proc. IEEE, Vol. 60, No. 7, 854-861 (1972).
15. M. M.Sondhi, Proc. IEEE, Vol. 60, No. 7, 842-853 (1972).
16. N. E. Nahi, Proc. IEEE, Vol. 60, No. 7, 872-877 (1972).
17. A. O. Aboutalib, L. M. Silverman, IEEE Trans. On Circuits And Systems T-CAS **75**, 278-286 (1975).
18. A. O. Aboutalib, M. S. Murphy, L.M. Silverman, IEEE Trans. Automat. Contr. AC **22**, 294-302 (1977).
19. Th. Hildebrand, Biol. Cybern. **36**, 229-234 (1980).
20. S. A. Rajala, R. J. P. de Figueiredo, IEEE Trans. On Acoustics, Speech and Signal Processing, Vol. ASSSP-29, No. 5, 1033-1042 (1981).
21. W. H. Zaagman, H. A. K. Mastebroek, R. R. de Ruyter van Steveninck, IEEE Trans, Syst. Man Cybern. SMC **13**, 900-906 (1983).

PATTERN CLASS DEGENERACY IN AN UNRESTRICTED STORAGE DENSITY MEMORY

Christopher L. Scofield, Douglas L. Reilly,
Charles Elbaum, Leon N. Cooper

Nestor, Inc., 1 Richmond Square, Providence, Rhode Island, 02906.

ABSTRACT

The study of distributed memory systems has produced a number of models which work well in limited domains. However, until recently, the application of such systems to real-world problems has been difficult because of storage limitations, and their inherent architectural (and for serial simulation, computational) complexity. Recent development of memories with unrestricted storage capacity and economical feedforward architectures has opened the way to the application of such systems to complex pattern recognition problems. However, such problems are sometimes underspecified by the features which describe the environment, and thus a significant portion of the pattern environment is often non-separable. We will review current work on high density memory systems and their network implementations. We will discuss a general learning algorithm for such high density memories and review its application to separable point sets. Finally, we will introduce an extension of this method for learning the probability distributions of non-separable point sets.

INTRODUCTION

Information storage in distributed content addressable memories has long been the topic of intense study. Early research focused on the development of correlation matrix memories [1, 2, 3, 4]. Workers in the field found that memories of this sort allowed storage of a number of distinct memories no larger than the number of dimensions of the input space. Further storage beyond this number caused the system to give an incorrect output for a memorized input.

Recent work on distributed memory systems has focused on single layer, recurrent networks. Hopfield [5, 6] introduced a method for the analysis of settling of activity in recurrent networks. This method defined the network as a dynamical system for which a global function called the 'energy' (actually a Liapunov function for the autonomous system describing the state of the network) could be defined. Hopfield showed that flow in state space is always toward the fixed points of the dynamical system if the matrix of recurrent connections satisfies certain conditions. With this property, Hopfield was able to define the fixed points as the sites of memories of network activity.

Like its forerunners, the Hopfield network is limited in storage capacity. Empirical study of the system found that for randomly chosen memories, storage capacity was limited to m ≤ 0.15N, where m is the number of memories that could be accurately recalled, and N is the dimensionality of the network (this has since been improved to m ≤ N, [7, 8]). The degradation of memory recall with increased storage density is directly related to the proliferation in the state space of unwanted local minima which serve as basins of flow.

UNRESTRICTED STORAGE DENSITY MEMORIES

Bachman et al. [9] have studied another relaxation system similar in some respects to the Hopfield network. However, in contrast to Hopfield, they have focused on defining a dynamical system in which the locations of the minima are explicitly known.

In particular, they have chosen a system with a Liapunov function given by

$$E = -1/L \sum_j Q_j |\mu - x_j|^{-L}, \tag{1}$$

where E is the total 'energy' of the network, $\mu(0)$ is a vector describing the initial network activity caused by a test pattern, and x_j, the site of the jth memory, for m memories in R^N. L is a parameter related to the network size. Then $\mu(0)$ relaxes to $\mu(T)$ = x_j for some memory j according to

$$\dot{\mu} = -\sum_j Q_j \, |\, \mu - x_j\,|^{-(L+2)} \, (\mu - x_j) \qquad (2)$$

This system is isomorphic to the classical electrostatic potential between a positive (unit) test charge, and negative charges Q_j at the sites x_j (for a 3-dimensional input space, and L = 1). The N-dimensional Coulomb energy function then defines exactly m basins of attraction to the fixed points located at the charge sites x_j. It can been shown that convergence to the closest distinct memory is guaranteed, *independent of the number of stored memories m*, for proper choice of N and L [9, 10].

Equation 1 shows that each cell receives feedback from the network in the form of a scalar

$$\sum_j Q_j \, |\, \mu - x_j\,|^{-L} . \qquad (3)$$

Importantly, this quantity is the same for all cells; it is as if a single virtual cell was computing the distance in activity space between the current state and stored states. The result of the computation is then broadcast to all of the cells in the network. A 2-layer feedforward network implementing such a system has been described elsewhere [10].

The connectivity for this architecture is of order m·N, where m is the number of stored memories and N is the dimensionality of layer 1. This is significant since the addition of a new memory m' = m + 1 will change the connectivity by the addition of N + 1 connections, whereas in the Hopfield network, addition of a new memory requires the addition of 2N + 1 connections.

An equilibrium feedforward network with similar properties has been under investigation for some time [11]. This model does not employ a relaxation procedure, and thus was not originally framed in the language of Liapunov functions. However, it is possible to define a similar system if we identify the locations of the 'prototypes' of this model as the locations in state space of potentials which satisfy the following conditions

$$E_j = -Q_j / R_o \quad \text{for } |\, \mu - x_j\,| < \lambda_j \qquad (4)$$

$$= 0 \quad \text{for } |\, \mu - x_j\,| > \lambda_j.$$

where R_o is a constant.

This form of potential is often referred to as the 'square-well' potential. This potential may be viewed as a limit of the N-dimensional Coulomb potential, in which the $1/R$ ($L = 1$) well is replaced with a square well (for which $L \gg 1$). Equation 4 describes an energy landscape which consists of plateaus of zero potential outside of wells with flat, zero slope basins. Since the landscape has only flat regions separated by discontinuous boundaries, the state of the network is always at equilibrium, and relaxation does not occur. For this reason, this system has been called an equilibrium model. This model, also referred to as the Restricted Coulomb Energy (RCE)[14] model, shares the property of unrestricted storage density.

LEARNING IN HIGH DENSITY MEMORIES

A simple learning algorithm for the placement of the wells has been described in detail elsewhere [11, 12].

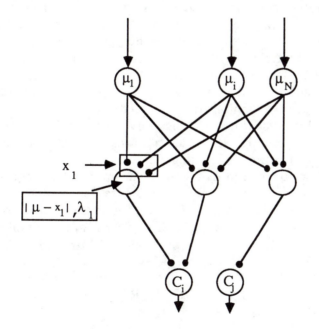

Figure1: 3-layer feedforward network. Cell i computes the quantity $|\mu - x_i|$ and compares to internal threshold λ_i.

Reilly et. al. have employed a three layer feedforward network (figure 1) which allows the generalization of a content addressable memory to a pattern classification memory. Because the locations of the minima are explicitly known in the equilibrium model, it is possible to dynamically program the energy function for an arbitrary energy landscape. This allows the construction of geographies of basins associated with the classes constituting the pattern environment. Rapid learning of complex, non-linear, disjoint, class regions is possible by this method [12, 13].

LEARNING NON-SEPARABLE CLASS REGIONS

Previous studies have focused on the acquisition of the geography and boundaries of non-linearly separable point sets. However, a method by which such high density models can acquire the probability distributions of non-separable sets has not been described.

Non-separable sets are defined as point sets in the state space of a system which are labelled with multiple class affiliations. This can occur because the input space has not carried all of the features in the pattern environment, or because the pattern set itself is not separable. Points may be degenerate with respect to the explicit features of the space, however they may have different probability distributions within the environment. This structure in the environment is important information for the identification of patterns by such memories in the presence of feature space degeneracies.

We now describe one possible mechanism for the acquisition of the probability distribution of non-separable points. It is assumed that all points in some region R of the state space of the network are the site of events $\mu(0, C_i)$ which are examples of pattern classes $C = \{C_1, ..., C_M\}$. A basin of attraction, $x_k(C_i)$, defined by equation 4, is placed at each site $\mu(0, C_i)$ unless

$$| \mu(0, C_i) - x_j(C_i) | < R_o, \qquad (5)$$

that is, unless a memory at x_j (of the class C_i) already contains $\mu(0, C_i)$. The initial values of Q_o and R_o at $x_k(C_i)$ are a constant for all sites x_j. Thus as events of the classes $C_1, ..., C_M$ occur at a particular site in R, multiple wells are placed at this location.

If a well $x_j(C_i)$ correctly covers an event $\mu(0, C_i)$, then the charge at that site (which defines the depth of the well) is incremented by a constant amount ΔQ_o. In this manner, the region R is covered with wells of all classes $\{C_1,..., C_M\}$, with the depth of well $x_j(C_i)$ proportional to the frequency of occurence of C_i at x_j.

The architecture of this network is exactly the same as that already described. As before, this network acquires a new cell for each well placed in the energy landscape. Thus we are able to describe the meaning of wells that overlap as the competition by multiple cells in layer 2 in firing for the pattern of activity in the input layer.

APPLICATIONS

This system has been applied to a problem in the area of risk assessment in mortgage lending. The input space consisted of feature detectors with continuous firing rates proportional to the values of 23 variables in the application for a mortgage. For this set of features, a significant portion of the space was non-separable.

Figures 2a and 2b illustrate the probability distributions of high and low risk applications for two of the features. It is clear that in this 2-dimensional subspace, the regions of high and low risk are non-separable but have different distributions.

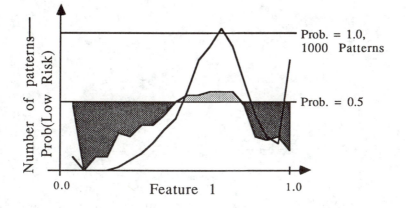

Figure 2a: Probability distribution for High and Low risk patterns for feature 1.

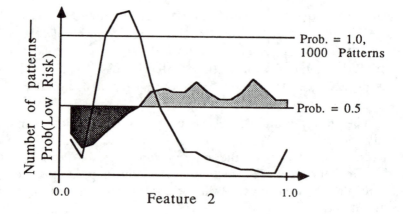

Figure 2b: Probability distribution for High and Low risk patterns for feature 2.

Figure 3 depicts the probability distributions acquired by the system for this 2-dimensional subspace. In this image, circle radius is proportional to the degree of risk: Small circles are regions of low risk, and large circles are regions of high risk.

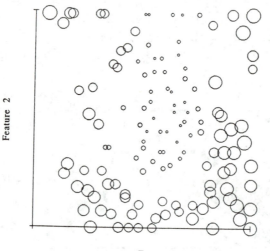

Figure 3: Probability distribition for Low and High risk. Small circles indicate low risk regions and large circles indicate high risk regions.

Of particular interest is the clear clustering of high and low risk regions in the 2-d map. Note that the regions are in fact non-linearly separable.

DISCUSSION

We have presented a simple method for the acquisition of probability distributions in non-separable point sets. This method generates an energy landscape of potential wells with depths that are proportional to the local probability density of the classes of patterns in the environment. These well depths set the probability of firing of class cells in a 3-layer feedforward network.

Application of this method to a problem in risk assessment has shown that even completely non-separable subspaces may be modeled with surprising accuracy. This method improves pattern classification in such problems with little additional computational burden.

This algorithm has been run in conjunction with the method described by Reilly et. al.[11] for separable regions. This combined system is able to generate non-linear decision surfaces between the separable zones, and approximate the probability distributions of the non-separable zones in a seemless manner. Further discussion of this system will appear in future reports.

Current work is focused on the development of a more general method for modelling the scale of variations in the distributions. Sensitivity to this scale suggests that the transition from separable to non-separable regions is smooth and should not be handled with a 'hard' threshold.

ACKNOWLEDGEMENTS

We would like to thank Ed Collins and Sushmito Ghosh for their significant contributions to this work through the development of the mortgage risk assessment application.

REFERENCES

[1] Anderson, J.A.: A simple neural network generating an interactive memory. Math. Biosci. **14**, 197-220 (1972).

[2] Cooper, L.N.: A possible organization of animal memory and learning. In: Proceedings of the Nobel Symposium on Collective Properties of Physical Systems, Lundquist, B., Lundquist, S. (eds.). (24), 252-264 London, New York: Academic Press 1973.

[3] Kohonen, T.: Correlation matrix memories. IEEE Trans. Comput. 21, 353-359 (1972).

[4] Kohonen, T.: Associative memory - a system-theoretical approach. Berlin, Heidelberg, New York: Springer 1977.

[5] Hopfield, J.J.: Neural networks and physical systems with emergent collective computational abilities. Proc. Natl. Acad. Sci. USA 79, 2554-2558 (April 1982).

[6] Hopfield, J.J.: Neurons with graded response have collective computational properties like those of two-state neurons. Proc. Natl. Acad. Sci. USA 81, 2088-3092 (May, 1984).

[7] Hopfield, J.J., Feinstein, D.I., Palmer, R.G.: 'Unlearning' has a stabilizing effect in collective memories. Nature 304, 158-159 (July 1983).

[8] Potter, T.W.: Ph.D. Dissertation in advanced technology, S.U.N.Y. Binghampton, (unpublished).

[9] Bachmann, C.M., Cooper, L.N., Dembo, A., Zeitouni, O.: A relaxation model for memory with high density storage. to be published in Proc. Natl. Acad. Sci. USA.

[10] Dembo, A., Zeitouni, O.: ARO Technical Report, Brown University, Center for Neural Science, Providence, R.I., (1987), also submitted to Phys. Rev. A.

[11] Reilly, D.L., Cooper, L.N., Elbaum, C.: A neural model for category learning. Biol. Cybern. 45, 35-41 (1982).

[12] Reilly, D.L., Scofield, C., Elbaum, C., Cooper, L.N.: Learning system architectures composed of multiple learning modules. to appear in Proc. First Int'l. Conf. on Neural Networks (1987).

[13] Rimey, R., Gouin, P., Scofield, C., Reilly, D.L.: Real-time 3-D object classification using a learning system. Intelligent Robots and Computer Vision, Proc. SPIE 726 (1986).

[14] Reilly, D.L., Scofield, C. L., Elbaum, C., Cooper, L.N: Neural Networks with low connectivity and unrestricted memory storage density. To be published.

A MEAN FIELD THEORY OF LAYER IV OF VISUAL CORTEX AND ITS APPLICATION TO ARTIFICIAL NEURAL NETWORKS*

Christopher L. Scofield
Center for Neural Science and Physics Department
Brown University
Providence, Rhode Island 02912
and
Nestor, Inc., 1 Richmond Square, Providence, Rhode Island, 02906.

ABSTRACT

A single cell theory for the development of selectivity and ocular dominance in visual cortex has been presented previously by Bienenstock, Cooper and Munro[1]. This has been extended to a network applicable to layer IV of visual cortex[2]. In this paper we present a mean field approximation that captures in a fairly transparent manner the qualitative, and many of the quantitative, results of the network theory. Finally, we consider the application of this theory to artificial neural networks and show that a significant reduction in architectural complexity is possible.

A SINGLE LAYER NETWORK AND THE MEAN FIELD APPROXIMATION

We consider a single layer network of ideal neurons which receive signals from outside of the layer and from cells within the layer (Figure 1). The activity of the ith cell in the network is

$$c_i = m_i \, d + \sum_j L_{ij} \, c_j. \qquad (1)$$

d is a vector of afferent signals to the network. Each cell receives input from n fibers outside of the cortical network through the matrix of synapses m_i. Intra-layer input to each cell is then transmitted through the matrix of cortico-cortical synapses L.

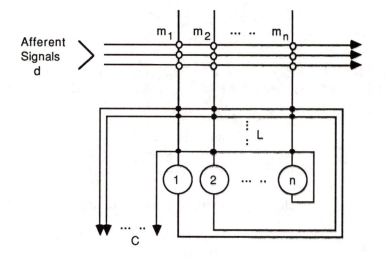

Figure 1: The general single layer recurrent network. Light circles are the LGN-cortical synapses. Dark circles are the (non-modifiable) cortico-cortical synapses.

We now expand the response of the ith cell into individual terms describing the number of cortical synapses traversed by the signal d before arriving through synapses L_{ij} at cell i. Expanding c_j in (1), the response of cell i becomes

$$c_i = m_i \, d + \sum_j L_{ij} \, m_j \, d + \sum_j L_{ij} \sum_k L_{jk} \, m_k \, d + \sum_j L_{ij} \sum_k L_{jk} \sum_n L_{kn} \, m_n \, d + ... \quad (2)$$

Note that each term contains a factor of the form

$$\sum_p L_{qp} \, m_p d.$$

This factor describes the first order effect, on cell q, of the cortical transformation of the signal d. The mean field approximation consists of estimating this factor to be a constant, independant of cell location

$$\sum_p L_{qp} \, m_p d = N \, \bar{m} d \, L_o = constant. \quad (3)$$

This assumption does not imply that each cell in the network is selective to the same pattern, (and thus that $m_i = m_j$). Rather, the assumption is that the vector sum is a constant

$$(\sum_p L_{qp} m_p) d = (N \bar{m} L_o) d.$$

This amounts to assuming that each cell in the network is surrounded by a population of cells which represent, on average, all possible pattern preferences. Thus the vector sum of the afferent synaptic states describing these pattern preferences is a constant independent of location.

Finally, if we assume that the lateral connection strengths are a function only of i-j then L_{ij} becomes a circular matrix so that

$$\sum_i L_{ij} = \sum_j L_{ji} = L_o = \text{constant}.$$

Then the response of the cell i becomes

$$c_i = m_i d + (L_o + L_o^2 + \dots) \bar{m} d . \tag{4}$$

$$= m_i d + (N L_o /(1 - L_o)) \bar{c}, \qquad \text{for } | L_o | < 1$$

where we define the spatial average of cortical cell activity $\bar{c} = \bar{m} d$, and N is the average number of intracortical synapses.

Here, in a manner similar to that in the theory of magnetism, we have replaced the effect of individual cortical cells by their average effect (as though all other cortical cells can be replaced by an 'effective' cell, figure 2). Note that we have retained all orders of synaptic traversal of the signal d.

Thus, we now focus on the activity of the layer after 'relaxation' to equilibrium. In the mean field approximation we can therefore write

$$= (m_i - \alpha) d \tag{5}$$

where the mean field

$$\alpha = a \bar{m}$$

with

$$a = N |L_o| (1 + |L_o|)^{-1},$$

and we asume that $L_0 < 0$ (the network is, on average, inhibitory).

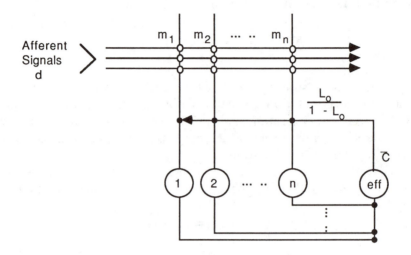

Figure 2: The single layer mean field network. Detailed connectivity between all cells of the network is replaced with a single (non-modifiable) synapse from an 'effective' cell.

LEARNING IN THE CORTICAL NETWORK

We will first consider evolution of the network according to a synaptic modification rule that has been studied in detail, for single cells, elsewhere[1, 3]. We consider the LGN - cortical synapses to be the site of plasticity and assume for maximum simplicity that there is no modification of cortico-cortical synapses. Then

$$\dot{m}_i = \phi(c_i, \bar{\bar{c}}_i)\, d \qquad (6)$$

$$\dot{L}_{ij} = 0.$$

In what follows $\bar{c}$ denotes the spatial average over cortical cells, while $\bar{c}_i$ denotes the time averaged activity of the ith cortical cell. The function ϕ has been discussed extensively elsewhere. Here we note that ϕ describes a function of the cell response that has both hebbian and anti-hebbian regions.

This leads to a very complex set of non-linear stochastic equations that have been analyzed partially elsewhere[2]. In general, the afferent synaptic state has fixed points that are stable and selective and unstable fixed points that are non-selective[1, 2]. These arguments may now be generalized for the network. In the mean field approximation

$$\dot{m}_i(\alpha) = \phi(c_i(\alpha), \bar{\bar{c}}_i(\alpha)) \, d = \phi[m_i(\alpha) - \alpha] \, d \qquad (7)$$

The mean field, α has a time dependent component $\bar{m}$. This varies as the average over all of the network modifiable synapses and, in most environmental situations, should change slowly compared to the change of the modifiable synapses to a single cell. Then in this approximation we can write

$$(\dot{m_i(\alpha)}-\alpha) = \phi[m_i(\alpha) - \alpha] \, d. \qquad (8)$$

We see that there is a mapping

$$m_i' <\longrightarrow m_i(\alpha) - \alpha \qquad (9)$$

such that for every $m_i(\alpha)$ there exists a corresponding (mapped) point m_i' which satisfies

$$\dot{m}_i' = \phi[m_i'] \, d,$$

the original equation for the mean field zero theory. It can be shown [2, 4] that for every fixed point of $m_i(\alpha = 0)$, there exists a corresponding fixed point $m_i(\alpha)$ with the same selectivity and stability properties. The fixed points are available to the neurons if there is sufficient inhibition in the network ($|L_o|$ is sufficiently large).

APPLICATION OF THE MEAN FIELD NETWORK TO LAYER IV OF VISUAL CORTEX

Neurons in the primary visual cortex of normal adult cats are sharply tuned for the orientation of an elongated slit of light and most are activated by stimulation of either eye. Both of these properties--orientation selectivity and binocularity--depend on the type of visual environment experienced during a critical period of early postnatal development. For example, deprivation of patterned input during this critical period leads to loss of orientation selectivity while monocular deprivation (MD) results in a dramatic shift in the ocular dominance of cortical neurons such that most will be responsive exclusively to the open eye. The ocular dominance shift after MD is the best known and most intensively studied type of visual cortical plasticity.

The behavior of visual cortical cells in various rearing conditions suggests that some cells respond more rapidly to environmental changes than others. In monocular deprivation, for example, some cells remain responsive to the closed eye in spite of the very large shift of most cells to the open eye. Singer et. al.[5] found, using intracellular recording, that geniculo-cortical synapses on inhibitory interneurons are more resistant to monocular deprivation than are synapses on pyramidal cell dendrites. Recent work suggests that the density of inhibitory GABAergic synapses in kitten striate cortex is also unaffected by MD during the cortical period [6, 7].

These results suggest that some LGN-cortical synapses modify rapidly, while others modify relatively slowly, with slow modification of some cortico-cortical synapses. Excitatory LGN-cortical synapses into excitatory cells may be those that modify primarily. To embody these facts we introduce two types of LGN-cortical synapses: those (m_i) that modify and those (z_k) that remain relatively constant. In a simple limit we have

$$\dot{m}_i = \phi(c_i, \bar{\bar{c}}_i)\, d$$

and

$$\dot{z}_k = 0.$$

(10)

We assume for simplicity and consistent with the above physiological interpretation that these two types of synapses are confined to two different classes of cells and that both left and right eye have similar synapses (both m_i or both z_k) on a given cell. Then, for binocular cells, in the mean field approximation (where binocular terms are in italics)

$$c_i(\alpha) = (m_i - \alpha)d = (m_i^l - \alpha^l)\cdot d^l + (m_i^r - \alpha^r)\cdot d^r$$

$$c_k(\alpha) = (z_k - \alpha)d = (z_k^l - \alpha^l)\cdot d^l + (z_k^r - \alpha^r)\cdot d^r,$$

where $d^{l(r)}$ are the explicit left (right) eye time averaged signals arriving form the LGN. Note that $\alpha^{l(r)}$ contain terms from modifiable and non-modifiable synapses:

$$\alpha^{l(r)} = a\, (\bar{m}^{l(r)} + \bar{z}^{l(r)}).$$

Under conditions of monocular deprivation, the animal is reared with one eye closed. For the sake of analysis assume that the right eye is closed and that only noise-like signals arrive at cortex from the right eye. Then the environment of the cortical cells is:

$$d = (d^j, n) \tag{12}$$

Further, assume that the left eye synapses have reached their selective fixed point, selective to pattern d^1. Then $(m_i^l,\, m_i^r) =$

$(m_i^{l*},\, x_i)$ with $|x_i| \ll |m_i^{l*}|$. Following the methods of BCM, a local

linear analysis of the ϕ - function is employed to show that for the closed eye

$$x_i = a\,(1 - \lambda a)^{-1}\bar{z}^r. \tag{13}$$

where $\lambda = N_m/N$ is the ratio of the number modifiable cells to the total number of cells in the network. That is, the asymptotic state of the closed eye synapses is a scaled function of the mean-field due to non-modifiable (inhibitory) cortical cells. The scale of this state is set not only by the proportion of non-modifiable cells, but in addition, by the averaged intracortical synaptic strength L_0.

Thus contrasted with the mean field zero theory the deprived eye LGN-cortical synapses do not go to zero. Rather they approach the constant value dependent on the average inhibition produced by the non-modifiable cells in such a way that the asymptotic output of the cortical cell is zero (it cannot be driven by the deprived eye). However lessening the effect of inhibitory synapses (e.g. by application of an inhibitory blocking agent such as bicuculine) reduces the magnitude of α so that one could once more obtain a response from the deprived eye.

We find, consistent with previous theory and experiment, that most learning can occur in the LGN-cortical synapse, for inhibitory (cortico-cortical) synapses need not modify. Some non-modifiable LGN-cortical synapses are required.

THE MEAN FIELD APPROXIMATION AND ARTIFICIAL NEURAL NETWORKS

The mean field approximation may be applied to networks in which the cortico-cortical feedback is a general function of cell activity. In particular, the feedback may measure the difference between the network activity and memories of network activity. In this way, a network may be used as a content addressable memory. We have been discussing the properties of a mean field network after equilibrium has been reached. We now focus on the detailed time dependence of the relaxation of the cell activity to a state of equilibrium.

Hopfield[8] introduced a simple formalism for the analysis of the time dependence of network activity. In this model, network activity is mapped onto a physical system in which the state of neuron activity is considered as a 'particle' on a potential energy surface. Identification of the pattern occurs when the activity 'relaxes' to a nearby minima of the energy. Thus minima are employed as the sites of memories. For a Hopfield network of N neurons, the intra-layer connectivity required is of order N^2. This connectivity is a significant constraint on the practical implementation of such systems for large scale problems. Further, the Hopfield model allows a storage capacity which is limited to m < N memories[8, 9]. This is a result of the proliferation of unwanted local minima in the 'energy' surface.

Recently, Bachmann et al.[10], have proposed a model for the relaxation of network activity in which memories of activity patterns are the sites of negative 'charges', and the activity caused by a test pattern is a positive test 'charge'. Then in this model, the energy function is the electrostatic energy of the (unit) test charge with the collection of charges at the memory sites

$$E = -1/L \sum_j Q_j \mid \mu - x_j \mid^{-L}, \qquad (14)$$

where $\mu(0)$ is a vector describing the initial network activity caused by a test pattern, and x_j, the site of the jth memory. L is a parameter related to the network size.

This model has the advantage that storage density is not restricted by the the network size as it is in the Hopfield model, and in addition, the architecture employs a connectivity of order $m \times N$. Note that at each stage in the settling of $\mu(t)$ to a memory (of network activity) x_j, the only feedback from the network to each cell is the scalar

$$\sum_j Q_j \, | \, \mu - x_j \, | \, - L \tag{15}$$

This quantity is an integrated measure of the distance of the current network state from stored memories. Importantly, this measure is the same for all cells; it is as if a single virtual cell was computing the distance in activity space between the current state and stored states. The result of the computation is then broadcast to all of the cells in the network. This is a generalization of the idea that the detailed activity of each cell in the network need not be fed back to each cell. Rather some global measure, performed by a single 'effective' cell is all that is sufficient in the feedback.

DISCUSSION

We have been discussing a formalism for the analysis of networks of ideal neurons based on a mean field approximation of the detailed activity of the cells in the network. We find that a simple assumption concerning the spatial distribution of the pattern preferences of the cells allows a great simplification of the analysis. In particular, the detailed activity of the cells of the network may be replaced with a mean field that in effect is computed by a single 'effective' cell.

Further, the application of this formalism to the cortical layer IV of visual cortex allows the prediction that much of learning in cortex may be localized to the LGN-cortical synaptic states, and that cortico-cortical plasticity is relatively unimportant. We find, in agreement with experiment, that monocular deprivation of the cortical cells will drive closed-eye responses to zero, but chemical blockage of the cortical inhibitory pathways would reveal non-zero closed-eye synaptic states.

692

Finally, the mean field approximation allows the development of single layer models of memory storage that are unrestricted in storage density, but require a connectivity of order m×N. This is significant for the fabrication of practical content addressable memories.

ACKNOWLEDGEMENTS

I would like to thank Leon Cooper for many helpful discussions and the contributions he made to this work.

*This work was supported by the Office of Naval Research and the Army Research Office under contracts #N00014-86-K-0041 and #DAAG-29-84-K-0202.

REFERENCES

[1] Bienenstock, E. L., Cooper, L. N & Munro, P. W. (1982) *J. Neuroscience* **2**, 32-48.
[2] Scofield, C. L. (1984) Unpublished Dissertation.
[3] Cooper, L. N, Munro, P. W. & Scofield, C. L. (1985) in *Synaptic Modification, Neuron Selectivity and Nervous System Organization*, ed. C. Levy, J. A. Anderson & S. Lehmkuhle, (Erlbaum Assoc., N. J.).
[4] Cooper, L. N & Scofield, C. L. (to be published) *Proc. Natl. Acad. Sci. USA*..
[5] Singer, W. (1977) **Brain Res. 134**, 508-000.
[6] Bear, M. F., Schmechel D. M., & Ebner, F. F. (1985) *J. Neurosci.* **5**, 1262-0000.
[7] Mower, G. D., White, W. F., & Rustad, R. (1986) *Brain Res.* **380**, 253-000.
[8] Hopfield, J. J. (1982) *Proc. Natl. Acad.* Sci. USA **79**, 2554-2558.
[9] Hopfield, J. J., Feinstein, D. I., & Palmer, R. G. (1983) *Nature* **304**, 158-159.
[10] Bachmann, C. M., Cooper, L. N, Dembo, A. & Zeitouni, O. (to be published) *Proc. Natl. Acad. Sci. USA*.

Teaching Artificial Neural Systems to Drive:
Manual Training Techniques for Autonomous Systems

J. F. Shepanski and S. A. Macy

TRW, Inc.
One Space Park, O2/1779
Redondo Beach, CA 90278

Abstract

We have developed a methodology for manually training autonomous control systems based on artificial neural systems (ANS). In applications where the rule set governing an expert's decisions is difficult to formulate, ANS can be used to extract rules by associating the information an expert receives with the actions he takes. Properly constructed networks imitate rules of behavior that permits them to function autonomously when they are trained on the spanning set of possible situations. This training can be provided manually, either under the direct supervision of a system trainer, or indirectly using a background mode where the network assimilates training data as the expert performs his day-to-day tasks. To demonstrate these methods we have trained an ANS network to drive a vehicle through simulated freeway traffic.

Introduction

Computational systems employing fine grained parallelism are revolutionizing the way we approach a number of long standing problems involving pattern recognition and cognitive processing. The field spans a wide variety of computational networks, from constructs emulating neural functions, to more crystalline configurations that resemble systolic arrays. Several titles are used to describe this broad area of research, we use the term artificial neural systems (ANS). Our concern in this work is the use of ANS for manually training certain types of autonomous systems where the desired rules of behavior are difficult to formulate.

Artificial neural systems consist of a number of processing elements interconnected in a weighted, user-specified fashion, the interconnection weights acting as memory for the system. Each processing element calculates an output value based on the weighted sum of its inputs. In addition, the input data is correlated with the output or desired output (specified by an instructive agent) in a training rule that is used to adjust the interconnection weights. In this way the network learns patterns or imitates rules of behavior and decision making.

The particular ANS architecture we use is a variation of Rummelhart et. al. [1] multi-layer perceptron employing the generalized delta rule (GDR). Instead of a single, multi-layer structure, our final network has a a multiple component or "block" configuration where one block's output feeds into another (see Figure 3). The training methodology we have developed is not tied to a particular training rule or architecture and should work well with alternative networks like Grossberg's adaptive resonance model[2].

The equations describing the network are derived and described in detail by Rumelhart et. al.[1]. In summary, they are:

Transfer function: $\quad o_j = (1 + e^{-S_j})^{-1}, \quad S_j = \sum_{i=0}^{n} w_{ji} o_i;$ $\qquad(1)$

Weight adaptation rule: $\quad \Delta w_{ji} = (1 - \alpha_{ji}) \eta_{ji} \delta_j o_i + \alpha_{ji} \Delta w_{ji}^{\text{previous}};$ $\qquad(2)$

Error calculation: $\quad \delta_j = o_j (1 - o_j) \sum_{k=1}^{m} \delta_k w_{kj},$ $\qquad(3)$

where o_j is the output of processing element j or a sensor input, w_{ji} is the interconnection weight leading from element i to j, n is the number of inputs to j, Δw is the adjustment of w, η is the training constant, α is the training "momentum," δ_j is the calculated error for element j, and m is the fanout of a given element. Element zero is a constant input, equal to one, so that w_{j0} is equivalent to the bias threshold of element j. The $(1 - \alpha)$ factor in equation (2) differs from standard GDR formulation, but it is useful for keeping track of the relative magnitudes of the two terms. For the network's output layer the summation in equation (3) is replaced with the difference between the desired and actual output value of element j.

These networks are usually trained by presenting the system with sets of input/output data vectors in cyclic fashion, the entire cycle of database presentation repeated dozens of times. This method is effective when the training agent is a computer operating in batch mode, but would be intolerable for a human instructor. There are two developments that will help real-time human training. The first is a more efficient incorporation of data/response patterns into a network. The second, which we are addressing in this paper, is a suitable environment wherein a man and ANS network can interact in training situation with minimum inconvenience or boredom on the human's part. The ability to systematically train networks in this fashion is extremely useful for developing certain types of expert systems including automatic signal processors, autopilots, robots and other autonomous machines. We report a number of techniques aimed at facilitating this type of training, and we propose a general method for teaching these networks.

System Development

Our work focuses on the utility of ANS for system control. It began as an application of Barto and Sutton's associative search network[3]. Although their approach was useful in a number of ways, it fell short when we tried to use it for capturing the subtleties of human decision-making. In response we shifted our emphasis from constructing goal functions for automatic learning, to methods for training networks using direct human instruction. An integral part of this is the development of suitable interfaces between humans, networks and the outside world or simulator. In this section we will report various approaches to these ends, and describe a general methodology for manually teaching ANS networks. To demonstrate these techniques we taught a network to drive a robot vehicle down a simulated highway in traffic. This application combines binary decision making and control of continuous parameters.

Initially we investigated the use of automatic learning based on goal functions[3] for training control systems. We trained a network-controlled vehicle to maintain acceptable following distances from cars ahead of it. On a graphics workstation, a one lane circular track was constructed and occupied by two vehicles: a network-controlled robot car and a pace car that varied its speed at random.. Input data to the network consisted of the separation distance and the speed of the robot vehicle. The values of a goal function were translated into desired output for GDR training. Output controls consisted of three binary decision elements: 1) accelerate one increment of speed, 2) maintain speed, and 3) decelerate one increment of speed. At all times the desired output vector had exactly one of these three elements active. The goal function was quadratic with a minimum corresponding to the optimal following distance. Although it had no direct control over the simulation, the goal function positively or negatively reinforced the system's behavior.

The network was given complete control of the robot vehicle, and the human trainer had no influence except the ability to start and terminate training. This proved unsatisfactory because the initial system behavior--governed by random interconnection weights--was very unstable. The robot tended to run over the car in front of it before significant training occurred. By carefully halting and restarting training we achieved stable system behavior. At first the following distance maintained by the robot car oscillated as if the vehicle was attached by a spring to the pace car. This activity gradually damped. After about one thousand training steps the vehicle maintained the optimal following distance and responded quickly to changes in the pace car's speed.

Constructing composite goal functions to promote more sophisticated abilities proved difficult, even ill-defined, because there were many unspecified parameters. To generate goal functions for these abilities would be similar to conventional programming--the type of labor we want to circumvent using ANS. On the other hand, humans are adept at assessing complex situations and making decisions based on qualitative data, but their "goal functions" are difficult if not impossible to capture analytically. One attraction of ANS is that it can imitate behavior based on these elusive rules without formally specifying them. At this point we turned our efforts to manual training techniques.

The initially trained network was grafted into a larger system and augmented with additional inputs: distance and speed information on nearby pace cars in a second traffic lane, and an output control signal governing lane changes. The original network's ability to maintain a safe following distance was retained intact. This grafting procedure is one of two methods we studied for adding new abilities to an existing system. (The second, which employs a block structure, is described below.) The network remained in direct control of the robot vehicle, but a human trainer instructed it when and when not to change lanes. His commands were interpreted as the desired output and used in the GDR training algorithm. This technique, which we call coaching, proved useful and the network quickly correlated its environmental inputs with the teacher's instructions. The network became adept at changing lanes and weaving through traffic. We found that the network took on the behavior pattern of its trainer. A conservative teacher produced a timid network, while an aggressive trainer produced a network that tended to cut off other automobiles and squeeze through tight openings. Despite its success, the coaching method of training did not solve the problem of initial network instability.

The stability problem was solved by giving the trainer direct control over the simulation. The system configuration (Figure 1), allows the expert to exert control or release it to the network. During initial training the expert is in the driver's seat while the network acts the role of

696

apprentice. It receives sensor information, predicts system commands, and compares its predictions against the desired output (ie. the trainer's commands). Figure 2 shows the data and command flow in detail. Input data is processed through different channels and presented to the trainer and network. Where visual and audio formats are effective for humans, the network uses information in vector form. This differentiation of data presentation is a limitation of the system; removing it is a task for future research. The trainer issues control commands in accordance with his assigned task while the network takes the trainer's actions as desired system responses and correlates these with the input. We refer to this procedure as master/apprentice training, network training proceeds invisibly in the background as the expert proceeds with his day to day work. It avoids the instability problem because the network is free to make errors without the adverse consequence of throwing the operating environment into disarray.

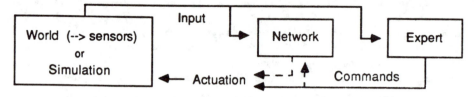

Figure 1. A scheme for manually training ANS networks. Input data is received by both the network and trainer. The trainer issues commands that are actuated (solid command line), or he coaches the network in how it ought to respond (broken command line).

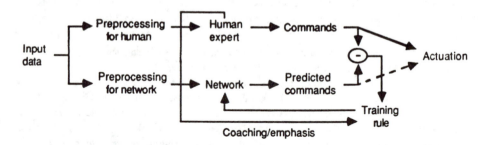

Figure 2. Data and command flow in the training system. Input data is processed and presented to the trainer and network. In master/apprentice training (solid command line), the trainer's orders are actuated and the network treats his commands as the system's desired output. In coaching, the network's predicted commands are actuated (broken command line), and the trainer influences weight adaptation by specifying the desired system output and controlling the values of training constants--his "suggestions" are not directly actuated.

Once initial, background training is complete, the expert proceeds in a more formal manner to teach the network. He releases control of the command system to the network in order to evaluate its behavior and weaknesses. He then resumes control and works through a series of scenarios designed to train the network out of its bad behavior. By switching back and forth between human and network control, the expert assesses the network's reliability and teaches correct responses as needed. We find master/apprentice training works well for behavior involving continuous functions, like steering. On the other hand, coaching is appropriate for decision functions, like when the car ought to pass. Our methodology employs both techniques.

The Driving Network

The fully developed freeway simulation consists of a two lane highway that is made of joined straight and curved segments which vary at random in length (and curvature). Several pace cars move at random speeds near the robot vehicle. The network is given the tasks of tracking the road, negotiating curves, returning to the road if placed far afield, maintaining safe distances from the pace cars, and changing lanes when appropriate. Instead of a single multi-layer structure, the network is composed of two blocks; one controls the steering and the other regulates speed and decides when the vehicle should change lanes (Figure 3). The first block receives information about the position and speed of the robot vehicle relative to other cars in its vicinity. Its output is used to determine the automobile's speed and whether the robot should change lanes. The passing signal is converted to a lane assignment based on the car's current lane position. The second block receives the lane assignment and data pertinent to the position and orientation of the vehicle with respect to the road. The output is used to determine the steering angle of the robot car.

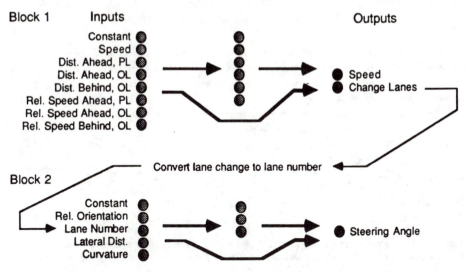

Figure 3. The two blocks of the driving ANS network. Heavy arrows indicate total interconnectivity between layers. PL designates the traffic lane presently occupied by the robot vehicle, OL refers to the other lane, curvature refers to the road, lane number is either 0 or 1, relative orientation and lateral distance refers to the robot car's direction and position relative to the road's direction and center line, respectively.

The input data is displayed in pictorial and textual form to the driving instructor. He views the road and nearby vehicles from the perspective of the driver's seat or overhead. The network receives information in the form of a vector whose elements have been scaled to unitary order, O(1). Wide ranging input parameters, like distance, are compressed using the hyperbolic tangent or logarithmic functions. In each block, the input layer is totally interconnected to both the output and a hidden layer. Our scheme trains in real time, and as we discuss later, it trains more smoothly with a small modification of the training algorithm.

Output is interpreted in two ways: as a binary decision or as a continuously varying parameter. The first simply compares the sigmoid output against a threshold. The second scales the output to an appropriate range for its application. For example, on the steering output element, a 0.5 value is interpreted as a zero steering angle. Left and right turns of varying degrees are initiated when this output is above or below 0.5, respectively.

The network is divided into two blocks that can be trained separately. Beside being conceptually easier to understand, we find this component approach is easy to train systematically. Because each block has a restricted, well-defined set of tasks, the trainer can concentrate specifically on those functions without being concerned that other aspects of the network behavior are deteriorating.

We trained the system from bottom up, first teaching the network to stay on the road, negotiate curves, change lanes, and how to return if the vehicle strayed off the highway. Block 2, responsible for steering, learned these skills in a few minutes using the master/apprentice mode. It tended to steer more slowly than a human but further training progressively improved its responsiveness.

We experimented with different training constants and "momentum" values. Large η values, about 1, caused weights to change too coarsely. η values an order of magnitude smaller worked well. We found no advantage in using momentum for this method of training, in fact, the system responded about three times more slowly when $\alpha = 0.9$ than when the momentum term was dropped. Our standard training parameters were $\eta = 0.2$, and $\alpha = 0.0$.

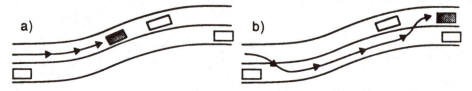

Figure 4. Typical behavior of a network-controlled vehicle (dark rectangle) when trained by
a) a conservative driver, and b) a reckless driver. Speed is indicated by the length of the arrows.

After Block 2 was trained, we gave steering control to the network and concentrated on teaching the network to change lanes and adjust speed. Speed control in this case was a continuous variable and was best taught using master/apprentice training. On the other hand, the binary decision to change lanes was best taught by coaching. About ten minutes of training were needed to teach the network to weave through traffic. We found that the network readily adapts the behavioral pattern of its trainer. A conservative trainer generated a network that hardly ever passed, while an aggressive trainer produced a network that drove recklessly and tended to cut off other cars (Figure 4).

Discussion

One of the strengths of expert systems based on ANS is that the use of input data in the decision making and control process does not have to be specified. The network adapts its internal weights to conform to input/output correlations it discovers. It is important, however, that data used by the human expert is also available to the network. The different processing of sensor data for man and network may have important consequences, key information may be presented to the man but not the machine.

This difference in data processing is particularly worrisome for image data where human ability to extract detail is vastly superior to our automatic image processing capabilities. Though we would not require an image processing system to understand images, it would have to extract relevant information from cluttered backgrounds. Until we have sufficiently sophisticated algorithms or networks to do this, our efforts at constructing expert systems which handle image data are handicapped.

Scaling input data to the unitary order of magnitude is important for training stability. This is evident from equations (1) and (2). The sigmoid transfer function ranges from 0.1 to 0.9 in approximately four units, that is, over an $O(1)$ domain. If system response must change in reaction to a large, $O(n)$ swing of a given input parameter, the weight associated with that input will be trained toward an $O(n^{-1})$ magnitude. On the other hand, if the same system responds to an input whose range is $O(1)$, its associated weight will also be $O(1)$. The weight adjustment equation does not recognize differences in weight magnitude, therefore relatively small weights will undergo wild magnitude adjustments and converge weakly. On the other hand, if all input parameters are of the same magnitude their associated weights will reflect this and the training constant can be adjusted for gentle weight convergence. Because the output of hidden units are constrained between zero and one, $O(1)$ is a good target range for input parameters. Both the hyperbolic tangent and logarithmic functions are useful for scaling wide ranging inputs. A useful form of the latter is

$$
x' = \begin{cases}
\beta[1+\ln(x/\alpha)] & \text{if } \alpha < x, \\
\beta x/\alpha & \text{if } -\alpha \leq x \leq \alpha, \\
-\beta[1+\ln(-x/\alpha)] & \text{if } x < -\alpha,
\end{cases}
\tag{4}
$$

where $\alpha > 0$ and defines the limits of the intermediate linear section, and β is a scaling factor. This symmetric logarithmic function is continuous in its first derivative, and useful when network behavior should change slowly as a parameter increases without bound. On the other hand, if the system should approach a limiting behavior, the tanh function is appropriate.

Weight adaptation is also complicated by relaxing the common practice of restricting interconnections to adjacent layers. Equation (3) shows that the calculated error for a hidden layer-given comparable weights, fanouts and output errors-will be one quarter or less than that of the

700

output layer. This is caused by the slope factor, $o_i(1-o_i)$. The difference in error magnitudes is not noticeable in networks restricted to adjacent layer interconnectivity. But when this constraint is released the effect of errors originating directly from an output unit has 4^d times the magnitude and effect of an error originating from a hidden unit removed d layers from the output layer. Compared to the corrections arising from the output units, those from the hidden units have little influence on weight adjustment, and the power of a multilayer structure is weakened. The system will train if we restrict connections to adjacent layers, but it trains slowly. To compensate for this effect we attenuate the error magnitudes originating from the output layer by the above factor. This heuristic procedure works well and facilitates smooth learning.

Though we have made progress in real-time learning systems using GDR, compared to humans-who can learn from a single data presentation-they remain relatively sluggish in learning and response rates. We are interested in improvements of the GDR algorithm or alternative architectures that facilitate one-shot or rapid learning. In the latter case we are considering least squares restoration techniques[4] and Grossberg and Carpenter's adaptive resonance models[3,5].

The construction of automated expert systems by observation of human personnel is attractive because of its efficient use of the expert's time and effort. Though the classic AI approach of rule base inference is applicable when such rules are clear cut and well organized, too often a human expert can not put his decision making process in words or specify the values of parameters that influence him. The attraction of ANS based systems is that imitations of expert behavior emerge as a natural consequence of their training.

References

1) D. E. Rumelhart, G. E. Hinton, and R. J. Williams, "Learning Internal Representations by Error Propagation," in *Parallel Distributed Processing: Explorations in the Microstructure of Cognition, Vol. I*, D. E. Rumelhart and J. L. McClelland (Eds.), chap. 8, (1986), Bradford Books/MIT Press, Cambridge

2) S. Grossberg, *Studies of Mind and Brain*, (1982), Reidel, Boston

3) A. Barto and R. Sutton, "Landmark Learning: An Illustration of Associative Search," *Biological Cybernetics,***42**, (1981), p. 1

4) A. Rosenfeld and A. Kak, *Digital Picture Processing, Vol. 1*, chap. 7, (1982), Academic Press, New York

5) G. A. Carpenter and S. Grossberg, "A Massively Parallel Architecture for a Self-organizing Neural Pattern Recognition Machine," *Computer Vision, Graphics and Image Processing*, **37**, (1987), p.54

DISCOVERING STRUCTURE FROM MOTION IN MONKEY, MAN AND MACHINE

Ralph M. Siegel[*]

The Salk Institute of Biology, La Jolla, Ca. 92037

ABSTRACT

The ability to obtain three-dimensional structure from visual motion is important for survival of human and non-human primates. Using a parallel processing model, the current work explores how the biological visual system might solve this problem and how the neurophysiologist might go about understanding the solution.

INTRODUCTION

Psychophysical experiments have shown that monkey and man are equally adept at obtaining three dimensional structure from motion[1]. In the present work, much effort has been expended mimicking the visual system. This was done for one main reason: the model was designed to help direct physiological experiments in the primate. It was hoped that if an approach for understanding the model could be developed, the approach could then be directed at the primate's visual system.

Early in this century, von Helmholtz[2] described the problem of extracting three-dimensional structure from motion:

> Suppose, for instance, that a person is standing still in a thick woods, where it is impossible for him to distinguish, except vaguely and roughly, in the mass of foliage and branches all around him what belongs to one tree and what to another, or how far apart the separate trees are, etc. But the moment he begins to move forward, everything disentangles itself, and immediately he gets an apperception of the material content of the woods and their relation to each other in space, just as if he were looking at a good stereoscopic view of it.

If the object moves, rather than the observer, the perception of three-dimensional structure from motion is still obtained. Object-centered structure from motion is examined in this report. Lesion studies in monkey have demonstrated that two extra-striate visual cortices called the middle temporal area (abbreviated MT

*Current address: Laboratory of Neurobiology, The Rockefeller University, 1230 York Avenue, New York, NY 10021

or V5) and the medial superior temporal area (MST)[3,4] are involved in obtaining structure from motion. The present model is meant to mimic the V5-MST part of the cortical circuitry involved in obtaining structure from motion. The model attempts to determine if the visual image corresponds to a three-dimensional object.

THE STRUCTURE FROM MOTION STIMULUS

The problem that the model solved was the same as that posed in the studies of monkey and man[1]. Structured and unstructured motion displays of a hollow, orthographically projected cylinder were computed (Figure 1). The cylinder rotates about its vertical axis. The unstructured stimulus was generated by shuffling the velocity vectors randomly on the display screen. The overall velocity and spatial distribution for the two displays are identical; only the spatial relationships have been changed in the unstructured stimulus. Human subjects report that the points are moving on the surface of a hollow cylinder when viewing the structured stimulus. With the unstructured stimulus, most subjects report that they have no sense of three-dimensional structure.

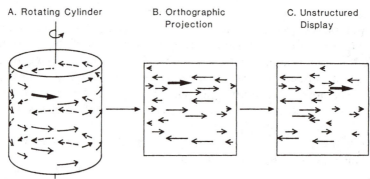

Figure 1. The structured and unstructured motion stimulus. A) "N" points are randomly placed on the surface of a cylinder. B) The points are orthographically projected. The motion gives a strong percept of a hollow cylinder. C) The unstructured stimulus was generated by shuffling the velocity vectors randomly on the screen.

FUNCTIONAL ARCHITECTURE OF THE MODEL

As with the primate subjects, the model was required to only indicate whether or not the display was structured. Subjects were not required to describe the shape, velocity or size of the cylinder. Thus the output cell* of the model signaled "1" if

---

*By cell, I mean a processing unit of the model which may correspond to a single neuron or group of neurons. The term neuron refers only to the actual wetware in the brain.

structured and "0" if not structured. This output layer corresponds to the cortical area MST of macaque monkey which appear to be sensitive to the global organization of the motion image[5]. It is not known if MST neurons will distinguish between structured and unstructured images.

The input to the model was based on physiological studies in the macaque monkey. Neurons in area V5 have a retinotopic representation of visual space[6,7].

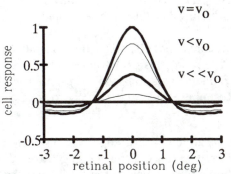

For each retinotopic location there is an encoding of a wide range of velocities[8]. Thus in the model's input representation, there were cells that represent different combinations of velocity and retinotopic spatial position. Furthermore motion velocity neurons in V5 have a center-surround opponent organization[9]. The width of the receptive fields was taken from the data of Albright et al.[8]. A typical receptive field of the model is shown in Figure 2.

Figure 2. The receptive field of an input layer cell. The optimal velocity is "v_o".

It was possible to determine what the activity of the input cells would be for the rotating cylinder given this representation. The activation pattern of the set of input cells was computed by convolving the velocity points with the difference of gaussians. The activity of the 100 input cells for an image of 20 points, with an angular velocity of 8°/sec is presented in Figure 3.

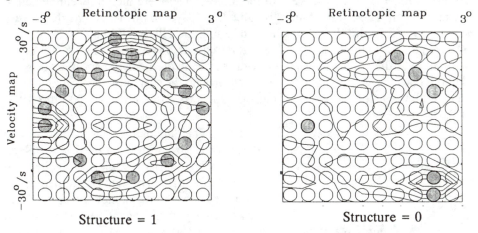

Figure 3. The input cell's activation pattern for a structured and unstructured stimulus. The circles correspond to the cells of the input layer. The contours were com-

puted using a linear interpolation between the individual cells. The horizontal axis corresponds to the position along the horizontal meridian. The vertical axis corresponds to the speed along the horizontal meridian. Thus activation of a cell in the upper right hand corner of the graph correspond to a velocity of 30^0/sec towards the right at a location of 3^0 to the right along the horizontal meridian.

Inspection of this input pattern suggested that the problem of detecting three-dimensional structure from motion may be reduced to a pattern recognition task. The problem was then: "Given a sparsely sampled input motion flow field, determine whether it corresponds best to a structured or unstructured object."

It was next necessary to determine the connections between the two input and output layers such that the model will be able to correctly signal structure or no structure (1 or 0) over a wide range of cylinder radii and rotational velocities. A parallel distributed network of the type used by Rosenberg and Sejnowski[10] provided the functional architecture (Figure 4).

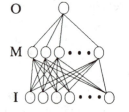

Figure 4. The parallel architecture used to extract structure from motion. The input layer (I), corresponding to area V5, mapped the position and speed along the horizontal axis. The output layer (O) corresponded to area MST that, it is proposed, signals structure or not. The middle layer (M) may exist in either V5 or MST.

The input layer of cells was fully connected to the middle layer of cells. The middle layer of cells represented an intermediate stage of processing and may be in either V5 or MST. All of the cells of the middle layer were then fully connected to the output cell. The inputs from cells of the lower layer to the next higher level were summed linearly and then "thresholded" using the Hill equation $X^3/(X^3+0.5^3)$. The weights between the layers were initially chosen between ± 1. The values of the weights were then adjusted using back-propagation methods (steepest descent) so that the network would "learn" to correctly predict the structure of the input image. The model learned to correctly perform the task after about 10,000 iterations (Figure 5).

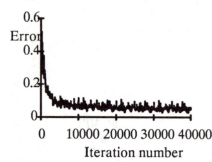

Figure 5. The "education" of the network to perform the structure from motion problem. The iteration number is plotted against the mean square error. The error is defined as the difference between the model's prediction and the known structure. The model was trained on a set of structured and unstructured cylinders with a wide range of radii, number of points, and rotational velocities.

PSYCHOPHYSICAL PERFORMANCE OF THE MODEL

The model's performance was comparable to that of monkey and man with respect to fraction of structure and number of points in the display (Figure 6). The model was indeed performing a global analysis as shown by allowing the model to view only a portion of the image. Like man and monkey, the model's performance suffers. Thus it appears that the model's performance was quite similar to known monkey and human psychophysics.

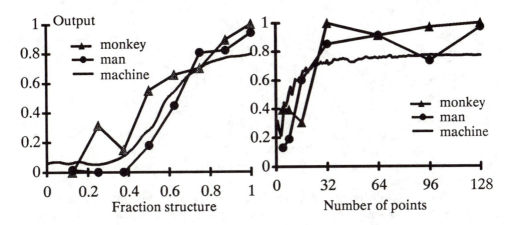

Figure 6. Psychophysical performance of the model. A. The effect of varying the fraction of structure. As the fraction of structure increase, the model's performance improves. Thirty repetitions were averaged for each value of structure for the model. The fraction of structure is defined as $(1-R_s/R_c)$, where R_s is the radius of shuffling of the motion vectors and R_c is the radius of the cylinder. The human and monkey data are taken from psychophysical studies[1].

HOW IS IT DONE?

The model has similar performance to monkey and man. It was next possible to examine this artificial network in order to obtain hints for studying the biological system. Following the approach of an electrophysiologist, receptive field maps for all the cells of the middle and output layers were made by activating individual input cells. The receptive field of some middle layer cells are shown in Figure 7. The layout of these maps are quite similar to that of Figure 4. However, now the activity of one cell in the middle layer is plotted as a function of the location and speed of a motion stimulus in the input layer. One could imagine that an electrode was placed in one of the cells of the middle layer while the experimentalist moved a bar about the

706

horizontal meridian with different locations and speeds. The activity of the cell is then plotted as a function of position and space.

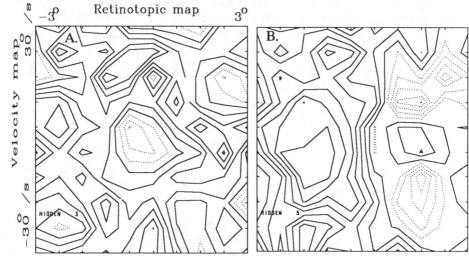

Figure 7. The activity of two different cells in the middle layer. Activity is plotted as a contour map as a function of horizontal position and speed. Dotted lines indicate inhibition.

These middle layer receptive field maps were interesting because they appear to be quite simple and symmetrical. In some, the inhibitory central regions of the receptive field were surrounded by excitatory regions (Figure 7A). Complementary cells were also found. In others, there are inhibitory bands adjacent to excitatory bands (Figure 7B). The above results suggest that neurons involved in extracting structure from motion may have relatively simple receptive fields in the spatial velocity domain. These receptive fields might be thought of as breaking the image down into component parts (i.e. a basis set). Correct recombination of these second order cells could then be used to detect the presence of a three-dimensional structure.

The output cell also had a simple receptive field again with interesting symmetries (Figure 8). However, the receptive field analysis is insufficient to indicate the role of the cell. Therefore in order to properly understand the "mean-

ing" of the cell's receptive field, it is necessary to use stimuli that are "real world relevant"- in this case the structure from motion stimuli. The output cell would give its maximal response only when a cylinder stimulus is presented.

Figure 8. The receptive field map of the output layer cell. Nothing about this receptive field structure indicates the cell is involved in obtaining structure from motion.

This work predicts that neurons in cortex involved in extracting structure from motion will have relatively simple receptive fields. In order to test this hypothesis, it will be necessary to make careful maps of these cells using small patches of motion (Figure 9). Known qualitative results in areas V5 and MST are consistent with, but do not prove, this hypothesis. As well, it will be necessary to use "relevant" stimuli (e.g. three-dimensional objects). If such simple receptive fields are indeed used in structure from motion, then support will be found for the idea that a simple cortical circuit (e.g. center-surround) can be used for many different visual analyses.

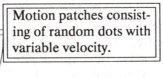

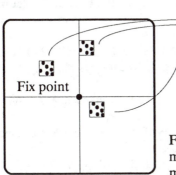

Fix point

Figure 9. It may be necessary to make careful maps of these neurons using small patches of motion, in order to observe the postulated simple receptive field properties of cortical neurons involved in extracting structure from motion. Such structures may not be apparent using hand moved bar stimuli.

DISCUSSION

In conclusion, it is possible to extract the three-dimensional structure of a rotating cylinder using a parallel network based on a similar functional architecture as found in primate cortex. The present model has similar psychophysics to monkey and man. The receptive field structures that underlie the present model are simple when viewed using a spatial-velocity representation. It is suggested that in order to understand how the visual system extracts structure from motion, quantitative spatial-velocity maps of cortical neurons involved need to be made. One also needs to use stimuli derived from the "real world" in order to understand how they may be used in visual field analysis. There are similarities between the shapes of the receptive fields involved in analyzing structure from motion and receptive fields in striate cortex[11]. It may be that similar cortical mechanisms and connections are used to perform different functions in different cortical areas. Lastly, this model demonstrates that the use of parallel architectures that are closely modeled on the cortical representation is a computationally efficient means to solve problems in vision. Thus as a final caveat, I would like to advise the creators of networks that solve ethologically realistic problems to use solutions that evolution has provided.

REFERENCES

1. R.M. Siegel and R.A. Andersen, Nature (Lond.) (1988).

2. H. von Helmholtz, Treatise on Physiological Optics (Dover Publications, N.Y., 1910), p. 297.

3. R.M. Siegel and R.A. Andersen, Soc. Neurosci. Abstr., 12, p. 1183 (1986).

4. R.M. Siegel and R.A. Andersen, Localization of function in extra-striate cortex: the effect of ibotenic acid lesions on motion sensitivity in Rhesus monkey, (in preparation).

5. K. Tanaka, K. Hikosaka, H. Saito, M. Yukie, Y. Fukada, and E. Iwai, J., Neurosci., 6, pp. 134-144 (1986).

6. S.M. Zeki, Brain Res., 35, pp. 528-532 (1971).

7. J.H.R. Maunsell and D.C. Van Essen, J. Neurophysiol., 49, pp. 1127-1147 (1983).

8. T.D. Albright, R. Desimone, and C.G. Gross, J. Neurophysiol., 51, pp. 16-31 (1984).

9. J. Allman, F. Miezen, and E. McGuinness, Ann. Rev. Neurosci., 8, pp. 407-430 (1985).

10. C.R. Rosenberg and T.J. Sejnowski, in: Reports of the Cognitive Neuropsychology Laboratory, John-Hopkins University (1986).

11. D.H. Hubel and T.N. Wiesel, Proc. R. Soc. Lond. B., 198, pp.1-59 (1977).

This work was supported by the Salk Institute for Biological Studies, The San Diego Supercomputer Center, and PHS NS07457-02.

TIME-SEQUENTIAL SELF-ORGANIZATION OF HIERARCHICAL NEURAL NETWORKS

Ronald H. Silverman
Cornell University Medical College, New York, NY 10021

Andrew S. Noetzel
Polytechnic University, Brooklyn, NY 11201

ABSTRACT

Self-organization of multi-layered networks can be realized by time-sequential organization of successive neural layers. Lateral inhibition operating in the surround of firing cells in each layer provides for unsupervised capture of excitation patterns presented by the previous layer. By presenting patterns of increasing complexity, in co-ordination with network self-organization, higher levels of the hierarchy capture concepts implicit in the pattern set.

INTRODUCTION

A fundamental difficulty in self-organization of hierarchical, multi-layered, networks of simple neuron-like cells is the determination of the direction of adjustment of synaptic link weights between neural layers not directly connected to input or output patterns. Several different approaches have been used to address this problem. One is to provide teaching inputs to the cells in internal layers of the hierarchy. Another is use of back-propagated error signals[1,2] from the uppermost neural layer, which is fixed to a desired output pattern. A third is the "competitive learning" mechanism,[3] in which a Hebbian synaptic modification rule is used, with mutual inhibition among cells of each layer preventing them from becoming conditioned to the same patterns.

The use of explicit teaching inputs is generally felt to be undesirable because such signals must, in essence, provide individual direction to each neuron in internal layers of the network. This requires extensive control signals, and is somewhat contrary to the notion of a self-organizing system.

Back-propagation provides direction for link weight modification of internal layers based on feedback from higher neural layers. This method allows true self-organization, but at the cost of specialized neural pathways over which these feedback signals must travel.

In this report, we describe a simple feed-forward method for self-organization of hierarchical neural networks. The method is a variation of the technique of competitive learning. It calls for successive neural layers to initiate modification of their afferent synaptic link weights only after the previous layer has completed its own self-organization. Additionally, the nature of the patterns captured can be controlled by providing an organized group of pattern sets which would excite the lowermost (input) layer of the network in concert with training of successive layers. Such a collection of pattern sets might be viewed as a "lesson plan."

MODEL

The network is composed of neuron-like cells, organized in hierarchical layers. Each cell is excited by variably weighted afferent connections from the outputs of the previous (lower) layer. Cells of the lowest layer take on the values of the input pattern. The cells themselves are of the McCulloch-Pitts type: they fire only after their excitation exceeds a threshold, and are otherwise inactive. Let $S_i(t)$ $\varepsilon\{0,1\}$ be the state of cell i at time t. Let w_{ij}, a real number ranging from 0 to 1, be the weight, or strength, of the synapse connecting cell i to cell j. Let e_{ij} be the local excitation of cell i at the synaptic connection from cell j. The excitation received along each synaptic connection is integrated locally over time as follows:

$$e_{ij}(t) = e_{ij}(t-1) + w_{ij}S_i(t) \tag{1}$$

Synaptic connections may, therefore be viewed as capacitive. The total excitation, E_j, is the sum of the local excitations of cell j.

$$E_j(t) = \sum_i e_{ij}(t) \tag{2}$$

The use of the time-integrated activity of a synaptic connection between two neurons, instead of the more usual instantaneous classification of neurons as "active" or "inactive", permits each synapse to provide a statistical measure of the activity of the input, which is assumed to be inherently stochastic. It also embodies the principle of learning based on locally available information and allows for implementations of the synapse as a capacitive element.

Over time, the total excitation of individual neurons on a give layer will increase. When excitation exceeds a threshold, θ, then the neuron fires, otherwise it is inactive.

$$S_j(t) = 1 \text{ if } E_j(t) > \theta \tag{3}$$
$$\text{else}$$
$$S_j(t) = 0$$

During a neuron's training phase, a modified Hebbian rule results in changes in afferent synaptic link weights such that, upon firing, synapses with integrated activity greater than mean activity are reinforced, and those with less than mean activity are weakened. More formally, if $S_j(t) = 1$ then the synapse weights are modified by

$$w_{ij}(t) = w_{ij}(t-1) + \text{sign}(e_{ij}(t) - \theta/n)k \cdot \text{sine}(\pi w_{ij}) \tag{4}$$

Here, n represents the fan-in to a cell, and k is a small, positive constant. The "sign" function specifies the direction of change and the "sine" function determines the magnitude of change. The sine curve provides the property that intermediate link weights are subject to larger modifications than weights near zero or saturation. This helps provide for stable end-states after learning.

Another effect of the integration of synaptic activity may be seen. A synapse of small weight is allowed to contribute to the firing of a cell (and hence have its weight incremented) if a series of patterns presented to the network consistently excite that synapse. The sequence of pattern presentations, therefore, becomes a factor in network self-organization.

Upon firing, the active cell inhibits other cells in its vicinity (lateral inhibition). This mechanism supports unsupervised, competitive learning. By preventing cells in the neighborhood of an active cell from modifying their afferent connections in response to a pattern, they are left available for capture of new patterns. Suppose there are n cells in a particular level. The lateral inhibitory mechanism is specified as follows:

$$\text{If } S_j(t) = 1 \text{ then}$$
$$e_{ik}(t) = 0 \quad \text{for all } i, \text{ for } k = (j-m)\bmod(n) \text{ to } (j+m)\bmod(n) \qquad (5)$$

Here, m specifies the size of a "neighborhood." A neighborhood significantly larger than a pattern set will result in a number of untrained cells. A neighborhood smaller than the pattern set will tend to cause cells to attempt to capture more than one pattern.

Schematic representations of an individual cell and the network organization are provided in Figures 1 and 2.

It is the pattern generator, or "instructor", that controls the form that network organization will take. The initial set of patterns are repeated until the first layer is trained. Next, a new pattern set is used to excite the lowermost (trained) level of the network, and so, induce training in the next layer of the hierarchy. Each of the patterns of the new set is composed of elements (or subpatterns) of the old set. The structure of successive pattern sets is such that each set is either a more complex combination of elements from the previous set (as words are composed of letters) or a generalization of some concept implicit in the previous set (such as line orientation).

Network organization, as described above, requires some exchange of control signals between the network and the instructor. The instructor requires information regarding firing of cells during training in order to switch to a new patterns appropriately. Obviously, if patterns are switched before any cells fire, learning will either not take place or will be smeared over a number of patterns. If a single pattern excites the network until one or more cells are fully trained, subsequent presentation of a non-orthogonal pattern could cause the trained cell to fire before any naive cell because of its saturated link weights. The solution is simply to allow gradual training over the full complement of the pattern set. After a few firings, a new pattern should be provided. After a layer has been trained, the instructor provides a control signal to that layer which permanently fixes the layer's afferent synaptic link weights.

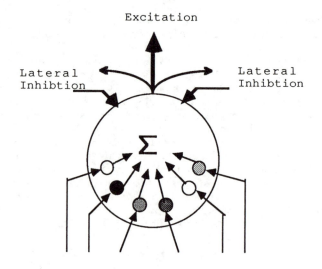

Excitation

Lateral
Inhibtion

Lateral
Inhibtion

Σ

Excitatory Inputs

Fig. 1. Schematic of neuron.
Shading of afferent synaptic connections
indicates variations in levels of local
time-integrated excitation.

Fig. 2. Schematic of network showing
lateral inhibition and forward excitation.
Shading of neurons, indicating degree of
training, indicates time-sequential
organization of successive neural layers.

SIMULATIONS

As an example, simulations were run in which a network was taught to differentiate vertical from horizontal line orientation. This problem is of interest because it represents a case in which pattern sets cannot be separated by a single layer of connections. This is so because the set of vertical (or horizontal) lines has activity at all positions within the input matrix.

Two variations were simulated. In the first simulation, the input was a 4x4 matrix. This was completely connected with unidirectional links to 25 cells. These cells had fixed inhibitory connections to the nearest five cells on either side (using a circular arrangement), and excited, using complete connectivity, a ring of eight cells, with inhibition over the nearest neighbor on either side.

Initially, all excitatory link weights were small, random numbers. Each pattern of the initial input consisted of a single active row or column in the input matrix. Active elements had, during any clock cycle, a probability of 0.5 of being "on", while inactive elements had a 0.05 probability of being "on."

After exposure to the initial pattern set, all cells on the first layer captured some input pattern, and all eight patterns had been captured by two or more cells.

The next pattern set consisted of two subsets of four vertical and four horizontal lines. The individual lines were presented until a few firings took place within the trained layer, and then another line from the same subset was used to excite the network. After the upper layer responed with a few firings, and some training occured, the other set was used to excite the network in a similar manner. After five cycles, all cells on the uppermost layer had become sensitive, in a postionally independent manner, to lines of a vertical or a horizontal orientation. Due to lateral inhibition, adjacent cells developed opposite orientation specificities.

In the second simulation, a 6x6 input matrix was connected to six cells, which were, in turn, connected to two cells. For this network, the lateral inhibitory range extended over the entire set of cells of each layer. The initial input set consisted of six patterns, each of which was a pair of either vertical lines or horizontal lines. After excitation by this set, each of the six middle level cells became sensitized to one of the input patterns. Next, the set of vertical and horizontal patterns were grouped into two subsets: vertical lines and horizontal lines. Individual patterns from one subset were presented until a cell, of the previously trained layer, fired. After one of the two cells on the uppermost layer fired, the procedure was repeated with the pattern set of opposite orientation. After 25 cycles, the two cells on the uppermost layer had developed opposite orientation specificities. Each of these cells was shown to be responsive, in a positionally independent manner, to any single line of appropriate orientation.

CONCLUSION

Competitive learning mechanisms, when applied sequentially to successive layers in a hierarchical structure, can capture pattern elements, at lower levels of the hierarchy, and their generalizations, or abstractions, at higher levels.

In the above mechanism, learning is externally directed, not by explicit teaching signals or back-propagation, but by provision of instruction sets consisting of patterns of increasing complexity, to be input to the lowermost layer of the network in concert with successive organization of higher neural layers.

The central difficulty of this method involves the design of pattern sets - a procedure whose requirements may not be obvious in all cases. The method is, however, attractive due to its simplicity of concept and design, providing for multi-level self-organization without direction by elaborate control signals.

Several research goals suggest themselves: 1) simplification or elimination of control signals, 2) generalization of rules for structuring of pattern sets, 3) extension of this learning principle to recurrent networks, and 4) gaining a deeper understanding of the role of time as a factor in network self-organization.

REFERENCES

1. D. E. Rumelhart and G.E. Hinton, Nature 323, 533 (1986).
2. K. A. Fukushima, Biol. Cybern. 55, 5 (1986).
3. D. E. Rumelhart and D. Zipser, Cog. Sci. 9, 75 (1985).

A COMPUTER SIMULATION OF CEREBRAL NEOCORTEX:
COMPUTATIONAL CAPABILITIES OF NONLINEAR NEURAL NETWORKS

Alexander Singer* and John P. Donoghue**

*Department of Biophysics, Johns Hopkins University, Baltimore, MD 21218 (to whom all correspondence should be addressed)

**Center for Neural Science, Brown University, Providence, RI 02912

716

A synthetic neural network simulation of cerebral neocortex was developed based on detailed anatomy and physiology. Processing elements possess temporal nonlinearities and connection patterns similar to those of cortical neurons. The network was able to replicate spatial and temporal integration properties found experimentally in neocortex. A certain level of randomness was found to be crucial for the robustness of at least some of the network's computational capabilities. Emphasis was placed on how synthetic simulations can be of use to the study of both artificial and biological neural networks.

A variety of fields have benefited from the use of computer simulations. This is true in spite of the fact that general theories and conceptual models are lacking in many fields and contrasts with the use of simulations to explore existing theoretical structures that are extremely complex (cf. MacGregor and Lewis, 1977). When theoretical superstructures are missing, simulations can be used to synthesize empirical findings into a system which can then be studied analytically in and of itself. The vast compendium of neuroanatomical and neurophysiological data that has been collected and the concomitant absence of theories of brain function (Crick, 1979; Lewin, 1982) makes neuroscience an ideal candidate for the application of synthetic simulations. Furthermore, in keeping with the spirit of this meeting, neural network simulations which synthesize biological data can make contributions to the study of artificial neural systems as general information processing machines as well as to the study of the brain. A synthetic simulation of cerebral neocortex is presented here and is intended to be an example of how traffic might flow on the two-way street which this conference is trying to build between artificial neural network modelers and neuroscientists.

The fact that cerebral neocortex is involved in some of the highest forms of information processing and the fact that a wide variety of neurophysiological and neuroanatomical data are amenable to simulation motivated the present development of a synthetic simulation of neocortex. The simulation itself is comparatively simple; nevertheless it is more realistic in terms of its structure and elemental processing units than most artificial neural networks.

The neurons from which our simulation is constructed go beyond the simple sigmoid or hard-saturation nonlinearities of most artificial neural systems. For example, because inputs to actual neurons are mediated by ion currents whose driving force depends on the membrane potential of the neuron, the amplitude of a cell's response to an input, i.e. the amplitude of the post-synaptic potential (PSP), depends not only on the strength of the synapse at which the input arrives, but also on the state of the neuron at the time of the input's arrival. This aspect of classical neuron electrophysiology has been implemented in our simulation (figure 1A), and leads to another important nonlinearity of neurons: namely, current shunting. Primarily effective as shunting inhibition, excitatory current can be shunted out an inhibitory synapse so that the sum of an inhibitory postsynaptic potential and an excitatory postsynaptic potential of equal amplitude does not result in mutual cancellation. Instead, interactions between the ion reversal potentials, conductance values, relative timing of inputs, and spatial locations of synapses determine the amplitude of the response in a nonlinear fashion (figure 1B) (see Koch, Poggio, and Torre, 1983 for a quantitative analysis). These properties of actual neurons have been ignored by most artificial neural network designers, though detailed knowledge of them has existed for decades and in spite of the fact that they can be used to implement complex computations (e.g. Torre and Poggio, 1978; Houchin, 1975).

The development of action potentials and spatial interactions within the model neurons have been simplified in our simulation. Action potentials involve preprogrammed fluctuations in the membrane potential of our neurons and result in an absolute and a relative refractory period. Thus, during the time a cell is firing a spike synaptic inputs are ignored, and immediately following an action potential the neuron is hyperpolarized. The modeling of spatial interactions is also limited since neurons are modeled primarily as spheres. Though the spheres can be deformed through control of a synaptic weight which modulates the amplitudes of ion conductances, detailed dendritic interactions are not simulated. Nonetheless, the fact that inhibition is generally closer to a cortical neuron's soma while excitation is more distal in a cell's dendritic tree is simulated through the use of stronger inhibitory synapses and relatively weaker excitatory synapses.

The relative strengths of synapses in a neural network define its connectivity. Though initial connectivity is random in many artificial networks, brains can be thought to contain a combination of randomness and fixed structure at distinct levels (Szentagothai, 1978). From a macroscopic perspective, all of cerebral neocortex might be structured in a modular fashion analogous to the way the barrel field of mouse somatosensory cortex is structured (Woolsey and Van der Loos, 1970). Though speculative, arguments for the existence of some sort of anatomical modularity over the entire cortex are gaining ground

718

(Mountcastle, 1978; Szentagothai, 1979; Shepherd, in press). Thus, inspired by the barrels of mice and by growing interest in functional units of 50 to 100 microns with on the order of 1000 neurons, our simulation is built up of five modules (60 cells each) with more dense local interconnections and fewer intermodular contacts. Furthermore, a wide variety of neuronal classification schemes have led us to subdivide the gross structure of each module so as to contain four classes of neurons: cortico-cortical pyramids, output pyramids, spiny stellate or local excitatory cells, and GABAergic or inhibirtory cells.

At this level of analysis, the impressed structure allows for control over a variety of pathways. In our simulation each class of neurons within a module is connected to every other class and intermodular connections are provided along pathways from cortico-cortical pyramids to inhibitory cells, output pyramids, and cortico-cortical pyramids in immediately adjacent modules. A general sense of how strong a pathway is can be inferred from the product of the number of synapses a neuron receives from a particular class and the strength of each of those synapses. The broad architecture of the simulation is further structured to emphasize a three step path: Inputs to the network impact most strongly on the spiny stellate cells of the module receiving the input; these cells in turn project to cortico-cortical pyramidal cells more strongly than they do to other cell types; and finally, the pathway from the cortico-cortical pyramids to the output pyramidal cells of the same module is also particularly strong. This general architecture (figure 2) has received empirical support in many regions of cortex (Jones, 1986).

In distinction to this synaptic architecture, a fine-grain connectivity is defined in our simulated network as well. At a more microscopic level, connectivity in the network is random. Thus, within the confines of the architecture described above, the determination of which neuron of a particular class is connected to which other cell in a target class is done at random. Two distinct levels of connectivity have, therefore, been established (figure 3). Together they provide a middle ground between the completely arbitrary connectivity of many artificial neural networks and the problem specific connectivities of other artificial systems. This distinction between gross synaptic architecture and fine-grain connectivity also has intuitive appeal for theories of brain development and, as we shall see, has non-trivial effects on the computational capabilities of the network as a whole.

With defintions for input integration within the local processors, that is within the neurons, and with the establishment of connectivity patterns, the network is complete and ready to perform as a computational unit. In order to judge the simulation's capabilities in some rough way, a qualitative analysis of its response to an input will suffice. Figure 4 shows the response of the network to an input composed of a small burst of action potentials arriving at a single module. The data is displayed as a raster in which time is mapped along the abscissa and all the cells of the network are arranged by module and cell class along the ordinate. Each marker on the graph represents a single action potential fired by the appropriate neuron at the indicated time. Qualitatively, what is of importance is the fact that the network does not remain unresponsive, saturate with activity in all neurons, or oscillate in any way. Of course, that the network behave this way was predetermined by the combination of the properties of the neurons with a judicious selection of synaptic weights and path strengths. The properties of the neurons were fixed from physiological data, and once a synaptic architecture was found which produced the results in figure 4, that too was fixed. A more detailed analysis of the temporal firing pattern and of the distribution of activity over the different cell classes might reveal important network properties and the relative importance of various pathways to the overall function. Such an analysis of the sensitivity of the network to different path strengths and even to intracellular parameters will, however, have to be postponed. Suffice it to say at this point that the network, as structured, has some nonzero, finite, non-oscillatory response which, qualitatively, might not offend a physiologist judging cortical activity.

Though the synaptic architecture was tailored manually and fixed so as to produce "reasonable" results, the fine-grain connectivity , i.e. the determination of exactly which cell in a class connects to which other cell, was random. An important property of artificial (and presumably biological) neural networks can be uncovered by exploiting the distinction between levels of connectivity described above. Before doing so, however, a detail of neural network design must be made explicit. Any network, either artificial or biological, must contend with the time it takes to communicate among the processing elements. In the brain, the time it takes for an action potential to travel from one neuron to another depends on the conduction velocity of the axon down which the spike is traveling and on the delay that occurs at the synapse connecting the cells. Roughly, the total transmission time from one cortical neuron to another lies between 1 and 5 milliseconds. In our simulation two paradigms were used. In one case, the transmission times between all neurons were standardized at 1 msec.* Alternatively, the transmission times were fixed at random, though admittedly unphysiological, values between 0.1 and 2 msec.

Now, if the time it takes for an action potential to travel from one neuron to another were fixed for all cells at 1 msec, different fine-grain connectivity patterns are found to produce entirely distinct network responses to the same input, in spite of the fact that the gross synaptic architecture remained constant. This was true no matter what particular synaptic architecture was used. If, on the other hand, one changes the transmission times so that they vary randomly between 0.1 and 2 msec, it becomes easy to find sets of synaptic strengths that were robust with respect to changes in the fine-grain connectivity. Thus, a wide search of path strengths failed to produce a network which was robust to changes in fine-grain connectivity in the case of identical transmission times, while a set of synaptic weights that produced robust responses was easy to find when the transmission times were randomized. Figure 5 summarizes this result. In the figure overall network activity is measured simply as the total number of action potentials generated by pyramidal cells during an experiment and robustness can be judged as the relative stability of this response. The abscissa plots distinct experiments using the same synaptic architecture with different fine-grain connectivity patterns. Thus, though the synaptic architecture remains constant, the different trials represent changes in which particular cell is connected to which other cell. The results show quite dramatically that the network in which the transmission times are randomly distributed is more robust with respect to changes in fine-grain connectivity than the network in which the transmission times are all 1 msec.

It is important to note that in either case, both when the network was robust and when changes of fine-grain connectivity produced gross changes in network output, the synaptic architectures produced outputs like that in figure 4 with some fine-grain connectivities. If the response of the network to an input can be considered the result of

---

* Because neurons receive varying amounts of input and because integration is performed by summating excitatory and inhibitory postsynaptic potentials in a nonlinear way, the time each neuron needs to summate its inputs and produce an action potential varies from neuron to neuron and from time to time. This then allows for asynchronous firing in spite of the identical transmission times.

some computation, figure 5 reveals that the *same* computational capability is not robust with respect to changes in fine-grain connectivity when transmission times between neurons are all 1 msec, but is more robust when these times are randomized. Thus, a single computational capability, viz. a response like that in figure 4 to a single input, was found to exist in networks with different synaptic architectures and different transmission time paradigms; this computational capability, however, varied in terms of its robustness with respect to changes in fine-grain connectivity when present in either of the transmission time paradigms.

A more complex computational capability emerged from the neural network simulation we have developed and described. If we label two neighboring modules C2 and C3, an input to C2 will suppress the response of C3 to a second input at C3 if the second input is delayed. A convenient way of representing this spatio-temporal integration property is given in figure 6. The ordinate plots the ratio of the normal response of one module (say C3) to the response of the module to the same input when an input to a neighboring module (say C2) preceds the input to the original module (C3). Thus, a value of one on the ordinate means the earlier spatially distinct input had no effect on the response of the module in which this property is being measured. A value less than one represents suppression, while values greater than one represent enhancement. On the abscissa, the interstimulus interval is plotted. From figure 6, it can be seen that significant suppression of the pyramidal cell output, mostly of the output pyramidal cell output, occurs when the inputs are separated by 10 to 30 msec. This response can be characterized as a sort of dynamic lateral inhibition since an input is suppressing the ability of a neighboring region to respond when the input pairs have a particular time course. This property could play a variety of role in biological and artificial neural networks. One role for this spatio-temporal integration property, for example, might be in detecting the velocity of a moving stimulus.

The emergent spatio-temporal property of the network just described was not explicitly built into the network. Moreover, no set of synaptic weights was able to give rise to this computational capability when transmission times were all set to 1 msec. Thus, in addition to providing robustness, the random transmission times also enabled a more complex property to emerge. The important factor in the appearances of both the robustness and the dynamic lateral inhibition was randomization; though it was implemented as randomly varying transmission times, random spontaneous activity would have played the same role. From the viewpoint, then, of the engineer designing artificial neural networks, the neural network presented here has instructional value in spite of the fact that it was designed to synthesize biological data. Specifically, it motivates the consideration of randomness as a design constraint.

From the prespective of the biologists attending this meeting, a simple fact will reveal the importance of synthetic simulations. The dynamic lateral inhibition presented in figure 6 is known to exist in rat somatosensory cortex (Simons, 1985). By deflecting the whiskers on a rat's face, Simons was able to stimulate individual barrels of the postero-medial somatosensory barrel field in combinations which revealed similar spatio-temporal interactions among the responses of the cortical neurons of the barrel field. The temporal suppression he reported even has a time course similar to that of the simulation. What the experiment did not reveal, however, was the class of cell in which suppression was seen; the simulation located most of the suppression in the output pyramidal cells. Hence, for a biologist, even a simple synthetic simulation like the one presented here can make definitive predictions. What differentiates the predictions made by synthetic simulations from those of more general artificial neural systems, of course, is that the strong biological foundations of synthetic simulations provide an easily grasped and highly relevant framework for both predictions and experimental verification.

One of the advertised purposes of this meeting was to "bring together neurobiologists, cognitive psychologists, engineers, and physicists with common interest in natural and artificial neural networks." Towards that end, synthetic computer simulations, i.e. simulations which follow known neurophysiological and neuroanatomical data as if they comprised a complex recipe, can provide an experimental medium which is useful for both biologists and engineers. The simulation of cerebral neocortex developed here has information regarding the role of randomness in the the robustness and presence of various computational capabilities as well as information regarding the value of distinct levels of connectivity to contribute to the design of artificial neural networks. At the same time, the synthetic nature of the network provides the biologist with an environment in which he can test notions of actual neural function as well as with a system which replicates known properties of biological systems and makes explicit predictions. Providing two-way interactions, synthetic simulations like this one will allow future generations of artificial neural networks to benefit from the empirical findings of biologists, while the slowly evolving theories of brain function benefit from the more generalizable results and methods of engineers.

References

Crick, F. H. C. (1979) Thinking about the brain, *Scientific American*, **241**:219 - 232.

Houchin, J. (1975) Direction specificity in cortical responses to moving stimuli -- a simple model. *Proceedings of the Physiological Society*, **247**:7 - 9.

Jones, E. G. (1986) Connectivity of primate sensory-motor cortex , in *Cerebral Cortex*, vol. 5, E. G. Jones and A. Peters (eds), Plenum Press, New York.

Koch, C., Poggio, T., and Torre, V. (1983) Nonlinear interactions in a dendritic tree: Localization, timing, and role in information processing. *Proceedings of the National Academy of Science, USA*, **80**:2799 - 2802.

Lewin, R. (1982) Neuroscientists look for theories, *Science*, **216**:507.

MacGregor, R.J. and Lewis, E.R. (1977) *Neural Modeling*, Plenum Press, New York.

Mountcastle, V. B. (1978) An organizing principle for cerebral function: The unit module and the distributed system, in *The Mindful Brain*, G. M. Edelman and V. B. Mountcastle (eds.), MIT Press, Cambridge, MA.

Shepherd, G.M. (in press) Basic circuit of cortical organization, in *Perspectives in Memory Research*, M.S. Gazzaniga (ed.), MIT Press, Cambridge, MA.
Simons, D. J. (1985) Temporal and spatial integration in the rat SI vibrissa cortex, *Journal of Neurophysiology*, **54**:615 - 635.

Szenthágothai, J. (1978) Specificity versus (quasi-) randomness in cortical connectivity, in *Architectonics of the Cerebral Cortex*, M. A. B. Brazier and H. Petsche (eds.), Raven Press, New York.

Szentágothai, J. (1979) Local neuron circuits in the neocortex, in *The Neurosciences. Fourth Study Program*, F. O. Schmitt and F. G. Worden (eds.), MIT Press, Cambridge, MA.

Torre, V. and Poggio, T. (1978) A synaptic mechanism possibly underlying directional selectivity to motion, *Proceeding of the Royal Society (London) B*, **202**:409 -416.

Woolsey, T.A. and Van der Loos, H. (1970) Structural organization of layer IV in the somatosensory region (SI) of mouse cerebral cortex, *Brain Research*, **17**:205-242.

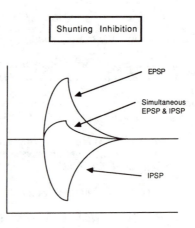

Figure 1A: Intracellular records of post-synaptic potentials resulting from single excitatory and inhibitory inputs to cells at different resting potentials.

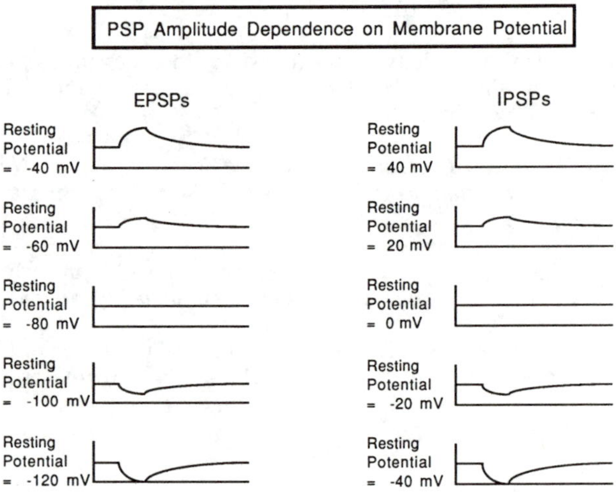

PSP Amplitude Dependence on Membrane Potential

EPSPs

Resting Potential = -40 mV

Resting Potential = -60 mV

Resting Potential = -80 mV

Resting Potential = -100 mV

Resting Potential = -120 mV

IPSPs

Resting Potential = 40 mV

Resting Potential = 20 mV

Resting Potential = 0 mV

Resting Potential = -20 mV

Resting Potential = -40 mV

Figure 1B: Illustration of the current shunting nonlinearity present in the model neurons. Though the simultaneous arrival of postsynaptic potentials of equal and opposite amplitude would result in no deflection in the membrane potential of a simple linear neuron model, a variety of factors contribute to the nonlinear response of actual neurons and of the neurons modeled in the present simulation.

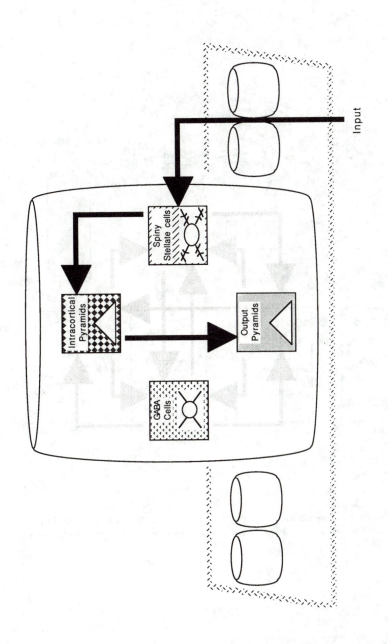

Figure 2: A schematic representation of the simulated cortical network. Five modules are used, each containing sixty neurons. Neurons are divided into four classes. Numerals within the caricatured neurons represent the number of cells in that particular class that are simulated. Though all cell classes are connected to all other classes, the pathway from input to spiny stellate to cortico-cortical pyramids to output pyramids is particularly strong.

726

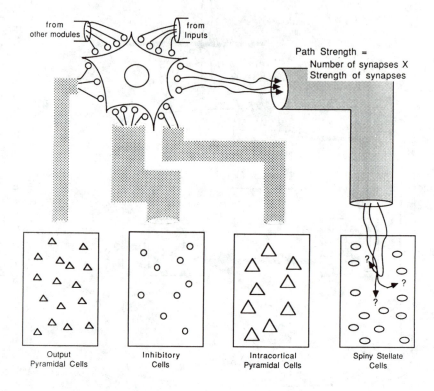

Figure 3: Two levels of connectivity are defined in the network. Gross synaptic architecture is defined among classes of cells. Fine-grain connectivity specifies which cell connects to which other cell and is determined at random.

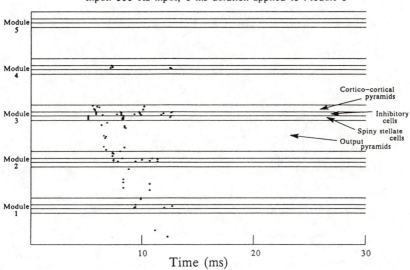

Figure 4: Sample response of the entire network to a small burst of action potentials delivered to module 3.

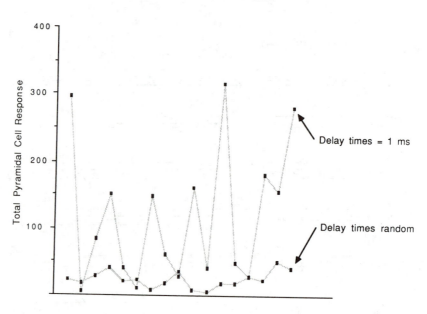

Robustness With Respect to Connectivity Pattern

Synaptic Architecture Constant

Individual Trials with Different Fine-grain Connectivity Patterns

Figure 5: Plot of an arbitrary activity measure (total spike activity in all pyramidal cells) versus various instatiations of the same connectional architecture. Along the abscissa are represented the different fine-grained patterns of connectivity within a fixed connectional architecture. In one case the conductance times between all cells was 1 msec and in the other case the times were selected at random from values between 0.1 msec and 2 msec. This experiment shows the greater overall stability produced by random conduction times.

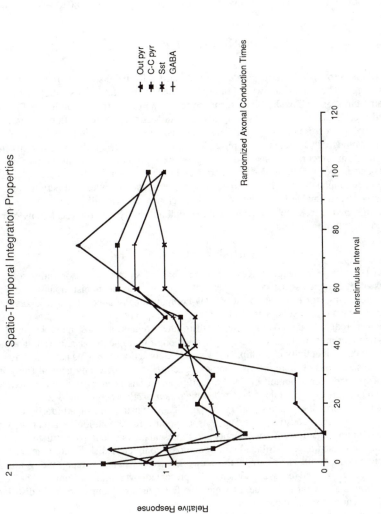

Figure 6: Spatio-temporal integration within the network. Plot of the time course of response suppression in the various cell classes. The ordinate plots the ratio of average cell activity (in terms of spikes) to a direct input after the presentation of an input to a neighboring module, and the average reponse to an input in the absence of prior input to an adjacent module. Values greater than one represent an enhancement of activity in response to the spatially distinct preceding input, while values less than one represent a suppression of the normal reponse. The abscissa plots the interstimulus interval. Note that the response suppression is most striking in only one class of cells.

Analysis of distributed representation of
constituent structure in connectionist systems

Paul Smolensky

Department of Computer Science, University of Colorado, Boulder, CO 80309–0430

Abstract

A general method, the *tensor product representation*, is described for the distributed representation of value/variable bindings. The method allows the fully distributed representation of symbolic structures: the roles in the structures, as well as the fillers for those roles, can be arbitrarily non-local. Fully and partially localized special cases reduce to existing cases of connectionist representations of structured data; the tensor product representation generalizes these and the few existing examples of fully distributed representations of structures. The representation saturates gracefully as larger structures are represented; it permits recursive construction of complex representations from simpler ones; it respects the independence of the capacities to generate and maintain multiple bindings in parallel; it extends naturally to continuous structures and continuous representational patterns; it permits values to also serve as variables; it enables analysis of the interference of symbolic structures stored in associative memories; and it leads to characterization of optimal distributed representations of roles and a recirculation algorithm for learning them.

Introduction

Any model of complex information processing in networks of simple processors must solve the problem of representing complex structures over network elements. Connectionist models of realistic natural language processing, for example, must employ computationally adequate representations of complex sentences. Many connectionists feel that to develop connectionist systems with the computational power required by complex tasks, distributed representations must be used: an individual processing unit must participate in the representation of multiple items, and each item must be represented as a pattern of activity of multiple processors. Connectionist models have used more or less distributed representations of more or less complex structures, but little if any general analysis of the problem of distributed representation of complex information has been carried out. This paper reports results of an analysis of a general method called the *tensor product representation*.

The language-based formalisms traditional in AI permit the construction of arbitrarily complex structures by piecing together constituents. The tensor product representation is a connectionist method of combining representations of constituents into representations of complex structures. If the constituents that are combined have distributed representations, completely distributed representations of complex structures can result: each part of the network is responsible for representing multiple constituents in the structure, and each constituent is represented over multiple units. The tensor product representation is a general technique, of which the few existing examples of fully distributed representations of structures are particular cases.

The tensor product representation rests on identifying natural counterparts within connectionist computation of certain fundamental elements of symbolic computation. In the present analysis, the problem of distributed representation of symbolic structures is characterized as the problem of taking complex structures with certain relations to their constituent symbols and mapping them into activity vectors—patterns of activation—with corresponding relations to the activity vectors representing their constituents. Central to the analysis is identifying a connectionist counterpart of *variable binding*: a method for binding together a distributed representation of a variable and a distributed representation of a value into a distributed representation of a variable/value binding—a representation which can co-exist on exactly the same network units with representations of other variable/value bindings, with limited confusion of which variables are bound to which values.

In summary, the analysis of the tensor product representation

(1)   provides a general technique for constructing fully distributed representations of arbitrarily complex structures;

(2)   clarifies existing representations found in particular models by showing what particular design decisions they embody;

(3)   allows the proof of a number of general computational properties of the representation;

(4)   identifies natural counterparts within connectionist computation of elements of symbolic computation, in particular, variable binding.

The recent emergence to prominence of the connectionist approach to AI raises the question of the relation between the nonsymbolic computation occurring in connectionist systems and the symbolic computation traditional in AI. The research reported here is part of an attempt to marry the two types of computation, to develop for AI a form of computation that adds crucial aspects of the power of symbolic computation to the power of connectionist computation: massively parallel soft constraint satisfaction. One way to marry these approaches is to implement serial symbol manipulation in a connectionist system[1,2]. The research described here takes a different tack. In a massively parallel system the processing of symbolic structures—for example, representations of parsed sentences—need not be limited to a series of elementary manipulations: indeed one would expect the processing to involve massively parallel soft constraint satisfaction. But in order for such processing to occur, a satisfactory answer must be found for the question: *How can symbolic structures, or structured data in general, be naturally represented in connectionist systems?* The difficulty here turns on one of the most fundamental problems for relating symbolic and connectionist computation: *How can variable binding be naturally performed in connectionist systems?*

This paper provides an overview of a lengthy analysis reported elsewhere[3] of a general connectionist method for variable binding and an associated method for representing structured data. The purpose of this paper is to introduce the basic idea of the method and survey some of the results; the reader is referred to the full report for precise definitions and theorems, more extended examples, and proofs.

The problem

Suppose we want to represent a simple structured object, say a sequence of elements, in a connectionist system. The simplest method, which has been used in many models, is to dedicate a network processing unit to each possible element in each possible position[4-9]. This is a *purely local representation*. One way of looking at the purely local representation is that the binding of constituents to the variables representing their positions is achieved by dedicating a separate unit to every possible binding, and then by activating the appropriate individual units.

Purely local representations of this sort have some advantages[10], but they have a number of serious problems. Three immediately relevant problems are these:

(1)   The number of units needed is *#elements* * *#positions*; most of these processors are inactive and doing no work at any given time.

(2)   The number of positions in the structures that can be represented has a fixed, rigid upper limit.

(3)   If there is a notion of similar elements, the representation does not take advantage of this: similar sequences do not have similar representations.

The technique of *distributed representation* is a well-known way of coping with the first and third problems[11-14]. If elements are represented as *patterns of activity* over a population of processing units, and if each unit can participate in the representation of many elements, then the number of elements that can be represented is much greater than the number of units, and similar elements can be represented by similar patterns, greatly enhancing the power of the network to learn and take advantage of generalizations.

Distributed representations of elements in structures (like sequences) have been successfully used in many models[1,4,5,15–18]. For each position in the structure, a *pool of units* is dedicated. The element occurring in that position is represented by a pattern of activity over the units in the pool.

Note that this technique goes only part of the way towards a truly distributed representation of the entire structure. While the *values* of the variables defining the roles in the structure are represented by distributed patterns instead of dedicated units, the *variables themselves* are represented by localized, dedicated pools. For this reason I will call this type of representation *semi-local.*

Because the representation of variables is still local, semi-local representations retain the second of the problems of purely local representations listed above. While the generic behavior of connectionist systems is to gradually overload as they attempt to hold more and more information, with dedicated pools representing role variables in structures, there is no loading at all until the pools are exhausted—and then there is complete saturation. The pools are essentially *registers,* and the representation of the structure as a whole has more of the characteristics of von Neumann storage than connectionist representation. A fully distributed connectionist representation of structured data would saturate gracefully.

Because the representation of variables in semi-local representations is local, semi-local representations also retain part of the third problem of purely local representations. Similar elements have similar representations *only if they occupy exactly the same role in the structure.* A notion of similarity of roles cannot be incorporated in the semi-local representation.

Tensor product binding

There is a way of viewing both the local and semi-local representations of structures that makes a generalization to fully distributed representations immediately apparent. Consider the following structure: strings of length no more than four letters. Fig. 1 shows a purely local representation and Fig. 2 shows a semi-local representation (both of which appeared in the letter-perception model of McClelland and Rumelhart[4,5]). In each case, the variable binding has been viewed in the same way. On the left edge is a set of imagined units which can represent an element in the structure—a filler of a role; these are the *filler units.* On the bottom edge is a set of imagined units which can represent a role: these are the *role units.* The remaining units are the ones really used to represent the structure: the *binding units.* They are arranged so that there is one for each pair of filler and role units.

In the purely local case, both the filler and the role are represented by a "pattern of activity" localized to a single unit. In the semi-local case, the filler is represented by a distributed pattern of activity but the role is still represented by a localized pattern. In either case, the binding of the filler to the role is achieved by a simple product operation: the activity of each binding unit is the product of the activities of the associated filler and role unit. In the vocabulary of matrix algebra, the activity representing the value/variable binding forms a matrix which is the *outer product* of the activity vector representing the value and the activity vector representing the variable. In the terminology of vector spaces, *the value/variable binding vector is the tensor product of the value vector and the variable vector.* This is what I refer to as the *tensor product representation* for variable bindings.

Since the activity vectors for roles in Figs. 1 and 2 consist of all zeroes except for a single activity of 1, the tensor product operation is utterly trivial. The local and semi-local cases are trivial special cases of a general binding procedure capable of producing completely distributed representations. Fig. 3 shows a distributed case designed for visual transparency. Imagine we are representing speech data, and have a sequence of values for the energy in a particular formant at successive times. In Fig. 3, distributed patterns are used to represent both the energy value and the variable to which it is bound: the position in time. The particular binding shown is of an energy value 2 (on a scale of 1–4) to the time 4. The peaks in the patterns indicate the value and variable being represented.

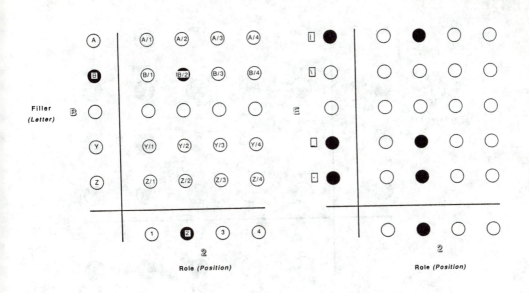

Fig. 1. Purely local representation of strings.   **Fig. 2.** Semi-local representation of strings.

If the patterns representing the value and variable being bound together are not as simple as those used in Fig. 3, the tensor product pattern representing the binding will not of course be particularly visually informative. Such would be the case if the patterns for the fillers and roles in a structure were defined with respect to a set of filler and role *features*: such distributed bindings have been used effectively by McClelland and Kawamoto[18] and by Derthick[19,20]. The extreme mathematical simplicity of the tensor product operation makes feasible an analysis of the general, fully distributed case.

Each binding unit in the tensor product representation corresponds to a pair of imaginary role and filler units. A binding unit can be readily interpreted semantically if its corresponding filler and role units can. The activity of the binding unit indicates that in the structure being represented an element is present which possesses the feature indicated by the corresponding filler unit *and* which occupies a role in the structure which possesses the feature indicated by the corresponding role unit. The binding unit thus detects a *conjunction* of a pair of filler and role features. (Higher-order conjunctions will arise later.)

A structure consists of multiple filler/role bindings. So far we have only discussed the representation of a single binding. In the purely local and semi-local cases, there are separate pools for different roles, and it is obvious how to combine bindings: simultaneously represent them in the separate pools. In the case of a fully distributed tensor product binding (eg., Fig. 3), each single binding is a pattern of activity that extends across the entire set of binding units. The simplest possibility for combining these patterns is simply to *add them up*; that is, to superimpose all the bindings on top of each other. In the special cases of purely local and semi-local representations, this procedure reduces trivially to simultaneously representing the individual fillers in the separate pools.

734

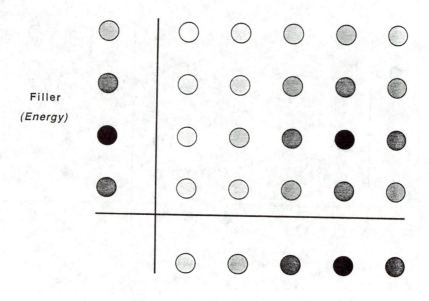

Filler

(Energy)

Role *(Time)*

Fig. 3. A visually transparent fully distributed tensor product representation.

The process of superimposing the separate bindings produces a representation of structures with the usual connectionist properties. If the patterns representing the roles are not too similar, the separate bindings can all be kept straight. It is not necessary for the role patterns to be non-overlapping, as they are in the purely local and semi-local cases; it is sufficient that the patterns be linearly independent. Then there is a simple operation that will correctly extract the filler for any role from the representation of the structure as a whole. If the patterns are not just linearly independent, but are also orthogonal, this operation becomes quite direct; we will get to it shortly. For now, the point is that simply superimposing the separate bindings is sufficient. If the role patterns are not too similar, the separate bindings do not interfere. The representation gracefully saturates if more and more roles are filled, since the role patterns being used lose their distinctness once their number approaches that of the role units.

Thus problem (2) listed above, shared by purely local and semi-local representations, is at last removed in fully distributed tensor product representations: they do not accomodate structures only up to a certain rigid limit, beyond which they are completely saturated; rather, they saturate gracefully. The third problem is also fully addressed, as similar roles can be represented by similar patterns in the tensor product representation and then generalizations both across similar fillers and across similar roles can be learned and exploited.

The definition of the tensor product representation of structured data can be summed up as follows:

(a) Let a set S of structured objects be given a *role decomposition*: a set of fillers, F, a set of roles R, and for each object s a corresponding set of bindings $\beta(s) = \{f/r : f \text{ fills role } r \text{ in } s\}$.

(b) Let a connectionist representation of the fillers F be given; f is represented by the activity vector **f**.

(c) Let a connectionist representation of the roles R be given; r is represented by the activity vector **r**.

(d)   Then the corresponding tensor product representation of s is $\sum_{f/r \in \beta(s)} \mathbf{f} \otimes \mathbf{r}$ (where $\otimes$ denotes the tensor product operation).

In the next section I will discuss a model using a fully distributed tensor product representation, which will require a brief consideration of role decompositions. I will then go on to summarize general properties of the tensor product representation.

Role decompositions

The most obvious role decompositions are *positional decompositions* that involve fixed position slots within a structure of pre-determined form. In the case of a string, such a role would be the ith position in the string; this was the decomposition used in the examples of Figs. 1 through 3. Another example comes from McClelland and Kawamoto's model[18] for learning to assign case roles. They considered sentences of the form *The N_1 V the N_2 with the N_3*; the four roles were the slots for the three nouns and the verb.

A less obvious but sometimes quite powerful role decomposition involves not fixed positions of elements but rather their *local context*. As an example, in the case of strings of letters, such roles might be $r_{x_y} = $ *is preceded by x and followed by y*, for various letters x and y.

Such a local context decomposition was used to considerable advantage by Rumelhart and McClelland in their model of learning the morphophonology of the English past tense[21]. Their structures were strings of phonetic segments, and the context decomposition was well-suited for the task because the generalizations the model needed to learn involved the transformations undergone by phonemes occurring in different local contexts.

Rumelhart and McClelland's representation of phonetic strings is an example of a fully distributed tensor product representation. The fillers were phonetic segments, which were represented by a pattern of phonetic features, and the roles were nearest-neighbor phonetic contexts, which were also represented as distributed patterns. The distributed representation of the roles was in fact itself a tensor product representation: the roles themselves have a constituent structure which can be further broken down through another role decomposition. The roles are indexed by a left and right neighbor; in essence, a string of two phonetic segments. This string too can be decomposed by a context decomposition; the filler can be taken to be the left neighbor, and the role can be indexed by the right neighbor. Thus the vowel $[i]$ in the word *week* is bound to the role r_{w_k}, and this role is in turn a binding of the filler $[w]$ in the sub-role r'_{k}. The pattern for $[i]$ is a vector $\mathbf{i}$ of phonetic features; the pattern for $[w]$ is another such vector of features $\mathbf{w}$, and the pattern for the sub-role r'_{k} is a third vector $\mathbf{k}$ consisting of the phonetic features of $[k]$. The binding for the $[i]$ in *week* is thus $\mathbf{i} \otimes (\mathbf{w} \otimes \mathbf{k})$. Each unit in the representation represents a *third-order* conjunction of a phonetic feature for a central segment together with two phonetic features for its left and right neighbors. [To get precisely the representation used by Rumelhart and McClelland, we have to take this tensor product representation of the roles (eg. r_{w_k}) and throw out a number of the binding units generated in this further decomposition; only certain combinations of features of the left and right neighbors were used. The distributed representation of letter triples used by Touretzky and Hinton[1] can be viewed as a similar third-order tensor product derived from nested context decompositions, with some binding units thrown away—in fact, all binding units off the main diagonal were discarded.]

This example illustrates how role decompositions can be iterated, leading to iterated tensor product representations. Whenever the fillers or roles of one decomposition are structured objects, they can themselves be further reduced by another role decomposition.

It is often useful to consider recursive role decompositions in which the fillers are the same type of object as the original structure. It is clear from the above definition that such a decomposition cannot be used to generate a tensor product representation. Nonetheless, recursive role decompositions *can* be used to relate the tensor product representation of complex structures to the tensor product representations of simpler structures. For example, consider Lisp binary tree structures built from a set A of atoms. A non-recursive decomposition uses A as the set of fillers, with each role being the

736

occupation of a certain position in the tree by an atom. From this decomposition a tensor product representation can be constructed. Then it can be seen that the operations *car*, *cdr*, and *cons* correspond to certain linear operators **car**, **cdr**, and **cons** in the vector space of activation vectors. Just as complex S-expressions can be constructed from atoms using *cons*, so their connectionist representations can be constructed from the simple representation of atoms by the application of **cons**. (This serial "construction" of the complex representation from the simpler ones is done by *the analyst*, not necessarily by the network; **cons** is a static, descriptive, mathematical operator—not necessarily a transformation to be carried out by a network.)

Binding and unbinding in connectionist networks

So far, the operation of binding a value to a variable has been described mathematically and pictured in Figs. 1–3 in terms of "imagined" filler units and role units. Of course, the binding operation can actually be performed in a network if the filler and role units are really there. Fig. 4 shows one way this can be done. The triangular junctions are Hinton's multiplicative connections[22]: the incoming activities from the role and filler units are multiplied at the junction and passed on to the binding unit.

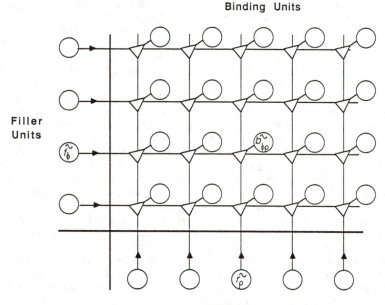

Fig. 4. A network for tensor product binding and unbinding.

"Unbinding" can also be performed by the network of Fig. 4. Suppose the tensor product representation of a structure is present in the binding units, and we want to extract the filler for a particular role. As mentioned above, this can be done accurately if the role patterns are linearly independent (and if each role is bound to only one filler). It can be shown that in this case, for each role there is a pattern of activity which, if set up on the role units, will lead to a pattern on the filler units that represents the corresponding filler. (If the role vectors are orthogonal, this pattern is the same as the role pattern.) As in Hinton's model[20], it is assumed here that the triangular junctions work in all directions, so that now they take the product of activity coming in from the binding and role units and pass it on to the filler units, which sum all incoming activity.

The network of Fig. 4 can bind one value/variable pair at a time. In order to build up the representation of an entire structure, the binding units would have to accumulate activity over an extended period of time during which all the individual bindings would be performed serially. Multiple bindings could occur in parallel if part of the apparatus of Fig. 4 were duplicated: this requires several copies of the sets of filler and role units, paired up with triangular junctions, all feeding into a single set of binding units.

Notice that in this scheme there are two *independent* capacities for parallel binding: the capacity to *generate* bindings in parallel, and the capacity to *maintain* bindings simultaneously. The former is determined by the degree of duplication of the filler/role unit sets (in Fig. 4, for example, the parallel generation capacity is 1). The parallel maintenance capacity is determined by the number of possible linearly independent role patterns, i.e. the number of role units in each set. It is logical that these two capacities should be independent, and in the case of the human visual and linguistic systems it seems that our maintenance capacity far exceeds our generation capacity[21]. Note that in purely local and semi-local representations, there is a separate pool of units dedicated to the representation of each role, so there is a tendency to suppose that the two capacities are equal. As long as a connectionist model deals with structures (like four-letter words) that are so small that the number of bindings involved is within the human parallel generation capacity, there is no harm done. But when connectionist models address the human representation of large structures (like entire scenes or discourses), it will be important to be able to maintain a large number of bindings even though the number that can be generated in parallel is much smaller.

Further properties and extensions

Continuous structures. It can be argued that underlying the connectionist approach is a fundamentally continuous formalization of computation[13]. This would suggest that a natural connectionist representation of structure would apply at least as well to continuous structures as to discrete ones. It is therefore of some interest that the tensor product representation applies equally well to structures characterized by a continuum of roles: a "string" extending through continuous time, for example, as in continuous speech. In place of a sum over a discrete set of bindings, $\sum_i \mathbf{f}_i \otimes \mathbf{r}_i$ we have an integral over a continuum of bindings: $\int_t \mathbf{f}(t) \otimes \mathbf{r}(t)\, dt$ This goes over exactly to the discrete case if the fillers are discrete step-functions of time.

Continuous patterns. There is a second sense in which the tensor product representation extends naturally to the continuum. If the patterns representing fillers and/or roles are continuous curves rather than discrete sets of activities, the tensor product operation is still well-defined. (Imagine Fig. 3 with the filler and role patterns being continuous peaked curves instead of discrete approximations; the binding pattern is then a continuous peaked two-dimensional surface.) In this case, the vectors $\mathbf{f}$ and/or $\mathbf{r}$ are members of infinite-dimensional function spaces; regarding them as patterns of activity over a set of processors would require an infinite number of processors. While this might pose some problems for computer simulation, the case where $\mathbf{f}$ and/of $\mathbf{r}$ are functions rather than finite-dimensional vectors is not particularly problematic analytically. And if a problem with a continuum of roles is being considered, it may be desirable to assume a continuum of linearly independent role vectors: this requires considering infinite-dimensional representations.

Values as variables. Treating both values and variables symmetrically as done in the tensor product representation makes it possible for the same entity to simultaneously serve both as a value and as a variable. In symbolic computation it often happens that the value bound to one variable is itself a variable which in turn has a value bound to it. In a semi-local representation, where variables are localized pools of units and values are patterns of activity in these pools, it is difficult to see how the same entity can simultaneously serve as both value and variable. In the tensor product representation, both values and variables are patterns of activity, and whether a pattern is serving as a "variable" or "value"—or both—might be merely a matter of descriptive preference.

738

Symbolic structures in associative memories. The mathematical simplicity of the tensor product representation makes it possible to characterize conditions under which a set of symbolic structures can be stored in an associative memory without interference. These conditions involve an interesting mixture of the numerical character of the associative memory and the discrete character of the stored data.

Learning optimal role patterns by recirculation. While the use of distributed patterns to represent *constituents* in structures is well-known, the use of such patterns to represent *roles* in structures poses some new challenges. In some domains, features for roles are familiar or easy to imagine; eg., features of semantic roles in a case-frame semantics. But it is worth considering the problem of distributed role representations in domain-independent terms as well. The patterns used to represent roles determine how information about a structure's fillers will be coded, and these role patterns have an effect on how much information can subsequently be extracted from the representation by connectionist processing. The challenge of making the most information available for such future extraction can be posed as follows. Assume enough apparatus has been provided to do all the variable binding in parallel in a network like that of Fig. 4. Then we can dedicate a set of role units to each role; the pattern for each role can be set up once and for all in one set of role units. Since the activity of the role units provide multipliers for filler values at the triangular junctions, we can treat these fixed role patterns as weights on the lines from the filler units to the binding units. The problem of finding good role patterns now becomes the problem of finding good weights for encoding the fillers into the binding units.

Now suppose that a second set of connections is used to try to extract all the fillers from the representation of the structure in the binding units. Let the weights on this second set of connections be chosen to minimize the mean-squared differences between the extracted filler patterns and the actual original filler patterns. Let a set of role vectors be called *optimal* if this mean-squared error is as small as possible.

It turns out that optimal role vectors can be characterized fairly simply both algebraically and geometrically (with the help of results from Williams[24]). Furthermore, having imbedded the role vectors as weights in a connectionist net, it is possible for the network to learn optimal role vectors by a fairly simple learning algorithm. The algorithm is derived as a gradient descent in the mean-squared error, and is what G. E. Hinton and J. L. McClelland (unpublished communication) have called a *recirculation algorithm*: it works by circulating activity around a closed network loop and training on the difference between the activities at a given node on successive passes.

Acknowledgements

This research has been supported by NSF grants IRI-8609599 and ECE-8617947, by the Sloan Foundation's computational neuroscience program, and by the Department of Computer Science and Institute of Cognitive Science at the University of Colorado at Boulder.

References

1. D. S. Touretzky & G. E. Hinton. *Proceedings of the International Joint Conference on Artificial Intelligence*, 238-243 (1985).
2. D. S. Touretzky. *Proceedings of the 8th Conference of the Cognitive Science Society*, 522–530 (1986).
3. P. Smolensky. Technical Report CU–CS–355–87, Department of Computer Science, University of Colorado at Boulder (1987).
4. J. L. McClelland & D. E. Rumelhart. *Psychological Review* 88, 375–407 (1981).
5. D. E. Rumelhart & J. L. McClelland. *Psychological Review* 89, 60–94 (1982).
6. M. Fanty. Technical Report 174, Department of Computer Science, University of Rochester (1985).
7. J. A. Feldman. *The Behavioral and Brain Sciences* 8, 265–289 (1985).
8. J. L. McClelland & J. L. Elman. In J. L. McClelland, D. E. Rumelhart, & the PDP Research Group, *Parallel distributed processing: Explorations in the microstructure of cognition. Vol. 2: Psychological and biological models.* Cambridge, MA: MIT Press/Bradford Books, 58–121 (1986).
9. T. J. Sejnowski & C. R. Rosenberg. *Complex Systems* 1, 145–168 (1987).
10. J. A. Feldman. Technical Report 189, Department of Computer Science, University of Rochester (1986).
11. J. A. Anderson & G. E. Hinton. In G. E. Hinton and J. A. Anderson, Eds., *Parallel models of associative memory.* Hillsdale, NJ: Erlbaum, 9–48 (1981).
12. G. E. Hinton, J. L. McClelland, & D. E. Rumelhart. In D. E. Rumelhart, J. L. McClelland, & the PDP Research Group, *Parallel distributed processing: Explorations in the microstructure of cognition. Vol. 1: Foundations.* Cambridge, MA: MIT Press/Bradford Books, 77–109 (1986).
13. P. Smolensky. *The Behavioral and Brain Sciences* 11(1) (in press).
14. P. Smolensky. In J. L. McClelland, D. E. Rumelhart, & the PDP Research Group, *Parallel distributed processing: Explorations in the microstructure of cognition. Vol. 2: Psychological and biological models.* Cambridge, MA: MIT Press/Bradford Books, 390–431 (1986).
15. G. E. Hinton. In Hinton, G.E. and Anderson, J.A., Eds., *Parallel models of associative memory.* Hillsdale, NJ: Erlbaum, 161–188 (1981).
16. M. S. Riley & P. Smolensky. *Proceedings of the Sixth Annual Conference of the Cognitive Science Society*, 286–292 (1984).
17. P. Smolensky. In D. E. Rumelhart, J. L. McClelland, & the PDP Research Group, *Parallel distributed processing: Explorations in the microstructure of cognition. Vol. 1: Foundations.* Cambridge, MA: MIT Press/Bradford Books, 194–281 (1986).
18. J. L. McClelland & A. H. Kawamoto. In J. L. McClelland, D. E. Rumelhart, & the PDP Research Group, *Parallel distributed processing: Explorations in the microstructure of cognition. Vol. 2: Psychological and biological models.* Cambridge, MA: MIT Press/Bradford Books, 272–326 (1986).
19. M. Derthick. *Proceedings of the National Conference on Artificial Intelligence*, 346–351 (1987).
20. M. Derthick. *Proceedings of the Annual Conference of the Cognitive Science Society*, 131–142 (1987).
21. D. E. Rumelhart & J. L. McClelland. In J. L. McClelland, D. E. Rumelhart, & the PDP Research Group, *Parallel distributed processing: Explorations in the microstructure of cognition. Vol. 2: Psychological and biological models.* Cambridge, MA: MIT Press/Bradford Books, 216–271 (1986)
22. G. E. Hinton. *Proceedings of the Seventh International Joint Conference on Artificial Intelligence*, 683–685 (1981).
23. A. M. Treisman & H. Schmidt. *Cognitive Psychology* 14, 107–141 (1982).
24. R. J. Williams. Technical Report 8501, Institute of Cognitive Science, University of California, San Diego (1985).

SPATIAL ORGANIZATION OF NEURAL NETWORKS:
A PROBABILISTIC MODELING APPROACH

A. Stafylopatis
M. Dikaiakos
D. Kontoravdis
National Technical University of Athens, Department of Electrical Engineering, Computer Science Division, 157 73 Zographos, Athens, Greece.

ABSTRACT

The aim of this paper is to explore the spatial organization of neural networks under Markovian assumptions, in what concerns the behaviour of individual cells and the interconnection mechanism. Space-organizational properties of neural nets are very relevant in image modeling and pattern analysis, where spatial computations on stochastic two-dimensional image fields are involved. As a first approach we develop a random neural network model, based upon simple probabilistic assumptions, whose organization is studied by means of discrete-event simulation. We then investigate the possibility of approximating the random network's behaviour by using an analytical approach originating from the theory of general product-form queueing networks. The neural network is described by an open network of nodes, in which customers moving from node to node represent stimulations and connections between nodes are expressed in terms of suitably selected routing probabilities. We obtain the solution of the model under different disciplines affecting the time spent by a stimulation at each node visited. Results concerning the distribution of excitation in the network as a function of network topology and external stimulation arrival pattern are compared with measures obtained from the simulation and validate the approach followed.

INTRODUCTION

Neural net models have been studied for many years in an attempt to achieve brain-like performance in computing systems. These models are composed of a large number of interconnected computational elements and their structure reflects our present understanding of the organizing principles of biological nervous systems. In the beginning, neural nets, or other equivalent models, were rather intended to represent the logic arising in certain situations than to provide an accurate description in a realistic context. However, in the last decade or so the knowledge of what goes on in the brain has increased tremendously. New discoveries in natural systems, make it now reasonable to examine the possibilities of using modern technology in order to synthesige systems that have some of the properties of real neural systems [8,9,10,11].

In the original neural net model developed in 1943 by McCulloch and Pitts [1,2] the network is made of many interacting components, known as the "McCulloch-Pitts cells" or "formal neurons", which are simple logical units with two possible states changing state according to a threshold function of their inputs. Related automata models have been used later for gene control systems (genetic networks) [3], in which genes are represented as binary automata changing state according to boolean functions of their inputs. Boolean networks constitute a more general model, whose dynamical behaviour has been studied extensively. Due to the large number of elements, the exact structure of the connections and the functions of individual components are generally unknown and assumed to be distributed at random. Several studies on these random boolean networks [5,6] have shown that they exhibit a surprisingly stable behaviour in what concerns their temporal and spatial organization. However, very few formal analytical results are available, since most studies concern statistical descriptions and computer simulations.

The temporal and spatial organization of random boolean networks is of particular interest in the attempt of understanding the properties of such systems, and models originating from the theory of stochastic processes [13] seem to be very useful. Spatial properties of neural nets are most important in the field of image recognition [12]. In the biological eye, a level-normalization computation is performed by the layer of horizontal cells, which are fed by the immediately preceding layer of photoreceptors. The horizontal cells take the outputs of the receptors and average them spatially, this average being weighted on a nearest-neighbor basis. This procedure corresponds to a mechanism for determining the brightness level of pixels in an image field by using an array of processing elements. The principle of local computation is usually adopted in models used for representing and generating textured images. Among the stochastic models applied to analyzing the parameters of image fields, the random Markov field model [7,14] seems to give a suitably structured representation, which is mainly due to the application of the markovian property in space. This type of modeling constitutes a promising alternative in the study of spatial organization phenomena in neural nets.

The approach taken in this paper aims to investigate some aspects of spatial organization under simple stochastic assumptions. In the next section we develop a model for random neural networks assuming boolean operation of individual cells. The behaviour of this model, obtained through simulation experiments, is then approximated by using techniques from the theory of queueing networks. The approximation yields quite interesting results as illustrated by various examples.

THE RANDOM NETWORK MODEL

We define a random neural network as a set of elements or cells, each one of which can be in one of two different states: firing or quiet. Cells are interconnected to form an NxN grid, where each grid point is occupied by a cell. We shall consider only connections between neighbors, so that each cell is connected to 4 among the other cells: two input and two output cells (the output of a cell is equal to its internal state and it is sent to its output cells which use it as one of their inputs). The network topology is thus specified

by its incidence matrix A of dimension MxM, where $M=N^2$. This matrix takes a simple form in the case of neighbor-connection considered here. We further assume a periodic structure of connections in what concerns inputs and outputs; we will be interested in the following two types of networks depending upon the period of reproduction for elementary square modules [5], as shown in Fig.1:
- Propagative nets (Period 1)
- Looping nets   (Period 2)

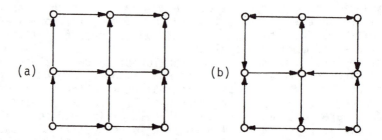

Fig.1. (a) Propagative connections, (b) Looping connections

At the edges of the grid, circular connections are established (modulo N), so that the network can be viewed as supported by a torus.
    The operation of the network is non-autonomous: changes of state are determined by both the interaction among cells and the influence of external stimulations. We assume that stimulations arrive from the outside world according to a Poisson process with parameter λ. Each arriving stimulation is associated with exactly one cell of the network; the cell concerned is determined by means of a given discrete probability distribution q_i ($1 \leq i \leq M$), considering an one-dimensional labeling of the M cells.
    The operation of each individual cell is asynchronous and can be described in terms of the following rules:
- A quiet cell moves to the firing state if it receives an arriving stimulation or if a boolean function of its inputs becomes true.
- A firing cell moves to the quiet state if a boolean function of its inputs becomes false.
- Changes of state imply a reaction delay of the cell concerned; these delays are independent identically distributed random variables following a negative exponential distribution with parameter ν.
According to these rules, the operation of a cell can be viewed as illustrated by Fig.2, where the horizontal axis represents time and the numbers 0,1,2 and 3 represent phases of an operation cycle. Phases 1 and 3, as indicated in Fig.2, correspond to reaction delays. In this sense, the quiet and firing states, as defined in the begining of this section, represent the aggregates of phases 0,1 and 2,3 respectively. External stimulations affect the receiving cell only when it is in phase 0; otherwise we consider that the stimulation is lost. In the same way, we assume that changes of the value of the input boolean function do not affect the operation of the cell during phases 1 and 3. The conditions are checked only at the end of the respective reaction delay.

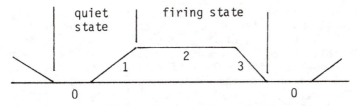

Fig.2. Changes of state for individual cells

The above defined model includes some features of the original McCulloch-Pitts cells [1,2] . In fact, it represents an asynchronous counterpart of the latter, in which boolean functions are considered instead of threshold functions. However, it can be shown that any McCulloch and Pitts' neural network can be implemented by a boolean network designed in an appropriate fashion [5]. In what follows, we will consider that the firing condition for each individual cell is determined by an "or" function of its inputs.

Under the assumptions adopted, the evolution of the network in time can be described by a continuous-parameter Markov process. However, the size of the state-space and the complexity of the system are such that no analytical solution is tractable. The spatial organization of the network could be expressed in terms of the steady-state probability distribution for the Markov process. A more useful representation is provided by the marginal probability distributions for all cells in the network, or equivalently by the probability of being in the firing state for each cell. This measure expresses the level of excitation for each point in the grid.

We have studied the behaviour of the above model by means of simulation experiments for various cases depending upon the network size, the connection type, the distribution of external stimulation arrivals on the grid and the parameters λ and γ. Some examples are illustrated in the last section, in comparison with results obtained using the approach discussed in the next section. The estimations obtained concern the probability of being in the firing state for all cells in the network. The simulation was implemented according to the "batched means" method; each run was carried out until the width of the 95% confidence interval was less that 10% of the estimated mean value for each cell, or until a maximum number of events had been simulated depending upon the size of the network.

THE ANALYTICAL APPROACH

The neural network model considered in the previous section exhibited the markovian property in both time and space. Markovianity in space, expressed by the principle of "neighbor-connections", is the basic feature of Markov random fields [7,14], as already discussed. Our idea is to attempt an approximation of the random neural network model by using a well-known model, which is markovian in time, and applying the constraint of markovianity in space. The model considered is an open queueing network, which belongs to the general class of queueing networks admitting a product-form solution [4]. In fact, one could distinguish several common features in the two network models.

A neural network, in general, receives information in the form of external stimulation signals and performs some computation on this information, which is represented by changes of its state. The operation of the network can be viewed as a flow of excitement among the cells and the spatial distribution of this excitement represents the response of the network to the information received. This kind of operation is particularly relevant in the processing of image fields. On the other hand, in queueing networks, composed of a number of service station nodes, customers arrive from the outside world and spend some time in the network, during which they more from node to node, waiting and receiving service at each node visited. Following the external arrival pattern, the interconnection of nodes and the other network parameters, the operation of the network is characterized by a distribution of activity among the nodes.

Let us now consider a queueing network, where nodes represent cells and customers represent stimulations moving from cell to cell following the topology of the network. Our aim is to define the network's characteristics in a way to match those of the neural net model as much as possible. Our queueing network model is completely specified by the following assumptions:

- The network is composed of $M=N^2$ nodes arranged on an NxN rectangular grid, as in the previous case. Interconnections are expressed by means of a matrix R of routing probabilities: r_{ij} ($1 \leq i, j \leq M$) represents the probability that a stimulation (customer) leaving node i will next visit node j. Since it is an open network, after visiting an arbitrary number of cells, stimulations may eventually leave the network. Let r_{i0} denote the probability of leaving the network upon leaving node i. In what follows, we will assume that r_{i0}=s for all nodes. In what concerns the routing probabilities r_{ij}, they are determined by the two interconnection schemata considered in the previous section (propagative and looping connections): each node i has two output nodes j, for which the routing probabilities are equally distributed. Thus, $r_{ij}=(1-s)/2$ for the two output nodes of i and equal to zero for all other nodes in the network.

- External stimulation arrivals follow a Poisson process with parameter λ and are routed to the nodes according to the probability distribution q_i ($1 \leq i \leq M$) as in the previous section.

- Stimulations receive a "service time" at each node visited. Service times are independent identically distributed random variables, which are exponentially distributed with parameter γ. The time spent by a stimulation at a node depends also upon the "service discipline" adopted. We shall consider two types of service disciplines according to the general queueing network model [4]: the first-come-first-served (FCFS) discipline, where customers are served in the order of their arrival to the node, and the infinite-server (IS) discipline, where a customer's service is immediately assumed upon arrival to the node, as if there were always a server available for each arriving customer (the second type includes no waiting delay). We will refer to the above two types of nodes as type 1 and type 2 respectively. In either case, all nodes of the network will be of the same type.

The steady-state solution of the above network is a straightforward application of the general BCMP theorem [4] according to the simple assumptions considered. The state of the system is described by the vector $(k_1,k_2,\ldots,k_M)$, where k_i is the number of customers present at node i. We first define the traffic intensity ρ_i for each node i as

$$\rho_i = \lambda e_i/\gamma \quad , \quad i = 1,2,\ldots,M \qquad (1)$$

where the quantities $\{e_i\}$ are the solution of the following set of linear equations:

$$e_i = q_i + \sum_{j=1}^{M} e_j r_{ji} \quad , \quad i = 1,2,\ldots,M \qquad (2)$$

It can be easily seen that, in fact, e_i represents the average number of visits a customer makes to node i during his sojourn in the network. The existence of a steady-state distribution for the system depends on the solution of the above set. Following the general theorem [4], the joint steady-state distribution takes the form of a product of independent distributions for the nodes:

$$p(k_1,k_2,\ldots,k_M) = p_1(k_1)p_2(k_2)\ldots p_M(k_M) \qquad (3)$$

where

$$p_i(k_i) = \begin{cases} (1-\rho_i)\rho_i^{k_i} & \text{(Type 1)} \\ e^{-\rho_i}\dfrac{\rho_i^{k_i}}{k_i!} & \text{(Type 2)} \end{cases} \qquad (4)$$

provided that the stability condition $\rho_i<1$ is satisfied for type 1 nodes.

The product form solution of this type of network expresses the idea of global and local balance which is characteristic of ergodic Markov processes. We can then proceed to deriving the desired measure for each node in the network; we are interested in the probability of being active for each node, which can be interpreted as the probability that at least one customer is present at the node:

$$P(k_i>0)=1-p_i(0) = \begin{cases} \rho_i & \text{(Type 1)} \\ 1-e^{-\rho_i} & \text{(Type 2)} \end{cases} \qquad (5)$$

The variation in space of the above quantity will be studied with respect to the corresponding measure obtained from simulation experiments for the neural network model.

NUMERICAL AND SIMULATION EXAMPLES

Simulations and numerical solutions of the queueing network model were run for different values of the parameters. The network sizes considered are relatively small but can provide useful information on the spatial organization of the networks. For both types of service discipline discussed in the previous section, the approach followed yields a very good approximation of the network's organization in most regions of the rectangular grid. The choice of the probability s of leaving the network plays a critical role in the beha-

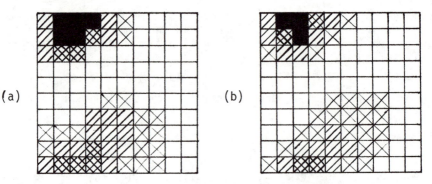

Fig.3. A 10x10 network with $\lambda=1$, $\gamma=1$ and propagative connections. External stimulations are uniformly distributed over a 3x3 square on the upper left corner of the grid. (a) simulation (b) Queueing network approach with s=0.05 and type 2 nodes.

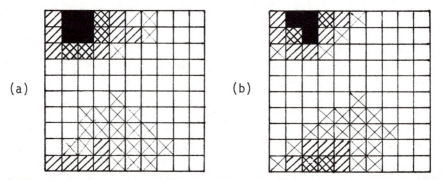

Fig.4. The network of Fig.3 with $\lambda=2$ (a) Simulation (b) Queueing network approach with s=0.08 and type 2 nodes.

viour of the queueing model, and must have a non-zero value in order for the network to be stable. Good results are obtained for very small values of s; in fact, this parameter represents the phenomenon of excitation being "lost" somewhere in the network. Graphical representations for various cases are shown in Figures 3-7. We have used a coloring of five "grey levels", defined by dividing into five segments the interval between the smallest and the largest value of the probability on the grid; the normalization is performed with respect to simulation results. This type of representation is less accurate than directly providing numerical values, but is more clear for describing the organization of the system. In each case, the results shown for the queueing model concern only one type of nodes, the one that best fits the simulation results, which is type 2 in the majority of cases examined. The graphical representation illustrates the structuring of the distribution of excitation on the grid in terms of functionally connected regions of high and low

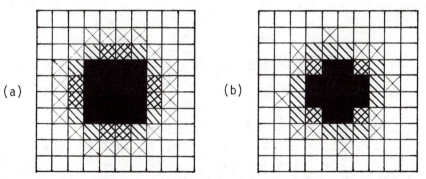

Fig.5. A 10x10 network with λ=1, γ=1 and looping connections. External stimulations are uniformly distributed over a 4x4 square on the center of the grid. (a) Simulation (b) Queueing network approach with s=0.07 and type 2 nodes.

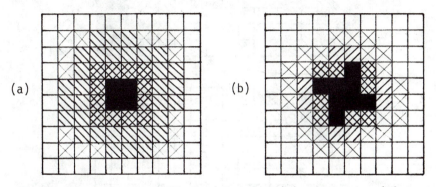

Fig.6. The network of Fig.5 with λ=0.5 (a) Simulation (b) Queueing network approach with s=0.03 and type 2 nodes.

excitation. We notice that clustering of nodes mainly follows the spatial distribution of external stimulations and is more sharply structured in the case of looping connections.

CONCLUSION

We have developed in this paper a simple continuous-time probabilistic model of neural nets in an attempt to investigate their spatial organization. The model incorporates some of the main features of the McCulloch-Pitts "formal neurons" model and assumes boolean operation of the elementary cells. The steady-state behaviour of the model was approximated by means of a queueing network model with suitably chosen parameters. Results obtained from the solution of the above approximation were compared with simulation results of the initial model, which validate the approximation. This simplified approach is a first step in an attempt to study the organiza-

748

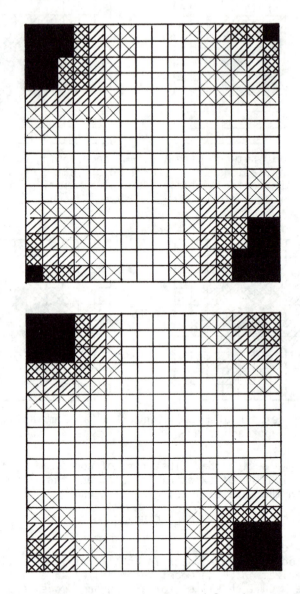

(a)

(b)

Fig.7. A 16x16 network with $\lambda=1$, $\gamma=1$ and looping connections. External stimulations are uniformly distributed over two 4x4 squares on the upper left and lower right corners of the grid. (a) Simulation (b) Queueing network approach with s=0.05 and type 1 nodes.

tional properties of neural nets by means of markovian modeling techniques.

REFERENCES

1. W. S. McCulloch, W. Pitts, "A Logical Calculus of the Ideas Immanent in Nervous Activity", Bull. of Math. Biophysics 5, 115-133 (1943).
2. M. L. Minsky, Computation: Finite and Infinite Machines (Prentice Hall, 1967).
3. S. Kauffman, "Behaviour of Randomly Constructed Genetic Nets", in Towards a Theoretical Biology, Ed. C. H. Waddington (Edinburgh University Press, 1970).
4. F. Baskett, K. M. Chandy, R. R. Muntz, F. G. Palacios, "Open, Closed and Mixed Networks of Queues with Different Classes of Customers", J. ACM, 22 (1975).
5. H. Atlan, F. Fogelman-Soulié, J. Salomon, G. Weisbuch, "Random Boolean Networks", Cyb. and Syst. 12 (1981).
6. F. Folgeman-Soulié, E. Goles-Chacc, G. Weisbuch, "Specific Roles of the Different Boolean Mappings in Random Networks", Bull. of Math. Biology, Vol.44, No 5 (1982).
7. G. R. Cross, A. K. Jain, "Markov Random Field Texture Models", IEEE Trans. on PAMI, Vol. PAMI-5, No 1 (1983).
8. E. R. Kandel, J. H. Schwartz, Principles of Neural Science, (Elsevier, N.Y., 1985).
9. J. J. Hopfield, D. W. Tank, "Computing with Neural Circuits: A Model", Science, Vol. 233, 625-633 (1986).
10. Y. S. Abu-Mostafa, D. Psaltis, "Optical Neural Computers", Scient. Amer., 256, 88-95 (1987).
11. R. P. Lippmann, "An Introduction to Computing with Neural Nets", IEEE ASSP Mag. (Apr. 1987).
12. C. A. Mead, "Neural Hardware for Vision", Eng. and Scie. (June 1987).
13. E. Gelenbe, A. Stafylopatis, "Temporal Behaviour of Neural Networks", IEEE First Intern. Conf. on Neural Networks, San Diego, CA (June 1987).
14. L. Onural, "A Systematic Procedure to Generate Connected Binary Fractal Patterns with Resolution-varying Texture", Sec. Intern. Sympt. on Comp. and Inform. Sciences, Istanbul, Turkey (Oct. 1987).

A DYNAMICAL APPROACH TO TEMPORAL PATTERN PROCESSING

W. Scott Stornetta
Stanford University, Physics Department, Stanford, Ca., 94305

Tad Hogg and B. A. Huberman
Xerox Palo Alto Research Center, Palo Alto, Ca. 94304

ABSTRACT

Recognizing patterns with temporal context is important for such tasks as speech recognition, motion detection and signature verification. We propose an architecture in which time serves as its own representation, and temporal context is encoded in the state of the nodes. We contrast this with the approach of replicating portions of the architecture to represent time.

As one example of these ideas, we demonstrate an architecture with capacitive inputs serving as temporal feature detectors in an otherwise standard back propagation model. Experiments involving motion detection and word discrimination serve to illustrate novel features of the system. Finally, we discuss possible extensions of the architecture.

INTRODUCTION

Recent interest in connectionist, or "neural" networks has emphasized their ability to store, retrieve and process patterns[1,2]. For most applications, the patterns to be processed are static in the sense that they lack temporal context.

Another important class consists of those problems that require the processing of temporal patterns. In these the information to be learned or processed is not a particular pattern but a sequence of patterns. Such problems include speech processing, signature verification, motion detection, and predictive signal processing[3-8].

More precisely, temporal pattern processing means that the desired output depends not only on the current input but also on those preceding or following it as well. This implies that two identical inputs at different time steps might yield different desired outputs depending on what patterns precede or follow them.

There is another feature characteristic of much temporal pattern processing. Here an entire sequence of patterns is recognized as a single distinct category, generating a single output. A typical example of this would be the need to recognize words from a rapidly sampled acoustic signal. One should respond only once to the appearance of each word, even though the word consists of many samples. Thus, each input may not produce an output.

With these features in mind, there are at least three additional issues which networks that process temporal patterns must address, above and beyond those that work with static patterns. The first is how to represent temporal context in the state of the network. The second is how to train at intermediate time steps before a temporal pattern is complete. The third issue is how to interpret the outputs during recognition, that is, how to tell when the sequence has been completed. Solutions to each of these issues require the construction of appropriate input and output representations. This paper is an attempt to address these issues, particularly the issue of representing temporal context in the state of the machine. We note in passing that the recognition of temporal sequences is distinct from the related problem of generating a sequence, given its first few members[9,10,11].

TEMPORAL CLASSIFICATION

With some exceptions[10,12], in most previous work on temporal problems the systems record the temporal pattern by replicating part of the architecture for each time step. In some instances input nodes and their associated links are replicated[3,4]. In other cases only the weights or links are replicated, once for each of several time delays[7,8]. In either case, this amounts to mapping the temporal pattern into a spatial one of much higher dimension before processing.

These systems have generated significant and encouraging results. However, these approaches also have inherent drawbacks. First, by replicating portions of the architecture for each time step the amount of redundant computation is significantly increased. This problem becomes extreme when the signal is sampled very frequently[4]. Next, by relying on replications of the architecture for each time step, the system is quite inflexible to variations in the rate at which the data is presented or size of the temporal window. Any variability in the rate of the input signal can generate an input pattern which bears little or no resemblance to the trained pattern. Such variability is an important issue, for example, in speech recognition. Moreover, having a temporal window of any fixed length makes it manifestly impossible to detect contextual effects on time scales longer than the window size. An additional difficulty is that a misaligned signal, in its spatial representation, may have very little resemblance to the correctly aligned training signal. That is, these systems typically suffer from not being translationally invariant in time.

Networks based on relaxation to equilibrium[11,13,14] also have difficulties for use with temporal problems. Such an approach removes any dependence on initial conditions and hence is difficult to reconcile directly with temporal problems, which by their nature depend on inputs from earlier times. Also, if a temporal problem is to be handled in terms of relaxation to equilibrium, the equilibrium points themselves must be changing in time.

A NON-REPLICATED, DYNAMIC ARCHITECTURE

We believe that many of the difficulties mentioned above are tied to the attempt to map an inherently dynamical problem into a static problem of higher dimension. As an alternative, we propose to represent the history of the inputs in the state of the nodes of a system, rather than by adding additional units. Such an approach to capturing temporal context shows some very immediate advantages over the systems mentioned above. First, it requires no replication of units for each distinct time step. Second, it does not fix in the architecture itself the window for temporal context or the presentation rate. These advantages are a direct result of the decision to let time serve as its own representation for temporal sequences, rather than creating additional spatial dimensions to represent time.

In addition to providing a solution to the above problems, this system lends itself naturally to interpretation as an evolving dynamical system. Our approach allows one to think of the process of mapping an evolving input into a discrete sequence of outputs (such as mapping continuous speech input into a sequence of words) as a dynamical system moving from one attractor to another[15].

As a preliminary example of the application of these ideas, we introduce a system that captures the temporal context of input patterns without replicating units for each time step. We modify the conventional back propagation algorithm by making the input units capacitive. In contrast to the conventional architecture in which the input nodes are used simply to distribute the signal to the next layer, our system performs an additional computation. Specifically, let X_i be the value computed by an input node at time t_i, and I_i be the input signal to this node at the same time. Then the node computes successive values according to

$$X_{i+1} = aI_{i+1} + dX_i \tag{1}$$

where a is an input amplitude and d is a decay rate. Thus, the result computed by an input unit is the sum of the current input value multiplied by a, plus a fractional part, d, of the previously computed value of the input unit. In the absence of further input, this produces an exponential decay in the activation of the input nodes. The value for d is chosen so that this decay reaches $1/e$ of its original value in a time τ characteristic of the time scale for the particular problem, i.e., $d = e^{-\tau r}$, where r is the presentation rate. The value for a is chosen to produce a specified maximum value for X, given by

$aI_{max}/(1-d)$. We note that Eq. (1) is equivalent to having a non-modifiable recurrent link with weight d on the input nodes, as illustrated in Fig. 1.

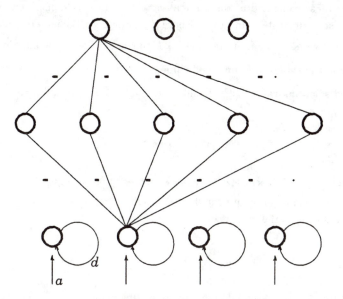

Fig. 1: Schematic architecture with capacitive inputs. The input nodes compute values according to Eq. (1). Hidden and output units are identical to standard back propagation nets.

The processing which takes place at the input node can also be thought of in terms of an infinite impulse response (IIR) digital filter. The infinite impulse response of the filter allows input from the arbitrarily distant past to influence the current output of the filter, in contrast to methods which employ fixed windows, which can be viewed in terms of finite impulse response (FIR) filters. The capacitive node of Fig. 1 is equivalent to pre-processing the signal with a filter with transfer function $a/(1-dz^{-1})$.

This system has the unique feature that a simple transformation of the parameters a and d allows it to respond in a near-optimal way to a signal which differs from the training signal in its rate. Consider a system initially trained at rate r with decay rate d and amplitude a. To make use of these weights for a different presentation rate, r', one simply adjusts the values a' and d' according to

$$d' = d^{r/r'} \qquad (2)$$

$$a' = a\,\frac{1-d'}{1-d} \qquad (3)$$

These equations can be derived by the following argument. The general idea is that the values computed by the input nodes at the new rate should be as close as possible to those computed at the original rate. Specifically, suppose one wishes to change the sampling rate from r to nr, where n is an integer. Suppose that at a time t_0 the computed value of the input node is X_0. If this node receives no additional input, then after m time steps, the computed value of the input node will be $X_0 d^m$. For the more rapid sampling rate, $X_0 d^m$ should be the value obtained after nm time steps. Thus we require

$$X_0 d^m = X_0 d'^{mn} \tag{4}$$

which leads to Eq. (2) because $n = r'/r$. Now suppose that an input I is presented m times in succession to an input node that is initially zero. After the mth presentation, the computed value of the input node is

$$aI \frac{1-d^m}{1-d} \tag{5}$$

Requiring this value to be equal to the corresponding value for the faster presentation rate after nm time steps leads to Eq. (3). These equations, then, make the computed values of the input nodes identical, independent of the presentation rate. Of course, this statement only holds exactly in the limit that the computed values of the input nodes change only infinitesimally from one time step to the next. Thus, in practice, one must insure that the signal is sampled frequently enough that the computed value of the input nodes is slowly changing.

The point in weight space obtained after initial training at the rate r has two desirable properties. First, it can be trained on a signal at one sampling rate and then the values of the weights arrived at can be used as a near-optimal starting point to further train the system on the same signal but at a different sampling rate. Alternatively, the system can respond to temporal patterns which differ in rate from the training signal, without any retraining of the weights. These factors are a result of the choice of input representation, which essentially present the same pattern to the hidden unit and other layers, independent of sampling rate. These features highlight the fact that in this system the weights to some degree represent the temporal pattern independent of the rate of presentation. In contrast, in systems which use temporal windows, the weights obtained after training on a signal at one sampling rate would have little or no relation to the desired values of the weights for a different sampling rate or window size.

EXPERIMENTS

As an illustration of this architecture and related algorithm, a three-layer, 15-30-2 system was trained to detect the leftward or rightward motion of a gaussian pulse moving across the field of input units with sudden changes in direction. The values of d and a were 0.7788 and 0.4424, respectively. These values were chosen to give a characteristic decay time of 4 time steps with a maximum value computed by the input nodes of 2.0. The pulse was of unit height with a half-width, σ, of 1.3. Figure 2 shows the input pulse as well as the values computed by the input nodes for leftward or rightward motion. Once trained at a velocity of 0.1 unit per sampling time, the velocity was varied over a wide range, from a factor of 2 slower to a factor of 2 faster as shown in Fig. 3. For small variations in velocity the system continued to correctly identify the type of motion. More impressive was its performance when the scaling relations given in Eqs. (2) and (3) were used to modify the amplitude and decay rate. In this case, acceptable performance was achieved over the entire range of velocities tested. This was without any additional retraining at the new rates. The difference in performance between the two curves also demonstrates that the excellent performance of the system is not an anomaly of the particular problem chosen, but characteristic of rescaling a and d according to Eqs. (2) and (3). We thus see that a simple use of capacitive links to store temporal context allows for motion detection at variable velocities.

A second experiment involving speech data was performed to compare the system's performance to the time-delay-neural-network of Watrous and Shastri[8]. In their work, they trained a system to discriminate between suitably processed acoustic signals of the words "no" and "go." Once trained on a single utterance, the system was able to correctly identify other samples of these words from the same speaker. One drawback of their approach was that the weights did not converge to a fixed point. We were therefore particularly interested in whether our system could converge smoothly and rapidly to a stable solution, using the same data, and yet generalize as well as theirs did. This experiment also provided an opportunity to test a solution to the intermediate step training problem.

The architecture was a 16-30-2 network. Each of the input nodes received an input signal corresponding to the energy (sampled every 2.5 milliseconds) as a function of time in one of 16 frequency channels. The input values were normalized to lie in the range 0.0 to 1.0. The values of d and a were 0.9944 and 0.022, respectively. These values were chosen to give a characteristic decay time comparable to the length of each word (they were nearly the same length), and a maximum value computed by the input nodes of 4.0. For an input signal that was part of the word "no", the training signal was (1.0, 0.0), while for the word "go" it was (0.0, 1.0). Thus the outputs that were compared to the training signal can be interpreted as evidence for one word or the other at each time step. The error shown in Fig. 4 is the sum of the squares of the difference between the desired outputs and the computed outputs for each time step, for both words, after training up to the number of iterations indicated along the x-axis.

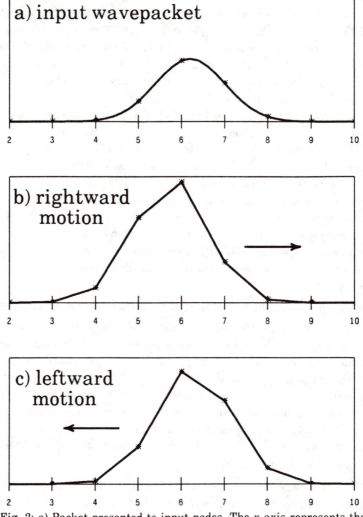

Fig. 2: a) Packet presented to input nodes. The x-axis represents the input nodes. b) Computed values from input nodes during rightward motion. c) Computed values during leftward motion.

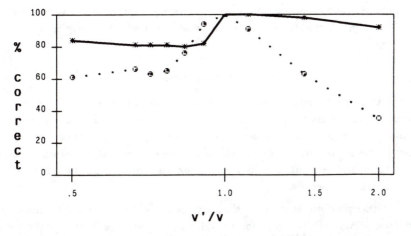

Fig. 3: Performance of motion detection experiment for various velocities. Dashed curve is performance without scaling and solid curve is with the scaling given in Eqs. (2) and (3).

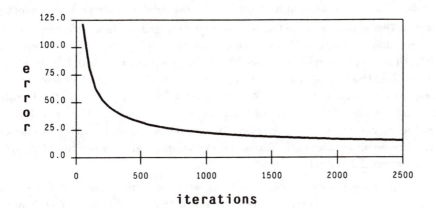

Fig. 4: Error in no/go discrimination as a function of the number of training iterations.

Evidence for each word was obtained by summing the values of the respective nodes over time. This suggests a mechanism for signaling the completion of a sequence: when this sum crosses a certain threshold value, the sequence (in this case, the word) is considered recognized. Moreover, it may be possible to extend this mechanism to apply to the case of connected speech: after a word is recognized, the sums could be reset to zero, and the input nodes reinitialized.

Once we had trained the system on a single utterance, we tested the performance of the resulting weights on additional utterances of the same speaker.

758

Preliminary results indicate an ability to correctly discriminate between "no" and "go." This suggests that the system has at least a limited ability to generalize in this task domain.

DISCUSSION

At a more general level, this paper raises and addresses some issues of representation. By choosing input and output representations in a particular way, we are able to make a static optimizer work on a temporal problem while still allowing time to serve as its own representation. In this broader context, one realizes that the choice of capacitive inputs for the input nodes was only one among many possible temporal feature detectors.

Other possibilities include refractory units, derivative units and delayed spike units. Refractory units would compute a value which was some fraction of the current input. The fraction would decrease the more frequently and recently the node had been "on" in the recent past. A derivative unit would have a larger output the more rapidly a signal changed from one time step to the next. A delayed spike unit might have a transfer function of the form $It^n e^{-\alpha t}$, where t is the time since the presentation of the signal. This is similar to the function used by Tank and Hopfield[7], but here it could serve a different purpose. The maximum value that a given input generated would be delayed by a certain amount of time. By similarly delaying the training signal, the system could be trained to recognize a given input in the context of signals not only preceding but also following it. An important point to note is that the transfer functions of each of these proposed temporal feature detectors could be rescaled in a manner similar to the capacitive nodes. This would preserve the property of the system that the weights contain information about the temporal sequence to some degree independent of the sampling rate.

An even more ambitious possibility would be to have the system train the parameters, such as d in the capacitive node case. It may be feasible to do this in the same way that weights are trained, namely by taking the partial of the computed error with respect to the parameter in question. Such a system may be able to determine the relevant time scales of a temporal signal and adapt accordingly.

ACKNOWLEDGEMENTS

We are grateful for fruitful discussions with Jeff Kephart and the help of Raymond Watrous in providing data from his own experiments. This work was partially supported by DARPA ISTO Contract # N00140-86-C-8996 and ONR Contract # N00014-82-0699.

1. D. Rumelhart, ed., *Parallel Distributed Processing*, (MIT Press, Cambridge, 1986).

2. J. Denker, ed., *Neural Networks for Computing*, AIP Conf. Proc.,151 (1986).

3. T. J. Sejnowski and C. R. Rosenberg, *NETtalk: A Parallel Network that Learns to Read Aloud*, Johns Hopkins Univ. Report No. JHU/EECS-86/01 (1986).

4. J.L. McClelland and J.L. Elman, in *Parallel Distributed Processing*, vol. II, p. 58.

5. W. Keirstead and B.A. Huberman, *Phys. Rev. Lett.* **56**, 1094 (1986).

6. A. Lapedes and R. Farber, *Nonlinear Signal Processing Using Neural Networks*, Los Alamos preprint LA-UR-87-2662 (1987).

7. D. Tank and J. Hopfield, *Proc. Nat. Acad. Sci.*, **84**, 1896 (1987).

8. R. Watrous and L. Shastri, *Proc. 9th Ann. Conf. Cog. Sci. Soc.*, (Lawrence Erlbaum, Hillsdale, 1987), p. 518.

9. P. Kanerva, *Self-Propagating Search: A Unified Theory of Memory*, Stanford Univ. Report No. CSLI-84-7 (1984).

10. M.I. Jordan, *Proc. 8th Ann. Conf. Cog. Sci. Soc.*, (Lawrence Erlbaum, Hillsdale, 1986), p. 531.

11. J. Hopfield, *Proc. Nat. Acad. Sci.*, **79**, 2554 (1982).

12. S. Grossberg, *The Adaptive Brain*, vol. II, ch. 6, (North-Holland, Amsterdam, 1987).

13. G. Hinton and T. J. Sejnowski, in *Parallel Distributed Processing*, vol. I, p. 282.

14. B. Gold, in *Neural Networks for Computing*, p. 158.

15. T. Hogg and B.A. Huberman, *Phys. Rev.* **A32**, 2338 (1985).

A NOVEL NET THAT LEARNS
SEQUENTIAL DECISION PROCESS

G.Z. SUN, Y.C. LEE and H.H. CHEN

Department of Physics and Astronomy
and
Institute for Advanced Computer Studies
UNIVERSITY OF MARYLAND,COLLEGE PARK,MD 20742

ABSTRACT

We propose a new scheme to construct neural networks to classify patterns. The new scheme has several novel features :

1. We focus attention on the important attributes of patterns in ranking order. Extract the most important ones first and the less important ones later.

2. In training we use the information as a measure instead of the error function.

3. A multi-perceptron-like architecture is formed auomatically. Decision is made according to the tree structure of learned attributes.

This new scheme is expected to self-organize and perform well in large scale problems.

1 INTRODUCTION

It is well known that two-layered perceptron with binary connections but no hidden units is unsuitable as a classifier due to its limited power [1]. It cannot solve even the simple *exclusive-or* problem. Two extensions have been proposed to remedy this problem. The first is to use higher order connections [2]. It has been demonstrated that high order connections could in many cases solve the problem with speed and high accuracy [3], [4]. The representations in general are more local than distributive. The main drawback is however the combinatorial explosion of the number of high-order terms. Some kind of heuristic judgement has to be made in the choice of these terms to be represented in the network.

A second proposal is the multi-layered binary network with hidden units [5]. These hidden units function as features extracted from the bottom input layer to facilitate the classification of patterns by the output units. In order to train the weights, learning algorithms have been proposed that back-propagate the errors from the visible output layer to the hidden layers for eventual adaptation to the desired values. The multi-layered networks enjoy great popularity in their flexibility.

However, there are also problems in implementing the multi-layered nets. Firstly, there is the problem of allocating the resources. Namely, how many hidden units would be optimal for a particular problem. If we allocate too many, it is not only wasteful but also could negatively affect the performance of the network. Since too many hidden units implies too many free parameters to fit specifically the training patterns. Their ability to generalize to noval test patterns would be adversely affected. On the other hand, if too few hidden units were allocated then the network would not have the power even to represent the trainig set. How could one judge beforehand how many are needed in solving a problem? This is similar to the problem encountered in the high order net in its choice of high order terms to be represented.

Secondly, there is also the problem of scaling up the network. Since the network represents a parallel or coorperative process of the whole system, each added unit would interact with every other units. This would become a serious problem when the size of our patterns becomes large.

Thirdly, there is no sequential communication among the patterns in the conventional network. To accomplish a cognitive function we would need the patterns to interact and communicate with each other as the human reasoning does. It is difficult to envision such an interacton in current systems which are basically input-output mappings.

2 THE NEW SCHEME

In this paper, we would like to propose a scheme that constructs a network taking advantages of both the parallel and the sequential processes.

We note that in order to classify patterns, one has to extract the intrinsic features, which we call attributes. For a complex pattern set, there may be a large number of attributes. But differnt attributes may have different ranking of importance. Instead of extracing them all simultaneously it may be wiser to extract them sequentially in order of its importance [6], [7]. Here the importance of an attribute is determined by its ability to partition the pattern set into sub-categories. A measure of this ability of a processing unit should be based on the extracted information. For simplicity, let us assume that there are only two categories so that the units have only binary output values 1 and 0 (but the input patterns may have analog representations). We call these units, including their connection weights to the input layer, *nodes*. For given connection weights, the patterns that are classified by a *node* as in category 1 may have their true classifications either 1 or 0. Similarly, the patterns that are classified by a *node* as in category 0 may also have their true classifications either 1 or 0. As a result, four groups of patterns are formed: (1,1), (0,0), (1,0), (0,1). We then need to judge on the efficiency of the *node* by its ability to split these patterns optimally. To do this we shall construct the impurity fuctions for the *node*. Before splitting, the impurity of the input patterns reaching the node is given by

$$I_b = -P_1^b \log P_1^b - P_0^b \log P_0^b \qquad (1)$$

where $P_1^b = N_1^b/N$ is the probability of being truely classified as in category 1, and $P_0^b = N_0^b/N$ is the probability of being truely classified as in category 0. After splitting, the patterns are channelled into two branches, the impurity becomes

$$I_a = -P_1^a \sum_{j=0,1} P(j,1) \log P(j,1) - P_0^a \sum_{j=0,1} P(j,0) \log P(j,0) \qquad (2)$$

where $P_1^a = N_1^a/N$ is the probability of being classified by the node as in category 1, $P_0^a = N_0^b/N$ is the probability of being classified by the node as in category 0, and $P(j,i)$ is the probability of a pattern, which should be in category j, but is classified by the node as in category i. The difference

$$\Delta I = I_b - I_a \qquad (3)$$

represents the decrease of the impurity at the node after splitting. It is the quantity that we seek to optimize at each node. The logarithm in the impurity function come from the information entropy of Shannon and Weaver. For all practical purpose, we found the optimization of (3) the same as maximizing the entropy [6]

$$S = \frac{N_1}{N}\left[(\frac{N_{01}}{N_1})^2 + (\frac{N_{11}}{N_1})^2\right] + \frac{N_0}{N}\left[(\frac{N_{00}}{N_0})^2 + (\frac{N_{10}}{N_0})^2\right] \qquad (4)$$

where N_i is the number of training patterns classified by the node as in category i, N_{ij} is the number of training patterns with true classification in category i but classified by the node as in category j. Later we shall call the terms in the first bracket S_1 and the second S_2. Obviously, we have

$$N_i = N_{0i} + N_{1i}, \qquad i = 0,1$$

After we trained the first unit, the training patterns were split into two branches by the unit. If the classificaton in either one of these two branches is pure enough, or equivalently either one of S_1 and S_2 is fairly close to 1, then we would terminate that branch (or branches) as a leaf of the decision tree, and classify the patterns as such. On the other hand, if either branch is not pure enough, we add additional node to split the pattern set further. The subsequent unit is trained with only those patterns channeled through this branch. These operations are repeated until all the branches are terminated as leaves.

3   LEARNING ALGORITHM

We used the stochastic gradient descent method to learn the weights of each node. The training set for each node are those patterns being channeled to this node. As stated in the previous section, we seek to maximize the entropy function S. The learning of the weights is therefore conducted through

$$\triangle W_j = \eta \frac{\partial S}{\partial W_j} \tag{5}$$

Where η is the learning rate. The gradient of S can be calculated from the following equation

$$\frac{\partial S}{\partial W_j} = \frac{1}{N}\Big[(1 - 2\frac{N_{01}^2}{N_1^2})\frac{\partial N_{11}}{\partial W_j} + (1 - 2\frac{N_{11}^2}{N_1^2})\frac{\partial N_{01}}{\partial W_j} +$$

$$(1 - 2\frac{N_{00}^2}{N_0^2})\frac{\partial N_{10}}{\partial W_j} + (1 - 2\frac{N_{10}^2}{N_0^2})\frac{\partial N_{00}}{\partial W_j}\Big] \tag{6}$$

Using analog units

$$O^r = \frac{1}{1 + exp(-\sum_j W_j I_j^r)} \tag{7}$$

we have

$$\frac{\partial O^r}{\partial W_j} = O^r(1 - O^r)I_j^r \tag{8}$$

Furthermore, let $A^r = 1$ or 0 being the true answer for the input pattern r , then

$$N_{ij} = \sum_{r=1}^{N}\Big[iA^r + (1 - i)(1 - A^r)\Big]\Big[jO^r + (1 - j)(1 - O^r)\Big] \tag{9}$$

Substituting these into equation (5), we get

$$\triangle W_j = 2\eta\sum_r\Big[2A^r(\frac{N_{11}}{N_1} - \frac{N_{10}}{N_0}) + \frac{N_{10}^2}{N_0^2} - \frac{N_{11}^2}{N_1^2}\Big]O^r(1 - O^r)I_j^r \tag{10}$$

In applying the formula (10),instead of calculating the whole summation at once, we update the weights for each pattern individually. Meanwhile we update N_{ij} in accord with equation (9).

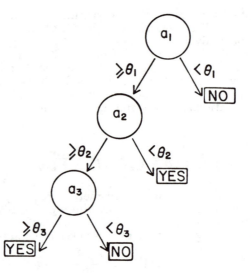

Figure 1: *The given classification tree, where θ_1, θ_2 and θ_3 are chosen to be all zeros in the numerical example.*

4 AN EXAMPLE

To illustrate our method, we construct an example which is itself a decision tree. Assuming there are three hidden variables a_1, a_2, a_3, a pattern is given by a ten-dimensional vector $I_1, I_2, \ldots, I_{10}$, constructed from the three hidden variables as follows

$$
\begin{aligned}
I_1 &= a_1 + a_3 & I_6 &= 2a_3 \\
I_2 &= 2a_1 - a_2 & I_7 &= a_3 - a_1 \\
I_3 &= a_3 - 2a_2 & I_8 &= 2a_1 + 3a_3 \\
I_4 &= a_1 + 2a_2 + 3a_3 & I_9 &= 4a_3 - 3a_1 \\
I_5 &= 5a_1 - 4a_4 & I_{10} &= 2a_1 + 2a_2 + 2a_3.
\end{aligned}
$$

A given pattern is classified as either 1 (yes) or 0 (no) according to the corresponding values of the hidden variables a_1, a_2, a_3. The actual decision is derived from the decision tree in *Fig*.1.

In order to learn this classification tree, we construct a training set of 5000 patterns generated by randomly chosen values a_1, a_2, a_3 in the interval -1 to +1. We randomly choose the initial weights for each node, and terminate

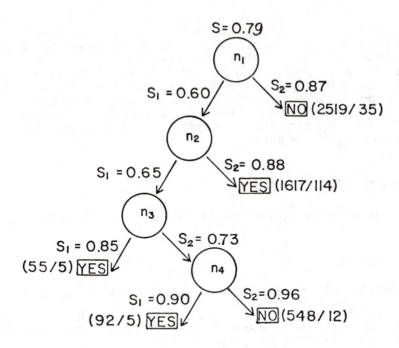

Figure 2: The learned classification tree structure

a branch as a leaf whenever the branch entropy is greater than 0.80. The entropy is started at $S = 0.65$, and terminated at its maximum value $S = 0.79$ for the first node. The two branches of this node have the entropy fuction valued at $S_1 = 0.61, S_2 = 0.87$ respectively. This corrosponds to 2446 patterns channeled to the first branch and 2554 to the second. Since $S_2 > 0.80$ we terminate the second branch. Among 2554 patterns channeled to the second branch there are 2519 patterns with true classification as *no* and 35 *yes* which are considered as errors. After completing the whole training process, there are totally four nodes automatically introduced. The final result is shown in a tree structure in $Fig.2$.

The total errors classified by the learned tree are 3.4 % of the 5000 trainig patterns. After trainig we have tested the result using 10000 novel patterns, the error among which is 3.2 %.

5   SUMMARY

We propose here a new scheme to construct neural network that can automatically learn the attributes sequentially to facilitate the classification of patterns according to the ranking importance of each attribute. This scheme uses information as a measure of the performance of each unit. It is self-organized into a presumably *optimal* structure for a specific task. The sequential learning procedure focuses attention of the network to the most important attribute first and then branches out' to the less important attributes. This strategy of searching for attributes would alleviate the scale up problem forced by the overall parallel back-propagation scheme. It also avoids the problem of resource allocation encountered in the high-order net and the multi-layered net. In the example we showed the performance of the new method is satisfactory. We expect much better performance in problems that demand large size of units.

6   acknowledgement

This work is partially supported by AFOSR under the grant 87-0388.

References

[1] M. Minsky and S. Papert, *Perceptron*, MIT Press Cambridge, Ma(1969).

[2] Y.C. Lee, G. Doolen, H.H. Chen, G.Z. Sun, T. Maxwell, H.Y. Lee and C.L. Giles, *Machine Learning Using A High Order Connection Netweork*, Physica **D22**,776-306 (1986).

[3] H.H. Chen, Y.C. Lee, G.Z. Sun, H.Y. Lee, T. Maxwell and C.L. Giles, *High Order Connection Model For Associate Memory*, AIP Proceedings Vol.151,p.86, Ed. John Denker (1986).

[4] T. Maxwell, C.L. Giles, Y.C. Lee and H.H. Chen, *Nonlinear Dynamics of Artificial Neural System*, AIP Proceedings Vol.151,p.299, Ed. John Denker(1986).

[5] D. Rummenlhart and J. McClelland, *Parallel Distributive Processing*, MIT Press(1986).

[6] L. Breiman, J. Friedman, R. Olshen, C.J. Stone, *Classification and Regression Trees*,Wadsworth Belmont, California(1984).

[7] J.R. Quinlan, *Machine Learning*, Vol.1 No.1(1986).

SELF-ORGANIZATION OF ASSOCIATIVE DATABASE
AND ITS APPLICATIONS

Hisashi Suzuki and Suguru Arimoto
Osaka University, Toyonaka, Osaka 560, Japan

ABSTRACT

An efficient method of self-organizing associative databases is proposed together with applications to robot eyesight systems. The proposed databases can associate any input with some output. In the first half part of discussion, an algorithm of self-organization is proposed. From an aspect of hardware, it produces a new style of neural network. In the latter half part, an applicability to handwritten letter recognition and that to an autonomous mobile robot system are demonstrated.

INTRODUCTION

Let a mapping $f : X \rightarrow Y$ be given. Here, X is a finite or infinite set, and Y is another finite or infinite set. A learning machine observes any set of pairs (x, y) sampled randomly from $X \times Y$. ($X \times Y$ means the Cartesian product of X and Y.) And, it computes some estimate $\hat{f} : X \rightarrow Y$ of f to make small, the estimation error in some measure.

Usually we say that: the faster the decrease of estimation error with increase of the number of samples, the better the learning machine. However, such expression on performance is incomplete. Since, it lacks consideration on the candidates of f of $\hat{f}$ assumed preliminarily. Then, how should we find out good learning machines? To clarify this conception, let us discuss for a while on some types of learning machines. And, let us advance the understanding of the self-organization of associative database.

· Parameter Type

An ordinary type of learning machine assumes an equation relating x's and y's with parameters being indefinite, namely, a structure of f. It is equivalent to define implicitly a set $\hat{F}$ of candidates of $\hat{f}$. ($\hat{F}$ is some subset of mappings from X to Y.) And, it computes values of the parameters based on the observed samples. We call such type a parameter type.

For a learning machine defined well, if $\hat{F} \ni f$, $\hat{f}$ approaches f as the number of samples increases. In the alternative case, however, some estimation error remains eternally. Thus, a problem of designing a learning machine returns to find out a proper structure of f in this sense.

On the other hand, the assumed structure of f is demanded to be as compact as possible to achieve a fast learning. In other words, the number of parameters should be small. Since, if the parameters are few, some $\hat{f}$ can be uniquely determined even though the observed samples are few. However, this demand of being proper contradicts to that of being compact. Consequently, in the parameter type, the better the compactness of the assumed structure that is proper, the better the learning machine. This is the most elementary conception when we design learning machines.

· Universality and Ordinary Neural Networks

Now suppose that a sufficient knowledge on f is given though f itself is unknown. In this case, it is comparatively easy to find out proper and compact structures of f. In the alternative case, however, it is sometimes difficult. A possible solution is to give up the compactness and assume an almighty structure that can cover various f's. A combination of some orthogonal bases of the infinite dimension is such a structure. Neural networks[1,2] are its approximations obtained by truncating finitely the dimension for implementation.

A main topic in designing neural networks is to establish such desirable structures of f. This work includes developing practical procedures that compute values of coefficients from the observed samples. Such discussions are flourishing since 1980 while many efficient methods have been proposed. Recently, even hardware units computing coefficients in parallel for speed-up are sold, e.g., ANZA, Mark III, Odyssey and Σ-1.

Nevertheless, in neural networks, there always exists a danger of some error remaining eternally in estimating f. Precisely speaking, suppose that a combination of the bases of a finite number can define a structure of f essentially. In other words, suppose that $\hat{F} \ni f$, or f is located near $\hat{F}$. In such case, the estimation error is none or negligible. However, if f is distant from $\hat{F}$, the estimation error never becomes negligible. Indeed, many researches report that the following situation appears when f is too complex. Once the estimation error converges to some value (> 0) as the number of samples increases, it decreases hardly even though the dimension is heighten. This property sometimes is a considerable defect of neural networks.

· Recursive Type

The recursive type is founded on another methodology of learning that should be as follows. At the initial stage of no sample, the set $\hat{F}_0$ (instead of notation $\hat{F}$) of candidates of $\hat{f}$ equals to the set of all mappings from X to Y. After observing the first sample $(x_1, y_1) \in X \times Y$, $\hat{F}_0$ is reduced to $\hat{F}_1$ so that $\hat{f}(x_1) = y_1$ for any $\hat{f} \in \hat{F}$. After observing the second sample $(x_2, y_2) \in X \times Y$, $\hat{F}_1$ is further reduced to $\hat{F}_2$ so that $\hat{f}(x_1) = y_1$ and $\hat{f}(x_2) = y_2$ for any $\hat{f} \in \hat{F}$. Thus, the candidate set $\hat{F}$ becomes gradually small as observation of samples proceeds. The $\hat{f}$ after observing i-samples, which we write $\hat{f}_i$, is one of the most likelihood estimation of f selected in $\hat{F}_i$. Hence, contrarily to the parameter type, the recursive type guarantees surely that $\hat{f}$ approaches to f as the number of samples increases.

The recursive type, if observes a sample (x_i, y_i), rewrites values $\hat{f}_{i-1}(\tilde{x})$'s to $\hat{f}_i(\tilde{x})$'s for some $\tilde{x}$'s correlated to the sample. Hence, this type has an architecture composed of a rule for rewriting and a free memory space. Such architecture forms naturally a kind of database that builds up management systems of data in a self-organizing way. However, this database differs from ordinary ones in the following sense. It does not only record the samples already observed, but computes some estimation of $f(x)$ for any $x \in X$. We call such database an associative database.

The first subject in constructing associative databases is how we establish the rule for rewriting. For this purpose, we adapt a measure called the dissimilarity. Here, a dissimilarity means a mapping $d : X \times X \to \{\text{reals} > 0\}$ such that for any $(x, \tilde{x}) \in X \times X$, $d(x, \tilde{x}) > 0$ whenever $f(x) \neq f(\tilde{x})$. However, it is not necessarily defined with a single formula. It is definable with, for example, a collection of rules written in forms of "if $\cdots$ then $\cdots$."

The dissimilarity d defines a structure of f locally in $X \times Y$. Hence, even though the knowledge on f is imperfect, we can reflect it on d in some heuristic way. Hence, contrarily to neural networks, it is possible to accelerate the speed of learning by establishing d well. Especially, we can easily find out simple d's for those f's which process analogically information like a human. (See the applications in this paper.) And, for such f's, the recursive type shows strongly its effectiveness.

We denote a sequence of observed samples by $(x_1, y_1), (x_2, y_2), \cdots$. One of the simplest constructions of associative databases $\hat{f}_i$ after observing i-samples $(i = 1, 2, \cdots)$ is as follows.

Algorithm 1. At the initial stage, let S_0 be the empty set. For every $i = 1, 2, \cdots$, let $\hat{f}_{i-1}(x)$ for any $x \in X$ equal some y^* such that $(x^*, y^*) \in S_{i-1}$ and

$$d(x, x^*) = \min_{(\tilde{x}, \tilde{y}) \in S_{i-1}} d(x, \tilde{x}) . \tag{1}$$

Furthermore, add (x_i, y_i) to S_{i-1} to produce S_i, i.e., $S_i = S_{i-1} \cup \{(x_i, y_i)\}$.

Another version improved to economize the memory is as follows.

Algorithm 2. At the initial stage, let S_0 be composed of an arbitrary element in $X \times Y$. For every $i = 1, 2, \cdots$, let $\hat{f}_{i-1}(x)$ for any $x \in X$ equal some y^* such that $(x^*, y^*) \in S_{i-1}$ and

$$d(x, x^*) = \min_{(\tilde{x}, \tilde{y}) \in S_{i-1}} d(x, \tilde{x}) .$$

Furthermore, if $\hat{f}_{i-1}(x_i) \neq y_i$ then let $S_i = S_{i-1}$, or add (x_i, y_i) to S_{i-1} to produce S_i, i.e., $S_i = S_{i-1} \cup \{(x_i, y_i)\}$.

In either construction, $\hat{f}_i$ approaches to f as i increases. However, the computation time grows proportionally to the size of S_i. The second subject in constructing associative databases is what addressing rule we should employ to economize the computation time. In the subsequent chapters, a construction of associative database for this purpose is proposed. It manages data in a form of binary tree.

SELF-ORGANIZATION OF ASSOCIATIVE DATABASE

Given a sample sequence $(x_1, y_1), (x_2, y_2), \cdots$, the algorithm for constructing associative database is as follows.

Algorithm 3.

Step 1(Initialization): Let $(x[\text{root}], y[\text{root}]) = (x_1, y_1)$. Here, $x[\cdot]$ and $y[\cdot]$ are variables assigned for respective nodes to memorize data. Furthermore, let $t = 1$.

Step 2: Increase t by 1, and put x_t in. After reset a pointer n to the root, repeat the following until n arrives at some terminal node, i.e., leaf.

Notations $\acute{n}$ and $\grave{n}$ mean the descendant nodes of n. If $d(x_t, x[\acute{n}]) \leq d(x_t, x[\grave{n}])$, let $n = \acute{n}$. Otherwise, let $n = \grave{n}$.

Step 3: Display $y[n]$ as the related information. Next, put y_t in. If $y[n] = y_t$, back to step 2. Otherwise, first establish new descendant nodes $\acute{n}$ and $\grave{n}$. Secondly, let

$$(x[\acute{n}], y[\acute{n}]) = (x[n], y[n]), \tag{2}$$
$$(x[\grave{n}], y[\grave{n}]) = (x_t, y_t). \tag{3}$$

Finally, back to step 2. Here, the loop of step 2–3 can be stopped at any time and also can be continued.

Now, suppose that gate elements, namely, artificial "synapses" that play the role of branching by d are prepared. Then, we obtain a new style of neural network with gate elements being randomly connected by this algorithm.

LETTER RECOGNITION

Recently, the vertical slitting method for recognizing typographic English letters[3], the elastic matching method for recognizing handwritten discrete English letters[4], the global training and fuzzy logic search method for recognizing Chinese characters written in square style[5], etc. are published. The self-organization of associative database realizes the recognition of handwritten continuous English letters.

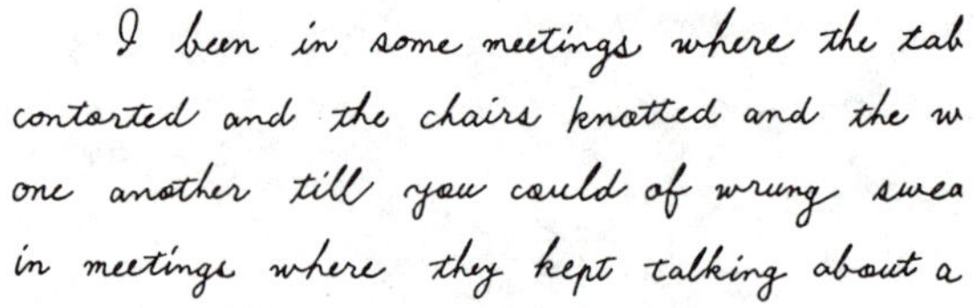

I been in some meetings where the tab
contorted and the chairs knotted and the w
one another till you could of wrung swea
in meetings where they kept talking about a

Fig. 1. Source document.

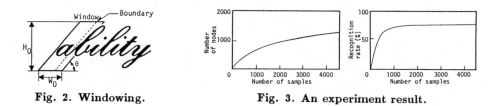

Fig. 2. Windowing.   **Fig. 3. An experiment result.**

An image scanner takes a document image (**Fig. 1**). The letter recognizer uses a parallelogram window that at least can cover the maximal letter (**Fig. 2**), and processes the sequence of letters while shifting the window. That is, the recognizer scans a word in a slant direction. And, it places the window so that its left vicinity may be on the first black point detected. Then, the window catches a letter and some part of the succeeding letter. If recognition of the head letter is performed, its end position, namely, the boundary line between two letters becomes known. Hence, by starting the scanning from this boundary and repeating the above operations, the recognizer accomplishes recursively the task. Thus the major problem comes to identifying the head letter in the window.

Considering it, we define the following.

- Regard window images as x's, and define X accordingly.

- For a $(x, \tilde{x}) \in X \times X$, denote by $\tilde{B}$ a black point in the left area from the boundary on window image $\tilde{x}$. Project each $\tilde{B}$ onto window image x. Then, measure the Euclidean distance δ between $\tilde{B}$ and a black point B on x being the closest to $\tilde{B}$. Let $d(x, \tilde{x})$ be the summation of δ's for all black points $\tilde{B}$'s on $\tilde{x}$ divided by the number of $\tilde{B}$'s.

- Regard couples of the "reading" and the position of boundary as y's, and define Y accordingly.

An operator teaches the recognizer in interaction the relation between window image and reading&boundary with algorithm 3. Precisely, if the recalled reading is incorrect, the operator teaches a correct reading via the console. Moreover, if the boundary position is incorrect, he teaches a correct position via the mouse.

Fig. 1 shows partially a document image used in this experiment. **Fig. 3** shows the change of the number of nodes and that of the recognition rate defined as the relative frequency of correct answers in the past 1000 trials. Specifications of the window are height = 20dot, width = 10dot, and slant angular = 68deg. In this example, the levels of tree were distributed in 6–19 at time 4000 and the recognition rate converged to about 74%. Experimentally, the recognition rate converges to about 60–85% in most cases, and to 95% at a rare case. However, it does not attain 100% since, e.g., "c" and "e" are not distinguishable because of excessive fluctuation in writing. If the consistency of the x, y-relation is not assured like this, the number of nodes increases endlessly (cf. **Fig. 3**). Hence, it is clever to stop the learning when the recognition rate attains some upper limit. To improve further the recognition rate, we must consider the spelling of words. It is one of future subjects.

OBSTACLE AVOIDING MOVEMENT

Various systems of camera type autonomous mobile robot are reported flourishingly[6-10]. The system made up by the authors (**Fig. 4**) also belongs to this category. Now, in mathematical methodologies, we solve usually the problem of obstacle avoiding movement as a cost minimization problem under some cost criterion established artificially. Contrarily, the self-organization of associative database reproduces faithfully the cost criterion of an operator. Therefore, motion of the robot after learning becomes very natural.

Now, the length, width and height of the robot are all about 0.7m, and the weight is about 30kg. The visual angle of camera is about 55deg. The robot has the following three factors of motion. It turns less than ±30deg, advances less than 1m, and controls speed less than 3km/h. The experiment was done on the passageway of width 2.5m inside a building which the authors' laboratories exist in (**Fig. 5**). Because of an experimental intention, we arrange boxes, smoking stands, gas cylinders, stools, handcarts, etc. on the passage way at random. We let the robot take an image through the camera, recall a similar image, and trace the route preliminarily recorded on it. For this purpose, we define the following.

- Let the camera face 28deg downward to take an image, and process it through a low pass filter. Scanning vertically the filtered image from the bottom to the top, search the first point C where the luminance changes excessively. Then, substitute all points from the bottom to C for white, and all points from C to the top for black (**Fig. 6**). (If no obstacle exists just in front of the robot, the white area shows the "free" area where the robot can move around.) Regard binary 32 × 32dot images processed thus as x's, and define X accordingly.

- For every $(x, \tilde{x}) \in X \times X$, let $d(x, \tilde{x})$ be the number of black points on the exclusive-or image between x and $\tilde{x}$.

- Regard as y's the images obtained by drawing routes on images x's, and define Y accordingly.

The robot superimposes, on the current camera image x, the route recalled for x, and inquires the operator instructions. The operator judges subjectively whether the suggested route is appropriate or not. In the negative answer, he draws a desirable route on x with the mouse to teach a new y to the robot. This operation defines implicitly a sample sequence of (x, y) reflecting the cost criterion of the operator.

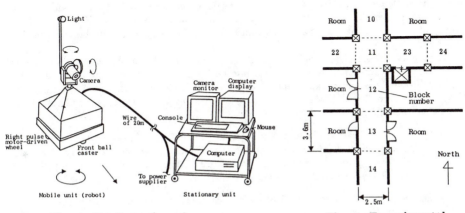

Fig. 4. Configuration of autonomous mobile robot system.

Fig. 5. Experimental environment.

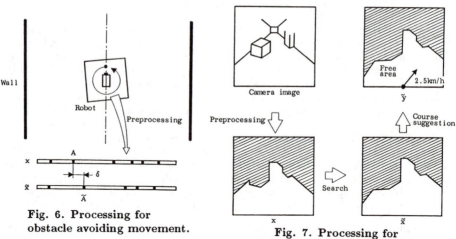

Fig. 6. Processing for
obstacle avoiding movement.

Fig. 7. Processing for
position identification.

We define the satisfaction rate by the relative frequency of acceptable suggestions of route in the past 100 trials. In a typical experiment, the change of satisfaction rate showed a similar tendency to **Fig. 3**, and it attains about 95% around time 800. Here, notice that the rest 5% does not mean directly the percentage of collision. (In practice, we prevent the collision by adopting some supplementary measure.) At time 800, the number of nodes was 145, and the levels of tree were distributed in 6–17.

The proposed method reflects delicately various characters of operator. For example, a robot trained by an operator O moves slowly with enough space against obstacles while one trained by another operator O' brushes quickly against obstacles. This fact gives us a hint on a method of printing "characters" into machines.

POSITION IDENTIFICATION

The robot can identify its position by recalling a similar landscape with the position data to a camera image. For this purpose, in principle, it suffices to regard camera images and position data as x's and y's, respectively. However, the memory capacity is finite in actual computers. Hence, we cannot but compress the camera images at a slight loss of information. Such compression is admittable as long as the precision of position identification is in an acceptable area. Thus, the major problem comes to find out some suitable compression method.

In the experimental environment (**Fig. 5**), juts are on the passageway at intervals of $3.6m$, and each section between adjacent juts has at most one door. The robot identifies roughly from a surrounding landscape which section itself places in. And, it uses temporarily a triangular surveying technique if an exact measure is necessary. To realize the former task, we define the following.

- Turn the camera to take a panorama image of 360deg. Scanning horizontally the center line, substitute the points where the luminance excessively changes for black and the other points for white (**Fig. 7**). Regard binary 360dot line images processed thus as x's, and define X accordingly.

- For every $(x, \tilde{x}) \in X \times X$, project each black point $\tilde{A}$ on $\tilde{x}$ onto x. And, measure the Euclidean distance δ between $\tilde{A}$ and a black point A on x being the closest to $\tilde{A}$. Let the summation of δ be S. Similarly, calculate $\tilde{S}$ by exchanging the roles of x and $\tilde{x}$. Denoting the numbers of A's and $\tilde{A}$'s respectively by n and $\tilde{n}$, define

$$d(x, \tilde{x}) = \frac{1}{2}\left(\frac{S}{n} + \frac{\tilde{S}}{\tilde{n}}\right). \tag{4}$$

- Regard positive integers labeled on sections as y's (cf. **Fig. 5**), and define Y accordingly.

In the learning mode, the robot checks exactly its position with a counter that is reset periodically by the operator. The robot runs arbitrarily on the passageways within 18m area and learns the relation between landscapes and position data. (Position identification beyond 18m area is achieved by crossing plural databases one another.) This task is automatic excepting the periodic reset of counter, namely, it is a kind of learning without teacher.

We define the identification rate by the relative frequency of correct recalls of position data in the past 100 trials. In a typical example, it converged to about 83% around time 400. At time 400, the number of levels was 202, and the levels of tree were distributed in 5–22. Since the identification failures of 17% can be rejected by considering the trajectory, no problem arises in practical use. In order to improve the identification rate, the compression ratio of camera images must be loosened. Such possibility depends on improvement of the hardware in the future.

Fig. 8 shows an example of actual motion of the robot based on the database for obstacle avoiding movement and that for position identification. This example corresponds to a case of moving from 14 to 23 in **Fig. 5**. Here, the time interval per frame is about 40sec.

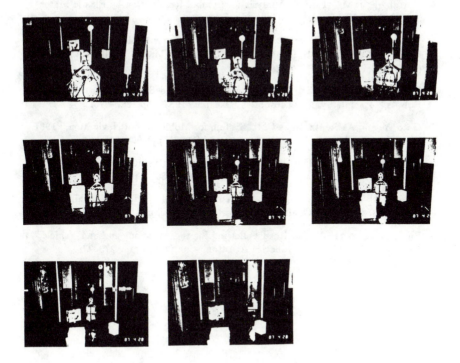

Fig. 8. Actual motion of the robot.

CONCLUSION

A method of self-organizing associative databases was proposed with the application to robot eyesight systems. The machine decomposes a global structure unknown into a set of local structures known and learns universally any input-output response. This framework of problem implies a wide application area other than the examples shown in this paper.

A defect of the algorithm 3 of self-organization is that the tree is balanced well only for a subclass of structures of f. A subject imposed us is to widen the class. A probable solution is to abolish the addressing rule depending directly on values of d and, instead, to establish another rule depending on the distribution function of values of d. It is now under investigation.

REFERENCES

1. Hopfield, J. J. and D. W. Tank, "Computing with Neural Circuit: A Model," Science **233** (1986), pp. 625–633.

2. Rumelhart, D. E. et al., "Learning Representations by Back-Propagating Errors," Nature **323** (1986), pp. 533–536.

3. Hull, J. J., "Hypothesis Generation in a Computational Model for Visual Word Recognition," IEEE Expert, **Fall** (1986), pp. 63–70.

4. Kurtzberg, J. M., "Feature Analysis for Symbol Recognition by Elastic Matching," IBM J. Res. Develop. **31-1** (1987), pp. 91–95.

5. Wang, Q. R. and C. Y. Suen, "Large Tree Classifier with Heuristic Search and Global Training," IEEE Trans. Pattern. Anal. & Mach. Intell. **PAMI 9-1** (1987) pp. 91–102.

6. Brooks, R. A. et al, "Self Calibration of Motion and Stereo Vision for Mobile Robots," 4th Int. Symp. of Robotics Research (1987), pp. 267–276.

7. Goto, Y. and A. Stentz, "The CMU System for Mobile Robot Navigation," 1987 IEEE Int. Conf. on Robotics & Automation (1987), pp. 99–105.

8. Madarasz, R. et al., "The Design of an Autonomous Vehicle for the Disabled," IEEE Jour. of Robotics & Automation **RA 2-3** (1986), pp. 117–125.

9. Triendl, E. and D. J. Kriegman, "Stereo Vision and Navigation within Buildings," 1987 IEEE Int. Conf. on Robotics & Automation (1987), pp. 1725–1730.

10. Turk, M. A. et al., "Video Road-Following for the Autonomous Land Vehicle," 1987 IEEE Int. Conf. on Robotics & Automation (1987), pp. 273–279.

A NEURAL–NETWORK SOLUTION TO THE CONCENTRATOR ASSIGNMENT PROBLEM

Gene A. Tagliarini
Edward W. Page

Department of Computer Science, Clemson University, Clemson, SC 29634-1906

ABSTRACT

Networks of simple analog processors having neuron–like properties have been employed to compute good solutions to a variety of optimization problems. This paper presents a neural–net solution to a resource allocation problem that arises in providing local access to the backbone of a wide–area communication network. The problem is described in terms of an energy function that can be mapped onto an analog computational network. Simulation results characterizing the performance of the neural computation are also presented.

INTRODUCTION

This paper presents a neural–network solution to a resource allocation problem that arises in providing access to the backbone of a communication network.[1] In the field of operations research, this problem was first known as the warehouse location problem and heuristics for finding feasible, suboptimal solutions have been developed previously.[2,3] More recently it has been known as the multifacility location problem[4] and as the concentrator assignment problem.[1]

THE HOPFIELD NEURAL NETWORK MODEL

The general structure of the Hopfield neural network model[5,6,7] is illustrated in Fig. 1. Neurons are modeled as amplifiers that have a sigmoid input/output curve as shown in Fig. 2. Synapses are modeled by permitting the output of any neuron to be connected to the input of any other neuron. The strength of the synapse is modeled by a resistive connection between the output of a neuron and the input to another. The amplifiers provide integrative analog summation of the currents that result from the connections to other neurons as well as connection to external inputs. To model both excitatory and inhibitory synaptic links, each amplifier provides both a normal output V and an inverted output $\overline{V}$. The normal outputs range between 0 and 1 while the inverting amplifier produces corresponding values between 0 and –1. The synaptic link between the output of one amplifier and the input of another is defined by a conductance T_{ij} which connects one of the outputs of amplifier j to the input of amplifier i. In the Hopfield model, the connection between neurons i and j is made with a resistor having a value $R_{ij} = 1/T_{ij}$. To provide an excitatory synaptic connection (positive T_{ij}), the resistor is connected to the normal output of

This research was supported by the U.S. Army Strategic Defense Command.

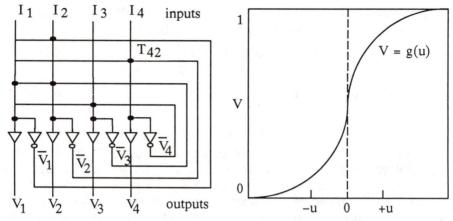

Fig. 1. Schematic for a simplified Hopfield network with four neurons.

Fig. 2. Amplifier input/output relationship

amplifier j. To provide an inhibitory connection (negative T_{ij}), the resistor is connected to the inverted output of amplifier j. The connections among the neurons are defined by a matrix T consisting of the conductances T_{ij}. Hopfield has shown that a symmetric T matrix ($T_{ij} = T_{ji}$) whose diagonal entries are all zeros, causes convergence to a stable state in which the output of each amplifier is either 0 or 1. Additionally, when the amplifiers are operated in the high–gain mode, the stable states of a network of n neurons correspond to the local minima of the quantity

$$E = (-1/2) \sum_{i=1}^{n} \sum_{j=1}^{n} T_{ij} V_i V_j - \sum_{i=1}^{n} V_i I_i \qquad (1)$$

where V_i is the output of the ith neuron and I_i is the externally supplied input to the ith neuron. Hopfield refers to E as the computational energy of the system.

THE CONCENTRATOR ASSIGNMENT PROBLEM

Consider a collection of n sites that are to be connected to m concentrators as illustrated in Fig. 3(a). The sites are indicated by the shaded circles and the concentrators are indicated by squares. The problem is to find an assignment of sites to concentrators that minimizes the total cost of the assignment and does not exceed the capacity of any concentrator. The constraints that must be met can be summarized as follows:

a) Each site i (i = 1, 2,..., n) is connected to exactly one concentrator; and

b) Each concentrator j (j = 1, 2,..., m) is connected to no more than k_j sites (where k_j is the capacity of concentrator j).

Figure 3(b) illustrates a possible solution to the problem represented in Fig. 3(a).

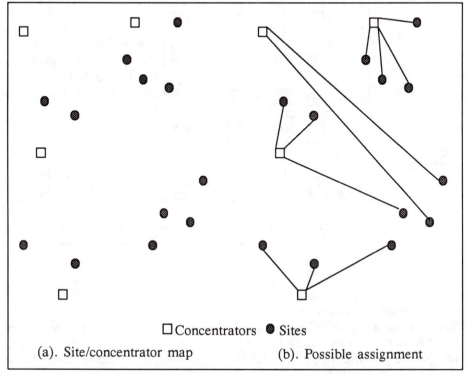

□ Concentrators   ● Sites

(a). Site/concentrator map   (b). Possible assignment

Fig. 3. Example concentrator assignment problem

If the cost of assigning site i to concentrator j is c_{ij} , then the total cost of a particular assignment is

$$\text{total cost} = \sum_{i=1}^{n} \sum_{j=1}^{m} x_{ij} \, c_{ij} \qquad (2)$$

where $x_{ij} = 1$ only if we actually decide to assign site i to concentrator j and is 0 otherwise. There are m^n possible assignments of sites to concentrators that satisfy constraint a). Exhaustive search techniques are therefore impractical except for relatively small values of m and n.

THE NEURAL NETWORK SOLUTION

This problem is amenable to solution using the Hopfield neural network model. The Hopfield model is used to represent a matrix of possible assignments of sites to concentrators as illustrated in Fig. 4. Each square corresponds

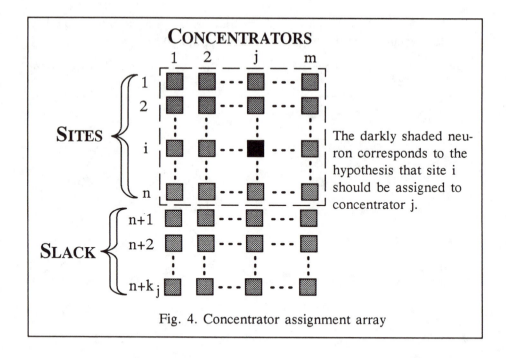

CONCENTRATORS

The darkly shaded neuron corresponds to the hypothesis that site i should be assigned to concentrator j.

Fig. 4. Concentrator assignment array

to a neuron and a neuron in row i and column j of the upper n rows of the array represents the hypothesis that site i should be connected to concentrator j. If the neuron in row i and column j is on, then site i should be assigned to concentrator j; if it is off, site i should not be assigned to concentrator j.

The neurons in the lower sub–array, indicated as "SLACK", are used to implement individual concentrator capacity constraints. The number of slack neurons in a column should equal the capacity (expressed as the number sites which can be accommodated) of the corresponding concentrator. While it is not necessary to assume that the concentrators have equal capacities, it was assumed here that they did and that their cumulative capacity is greater than or equal to the number of sites.

To enable the neurons in the network illustrated above to compute solutions to the concentrator problem, the network must realize an energy function in which the lowest energy states correspond to the least cost assignments. The energy function must therefore favor states which satisfy constraints a) and b) above as well as states that correspond to a minimum cost assignment. The energy function is implemented in terms of connection strengths between neurons. The following section details the construction of an appropriate energy function.

THE ENERGY FUNCTION

Consider the following energy equation:

$$E = A \sum_{i=1}^{n} \left(\sum_{j=1}^{m} V_{ij} - 1 \right)^2 + B \sum_{j=1}^{m} \left(\sum_{i=1}^{n+k_j} V_{ij} - k_j \right)^2 \qquad (3)$$

$$+ C \sum_{j=1}^{m} \sum_{i=1}^{n+k_j} V_{ij} \left(1 - V_{ij} \right)$$

where V_{ij} is the output of the amplifier in row i and column j of the neuron matrix, m and n are the number of concentrators and the number of sites respectively, and k_j is the capacity of concentrator j.

The first term will be minimum when the sum of the outputs in each row of neurons associated with a site equals one. Notice that this term influences only those rows of neurons which correspond to sites; no term is used to coerce the rows of slack neurons into a particular state.

The second term of the equation will be minimum when the sum of the outputs in each column equals the capacity k_j of the corresponding concentrator. The presence of the k_j slack neurons in each column allows this term to enforce the concentrator capacity restrictions. The effect of this term upon the upper sub-array of neurons (those which correspond to site assignments) is that no more than k_j sites will be assigned to concentrator j. The number of neurons to be turned on in column j is k_j; consequently, the number of neurons turned on in column j of the assignment sub-array will be less than or equal to k_j .

The third term causes the energy function to favor the "zero" and "one" states of the individual neurons by being minimum when all neurons are in one or the other of these states. This term influences all neurons in the network.

In summary, the first term enforces constraint a) and the second term enforces constraint b) above. The third term guarantees that a choice is actually made; it assures that each neuron in the matrix will assume a final state near zero or one corresponding to the x_{ij} term of the cost equation (Eq. 2).

After some algebraic re-arrangement, Eq. 3 can be written in the form of Eq. 1 where

$$T_{ij,kl} = \begin{cases} A * \delta(i,k) * (1-\delta(j,l)) + B * \delta(j,l) * (1-\delta(i,k)), & \text{if } i \leq n \text{ and } k \leq n \\ C * \delta(j,l) * (1-\delta(i,k)), & \text{if } i > n \text{ or } k > n. \end{cases} \qquad (4)$$

Here quadruple subscripts are used for the entries in the matrix T. Each entry indicates the strength of the connection between the neuron in row i and column j and the neuron in row k and column l of the neuron matrix. The function delta is given by

$$\delta(i , j) = \begin{cases} 1, \text{ if } i = j \\ 0, \text{ otherwise.} \end{cases} \tag{5}$$

The A and B terms specify inhibitions within a row or a column of the upper sub-array and the C term provides the column inhibitions required for the neurons in the sub-array of slack neurons.

Equation 3 specifies the form of a solution but it does not include a term that will cause the network to favor minimum cost assignments. To complete the formulation, the following term is added to each $T_{ij,kl}$:

$$\frac{D * \delta(j , l) * (1 - \delta(i , k))}{(\text{cost}[i , j] + \text{cost}[k , l])}$$

where cost[i , j] is the cost of assigning site i to concentrator j. The effect of this term is to reduce the inhibitions among the neurons that correspond to low cost assignments. The sum of the costs of assigning both site i to concentrator j and site k to concentrator l was used in order to maintain the symmetry of T.

The external input currents were derived from the energy equation (Eq.3) and are given by

$$I_{ij} = \begin{cases} 2 * k_j, \text{ if } i \leq n \\ 2 * k_j - 1, \text{ otherwise.} \end{cases} \tag{6}$$

This exemplifies a technique for combining external input currents which arise from combinations of certain basic types of constraints.

AN EXAMPLE

The neural network solution for a concentrator assignment problem consisting of twelve sites and five concentrators was simulated. All sites and concentrators were located within the unit square on a randomly generated map.

For this problem, it was assumed that no more than three sites could be assigned to a concentrator. The assignment cost matrix and a typical assignment resulting from the simulation are shown in Fig. 5. It is interesting to notice that the network proposed an assignment which made no use of concentrator 2.

Because the capacity of each concentrator k_j was assumed to be three sites, the external input current for each neuron in the upper sub-array was

$$I_{ij} = 6$$

while in the sub-array of slack neurons it was

$$I_{ij} = 5.$$

The other parameter values used in the simulation were

$$A = B = C = -2$$

and

$$D = 0.1 .$$

781

CONCENTRATORS

SITES	1	2	3	4	5
A	.47	.28	.55	(.12)	.46
B	.72	.75	(.33)	.40	.63
C	.95	.71	(.31)	.39	.92
D	.88	.78	(.06)	.38	.82
E	.31	.62	.81	.56	(.21)
F	.25	.51	.76	.46	(.16)
G	.17	.39	.77	.41	(.11)
H	(.66)	.81	.54	.52	.56
I	.60	.67	.44	(.36)	.51
J	(.58)	.84	.76	.66	.48
K	.42	.33	.55	(.15)	.38
L	(.19)	.60	1.05	.71	.18

Fig. 5. The concentrator assignment cost matrix with choices circled.

Since this choice of parameters results in a T matrix that is symmetric and whose diagonal entries are all zeros, the network will converge to the minima of Eq. 3. Furthermore, inclusion of the term which is weighted by the parameter D causes the network to favor minimum cost assignments.

To evaluate the performance of the simulated network, an exhaustive search of all solutions to the problem was conducted using a backtracking algorithm. A frequency distribution of the solution costs associated with the assignments generated by the exhaustive search is shown in Fig. 6. For comparison, a histogram of the results of one hundred consecutive runs of the neural–net simulation is shown in Fig. 7. Although the neural–net simulation did not find a global minimum, ninety–two of the one hundred assignments which it did find were among the best 0.01% of all solutions and the remaining eight were among the best 0.3%.

CONCLUSION

Neural networks can be used to find good, though not necessarily optimal, solutions to combinatorial optimization problems like the concentrator

782

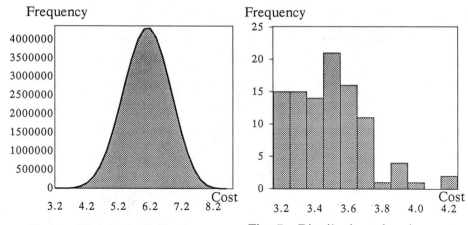

Fig. 6. Distribution of assignment costs resulting from an exhaustive search of all possible solutions.

Fig. 7. Distribution of assignment costs resulting from 100 consecutive executions of the neural net simulation.

assignment problem. In order to use a neural network to solve such problems, it is necessary to be able to represent a solution to the problem as a state of the network. Here the concentrator assignment problem was successfully mapped onto a Hopfield network by associating each neuron with the hypothesis that a given site should be assigned to a particular concentrator. An energy function was constructed to determine the connections that were needed and the resulting neural network was simulated.

While the neural network solution to the concentrator assignment problem did not find a globally minimum cost assignment, it very effectively rejected poor solutions. The network was even able to suggest assignments which would allow concentrators to be removed from the communication network.

REFERENCES

1. A. S. Tanenbaum, Computer Networks (Prentice–Hall: Englewood Cliffs, New Jersey, 1981), p. 83.

2. E. Feldman, F. A. Lehner and T. L. Ray, Manag. Sci. V12, 670 (1966).

3. A. Kuehn and M. Hamburger, Manag. Sci. V9, 643 (1966).

4. T. Aykin and A. J. G. Babu, J. of the Oper. Res. Soc. V38, N3, 241 (1987).

5. J. J. Hopfield, Proc. Natl. Acad. Sci. U. S. A., V79, 2554 (1982).

6. J. J. Hopfield and D. W. Tank, Bio. Cyber. V52, 141 (1985).

7. D. W. Tank and J. J. Hopfield, IEEE Trans. on Cir. and Sys. CAS–33, N5, 533 (1986).

USING NEURAL NETWORKS TO IMPROVE COCHLEAR IMPLANT SPEECH PERCEPTION

Manoel F. Tenorio
School of Electrical Engineering
Purdue University
West Lafayette, IN 47907

ABSTRACT

An increasing number of profoundly deaf patients suffering from sensorineural deafness are using cochlear implants as prostheses. After the implant, sound can be detected through the electrical stimulation of the remaining peripheral auditory nervous system. Although great progress has been achieved in this area, no useful speech recognition has been attained with either single or multiple channel cochlear implants.

Coding evidence suggests that it is necessary for any implant which would effectively couple with the natural speech perception system to simulate the temporal dispersion and other phenomena found in the natural receptors, and currently not implemented in any cochlear implants. To this end, it is presented here a computational model using artificial neural networks (ANN) to incorporate the natural phenomena in the artificial cochlear.

The ANN model presents a series of advantages to the implementation of such systems. First, the hardware requirements, with constraints on power, size, and processing speeds, can be taken into account together with the development of the underlining software, before the actual neural structures are totally defined. Second, the ANN model, since it is an abstraction of natural neurons, carries the necessary ingredients and is a close mapping for implementing the necessary functions. Third, some of the processing, like sorting and majority functions, could be implemented more efficiently, requiring only local decisions. Fourth, the ANN model allows function modifications through parametric modification (no software recoding), which permits a variety of fine-tuning experiments, with the opinion of the patients, to be conceived. Some of those will permit the user some freedom in system modification at real-time, allowing finer and more subjective adjustments to fit differences on the condition and operation of individual's remaining peripheral auditory system.

1. INTRODUCTION

The study of the model of sensory receptors can be carried out either via trying to understand how the natural receptors process incoming signals and build a representation code, or via the construction of artificial replacements. In the second case, we are interested in to what extent those artificial counterparts have the ability to replace the natural receptors.

Several groups are now carrying out the design of artificial sensors. Artificial cochleas seem to have a number of different designs and a tradition of experiments. These make them now available for widespread use as prostheses for patients who have sensorineural deafness caused by hair cell damage.

Although surgery is required for such implants, their performance has reached a level of maturity to induce patients to seek out these devices voluntarily. Unfortunately, only partial acoustic information is obtained by severely deaf patients with cochlear prosthesis. Useful patterns for speech communication are not yet fully recognizable through auditory prostheses. This problem with artificial receptors is true for both single implants, that stimulate large sections of the cochlea with signals that cover a large portion of the spectrum [4,5], and multi channel implants, that stimulate specific regions of the cochlea with specific portions of the auditory spectrum [3,13].

In this paper, we tackle the problem of artificial cochlear implants through the used of neurocomputing tools. The receptor model used here was developed by Gerald Wasserman of the Sensory Coding Laboratory, Department of Psychological Sciences, Purdue University [20], and the implants were performed by Richard Miyamoto of the Department of Otolaryngology, Indiana University Medical School [11].

The idea is to introduce with the cochlear implant, the computation that would be performed otherwise by the natural receptors. It would therefore be possible to experimentally manipulate the properties of the implant and measure the effect of coding variations on behavior. The model was constrained to be portable, simple to implant, fast enough computationally for on-line use, and built with a flexible paradigm, which would allow for modification of the different parts of the model, without having to reconstruct it entirely. In the next section, we review parts of the receptor model, and discuss the block diagram of the implant. Section 3 covers the limitations associated with the technique, and discusses the results obtained with a single neuron and one feedback loop. Section 4 discusses the implementations of these models using feedforward neural networks, and the computational advantages for doing so.

2. COCHLEAR IMPLANTS AND THE NEURON MODEL

Although patients cannot reliably recognize randomly chosen spoken words to them (when implanted with either multichannel or single channel devices), this is not to say that no information is extracted from speech. If the vocabulary is reduced to a limited set of words, patients perform significantly better than chance, at associating the word with a member of the set.

For these types of experiments, single channel implants correspond to reported performance of 14% to 20% better than chance, with 62% performance being the highest reported. For multiple channels, performances of 95% were reported. So far no one has investigated the differences in performance between the two types of implants. Since the two implants have so many differences, it is difficult to point out the cause for the better performance in the multiple channel case.

The results of such experiments are encouraging, and point to the fact that cochlea implants need only minor improvement to be able to mediate ad-lib speech perception successfully. Sensory coding studies have suggested a solution to the implant problem, by showing that the representation code generated by the sensory system is task dependent. This evidence came from comparison of intracellular recordings taken from a single receptor of intact subjects.

This coding evidence suggests that the temporal dispersion (time integration) found in natural receptors would be a necessary part of any cochlear implant. Present cochlear implants have no dispersion at all. Figure 2 shows the block diagram for a representative cochlear implant, the House-Urban stimulator. The acoustic signal is picked up by the microphone, which sends it to an AM oscillator. This modulation step is necessary to induce an electro-magnetic coupling between the external and internal coil. The internal coil has been surgically implanted, and it is connected to a pair of wires implanted inside and outside the cochlea.

Just incorporating the temporal dispersion model to an existing device would not replicate the fact that in natural receptors, temporal dispersion appears in conjunction to other operations which are strongly non linear. There are operations like selection of a portion of the spectrum, rectification, compression, and time-dispersion to be considered.

In figure 3, a modified implant is shown, which takes into consideration some of these operations. It is depicted as a single-channel implant, although the ultimate goal is to make it multichannel. Details of the operation of this device can be found elsewhere [21]. Here, it is important to mention that the implant would also have a compression/rectification function, and it would receive a feedback from the integrator stage in order to control its gain.

3. CHARACTERISTICS AND RESULTS OF THE IMPLANTS

The above model has been implemented as an off-line process, and then the patients were exposed to a preprocessed signal which emulated the operation of the device. It is not easy to define the amount of feedback needed in the system or the amount of time dispersion. It could also be that these parameters are variable across different conditions. Another variance in the experiment is the amount of damage (and type) among different individuals. So, these parameters have to be determined clinically.

The coupling between the artificial receptor and the natural system also presents problems. If a physical connection is used, it increases the risk of infections. When inductive methods are used, the coupling is never ideal. If portability and limited power is of concern in the implementation, then the limited energy available for coupling has to be used very effectively.

The computation of the receptor model has to be made in a way to allow for fast implementation. The signal transformation is to be computed on-line. Also, the results from clinical studies should be able to be incorporated fairly easily without having to reengineer the implant.

Now we present the results of the implementation of the transfer function of figure 4. Patients, drawn from a population described elsewhere [11,12,14], were given spoken sentences processed off-line, and simultaneously presented with a couple of words related to the context. Only one of them was the correct answer. The patient had two buttons, one for each alternative; he/she was to press the button which corresponded to the correct alternative. The results are shown in the tables below.

Patient 1 (Average of the population)

Percentage of correct alternatives

Dispersion		
No disp.	67%	
0.1 msec	78%	
0.3 msec	85%	Best performance

1 msec	76%
3 msec	72%

Table I: Phoneme discrimination in a two-alternate task.

Patient 2

Percentage of correct alternatives

Dispersion		
No disp.	50%	
1.0 msec	76%	Best performance

Table II: Sentence comprehension in a two-alternative task.

There were quite a lot of variations in the performance of the different patients, some been able to perform better at different dispersion and compression amounts than the average of the population. Since one cannot control the amount of damage in the system of each patient or differences in individuals, it is hard to predict the ideal values for a given patient. Nevertheless, the improvements observed are of undeniable value in improving speech perception.

4. THE NEUROCOMPUTING MODEL

In studying the implementation of such a system for on-line use, yet flexible enough to produce a carry-on device, we look at feedforward neurocomputer models as a possible answer. First, we wanted a model that easily produced a parallel implementation, so that the model could be expanded in a multichannel environment without compromising the speed of the system. Figure 5 shows the initial idea for the implementation of the device as a Single Instruction Multiple Data (SIMD) architecture.

The implant would be similar to the one described in Figure 4, except that the transfer function of the receptor would be performed by a two layer feed forward network (Figure 6). Since there is no way of finding out the values of compression and dispersion apart from clinical trials, or even if these values do change in certain conditions, we need to create a structure that is flexible enough to modify the program structure by simple manipulation of parameters. This is also the same problem we would face when trying to expand the system to a multichannel implant. Again, neuromorphic models provided a nice paradigm in which the dataflow and the function of the program could be altered by simple parameter (weight) change.

For this first implementation we chose to use the no-contact inductive coupling method. The drawback of this method is that all the information has to be compressed in a single channel for reliable transmission and cross talk elimination.

Since the inductive coupling of the implant is critical at every cycle, the most relevant information must be picked out of the processed signal. This information is then given all the available energy, and after all the coupling loss, it should be sufficient to provide for speech pattern discrimination. In a multichannel setting, this corresponds to doing a sorting of all the n signals in the channels, selecting the m highest signals, and adding them up for modulation. In a naive single processor implementation, this could correspond to n^2 comparisons, and in a multiprocessor implementation, $\log(n)$ comparisons. Both are dependent on the number of signals to be sorted.

We needed a scheme in which the sorting time would be constant with the number of channels, and would be easily implementable in analog circuitry, in case this became a future route. Our scheme is shown in Figure 7. Each channel is connected to a threshold element, whose threshold can be varied externally. A monotonically decreasing function scans the threshold values, from the highest possible value of the output to the lowest. The output of these elements will be high corresponding to the values that are the highest first. These output are summed with a quasi-integrator with threshold set to m. This element, when high, disables the scanning functions; and it corresponds to having found the m highest signals. This sorting is independent of the number of channels.

The output of the threshold units are fed into sigma-pi units which gates the signals to be modulated. The output of these units are summed and correspond to the final processed signal (Figure 8).

The user has full control of the characteristics of this device. The number of channels can be easily altered; the number of components allowed in the modulation can be changed; the amount of gain, rectification-compression, and dispersion of each channel can also be individually controlled. The entire system is easily implementable in analog integrated circuits, once the clinical tests have determine the optimum operational characteristics.

5. CONCLUSION

We have shown that the study of sensory implants can enhance our understanding of the representation schemes used for natural sensory receptors. In particular, implants can be enhanced significantly if the effects of the sensory processing and transfer functions are incorporated in the model.

We have also shown that neuromorphic computing paradigm provides a parallel and easily modifiable framework for signal processing structures, with advantages that perhaps cannot be offered by other technology.

We will soon start the use of the first on-line portable model, using a single processor. This model will provide a testbed for more extensive clinical trials of the implant. We will then move to the parallel implementation, and from there, possibly move toward analog circuitry implementation.

Another route for the use of neuromorphic computing in this domain is possibly the use of sensory recordings from healthy animals to train self-organizing adaptive learning networks, in order to design the implant transfer functions.

REFERENCES

[1] Bilger, R.C.; Black, F.O.; Hopkinson, N.T.; and Myers, E.N., "Implanted auditory prosthesis: An evaluation of subjects presently fitted with cochlear implants," *Otolaryngology,* 1977, Vol. 84, pp. 677-682.

[2] Bilger, R.C.; Black, F.O.; Hopkinson, N.T.; Myers, E.N.; Payne, J.L.; Stenson, N.R.; Vega, A.; and Wolf, R.V., "Evaluation of subjects presently fitted with implanted auditory prostheses," *Annals of Otology, Rhinology, and Laryngology,* 1977, Vol. 86(Supp. 38), pp. 1-176.

[3]   Eddington, D.K.; Dobelle, W.H.; Brackmann, D.E.; Mladejovsky, M.G.;
        and Parkin, J., "Place and periodicity pitch by stimulation of multiple
        scala tympani electrodes in deaf volunteers," *American Society for
        Artificial Internal Organs, Transactions,* 1978, Vol. 24, pp. 1-5.

[4]   House, W.F.; Berliner, K.; Crary, W.; Graham, M.; Luckey, R.; Norton,
        N.; Selters, W.; Tobin, H.; Urban, J.; and Wexler, M., "Cochlear
        implants," *Annals of Otology, Rhinology and Laryngology,* 1976, Vol.
        85(Supp. 27), pp. 1-93.

[5]   House, W.F. and Urban, J., "Long term results of electrode implanta-
        tion and electronic stimulation of the cochlea in man," *Annals of Otol-
        ogy, Rhinology and Laryngology,* 1973, Vol. 82, No. 2, pp. 504-517.

[6]   Ifukube, T. and White, R.L., "A speech processor with lateral inhibi-
        tion for an eight channel cochlear implant and its evaluation," *IEEE
        Trans. on Biomedical Engineering,* November 1987, Vol. BME-34, No.
        11.

[7]   Kong, K.-L., and Wasserman, G.S., "Changing response measures
        alters temporal summation in the receptor and spike potentials of the
        Limulus lateral eye," *Sensory Processes,* 1978, Vol. 2, pp. 21-31. (a)

[8]   Kong, K.-L., and Wasserman, G.S., "Temporal summation in the recep-
        tor potential of the *Limulus* lateral eye: Comparison between retinula
        and eccentric cells," *Sensory Processes,* 1978, Vol. 2, pp. 9-20. (b)

[9]   Michelson, R.P., "The results of electrical stimulation of the cochlea in
        human sensory deafness," *Annals of Otology, Rhinology and Laryngol-
        ogy,* 1971, Vol. 80, pp. 914-919.

[10]   Mladejovsky, M.G.; Eddington, D.K.; Dobelle, W.H.; and Brackmann,
        D.E., "Artificial hearing for the deaf by cochlear stimulation: Pitch
        modulation and some parametric thresholds," *American Society for
        Artificial Internal Organs, Transactions,* 1974, Vol. 21, pp. 1-7.

[11]   Miyamoto, R.T.; Gossett, S.K.; Groom, G.L.; Kienle, M.L.; Pope, M.L.;
        and Shallop, J.K., "Cochlear implants: An auditory prosthesis for the
        deaf," *Journal of the Indiana State Medical Association,* 1982, Vol. 75,
        pp. 174-177.

[12]   Miyamoto, R.T.; Myres, W.A.; Pope, M.L.; and Carotta, C.A.,
        "Cochlear implants for deaf children," *Laryngoscope,* 1986, Vol. 96, pp.
        990-996.

[13]   Pialoux, P.; Chouard, C.H.; Meyer, B.; and Fugain, C., "Indications
        and results of the multichannel cochlear implant," *Acta Otolaryngology,*
        1979, Vo.. 87, pp. 185-189.

[14]   Robbins, A.M.; Osberger, M.J.; Miyamoto, R.T.; Kienle, M.J.; and
        Myres, W.A., "Speech-tracking performance in single-channel cochlear
        implant subjects," *Journal of Speech and Hearing Research,* 1985, Vol.
        28, pp. 565-578.

[15]   Russell, I.J. and Sellick, P.M., "The tuning properties of cochlear hair
        cells," in E.F. Evans and J.P. Wilson (eds.), *Psychophysics and Physiol-
        ogy of Hearing,* London: Academic Press, 1977.

[16]   Wasserman, G.S., "*Limulus* psychophysics: Temporal summation in the
        ventral eye," *Journal of Experimental Psychology: General,* 1978, Vol.
        107, pp. 276-286.

[17] Wasserman, G.S., "*Limulus* psychophysics: Increment threshold," *Perception & Psychophysics,* 1981, Vol. 29, pp. 251-260.

[18] Wasserman, G.S.; Felsten, G.; and Easland, G.S., "Receptor saturation and the psychophysical function," *Investigative Ophthalmology and Visual Science,* 1978, Vol. 17, p. 155 (Abstract).

[19] Wasserman, G.S.; Felsten, G.; and Easland, G.S., "The psychophysical function: Harmonizing Fechner and Stevens," *Science,* 1979, Vol. 204, pp. 85-87.

[20] Wasserman, G.S., "Cochlear implant codes and speech perception in profoundly deaf," *Bulletin of Psychonomic Society,* Vol. (18)3, 1987.

[21] Wasserman, G.S.; Wang-Bennett, L.T.; and Miyamoto, R.T., "Temporal dispersion in natural receptors and pattern discrimination mediated by artificial receptor," *Proc. of the Fechner Centennial Symposium,* Hans Buffart (Ed.), Elsevier/North Holland, Amsterdam, 1987.

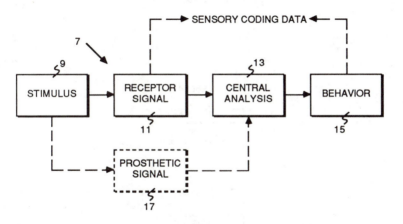

Fig. 1. Path of Natural and Prosthetic Signals.

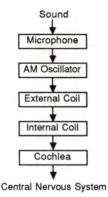

Fig. 2. The House-Urban Cochlear Implant.

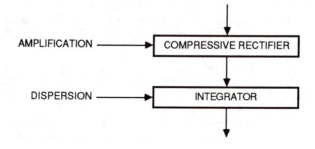

Fig. 3. Receptor Model

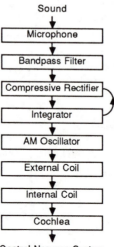

Fig. 4. Modified Implant Model.

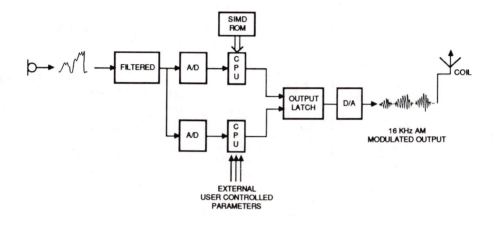

Fig. 5. Initial Concept for a SIMD Architecture.

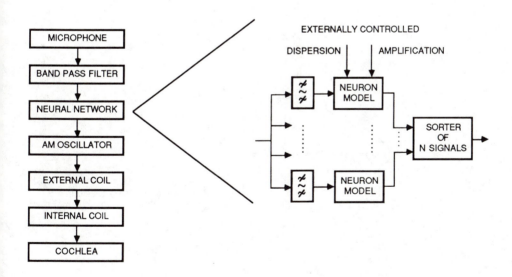

Fig. 6. Feedforward Neuron Model Implant.

SORTER OF n SIGNALS IN O(1)

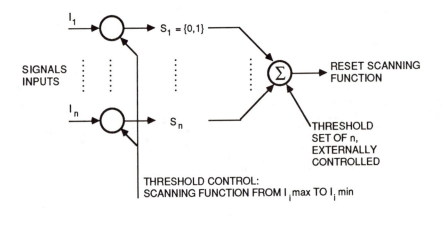

Fig. 7. Signal Sorting Circuit.

SIGNAL SELECTORS

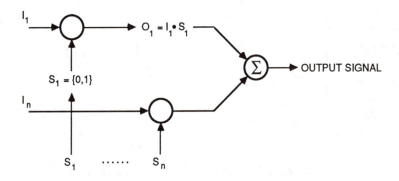

Fig. 8. Sigma-Pi Units for Signal Composition.

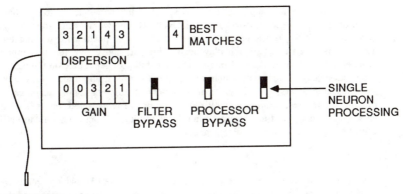

Fig. 9. Parameter Controls for Clinical Studies.

A 'Neural' Network that Learns to Play Backgammon

G. Tesauro

Center for Complex Systems Research, University of Illinois
at Urbana-Champaign, 508 S. Sixth St., Champaign, IL 61820

T. J. Sejnowski

Biophysics Dept., Johns Hopkins University, Baltimore, MD 21218

ABSTRACT

We describe a class of connectionist networks that have learned to play back-gammon at an intermediate-to-advanced level. The networks were trained by a supervised learning procedure on a large set of sample positions evaluated by a human expert. In actual match play against humans and conventional computer programs, the networks demonstrate substantial ability to generalize on the basis of expert knowledge. Our study touches on some of the most important issues in network learning theory, including the development of efficient coding schemes and training procedures, scaling, generalization, the use of real-valued inputs and outputs, and techniques for escaping from local minima. Practical applications in games and other domains are also discussed.

INTRODUCTION

A potentially quite useful testing ground for studying issues of knowledge representation and learning in networks can be found in the domain of game playing. Board games such as chess, go, backgammon, and Othello entail considerable sophistication and complexity at the advanced level, and mastery of expert concepts and strategies often takes years of intense study and practice for humans. However, the complexities in board games are embedded in relatively "clean" structured tasks with well-defined rules of play, and well-defined criteria for success and failure. This makes them amenable to automated play, and in fact most of these games have been extensively studied with conventional computer science techniques. Thus, direct comparisons of the results of network learning can be made with more conventional approaches.

In this paper, we describe an application of network learning to the game of backgammon. Backgammon is a difficult board game which appears to be well-suited to neural networks, because the way in which moves are selected is primarily on the basis of pattern-recognition or "judgemental" reasoning, as opposed to explicit "look-ahead," or tree-search computations. This is due to the probabilistic dice rolls in backgammon, which greatly expand the branching factor at each ply in the search (to over 400 in typical positions).

Our learning procedure is a supervised one[1] that requires a database of positions and moves that have been evaluated by an expert "teacher." In contrast, in an unsupervised procedure[2-4] learning would be based on the consequences of a given move (e.g., whether it led to a won or lost position), and explicit teacher instructions would not be required. However, unsupervised learning procedures thus far have been much less efficient at reaching high levels of performance than supervised learning procedures. In part, this advantage of supervised learning can be traced to the higher quantity and quality of information available from the teacher.

Studying a problem of the scale and complexity of backgammon leads one to confront important general issues in network learning. Amongst the most important are scaling and generalization. Most of the problems that have been examined with connectionist learning algorithms are relatively small scale and it is not known how well they will perform on much larger problems. Generalization is a key issue in learning to play backgammon since it is estimated that there are 10^{20} possible board positions, which is far in excess of the number of examples that can be provided during training. In this respect our study is the most severe test of generalization in any connectionist network to date.

We have also identified in this study a novel set of special techniques for training the network which were necessary to achieve good performance. A training set based on naturally occurring or random examples was not sufficient to bring the network to an advanced level of performance. Intelligent data-base design was necessary. Performance also improved when noise was added to the training procedure under some circumstances. Perhaps the most important factor in the success of the network was the method of encoding the input information. The best performance was achieved when the raw input information was encoded in a conceptually significant way, and a certain number of pre-computed features were added to the raw information. These lessons may also be useful when connectionist learning algorithms are applied to other difficult large-scale problems.

NETWORK AND DATA BASE SET-UP

Our network is trained to *select* moves (i.e. to produce a real-valued score for any given move), rather than to *generate* them. This avoids the difficulties of having to teach the network the concept of move legality. Instead, we envision our network operating in tandem with a preprocessor which would take the board position and roll as input, and produce all legal moves as output. The network would be trained to score each move, and the system would choose the move with the highest network score. Furthermore, the network is trained to produce relative scores for each move, rather than an absolute evaluation of each final position. This approach would have greater sensitivity in distinguishing between close alternatives, and corresponds more closely to the way humans actually evaluate moves.

The current data base contains a total of 3202 board positions, taken from various sources[5]. For each position there is a dice roll and a set of legal moves of that roll from that position. The moves receive commentary from a human expert in the form of a relative score in the range [-100,+100], with +100 representing the best possible move and -100 representing the worst possible move. One of us (G.T.) is a strong backgammon player, and played the role of human expert in entering these scores. Most of the moves in the data base were not scored, because it is not feasible for a human expert to comment on all possible moves. (The handling of these unscored lines of data in the training procedure will be discussed in the following section.)

An important result of our study is that in order to achieve the best performance, the data base of examples must be intelligently designed, rather than haphazardly accumulated. If one simply accumulates positions which occur in actual game play, for example, one will find that certain principles of play will appear over and over again in these positions, while other important principles may be used only rarely. This causes problems for the network, as it tends to "overlearn" the commonly used principles, and not learn at all the rarely used principles. Hence it is necessary to have both an intelligent selection mechanism to reduce the number of over-represented situations, and an intelligent design mechanism to enhance the number of examples which illustrate under-represented situations. This process is described in more detail elsewhere[5].

We use a deterministic, feed-forward network with an input layer, an output layer, and either one or two layers of hidden units, with full connectivity between adjacent layers. (We have tried a number of experiments with restricted receptive fields, and generally have not found them to be useful.) Since the desired output of the network is a single real value, only one output unit is required.

The coding of the input patterns is probably the most difficult and most important design issue. In its current configuration the input layer contains 459 input units. A location-based representation scheme is used, in which a certain number of input units are assigned to each of the 26 locations (24 basic plus White and Black bar) on the board. The input is inverted if necessary so that the network always sees a problem in which White is to play.

An example of the coding scheme used until very recently is shown in Fig. 1. This is essentially a unary encoding of the number of men at each board location, with a few exceptions as indicated in the diagram. This representation scheme worked fairly well, but had one peculiar problem in that after training, the network tended to prefer piling large numbers of men on certain points, in particular White's 5 point (the 20 point in the 1-24 numbering scheme). Fig. 2 illustrates an example of this peculiar behavior. In this position White is to play 5-1. Most humans would play 4-5,4-9 in this position; however, the network chose the move 4-9,19-20. This is actually a bad move, because it reduces White's chances of making further points in his inner board. The fault lies not with the data base used to train the network, but rather with the representation scheme used. In Fig. 1a, notice that unit 12 is turned on whenever the final position is a point, and the number of men is different from the initial position. For the 20 point in particular, this unit will develop strong excitatory weights due to cases in which the initial position is not a point (i.e., the move makes the point). The 20 point is such a valuable point to make that the excitation produced by turning unit 12 on might overwhelm the inhibition produced by the poor distribution of builders.

Figure 1-- Two schemes used to encode the raw position information in the network's input. Illustrated in each case is the encoding of two White men present before the move, and three White men present after the move. (a) An essentially unary coding of the number of men at a particular board location. Units 1-10 encode the initial position, units 11-16 encode the final position if there has been a change from the initial position. Units are turned on in the cases indicated on top of each unit, e.g., unit 1 is turned on if there are 5 or more Black men present, etc.. (b) A superior coding scheme with more units used to characterize the type of transition from initial to final position. An up arrow indicates an increase in the number of men, a down arrow indicates a decrease. Units 11-15 have conceptual interpretations: 11="clearing," 12="slotting," 13="breaking," 14="making," 15="stripping" a point.

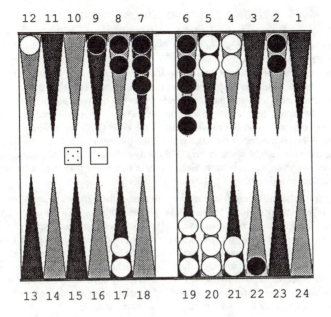

12 11 10 9 8 7   6 5 4 3 2 1

13 14 15 16 17 18   19 20 21 22 23 24

Figure 2-- A sample position illustrating a defect of the coding scheme in Fig. 1a. White is to play 5-1. With coding scheme (1a), the network prefers 4-9, 19-20. With coding scheme (1b), the network prefers 4-9, 4-5. The graphic display was generated on a Sun Microsystems workstation using the Gammontool program.

In conceptual terms, humans would say that unit 12 participates in the representation of two different concepts: the concept of *making* a point, and the concept of *changing* the number of men occupying a made point. These two concepts are unrelated, and there is no point in representing them with a common input unit. A superior representation scheme in which these concepts are separated is shown in Fig. 1b: In this representation unit 13 is turned on only for moves which make the point. Other moves which change the number of men on an already-made point do not activate unit 13, and thus do not receive any undeserved excitation. With this representation scheme the network no longer tends to pile large numbers of men on certain points, and its overall performance is significantly better.

In addition to this representation of the raw board position, we also utilize a number of input units to represent certain "pre-computed" features of the raw input. The principal goal of this study has been to investigate network learning, rather than simply to obtain high performance, and thus we have resisted the temptation of including sophisticated hand-crafted features in the input encoding. However, we have found that a few simple features are needed in practice to obtain minimal standards of competent play. With only "raw" board information, the order of the desired computation (as defined by Minsky and Papert[6]) is probably quite high, and the number of examples needed to learn such a difficult computation might be intractably large. By giving the network "hints" in the form of pre-computed features, this reduces the order of the computation, and thus might make more of the problem learnable in a tractable number of examples.

TRAINING AND TESTING PROCEDURES

To train the network, we have used the standard "back-propagation" learning algorithm[7-9] for modifying the connections in a multilayer feed-forward network. (A detailed discussion of learning parameters, etc., is provided elsewhere[5].) However, our procedure differs from the standard procedure due to the necessity of dealing with the large number of uncommented moves in the data base. One solution would be simply to avoid presenting these moves to the network. However, this would limit the variety of input patterns presented to the network in training, and certain types of inputs probably would be eliminated completely. The alternative procedure which we have adopted is to skip the uncommented moves most of the time (75% for ordinary rolls and 92% for double rolls), and the remainder of the time present the pattern to the network and generate a random teacher signal with a slight negative bias. This makes sense, because if a move has not received comment by the human expert, it is more likely to be a bad move than a good move. The random teacher signal is chosen uniformly from the interval [-65,+35].

We have used the following four measures to assess the network's performance after it has been trained: (i) performance on the training data, (ii) performance on a set of test data (1000 positions) which was not used to train the network, (iii) performance in actual game play against a conventional computer program (the program *Gammontool* of Sun Microsystems Inc.), and (iv) performance in game play against a human expert (G.T.). In the first two measures, we define the performance as the fraction of positions in which the network picks the correct move, i.e., those positions for which the move scored highest by the network agrees with the choice of the human expert. In the latter two measures, the performance is defined simply as the fraction of games won, without considering the complications of counting gammons or backgammons.

QUANTITATIVE RESULTS

A summary of our numerical results as measured by performance on the training set and against Gammontool is presented in Table 1. The best network that we have produced so far appears to defeat Gammontool nearly 60% of the time. Using this as a benchmark, we find that the most serious decrease in performance occurs by removing all pre-computed features from the input coding. This produces a network which wins at most about 41% of the time. The next most important effect is the removal of noise from the training procedure; this results in a network which wins 45% of the time. Next in importance is the presence of hidden units; a network without hidden units wins about 50% of the games against Gammontool. In contrast, effects such as varying the exact number of hidden units, the number of layers, or the size of the training set, results in only a few (1-3) percentage point decrease in the number of games won.

Also included in Table 1 is the result of an interesting experiment in which we removed our usual set of pre-computed features and substituted instead the individual terms of the Gammontool evaluation function. We found that the resulting network, after being trained on our expert training set, was able to defeat the Gammontool program by a small margin of 54 to 46 percent. The purpose of this experiment was to provide evidence of the usefulness of network learning as an adjunct to standard AI techniques for hand-crafting evaluation functions. Given a set of features to be used in an evaluation function which have been designed, for example, by interviewing a human expert, the problem remains as to how to "tune" these features, i.e., the relative weightings to associate to each feature, and at an advanced level, the context in which each feature is relevant. Little is known in general about how to approach this problem, and often the human programmer must resort to painstaking trial-and-error tuning by hand. We claim that network learning is a powerful, general-purpose, automated method of approaching this problem, and has the potential to produce a tuning which is superior to those produced by humans, given a data base of sufficiently high quality, and a suitable scheme for encoding the features. The result of our experiment provides evidence to support this claim, although it is not firmly established since we do not have highly accurate statistics, and we do not know how much human effort went into the tuning of the Gammontool evaluation function. More conclusive evidence would be provided if the experiment were repeated with a more sophisticated program such as Berliner's BKG[10], and similar results were obtained.

Network size	Training cycles	Perf. on test set	Perf. vs. Gammontool	Comments
(a) 459-24-24-1	20	.540	.59 ± .03	
(b) 459-24-1	22	.542	.57 ± .05	
(c) 459-24-1	24	.518	.58 ± .05	1600 posn. D.B.
(d) 459-12-1	10	.538	.54 ± .05	
(e) 410-24-12-1	16	.493	.54 ± .03	Gammontool features
(f) 459-1	22	.485	.50 ± .03	No hidden units
(g) 459-24-12-1	10	.499	.45 ± .03	No training noise
(h) 393-24-12-1	12	.488	.41 ± .02	No features

Table 1-- Summary of performance statistics for various networks. (a) The best network we have produced, containing two layers of hidden units, with 24 units in each layer. (b) A network with only one layer of 24 hidden units. (c) A network with 24 hidden units in a single layer, trained on a training set half the normal size. (d) A network with half the number of hidden units as in (b). (e) A network with features from the Gammontool evaluation function substituted for the normal features. (f) A network without hidden units. (g) A network trained with no noise in the training procedure. (h) A network with only a raw board description as input.

QUALITATIVE RESULTS

Analysis of the weights produced by training a network is an exceedingly difficult problem, which we have only been able to approach qualitatively. In Fig. 3 we present a diagram showing the connection strengths in a network with 651 input units and no hidden units. The figure shows the weights from each input unit to the output unit. (For purposes of illustration, we have shown a coding scheme with more units than normal to explicitly represent the transition from initial to final position.) Since the weights go directly to the output, the corresponding input units can be clearly interpreted as having either an overall excitatory or inhibitory effect on the score produced by the network.

A great deal of columnar structure is apparent in Fig. 3. This indicates that the network has learned that a particular number of men at a given location, or a particular type of transition at a given location, is either good or bad independent of the exact location on the board where it occurs. Furthermore, we can see the importance of each of the pre-computed features in the input coding. The most significant features seem to be the number of points made in the network's inner board, and the total blot exposure.

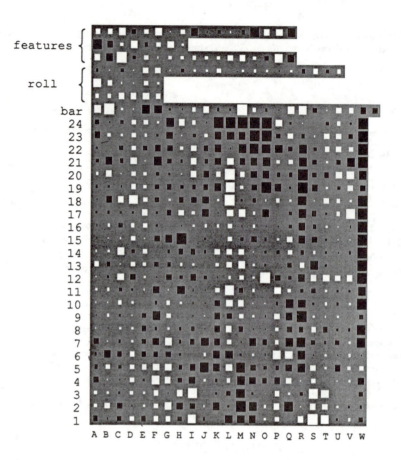

Figure 3-- A Hinton diagram for a network with 651 input units and no hidden units. Small squares indicate weights from a particular input unit to the output unit. White squares indicate positive weights, and black squares indicate negative weights. Size of square indicates magnitude of weight. First 24 rows from bottom up indicate raw board information. Letting x=number of men before the move and y=number of men after the move, the interpretations of columns are as follows: A: x<=-5; B: x=-4; C: x=-3; D: x<=-2; E: x=-1; F: x=1; G: x>=2; H: x=3; I: x=4; J: x>=5; K: x<1 & y=1; L: x<2 & y>=2; M: x<3 & y=3; N: x<4 & y=4; O: x<y & y>=5; P: x=1 & y=0; Q: x>=2 & y=0; R: x>=2 & y=1; S: x>=3 & y=2; T: x>=4 & y=3; U: x>=5 & y=4; V: x>y & y>=5; W: prob. of a White blot at this location being hit (pre-computed feature). The next row encodes number of men on White and Black bars. The next 3 rows encode roll information. Remaining rows encode various pre-computed features.

Much insight into the basis for the network's judgement of various moves has been gained by actually playing games against it. In fact, one of the most revealing tests of what the network has and has not learned came from a 20-game match played by G.T. against one of the latest generation of networks with 48 hidden units. (A detailed description of the match is given in Ref. 11.) The surprising result of this match was that the network actually won, 11 games to 9. However, a detailed analysis of the moves played by the network during the match indicates that the network was extremely lucky to have won so many games, and could not reasonably be expected to continue to do so well over a large number of games. Out of the 20 games played, there were 11 in which the network did not make any serious mistakes. The network won 6 out of these 11 games, a result which is quite reasonable. However, in 9 of the 20 games, the network made one or more serious (i.e. potentially fatal) "blunders." The seriousness of these mistakes would be equivalently to dropping a piece in chess. Such a mistake is nearly always fatal in chess against a good opponent; however in backgammon there are still chances due to the element of luck involved. In the 9 games in which the network blundered, it did manage to survive and win 5 of the games due to the element of luck. (We are assuming that the mistakes made by the human, if any, were only minor mistakes.) It is highly unlikely that this sort of result would be repeated. A much more likely result would be that the network would win only one or two of the games in which it made a serious error. This would put the network's expected performance against expert or near-expert humans at about the 35-40% level. (This has also been confirmed in play against other networks.)

We find that the network does act as if it has picked up many of the global concepts and strategies of advanced play. The network has also learned many important tactical elements of play at the advanced level. As for the specific kinds of mistakes made by the network, we find that they are not at all random, senseless mistakes, but instead fall into clear, well-defined conceptual categories, and furthermore, one can understand the reasons why these categories of mistakes are made. We do not have space here to describe these in detail, and refer the reader instead to Ref. 5.

To summarize, qualitative analysis of the network's play indicates that it has learned many important strategies and tactics of advanced backgammon. This gives the network very good overall performance in typical positions. However, the network's worst case performance leaves a great deal to be desired. The network is capable of making both serious, obvious, "blunders," as well more subtle mistakes, in many different types of positions. Worst case performance is important, because the network must make long sequences of moves throughout the course of a game without any serious mistakes in order to have a reasonable chance of winning against a skilled opponent. The prospects for improving the network's worst case performance appear to be mixed. It seems quite likely that many of the current "blunders" can be fixed with a reasonable number of hand-crafted examples added to the training set. However, many of the subtle mistakes are due to a lack of very sophisticated knowledge, such as the notion of timing. It is difficult to imagine that this kind of knowledge could be imparted to the network in only a few examples. Probably what is required is either an intractably large number of examples, or a major overhaul in either the pre-computed features or the training paradigm.

DISCUSSION

We have seen from both quantitative and qualitative measures that the network has learned a great deal about the general principles of backgammon play, and has not simply memorized the individual positions in the training set. Quantitatively, the measure of game performace provides a clear indication of the network's ability to generalize, because apart from the first couple of moves at the start of each game, the network must operate entirely on generalization. Qualitatively, one can see after playing several games against the network that there are certain characteristic kinds of positions in which it does well, and other kinds of positions in which it systematically makes well-defined types of mistakes. Due to the network's frequent "blunders," its overall level of play is only intermediate level, although it probably is somewhat better than the average intermediate-level player. Against the intermediate-level program Gammontool, our best network wins almost 60% of the games. However, against a human expert the network would only win about 35-40% of the time. Thus while the network does not play at expert level, it is sufficiently good to give an expert a hard time, and with luck in its favor can actually win a match to a small number of games.

Our simple supervised learning approach leaves out some very important sources of information which are readily available to humans. The network is never told that the underlying topological structure of its input space really corresponds to a one-dimensional spatial structure; all it knows is that the inputs form a 459-dimensional hypercube. It has no idea of the object of the game, nor of the sense of temporal causality, i.e. the notion that its actions have consequences, and how those consequences lead to the achievement of the objective. The teacher signal only says whether a given move is good or bad, without giving any indication as to what the teacher's reasons are for making such a judgement. Finally, the network is only capable of scoring single moves in isolation, without any idea of what other moves are available. These sources of knowledge are essential to the ability of humans to play backgammon well, and it seems likely that some way of incorporating them into the network learning paradigm will be necessary in order to achieve further substantial improvements in performance.

There are a number of ways in which these additional sources of knowledge might be incorporated, and we shall be exploring some of them in future work. For example, knowledge of alternative moves could be introduced by defining a more sophisticated error signal which takes into account not only the network and teacher scores for the current move, but also the network and teacher scores for other moves from the same position. However, the more immediate plans involve a continuation of the existing strategies of hand-crafting examples and coding scheme modifications to eliminate the most serious errors in the network's play. If these errors can be eliminated, and we are confident that this can be achieved, then the network would become substantially better than any commercially available program, and would be a serious challenge for human experts. We would expect 65% performance against Gammontool, and 45% performance against human experts.

Some of the results of our study have implications beyond backgammon to more general classes of difficult problems. One of the limitations we have found is that substantial human effort is required both in the design of the coding scheme and in the design of the training set. It is not sufficient to use a simple coding scheme and random training patterns, and let the automated network learning procedure take care of everything else. We expect this to be generally true when connectionist learning is applied to difficult problem domains.

On the positive side, we foresee a potential for combining connectionist learning techniques with conventional AI techniques for hand-crafting knowledge to make significant progress in the development of intelligent systems. From the practical point of view, network learning can be viewed as an "enhancer" of traditional techniques, which might produce systems with superior performance. For this particular application, the obvious way to combine the two approaches is in the use of pre-computed features in the input encoding. Any set of hand-crafted features used in a conventional evaluation function could be encoded as discrete or continuous activity levels of input units which represent the current board state along with the units representing the raw information. Given a suitable encoding scheme for these features, and a training set of sufficient size and quality (i.e., the scores in the training set should be better than those of the original evaluation function), it seems possible that the resulting network could outperform the original evaluation function, as evidenced by our experiment with the Gammontool features.

Network learning might also hold promise as a means of achieving the long-sought goal of automated feature discovery[2]. Our network certainly appears to have learned a great deal of knowledge from the training set which goes far beyond the amount of knowledge that was explicitly encoded in the input features. Some of this knowledge (primarily the lowest level components) is apparent from the weight diagram when there are no hidden units (Fig. 3). However, much of the network's knowledge remains inaccessible. What is needed now is a means of disentangling the novel features discovered by the network from either the patterns of activity in the hidden units, or from the massive number of connection strengths which characterize the network. This is one our top priorities for future research, although techniques for such "reverse engineering" of parallel networks are only beginning to be developed[12].

ACKNOWLEDGEMENTS

This work was inspired by a conference on "Evolution, Games and Learning" held at Los Alamos National Laboratory, May 20-24, 1985. We thank Sun Microsystems Inc. for providing the source code for their Gammontool program, Hans Berliner for providing some of the positions used in the data base, Subutai Ahmad for writing the weight display graphics package, Bill Bogstad for assistance in programming the back-propagation simulator, and Bartlett Mel, Peter Frey, and Scott Kirkpatrick for critical reviews of the manuscript. G.T. was supported in part by the National Center for Supercomputing Applications. T.J.S. was supported by a NSF Presidential Young Investigator Award, and by grants from the Seaver Institute and the Lounsbury Foundation.

REFERENCES

1. D. E. Rumelart and J. L. McClelland, eds., *Parallel Distributed Processing: Explorations in the Microstructure of Cognition*, Vols. 1 and 2 (Cambridge: MIT Press, 1986).

2. A. L. Samuel, "Some studies in machine learning using the game of checkers." *IBM J. Res. Dev.* 3, 210--229 (1959).

3. J. H. Holland, "Escaping brittleness: the possibilities of general-purpose learning algorithms applied to parallel rule-based systems." In: R. S. Michalski et al. (eds.), *Machine learning: an artificial intelligence approach, Vol. II* (Los Altos CA: Morgan-Kaufman, 1986).

4. R. S. Sutton, "Learning to predict by the methods of temporal differences," GTE Labs Tech. Report TR87-509.1 (1987).

5. G. Tesauro and T. J. Sejnowski, "A parallel network that learns to play backgammon." Univ. of Illinois at Urbana-Champaign, Center for Complex Systems Research Technical Report (1987).

6. M. Minsky and S. Papert, *Perceptrons* (Cambridge: MIT Press, 1969).

7. D. E. Rumelhart, G. E. Hinton, and R. J. Williams, "Learning representations by back-propagating errors." *Nature* 323, 533--536 (1986).

8. Y. Le Cun, "A learning procedure for asymmetric network." *Proceedings of Cognitiva (Paris)* 85, 599--604 (1985).

9. D. B. Parker, "Learning-logic." MIT Center for Computational Research in Economics and Management Science Tech. Report TR-47 (1985).

10. H. Berliner, "Backgammon computer program beats world champion." *Artificial Intelligence* 14, 205--220 (1980).

11. G. Tesauro, "Neural network defeats creator in backgammon match." Univ. of Illinois at Urbana-Champaign, Center for Complex Systems Research Technical Report (1987).

12. C. R. Rosenberg, "Revealing the structure of NETtalk's internal representations." Proceedings of the Ninth Annual Conference of the Cognitive Science Society (Hillsdale, NJ: Lawrence Erlbaum Associates, 1987).

INTRODUCTION TO A SYSTEM FOR IMPLEMENTING NEURAL NET CONNECTIONS ON SIMD ARCHITECTURES

Sherryl Tomboulian

Institute for Computer Applications in Science and Engineering
NASA Langley Research Center, Hampton VA 23665

ABSTRACT

Neural networks have attracted much interest recently, and using parallel architectures to simulate neural networks is a natural and necessary application. The SIMD model of parallel computation is chosen, because systems of this type can be built with large numbers of processing elements. However, such systems are not naturally suited to generalized communication. A method is proposed that allows an implementation of neural network connections on massively parallel SIMD architectures. The key to this system is an algorithm that allows the formation of arbitrary connections between the "neurons". A feature is the ability to add new connections quickly. It also has error recovery ability and is robust over a variety of network topologies. Simulations of the general connection system, and its implementation on the Connection Machine, indicate that the time and space requirements are proportional to the product of the average number of connections per neuron and the diameter of the interconnection network.

INTRODUCTION

Neural Networks hold great promise for biological research, artificial intelligence, and even as general computational devices. However, to study systems in a realistic manner, it is highly desirable to be able to simulate a network with tens of thousands or hundreds of thousands of neurons. This suggests the use of parallel hardware. The most natural method of exploiting parallelism would have each processor simulating a single neuron.

Consider the requirements of such a system. There should be a very large number of processing elements which can work in parallel. The computation that occurs at these elements is simple and based on local data. The processing elements must be able to have connections to other elements. All connections in the system must be able to be traversed in parallel. Connections must be added and deleted dynamically.

Given current technology, the only type of parallel model that can be constructed with tens of thousands or hundreds of thousands of processors is an SIMD architecture. In exchange for being able to build a system with so many processors, there are some inherent limitations. SIMD stands for single instruction multiple data[1] which means that all processors can work in parallel, but they must do exactly the same thing at the same time. This machine model is sufficient for the computation required within a neuron, however in such a system it is difficult to implement arbitrary connections between neurons. The Connection Machine[2] provides such a model, but uses a device called the router

This work was supported by the National Aeronautics and Space Administration under NASA Constract No. NAS1-18010-7 while the author was in residence at ICASE.

to deliver messages. The router is a complex piece of hardware that uses significant chip area, and without the additional hardware for the router, a machine could be built with significantly more processors. Since one of the objectives is to maximize the number of "neurons" it is desirable to eliminate the extra cost of a hardware router and instead use a software method.

Existing software algorithms for forming connections on SIMD machines are not sufficient for the requirements of a neural networks. They restrict the form of graph (neural network) that can be embedded to permutations[3,4] or sorts[5,6 combined with 7], the methods are network specific, and adding a new connection is highly time consuming.

The software routing method presented here is a unique algorithm which allows arbitrary neural networks to be embedded in machines with a wide variety of network topologies. The advantages of such an approach are numerous: A new connection can be added dynamically in the same amount of time that it takes to perform a parallel traversal of all connections. The method has error recovery ability in case of network failures. This method has relationships with natural neural models. When a new connection is to be formed, the two neurons being connected are activated, and then the system forms the connection without any knowledge of the "address" of the neuron-processors and without any instruction as to the method of forming the connecting path. The connections are entirely distributed; a processor only knows that connections pass through it – it doesn't know a connection's origin or final destination.

Some neural network applications have been implemented on massively parallel architectures, but they have run into restrictions due to communication. An implementation on the Connection Machine[8] discovered that it was more desirable to cluster processors in groups, and have each processor in a group represent one connection, rather than having one processor per neuron, because the router is designed to deliver one message at a time from each processor. This approach is contrary with the more natural paradigm of having one processor represent a neuron. The MPP [9], a massively parallel architecture with processors arranged in a mesh, has been used to implement neural nets[10], but because of a lack of generalized communication software, the method for edge connections is a regular communication pattern with all neurons within a specified distance. This is not an unreasonable approach, since within the brain neurons are usually locally connected, but there is also a need for longer connections between groups of neurons. The algorithms presented here can be used on both machines to facilitate arbitrary connections with an irregular number of connections at each processor.

MACHINE MODEL

As mentioned previously, since we desire to build a system with an large number of processing elements, the only technology currently available for building such large systems is the SIMD architecture model. In the SIMD model there is a single control unit and a very large number of slave processors that can execute the same instruction stream simultaneously. It is possible to disable some processors so that only some execute an instruction, but it is not possible to have two processor performing different instructions at the same time. The processors have exclusively local memory which is small (only a few thousand bits), and they have no facilities for local indirect addressing. In this scheme an *instruction* involves both a particular operation code *and* the local memory address. All processors must do this same thing to the same areas of their local memory at the same time.

The basic model of computation is bit-serial – each instruction operates on a bit at a time. To perform multiple bit operations, such as integer addition, requires several instructions. This model is chosen because it requires less hardware logic, and so would allow a machine to be built with a larger number of processors than could otherwise be achieved with a standard word-oriented approach. Of course, the algorithms presented here will also work for machines with more complex instruction abilities; the machine model described satisfies the minimal requirements.

An important requirement for connection formation is that the processors are connected in some topology. For instance, the processors might be connected in a grid so that each processor has a North, South, East, and West neighbor. The methods presented here work for a wide variety of network topologies. The requirements are: (1) there must be some path between any two processors; (2) every neighbor link must be bi-directional, i.e. if A is a neighbor of B, then B must be a neighbor of A; (3) the neighbor relations between processors must have a consistent invertible labeling. A more precise definition of the labeling requirements can be found in [11]. It suffices that most networks [12], including grid, hypercube, cube connected cycles[13], shuffle exchange[14], and mesh of trees[15] are admissible under the scheme. Additional requirements are that the processors be able to read from or write to their neighbors' memories, and that at least one of the processors acts as a serial port between the processors and the controller.

COMPUTATIONAL REQUIREMENTS

The machine model described here is sufficient for the computational requirements of a neuron. Adopt the paradigm that each processor represents one neuron. While several different models of neural networks exist with slightly different features, they are all fairly well characterized by computing a sum or product of the neighbors values, and if a certain threshold is exceeded, then the processor neuron will *fire*, i.e. activate other neurons. The machine model described here is more efficient at boolean computation, such as described by McCulloch and Pitts[16], since it is bit serial. Neural net models using integers and floating point arithmetic [17,18] will also work but will be somewhat slower since the time for computation is proportional to the number of bits of the operands.

The only computational difficulty lies in the fact that the system is SIMD, which means that the processes are synchronous. For some neural net models this is sufficient[18] however others require asynchronous behavior [17]. This can easily be achieved simply by turning the processors on and off based on a specified probability distribution. (For a survey of some different neural networks see [19]).

CONNECTION ASSUMPTIONS

Many models of neural networks assume fully connected systems. This model is considered unrealistic, and the method presented here will work better for models that contain more sparsely connected systems. While the method will work for dense connections, the time and space required is proportional to the number of edges, and becomes prohibitively expensive.

Other than the sparse assumptions, there are no restrictions to the topological form of the network being simulated. For example, multiple layered systems, slightly irregular structures, and completely random connections are all handled easily. The system does function better if there is locality in the neural network. These assumptions seem to fit the biological model of neurons.

THE CONNECTION FORMATION METHOD

A fundamental part of a neural network implementation is the realization of the connections between neurons. This is done using a software scheme first presented in [11,20]. The original method was intended for realizing directed graphs in SIMD architectures. Since a neural network is a graph with the neurons being vertices and the connections being arcs, the method maps perfectly to this system. Henceforth the terms neuron and vertex and the terms arc and connection will be used interchangeably.

The software system presented here for implementing the connections has several parts. Each processor will be assigned exactly one neuron. (Of course some processors may be "free" or unallocated, but even "free" processor participate in the routing process.) Each connection will be realized as a path in the topology of processors. A labeling of these paths in time and space is introduced which allows efficient routing algorithms and a set-up strategy is introduced that allows new connections to be added quickly.

The standard computer science approach to forming the connection would be to store the addresses of the processors to which a given neuron is connected. Then, using a routing algorithm, messages could be passed to the processors with the specified destination. However, the SIMD architecture does not lend itself to standard message passing schemes because processors cannot do indirect addressing, so buffering of values is difficult and costly.

Instead, a scheme is introduced which is closer to the natural neuron-synapse structures. Instead of having an address for each connection, the connection is actually represented as a fixed path between the processors, using time as a virtual dimension. The path a connection takes through the network of processors is statically encoded in the local memories of the neurons that it passes through. To achieve this, the following data structures will be resident at each processor.

```
ALLOCATED ---- boolean flag indicating
    whether this processor is assigned
    a vertex (neuron) in the graph
VERTEX LABEL --- label of graph vertex  (neuron)
HAS_NEIGHBOR[1..neighbor_limit] flag
    indicating the existence of neighbors
SLOTS[1..T] OF      arc path information
    START---------new arc starts here
    DIRECTION------direction to send
                    {1..neighbor_limit,FREE}
    END----------arc ends here
    ARC LABEL-----label of arc
```

The ALLOCATED and VERTEX LABEL field indicates that the processor has been assigned a vertex in the graph (neuron). The HAS NEIGHBOR field is used to indicate whether a physical wire exists in the particular direction; it allows irregular network topologies and boundary conditions to be supported. The SLOTS data structure is the key to realizing the connections. It is used to instruct the processor where to send a message and to insure that paths are constructed in such a way that no collisions will occur.

SLOTS is an array with T elements. The value T is called the time quantum. Traversing all the edges of the embedded graph in parallel will take a certain amount of time since messages must be passed along through a sequence of neighboring processors. Forming these parallel connections will be considered an uninterruptable operation which will take T steps. The SLOTS array is used to tell the processors what they should do on each relative time position within the time quantum.

One of the characteristics of this algorithm is that a fixed path is chosen to represent the connection between two processors, and once chosen it is never changed. For example, consider the grid below.

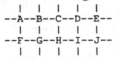

Fig. 1. Grid Example

If there is an arc between A and H, there are several possible paths: East-East-South, East-South-East, and South-East-East. Only one of these paths will be chosen between A and H, and that same path will always be used. Besides being invariant in space, paths are also invariant in time. As stated above, traversal is done within a time quantum T. Paths do no have to start on time 1, but can be scheduled to start at some relative offset within the time quantum. Once the starting time for the path has been fixed, it is never changed. Another requirement is that a message can not be buffered, it must proceed along the specified directions without interruption. For example, if the path is of length 3 and it starts at time 1, then it will arrive at time 4. Alternatively, if it starts at time 2 it will arrive at time 5. Further, it is necessary to place the paths so that no collisions occur; that is, no two paths can be at the same processor at the same instant in time. Essentially time adds an extra dimension to the topology of the network, and within this space-time network all data paths must be non-conflicting. The rules for constructing paths that fulfill these requirements are listed below.

- At most one connection can enter a processor at a given time, and at most one connection can leave a processor at a given time. It is possible to have both one coming and one going at the same time. Note that this does not mean that a processor can have only one connection; it means that it can have only one connection during any one of the T time steps. It can have as many as T connections going through it.

- Any path between two processors (u,v) representing a connection must consist of steps at contiguous times. For example, if the path from processor u to processor v is u,f,g,h,v , then if the arc from u-f is assigned time 1, f-g must have time 2, g-h time 3, and h-v time 4. Likewise if u-f occurs at time 5, then arc h-v will occur time 8.

When these rules are used when forming paths, the SLOTS structure can be used to mark the paths. Each path goes through neighboring processors at successive time steps. For each of these time steps the DIRECTION field of the SLOTS structure is marked, telling the processor which direction it should pass a message if it receives it on that time. SLOTS serves both to instruct the processors how to send messages, and to indicate that a processor is busy at a certain time slot so that when new paths are constructed it can be guaranteed that they won't conflict with current paths.

Consider the following example. Suppose we are given the directed graph with vertices A,B,C,D and edges A − > C, B − > C,B − > D, and D − > A. This is to be done where A,B,C, and D have been assigned to successive elements of a linear array. (A linear array in not a good network for this scheme, but is a convenient source of examples.)

Logical Connections

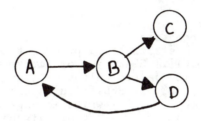

Fig. 2. Graph Example

A,B,C,D are successive members in a linear array

```
1---2---3---4
A---B---C---D
```

First, A ->C can be completed with the map East-East, so
Slots[A][1].direction = E, Slots[B][2].direction=E,
Slots[C][2].end = 1 .

B->C can be done with the map East, it can start at time 1,
since Slots[B][1].direction and Slots[C][1].end are free.

B->D goes through C then to D, its map is East-East. B is
occupied at time 1 and 2. It is free at time 3,
so Slots[B][3].direction = E, Slots[C][4].direction = E,
Slots[D][4].end = 1.

D->A must go through C,B,A. using map West-West-West.
D is free on time 1, C is free on time 2, but B is occupied
on time 3. D is free on time 2, but C is occupied on time 3.
It can start from D at time 3, Slots[D][3].direction = W,
Slots[C][4].direction = W, Slots[B][5].direction = W,
Slots[A][5].end=1

Every processor acts as a conduit for its neighbors messages. No processor knows where any message is going to or coming from, but each processor knows what it must do to establish the local connections.

The use of contiguous time slots is vital to the correct operation of the system. If all edge-paths are established according to the above rules, there is a simple method for making the connections. The paths have been restricted so that there will be no collisions, and paths' directions use consecutive time slots. Hence if all arcs at time i send a message to their neighbors, then each processor is guaranteed no more than 1 message coming to it. The end of a path is specified by setting a separate bit that is tested after each message is received. A separate start bit indicates when a path starts. The start bit is needed because the SLOTS array just tells the processors where to send a message, regardless of how that message arrived. The start array indicates when a message originates, as opposed to arriving from a neighbor.

The following algorithm is basic to the routing system.

```
for i = time 1 to T
      FORALL processors
      /* if an arc starts or is passing through at this time*/
            if SLOT[i].START = 1 or active = 1
                  for j=1 to neighbor-limit
                        if SLOT[i].direction= j
                              write message bit to in-box
                                    of neighbor j;
                  set active = 0;
                  FORALL processor that just received a message
                  if end[i]
                        move in-box to message-destination;
                  else
                        move in-box to out-box;
                        set active bit = 1;
```

This code follows the method mentioned above. The time slots are looped through and the messages are passed in the appropriate directions as specified in the SLOTS array. Two bits, in-box and out-box, are used for message passing so that an out-going message won't be overwritten by an in-coming message before it gets transferred. The inner loop *for j = 1 to neighbor limit* checks each of the possible neighbor directions and sends the message to the correct neighbor. For instance, in a grid the neighbor limit is 4, for North, South, East, and West neighbors. The time complexity of data movement is O(T times neighbor-limit).

SETTING UP CONNECTIONS

One of the goals in developing this system was to have a method for adding new connections quickly. Paths are added so that they don't conflict with any previously constructed path. Once a path is placed it will not be re-routed by the basic placement algorithm; it will always start at the same spot at the same time. The basic idea of the method for placing a connection is to start from the source processor and in parallel examine all possible paths outward from it that do not conflict with pre-established paths and which adhere to the sequential time constraint. As the trial paths are flooding the system, they are recorded in temporary storage. At the end of this deluge of trial paths *all* possible paths will have been examined. If the destination processor has been reached, then a path exists under the current time-space restrictions. Using the stored information a path can be backtraced and recorded in the SLOTS structure. This is similar to the Lee-Moore routing algorithm[21,22] for finding a path in a system, but with the sequential time restriction.

For example, suppose that the connection (u,v) is to be added. First it is assumed that processors for u and v have already been determined, otherwise (as a simplification) assume a random allocation from a pool of free processors. A parallel breadth-first search will be performed starting from the source processor. During the propagation phase a processor which receives a message checks its SLOTS array to see if they are busy on that time step, if not it will propagate to its neighbors on the next time step. For instance, suppose a trial path starts at time 1 and moves to a neighboring processor, but that neighbor is already busy at time 1 (as can be seen by examining the DIRECTION-SLOT.) Since a path that would go through this neighbor at this time is not legal, the trial path would commit suicide, that is, it stops propagating itself. If the processor slot for time 2 was free, the trial path would attempt to propagate to all of its' neighbors at time 3.

Using this technique paths can be constructed with essentially no knowledge of the relative locations of the "neurons" being connected or the underlying topology. Variations on the outlined method, such as choosing the shortest path, can improve the choice of paths with very little overhead. If the entire network were known ahead of time, an off-line method could be used to construct the paths more efficiently; work on off-line methods is underway. However, the simple elegance of this basic method holds great appeal for systems that change slowly over time in unpredictable ways.

PERFORMANCE

Adding an edge (assuming one can be added), deleting any set of edges, or traversing all the edges in parallel, all have time complexity $O(T \times neighbor - limit)$. If it is assumed that neighbor limit is a small constant then the complexity is $O(T)$. Since T is related both to the time and space needed, it is a crucial factor in determining the value of the algorithms presented. Some analytic bounds on T were presented in[11], but it is difficult to get a tight bound on T for general interconnection networks and dynamically changing graphs. A simulator was constructed to examine the behavior of the algorithms. Besides the simulated data, the algorithms mentioned were actually implemented for the Connection Machine. The data produced by the simulator is consistent with that produced by the real machine. The major result is that the size of T appears proportional to the average degree of the graph times the diameter of the interconnection network[20].

812

FURTHER RESEARCH

This paper has been largely concerned with a system that can realize the connections in a neural network when the two neurons to be joined have been activated. The tests conducted have been concerned with the validity of the method for implementing connections, rather than with a full simulation of a neural network. Clearly this is the next step.

A natural extension of this method is a system which can form its own connections based solely on the activity of certain neurons, without having to explicitly activate the source and destination neurons. This is an exciting avenue, and further results should be forthcoming.

Another area of research involves the formation of branching paths. The current method takes an arc in the neural network and realizes it as a unique path in space-time. A variation that has similarities to dendritic structure would allow a path coming from a neuron to branch and go to several target neurons. This extension would allow for a much more economical embedding system. Simulations are currently underway.

CONCLUSIONS

A method has been outlined which allows the implementation of neural nets connections on a class of parallel architectures which can be constructed with very large numbers of processing elements. To economize on hardware so as to maximize the number of processing element buildable, it was assumed that the processors only have local connections; no hardware is provided for communication. Some simple algorithms have been presented which allow neural nets with arbitrary connections to be embedded in SIMD architectures having a variety of topologies. The time for performing a parallel traversal and for adding a new connection appears to be proportional to the diameter of the topology times the average number of arcs in the graph being embedded. In a system where the topology has diameter $O(logN)$, and where the degree of the graph being embedded is bounded by a constant, the time is apparently $O(logN)$. This makes it competitive with existing methods for SIMD routing, with the advantages that there are no apriori requirements for the form of the data, and the topological requirements are extremely general. Also, with our approach new arcs can be added without reconfiguring the entire system. The simplicity of the implementation and the flexibility of the method suggest that it could be an important tool for using SIMD architectures for neural network simulation.

BIBLIOGRAPHY

1. M.J. Flynn, "Some computer organizations and their effectiveness", IEEE Trans Comput., vol C-21, no.9, pp. 948-960.
2. W. Hillis, "The Connection Machine", MIT Press, Cambridge, Mass, 1985.
3. D. Nassimi, S. Sahni, "Parallel Algorithms to Set-up the Benes Permutation Network", Proc. Workshop on Interconnection Networks for Parallel and Distributed Processing, April 1980.
4. D. Nassimi, S. Sahni, "Benes Network and Parallel Permutation Algorithms", IEEE Transactions on Computers, Vol C-30, No 5, May 1981.
5. D. Nassimi, S. Sahni, "Parallel Permutation and Sorting Algorithms and a

New Generalized Connection Network", *JACM*, Vol. 29, No. 3, July 1982 pp. 642-667

6. K.E. Batcher, "Sorting Networks and their Applications", The Proceedings of AFIPS 1968 SJCC, 1968, pp. 307-314.

7. C. Thompson, "Generalized connection networks for parallel processor inter-communication", *IEEE Tran. Computers*, Vol C, No 27, Dec 78, pp. 1119-1125.

8. Nathan H. Brown, Jr., "Neural Network Implementation Approaches for the Connection Machine", presented at the 1987 conference on Neural Information Processing Systems – Natural and Synthetic.

9. K.E. Batcher, "Design of a massively parallel processor", *IEEE Trans on Computers*, Sept 1980, pp. 836-840.

10. H.M. Hastings, S. Waner, "Neural Nets on the MPP", Frontiers of Massively Parallel Scientific Computation, NASA Conference Publication 2478, NASA Goddard Space Flight Center, Greenbelt Maryland, 1986.

11. S. Tomboulian, "A System for Routing Arbitrary Communication Graphs on SIMD Architectures", Doctoral Dissertation, Dept of Computer Science, Duke University, Durham NC.

12. T. Feng, "A Survey of Interconnection Networks", *Computer*, Dec 1981, pp.12-27.

13. F. Preparata and J. Vuillemin, "The Cube Connected Cycles: a Versatile Network for Parallel Computation", *Comm. ACM*, Vol 24, No 5 May 1981, pp. 300-309.

14. H. Stone, "Parallel processing with the perfect shuffle", *IEEE Trans. Computers*, Vol C, No 20, Feb 1971, pp. 153-161.

15. T. Leighton, "Parallel Computation Using Meshes of Trees", *Proc. International Workshop on Graph Theory Concepts in Computer Science*, 1983.

16. W.S. McCulloch, and W. Pitts, "A Logical Calculus of the Ideas Imminent in Nervous Activity," *Bulletin of Mathematical Biophysics*, Vol 5, 1943, pp.115-133.

17. J.J. Hopfield, "Neural networks and physical systems with emergent collective computational abilities", *Proc. Natl. Aca. Sci.*, Vol 79, April 1982, pp. 2554-2558.

18. T. Kohonen, "Self-Organization and Associative Memory, Springer-Verlag, Berlin , 1984.

19. R.P. Lippmann, "An Introduction to Computing with Neural Nets", *IEEE AASP*, April 1987, pp. 4-22.

20. S. Tomboulian, "A System for Routing Directed Graphs on SIMD Architectures", ICASE Report No. 87-14, NASA Langley Research Center, Hampton, VA.

21. C.Y. Lee, "An algorithm for path connections and its applications", *IRE Trans Elec Comput*, Vol. EC-10, Sept. 1961, pp. 346-365.

22. E. F. Moore, "Shortest path through a maze", *Annals of Computation Laboratory*, vol. 30. Cambridge, MA: Harvard Univ. Press, 1959, pp.285-292.

NEUROMORPHIC NETWORKS BASED ON SPARSE OPTICAL ORTHOGONAL CODES

Mario P. Vecchi and Jawad A. Salehi

Bell Communications Research

435 South Street

Morristown, NJ 07960-1961

Abstract

A family of neuromorphic networks specifically designed for communications and optical signal processing applications is presented. The information is encoded utilizing sparse Optical Orthogonal Code sequences on the basis of unipolar, binary $(0,1)$ signals. The generalized synaptic connectivity matrix is also unipolar, and clipped to binary $(0,1)$ values. In addition to high-capacity associative memory, the resulting neural networks can be used to implement general functions, such as code filtering, code mapping, code joining, code shifting and code projecting.

1 Introduction

Synthetic neural nets[1,2] represent an active and growing research field. Fundamental issues, as well as practical implementations with electronic and optical devices are being studied. In addition, several learning algorithms have been studied, for example stochastically adaptive systems[3] based on many-body physics optimization concepts[4,5].

Signal processing in the optical domain has also been an active field of research. A wide variety of non-linear all-optical devices are being studied, directed towards applications both in optical computating and in optical switching. In particular, the development of Optical Orthogonal Codes (OOC)[6] is specifically interesting to optical communications applications, as it has been demonstrated in the context of Code Division Multiple Access (CDMA)[7].

In this paper we present a new class of neuromorphic networks, specifically designed for optical signal processing and communications, that encode the information in *sparse* OOC's. In Section 2 we review some basic concepts. The new neuromorphic networks are defined in Section 3, and their associative memory properties are presented in Section 4. In Section 5 other general network functions are discussed. Concluding remarks are given in Section 6.

2 Neural Networks and Optical Orthogonal Codes

2.1 Neural Network Model

Neural network are generally based on multiply-threshold-feedback cycles. In the Hopfield model[2], for instance, a connectivity $\overline{T}$ matrix stores the M different memory elements, labeled m, by the sum of outer products,

$$T_{ij} = \sum_m^M u_i^m u_j^m \ ; \ \ i,j = 1,2...N \tag{1}$$

where the state vectors $\underline{u}^m$ represent the memory elements in the bipolar $(-1, 1)$ basis. The diagonal matrix elements in the Hopfield model are set to zero, $T_{ii} = 0$.

For a typical memory recall cycle, an input vector $\underline{v}^{in}$, which is close to a particular memory element $m = k$, multiplies the $\overline{T}$ matrix, such that the output vector $\underline{v}^{out}$ is given by

$$\hat{v}_i^{out} = \sum_{j=1}^{N} T_{ij} v_j^{in} \; ; \; i, j = 1, 2 ... N \tag{2}$$

and can be seen to reduce to

$$\hat{v}_i^{out} \approx (N - 1) u_i^k + \sqrt{(N - 1)(M - 1)} \tag{3}$$

for large N and in the case of randomly coded memory elements $\underline{u}^m$.

In the Hopfield model, each output $\hat{\underline{v}}^{out}$ is passed through a thresholding stage around zero. The thresholded output signals are then fed back, and the multiply and threshold cycle is repeated until a final stable output $\underline{v}^{out}$ is obtained. If the input $\underline{v}^{in}$ is sufficiently close to $\underline{u}^k$, and the number of state vectors is small (i.e. $M \ll N$), the final output will converge to memory element $m = k$, that is, $\underline{v}^{out} \to \underline{u}^k$. The associative memory property of the network is thus established.

2.2 Optical Orthogonal Codes

The OOC sequences have been developed[6,7] for optical CDMA systems. Their properties have been specifically designed for this purpose, based on the following two conditions: each sequence can be easily distinguished from a shifted version of itself, and each sequence can be easily distinguished from any other shifted or unshifted sequence in the set. Mathematically, the above two conditions are expressed in terms of auto- and crosscorrelation functions. Because of the non-negative nature of optical signals[1], OOC are based on unipolar $(0, 1)$ signals[7].

In general, a family of OOC is defined by the following parameters:

- F, the *length* of the code,

- K, the *weight* of the code, that is, the number of 1's in the sequence,

- λ_a, the auto-correlation value for all possible shifts, other than the zero shift,

- λ_c, the cross-correlation value for all possible shifts, including the zero shift.

For a given code length F, the maximum number of distinct sequences in a family of OOC depends on the chosen parameters, that is, the weight of the code K and the allowed overlap λ_a and λ_c. In this paper we will consider OOC belonging to the *minimum overlap* class, $\lambda_a = \lambda_c = 1$.

---

[1] We refer to optical *intensity* signals, and not to detection systems sensitive to phase information.

3   Neuromorphic Optical Networks

Our neuromorphic networks are designed to take full advantage of the properties of the OOC. The connectivity matrix $\overline{T}$ is defined as a sum of outer products, by analogy with (1), but with the following important modifications:

1. The memory vectors are defined by the sequences of a given family of OOC, with a basis given by the unipolar, binary pair $(0, 1)$. The dimension of the sparse vectors is given by the length of the code F, and the maximum number of available items depends on the chosen family of OOC.

2. All of the matrix elements T_{ij} are clipped to unipolar, binary $(0, 1)$ values, resulting in a sparse and simplified connectivity matrix, without any loss in the functional properties defined by our neuromorphic networks.

3. The diagonal matrix elements T_{ii} are *not* set to zero, as they reflect important information implicit in the OOC sequences.

4. The threshold value is *not* zero, but it is chosen to be equal to K, the weight of the OOC.

5. The connectivity matrix $\overline{T}$ is generalized to allow for the possibility of a variety of outer product options: self-outer products, as in (1), for associative memory, but also cross-outer products of different forms to implement various other system functions.

A simplified schematic diagram of a possible optical neuromorphic processor is shown in Figure 1. This implementation is equivalent to an incoherent optical matrix-vector multiplier[8], with the addition of nonlinear functions. The input vector is clipped using an optical hard-limiter with a threshold setting at 1, and then it is anamorphically imaged onto the connectivity mask for $\overline{T}$. In this way, the ith pixel of the input vector is imaged onto the ith column of the $\overline{T}$ mask. The light passing through the mask is then anamorphically imaged onto a line of optical threshold elements with a threshold setting equal to K, such that the jth row is imaged onto the jth threshold element.

4   Associative Memory

The associative memory function is defined by a connectivity matrix $\overline{T}^{MEM}$ given by:

$$T_{ij}^{MEM} = \mathcal{G}\left\{\sum_{m}^{M} x_i^m x_j^m\right\} \quad ; \quad i,j = 1, 2 ... F \tag{4}$$

where each memory element $\underline{x}^m$ corresponds to a given sequence of the OOC family, with code length F. The matrix elements of $\overline{T}^{MEM}$ are all clipped, unipolar values, as indicated by the function $\mathcal{G}\{\}$, such that,

$$\mathcal{G}\{\zeta\} = \begin{cases} 1 & if \; \zeta \geq 1 \\ 0 & if \; \zeta < 1 \end{cases} \tag{5}$$

We will now show that an input vector $\underline{x}^k$, which corresponds to memory element $m = k$, will produce a stable output (equal to the wanted memory vector) in a *single pass* of the multiply and threshold process.

The *multiplication* can be written as:

$$\hat{v}_i^{out} = \sum_j^F T_{ij}^{MEM} x_j^k \tag{6}$$

We remember that the non-linear clipping function $\mathcal{G}\{\}$ is to be applied *first* to obtain $\overline{T}^{MEM}$. Hence,

$$\hat{v}_i^{out} = \sum_j^F x_j^k \; \mathcal{G}\left\{ x_i^k x_j^k + \sum_{m \neq k}^M x_i^m x_j^m \right\} \tag{7}$$

For $x_i^k = 0$, only the second term in (7) contributes, and the pseudo-orthogonality properties of the OOC allow us to write:

$$\sum_{j \neq i}^F x_j^k \; \mathcal{G}\left\{ \sum_{m \neq k}^M x_i^m x_j^m \right\} \leq \lambda_c, \tag{8}$$

where the cross-correlation value is $\lambda_c < K$.

For $x_i^k = 1$, we again consider the properties of the OOC to obtain for the first term of (7):

$$\sum_j^F x_j^k x_i^k x_j^k = K x_i^k, \tag{9}$$

where K is the weight of the OOC.

Therefore, the result of the multiplication operation given by (7) can be written as:

$$\hat{v}_i^{out} = K x_i^k + \left[\begin{array}{c} value\ strictly \\ less\ than\ K \end{array} \right] \tag{10}$$

The *thresholding* operation follows, around the value K as explained in Section 3. That is, (10) is thresholded such that:

$$v_i^{out} = \left\{ \begin{array}{ll} 1 & \text{if } \hat{v}_i^{out} \geq K \\ 0 & \text{if } \hat{v}_i^{out} < K, \end{array} \right. \tag{11}$$

hence, the final output at the end of a *single pass* will be given by: $v_i^{out} = x_i^k$.

The result just obtained can be extended to demonstrate the single pass convergence when the input vector is close, but not necessarily equal, to a stored memory element. We can draw the following conclusions regarding the properties of our neuromorphic networks based on OOC:

- For any given input vector $\underline{v}^{in}$, the single pass output will correspond to the memory vector $\underline{x}^m$ which has the smallest Hamming distance to the input.

- If the input vector $\underline{v}^{in}$ is missing a single 1-element from the K 1's of an OOC, the single pass output will be the null or zero vector.

- If the input vector $\underline{v}^{in}$ has the same Hamming distance to two (or more) memory vectors $\underline{x}^m$, the single pass output will be the logical sum of those memory vectors.

The ideas just discussed were tested with a computer simulation. An example of associative memory is shown in Table 1, corresponding to the OOC class of length $F = 21$ and weight $K = 2$. For this case, the maximum number of independent sequences is $M = 10$. The connectivity matrix $\overline{T}^{MEM}$ is seen in Table 1, where one can clearly appreciate the simplifying features of our model, both in terms of the sparsity and of the unipolar, clipped values of the matrix elements. The computer simulations for this example are shown in Table 2. The input vectors $\underline{a}$ and $\underline{b}$ show the error-correcting memory recovery properties. The input vector $\underline{c}$ is equally distant to memory vectors $\underline{x}^3$ and $\underline{x}^8$, resulting in an output which is the sum $(\underline{x}^3 \oplus \underline{x}^8)$. And finally, input vector $\underline{d}$ is closest to $\underline{x}^1$, but one 1 is missing, and the output is the zero vector. The mask in Figure 1 shows the optical realization of the Table 1, where the transparent pixels correspond to the 1's and the opaque pixels to the 0's of the connectivity matrix $\overline{T}^{MEM}$.

It should be pointed out that the capacity of our network is significant. From the previous example, the capacity is seen to be $\approx F/2$ for *single pass* memory recovery. This result compares favorably with the capacity of a Hopfield model[9], of $\approx F/4\ln F$.

5   General Network Functions

Our neuromorphic networks, based on OOC, can be generalized to perform functions other than associative memory storage by constructing non-symmetrical connectivity matrices. The single pass convergence of our networks avoids the possibility of limit-cycle oscillations. We can write in general:

$$T_{ij} = \mathcal{G}\left\{ \sum_{m=1}^{M} y_i^m x_j^m \right\},\tag{12}$$

where each pair defined by m includes two vectors y^m and $\underline{x}^m$, which are *not necessarily equal.* The clipping function $\mathcal{G}\{\}$ insures that all matrix elements are binary (0,1) values. The possible choice of vector pairs is not completely arbitrary, but there is a wide variety of functions that can be implemented for each family of OOC. We will now discuss some of the applications that are of particular interest in optical communication systems.

5.1   Code Filtering (CDMA)

Figure 2 shows an optical CDMA network in a star configuration. M nodes are inter-connected with optical fibers to a passive MxM star coupler that broadcasts the optical signals. At each node there is a data encoder that maps each bit of information to the OOC sequence corresponding to the user for which the transmission is intended. In addition, each node has a filter and decoder that recognizes its specific OOC sequence. The optical transmission rate has been expanded by a factor F corresponding to the length of the OOC sequence. Within the context of a CDMA communication system[7], the filter or decoder must perform the function of recognizing a specific OOC sequence in the presence of other interfering codes sent on the common transmission medium.

We can think, then, of one of our neuromorphic networks as a filter, placed at a given receiver node, that will recognize the specific code that it was programmed for.

We define for this purpose a connectivity matrix as

$$T_{ij}^{CDMA} = x_i^k x_j^k \;\; ; \;\; i, j = 1, 2 \ldots F, \tag{13}$$

where only one vector $\underline{x}^k$ is stored at each node. This symmetric, clipped connectivity matrix will give an output equal to $\underline{x}^k$ whenever the input contains this vector, and a null or zero output vector otherwise. It is clear by comparing (13) with (4) that the CDMA filtering matrix is equivalent to an associative memory matrix with only one item imprinted in the memory. Hence the discussion of Section 4 directly applies to the understanding of the behaviour of $\overline{T}^{CDMA}$.

In order to evaluate the performance of our neuromorphic network as a CDMA filter, computer simulations were performed. Table 3 presents the $\overline{T}^{CDMA}$ matrix for a particular node defined by $\underline{x}^k$ of a CDMA system based on the OOC family $F = 21$, $K = 2$. The total number of distinct codes for this OOC family is $M = 10$, hence there are 9 additional OOC sequences that interfere with $\underline{x}^k$, labeled in Table 3 $\underline{x}^1$ to $\underline{x}^9$.

The performance was simulated by generating random composite sequences from the set of codes $\underline{x}^1$ to $\underline{x}^9$ arbitrarily shifted. All inputs are unipolar and clipped $(0,1)$ signals. The results presented in Table 4 give examples of our simulation for the $\overline{T}^{CDMA}$ matrix shown in Table 3. The input $\underline{a}$ is the (logical) sum of a 1-bit (vector $\underline{x}^k$), plus interfering signals from arbitrarily shifted sequences of $\underline{x}^2$, $\underline{x}^3$, $\underline{x}^4$, $\underline{x}^6$ and $\underline{x}^9$. The output of the neuromorphic network is seen to recover accurately the desired vector $\underline{x}^k$. The input vector $\underline{b}$ contains a 0-bit (null vector), plus the shifted sequences of $\underline{x}^1$, $\underline{x}^2$, $\underline{x}^3$, $\underline{x}^6$, $\underline{x}^7$ and $\underline{x}^8$, and we see that the output correctly recovers a 0-bit.

As discussed in Section 4, our neuromorphic network will always correctly recognize a 1-bit (vector $\underline{x}^k$) presented to its input. On the other hand [2], there is the possibility of making an error when a 0-bit is sent, and the interfering signals from other nodes happen to generate the chip positions of $\underline{x}^k$. This case is shown by input vector $\underline{c}$ of Table 4, which contains a 0-bit (null vector), plus shifted sequences of $\underline{x}^2$, $\underline{x}^3$, $\underline{x}^4$, $\underline{x}^5$, $\underline{x}^6$, $\underline{x}^7$ and $\underline{x}^8$ in such a way that the output is erroneously given as a 1-bit. The properties of the OOC sequences are specifically chosen to minimize these errors[7], and the statistical results of our simulation are also shown in Table 4. It is seen that, as expected, when a 1-bit is sent it is always correctly recognized. On the other hand, when 0-bits are sent, occasional errors occur. Our simulation, yields an overall bit error rate (BER) of $BER_{sim} = 5.88\%$, as shown in Table 4.

These results can be compared with theoretical calculations[7] which yield an estimate for the BER for the CDMA system described:

$$BER_{calc} \approx \frac{1}{2} \prod_{k=0}^{K-1} \left[1 - q^{M-1-k} \right], \tag{14}$$

where $q \equiv 1 - \frac{K}{2F}$. For the example of the OOC family $F = 21$, $K = 2$, with $M = 10$, the above expression yields $BER_{calc} \approx 5.74\%$.

---

[2]Our channel can be described, then, as a binary Z-channel between each two nodes dynamically establishing a communication path

It is seen, therefore, that our neuromorphic network approaches the minimum possible BER for a given family of OOC. In fact, the results obtained using our $\overline{T}^{CDMA}$ are equivalent CDMA detection scheme based on "optical-AND-gates"[10], which corresponds to the limiting BER determined by the properties of the OOC themselves[3]. The optical mask corresponding to the code filtering function is shown in Figure 3.

5.2 Other Functions

As a first example of a non-symmetric $\overline{T}$ matrix, let us consider the function of *mapping* an input code to a corresponding different output code. We define our mapping matrix as:

$$T_{ij}^{MAP} = \mathcal{G}\left\{\sum_m y_i^m x_j^m\right\} \quad ; \quad i,j = 1,2...F, \tag{15}$$

where an input vector $\underline{x}^m$ will produce a *different* output vector code $\underline{y}^m$.

The function of code *joining* is defined by a transfer function that takes a given input code and produces at the output a chosen combination of two or more codes. This function is performed by expressing the general matrix given by 12 as follows:

$$T_{ij}^{JOIN} = \mathcal{G}\left\{\sum_m (y_i^m + w_i^m + ...)x_j^m\right\} \quad ; \quad i,j = 1,2...F, \tag{16}$$

where an input vector $\underline{x}^m$ will result in an output that joins several vector codes ($\underline{y}^m \oplus \underline{w}^m \oplus ...$).

The code *shifting* matrix $\overline{T}^{SHIFT}$ will allow for the shift of a given code sequence, such that both input and output correspond to the same code, but shifted with respect to itself. That is,

$$T_{ij}^{SHIFT} = \mathcal{G}\left\{\sum_m x(s)_i^m x(0)_j^m\right\} \quad ; \quad i,j = 1,2...F, \tag{17}$$

where we have indicated an unshifted code sequence by $\underline{x}(0)^m$, and its corresponding output pair as a shifted version of itself $\underline{x}(s)^m$.

The code *projecting* function corresponds to processing an input vector that contains the logical sum of several codes, and projecting at the output a selected single code sequence. The corresponding matrix $\overline{T}^{PROJ}$ is given by:

$$T_{ij}^{PROJ} = \mathcal{G}\left\{\sum_m x_i^m(y_j^m + w_j^m + ...)\right\} \quad ; \quad i,j = 1,2...F, \tag{18}$$

where each input vector ($\underline{y}^m \oplus \underline{w}^m \oplus ...$) will project at the output to a single code $\underline{x}^m$. In general, the resulting output code sequence $\underline{x}^m$ could correspond to a code *not necessarely* contained in the input vector.

The performance and error correcting properties of these, and other, general functions follow a similar behaviour as discussed in Section 4.

---

[3]The BER for the OOC family shown in this example are far too large for a useful CDMA communications system. Our choice intended to show computer simulated results within a reasonable computation time.

6   Conclusions

The neuromorphic networks presented, based on sparse Optical Orthogonal Code (OOC) sequences, have been shown to have a number of attractive properties. The unipolar, clipped nature of the synaptic connectivity matrix simplifies the implementation. The single pass convergence further allows for general network functions that are expected to be of particular interest in communications and signal processing systems.

The coding of the information, based on OOC, has also been shown to result in high capacity associative memories. The combination of efficient associative memory properties, plus a variety of general network functions, also suggests the possible application of our neuromorphic networks in the implementation of computational functions based on optical symbolic substitution.

The family of neuromorphic networks discussed here emphasizes the importance of understanding the general properties of non-negative systems based on sparse codes[11]. It is hoped that our results will stimulate further work on the fundamental relationship between coding, or representations, and the information processing properties of neural nets.

Acknowledgement

We thank J. Y. N. Hui and J. Alspector for many useful discussions, and C. A. Brackett for his support and encouragement of this research.

References

[1] S. Grossberg. In K. Schmitt, editor, *Delay and Functional-Differential Equations and Their Applications*, page 121, Academic Press, New York, NY, 1972.

[2] J. J. Hopfield. Neural Networks and Physical Systems with Emergent Collective Computational Abilities. *Proc. Nat. Acad. Sci. USA*, 79:2254, 1982.

[3] D. H. Ackley, G. E. Hinton, and T. J. Sejnowski. A Learning Algorithm for Boltzmann Machines. *Cogn. Sci.*, 9:147, 1985.

[4] S. Kirkpatrick, C. D. Gelatt, and M. P. Vecchi. Optimization by Simulated Annealing. *Science*, 220:671, 1983.

[5] M. P. Vecchi and S. Kirkpatrick. Global Wiring by Simulated Annealing. *IEEE Trans. CAD of Integrated Circuits and Systems*, CAD-2:215, 1983.

[6] F. R. K. Chung, J. A. Salehi, and V. K. Wei. Optical Orthogonal Codes: Design, Analysis and Applications. In *IEEE International Symposium on Information Theory, Catalog No. 86CH2374-7*, 1986. Accepted for publication in IEEE Trans. on Information Theory.

[7] J. A. Salehi and C. A. Brackett. Fundamental Principles of Fiber Optics Code Division Multiple Access. In *IEEE International Conference on Communications*, 1987.

[8] N. H. Farhat, D. Psaltis, A. Prata, and E. Paek. Optical Implementation of the Hopfield Model. *Appl. Opt.*, 24:1469, 1985.

[9] R. J. McEliece, E. C. Posner, E. R. Rodemich, and S. S. Venkatesh. The Capacity of Hopfield Associative Memory. *IEEE Trans. on Information Theory*, IT-33:461, 1987.

[10] J. A. Salehi. Principles and Applications of Optical AND Gates in Fiber Optics Code Division Multiple Access Networks. *In preparation*, 1987.

[11] G. Palm. Technical comments. *Science*, 235:1226, 1987.

Table 1: Associative Memory. Example showing storage of 10 distinct code sequences corresponding to the chosen OOC family.

Table 3: Code Filtering (CDMA).

Table 1 — OOC Family: $F = 31$, $K = 3$

Code Vectors:

g^1	1 1 0 0 0 0 0 0 0 0 0 0 0 0 0 0 0 0 0 0 0
g^2	0 0 1 0 1 0 0 0 0 0 0 0 0 0 0 0 0 0 0 0 0
g^3	0 0 0 1 0 0 1 0 0 0 0 0 0 0 0 0 0 0 0 0 0
g^4	0 0 0 0 0 1 0 0 1 0 0 0 0 0 0 0 0 0 0 0 0
g^5	0 0 0 0 0 0 1 0 0 0 1 0 0 0 0 0 0 0 0 0 0
g^6	0 0 0 0 0 0 0 1 0 0 0 0 1 0 0 0 0 0 0 0 0
g^7	0 0 0 0 0 0 0 0 1 0 0 0 0 0 1 0 0 0 0 0 0
g^8	0 0 0 0 0 0 0 0 0 1 0 0 0 0 0 0 1 0 0 0 0
g^9	0 0 0 1 0 0 0 0 0 0 1 0 0 0 0 0 0 0 0 0 0
g^{10}	0 0 0 0 1 0 0 0 0 0 0 0 1 0 0 0 0 0 0 0 0

Connectivity Matrix T^{MEM}:

(binary connectivity matrix)

Table 3 — OOC Family: $F = 31$, $K = 3$

Imprinted Code for Node k:

g^k	0 0 0 0 0 0 0 1 0 0 0 0 1 0 0 0 0 0 0 0 0

Connectivity Matrix T^{CDMA}:

(binary connectivity matrix)

Remaining 9 Interfering Codes:

g^1	1 1 0 0 0 0 0 0 0 0 0 0 0 0 0 0 0 0 0 0 0
g^2	0 0 1 0 1 0 0 0 0 0 0 0 0 0 0 0 0 0 0 0 0
g^3	0 0 0 1 0 0 1 0 0 0 0 0 0 0 0 0 0 0 0 0 0
g^4	0 0 0 0 0 1 0 0 1 0 0 0 0 0 0 0 0 0 0 0 0
g^5	0 0 0 0 0 0 1 0 0 0 1 0 0 0 0 0 0 0 0 0 0
g^6	0 0 0 0 0 0 0 1 0 0 0 0 1 0 0 0 0 0 0 0 0
g^7	0 0 0 0 0 0 0 0 1 0 0 0 0 0 1 0 0 0 0 0 0
g^8	0 0 0 1 0 0 0 0 0 0 1 0 0 0 0 0 0 0 0 0 0
g^9	0 0 0 0 1 0 0 0 0 0 0 0 1 0 0 0 0 0 0 0 0

Table 2: Associative Memory, Performance Simulation. Single pass convergence, based on the T^{MEM} matrix given in Table 1. For each input, the Hamming distance to the *closest* stored memory vector is given.

Table 2 — OOC Family: $F = 31$, $K = 3$

Input Vector:

a	1 1 0 1 0 0 0 0 0 1 0 0 0 0 0 1 0 0 0 0 0
	Hamming distance from $g^1 = 4$

Output Vector:

g^1	1 1 0 0 0 0 0 0 0 0 0 0 0 0 0 0 0 0 0 0 0

Input Vector:

b	1 0 0 1 0 0 1 0 1 0 0 0 1 0 1 0 0 1 1 0 0
	Hamming distance from $g^3 = 6$

Output Vector:

g^3	0 0 0 1 0 0 1 0 0 0 0 0 0 0 0 0 0 0 0 0 0

Input Vector:

c	1 0 0 1 0 0 1 0 0 1 0 1 0 0 0 1 0 0 0 1 0
	Hamming distance from $g^? = 5$
	Hamming distance from $g^? = 5$

Output Vector:

$g^? \oplus g^?$	0 0 0 1 0 0 1 0 0 0 0 1 0 0 0 0 0 0 0 1 0

Input Vector:

d	0 1 0 0 0 0 0 0 0 0 0 0 0 0 0 0 0 0 0 0 0
	Hamming distance from $g^1 = 1$

Output Vector:

$zero$	0 0 0 0 0 0 0 0 0 0 0 0 0 0 0 0 0 0 0 0 0

Table 4: Code Filtering (CDMA), Performance Simulation. Single pass convergence, based on the T^{CDMA} matrix given in Table 3. Each input is specified by the intended message bit (1-bit = g^k; 0-bit = zero vector), plus the indicated interfering codes, arbitrarily shifted with respect to the message bit.

Table 4 — OOC Family: $F = 31$, $K = 3$

Input Vector = $g^k \oplus [g^1 \oplus g^3 \oplus g^5 \oplus g^8]$:

a	1 0 0 1 1 0 1 0 1 0 0 1 1 1 1 0 0 0 0 0 0

Output Vector: *correct*

g^k	0 0 0 0 0 0 0 1 0 0 0 0 1 0 0 0 0 0 0 0 0

Input Vector = $[zero] \oplus [g^1 \oplus g^3 \oplus g^5 \oplus g^8]$:

b	1 0 0 0 1 0 1 0 0 0 0 1 1 1 0 0 1 1 0 0 0

Output Vector: *correct*

$[zero]$	0 0 0 0 0 0 0 0 0 0 0 0 0 0 0 0 0 0 0 0 0

Input Vector = $[zero] \oplus [g^1 \oplus g^3 \oplus g^5 \oplus g^7 \oplus g^8 \oplus g^9]$:

c	0 0 0 1 1 0 1 1 1 1 0 1 1 1 0 0 1 1 0 1 0

Output Vector: *error*

g^k	0 0 0 0 0 0 0 1 0 0 0 0 1 0 0 0 0 0 0 0 0

Statistical Behaviour

	1-bit	0-bit	Total
Samples	50037	50388	100425
Errors	0	5901	5901
BER %	0.00%	11.71%	5.88%

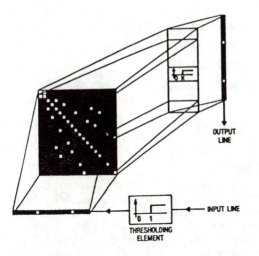

Figure 1:
Schematic diagram of an optical neuromorphic processor using sparse Optical Orthogonal Codes. Notice the absence of feedback because of the single-pass convergence. The mask shown represents the realization of the content-addressable memory of Table 1.

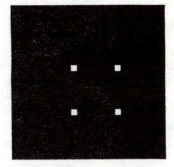

Figure 3:
Optical realization of a code filtering (CDMA) mask of Table 3. The 1's are represented by the transparent pixels, and the 0's by the opaque pixels.

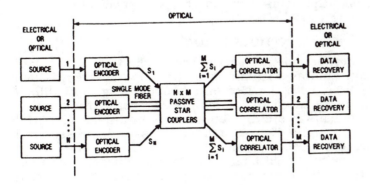

Figure 2:
Schematic diagram of a CDMA communications system over an Optical Fiber interconnection network. Each node represents one of the M possible distinct users in the system.

SYNCHRONIZATION IN NEURAL NETS

Jacques J. Vidal

University of California Los Angeles, Los Angeles, Ca. 90024

John Haggerty*

ABSTRACT

The paper presents an artificial neural network concept (the Synchronizable Oscillator Networks) where the instants of individual firings in the form of point processes constitute the only form of information transmitted between joining neurons. This type of communication contrasts with that which is assumed in most other models which typically are continuous or discrete value-passing networks. Limiting the messages received by each processing unit to time markers that signal the firing of other units presents significant implementation advantages.

In our model, neurons fire spontaneously and regularly in the absence of perturbation. When interaction is present, the scheduled firings are advanced or delayed by the firing of neighboring neurons. Networks of such neurons become global oscillators which exhibit multiple synchronizing attractors. From arbitrary initial states, energy minimization learning procedures can make the network converge to oscillatory modes that satisfy multi-dimensional constraints Such networks can directly represent routing and scheduling problems that consist of ordering sequences of events.

INTRODUCTION

Most neural network models derive from variants of Rosenblatt's original perceptron and as such are value-passing networks. This is the case in particular with the networks proposed by Fukushima[1], Hopfield[2], Rumelhart[3] and many others. In every case, the inputs to the processing elements are either binary or continuous amplitude signals which are weighted by synaptic gains and subsequently summed (integrated). The resulting activation is then passed through a sigmoid or threshold filter and again produce a continuous or quantized output which may become the input to other neurons. The behavior of these models can be related to that of living neurons even if they fall considerably short of accounting for their complexity. Indeed, it can be observed with many real neurons that action potentials (spikes) are fired and propagate down the axonal branches when the internal activation reaches some threshold and that higher

John Haggerty is with Interactive Systems Los angeles
3030 W. 6th St. LA, Ca. 90020

input rates levels result in more rapid firing. Behind these traditional models, there is the assumption that the average frequency of action potentials is the carrier of information between neurons. Because of integration, the firings of individual neurons are considered effective only to the extent to which they contribute to the average intensities It is therefore assumed that the activity is simply "frequency coded". The exact timing of individual firing is ignored.

This view however does not cover some other well known aspects of neural communication. Indeed, the precise timing of spike arrivals can make a crucial difference to the outcome of some neural interactions. One classic example is that of pre-synaptic inhibition, a widespread mechanism in the brain machinery. Several studies have also demonstrated the occurrence and functional importance of precise timing or phase relationship between cooperating neurons in local networks[4, 5] .

The model presented in this paper contrasts with the ones just mentioned in that in the networks each firing is considered as an individual output event. On the input side of each node, the firing of other nodes (the presynaptic neurons) either delay (inhibit) or advance (excite) the node firing. As seen earlier, this type of neuronal interaction which would be called phase-modulation in engineering systems, can also find its rationale in experimental neurophysiology. Neurophysiological plausibility however is not the major concern here. Rather, we propose to explore a potentially useful mechanism for parallel distributed computing. The merit of this approach for artificial neural networks is that digital pulses are used for internode communication instead of analog voltages. The model is particularly well suited to the time-ordering and sequencing found in a large class of routing and trajectory control problems.

NEURONS AS SYNCHRONIZABLE OSCILLATORS:

In our model, the processing elements (the "neurons") are relaxation oscillators with built-in self-inhibition. A relaxation oscillator is a dynamic system that is capable of accumulating potential energy until some threshold or breakdown point is reached. At that point the energy is abruptly released, and a new cycle begins.

The description above fits the dynamic behavior of neuronal membranes. A richly structured empirical model of this behavior is found in the well-established differential formulation of Hodgkin and Huxley[6] and in a simplified version given by Fitzhugh[7]. These differential equations account for the foundations of neuronal activity and are also capable of representing subthreshold behavior and the refractoriness that follows each firing. When the membrane potential enters the critical region, an abrupt depolarization, i.e., a collapse of the potential difference across the membrane occurs followed by a somewhat slower recovery. This brief electrical shorting of the membrane is called the action potential or "spike" and constitutes the output event for the neuron. If the causes for the initial depolarization are maintained, oscillation ("limit-cycles") develops, generating multiple firings. Depending on input level and membrane parameters, the oscillation can be limited to a single spike, or may produce an oscillatory burst, or even continually sustained activity.

The present model shares the same general properties but uses the much simpler description of relaxation oscillator illustrated on Figure 1.

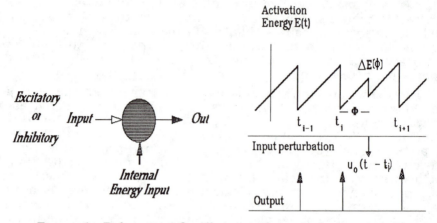

Figure 1 *Relaxation Oscillator with perturbation input*

Firing occurs when the energy level E(t) reaches some critical level E_c. Assuming a constant rate of energy influx a, firing will occur with the natural period

$$T = \frac{E_c}{a}$$

When pre-synaptic pulses impinge on the course of energy accumulation, the firing schedule is disturbed. Letting t_o represent the instant of the last firing of the cell and t_j, $(j = 1,2,...J)$, the intants of impinging arrivals from other cells:

$$E(t - t_o) = a(t - t_o) + \Sigma\ w_j..u_o(t - t_i)\ ;\ E \le E_c$$

where $u_o(t)$ represents the unit impulse at $t=0$.

The dramatic complexity of synchronization dynamics can be appreciated by considering the simplest possible case, that of a master slave interaction between two regularly firing oscillator units A and B, with natural periods T_A and T_B. At the instants of firing, unit A unidirectionally sends a spike signal to unit B which is received at some interval Φ measured from the last time B fired.

Upon reception the spike is transformed into a quantum of energy ΔE which depends upon the post-firing arrival time Φ. The relationship $\Delta E(\Phi)$ can be shaped to represent refractoriness and other post-spike properties. Here it is assumed to be a simple ramp function. If the interaction is inhibitory, the consequence of this arrival is that the next firing of unit B is delayed (with respect to what its schedule would have been in absence of perturbation) by some positive interval δ (Figure 2). Because of the shape of $\Delta E(\Phi)$, the delaying action, nil immediately after firing, becomes longer for impinging pre-synaptic spikes that arrive later in the interval. If the interaction is excitatory, the delay is negative, i.e. a shortening of the natural firing interval. Under very general assumptions regarding the function $\Delta E(\Phi)$, B will tend to synchronize to A. Within a given range of coupling gains, the phase Φ will self-adjust until equilibrium is achieved. With a given $\Delta E(\Phi)$, this equilibrium corresponds to a distribution of maximum entropy, i.e., to the point where both cells receive the same amouint of activation, during their common cycle.

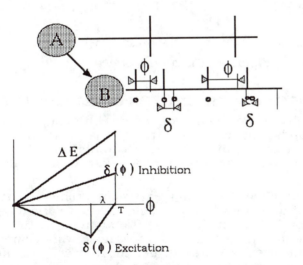

Figure 2 *Relationship between phase and delay when input efficiently increases linearly in the after-spike interval*

The synchronization dynamics presents an attractor for each rational frequency pair. To each ratio is associated a range of stability but only the ratios of lowest cardinality have wide zones of phase-locking (Figure 3). The wider stability zones correspond to a one to one ratio between f_A and f_B (or between their inverses T_A and T_B). Kohn and Segundo have demonstrated that such phase locking occurs in living invertebrate neurons and pointed out the paradoxical nature of phase-locked inhibition which, within each stability region, takes the appearence of excitation since small increases in input firing rate will locally result in increased output rates [8, 5].

The areas between these ranges of stability have the appearance of unstable transitions but in fact, as recently pointed out by Bak[9], form an infinity of locking steps known as the Devil's Staircase, corresponding to the infinity of intermediate rational pairs (figure 3). Bak showed that the staircase is self-similar under scaling and that the transitions form a fractal Cantor set with a fractal dimension which is a universal constant of dynamic systems.

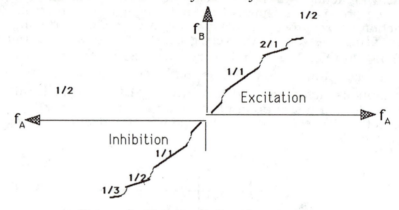

Figure 3 *Unilateral Synchronization:*

CONSTRAINT SATISFACTION IN OSCILLATOR NETWORKS

The global synchronization of an interconnected network of mutually phase-locking oscillators is a constraint satisfaction problem. For each synchronization equilibrium, the nodes fire in interlocked patterns that organize inter-spike intervals into integer ratios.

The often cited "Traveling Salesman Problem", the archetype for a class of important "hard" problems, is a special case when the ratio must be 1/1; all nodes must fire at the same frequency. Here the equilibrium condition is that every node will accumulate the the same amount of energy during the global cycle. Furthermore, the firings must be ordered along a minimal path.

Using stochastic energy minimization and simulated annealing, the first simulations have demonstrated the feasibility of the approach with a limited number of nodes. The TSP is isomorphic to many other sequencing problems which involve distributed constraints.and fall into the oscillator array neural net paradigm in a particularly natural way. Work is being pursued to more rigorously establish the limits of applicability of the model..

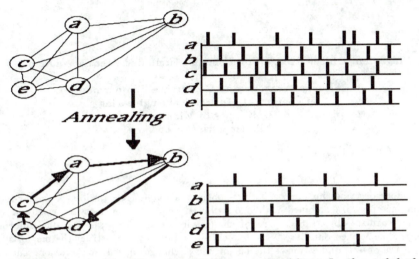

Figure 4. *The Traveling Salesman Problem: In the global oscillation of minimal energy each node is constrained to fire at the same rate in the order corresponding to the minimal path.*

ACKNOWLEDGEMENT

Research supported in part by Aerojet Electro-Systems under the Aerojet-UCLA Cooperative Research Master Agreement No. D841211, and by NASA NAG 2-302.

REFERENCES

1. K. Fukushima, Biol. Cybern. **20**, 121 (1975).
2. J.J. Hopfield, Proc. Nat. Acad. Sci. **79**, 2556 (1982).
3. D.E. Rumelhart, G.E. Hinton, and R.J. Williams, *Parallel Distributed Processing: Explorations in the Microstructure of Cognition,* (MIT Press, Cambridge, MA., 1986) p. 318.
4. J.P. Segundo, G.P. Moore, N.J. Stensaas, and T.H. Bullock, J. Exp. Biol. **40**, 643, (1963).
5. J.P. Segundo and A.F. Kohn, Biol Cyber **40**, 113 (1981).
6. A.L. Hodgkin and A.F. Huxley, J. Physiol. **117**, 500 (1952).
7. Fitzhugh, Biophysics J., **1**, 445 (1961).
8. A.F. Kohn, A. Freitas da Rocha, and J.P. Segundo, Biol. Cybern. **41**, 5 (1981).
9. P. Bak, Phys. Today (Dec 1986) p. 38 .
10. J. Haggerty and J.J. Vidal, UCLA BCI Report, 1975.

Invariant Object Recognition Using a Distributed Associative Memory

Harry Wechsler and George Lee Zimmerman
Department of Electrical Engineering
University of Minnesota
Minneapolis, MN 55455

Abstract

This paper describes an approach to 2-dimensional object recognition. Complex-log conformal mapping is combined with a distributed associative memory to create a system which recognizes objects regardless of changes in rotation or scale. Recalled information from the memorized database is used to classify an object, reconstruct the memorized version of the object, and estimate the magnitude of changes in scale or rotation. The system response is resistant to moderate amounts of noise and occlusion. Several experiments, using real, gray scale images, are presented to show the feasibility of our approach.

Introduction

The challenge of the visual recognition problem stems from the fact that the projection of an object onto an image can be confounded by several dimensions of variability such as uncertain perspective, changing orientation and scale, sensor noise, occlusion, and non-uniform illumination. A vision system must not only be able to sense the identity of an object despite this variability, but must also be able to characterize such variability -- because the variability inherently carries much of the valuable information about the world. Our goal is to derive the functional characteristics of image representations suitable for invariant recognition using a distributed associative memory. The main question is that of finding appropriate transformations such that interactions between the internal structure of the resulting representations and the distributed associative memory yield invariant recognition. As Simon [1] points out, all mathematical derivation can be viewed simply as a change of representation, making evident what was previously true but obscure. This view can be extended to all problem solving. Solving a problem then means transforming it so as to make the solution transparent.

We approach the problem of object recognition with three requirements: classification, reconstruction, and characterization. Classification implies the ability to distinguish objects that were previously encountered. Reconstruction is the process by which memorized images can be drawn from memory given a distorted version exists at the input. Characterization involves extracting information about how the object has changed from the way in which it was memorized. Our goal in this paper is to discuss a system which is able to recognize memorized 2-dimensional objects regardless of geometric distortions like changes in scale and orientation, and can characterize those transformations. The system also allows for noise and occlusion and is tolerant of memory faults.

The following sections, Invariant Representation and Distributed Associative Memory, respectively, describe the various components of the system in detail. The Experiments section presents the results from several experiments we have performed on real data. The paper concludes with a discussion of our results and their implications for future research.

1. Invariant Representation

The goal of this section is to examine the various components used to produce the vectors which are associated in the distributed associative memory. The block diagram which describes the various functional units involved in obtaining an invariant image representation is shown in Figure 1. The image is complex-log conformally mapped so that rotation and scale changes become translation in the transform domain. Along with the conformal mapping, the image is also filtered by a space variant filter to reduce the effects of aliasing. The conformally mapped image is then processed through a Laplacian in order to solve some problems associated with the conformal mapping. The Fourier transform of both the conformally mapped image and the Laplacian processed image produce the four output vectors. The magnitude output vector $|\bullet|_1$ is invariant to linear transformations of the object in the input image. The phase output vector Φ_2 contains information concerning the spatial properties of the object in the input image.

1.1 Complex-Log Mapping and Space Variant Filtering

The first box of the block diagram given in Figure 1 consists of two components: Complex-log mapping and space variant filtering. Complex-log mapping transforms an image from rectangular coordinates to polar exponential coordinates. This transformation changes rotation and scale into translation. If the image is mapped onto a complex plane then each pixel (x,y) on the Cartesian plane can be described mathematically by $z = x + jy$. The complex-log mapped points w are described by

$$w = \ln(z) = \ln(|z|) + j\theta_z \qquad (1)$$

where $|z| = (x^2 + y^2)^{\frac{1}{2}}$ and $\theta_z = \tan^{-1}(y/x)$.

Our system sampled 256x256 pixel images to construct 64x64 complex-log mapped images. Samples were taken along radial lines spaced 5.6 degrees apart. Along each radial line the step size between samples increased by powers of 1.08. These numbers are derived from the number of pixels in the original image and the number of samples in the complex-log mapped image. An excellent examination of the different conditions involved in selecting the appropriate number of samples for a complex-log mapped image is given in [2]. The non-linear sampling can be split into two distinct parts along each radial line. Toward the center of the image the samples are dense enough that no anti-aliasing filter is needed. Samples taken at the edge of the image are large and an anti-aliasing filter is necessary. The image filtered in this manner has a circular region around the center which corresponds to an area of highest resolution. The size of this region is a function of the number of angular samples and radial samples. The filtering is done, at the same time as the sampling, by convolving truncated Bessel functions with the image in the space domain. The width of the Bessel functions main lobe is inversely proportional to the eccentricity of the sample point.

A problem associated with the complex-log mapping is sensitivity to center misalignment of the sampled image. Small shifts from the center causes dramatic distortions in the complex-log mapped image. Our system assumes that the object is centered in the image frame. Slight misalignments are considered noise. Large misalignments are considered as translations and could be accounted for by changing the gaze in such a way as to bring the object into the center of the frame. The decision about what to bring into the center of the frame is an active function and should be determined by the task. An example of a system which could be used to guide the translation process was developed by Anderson and Burt [3]. Their pyramid system analyzes the input image at different tem-

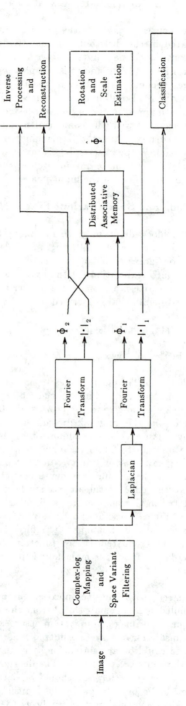

Figure 1. Block Diagram of the System.

poral and spatial resolution levels. Their smart sensor was then able to shift its fixation such that interesting parts of the image (ie. something large and moving) was brought into the central part of the frame for recognition.

1.2 Fourier Transform

The second box in the block diagram of Figure 1 is the Fourier transform. The Fourier transform of a 2-dimensional image f(x,y) is given by

$$F(u,v) = \int\limits_{-\infty}^{\infty} \int\limits_{-\infty}^{\infty} f(x,y)e^{-j(ux+vy)} \, dx \, dy \qquad (2)$$

and can be described by two 2-dimensional functions corresponding to the magnitude $|F(u,v)|$ and phase $\Phi_F(u,v)$. The magnitude component of the Fourier transform which is invariant to translation, carries much of the contrast information of the image. The phase component of the Fourier transform carries information about how things are placed in an image. Translation of f(x,y) corresponds to the addition of a linear phase component. The complex-log mapping transforms rotation and scale into translation and the magnitude of the Fourier transform is invariant to those translations so that $|\bullet|_1$ will not change significantly with rotation and scale of the object in the image.

1.3 Laplacian

The Laplacian that we use is a difference-of-Gaussians (DOG) approximation to the $\nabla^2 G$ function as given by Marr [4].

$$\nabla^2 G = \frac{1}{\pi\sigma^4} [1 - r^2/2\sigma^2] \, e^{\{-r^2/2\sigma^2\}} \qquad (3)$$

The result of convolving the Laplacian with an image can be viewed as a two step process. The image is blurred by a Gaussian kernel of a specified width σ. Then the isotropic second derivative of the blurred image is computed. The width of the Gaussian kernel is chosen such that the conformally mapped image is visible -- approximately 2 pixels in our experiments. The Laplacian sharpens the edges of the object in the image and sets any region that did not change much to zero. Below we describe the benefits from using the Laplacian.

The Laplacian eliminates the stretching problem encountered by the complex-log mapping due to changes in object size. When an object is expanded the complex-log mapped image will translate. The pixels vacated by this translation will be filled with more pixels sampled from the center of the scaled object. These new pixels will not be significantly different than the displaced pixels so the result looks like a stretching in the complex-log mapped image. The Laplacian of the complex-log mapped image will set the new pixels to zero because they do not significantly change from their surrounding pixels. The Laplacian eliminates high frequency spreading due to the finite structure of the discrete Fourier transform and enhances the differences between memorized objects by accentuating edges and de-emphasizing areas of little change.

2. Distributed Associative Memory (DAM)

The particular form of distributed associative memory that we deal with in this paper is a memory matrix which modifies the flow of information. Stimulus vectors are associated with response vectors and the result of this association is spread over the entire memory space. Distributing in this manner means that information about a small portion of the association can be found in a large area of the memory. New associations are placed over the older ones and are allowed to interact. This means that the size of the memory matrix stays the same regardless of the number of associations that have been memorized. Because the associations are allowed to interact with each other an implicit representation of structural relationships and contextual information can develop, and as a consequence a very rich level of interactions can be captured. There are few restrictions on what vectors can be associated there can exist extensive indexing and cross-referencing in the memory. Distributed associative memory captures a distributed representation which is context dependent. This is quite different from the simplistic behavioral model [5].

The *construction* stage assumes that there are n pairs of m-dimensional vectors that are to be associated by the distributed associative memory. This can be written as

$$M\vec{s_i} = \vec{r_i} \quad \text{for } i = 1,...,n \tag{4}$$

where $\vec{s_i}$ denotes the ith stimulus vector and $\vec{r_i}$ denotes the ith corresponding response vector. We want to construct a memory matrix M such that when the kth stimulus vector $\vec{s_k}$ is projected onto the space defined by M the resulting projection will be the corresponding response vector $\vec{r_k}$. More specifically we want to solve the following equation:

$$MS = R \tag{5}$$

where $S = [\vec{s_1} \mid \vec{s_2} \mid ... \mid \vec{s_n}]$ and $R = [\vec{r_1} \mid \vec{r_2} \mid ... \mid \vec{r_n}]$. A unique solution for this equation does not necessarily exist for any arbitrary group of associations that might be chosen. Usually, the number of associations n is smaller than m, the length of the vector to be associated, so the system of equations is underconstrained. The constraint used to solve for a unique matrix M is that of minimizing the square error, $\|MS - R\|^2$, which results in the solution

$$M = RS^+ \tag{6}$$

where S^+ is known as the Moore-Penrose generalized inverse of S [6].

The *recall* operation projects an unknown stimulus vector $\tilde{s}$ onto the memory space M. The resulting projection yields the response vector $\tilde{r}$

$$\tilde{r} = M\tilde{s} \tag{7}$$

If the memorized stimulus vectors are independent and the unknown stimulus vector $\tilde{s}$ is one of the memorized vectors $\vec{s_k}$, then the recalled vector will be the associated response vector $\vec{r_k}$. If the memorized stimulus vectors are dependent, then the vector recalled by one of the memorized stimulus vectors will contain the associated response vector and some *crosstalk* from the other stored response vectors.

The recall can be viewed as the weighted sum of the response vectors. The recall begins by assigning weights according to how well the unknown stimulus vector matches with the memorized stimulus vector using a linear least squares classifier. The response vectors are multiplied by the weights and summed together to build the recalled response vector. The recalled response vector is usually dominated by the memorized response vector that is closest to the unknown stimulus vector.

Assume that there are n associations in the memory and each of the associated stimulus and response vectors have m elements. This means that the memory matrix has m^2 elements. Also assume that the noise that is added to each element of a memorized stimulus vector is independent, zero mean, with a variance of σ_i^2. The recall from the memory is then

$$\vec{r} = \vec{r}_k + \vec{v}_o = M(\vec{s}_k + \vec{v}_i) = \vec{r}_k + M\vec{v}_i \tag{8}$$

where $\vec{v}_i$ is the input noise vector and $\vec{v}_o$ is the output noise vector. The ratio of the average output noise variance to the average input noise variance is

$$\sigma_o^2/\sigma_i^2 = \frac{1}{m}\mathrm{Tr}[MM^T] \tag{9}$$

For the autoassociative case this simplifies to

$$\sigma_o^2/\sigma_i^2 = \frac{n}{m} \tag{10}$$

This says that when a noisy version of a memorized input vector is applied to the memory the recall is improved by a factor corresponding to the ratio of the number of memorized vectors to the number of elements in the vectors. For the heteroassociative memory matrix a similar formula holds as long as n is less than m [7].

$$\sigma_o^2/\sigma_i^2 = \frac{1}{m}\mathrm{Tr}[RR^T]\mathrm{Tr}[(S^TS)^{-1}] \tag{11}$$

Fault tolerance is a byproduct of the distributed nature and error correcting capabilities of the distributed associative memory. By distributing the information, no single memory cell carries a significant portion of the information critical to the overall performance of the memory.

3. Experiments

In this section we discuss the result of computer simulations of our system. Images of objects are first preprocessed through the subsystem outlined in section 2. The output of such a subsystem is four vectors: $|\bullet|_1$, Φ_1, $|\bullet|_2$, and Φ_2. We construct the memory by associating the stimulus vector $|\bullet|_1$ with the response vector Φ_2 for each object in the database. To perform a recall from the memory the unknown image is preprocessed by the same subsystem to produce the vectors $|\bullet|_1$, Φ_1, $|\bullet|_2$, and Φ_2. The resulting stimulus vector $|\bullet|_1$ is projected onto the memory matrix to produce a response vector which is an estimate of the memorized phase Φ_2. The estimated phase vector Φ_2 and the magnitude $|\bullet|_1$ are used to reconstruct the memorized object. The difference between the estimated phase Φ_2 and the unknown phase Φ_2 is used to estimate the amount of rotation and scale experienced by the object.

The database of images consists of twelve objects: four keys, four mechanical parts, and four leaves. The objects were chosen for their essentially two-dimensional structure. Each object was photographed using a digitizing video camera against a black background. We emphasize that all of the images used in creating and testing the recognition system were taken at different times using various camera rotations and distances. The images are digitized to 256x256, eight bit quantized pixels, and each object covers an area of about 40x40 pixels. This small object size relative to the background is necessary due to the non-linear sampling of the complex-log mapping. The objects were centered within the frame by hand. This is the source of much of the noise and could have been done automatically using the object's center of mass or some other criteria determined by the task. The orientation of each memorized object was arbitrarily chosen such that their major axis was vertical. The 2-dimensional images that are the output from the invariant representation subsystem are scanned horizontally to form the vectors for memorization. The database used for these experiments is shown in Figure 2.

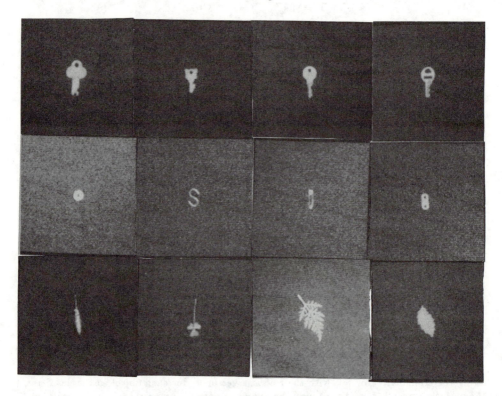

Figure 2. The Database of Objects Used in the Experiments

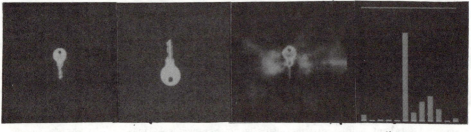

a) Original   b) Unknown   c) Recall: rotated 135°   d) Memory:6

SNR: -3.37 Db

Figure 3. Recall Using a Rotated and scaled key

The first example of the operation of our system is shown in Figure 3. Figure 3a) is the image of one of the keys as it was memorized. Figure 3b) is the unknown object presented to our system. The unknown object in this case is the same key that has been rotated by 180 degrees and scaled. Figure 3c) is the recalled, reconstructed image. The rounded edges of the recalled image are artifacts of the complex-log mapping. Notice that the reconstructed recall is the unrotated memorized key with some noise caused by errors in the recalled phase. Figure 3d) is a histogram which graphically displays the classification vector which corresponds to $S^+\bar{s}$. The histogram shows the interplay between the memorized images and the unknown image. The "6" on the bargraph indicates which of the twelve classes the unknown object belongs. The histogram gives a value which is the best linear estimate of the image relative to the memorized objects. Another measure, the signal-to-noise ratio (SNR), is given at the bottom of the recalled image. SNR compares the variance of the ideal recall after processing with the variance of the difference between the ideal and actual recall. This is a measure of the amount of noise in the recall. The SNR does not carry much information about the quality of the recall image because the noise measured by the SNR is due to many factors such as misalignment of the center, changing reflections, and dependence between other memorized objects -- each affecting quality in a variety of ways. Rotation and scale estimates are made using a vector D corresponding to the difference between the unknown vector Φ_2 and the recalled vector Φ_2. In an ideal situation D will be a plane whose gradient indicates the exact amount of rotation and scale the recalled object has experienced. In our system the recalled vector Φ_2 is corrupted with noise which means rotation and scale have to be estimated. The estimate is made by letting the first order difference D at each point in the plane vote for a specified range of rotation or scale.

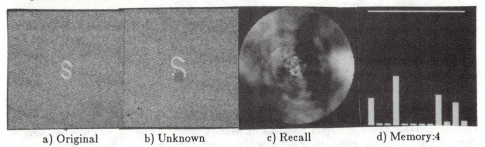

a) Original   b) Unknown   c) Recall   d) Memory:4

Figure 4 Recall Using Scaled and Rotated "S" with Occlusion

Figure 4 is an example of occlusion. The unknown object in this case is an "S" curve which is larger and slightly tilted from the memorized "S" curve. A portion of the bottom curve was occluded. The resulting reconstruction is very noisy but has filled in the missing part of the bottom curve. The noisy recall is reflected in both the SNR and the interplay between the memories shown by the histogram.

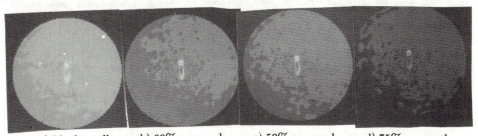

a) Ideal recall   b) 30% removed   c) 50% removed   d) 75% removed

Figure 5. Recall for Memory Matrix Randomly Set to Zero

Figure 5 is the result of *randomly* setting the elements of the memory matrix to zero. Figure 5a) shows is the ideal recall. Figure 5b) is the recall after 30 percent of the memory matrix has been set to zero. Figure 5c) is the recall for 50 percent and Figure 5d) is the recall for 75 percent. Even when 90 percent of the memory matrix has been set to zero a faint outline of the pin could still be seen in the recall. This result is important in two ways. First, it shows that the distributed associative memory is robust in the presence of noise. Second, it shows that a completely connected network is not necessary and as a consequence a scheme for data compression of the memory matrix could be found.

4. Conclusion

In this paper we demonstrate a computer vision system which recognizes 2-dimensional objects invariant to rotation or scale. The system combines an invariant representation of the input images with a distributed associative memory such that objects can be classified, reconstructed, and characterized. The distributed associative memory is resistant to moderate amounts of noise and occlusion. Several experiments, demonstrating the ability of our computer vision system to operate on real, grey scale images, were presented.

Neural network models, of which the distributed associative memory is one example, were originally developed to simulate biological memory. They are characterized by a large number of highly interconnected simple processors which operate in parallel. An excellent review of the many neural network models is given in [8]. The distributed associative memory we use is linear, and as a result there are certain desirable properties which will not be exhibited by our computer vision system. For example, feedback through our system will not improve recall from the memory. Recall could be improved if a non-linear element, such as a sigmoid function, is introduced into the feedback loop. Non-linear neural networks, such as those proposed by Hopfield [9] or Anderson et. al. [10], can achieve this type of improvement because each memorized pattern is associated with stable points in an energy space. The price to be paid for the introduction of non-linearities into a memory system is that the system will be difficult to analyze and can be unstable. Implementing our computer vision system using non-linear distributed associative memory is a goal of our future research.

We are presently extending our work toward 3-dimensional object recognition. Much of the present research in 3-dimensional object recognition is limited to polyhedral, non-occluded objects in a clean, highly controlled environment. Most systems are edge based and use a generate-and-test paradigm to estimate the position and orientation of recognized objects. We propose to use an approach based on characteristic views [11] or aspects [12] which suggests that the infinite 2-dimensional projections of a 3-dimensional object can be grouped into a finite number of topological equivalence classes. An efficient 3-dimensional recognition system would require a parallel indexing method to search for object models in the presence of geometric distortions, noise, and occlusion. Our object recognition system using distributed associative memory can fulfill those requirements with respect to characteristic views.

References

[1] Simon, H. A., (1984), **The Science of the Artificial (2nd ed.),** MIT Press.
[2] Massone, L., G. Sandini, and V. Tagliasco (1985), "Form-invariant" topological mapping strategy for 2D shape recognition, **CVGIP,** 30, 169-188.
[3] Anderson, C. H., P. J. Burt, and G. S. Van Der Wal (1985), Change detection and tracking using pyramid transform techniques, **Proc. of the SPIE Conference on Intelligence, Robots, and Computer Vision,** Vol. 579, 72-78.

[4] Marr, D. (1982), **Vision,** W. H. Freeman, 1982.

[5] Hebb, O. D. (1949), **The Organization of Behavior,** New York: Wiley.

[6] Kohonen, T. (1984), **Self-Organization and Associative-Memories,** Springer-Verlag.

[7] Stiles, G. S. and D. L. Denq (1985), On the effect of noise on the Moore-Penrose generalized inverse associative memory, **IEEE Trans. on PAMI,** 7, 3, 358-360.

[8] McClelland, J. L., and D. E. Rumelhart, and the PDP Research Group (Eds.) (1986), **Parallel Distributed, Processing,** Vol. 1, 2, MIT Press.

[9] Hopfield, J. J. (1982), Neural networks and physical systems with emergent collective computational abilities, **Proc. Natl. Acad. Sci. USA,** 79, April 1982.

[10] Anderson, J. A., J. W. Silverstein, S. A. Ritz, and R. S. Jones (1977), Distinctive features, categorical perception, and probability learning: some applications of a neural model, **Psychol. Rev.,** 84,413-451.

[11] Chakravarty, I., and H. Freeman (1982), Characteristic views as a basis for 3-D object recognition, **Proc. SPIE on Robot Vision,** 336, 37-45.

[12] Koenderink, J. J., and A. J. Van Doorn (1979), Internal representation of solid shape with respect to vision, **Biol. Cybern.,** 32,4,211-216.

LEARNING IN NETWORKS OF
NONDETERMINISTIC ADAPTIVE LOGIC ELEMENTS

Richard C. Windecker*

AT&T Bell Laboratories, Middletown, NJ 07748

ABSTRACT

This paper presents a model of nondeterministic adaptive automata that are constructed from simpler nondeterministic adaptive information processing elements. The first half of the paper describes the model. The second half discusses some of its significant adaptive properties using computer simulation examples. Chief among these properties is that network aggregates of the model elements can adapt appropriately when a *single* reinforcement channel provides the same positive or negative reinforcement signal to *all* adaptive elements of the network at the same time. This holds for multiple-input, multiple-output, multiple-layered, combinational and sequential networks. It also holds when some network elements are "hidden" in that their outputs are not directly seen by the external environment.

INTRODUCTION

There are two primary motivations for studying models of adaptive automata constructed from simple parts. First, they let us learn things about real biological systems whose properties are difficult to study directly: We form a hypothesis about such systems, embody it in a model, and then see if the model has reasonable learning and behavioral properties. In the present work, the hypothesis being tested is: that much of an animal's behavior as determined by its nervous system is intrinsically *non*deterministic; that learning consists of *incremental* changes in the probabilities governing the animal's behavior; and that this is a consequence of the animal's nervous system consisting of an aggregate of information processing elements some of which are individually *non*deterministic and adaptive. The second motivation for studying models of this type is to find ways of building machines that can *learn* to do (artificially) intelligent and practical things. This approach has the potential of complementing the currently more developed approach of programming intelligence into machines.

We do not assert that there is necessarily a one-to-one correspondence between real physiological neurons and the postulated model information processing elements. Thus, the model may be loosely termed a "neural network model," but is more accurately described as a model of adaptive automata constructed from simple adaptive parts.

---

* The main ideas in this paper were conceived and initially developed while the author was at the University of Chiang Mai, Thailand (1972-73). The ideas were developed further and put in a form consistent with existing switching and automata theory during the next four years. For two of those years, the author was at the University of Guelph, Ontario, supported of National Research Council of Canada Grant #A6983.

It almost certainly has to be a property of any acceptable model of animal learning that a *single* reinforcement channel providing reinforcement to *all* the adaptive elements in a network (or subnetwork) can effectively cause that network to adapt appropriately. Otherwise, methods of providing separate, specific reinforcement to all adaptive elements in the network must be postulated. Clearly, the environment reinforces an animal as a whole and the *same* reinforcement mechanism can cause the animal to adapt to *many* types of situation. Thus, the reinforcement system is non-specific to particular adaptive elements and particular behaviors. The model presented here has this property.

The model described here is a close cousin to the family of models recently described by Barto and coworkers [1-4]. The most significant difference are: 1) In the present model, we define the timing discipline for networks of elements more explicitly and completely. This particular timing discipline makes the present model consistent with a nondeterministic extension of switching and automata theory previously described [5]. 2) In the present model, the reinforcement algorithm that adjusts the weights is kept very simple. With this algorithm, positive and negative reinforcement have symmetric and opposite effects on the weights. This ensures that the logical signals are symmetric opposites of each other. (Even small differences in the reinforcement algorithm can make both subtle as well as profound differences in the behavior of the model.) We also allow, null, or zero, reinforcement.

As in the family of models described by Barto, networks constructed within the present model can get "stuck" at a suboptimal behavior during learning and therefore not arrive at the optimal adapted state. The complexity of the Barto reinforcement algorithm is designed partly to overcome this tendency. In the present work, we emphasize the use of *training strategies* when we wish to ensure that the network arrives at an optimal state. (In nature, it seems likely that getting "stuck" at suboptimal behavior is common.) In all networks studied so far, it has been easy to find strategies that prevent the network from getting stuck.

The chief contributions of the present work are: 1) The establishment of a close connection between these types of models and ordinary, nonadaptive, switching and automata theory [5]. This makes the wealth of knowledge in this area, especially network synthesis and analysis methods, readily applicable to the study of adaptive networks. 2) The experimental demonstration that *sequential* ("recurrent") nondeterministic adaptive networks can adapt appropriately. Such networks can learn to produce outputs that depend on the recent *sequence* of past inputs, not just the current inputs. 3) The demonstration that the use of *training strategies* can not only prevent a network from getting stuck, but may also result in more rapid learning. Thus, such strategies may be able to compensate, or even more than compensate, for reduced complexity in the model itself.

References 2-4 and 6 provide a comprehensive background and guide to the literature on both deterministic and nondeterministic adaptive automata including those constructed from simple parts and those not.

THE MODEL ADAPTIVE ELEMENT

The model adaptive element postulated in this work is a nondeterministic, adaptive generalization of threshold logic [7]. Thus, we call these elements Nondeterministic Adaptive Threshold-logic gates (NATs). The output chosen by a NAT at any given time is *not* a function of its inputs. Rather, it is chosen by a stochastic process according to certain probabilities. It is these *probabilities* that are a function of the inputs.

A NAT is like an ordinary logic gate in that it accepts logical inputs that are two-valued and produces a logical output that is two-valued. We let these values be

+1 and −1. A NAT also has a timing input channel and a reinforcement input channel. The NAT operates on a three-part cycle: 1) Logical input signals are changed and remain constant. 2) A timing signal is received and the NAT selects a new output based on the inputs at that moment. The new output remains constant. 3) A reinforcement signal is received and the weights are incremented according to certain rules.

Let N be the number of logical input channels, let x_i represent the ith input signal, and let z be the output. The NAT has within it $N+1$ "weights," $w_0, w_1, ..., w_N$. The weights are confined to integer values. For a given set of inputs, the gate calculates the quantity W:

$$W = w_0 + w_1 x_1 + w_2 x_2 + w_3 x_3 + ... + w_N x_N = w_0 + \vec{W} \cdot \vec{X} \qquad (1)$$

Then the probability that output $z = +1$ is chosen is:

$$P(z = +1) = \frac{1}{\sqrt{2\pi}\,\sigma} \int_{-\infty}^{W} e^{-\frac{x^2}{2\sigma^2}}\, dx = \frac{1}{\sqrt{\pi}} \int_{-\infty}^{W/\sqrt{2}\sigma} e^{-\varsigma^2}\, d\varsigma \qquad (2)$$

where $\varsigma = x/\sqrt{2}\sigma$. (An equivalent formulation is to let the NAT generate a random number, w_σ, according to the normal distribution with mean zero and variance σ^2. Then if $W > -w_\sigma$, the gate selects the output $z = +1$. If $W < -w_\sigma$, the gate selects output $z = -1$. If $W = -w_\sigma$, the gate selects output −1 or +1 with equal probability.)

Reinforcement signals, R, may have one of *three* values: +1, −1, and 0 representing positive, negative, and no reinforcement, respectively. If +1 reinforcement is received, each weight is incremented by one in the direction that makes the current output, z, *more likely* to occur in the future when the same inputs are applied; if −1 reinforcement is received, each weight is incremented in the direction that makes the current output *less likely*; if 0 reinforcement is received, the weights are not changed. These rules may be summarized: $\Delta w_0 = zR$ and $\Delta w_i = x_i zR$ for $i > 0$.

NATs operate in discrete time because if the NAT can choose output +1 or −1, depending on a stochastic process, it has to be told *when* to select a new output. It cannot "run freely," or it could be constantly changing output. Nor can it change output only when its inputs change because it may need to select a new output even when they do not change.

The normal distribution is used for heuristic reasons. If a real neuron (or an aggregate of neurons) uses a stochastic process to produce nondeterministic behavior, it is likely that process can be described by the normal distribution. In any case, the *exact* relationship between $P(z = +1)$ and W is not critical. What *is* important is that $P(z = +1)$ be monotonically increasing in W, go to 0 and 1 asymptotically as W goes to $-\infty$ and $+\infty$, respectively, and equal 0.5 at $W = 0$.

The parameter σ is adjustable. We use 10 in the computer simulation experiments described below. Experimentally, values near 10 work reasonably well for networks of NATs having few inputs. Note that as σ goes to zero, the behavior of a NAT approximates that of an ordinary deterministic adaptive threshold logic gate with the difference that the output for the case $W = 0$ is not arbitrary: The NAT will select output +1 or −1 with equal probability.

Note that for *all* values of W, the probabilities are greater than zero that either +1 or −1 will be chosen, although for large values of W (relative to σ) for all practical purposes, the behavior is deterministic. There are many values of the weights that cause the NAT to *approximate* the behavior of a deterministic threshold logic gate. For the same reasons that deterministic threshold logic gates cannot realize all 2^{2^N} functions of N variables [7], so a NAT cannot learn to approximate *any* deterministic function; only the threshold logic functions.

Note also that when the weights are near zero, a NAT adapts most rapidly when both positive and negative reinforcement are used in approximately equal amounts. As the NAT becomes more likely to produce the appropriate behavior, the opportunity to use negative reinforcement decreases while the opportunity to use positive reinforcement increases. This means that a NAT cannot learn to (nearly) always select a certain output if negative reinforcement alone is used. Thus, positive reinforcement has an important role in this model. (In most deterministic models, positive reinforcement is not useful.)

Note further that there is no hysteresis in NAT learning. For a given configuration of inputs, a $+1$ output followed by a $+1$ reinforcement has exactly the same effect on all the weights as a -1 output followed by a -1 reinforcement. So the *order* of such events has no effect on the final values of the weights.

Finally, if only *negative* reinforcement is applied to a NAT, independent of output, for a particular combination of inputs, the weights will change in the direction that makes W tend toward zero and once there, follow a random walk centered on zero. (The further W is from zero, the more likely its next step will be toward zero.) If all possible input combinations are applied with more or less equal probability, *all* the weights will tend toward zero and then follow random walks centered on zero. In this case, the NAT will select $+1$ or -1 with more or less equal probability without regard to its inputs.

NETWORKS

NATs may be connected together in networks (NAT-nets). The inputs to a NAT in such a network can be selected from among: 1) the set of inputs to the entire network, 2) the set of outputs from other NATs in the network, and 3) its own output. The outputs of the network may be chosen from among: 1) the inputs to the network as a whole, and 2) the outputs of the various NATs in the network.

Following Ref. 5, we impose a timing discipline on a NAT-net. The network is organized into layers such that each NAT belongs to one layer. Letting L be the number of layers, the network operates as follows: 1) All NATs in a given layer receive timing signals at the same time and select a new output at the same time. 2) Timing signals are received by the different layers, in sequence, from 1 to L. 3) Inputs to the network as a whole are levels that may change only *before* Layer 1 receives its timing signal. Similarly, outputs from the network as a whole are available to the environment only *after* Layer L has received its timing signal. Reinforcement to the network as a whole is accepted only *after* outputs are made available to the environment. The same reinforcement signal is distributed to *all* NATs in the network at the same time.

With these rules, NAT-nets operate through a sequence of timing cycles. In each cycle: 1) Network inputs are changed. 2) Layers 1 through L select new outputs, in sequence. 3) Network outputs are made available to the environment. 4) Reinforcement is received from the environment. We call each such cycle a "trial" and a sequence of such trials is a "session."

This model is very general. If, for each gate, inputs are selected only from among the inputs to the network as a whole and from the outputs of gates in layers preceding it in the timing cycle, then the network is *combinational*. In this case, the probability of the network producing a given output configuration is a function of the inputs at the start of the timing cycle. If at least one NAT has one input from a

NAT in the same layer or from a subsequent layer in the timing cycle, then the network is *sequential*. In this case, the network may have "internal states" that allow it to remember information from one cycle to the next. Thus, the probabilities governing its choice of outputs may depend on inputs in previous cycles. So sequential NAT-nets may have short-term memory embodied in internal states and long-term memory embodied in the weights. In Ref. 5, we showed that sequential networks can be constructed by adding feedback paths to combinational networks and any sequential network can be put in this standard form.

In information-theoretic terms: 1) A NAT-net with no inputs and some outputs is an "information source." 2) A NAT-net with both inputs and outputs is an information "channel." 3) A combinational NAT-net is "memory-less" while a sequential NAT-net has memory. In this context, note that a NAT-net may operate in an environment that is either deterministic or nondeterministic. Both the logical and the reinforcement inputs can be selected by stochastic processes. Note also that nondeterministic and deterministic elements as well as adaptive and nonadaptive elements can be combined in one network. (It may be that the decision-making parts of an animal's nervous system are nondeterministic and adaptive while the information transmitting parts (sensory data-gathering and the motor output parts) are deterministic and nonadaptive.)

One capability that combinational NAT-nets possess is that of "pattern recognizers." A network having many inputs and one or a few outputs can "recognize" a small subset of the potential input patterns by producing a particular output pattern with high probability when a member of the recognized subset appears and a different output pattern otherwise. In practice, the number of possible input patterns may be so large that we cannot present them all for training purposes and must be content to train the network to recognize one subset by distinguishing it (with different output pattern) from another subset. In this case, if a pattern is subsequently presented to the network that has not been in one of the training sets, the probabilities governing its output may approach one or zero, but may well be closer to 0.5. The exact values will depend on the details of the training period. If the new pattern is similar to those in one of the training sets, the NAT-net will often have a high probability of producing the same output as for that set. This *associative* property is the analog of the well known associative property in deterministic models. If the network lacks sufficient complexity for the separation we wish to make, then it cannot be trained. For example, a single *N*-input NAT cannot be trained to recognize *any* arbitrary set of input patterns by selecting the +1 output when one of them is presented and −1 otherwise. It can only be trained to make separations that correspond to threshold functions.

A combinational NAT-net can also produce patterns. By analogy with a pattern recognizer, a NAT-net with none or a few inputs and a larger number of outputs can learn for each input pattern to produce a particular subset of the possible output patterns. Since the mapping may be few-to-many, instead of many-to-few, the goal of training in this case may or may not be to have the network approximate deterministic behavior. Clearly, the distinction between pattern recognizers and pattern producers is somewhat arbitrary: in general, NAT-nets are pattern transducers that map subsets of input patterns into subsets of output patterns. A *sequential* network can "recognize" patterns in the time-sequence of network inputs and produce patterns in the time-sequence of outputs.

SIMULATION EXPERIMENTS

In this Section, we discuss computer simulation results for three types of multiple-element networks. For two of these types, certain strategies are used to train the networks. In general, these strategies have two parts that alternate, as needed. The first part is a general scheme for providing network inputs and reinforcement that tends to train all elements in the network in the desired direction. The second part is substituted temporarily when it becomes apparent that the network is getting stuck in some suboptimal behavior. It is focussed on getting the network unstuck. The strategies used here are intuitive. In general, there appear to be many strategies that will lead the network to the desired behavior. While we have made some attempt to find strategies that are reasonably efficient, it is very unlikely that the ones used are optimal. Finally, these strategies have been tested in hundreds of training sessions. Although they worked in all such sessions, there may be some (depending on the sequence of random numbers generated) in which they would not work.

In describing the networks simulated, Figs. 1-3, we use the diagramatic conventions defined in Ref. 5: We put all NATs in the same layer in a vertical line, with the various layers arranged from left to right in their order in the timing cycle. Inputs to the entire network come in from the left; outputs go out to the right. Because the timing cycle is fixed, we omit the timing inputs in these figures. For similar reasons, we also omit the reinforcement inputs.

In the simulations described here, the weights in the NATs start at zero making the network outputs completely random in the sense that on any given trial, all outputs are equally likely to occur, independent of past or present inputs. As learning proceeds, some or all the weights become large, so that the NAT-net's selection of outputs is strongly influenced by some or all of its inputs and internal connections. (Note that if the weights do not start at zero, they can be driven close to zero by using negative reinforcement.) In general, the optimum behavior toward which the network adapts is deterministic. However, because the probabilities are never identically equal to zero or one, we apply an arbitrary criterion and say that a NAT-net has learned the appropriate behavior when that criterion is satisfied. In real biological systems, we cannot know the weights or the exact probabilities governing the behavior of the individual adaptive elements. Therefore, it is appropriate to use a criterion based on observable behavior. For example, the criterion might be that the network selects the correct response (and continues to receive appropriate reinforcement) 25 times in a row.

Note that NAT-nets can adapt appropriately when the environment is not deliberately trying to make the them behave in a particular way. For example, the environment may provide inputs according to some (not necessarily deterministic) pattern and there may be some independent mechanism that determines whether the NAT-net is responding appropriately or not and provides the reinforcement accordingly. One paradigm for this situation is a game in which the NAT-net and the environment are players. The reinforcement scheme is simple: if, according to the rules of the game, the NAT-net wins a play (= trial) of the game, reinforcement is $+1$, if it loses, -1.

For a NAT-net to adapt appropriately in this situation, the game must consist of a series of similar plays. If the game is competitive, the best strategy a given player has depends on how much information he has about the opponent and vice versa. If a player assumes that his opponent is all-knowing, then his best strategy is to minimize his maximum loss and this often means playing at random, or a least according to certain probabilities. If a player knows a lot about how his opponent plays, his best strategy may be to maximize gain. This often means playing according to some deterministic strategy.

The example networks described here are special cases of three types: pattern producing (combinational multiple-output) networks, pattern recognizing (combinational multiple-input, multiple-layered, few-output) networks, and game playing (sequential) networks. The associative properties of NATs and NAT-nets are not emphasized here because they are analogous to the well known associative properties of other related models.

A Class of Simple Pattern Producing Networks

A simple class of pattern producing networks consists of the single-layer type shown in Fig. 1. Each of M NATs in such a network has no inputs, only an output. As a consequence, each has only one weight, w_0. The network is a simple, adaptive, information source.

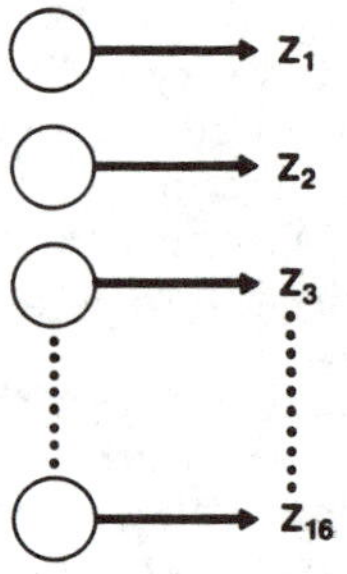

Consider first the case in which the network contains only one NAT and we wish to train it to always produce a simple "pattern," +1. We give positive reinforcement when it selects +1 and negative reinforcement otherwise. If w_0 starts at 0, it will quickly grow large making the probability of selecting +1 approach unity. The criterion we use for deciding that the network is trained is that it produce a string of 25 correct outputs. Table I

Fig. 1. A Simple Pattern Producing Network

shows that in 100 sessions, this one-NAT network selected +1 output for the next 25 trials starting, on average, at trial 13.

Next consider a network with two NATs. They can produce four different output patterns. If both weights are 0, they will produce each of the patterns with equal probability. But they can be trained to produce one pattern (nearly) all the time. If we wish to train this subnetwork to produce the pattern (in vector notation) [+1 +1], one strategy is to give no reinforcement if it produces patterns [-1 +1] or [+1 -1], give it positive reinforcement if it produces [+1 +1] and negative reinforcement if it produces [-1 -1]. Table I shows that in 100 sessions, this network learned to produce the desired pattern (by producing a string of 25 correct outputs) in about 25 trials. Because we initially gave reinforcement only about 50% of the time, it took longer to train two NATS than one.

M	Min	Ave	Max
1	1	13	26
2	8	25	43
4	18	35	60
8	44	70	109
16	49	115	215

Table I. Training Times For Networks Per Fig. 1.

Next, consider the 16-NAT network in Fig. 1. Now there are 2^{16} possible patterns the network can produce. When all the weights are zero, each has probability 2^{-16} of being produced. An *ineffective* strategy for training this network is to provide positive reinforcement when the desired pattern is produced, negative reinforcement when its opposite is produced, and zero reinforcement otherwise. A better strategy is to focus on one output of the network at a time, training each NAT separately (as above) to have a high probability of producing the desired output. Once all are trained to a relatively high level, the network as a whole has a reasonable chance of producing exactly the correct output. Now we can provide positive reinforcement when it does and no reinforcement otherwise. With this two-stage hybrid strategy, the network will soon meet the training criterion. The time it takes to train a network of M elements with a strategy of this type is roughly proportional to M, not $2^{(M-1)}$, as for the first strategy.

A still more efficient strategy is to alternate between a general substrategy and a substrategy focussed on keeping the network from getting "stuck." One effective general substrategy is to give positive reinforcement when more than half of the NATs select the desired output, negative reinforcement when less than half select the desired output, and no reinforcement when exactly half select the desired output. This substrategy starts out with approximately equal amounts of positive and negative reinforcement being applied. Soon, the network selects more than half of the outputs correctly more and more of the time. Unfortunately, there will tend to be a minority subset with low probability of selecting the correct output. At this stage, we must recognize this subset and switch to a substrategy that focuses on the elements of this subset following the strategy for one or two elements, above. When all NATs have a sufficiently high probability of selecting the desired output, training can conclude with the first substrategy.

The strategies used to obtain the results for $M = 4, 8$, and 16 in Table I were slightly more complicated variants of this two-part strategy. In all of them, a running average was kept of the number of right responses given by each NAT. Letting C_i be the "correct" output for z_i, the running average after the tth trial, $A_i(t)$, is:

$$A_i(t) = BA_i(t - 1) + C_i z_i(t) \tag{3}$$

where B is a fraction generally in the range 0.75 to 0.9. If $A_i(t)$ for a particular i gets too far below the combined average for all i, then training focuses on the ith element until its average improves. The significance of the results given in Table I is not the details of the strategies used, nor how close the training times may be to the optimum. Rather, it is the demonstration that training strategies *exist* such that the training time grows significantly more slowly than in proportion to M.

A Simple Pattern Recognizing Network

As mentioned above, there are fewer threshold logic functions of N variables (for $N > 1$) than the total possible functions. For $N = 2$, there are 14. The remining two are the "exclusive or" (XOR) and its complement. Multi-layered networks are needed to realize these functions, and an important test of any adaptive network model is its ability to learn XOR. The

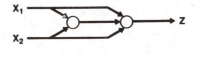

Fig. 2. A Two-Element Network That Learns XOR

network in Fig. 2 is one of the simplest networks capable of learning this function. Table II gives the results of 100 training sessions with this network. The strategy used to obtain these results again had two parts. The general part consisted of supplying each of the four possible input patterns to the network in rotation, giving appropriate reinforcement each trial. The second part involved

Network	Function	Min	Ave	Max
Fig. 2	OR	18	57	106
Fig. 2	XOR	218	681	1992
Ref. 2	XOR	~700	~3500	~14,300
Ref. 8	XOR	2232	-	-

Table II. Training Times For The Network In Fig. 2.

keeping a running average (similar to Eq. (3)) of the responses of the network by input combination. When the average for one combination fell significantly behind the average for all, training was focused on just that combination until performance improved. The criterion used for deciding when training was complete was a sequence of 50 correct responses (for all input patterns together).

For comparison, Table II shows results for the same network trained to realize the normal OR function. Also shown for comparison are numbers taken from Refs. 2 and 8 for the equivalent network in those different models. These are nondeterministic and deterministic models, respectively. The numbers from Ref. 2 are not exactly comparable with the present results for several reasons. These include: 1) The criterion for judging when the task was learned was not the same; 2) In Ref. 2, the "wrong" reinforcement was deliberately applied 10% of the time to test learning in this situation; 3) Neither model was optimized for the particular task at hand. Nonetheless, if these (and other) differences were taken into account, it is likely that the NAT-net would have learned the XOR function significantly faster.

The significance of the present results is that they suggest that the use of a training strategy can not only prevent a network from getting stuck, but may also facilitate more rapid learning. Thus, such strategies can compensate, or more than compensate, for reduced complexity in the reinforcement algorithm.

A Simple Game-Playing Network

Here, we consider NAT-nets in the context of the game of "matching pennies." In this game, each player has a stack of pennies. At each play of the game, each player places one of his pennies, heads up or heads down, but covered, in front of him. Each player uncovers his penny at the same time. If they match, player A adds both to his stack, otherwise, player B takes both.

Game theory says that the strategy of each player that minimizes his maximum loss is to play heads and tails at random. Then A cannot predict B's behavior and at best can win 50% of the time and likewise for B with respect to A. This is a conservative strategy on the part of each player because each assumes that the other has (or can derive through a sequence of plays), and can use, information about the other player's strategy. Here, we make the different assumption that: 1) Player B does not play at random, 2) Player B has no information about A's strategy, and 3) Player B is incapable of inferring any information about A through a sequence of plays and in any event is incapable of changing its strategy. Then, if A has no information about B's pattern of playing at the start of the game, A's best course of action is to try to infer a non-random pattern in B's playing through a sequence of plays and subsequently take advantage of that knowledge to win more often than 50% of the time. An adaptive NAT-net, as A, can adapt appropriately in situations of this type. For example, suppose a single NAT of the type in Fig. 1 plays A, where $+1$ output means heads, -1 output means tails. A third agent supplies reinforcement $+1$ if the NAT wins a play, -1 otherwise. Suppose B plays heads with 0.55 probability and tails with 0.45 probability. Then A will learn over time to play heads 100% of the time and thereby maximize its total winnings by winning 55% of the time.

A more complicated situation is the following. Suppose B repeats its own move *two* plays ago 80% of the time, and plays the opposite 20% of the time. A NAT-net with the potential to adapt to this strategy and win 80% of the time is shown in Fig. 3. This is a sequential network shown in the standard form of a combinational network (in the dotted rectangle) plus a feedback path. The input to the network at time t is B's play at $t - 1$. The output is A's move. The top NAT selects its output at time t based partly on the bottom NAT's output at time $t - 1$. The bottom NAT selects its output at $t - 1$ based on its input at that time which is B's output at $t - 2$. Thus, the network as a whole can learn to select its output based on B's play two time increments past. Simulation of 100 sessions resulted in the network learning to do this 98 times. On average, it took 468 plays (Min 20, max 4137) to reach the point at which the network repeated B's move two times past on the next 50 plays. For two sessions the network got stuck (for an unknown number of plays greater than 25,000) playing the opposite of B's last move or always playing tails. (The first two-part strategy found that trains the network to repeat B's output two time increments past without getting stuck (not in the game-playing context) took an average of 260 trials (Min 25, Max 1943) to meet the training criterion.)

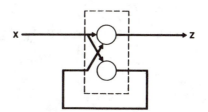
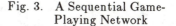

Fig. 3.   A Sequential Game-Playing Network

The significance of these results is that a *sequential* NAT-net can learn to produce appropriate behavior. Note that hidden NATs contributed to appropriate behavior for both this network and the one that learned XOR, above.

CONCLUDING REMARKS

The examples above have been kept simple in order to make them readily understandable. They are not exhaustive in the sense of covering all possible types of situations in which NAT-nets can adapt appropriately. Nor are they definitive in the sense of proving generally and in what situations NAT-nets can adapt appropriately. Rather, they are illustrative in the sense of demonstrating a variety of significant adaptive abilities. They provide an existence proof that NAT-nets can adapt appropriately and relatively easily in a wide variety of situations.

The fact that *non*deterministic models can learn when the same reinforcement is applied to all adaptive elements, while deterministic models generally cannot, supports the hypothesis that animal nervous systems may be (partly) nondeterministic. Experimental characterization of how animal learning does, or does not get "stuck," as a function of learning environment or training strategy, would be a useful test of the ideas presented here.

REFERENCES

1. Barto, A. G., "Game-Theoretic Cooperativity in Networks of Self-Interested Units," pp. 41-46 in Neural Networks for Computing, J. S. Denker, Ed., AIP Conference Proceedings 151, American Institute of Physics, New York, 1986.
2. Barto, A. G., Human Neurobiology, *4*, 229-256, 1985.
3. Barto, A. G., R. S. Sutton, and C. W. Anderson, IEEE Transactions on Systems, Man, and Cybernetics, SMC-13, No. 5, 834-846, 1983.
4. Barto, A. G., and P. Anandan, IEEE Transactions on Systems, Man, and Cybernetics, SMC-15, No. 3, 360-375, 1985.
5. Windecker, R. C., Information Sciences, *16*, 185-234 (1978).
6. Rumelhart, D. E., and J. L. McClelland, Parallel Distributed Processing, MIT Press, Cambridge, 1986.
7. Muroga, S., Threshold Logic And Its Applications, Wiley-Interscience, New York, 1971.
8. Rumelhart, D. E., G. E. Hinton, and R. J. Williams, Chapter 8 in Ref. 6.

Strategies for Teaching Layered Networks Classification Tasks

Ben S. Wittner [1] and John S. Denker
AT&T Bell Laboratories
Holmdel, New Jersey 07733

Abstract

There is a widespread misconception that the delta-rule is in some sense guaranteed to work on networks without hidden units. As previous authors have mentioned, there is no such guarantee for classification tasks. We will begin by presenting explicit counter-examples illustrating two different interesting ways in which the delta rule can fail. We go on to provide conditions which do guarantee that gradient descent will successfully train networks without hidden units to perform two-category classification tasks. We discuss the generalization of our ideas to networks with hidden units and to multi-category classification tasks.

The Classification Task

Consider networks of the form indicated in figure 1. We discuss various methods for training such a network, that is for adjusting its weight vector, $\mathbf{w}$. If we call the input $\mathbf{v}$, the output is $g(\mathbf{w} \cdot \mathbf{v})$, where g is some function.

The classification task we wish to train the network to perform is the following. Given two finite sets of vectors, F_1 and F_2, output a number greater than zero when a vector in F_1 is input, and output a number less than zero when a vector in F_2 is input. Without significant loss of generality, we assume that g is odd (i.e. $g(-s) = -g(s)$). In that case, the task can be reformulated as follows. Define [2]

$$F := F_1 \cup \{-\mathbf{v} \text{ such that } \mathbf{v} \in F_2\} \tag{1}$$

and output a number greater than zero when a vector in F is input. The former formulation is more natural in some sense, but the later formulation is somewhat more convenient for analysis and is the one we use. We call vectors in F, *training vectors*.

A Class of Gradient Descent Algorithms

We denote the solution set by

$$W := \{\mathbf{w} \text{ such that } g(\mathbf{w} \cdot \mathbf{v}) > 0 \text{ for all } \mathbf{v} \in F\}, \tag{2}$$

---

[1] Currently at NYNEX Science and Technology, 500 Westchester Ave., White Plains, NY 10604
[2] We use both $A := B$ and $B =: A$ to denote "A is by definition B".

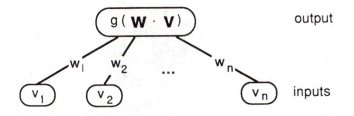

Figure 1: a simple network

and we are interested in rules for finding some weight vector in W. We restrict our attention to rules based upon gradient descent down error functions $E(\mathbf{w})$ of the form

$$E(\mathbf{w}) = \sum_{\mathbf{v} \in F} h(\mathbf{w} \cdot \mathbf{v}). \tag{3}$$

The delta-rule is of this form with

$$h(\mathbf{w} \cdot \mathbf{v}) = h_\delta(\mathbf{w} \cdot \mathbf{v}) := \frac{1}{2}(b - g(\mathbf{w} \cdot \mathbf{v}))^2 \tag{4}$$

for some positive number b called the *target* (Rumelhart, McClelland, et al.). We call the delta rule error function E_δ.

Failure of Delta-rule Using Obtainable Targets

Let g be any function that is odd and differentiable with $g'(s) > 0$ for all s. In this section we assume that the target b is in the range of g. We construct a set F of training vectors such that even though W is not empty, there is a local minimum of E_δ not located in W. In order to facilitate visualization, we begin by assuming that g is linear. We will then indicate why the construction works for the nonlinear case as well. We guess that this is the type of counter-example alluded to by Duda and Hart (p. 151) and by Minsky and Papert (p. 15).

The input vectors are two dimensional. The arrows in figure 2 represent the training vectors in F and the shaded region is W. There is one training vector, $\mathbf{v}^1$, in the second quadrant, and all the rest are in the first quadrant. The training vectors in the first quadrant are arranged in pairs symmetric about the ray R and ending on the line L. The line L is perpendicular to R, and intersects R at unit distance from the origin. Figure 2 only shows three of those symmetric pairs, but to make this construction work we might need many. The point $\mathbf{p}$ lies on R at a distance of $g^{-1}(b)$ from the origin.

We first consider the contribution to E_δ due to any single training vector, $\mathbf{v}$. The contribution is

$$(1/2)(b - g(\mathbf{w} \cdot \mathbf{v}))^2, \tag{5}$$

and is represented in figure 3 in the z-direction. Since g is linear and since b is in the

852

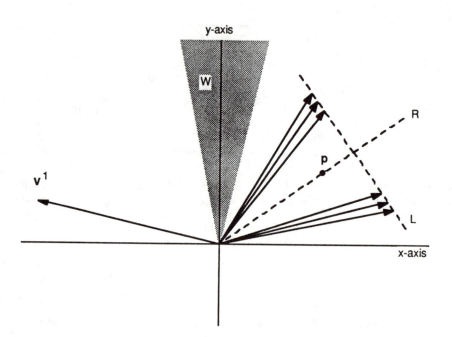

Figure 2: Counter-example for obtainable targets

range of g, the contribution is a quadratic trough with bottom on a line in the xy-plane and perpendicular to $\mathbf{v}$. If $\mathbf{v}$ is one of the training vectors in the first quadrant, then the point $\mathbf{p}$ lies along the bottom line of the trough.

Now we consider the contribution to the error function due to one of the symmetric pairs. It is the sum of two quadratic troughs with bottom lines intersecting only at the point $\mathbf{p}$. So it is a quadratic bowl with bottom point at $\mathbf{p}$.

Next we consider the contribution to E_δ due to all the training vectors in the first quadrant. It is a sum of quadratic bowls, all with bottom at $\mathbf{p}$. So it is itself a quadratic bowl with bottom at $\mathbf{p}$ and it can be made arbitrarily steep by having arbitrarily many of the symmetric pairs. Let us call this contribution E_0.

We denote by E_1 the contribution to E_δ due to $\mathbf{v}^1$. E_1 is a quadratic trough and E_0 is a quadratic bowl, so $E_\delta = E_0 + E_1$ is a quadratic bowl with a single minimum. That minimum is closer to W than is $\mathbf{p}$, but if the bowl defined by E_0 is sufficiently steep, the minimum will be sufficiently near $\mathbf{p}$ so as to not be in W. Q.E.D.

Since E_δ is a quadratic function of $\mathbf{w}$, it is easy to compute directly the zeroes of its gradient. In this way, we have confirmed the conclusion of the conceptual argument presented above.

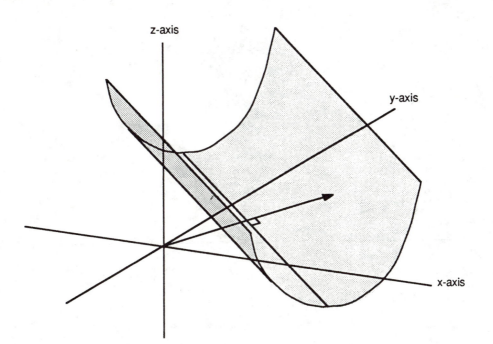

Figure 3: Error surface

We now remove the assumption that g is linear. The key observation is that

$$dh_\delta/ds \equiv h_\delta'(s) = (b - g(s))(-g'(s)) \tag{6}$$

still only has a single zero at $g^{-1}(b)$ and so $h(s)$ still has a single minimum at $g^{-1}(b)$. The contribution to E_δ due to the training vectors in the first quadrant therefore still has a global minimum on the xy-plane at the point $\mathbf{p}$. So, as in the linear case, if there are enough symmetric pairs of training vectors in the first quadrant, the value of E_0 at $\mathbf{p}$ can be made arbitrarily lower than the value along some circle in the xy-plane centered around $\mathbf{p}$, and $E_\delta = E_0 + E_1$ will have a local minimum arbitrarily near $\mathbf{p}$. Q.E.D.

Failure of Delta-rule Using Unobtainable Targets

We now consider the case where the target b is *greater* than any number in the range of g. The kind of counter-example presented in the previous section no longer exists, but we will show that for some choices of g, including the traditional choices, the delta rule can still fail. Specifically, we construct a set F of training vectors such that even though W is not empty, for some choices of initial weights, the path traced out by going down the gradient of E_δ never enters W.

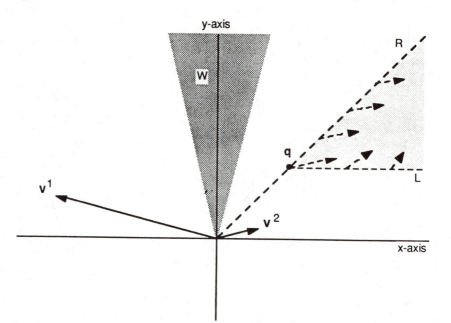

Figure 4: Counter-example for unobtainable targets

We suppose that g has the following property. There exists a number $r > 0$ such that

$$\lim_{s \to \infty} \frac{h_\delta'(-rs)}{h_\delta'(s)} = 0. \tag{7}$$

An example of such a g is

$$g(s) = \tanh(s) = \frac{2}{1 + e^{-2s}} - 1, \tag{8}$$

for which any r greater than 1 will do.

The solid arrows in figure 4 represent the training vectors in F and the more darkly shaded region is W. The set F has two elements,

$$\mathbf{v}^1 = \begin{bmatrix} -2 \\ 1 \end{bmatrix} \quad \text{and} \quad \mathbf{v}^2 = \left(\frac{1}{r}\right)\left(\frac{1}{3}\right)\begin{bmatrix} 2 \\ 1 \end{bmatrix} \tag{9}$$

The dotted ray, R lies on the diagonal $\{y = x\}$.

Since

$$E_\delta(\mathbf{w}) = h_\delta(\mathbf{w} \cdot \mathbf{v}^1) + h_\delta(\mathbf{w} \cdot \mathbf{v}^2), \tag{10}$$

the gradient descent algorithm follows the vector field

$$-\nabla E(\mathbf{w}) = -h_\delta'(\mathbf{w} \cdot \mathbf{v}^1)\mathbf{v}^1 - h_\delta'(\mathbf{w} \cdot \mathbf{v}^2)\mathbf{v}^2. \tag{11}$$

The reader can easily verify that for all $\mathbf{w}$ on R,

$$\mathbf{w} \cdot \mathbf{v}^1 = -r\mathbf{w} \cdot \mathbf{v}^2. \tag{12}$$

So by equation (7), if we constrain $\mathbf{w}$ to move along R,

$$\lim_{\mathbf{w} \to \infty} \frac{-h_\delta'(\mathbf{w} \cdot \mathbf{v}^1)}{-h_\delta'(\mathbf{w} \cdot \mathbf{v}^2)} = 0. \tag{13}$$

Combining equations (11) and (13) we see that there is a point $\mathbf{q}$ somewhere on R such that beyond $\mathbf{q}$, $-\nabla E(\mathbf{w})$ points into the region to the right of R, as indicated by the dotted arrows in figure 4.

Let L be the horizontal ray extending to the right from $\mathbf{q}$. Since for all s,

$$g'(s) > 0 \quad \text{and} \quad b > g(s), \tag{14}$$

we get that

$$-h_\delta'(s) = (b - g(s))g'(s) > 0. \tag{15}$$

So since both $\mathbf{v}^1$ and $\mathbf{v}^2$ have a positive y-component, $-\nabla E(\mathbf{w})$ also has a positive y-component for all $\mathbf{w}$. So once the algorithm following $-\nabla E$ enters the region above L and to the right of R (indicated by light shading in figure 4), it never leaves. Q.E.D.

Properties to Guarantee Gradient Descent Learning

In this section we present three properties of an error function which guarantee that gradient descent will not fail to enter a non-empty W.

We call an error function of the form presented in equation (3) *well formed* if h is differentiable and has the following three properties.

1. For all s, $-h'(s) \geq 0$ (i.e. h does not push in the wrong direction).

2. There exists some $\epsilon > 0$ such that $-h'(s) \geq \epsilon$ for all $s \leq 0$ (i.e. h keeps pushing if there is a misclassification).

3. h is bounded below.

Proposition 1 *If the error function is well formed, then gradient descent is guaranteed to enter W, provided W is not empty.*

The proof proceeds by contradiction. Suppose for some starting weight vector the path traced out by gradient descent never enters W. Since W is not empty, there is some non-zero $\mathbf{w}^*$ in W. Since F is finite,

$$\lambda := \min\{\mathbf{w}^* \cdot \mathbf{v} \text{ such that } \mathbf{v} \in F\} > 0. \tag{16}$$

Let $\mathbf{w}(t)$ be the path traced out by the gradient descent algorithm. So

$$\mathbf{w}'(t) = -\nabla E(\mathbf{w}(t)) = \sum_{\mathbf{v} \in F} -h'(\mathbf{w}(t) \cdot \mathbf{v})\mathbf{v} \qquad \text{for all } t. \tag{17}$$

Since we are assuming that at least one training vector is misclassified at all times, by properties 1 and 2 and equation (17),

$$\mathbf{w}^* \cdot \mathbf{w}'(t) \geq \epsilon\lambda \qquad \text{for all } t. \tag{18}$$

So

$$|\mathbf{w}'(t)| \geq \epsilon\lambda/|\mathbf{w}^*| =: \xi > 0 \qquad \text{for all } t. \tag{19}$$

By equations (17) and (19),

$$dE(\mathbf{w}(t))/dt = \nabla E \cdot \mathbf{w}'(t) = -\mathbf{w}'(t) \cdot \mathbf{w}'(t) \leq -\xi^2 < 0 \qquad \text{for all } t. \tag{20}$$

This means that

$$E(\mathbf{w}(t)) \to -\infty \quad \text{as} \quad t \to \infty. \tag{21}$$

But property 3 and the fact that F is finite guarantee that E is bounded below. This contradicts equation (21) and finishes the proof.

Consensus and Compromise

So far we have been concerned with the case in which F is separable (i.e. W is not empty). What kind of behavior do we desire in the non-separable case? One might hope that the algorithm will choose weights which produce correct results for as many of the training vectors as possible. We suggest that this is what gradient descent using a well formed error function does.

From investigations of many well formed error functions, we suspect the following well formed error function is representative. Let $g(s) = s$, and for some $b > 0$, let

$$h(s) = \begin{cases} (b - s)^2 & \text{if } s \leq b; \\ 0 & \text{otherwise.} \end{cases} \tag{22}$$

In all four frames of figure 5 there are three training vectors. Training vectors 1 and 2 are held fixed while 3 is rotated to become increasingly inconsistent with the others. In frames (i) and (ii) F is separable. The training set in frame (iii) lies just on the border between separability and non-separability, and the one in frame (iv) is in the interior of

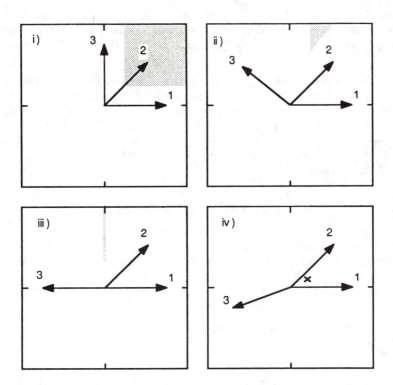

Figure 5: The transition between seperability and non-seperability

the non-separable regime. Regardless of the position of vector 3, the global minimum of the error function is the only minimum.

In frames (i) and (ii), the error function is zero on the shaded region and the shaded region is contained in W. As we move training vector number 3 towards its position in frame (iii), the situation remains the same except the shaded region moves arbitrarily far from the origin. At frame (iii) there is a discontinuity; the region on which the error function is at its global minimum is now the one-dimensional ray indicated by the shading. Once training vector 3 has moved into the interior of the non-separable regime, the region on which the error function has its global minimum is a point closer to training vectors 1 and 2 than to 3 (as indicated by the "x" in frame (iv)).

If all the training vectors can be satisfied, the algorithm does so; otherwise, it tries to satisfy as many as possible, and there is a discontinuity between the two regimes. We summarize this by saying that it finds a consensus if possible, otherwise it devises a compromise.

Hidden Layers

For networks with hidden units, it is probably impossible to prove anything like proposition 1. The reason is that even though property 2 assures that the top layer of weights

858

gets a non-vanishing error signal for misclassified inputs, the lower layers might still get a vanishingly weak signal if the units above them are operating in the saturated regime.

We believe it is nevertheless a good idea to use a well formed error function when training such networks. Based upon a probabilistic interpretation of the output of the network, Baum and Wilczek have suggested using an entropy error function (we thank J.J. Hopfield and D.W. Tank for bringing this to our attention). Their error function is well formed. Levin, Solla, and Fleisher report simulations in which switching to the entropy error function from the delta-rule introduced an order of magnitude speed-up of learning for a network with hidden units.

Multiple Categories

Often one wants to classify a given input vector into one of many categories. One popular way of implementing multiple categories in a feed-forward network is the following. Let the network have one output unit for each category. Denote by $o_j^{\mathbf{v}}(\mathbf{w})$ the output of the j-th output unit when input $\mathbf{v}$ is presented to the network having weights $\mathbf{w}$. The network is considered to have classified $\mathbf{v}$ as being in the k-th category if

$$o_k^{\mathbf{v}}(\mathbf{w}) > o_j^{\mathbf{v}}(\mathbf{w}) \quad \text{for all } j \neq k. \tag{23}$$

The way such a network is usually trained is the generalized delta-rule (Rumelhart, McClelland, et al.). Specifically, denote by $c(\mathbf{v})$ the desired classification of $\mathbf{v}$ and let

$$b_j^{\mathbf{v}} := \begin{cases} b & \text{if } j = c(\mathbf{v}); \\ -b & \text{otherwise,} \end{cases} \tag{24}$$

for some target $b > 0$. One then uses the error function

$$E(\mathbf{w}) := \sum_{\mathbf{v}} \sum_j \left(b_j^{\mathbf{v}} - o_j^{\mathbf{v}}(\mathbf{w}) \right)^2. \tag{25}$$

This formulation has several bothersome aspects. For one, the error function is not will formed. Secondly, the error function is trying to adjust the outputs, but what we really care about is the *differences* between the outputs. A symptom of this is the fact that the change made to the weights of the connections to any output unit does not depend on any of the weights of the connections to any of the other output units.

To remedy this and also the other defects of the delta rule we have been discussing, we suggest the following. For each $\mathbf{v}$ and j, define the relative coordinate

$$\beta_j^{\mathbf{v}}(\mathbf{w}) := o_{c(\mathbf{v})}^{\mathbf{v}}(\mathbf{w}) - o_j^{\mathbf{v}}(\mathbf{w}). \tag{26}$$

What we really want is all the β to be positive, so use the error function

$$E(\mathbf{w}) := \sum_{\mathbf{v}} \sum_{j \neq c(\mathbf{v})} h\left(\beta_j^{\mathbf{v}}(\mathbf{w})\right) \tag{27}$$

for some well formed h. In the simulations we have run, this does not always help, but sometimes it helps quite a bit.

We have one further suggestion. Property 2 of a well formed error function (and the fact that derivatives are continuous) means that the algorithm will not be completely satisfied with positive β; it will try to make them greater than zero by some non-zero margin. That is a good thing, because the training vectors are only representatives of the vectors one wants the network to correctly classify. Margins are critically important for obtaining robust performance on input vectors not in the training set. The problem is that the margin is expressed in meaningless units; it makes no sense to use the same numerical margin for an output unit which varies a lot as is used for an output unit which varies only a little. We suggest, therefore, that for each j and $\mathbf{v}$, keep a running estimate of $\sigma_j^{\mathbf{v}}(\mathbf{w})$, the variance of $\beta_j^{\mathbf{v}}(\mathbf{w})$, and replace $\beta_j^{\mathbf{v}}(\mathbf{w})$ in equation (27) by

$$\beta_j^{\mathbf{v}}(\mathbf{w})/\sigma_j^{\mathbf{v}}(\mathbf{w}). \tag{28}$$

Of course, when beginning the gradient descent, it is difficult to have a meaningful estimate of $\sigma_j^{\mathbf{v}}(\mathbf{w})$ because $\mathbf{w}$ is changing so much, but as the algorithm begins to converge, your estimate can become increasingly meaningful.

References

1. David Rumelhart, James McClelland, and the PDP Research Group, *Parallel Distributed Processing*, MIT Press, 1986

2. Richard Duda and Peter Hart, *Pattern Classification and Scene Analysis*, John Wiley & Sons, 1973.

3. Marvin Minsky and Seymour Papert, "On Perceptrons", Draft, 1987.

4. Eric Baum and Frank Wilczek, these proceedings.

5. Esther Levin, Sara A. Solla, and Michael Fleisher, private communications.

A METHOD FOR THE DESIGN OF STABLE LATERAL INHIBITION
NETWORKS THAT IS ROBUST IN THE PRESENCE
OF CIRCUIT PARASITICS

J.L. WYATT, Jr and D.L. STANDLEY
Department of Electrical Engineering and Computer Science
Massachusetts Institute of Technology
Cambridge, Massachusetts 02139

ABSTRACT

In the analog VLSI implementation of neural systems, it is
sometimes convenient to build lateral inhibition networks by using
a locally connected on-chip resistive grid. A serious problem
of unwanted spontaneous oscillation often arises with these
circuits and renders them unusable in practice. This paper reports
a design approach that guarantees such a system will be stable,
even though the values of designed elements and parasitic elements
in the resistive grid may be unknown. The method is based on a
rigorous, somewhat novel mathematical analysis using Tellegen's
theorem and the idea of Popov multipliers from control theory. It
is thoroughly practical because the criteria are local in the sense
that no overall analysis of the interconnected system is required,
empirical in the sense that they involve only measurable frequency
response data on the individual cells, and robust in the sense that
unmodelled parasitic resistances and capacitances in the interconnection network cannot affect the analysis.

I. INTRODUCTION

The term "lateral inhibition" first arose in neurophysiology to
describe a common form of neural circuitry in which the output of
each neuron in some population is used to inhibit the response of
each of its neighbors. Perhaps the best understood example is the
horizontal cell layer in the vertebrate retina, in which lateral
inhibition simultaneously enhances intensity edges and acts as an
automatic gain control to extend the dynamic range of the retina
as a whole[1]. The principle has been used in the design of artificial
neural system algorithms by Kohonen[2] and others and in the electronic
design of neural chips by Carver Mead et. al.[3,4].

In the VLSI implementation of neural systems, it is convenient
to build lateral inhibition networks by using a locally connected
on-chip resistive grid. Linear resistors fabricated in, e.g.,
polysilicon, yield a very compact realization, and nonlinear
resistive grids, made from MOS transistors, have been found useful
for image segmentation.[4,5] Networks of this type can be divided into
two classes: feedback systems and feedforward-only systems. In the
feedforward case one set of amplifiers imposes signal voltages or

currents on the grid and another set reads out the resulting response
for subsequent processing, while the same amplifiers both "write" to
the grid and "read" from it in a feedback arrangement. Feedforward
networks of this type are inherently stable, but feedback networks
need not be.

A practical example is one of Carver Mead's retina chips[3] that
achieves edge enhancement by means of lateral inhibition through a
resistive grid. Figure 1 shows a single cell in a continuous-time
version of this chip. Note that the capacitor voltage is affected
both by the local light intensity incident on that cell and by the
capacitor voltages on neighboring cells of identical design. Any
cell drives its neighbors, which drive both their distant neighbors
and the original cell in turn. Thus the necessary ingredients for
instability--active elements and signal feedback--are both present
in this system, and in fact the continuous-time version oscillates
so badly that the original design is scarcely usable in practice
with the lateral inhibition paths enabled.[6] Such oscillations can

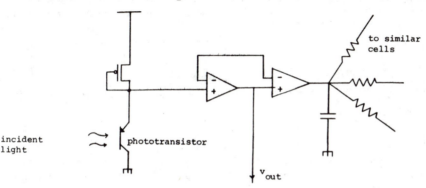

Figure 1. This photoreceptor and signal processor circuit, using two
MOS transconductance amplifiers, realizes lateral inhibition by
communicating with similar units through a resistive grid.

readily occur in any resistive grid circuit with active elements and
feedback, even when each individual cell is quite stable. Analysis
of the conditions of instability by straightforward methods appears
hopeless, since any repeated array contains many cells, each of
which influences many others directly or indirectly and is influenced
by them in turn, so that the number of simultaneously active feedback loops is enormous.

This paper reports a practical design approach that rigorously
guarantees such a system will be stable. The very simplest version
of the idea is intuitively obvious: design each individual cell so
that, although internally active, it acts like a passive system as
seen from the resistive grid. In circuit theory language, the
design goal here is that each cell's output impedance should be a
positive-real[7] function. This is sometimes not too difficult in
practice; we will show that the original network in Fig. 1 satisfies
this condition in the absence of certain parasitic elements. More
important, perhaps, it is a condition one can verify experimentally

by frequency-response measurements.

It is physically apparent that a collection of cells that appear passive at their terminals will form a stable system when interconnected through a passive medium such as a resistive grid. The research contributions, reported here in summary form, are i) a demonstration that this passivity or positive-real condition is much stronger than we actually need and that weaker conditions, more easily achieved in practice, suffice to guarantee stability of the linear network model, and ii) an extension of i) to the *nonlinear* domain that furthermore rules out large-signal oscillations under certain conditions.

II.   FIRST-ORDER LINEAR ANALYSIS OF A SINGLE CELL

We begin with a linear analysis of an elementary model for the circuit in Fig. 1. For an initial approximation to the output admittance of the cell we simplify the topology (without loss of relevant information) and use a naive' model for the transconductance amplifiers, as shown in Fig. 2.

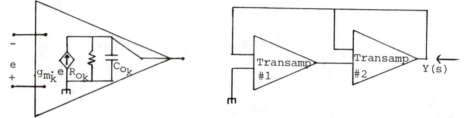

Figure 2.   Simplified network topology and transconductance amplifier model for the circuit in Fig. 1. The capacitor in Fig. 1 has been absorbed into C_{o_2}.

Straightforward calculations show that the output admittance is given by

$$Y(s) = [g_{m_2} + R_{o_2}^{-1} + s\, C_{o_2}] + \frac{g_{m_1}g_{m_2}R_{o_1}}{(1 + s\, R_{o_1}C_{o_1})} \; . \qquad (1)$$

This is a positive-real, i.e., passive, admittance since it can always be realized by a network of the form shown in Fig. 3, where $R_1 = (g_{m_2} + R_{o_2}^{-1})^{-1}$, $R_2 = (g_{m_1}g_{m_2}R_{o_1})^{-1}$, and $L = C_{o_1}/g_{m_1}g_{m_2}$.

Although the original circuit contains no inductors, the realization has both capacitors and inductors and thus is capable of damped oscillations. Nonetheless, *if* the transamp model in Fig. 2 were perfectly accurate, no network created by interconnecting such cells through a resistive grid (with parasitic capacitances) could exhibit sustained oscillations. For element values that may be typical in practice, the model in Fig. 3 has a lightly damped resonance around 1 KHz with a $Q \approx 10$. This disturbingly high Q suggests that the cell will be highly sensitive to parasitic elements not captured by the simple models in Fig. 2. Our preliminary

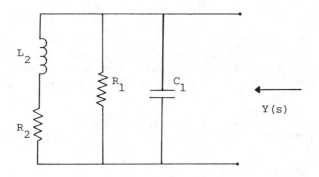

Figure 3. Passive network realization of the output admittance (eq. (1) of the circuit in Fig. 2.

analysis of a much more complex model extracted from a physical circuit layout created in Carver Mead's laboratory indicates that the output impedance will not be passive for all values of the trans-amp bias currents. But a definite explanation of the instability awaits a more careful circuit modelling effort and perhaps the design of an on-chip impedance measuring instrument.

III. POSITIVE-REAL FUNCTIONS, θ-POSITIVE FUNCTIONS, AND STABILITY OF LINEAR NETWORK MODELS

In the following discussion $s = \sigma+j\omega$ is a complex variable, $H(s)$ is a rational function (ratio of polynomials) in s with real coefficients, and we assume for simplicity that $H(s)$ has no pure imaginary poles. The term *closed right half plane* refers to the set of complex numbers s with $\text{Re}\{s\} \geq 0$.

Def. 1

The function $H(s)$ is said to be *positive-real* if a) it has no poles in the right half plane and b) $\text{Re}\{H(j\omega)\} \geq 0$ for all ω.

If we know at the outset that $H(s)$ has no right half plane poles, then Def. 1 reduces to a simple graphical criterion: $H(s)$ is positive-real if and only if the *Nyquist diagram* of $H(s)$ (i.e. the plot of $H(j\omega)$ for $\omega \geq 0$, as in Fig. 4) lies entirely in the closed right half plane.
Note that positive-real functions are necessarily stable since they have no right half plane poles, but stable functions are not necessarily positive-real, as Example 1 will show.
A deep link between positive real functions, physical networks and passivity is established by the classical result[7] in linear circuit theory which states that $H(s)$ is positive-real if and only if it is possible to synthesize a 2-terminal network of positive linear resistors, capacitors, inductors and ideal transformers that has $H(s)$ as its driving-point impedance or admittance.

<u>Def. 2</u>

The function H(s) is said to be θ-positive for a particular value of θ (θ ≠ 0, θ ≠ π), if a) H(s) has no poles in the right half plane, and b) the Nyquist plot of H(s) lies strictly to the right of the straight line passing through the origin at an angle θ to the real positive axis.

Note that every θ-positive function is stable and any function that is θ-positive with θ = π/2 is necessarily positive-real.

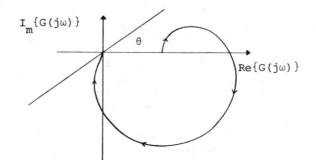

Figure 4. Nyquist diagram for a function that is θ-positive but not positive-real.

<u>Example 1</u>

The function

$$G(s) = \frac{(s+1)(s+40)}{(s+5)(s+6)(s+7)} \tag{2}$$

is θ-positive (for any θ between about 18° and 68°) and stable, but it is not positive-real since its Nyquist diagram, shown in Fig. 4, crosses into the left half plane.

The importance of θ-positive functions lies in the following observations: 1) an interconnection of passive linear resistors and capacitors and cells with stable linear impedances can result in an unstable network, b) such an instability cannot result if the impedances are also positive-real, c) θ-positive impedances form a larger class than positive-real ones and hence θ-positivity is a less demanding synthesis goal, and d) Theorem 1 below shows that such an instability cannot result if the impedances are θ-positive, even if they are not positive-real.

<u>Theorem 1</u>

Consider a linear network of arbitrary topology, consisting of any number of passive 2-terminal resistors and capacitors of arbitrary value driven by any number of active cells. If the output impedances of all the active cells are θ-positive for some common θ, $0<\theta\leq\frac{\pi}{2}$, then the network is stable.

The proof of Theorem 1 relies on Lemma 1 below.

Lemma 1

If $H(s)$ is θ-positive for some fixed θ, then for all s_o in the closed first quadrant of the complex plane, $H(s_o)$ lies strictly to the right of the straight line passing through the origin at an angle θ to the real positive axis, i.e., $\text{Re}\{s_o\} \geq 0$ and $\text{Im}\{s_o\} \geq 0 \Rightarrow \theta-\pi < \angle\, H(s_o) < \theta$.

Proof of Lemma 1 (Outline)

Let d be the function that assigns to each s in the closed right half plane the perpendicular distance $d(s)$ from $H(s)$ to the line defined in Def. 2. Note that $d(s)$ is harmonic in the closed right half plane, since H is analytic there. It then follows, by application of the maximum modulus principle[8] for harmonic functions, that d takes its minimum value on the boundary of its domain, which is the imaginary axis. This establishes Lemma 1.

Proof of Theorem 1 (Outline)

The network is unstable or marginally stable if and only if it has a natural frequency in the closed right half plane, and s_o is a natural frequency if and only if the network equations have a nonzero solution at s_o. Let $\{I_k\}$ denote the complex branch currents of such a solution. By Tellegen's theorem[9] the sum of the complex powers absorbed by the circuit elements must vanish at such a solution, i.e.,

$$\sum_{\text{resistances}} R_k|I_k|^2 + \sum_{\text{capacitances}} |I_k|^2/s_o C_k + \sum_{\substack{\text{cell} \\ \text{terminal pairs}}} Z_k(s_o)|I_k|^2 = 0,$$

$$(3)$$

where the second term is deleted in the special case $s_o=0$, since the complex power into capacitors vanishes at $s_o=0$.

If the network has a natural frequency in the closed right half plane, it must have one in the closed first quadrant since natural frequencies are either real or else occur in complex conjugate pairs. But (3) cannot be satisfied for any s_o in the closed first quadrant, as we can see by dividing both sides of (3) by $\sum_k |I_k|^2$, where the sum is taken over all network branches. After this division, (3) asserts that zero is a convex combination of terms of the form R_k, terms of the form $(C_k s_o)^{-1}$, and terms of the form $Z_k(s_o)$. Visualize where these terms lie in the complex plane: the first set lies on the real positive axis, the second set lies in the closed 4-th quadrant since s_o lies in the closed 1st quadrant by assumption, and the third set lies to the right of a line passing through the origin at an angle θ by Lemma 1. Thus all these terms lie strictly to the right of this line, which implies that no convex combination of them can equal zero. Hence the network is stable!

IV. STABILITY RESULT FOR NETWORKS WITH NONLINEAR
RESISTORS AND CAPACITORS

The previous result for linear networks can afford some limited insight into the behavior of nonlinear networks. First the nonlinear equations are linearized about an equilibrium point and Theorem 1 is applied to the linear model. If the linearized model is stable, then the equilibrium point of the original nonlinear network is *locally stable*, i.e., the network will return to that equilibrium point *if the initial condition is sufficiently near it*. But the result in this section, in contrast, applies to the full nonlinear circuit model and allows one to conclude that in certain circumstances the network cannot oscillate *even if* the initial state is *arbitrarily far from* the equilibrium point.

Def. 3

A function H(s) as described in Section III is said to satisfy the *Popov criterion*[10] if there exists a real number r>0 such that Re{(1+jωr) H(jω)} $\geq$ 0 for all ω.

Note that positive real functions satisfy the Popov criterion with r=0. And the reader can easily verify that G(s) in Example 1 satisfies the Popov criterion for a range of values of r. The important effect of the term (1+jωr) in Def. 3 is to rotate the Nyquist plot counterclockwise by progressively greater amounts up to 90° as ω increases.

Theorem 2

Consider a network consisting of nonlinear 2-terminal resistors and capacitors, and cells with linear output impedances $Z_k(s)$. Suppose

i) the resistor curves are characterized by continuously differentiable functions $i_k = g_k(v_k)$ where $g_k(0) = 0$ and $0 < g_k'(v_k) < G < \infty$ for all values of k and v_k,

ii) the capacitors are characterized by $i_k = C_k(v_k)\dot{v}_k$ with $0 < C_1 < C_k(v_k) < C_2 < \infty$ for all values of k and $\dot{v}_k$,

iii) the impedances $Z_k(s)$ have no poles in the closed right half plane and all satisfy the Popov criterion for some common value of r.

If these conditions are satisfied, then the network is stable in the sense that, for any initial condition,

$$\int_0^\infty \left(\sum_{\text{all branches}} i_k^2(t) \right) dt < \infty . \tag{4}$$

The proof, based on Tellegen's theorem, is rather involved. It will be omitted here and will appear elsewhere.

ACKNOWLEDGEMENT

We sincerely thank Professor Carver Mead of Cal Tech for enthusiastically supporting this work and for making it possible for us to present an early report on it in this conference proceedings. This work was supported by Defense Advanced Research Projects Agency (DoD), through the Office of Naval Research under ARPA Order No. 3872, Contract No. N00014-80-C-0622 and Defense Advanced Research Projects Agency (DARPA) Contract No. N00014-87-K-0825.

REFERENCES

1. F.S. Werblin, "The Control of Sensitivity on the Retina," Scientific American, Vol. 228, no. 1, Jan. 1983, pp. 70-79.
2. T. Kohonen, Self-Organization and Associative Memory, (vol. 8 in the Springer Series in Information Sciences), Springer Verlag, New York, 1984.
3. M.A. Sivilotti, M.A. Mahowald, and C.A. Mead, "Real Time Visual Computations Using Analog CMOS Processing Arrays," Advanced Research in VLSI - Proceedings of the 1987 Stanford Conference, P. Losleben, ed., MIT Press, 1987, pp. 295-312.
4. C.A. Mead, Analog VLSI and Neural Systems, Addison-Wesley, to appear in 1988.
5. J. Hutchinson, C. Koch, J. Luo and C. Mead, "Computing Motion Using Analog and Binary Resistive Networks," submitted to IEEE Transactions on Computers, August 1987.
6. M. Mahowald, personal communication.
7. B.D.O. Anderson and S. Vongpanitlerd, Network Analysis and Synthesis - A Modern Systems Theory Approach, Prentice-Hall, Englewood Cliffs, NJ., 1973.
8. L.V. Ahlfors, Complex Analysis, McGraw-Hill, New York, 1966, p. 164.
9. P. Penfield, Jr., R. Spence, and S. Duinker, Tellegen's Theorem and Electrical Networks, MIT Press, Cambridge, MA, 1970.
10. M. Vidyasagar, Nonlinear Systems Analysis, Prentice-Hall, Englewood Cliffs, NJ, 1970, pp. 211-217.

AUTHOR INDEX